Symbol	Meaning*
H_1	Alternative hypothesis (8.2)
H_i	ith hypothesis (3.6)
h	Number of parameters estimated from sample (9.7)
I	Irregular variation (13.2)
I_1	Price index for year 1 (14.2) or quantity index for year 1 (14.8)
k	Number of independent variables in a regression (12.5)
k	Number of population means being compared (10.6)
k	Number of treatments (10.8)
l	Width of class interval containing the median (2.4)
L_m	Lower limit of class interval containing the median (2.4)
λ	Mean number of events (arrivals in Appendix 5.1) per unit of time (5.11)
M	Median (2.4)
M	Number of times an experiment occurs (3.2)
M	Monetary gain (15.6)
M_0	Population median if null hypothesis is true (9.10)
M_i	Dummy variable for ith month (13.10)
m	Number of times a particular outcome occurs (3.2)
m	Mean number of customers that can be serviced per unit of time (Appendix 5.1)
m	Number of commodities in price index (14.2)
μ	Population mean (2.4)
μ_o	Population mean if null hypothesis is true (8.4)
μ_1, μ_2	Means of populations 1 and 2 (7.7)
μ_i	Mean of ith population (10.5)
μ_p	Mean of prior distribution of the mean (Appendix 7.1)
$\mu_{y.x}$	Mean of Y, given the value of X (11.4)
$\tilde{\mu}$	Bayesian estimate of the population mean (Appendix 7.1)
N	Number of observations in the population (2.4)
n	Number of observations in the sample (2.4)
n	Number of Bernoulli trials (4.7)
n	Number of blocks (10.8)
n_1,n_2	Sample sizes from populations 1 and 2 (7.7)
n_1,n_2	Numbers of each type in runs test (9.12)
υ	Number of degrees of freedom of the χ^2 distribution (9.3)
υ_1, υ_2	Numbers of degrees of freedom of the F distribution (10.4)
P	Population proportion (7.4)
P	Probability of success on each Bernoulli trial (4.7)
P_0	Population proportion if null hypothesis is true (8.5)
P_1, P_2	Population proportions in populations 1 and 2 (7.7)
$P(A)$	Probability of event A (3.4)
$P(A \text{ or } B)$	Probability of event A, or event B, or both (3.4)
$P(A \text{ and } B)$	Probability of both event A and event B (3.4)
$P(A\|B)$	Probability of event A, given that event B occurs (3.5)
$P(x)$	Probability that random variable X equals x (4.3)
$P(\text{not } A)$	Probability that event A does not occur (3.4)
P_w	Probability that a newly arrived customer will have to wait (Appendix 5.1)
P_i^o	Specified value of ith probability (9.2)

(Glossary continues inside the back cover)

*Number in parenthesis indicates the section where the symbol is introduced.

STATISTICS FOR BUSINESS AND ECONOMICS

Methods and Applications

To accompany the text:

Statistics for Business and Economics: Problems, Exercises, and Case Studies

Statistics for Business and Economics: Readings and Cases

STATISTICS FOR BUSINESS AND ECONOMICS

Methods and Applications

EDWIN MANSFIELD
UNIVERSITY OF PENNSYLVANIA

W • W • Norton & Company
NEW YORK LONDON

W. W. Norton & Company, Inc., 500 Fifth Avenue, New York, N.Y. 10110
W. W. Norton & Company Ltd., 25 New Street Square, London EC4A 3NT

Published simultaneously in Canada by George J. McLeod Limited, Toronto.
Printed in the United States of America.

Library of Congress Cataloging in Publication Data
Mansfield, Edwin.
Statistics for business and economics.
Includes index.
1. Statistics. I. Title.
HA29.M2463 1980 519.5 79-28719

3 4 5 6 7 8 9 0

ISBN 0-393-95057-3

To Edward D. Gruen and the late Frank A. Hanna, who encouraged and helped me to enter this field.

Contents

Chapter 8 Hypothesis Testing 245

Chapter 9 χ^2 (Chi-Square) Tests and Nonparametric Techniques 291

Chapter 10 Experimental Design and the Analysis of Variance 324

Preface

This book is written for first courses in statistics for undergraduate and graduate students in business administration, public administration, and economics. During the past twenty years, I have taught such courses at the Graduate School of Industrial Administration at Carnegie-Mellon University, at Harvard University, and at the University of Pennsylvania. Some of these courses have been designed primarily for business students, while others have been primarily for economics majors. Despite the fact that many good books exist in this field, I felt the need in these courses for a text that contained a fuller account of how statistical methods are used in the real world—a text that was broader in scope and more flexible than those currently available. Gradually, building on materials used in the classroom, the present volume evolved.

There are several features of this book that, in my opinion, differentiate it from others. First, there is a continual emphasis on the practical applications of statistics. In each chapter, there are case studies indicating how statistical techniques have been used by business firms and government agencies in solving actual problems. At the end of each chapter there is a case based on the actual experience of a particular firm or government agency. These cases, ranging from NASA's problems with the reliability of the Apollo space mission to Dow Chemical's sampling for clerical errors, are presented for solution by the student. In my judgment, both the case studies and the end-of-chapter cases add significantly to the effectiveness of the text because students are motivated to learn material much more thoroughly when they are convinced that it has important practical uses.

Second, this book provides a broader menu of topics than is included in most other texts at this level. I have included optional sections and appendixes dealing with basic aspects of econometrics, queuing theory, simulation, serial correlation, the analysis of residuals, and other topics frequently neglected in books of this sort. These are areas in which statistics has found important applications which many instructors would like to touch on. Also, more attention is paid to the role of the computer (particularly in regression analysis) here than in most other books at this level. An attempt is made as well to include a fuller treatment of modern decision theory than in most other texts.

Third, if used in a one-semester (or one-quarter) statistics course this book provides more flexibility than most other texts. It is designed so that several quite different types of one-semester courses can be taught from it: (1) for instructors who want to provide a relatively broad survey of the basic techniques of statistical inference, the outline in Table 1 may be useful; (2) for those who want to emphasize decision theory, the outline in Table 2 may be preferred; (3) for a course emphasizing regression and economic statistics, the outline in Table 3 may be more appropriate.[1] Moreover, the outlines in Tables 1, 2, and 3 are by no means the only choices. I have placed many topics in optional sections and appendixes, thus enabling the instructor to put together a course that fits his or her special needs.[2]

Table 1 Suggested Outline for a One-Semester Course Providing a Relatively Broad Survey of the Basic Techniques of Statistical Inference

Chapter 1
Chapter 2 (including Appendix 2.1)
Chapter 3
Chapter 4 (including Appendix 4.1)
Chapter 5
Chapter 6
Chapter 7
Chapter 8
Chapter 9 (probably omitting some sections)
Chapter 10 (probably omitting some sections)
Chapter 11
Chapter 15 (optional)

Table 2 Suggested Outline for a One-Semester Course Emphasizing Decision Theory and Management Science

Chapter 1
Chapter 2 (including Appendix 2.1)
Chapter 3
Chapter 4 (including Appendix 4.1)
Chapter 5 (including Appendix 5.1)
Chapter 6 (including Appendix 6.1)
Chapter 7 (including Appendix 7.1)
Chapter 8
Chapter 11
Chapter 15
Chapter 16

1. As indicated in Table 3, instructors who wish to skip the Poisson distribution can easily do so.
2. In addition, some instructors prefer to take up basic aspects of simple regression and correlation relatively early in the course. This can easily be done (see footnote 8 of Chapter 11). Also, some instructors prefer to take up decision theory early in the course. This, too, can easily be done.

Table 3 Suggested Outline for a One-Semester Course Emphasizing Regression and Economic Statistics

Chapter 1
Chapter 2 (including Appendix 2.1)
Chapter 3
Chapter 4 (including Appendix 4.1)
Chapter 5 (omitting Sections 5.10–5.12)
Chapter 6
Chapter 7
Chapter 8
Chapter 11
Chapter 12
Chapter 13 (optional)
Chapter 14
Chapter 15 (optional)

Fourth, I have tried to assemble an unusually abundant and varied selection of problems, problem sets, and supplementary materials for use by both the student and the instructor. In addition to the case studies at the ends of the chapters, each chapter contains approximately 30 problems. Besides the latter, *Statistics for Business and Economics: Problems, Exercises, and Case Studies*, which accompanies this text, includes numerous additional problems and questions as well as a detailed and realistic case study pertaining to each chapter. (There is also a *Solutions Manual* for instructors, which includes solutions to all problems and cases in the text.) Finally, *Statistics for Business and Economics: Readings and Cases* has been designed to supplement the basic text. These readings illustrate and help to illuminate the various topics in the text.

No mathematical background beyond high school algebra is required for an understanding of *Statistics for Business and Economics*. Mathematical derivations are generally not included in the text. The emphasis here is on providing students with solid and effective evidence concerning the power and applicability of modern statistical methods, on making sure that they can use these techniques, and on indicating the assumptions underlying these techniques, as well as their limitations. To accomplish these purposes, it is neither necessary nor appropriate to deluge students with mathematics.

In writing this book I have benefited from the comments and suggestions of many colleagues and students. Particular thanks go to the following teachers who have commented in detail on all or most of the manuscript: Gordon Antelmen, University of Chicago; Wallace Blischke, University of Southern California; Warren Boe, University of Iowa; Judd Hammock, California State University at Los Angeles; Arthur Hoerl, University of Delaware; D. L. Marx, Louisiana State University; Robert B. Miller, University of Wisconsin; Don R. Robinson, Illinois State University; Stanley Steinkamp, University of Illinois; Robert J. Thornton, Lehigh University;

Bruce Vavrichek, University of Maryland; Albert Wolinsky, California State University at Los Angeles; Gordon P. Wright, Tuck School, Dartmouth College; and Arnold Barnett, Massachusetts Institute of Technology. I also thank Donald S. Lamm, who did his usual good job with the editorial and publishing aspects of the work.

I am indebted to the Biometrika Trustees for allowing me to reproduce material from the *Biometrika Tables for Statisticians*; to Professors J. Durbin and G. S. Watson for permission to reproduce their tables; and to the Rand Corporation for permission to reproduce material from its tables of random numbers. Further, I am grateful to the literary executor of the late Sir Ronald A. Fisher, F.R.S.; to Dr. Frank Yates, F.R.S.; and to Longman Group Ltd., London, for permission to reprint part of Table 3 from their book *Statistical Tables for Biological, Agricultural, and Medical Research* (6th edition, 1974). Finally, special thanks go to my wife, Lucile, and to my two children who helped in countless ways with the preparation of this book.

Philadelphia, 1980 **E.M.**

STATISTICS FOR BUSINESS AND ECONOMICS

Methods and Applications

1

Introduction to Statistics

1.1 The Field of Statistics

Statistics is a field that contains powerful techniques for solving important business and economic problems. It is no more possible for a business executive or an economist to function without a knowledge of statistics than for a physicist to function without a knowledge of mathematics. To cite just one example of the uses of statistics, consider the decennial census which gathers basic data concerning the number, age distribution, and sex structure of the American people as well as changes in the occupational, educational, and industrial composition of the labor force. Knowledge derived from this census serves a multitude of purposes in both industry and government.

Basically, the field of statistics has two aspects: descriptive statistics and analytical statistics. *Descriptive statistics* is concerned with summarizing and describing a given set of data. For example, after a decennial census is conducted by the Bureau of the Census, government statisticians must summarize and describe the results. Since the amount of raw data gathered from the American people by the Census Bureau is immense, it is necessary to condense and interpret this information to make it meaningful and useful. Because this process may distort the results if not carefully handled, an important task of descriptive statistics is to limit such distortions.

Analytical statistics consists of a host of techniques that help decision makers to arrive at rational decisions under uncertainty. Generally, the data available to a business executive, economist, government official, or other such decision maker are incomplete, in the sense that they pertain to only part of the population in which the decision maker is interested. For example, firms continually have to test whether the materials they receive from suppliers conform to specifications of quality and performance. Since it is frequently too expensive or even impossible (in cases where the only effective tests destroy the materials) to test all incoming materials, firms must base their decisions on testing only a sample. A central function of analytical statistics is to specify ways in which decisions based on incomplete information of this sort can be made as effectively as possible.

It is important to note at the outset that statistics is concerned with *the ways in which* data should be gathered (and *whether* data should be gathered at all), as well as with *how* a particular set of data should be analyzed once it has been collected. *What one can legitimately conclude from a particular set of data depends on how the data have been collected.* If a market-research firm were to collect data concerning the tippling habits of the American people by sampling the inhabitants of nursing homes run by the Temperance Union, you would pay no attention to its results—and rightly so. But as we shall see, far less apparent—and thus more dangerous—mistakes can be made in data gathering, with unfortunate and costly results both to those who collect and analyze the data and those who use them.

1.2 The Design of Sample Surveys

The previous section mentioned two terms, *population* and *sample,* which must now be defined. *The* ***population*** *consists of the total collection of observations or measurements that are of interest to the statistician or decision maker.* It is difficult to give a more specific definition at this point, but in subsequent sections we shall provide many concrete examples of populations. *A* ***sample*** *is a subset of measurements taken from the population in which the statistician is interested.* As pointed out in the previous section, firms, government agencies, and other organizations continually draw samples to obtain needed information because it would be too expensive and time-consuming to try to obtain complete data concerning all relevant units. Thus, the federal government's unemployment statistics are based on a sample of families rather than on a complete count of all workers without jobs and looking for work. Television ratings, which have so strong an influence over which programs survive and which are canceled, are based on the TV-viewing decisions of a sample of families. It is easy to find examples of the use of sampling in all areas of industrial and government activity.

To illustrate in greater detail the use of sampling techniques by business firms, consider the case of the Chesapeake and Ohio Railroad, which has used sampling for many years to determine the amount of revenue it is owed on interline shipments (that is, shipments that go in part over other railroads).[1] To determine how much it is owed for interline shipments, the Chesapeake and Ohio could inspect the waybill of every shipment going over it and another railroad. (A waybill is the document describing the goods being shipped, the route, and the total charges for that particular shipment of freight.) Instead, the Chesapeake and Ohio has concluded that it is more practical to sample the waybills and rely on the results of the sample. Since fewer waybills must be inspected, the reduction in accuracy due to sampling is more than outweighed by the savings in clerical expense.

1. See J. Neter, "How Accountants Save Money by Sampling," in J. Tanur, *Statistics: A Guide to the Unknown* (San Francisco: Holden-Day, 1972). More accurately, the railroad is interested in interline, less-than-carload, freight shipments. To avoid awkward language, we refer to these simply as interline shipments.

Specifically, the railroad has used the following procedure. First, it has divided all waybills issued during a six-month period into five groups: (1) those with total freight charges of $5 or less; (2) those with charges of $5.01 to $10; (3) those with charges of $10.01 to $20; (4) those with charges of $20.01 to $40; and (5) those with charges of over $40. Then, the railroad has chosen a sample of the waybills in each group and has determined for each waybill in the sample the amount it is owed on interline shipments. From each sample the Chesapeake and Ohio has estimated the total amount it is owed on interline shipments in this group. Finally, the railroad has added the estimated totals for the five groups to estimate the total amount it is owed on *all* interline shipments.

How accurate has this sampling procedure been? Table 1.1 shows the results of an experiment that the Chesapeake and Ohio carried out to help answer this question. For a six-month period, the railroad examined all waybills to determine the actual amount it was owed for such shipments. As Table 1.1 indicates, this actual amount differed by only about 1/10 of 1 percent from the amount estimated from the sample. The sample cost the railroad only about $1,000, whereas the complete examination of all waybills cost $5,000. Thus, the railroad saved $4,000 by sampling while sacrificing only a negligible amount of accuracy. In this case, the savings due to the use of sampling techniques were of moderate, but by no means trivial, size; in other cases, the savings have turned out to be much greater. This does not mean that all sampling procedures are as accurate as for the Chesapeake and Ohio Railroad or that it is always better to sample than to carry out a complete count. What this example does illustrate is that, *if correct statistical principles are known and observed,* it is generally possible to design samples that will be sufficiently accurate for the purposes at hand. In some cases, pinpoint accuracy is required; in other cases, rough estimates will do. As we shall see in subsequent chapters, the nature of any given sampling procedure should depend on the accuracy required as well as on the type of population from which the sample is drawn.

TABLE 1.1

Characteristics and Accuracy of Chesapeake and Ohio's Sample of Waybills

Total number of waybills	22,984
Number of waybills sampled	2,072
Cost of examining all waybills (dollars)	5,000
Cost of sample (dollars)	1,000
Total amount due Chesapeake and Ohio on basis of examination of all waybills (dollars)	64,651
Total amount due Chesapeake and Ohio on basis of sample (dollars)	64,568
Percentage error in sample result	1/10 of 1%

SOURCE: J. Neter, "How Accountants Save Money by Sampling," in J. Tanur, *Statistics: A Guide to the Unknown* (San Francisco: Holden-Day, 1972).

TYPES OF POPULATIONS AND SAMPLES

At this point, let's return to the concept of the population. What was the population in the Chesapeake and Ohio case? It was the entire set of waybills; or more accurately, it was the set of amounts owed the railroad on interline shipments on these waybills. For the sake of simplicity, suppose that there are only 20 waybills and that the amount owed the railroad for interline shipments on each waybill is as shown in Table 1.2; then the population consists of the 20 numbers in Table 1.2.

TABLE 1.2
Hypothetical Population of Amounts (Dollars) Owed the Chesapeake and Ohio Railroad on Interline Shipments, 20 Waybills

2.10	4.05	2.91	2.88	3.31
2.70	3.05	3.06	2.94	3.23
2.62	3.04	3.20	2.73	3.02
3.19	3.20	3.40	2.56	2.99

Using a quite different example, suppose that the U.S. Department of Labor wants to estimate the percentage of unemployed individuals in the labor force 16 years old and over in a particular small town. In this case, what is the population? It is a set of observations indicating whether each person in the labor force 16 years old or over does or does not have a job. To simplify, suppose that there are in this town only 25 individuals in the labor force of age 16 or over, and that the employment status of each one is as shown in Table 1.3. (That is, the first person's employment status is given by the first word in the first column, the second person's employment status is given by the second word in the first column, and so on, until the last person's employment status is given by the last word in the fifth column.) Then the observations in Table 1.3 would constitute the population.

TABLE 1.3
Hypothetical Population of Whether or Not Employed, 25 People

Unemployed	Employed	Employed	Employed	Employed
Employed	Employed	Unemployed	Employed	Employed
Employed	Employed	Employed	Employed	Employed
Employed	Employed	Employed	Employed	Employed
Employed	Employed	Employed	Employed	Employed

When carrying out sample surveys, *a listing of all the elements or units in the population is often called a* **frame**. In the example of the Chesapeake and Ohio Railroad where the waybills were numbered consecutively, the frame is the list of waybill numbers. In the case of the unemployment rate, a listing of all persons in the labor force 16 years old or over would be a frame. *A* **census** *is a survey that attempts to include all the elements or units in the frame.* For example, the Census Bureau's decennial Census of Population attempts to include each person in the United States. If the frame is very large, it is generally impossible to include all the elements or units; nevertheless, surveys that do so with reasonable success are called censuses.

Populations are of various types. For example, some consist of *quantitative* information, whereas others consist of *qualitative* information. In the Chesapeake and Ohio example, the population consists of quantitative measurements; that is, ones that can be expressed numerically. Each observation or measurement in this population is the amount on a particular waybill that is owed the railroad for interline shipments. Each such amount is a number (which can vary from zero to as big as you like). In the unemployment case, the population consists of qualitative (non-numerical) information. Each observation or measurement in this population is a statement of whether or not the person in question has a job. Other qualitative characteristics of a person include his or her sex and education.

Populations can also be *finite* or *infinite*. If a population contains a finite number of members, it is called a *finite population*. Some finite populations, such as the hypothetical ones in Tables 1.2 and 1.3, contain relatively few members; others have a very large number of members. If a population contains an unlimited number of members, it is called an *infinite population*. Typically, infinite populations are conceptual constructs. In flipping a coin again and again, the sequence of heads and tails is an infinite population if the flipping is conceived of as continuing indefinitely. In some cases, as we shall see in subsequent chapters, it is convenient to treat a finite population as if it were infinite if the size of the sample is a small proportion of the size of the population.

As we have already stated, a sample is a subset of measurements taken from the population.[2] For example, suppose that we choose the waybills in the first column of Table 1.2 as our sample; then the sample consists of $2.10, $2.70, $2.62, and $3.19. Or suppose that we choose as a sample the individuals in the third column in Table 1.3; then the sample consists of "Employed," "Unemployed," "Employed," "Employed," and "Employed." A sample can be drawn in a variety of ways. Among those discussed in subsequent chapters are simple random sampling, stratified random sampling, systematic sampling, and cluster sampling. In order to use sample data intelligently it is essential that you have some familiarity with each of these methods.

1.3 The Design of Experiments

Unlike the case of the Chesapeake and Ohio Railroad, many statistical investigations are not aimed simply at estimating an unknown total or an unknown percentage. Instead, their purpose is to estimate the effect of one or more factors on some dependent variable. For example, the Exxon Corporation may be interested in estimating how much effect a 1 percent reduction in price will have on the quantity demanded of its gasoline. Or the Federal Reserve Board may be interested in estimating the effect on the prime interest

2. Even if a subset of measurements is taken from a population in which the statistician or decision maker is *not* interested, it is a sample; but it is a sample from the wrong population.

rate of a 3 percent increase in the money supply over a particular period of time. In many cases, it is impossible for firms or agencies to carry out experiments to help estimate these effects, either because the factors are beyond their control or because it would be too expensive or risky to experiment in this way. In such cases, statisticians have devised techniques for estimating these effects as accurately as possible from historical data.

One important limitation of historical studies of this kind is the difficulty in controlling the effects of factors other than those in which the investigator is interested. For example, the relationship between the price and the quantity demanded of, say, gasoline is likely to depend, among other things, on the level of consumer income and the prices of competing and complementary products (all of which are continually changing). Unless the effects of these other factors can be held constant, there is no meaningful way to estimate the price-demand relationship. As we shall see in later chapters, statisticians have devised techniques to help reduce—and in some cases, completely overcome—this limitation, but it should be noted that special problems are encountered in analyzing nonexperimental data of this sort.

In recent years, there have been increasing uses of direct and conscious experimentation in business and economics, ranging from experiments investigating consumers' responses to advertising to experiments investigating the response of the poor to cash subsidies. Imperial Chemical Industries (ICI), the huge British chemical firm, for instance, carried out the following experiment to estimate the effect of a chlorinating agent on the abrasion resistance of a certain type of rubber.[3] Each of 10 pieces of this rubber was cut in half, and one half of each piece was treated with the chlorinating agent, while the other half was untreated. Then the abrasion resistance of each half-piece was evaluated on a machine, and the difference between the abrasion resistance of the treated half-piece and the untreated half-piece was computed. Table 1.4 shows the 10 differences (one corresponding to each of the pieces in the sample).

These 10 differences can be viewed as a sample of 10 from the infinite population of differences which would result if pieces of rubber were subjected to this test indefinitely. Viewed in this way, this sample enabled ICI to estimate the average effect of the chlorinating agent on the abrasion resistance of the rubber. Also, it enabled ICI to *test certain hypotheses* about this effect. For example, some of ICI's personnel were interested in testing the hypothesis that on the average, this chlorinating agent had no effect on the abrasion resistance of the rubber. As we shall see in subsequent chapters, statistical methods can be used to test this hypothesis, based on the sample in Table 1.4.

Just as it is not easy to design a sample survey effectively, so it is not easy to create the proper design for an experiment. The time to worry about how an experiment should be designed is before, not after, the experiment is carried out. Too often, a firm or government agency finds that it cannot draw useful conclusions from an experiment because the experiment was improperly

3. See O. Davies, *The Design and Analysis of Industrial Experiments* (London: Oliver and Boyd, 1956), p. 13.

TABLE 1.4
Difference in Abrasion Resistance (Treated Part Minus Untreated Part), 10 Pieces of Rubber

Piece	Difference[a]
1	2.6
2	3.1
3	−0.2
4	1.7
5	0.6
6	1.2
7	2.2
8	1.1
9	−0.2
10	0.6

SOURCE: O. Davies, *The Design and Analysis of Industrial Experiments* (London: Oliver and Boyd, 1956), p. 13.
[a]The units in which these differences are expressed are given in the source and need not concern us here.

designed. *Before carrying out an experiment (or a sample survey) it is essential that the objectives of the experiment (or sample survey) be defined precisely.* Without such a statement of objectives, it is impossible to formulate a design that will obtain the desired information at reasonable cost. With such a statement of objectives, the statistician can provide useful and time-tested guidance for conducting the experiment and obtaining the desired information at minimum or close-to-minimum cost.

1.4 The Role of Probability in Statistics

As we have seen, a central feature of analytical statistics is its use of information concerning a sample to make inferences about the nature of the population from which the sample is drawn. For example, the Chesapeake and Ohio used sample data to make an inference concerning the total amount on all waybills it was owed for interline shipments. And Imperial Chemical Industries used sample data to make an inference concerning the true effect of a chlorinating agent on the abrasion resistance of rubber. Any conclusion based on a sample must be subject to a certain amount of uncertainty. Thus, even if ICI finds that the chlorinating agent increases abrasion resistance for 150 pieces of rubber, there is always a chance that these results are somehow a fluke and that, if the sample size were increased to 1,000 or to 10,000, the results would be reversed.

How much certainty can you attach to a particular statement about a population that is based on the results of a sample? Intuitively, you are likely to feel that the answer depends on the size of the sample. The bigger the sample, the more confidence you are likely to have in the sample results; the smaller the sample, the less confidence you are likely to have in the sample results. (Subsequent chapters will show that the sample size is indeed one determinant of how much confidence you can put in a sample result, but it is by no means the only determinant.) Why do you feel that increases in sample size increase the amount of confidence you should have in the sample results? Because, as the sample size increases, the ***probability*** that the sample result departs greatly from the corresponding result for the entire population becomes smaller and smaller. Thus, as the Chesapeake and Ohio takes a bigger and bigger sample of its waybills, it seems logical that we can place more and more confidence in the sample estimate of the total amount it is owed on interline shipments.

But what exactly do we mean by a probability? And how can we measure a particular probability? These questions are important to the statistician and to the user of statistics alike. To go beyond vague, intuitive notions about the degree of accuracy of a particular sample result we must draw on probability theory, a branch of mathematics distinct from, but closely related to, statistics. (To delve deeply into probability theory, one needs a considerable mathematical background. However, no mathematics beyond elementary high-school algebra is needed to understand the elements of probability theory required for an introductory course in business and economics statistics.) Until we present a more adequate definition of a probability in Chapter 3, we shall treat probability intuitively. That is, if a particular event has a 50-50 chance of occurring, we shall say that the probability of its occurring is 0.5. Similarly, if a particular event has one chance in four of occurring, we shall say that the probability of its occurring is 0.25. Or if a particular event has one chance in five of occurring, we shall say that the probability of its occurring is 0.2.

1.5 Decision Making under Uncertainty

As pointed out previously, statistics is meant to promote more rational decision making under uncertainty. To see how important and far-reaching this objective is, recall that practically all decisions are made under conditions of uncertainty, because it is seldom possible for the decision maker to forecast accurately the consequence of each alternative course of action. If you must decide whether to accept a job with a New York bank or a Los Angeles oil firm, there is no way for you to forecast accurately the ultimate consequences of accepting either offer. This is a decision you must make under uncertainty; and as with all decisions that must be made under uncertainty, you will be forced to gamble. Under conditions of uncertainty, even if an individual makes the best possible decisions, some will turn out to be wrong when judged with the advantage of hindsight. (The lively traffic through the divorce courts bears witness to this fact.)

A decision made by Maxwell House, the nation's largest producer of coffee, illustrates the sorts of decisions that modern statistical techniques can help to analyze. Maxwell House, together with the American Can Company, developed a new kind of keyless coffee container based on the tear-strip

TABLE 1.5
Changes in Maxwell House's Profit Corresponding to Selected Changes in Its Market Share

Price held constant		Price increased by 2 cents	
Change in market share (percentage points)	Change in profit (thousands of dollars)	Change in market share (percentage points)	Change in profit (thousands of dollars)
+2.8	+4,104	+2.5	+11,939
+1.0	-591	+1.0	+6,489
0	-840	0	+2,856
-0.6	-1,218	-1.5	-1,050

SOURCE: J. Newman, *Management Applications of Decision Theory* (New York: Harper and Row, 1971).

opening principle. One important decision that had to be made before introducing this new can was whether or not to raise the price of coffee in it by 2 cents per pound. (The quick-strip can was expected to cost an average of 0.7 cents more than the older container.) According to one analyst who studied this decision,[4] if Maxwell House raised its price by 2 cents per pound, it might have been reasonable to expect that each of the following consequences was equally likely to occur: (1) its market share would decline by about 1.5 percentage points; (2) its market share would remain constant; (3) its market share would increase by 1.0 percentage points; and (4) its market share would increase by 2.5 percentage points. The change in Maxwell House's profits corresponding to each of these changes in its market share is given in Table 1.5.

According to the same analyst, if Maxwell House did not raise its price, it might have been reasonable to expect that (1) the probability was 0.1 that its market share would decline by 0.6 percentage points; (2) the probability was 0.2 that its market share would remain constant; (3) the probability was 0.5 that its market share would increase by 1.0 percentage points; and (4) the probability was 0.2 that its market share would increase by 2.8 percentage points. The change in Maxwell House's profits corresponding to each of these market-share changes is provided in Table 1.5.

Based on this welter of facts and estimates, what should Maxwell House have done? Should it have increased the coffee price or not? Given the numbers in the two previous paragraphs and in Table 1.5, you probably feel uncomfortable for at least two reasons. First, you probably wonder how such probabilities can be derived. Second, you probably wonder how data of this sort can be used to help guide decision making. Both questions will be answered in due time. (Indeed, this case will be taken up in detail in a later

4. See J. Newman, *Management Applications of Decision Theory* (New York: Harper and Row, 1971)

chapter.) For now, we simply want to introduce you to this class of problem because it is an important type that statistical methods can help to solve.

1.6 Bias and Error

Because the field of statistics attempts to make inferences from a sample concerning a population, statistics must necessarily be concerned with ***error***. Any result based on a sample is likely to depart in some measure from the corresponding result for the population. For example, in Table 1.1, we saw that the Chesapeake and Ohio's sample result was in error by $83. (The total amount due the railroad on the basis of an examination of all waybills was $64,651, while the total amount due the railroad on the basis of the sample was $64,568.) Were the Chesapeake and Ohio to take another such sample it, too, would probably be in error, perhaps by more than $83, perhaps by less. Similarly, the results of ICI's sample (shown in Table 1.4) are almost certainly in error to some extent. That is, if ICI were to test 10,000 pieces of rubber rather than 10 pieces, the results would probably differ from those in Table 1.4. To repeat, all results based on a sample are likely to be in error, since the sample does not contain information concerning all items in the population.

The error in a particular sample result consists of two parts: ***experimental or sampling error*** (sometimes called ***random error***) and ***bias***. Experimental or sampling errors occur because of a large number of uncontrolled factors, which we can subsume under the term *chance*. For example, if the Chesapeake and Ohio draws 10 samples of waybills (each sample containing 2,074 waybills) at random[5] from the population of 22,984 waybills in Table 1.1, the results will differ from one sample to another. Why? Because of the luck of the draw. Or in the case of ICI's experiment, the average difference in abrasion resistance between treated and untreated half-pieces will differ from one sample of 10 pieces of rubber to another sample of 10 pieces of rubber. Why? Because the instruments that measure abrasion resistance contain small errors, because human beings make mistakes in reading these instruments, because the pieces of rubber sometimes are not homogeneous, and so on. Errors of this kind are experimental or sampling errors.

The essential characteristic of experimental or sampling errors is that they can reasonably be expected to cancel each other out over a period of time or over a large number of experiments or samples. Many uncontrolled factors operate to cause the results of a sample to be in error, sometimes on the high side and sometimes on the low side. However, if the sample or the experiment is repeated a large number of times the errors tend to offset each other when the results are averaged. In contrast, *bias consists of a systematic and persistent type of error which will not tend to cancel out.* For example, the Chesapeake and Ohio, instead of choosing waybills at random, might select its samples in such

5. The concept of drawing a sample at random is more sophisticated and technical than can be appreciated or explained adequately at this point. For present purposes, it is sufficient to regard *at random* as meaning that each unit in the population has the same chance of being drawn in the sample. Much more will be said about this in later chapters.

a way that waybills containing relatively large amounts owed the railroad on interline shipments have a higher chance of being picked than waybills containing relatively small amounts owed the railroad on interline shipments. If this were the case, there would be a systematic and persistent tendency for the sample estimates to be too high, and this error would not tend to cancel out if the sample were repeated many times and the results were averaged.

To summarize, an important reason for distinguishing between experimental or sampling error and bias is that *increases in a sample's size tend to reduce experimental or sampling errors, while such increases do not reduce bias.* Thus, although (as pointed out previously) the accuracy of the results of a sample tends to increase with the sample's size, not all errors can be eliminated by increasing sample size. If there are serious biases, the results may be considerably in error even if the sample is huge.

TWO MAJOR CAUSES OF BIAS

Bias can be particularly dangerous when, like termites or dry rot, its presence is undetected. Two types of methodological errors are commonly causes of serious bias. The first arises when *a sample is taken from a population that differs in an important way from the population that the statistician or decision maker is really interested in.* A classic case of this sort occurred in the 1930s when the *Literary Digest* sampled voters to predict whether the Democratic candidate, Franklin D. Roosevelt, or Alfred M. Landon, his Republican opponent, would win the 1936 presidential election. The *Literary Digest* sent questionnaires to about 10 million people. Although over 2 million questionnaires were filled out and returned to the *Digest*, the results, which predicted that Landon would win, were disastrously inaccurate. (Rather than winning, Landon turned in a losing performance not to be equaled for decades.) The basic reason for this statistical debacle was that the *Literary Digest* sampled the wrong population.

From which population did the *Digest* draw its sample? Not from all voters, but only from those who owned a telephone or a car. Since poorer voters often did not have telephones or cars in 1936, and since the poor tended to be much more heavily for Roosevelt than for Landon, this meant that the population sampled by the *Digest* was much more heavily Republican than the total population of voters. Further, more affluent and better-educated people are more likely to fill out and return questionnaires like the *Digest*'s, while poorer and less educated people are less likely to respond. This fact, too, meant that the *Digest*'s sample was more likely to contain Landon supporters than Roosevelt supporters. The result? Although the *Literary Digest*'s sample survey was less lethal, it was to statistics what the *Titanic* (the famous British ocean liner that sank in the North Atlantic in 1912, with over 1,500 lives lost) was to ocean navigation. (As for the *Literary Digest* itself, it subsequently folded.) The moral? Be sure that the population from which a sample is drawn is the population in which you are (or should be) interested.[6]

6. Of course, one should not be overly dogmatic about this. Sometimes it is not possible to obtain an adequate frame for the population one is interested in, and the only possibility is to do the best

A second methodological mistake which can be the source of serious bias arises when *the effect of the variable one wants to measure is mixed up inextricably with the effect of some other factor*. In this case, if one finds that the variable in question seems to have a certain effect, this estimated effect may be biased because it also reflects the effect of another factor. Table 1.6 presents part of the results of a nationwide test of the effectiveness of the Salk antipolio vaccine.[7] These data seem perfectly adequate to measure the effects of the vaccine on the incidence of polio in children, but in fact they contain a major bias because only those second-graders who received their parents' permission were given the vaccine, whereas all first- and third-graders were used to indicate what the incidence of polio would be without the vaccine. The bias arises because the incidence of polio was substantially lower among nonvaccinated children who did not receive permission than among those who did. Why? Because children of higher-income parents were more likely than children of lower-income parents to receive permission—and they were also more likely to contract polio. (Surprising as it may seem, children who grow up in less hygienic conditions are less likely to contract this disease!)

TABLE 1.6
Incidence of Polio in Two Groups of Children

School grade	Treatment	Number of children	Number afflicted with polio	Number afflicted per 100,000 children
Second	Salk vaccine	222,000	38	17
First and third	No vaccine	725,000	330	46

SOURCE: W. S. Youden, "Chance, Uncertainty, and Truth in Science," *Journal of Quality Technology*, 1972.

Table 1.7 shows the results of a more adequate experimental design in which this bias is eliminated. In this sample, children who received permission are differentiated from those who did not. Those with permission were assigned at random to either the group receiving the vaccine or the group receiving the placebo (something similar in appearance to the vaccine, but of no medical significance). As you can see, the results indicate that the reduction in the number afflicted with polio (per 100,000 children) due to the vaccine is much larger than in Table 1.6. Biases of the sort contained in Table 1.6 and in the *Literary Digest* survey can result in distorted information which can lead business executives and government officials to make costly and embarrassing mistakes. It is worth taking some trouble to avoid the methodological errors that result in these biases—unless, of course, you don't mind betting on a Landon for president.

---

one can with a frame which is only approximately what one would like to use. If this frame is close enough to what is really needed, little harm will result. Sometimes, however, it is not close enough, even though the analyst believes it to be.

7. See W. S. Youden, "Chance, Uncertainty, and Truth in Science," *Journal of Quality Technology*, 1972.

TABLE 1.7
Modified Experiment with Salk Vaccine

Permission given	Treatment	Number of children	Number afflicted with polio	Number afflicted per 100,000 children
Yes	Salk vaccine	201,000	33	16
Yes	Placebo	201,000	115	57
No	None	339,000	121	36

SOURCE: See Table 1.6.

EXERCISES

1.1 The mayor of a seaside vacation town must decide whether or not to support an effort to close down one of the town's elementary schools. To determine the extent to which public opinion favors such an action, he hires several interviewers to ask individuals selected at random on the town's boardwalk from 10 A.M. until noon each day during the second week in August whether they feel that an elementary school with less than 200 students should remain in operation.
(a) What is the population from which this sample is drawn?
(b) Is this population finite or infinite?
(c) Does this population consist of qualitative or quantitative measurements?
(d) What deficiencies can you see in this sample design? What possible biases are likely to exist in the results?
(e) What improvements in the sample design would you suggest?

1.2 Another seaside vacation resort is the scene of considerable controversy over whether or not bars should be allowed to stay open after midnight. The local newspaper, which favors the existing arrangements whereby bars must close at midnight, points out that when a neighboring community allowed bars to stay open after midnight the crime rate increased.
(a) What are the weaknesses in the newspaper's argument?
(b) Do you think that an experiment could be run to resolve this type of controversy? If so, what sort of an experiment should it be?
(c) Do you think that an analysis of data concerning past changes in crime rates in this and other communities might help resolve this controversy? If so, what kind of data should be examined, and how might it be analyzed?

1.3 A major private university mails a questionnaire to a sample of 10 percent of its alumni. This questionnaire asks whether the recipient of the questionnaire is for or against the university's policies with respect to curriculum, faculty hiring, coed dormitories, and a variety of other topics. About 30 percent of the questionnaires are filled out and returned. Do you think that there is likely to be bias as well as sampling error in the results? Explain.

1.4 Based on the information given in Section 1.5, what was the probability that Maxwell House's profit would increase by $11,939,000 if it increased the price of coffee by 2 cents per pound? What was the probability that its profit would decrease by $1,050,000 under these circumstances?

1.5 To determine the relative safety of today's aircraft and those of 40 years ago, a researcher compares the number of aircraft accidents in 1980 with that in 1940. Finding that the number of accidents has increased, the researcher concludes that today's airplanes are not as safe as the earlier ones. Do you agree? Why, or why not?

1.7 The Frequency Distribution

At the beginning of this chapter, we pointed out that the field of statistics can be divided into two parts: descriptive statistics and analytical statistics. One of the central tools of descriptive statistics is the frequency distribution. In this section, we describe the nature of a frequency distribution, how it can be represented graphically, and some guidelines that may prove useful in constructing frequency distributions.

CLASS INTERVALS

Although one seldom is provided with the entire population in which he or she is interested, let's start by assuming that you undertake a study for which you do have the entire population. Specifically, you are interested in determining the profitability of the 23 petroleum firms in Table 1.8 in 1973. The numbers in the table constitute the population. Given these numbers, how can we summarize and describe them? (That is, how can we put them into a more compact and comprehensible form?) The need to condense and summarize the data is not as great in this example as in many other cases, since there are only 23 numbers, in contrast to the hundreds or thousands of numbers in some other populations. However, this example can illustrate the essential principles on which we shall focus.

As a first step toward summarizing these data it is convenient to establish certain ***class intervals***, which we define as *classes or ranges of values that the observations or measurements can assume.* For example, Table 1.9 shows eight class intervals for these profit rates, the first being 6.95–8.95 percent, the second 8.95–10.95 percent, and so on. In other words, all values greater than or equal to 6.95 percent and less than 8.95 percent are included in the first class interval, and all values greater than or equal to 8.95 percent and less than 10.95 percent are included in the second class interval, and so on. Each class interval must have a ***lower limit*** and an ***upper limit***. An observation or measurement falls in a particular class interval if it is greater than or equal to this class interval's lower limit and less than its upper limit. The lower limit of the first class interval is 6.95, the lower limit of the second class interval is 8.95, and so on. The upper limit of the first class interval is 8.95, the upper limit of the second class interval is 10.95, and so on. The ***width*** of a class interval is equal to its upper limit minus its lower limit.

Given these class intervals, the next step is to determine how many observations or measurements in the population fall into each class interval. Using the data in Table 1.8, we find that two firms fall into the first class interval in Table 1.9, five fall into the second class interval, four firms fall into the third class interval, and so on. The resulting data, shown in Table 1.9, are

TABLE 1.8
Population of Profit Rates of 23 Major Petroleum Companies, 1973

Firm	Profit rate[a] (percent)
Exxon	17.8
Texaco	16.2
Mobil	14.9
Gulf	14.4
Standard Oil of California	14.5
Standard Oil of Indiana	12.4
Shell	10.7
Continental	13.4
Atlantic Richfield	8.7
Occidental	9.0
Phillips	11.7
Union	10.5
Sun	11.9
Ashland	15.5
Cities Service	9.6
Getty	9.0
Marathon	16.2
Standard Oil of Ohio	7.9
Pennzoil	13.3
Kerr McGee	11.2
Murphy	22.3
Commonwealth	17.7
American Petrofina	13.4

SOURCE: *Fortune*, 1974.

[a] *Profit rate* is defined here as net income as a percent of stockholders' equity.

TABLE 1.9
Frequency Distribution of Population of 1973 Profit Rates of 23 Major Petroleum Firms

Profit rate	Number of firms
6.95 and under 8.95 percent	2
8.95 and under 10.95 percent	5
10.95 and under 12.95 percent	4
12.95 and under 14.95 percent	6
14.95 and under 16.95 percent	3
16.95 and under 18.95 percent	2
18.95 and under 20.95 percent	0
20.95 and under 22.95 percent	1
Total	23

called a *frequency distribution. A* ***frequency distribution*** *shows the number of observations or measurements that are included in each of the class intervals.* Thus, a frequency distribution, if properly formulated, can summarize a set of numbers very effectively. For example, the frequency distribution in Table 1.9 makes it much easier to see how profitable the 23 petroleum firms were in 1973. Clearly, only a few of them (three) had 1973 profit rates greater than or equal to 16.95 percent, and only a few (two) had 1973 profit rates of less than 8.95 percent. The bulk of the firms had profit rates between 8.95 and 16.95 percent. This frequency distribution shows the salient features of the raw data in Table 1.8.

If the population consists of qualitative measurements, the classes in the frequency distribution are not ranges of numerical values, but possible qualitative observations. For example, if the population consists of observations concerning whether various people in a certain area are employed or unemployed, the frequency distribution shows the number of persons in the relevant area who are employed and the number unemployed. Table 1.10 constitutes such a frequency distribution for the United States in December 1978. Often, frequency distributions of this sort are presented graphically in the form of ***bar charts*** like that in Figure 1.1. Each bar in Figure 1.1 has a length—and thus an area, since the width of each bar is the same—representing the number of people employed or unemployed.

TABLE 1.10

Frequency Distribution of Employment Status of People in the Civilian Labor Force, United States, December 1978

Employment status	Number of persons
Employed	95,855,000
Unemployed	6,012,000
Total	101,867,000

SOURCE: *Economic Report of the President,* January 1979.

GRAPHICAL REPRESENTATIONS OF FREQUENCY DISTRIBUTIONS

When a population consists of quantitative measurements (such as the profit rates of the 23 oil firms), a useful way to represent the population's frequency distribution is to construct a ***histogram***. The histogram has a horizontal axis and a vertical axis, as shown in Figure 1.2. The horizontal axis is scaled to display all values of the measurements in the population (in this case the profit rates of the 23 oil firms). The horizontal axis is divided into segments corresponding to the class intervals, and a vertical bar is erected on each of these segments. In Figure 1.2 the segment from 6.95 percent to 8.95 percent on the horizontal axis is the first class interval, and on this segment a vertical bar is erected. Similarly, a vertical bar is erected on the segment of the horizontal axis corresponding to each of the other class intervals.

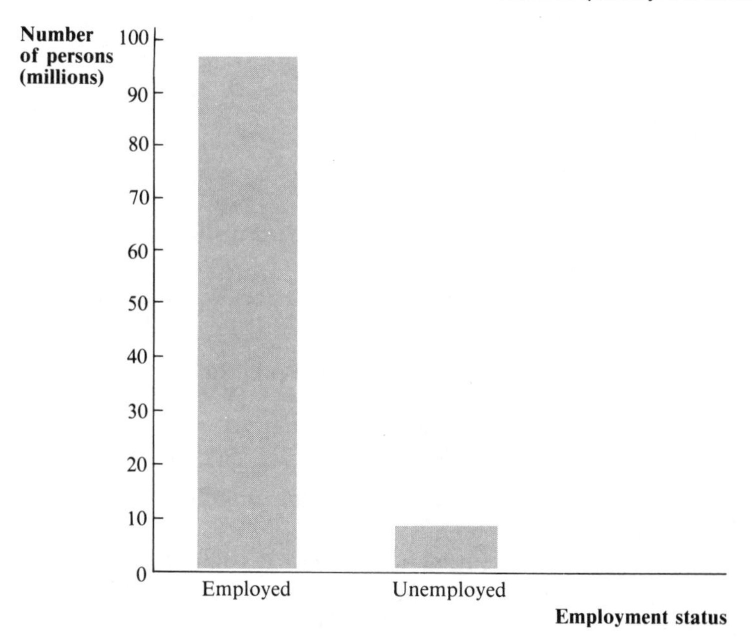

FIGURE 1.1 Bar Chart Showing Frequency Distribution of Employment Status of People in the Civilian Labor Force, December 1978

A histogram's vertical axis shows the number of observations (measurements) in the population which fall into any particular class interval. Thus, the height of the vertical bar for each class interval on the horizontal axis equals the number of observations (measurements) in the population that fall within this class interval. For example, in Figure 1.2 the height of the vertical bar for the first class interval (from 6.95 percent to 8.95 percent on the horizontal axis) is 2 units because the profit rates of 2 out of 23 firms fall into this class interval. Clearly, the histogram in Figure 1.2 is a simple and effective way to present the frequency distribution in Table 1.9 in graphical form. (If

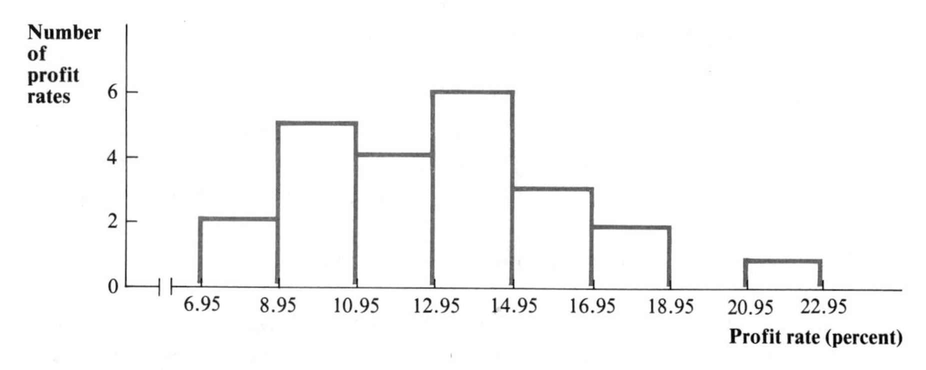

FIGURE 1.2 Histogram of Population of 1973 Profit Rates of 23 Major Petroleum Firms

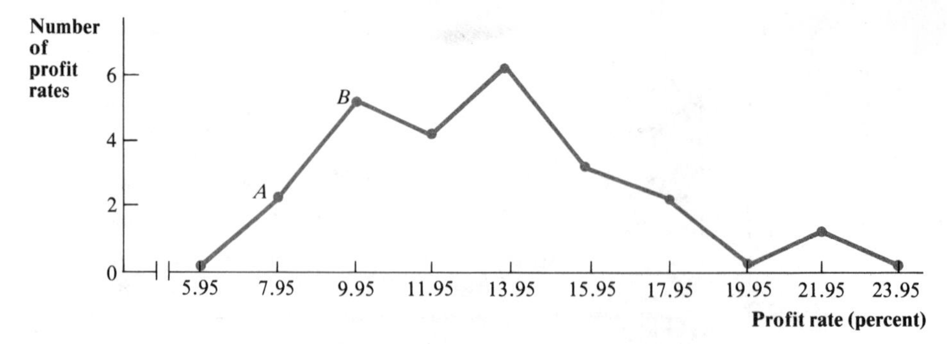

FIGURE 1.3 Frequency Polygon of Population of 1973 Profit Rates of 23 Major Petroleum Firms

the class intervals in a frequency distribution have unequal widths, the above instructions for constructing a histogram should be modified somewhat, as we shall see in Chapter 2.)

An alternative way of representing a frequency distribution graphically is to construct a *frequency polygon.* Like the histogram, the frequency polygon has a horizontal axis showing possible values of the measurements in the population, and a vertical axis showing the number of observations or measurements for each particular class interval. To construct a frequency polygon, we first find the *class mark* for each class interval. (The class mark is the point in a particular class interval midway between the interval's upper and lower limit.) We then plot the number of observations or measurements within a particular class interval against the class mark for that interval. In Figure 1.3, if we plot the number of profit rates of the oil firms in the first class interval against the class mark for this interval (7.95 percent), we get point A. If we plot the number of profit rates of the oil firms in the second class interval against the class mark for this interval (9.95 percent), we get point B. When we connect all resulting points, the resulting geometric figure is a frequency polygon. (Since there are zero firms in the 4.95–6.95 percent class interval and in the 22.95–24.95 percent class interval, the points for these class intervals lie on the horizontal axis.) Like the histogram, the frequency polygon is useful for portraying a frequency distribution graphically.

GUIDELINES FOR CONSTRUCTING FREQUENCY DISTRIBUTIONS

The purpose of constructing a frequency distribution is *to condense the mass of data contained in an array of population values into a table that is more readily appraised and comprehended.* To do this, the observations in the population are grouped into class intervals, as we have seen. There are no simple, cut-and-dried rules to insure that one will construct a proper frequency distribution. Moreover, there is no single frequency distribution which is the best representation of a given set of data; ordinarily, a number of different frequency distributions will do well enough. But in constructing a frequency distribution, one would be wise to consider the following guidelines.

Definition of Class Intervals. In defining class intervals, it is essential that each measurement in the population fall into some class interval. Thus, it would be wrong to define class intervals in Table 1.9 so that profit rates above 20 percent fell into no class interval. Such a definition of class intervals would result in a frequency distribution which distorts, at least to some extent, the nature of the population. Equally important, none of the class intervals should overlap. Thus, in Table 1.9 it would be wrong to define one class interval as 6.95–8.95 percent and another as 7.95–9.95 percent. If class intervals were defined in this way, it would be possible for a given measurement in the population to fall into more than one class interval.

Width of Class Intervals. In general, statisticians prefer to construct class intervals of *equal* widths because this makes it easier to compare the number of observations in different class intervals and to carry out a variety of computations which are described in the next chapter. However, it is sometimes not practical to use class intervals of equal widths. For example, in constructing a frequency distribution of the annual incomes of American families, we might want the width of class intervals to be $2,000 up to an income level of $20,000, and to be $5,000 from an income level of $20,000 to $30,000. Families with incomes of $30,000 or more might be included in a single class interval with no finite upper limit. (Such a class interval is called *open-ended*.) In this case the reason for preferring class intervals of unequal widths is that the bulk of the nation's families have incomes of under $20,000, and it is important in this income range to have the width of class intervals small enough so that the frequency distribution portrays the distribution of income with a reasonable degree of accuracy. However, it would be inappropriate to use the same width of class interval for the very affluent, since so many class intervals would result that the frequency distribution would be very unwieldy. For example, if $2,000 were used as the width of each class interval for incomes ranging from zero to $1 million, there would be 500 intervals!

Number of Class Intervals. Although there are no hard-and-fast rules for how many class intervals to use, it is generally recommended that there be no fewer than 5 and no more than 20. If data are lumped into too few class intervals, there is a major loss of information because much of the variation cannot be determined from the frequency distribution. However, if too many class intervals are used, the frequency distribution does little more than reproduce the array of data. (If there are enough class intervals, each interval may contain only 1 or zero observations.) What is needed is something in-between, and the answer is inevitably a matter of judgment.

Relationship between Number and Width of Class Intervals. When deciding on the number and width of class intervals to use, it is important to realize that interval width is related to the number of class intervals one picks. If all class intervals are to have equal widths, the following formula provides a reasonable estimate of how wide they should be:

$$\text{Width of class interval} = \frac{\text{Largest value} - \text{Smallest value}}{\text{Number of class intervals}},$$

where "Largest value" is the largest value of an observation in the population, and "Smallest value" is the smallest value of an observation in the population. What this formula does is indicate how wide each class interval must be in order to cover the distance between the largest and smallest values, given the chosen number of class intervals. Thus, if we want nine class intervals of equal widths for the data in Table 1.8,

$$\text{Width of class interval} = \frac{22.3 - 7.9}{9} = 1.60,$$

since the largest profit rate is 22.3 percent and the smallest is 7.9 percent. According to this result, the width of the class interval should be about 1.6 percentage points.

Position of Class Intervals. If possible, one should try to set up class intervals so that the midpoint or class mark of each interval is close to the average of the observations included in the class interval. For example, suppose that the data to be summarized are monetary amounts that tend to concentrate on multiples of $1 ($1; $2; $3; and so on). Then it would be better to use class intervals with multiples of $1 as class marks (class intervals of $0.50 and under $1.50; $1.50 and under $2.50; and so on) than ones with class marks that are quite different from the actual concentration of observations in the class intervals. The reason is that in calculating averages and other summary measures from frequency distributions, we assume that the class mark is representative of the values in each class interval. (More will be said about this in the next chapter.)

THE CUMULATIVE FREQUENCY DISTRIBUTION

In some cases, statisticians and decision makers are interested in the number of measurements in the population that lie below or above a certain value. Thus, an analyst in the oil industry might want to know how many firms in Table 1.8 had profit rates below 10.95 percent in 1973. In cases of this sort, it is useful to construct a cumulative frequency distribution, showing *the number of measurements in the population that are less than particular values.* Table 1.11 shows the cumulative frequency distribution for the profit rates of the oil firms in Table 1.8. The difference between the *cumulative* frequency distribution in Table 1.11 and the *ordinary* frequency distribution in Table 1.9 is that the cumulative frequency distribution shows the number of oil firms with profit rates *less than* those given in Table 1.11, whereas the ordinary frequency distribution in Table 1.9 shows the number of oil firms in *each* of the class intervals.

If one has the frequency distribution for a set of data, it is easy to construct a cumulative frequency distribution. To construct a cumulative frequency distribution for the profit rates of oil firms in Table 1.9, we proceed as follows. The number of firms with profit rates of less than 8.95 percent is 2

TABLE 1.11
Cumulative Frequency Distribution of Population of 1973 Profit Rates of 23 Major Petroleum Firms

Profit rate	Number of firms
Less than 6.95 percent	0
Less than 8.95 percent	2
Less than 10.95 percent	7
Less than 12.95 percent	11
Less than 14.95 percent	17
Less than 16.95 percent	20
Less than 18.95 percent	22
Less than 20.95 percent	22
Less than 22.95 percent	23

(the figure in the lowest class interval in Table 1.9); the number of firms with profit rates of less than 10.95 percent is 7 (the sum of the figures in the lowest two class intervals); the number of firms with profit rates of less than 12.95 percent is 11 (the sum of the figures in the lowest three class intervals); and so on. To obtain the number of firms with profit rates of less than the upper limit of a particular class interval, we add up (or *cumulate,* which accounts for the name of the cumulative frequency distribution) the number of firms in this and all lower class intervals.

Just as the ordinary frequency distribution can be portrayed graphically, so can the cumulative frequency distribution. To construct such a graph, one needs only to plot the number of observations in the population with values

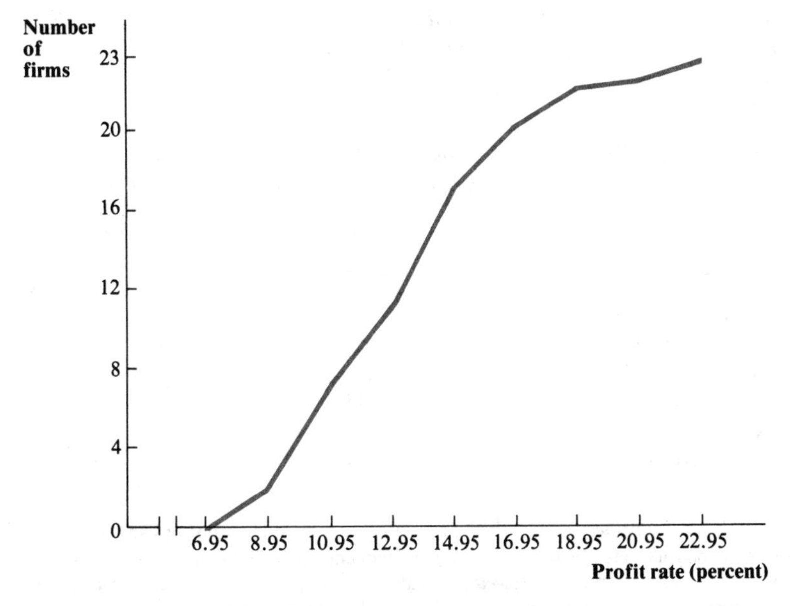

FIGURE 1.4 Ogive Showing Cumulative Frequency Distribution of Population of 1973 Profit Rates of 23 Major Petroleum Firms

less than a certain number against the number itself. For example, since there are two oil firms with profit rates of less than 8.95 percent, we would plot 2 on the vertical axis against 8.95 percent on the horizontal axis. Since there are seven oil firms with profit rates of less than 10.95 percent, we would plot 7 on the vertical axis against 10.95 percent on the horizontal axis. The results of plotting and connecting all such points with straight lines are shown in Figure 1.4. The resulting curve is called an *ogive.* Such curves are encountered frequently in business and economic statistics.[8]

FREQUENCY DISTRIBUTIONS FOR SAMPLE DATA

Finally, in setting up an actual frequency distribution, we usually do not have all the measurements in the population. If, as is usually the case, we have data concerning only a sample from the population, a frequency distribution (constructed according to the principles discussed above) is a useful way to describe and summarize the data. For example, take the case of the profit rates of the 23 oil firms. Regardless of whether the data in Table 1.8 are viewed as an entire population or as a sample from a larger population, it is useful to formulate a frequency distribution to summarize the data, and the same principles apply. Thus, our discussion in previous parts of this section applies to both sample and population data.

However, two important points should be recognized. First, *the frequency distribution of a sample is different from the frequency distribution of the population from which the sample is drawn.* Because the sample contains only a portion of the measurements in the population, the two frequency distributions are different and should not be confused. Second, *although we ordinarily do not have the data to construct a frequency distribution for the entire population, such a frequency distribution exists or can be imagined. It is important that we be able to visualize and think about this frequency distribution, since the questions that statistical investigations attempt to answer often are questions about this frequency distribution.*

EXERCISES

1.6 According to the Federal Trade Commission (FTC), the profit after taxes (as a percent of sales) of each manufacturing industry in the United States in 1974 was as follows:

Industry	Profit (as percent of sales)
Transportation equipment	3.9
Electrical equipment	4.3
Machinery (nonelectrical)	5.6
Fabricated metal products	4.0
Primary iron and steel	4.1
Primary nonferrous metals	5.4

(*table continues*)

8. In this section we have considered "less than" cumulative frequency distributions and ogives. It is also possible to construct "greater than" cumulative frequency distributions and ogives, which show the number of measurements that *exceed* particular values. To obtain the number of firms with profit rates *greater than* the lower limit of a particular class interval, we add up the number of firms in this and all higher class intervals.

Industry	Profit (as percent of sales)
Stone, clay, and glass	4.8
Instruments	8.4
Other durable goods	5.0
Food and kindred products	2.6
Tobacco	5.8
Textiles	2.8
Paper	5.4
Printing	4.8
Chemicals	6.8
Petroleum and coal	7.6
Rubber	4.0
Other nondurable products	2.1

(a) Construct a frequency distribution of the profit rates of all manufacturing industries.
(b) Construct a histogram based on these data.
(c) Construct a frequency polygon based on these data.

1.7 According to the Department of Agriculture, the total net income per farm in each of 10 midwestern states in 1973 was:

State	Income per farm (dollars)
Illinois	13,224
Indiana	11,282
Iowa	19,685
Kansas	17,018
Michigan	6,171
Minnesota	19,456
Nebraska	17,790
North Dakota	35,631
Ohio	5,212
South Dakota	22,928

(a) Are there too few states to construct a frequency distribution of their incomes per farm?
(b) If not, what frequency distribution would you suggest?

1.8 A supermarket inspects the work of one of its clerks at the checkout counter. For each customer, it determines what the customer's bill should be and what this clerk calculates it to be. For 50 customers the frequency distribution of the difference between the latter and the former is as follows:

Error	Number of customers
– $1.00 and under – $0.75	1
– $0.75 and under – $0.50	2
– $0.50 and under – $0.25	4
– $0.25 and under – $0.00	30
$0.00 and under $0.25	6
$0.25 and under $0.50	2
$0.50 and under $0.75	2
$0.75 and under $1.00	2
$1.00 and under $1.25	1

(a) What are the class intervals in this frequency distribution?
(b) Are the class intervals of equal width? (That is, is the difference between the upper and lower limit the same for each class interval?)
(c) If the error in a particular customer's bill equals −$0.25, into which class interval would this item fall?

1.9 Given only the frequency distribution in Table 1.9 (not the original data in Table 1.8), which of the following questions would you be able to answer:
(a) How many of these petroleum firms had profit rates equal to 18.95 percent or more?
(b) How many had profit rates of less than 12.95 percent?
(c) How many had profit rates of less than 9.95 percent?
(d) How many had profit rates of 11.95 percent or more?
(e) How many had profit rates of at least 8.95 percent but less than 20.95 percent?
(f) How many had profit rates of at least 11.95 percent but less than 14.95 percent?

1.10 If the width of each class interval in Table 1.9 were cut in half, how many of the questions in Exercise 1.9 above would you be able to answer? What are some of the advantages of relatively narrow class intervals? What are some of their disadvantages?

1.11 Construct a cumulative frequency distribution for the data in Exercise 1.8.

1.12 Plot the ogive for the cumulative frequency distribution for the data in Exercise 1.8.

CHAPTER REVIEW

1. The field of statistics consists of two parts, descriptive statistics and analytical statistics. *Descriptive statistics* is concerned with summarizing and describing a set of data. *Analytical statistics* consists of techniques which help decision makers come to rational decisions under uncertainty. Statistics is concerned with whether data should be gathered at all, with how data should be gathered, and with how a particular set of data should be analyzed once it has been collected.

2. Firms, government agencies, and other organizations are continually engaged in *sampling* in order to obtain needed information because it would be too expensive and time-consuming to try to obtain complete data concerning all relevant units. A *population* consists of the total collection of observations or measurements that are of interest to the statistician or decision maker in solving a particular problem. A population can consist of quantitative or qualitative information, and it may be finite or infinite. A *sample* is a subset of measurements taken from the population.

3. In recent years, there has been increasing utilization of direct and conscious experimentation in business and economics. Just as statistics is useful in designing a sample survey, it is also useful in designing an experiment. In both a survey and an experiment the objective should be to obtain the desired information at minimum cost. To promote this objective, it is important that the purposes of the experiment or sample survey be defined precisely before data are collected.

4. Modern statistical techniques are useful in promoting more rational decisions under uncertainty. Almost all decisions are made under uncertainty because it is seldom possible for the decision maker to forecast accurately the

consequence of each alternative course of action. Statisticians use the concept of *probability* to measure the amount of confidence that one can have in various sample results.

5. Because statistics attempts to make inferences from a sample concerning a population, it must be concerned with error. After all, any sample result is likely to depart in some measure from the corresponding result for the total population. The error in any particular sample result is composed of two parts: *experimental or sampling error*, and *bias*. Experimental or sampling error is due to a large number of uncontrolled factors which we subsume under the shorthand expression *chance. Bias* consists of a persistent, systematic sort of error. Increases in a sample's size tend to reduce experimental or sampling error, but not bias.

6. To summarize a body of data, it is useful to construct a *frequency distribution,* which is a table showing the number of measurements or observations that fall into each of a number of class intervals. To establish a frequency distribution, one must set up certain well-defined class intervals, each interval being defined by a lower limit and an upper limit. Frequency distributions are often presented in graphical as well as tabular form. A *histogram* is composed of a series of bars or rectangles; the bottom of each bar is the line segment on the horizontal axis corresponding to the interval from the class interval's lower limit to its upper limit. The area of each bar is proportional to the number of cases in the class. A *frequency polygon* is another type of graphical representation of a frequency distribution.

2

Summary and Description of Data

2.1 Introduction

As pointed out in Chapter 1, descriptive statistics is concerned with the summary and description of a body of data. While this concise definition may make descriptive statistics seem cut-and-dried and perhaps a bit dull, this is by no means true. The proper summary of a body of data involves much more than arithmetic. It entails avoiding a variety of pitfalls (many of which are discussed in this chapter) that can lead the unwary analyst or decision maker to false conclusions. Data can also be distorted intentionally by unscrupulous individuals and firms to mislead others. To avoid being misled, a knowledge of descriptive statistics is essential.

2.2 Types of Summary Measures

In the previous chapter we showed how a frequency distribution can be constructed to summarize and describe a set of data. However, in many situations where a frequency distribution would be too detailed and cumbersome, a few summary measures can present concisely the salient features of the data. Although summary measures provide much less information than the frequency distribution, in many situations the lack of a certain amount of information is not crucial and the greater conciseness of the summary measures makes them more useful than the frequency distribution. In general, two types of summary measures are most frequently used: measures of central tendency, and measures of dispersion.

Measures of Central Tendency. Often one wants a single number to represent the "average level" of a set of data. In other words, a number is needed that will indicate where the frequency distribution is centered. This number should tell us what a "typical" value of the measurements might be. To illustrate, let's return to the data concerning the 1973 profit rates of 23 major oil firms, which are shown in Table 2.1. If these data were collected to determine whether the oil companies were making much higher profits than

TABLE 2.1
Population of Profit Rates of 23 Major Petroleum Companies, 1973

Firm	Profit rate[a] (percent)
Exxon	17.8
Texaco	16.2
Mobil	14.9
Gulf	14.4
Standard Oil of California	14.5
Standard Oil of Indiana	12.4
Shell	10.7
Continental	13.4
Atlantic Richfield	8.7
Occidental	9.0
Phillips	11.7
Union	10.5
Sun	11.9
Ashland	15.5
Cities Service	9.6
Getty	9.0
Marathon	16.2
Standard Oil of Ohio	7.9
Pennzoil	13.3
Kerr McGee	11.2
Murphy	22.3
Commonwealth	17.7
American Petrofina	13.4

SOURCE: *Fortune*, 1974.
[a]*Profit rate* is defined here as net income as a percent of stockholders' equity.

firms in other industries, it might be sensible to ask "What is the typical level of the 1973 profit rates of these 23 oil firms?" As we shall see, there are several types of averages or measures of central tendency that can be used to help answer this question; and the choice among these depends on the purposes of the investigator and the nature of the data.

Measures of Dispersion. In addition to knowing the "average level" of a set of data, it is important to know the degree to which the individual measurements vary about this average. In other words, we need to know whether a frequency distribution is tightly packed around its average or whether there is a great deal of scatter about it. In the case of the oil firms in Table 2.1, an important question is "Regardless of what the average level of the 1973 profit rates of these firms may be, to what extent do their profit rates vary?" Statisticians have devised a number of measures of dispersion, which will be described in subsequent sections. As in the case of measures of central tendency, the choice among measures of dispersion depends on the purposes of the investigator and the nature of the data.

2.3 Parameters and Statistics

If we have all the measurements in a given *population*, we can calculate summary measures for that population as a whole. Such summary measures are called **parameters.** For example, if we calculate a particular kind of average of the profit rates in Table 2.1, the resulting average is a parameter since it is calculated from all the measurements in the relevant population. (Recall from Chapter 1 that these 23 observations are regarded as the population.) Or if we calculate a particular kind of measure of dispersion (again using the profit rates in Table 2.1), the result is a parameter since the calculation is based on all the measurements in the relevant population. As has been pointed out, we seldom have all the measurements in an entire population, but this does not mean we are not interested in the parameters of the population. On the contrary, *much of analytical statistics is designed to draw inferences from a sample concerning the value of a population parameter.*

If we have only *sample* data, we can calculate summary measures for the sample; such summary measures are called **statistics.** For example, if the Chesapeake and Ohio Railroad's statisticians calculated, from the sample of waybills in Chapter 1, a particular kind of average of the amounts owed the railroad on interline shipments, this resulting average is a statistic since it is calculated from the *measurements in a sample.* Or if the railroad's statisticians calculate a particular kind of measure of dispersion, using the amounts owed the railroad on interline shipments in the same sample of waybills, the result is also a statistic since the calculation is based on the *measurements in a sample.* As we shall see in subsequent chapters, a statistic from a sample is often used to estimate the analogous parameter of the entire population from which the sample is drawn. Thus, the average amount owed the Chesapeake and Ohio on interline shipments in the sample of waybills was used to estimate the average amount owed the railroad on such shipments in the population of *all* waybills.

2.4 Measures of Central Tendency

THE ARITHMETIC MEAN

There are several important types of measures of central tendency. The one used most frequently is the arithmetic mean. Like Molière's character in *Le Bourgeois Gentilhomme* who was surprised to learn that he had been speaking prose all his life, you may be surprised to learn that you have been using the arithmetic mean for a long time (although you probably have not been calling it by that name). *The* ***arithmetic mean*** *is the sum of the numbers included in the relevant set of data divided by the number of such numbers.* Let N denote how many numbers there are in a population; thus in Table 2.1, $N = 23$. If we order these numbers from 1 to N, X_1 being the first number, X_2 being the second number, and so on up to X_N, which is the Nth number, then the population mean is

$$\mu = \frac{X_1 + X_2 + X_3 + \ldots + X_N}{N}, \tag{2.1}$$

where μ is the Greek letter *mu*. In particular, if we order the numbers in Table 2.1 from the top down, X_1 being 17.8, X_2 being 16.2, and so forth, with X_{23} being 13.4,[1] then in the case of this population

$$\mu = \frac{17.8 + 16.2 + 14.9 + 14.4 + \ldots + 17.7 + 13.4}{23}$$

$$= \frac{302.2}{23}$$

$$= 13.1.$$

Consequently, the arithmetic mean of the population turns out to be 13.1 percent.

If the arithmetic mean is calculated for a sample rather than for the whole population, it is designated as $\bar{X}$ rather than μ. Whereas N stands for the number of measurements in the population, n stands for the number of measurements in the sample. Thus, the sample mean is defined as

$$\bar{X} = \frac{X_1 + X_2 + X_3 + \ldots + X_n}{n}. \tag{2.2}$$

This expression is sometimes written in the following form:

$$\bar{X} = \frac{\sum_{i=1}^{n} X_i}{n},$$

where Σ is the mathematical *summation sign*. What does ΣX_i mean? It means that the numbers to the right of the summation sign (that is, the values of X_i) should be summed from the lower limit on i (which is given below the Σ sign) to the upper limit on i (which is given above the Σ sign). Thus, in this case it means that X_i is to be summed from $i = 1$ to $i = n$. In other words, ΣX_i means the same as $X_1 + X_2 + \ldots + X_n$. (For further discussion of the summation sign and its uses, see Appendix 2.1.)

Whether the set of data is a sample or a whole population, it is sometimes necessary to calculate the arithmetic mean from grouped data —that is, from a frequency distribution. For example, the 1973 profit rates of 15 chemical firms presented in Table 2.2 can be regarded as a sample from the population of 1973 profit rates of all American chemical firms. (Why are they a sample, whereas the profit rates of the 23 oil firms are a population? Because they are a subset of the profit rates of chemical firms in which we are

1. It is important to note that this is only one of many orderings that could be used. For example, the firms (whose profit rates are included in Table 2.1) could be arranged in alphabetical order. This would do just as well.

TABLE 2.2
Frequency Distribution of Sample of 15 Profit Rates of Major Chemical Firms, 1973

Profit rate[a]	Number of firms
9 and under 11 percent	3
11 and under 13 percent	3
13 and under 15 percent	5
15 and under 17 percent	3
17 and under 19 percent	1
Total	15

SOURCE: *Fortune,* 1974.
[a]As in Table 2.1, the *profit rate* of a firm is defined as its net income as a percentage of stockholders' equity.

interested, whereas in the oil industry we are interested only in the profit rates of the 23 firms.) Since the profit rates of individual firms are not given in Table 2.2, we cannot use equation (2.2) to calculate $\bar{X}$. However, we can approximate the sum of the X_i by assuming that *the midpoint (class mark) of each class interval can be used to represent the value of the measurements in that class interval.* Thus, the sample mean can be approximated by

$$\bar{X} = \frac{f_1X'_1 + f_2X'_2 + \ldots + f_kX'_k}{n} = \frac{\sum_{j=1}^{k} f_jX'_j}{n}, \tag{2.3}$$

where f_1 is the number of measurements in the first class interval, X'_1 is the midpoint of the first class interval, f_2 is the number of measurements in the second class interval, X'_2 is the midpoint of the second class interval, and so on.

Applying equation (2.3) to the data in Table 2.2, we find that the sample mean can be approximated by

$$\bar{X} = \frac{3(10) + 3(12) + 5(14) + 3(16) + 1(18)}{15}$$

$$= \frac{202}{15} = 13.47.$$

Thus, the sample mean is about 13.5 percent. If the mean of the measurements in each class interval is close to the midpoint of the class interval, this approximation should entail only a small amount of error.[2] Even if data are available for all the measurements, calculating the mean from a frequency distribution of the data may be easier and less expensive.

2. At this point, you should understand more fully why we recommended in the previous chapter that class intervals be constructed so that the midpoint of each class interval is close to the average of the observations included in it. If this is done, the error in using equation (2.3) will be very small.

THE WEIGHTED ARITHMETIC MEAN

In some cases the measurements in a sample or a population should not be weighted equally, as in equations (2.1) and (2.2). Since some chemical or oil firms are much bigger than others, it might be argued that a firm's profit rate should be weighted according to its size in determining the average level of profit rates. If w_i is the weight attached to the ith measurement in a sample, the weighted arithmetic mean is

$$\bar{X}_w = \frac{\sum_{i=1}^{n} w_i X_i}{\sum_{i=1}^{n} w_i}. \tag{2.4}$$

For example, suppose that we have a sample of three firms' profit rates: 10 percent, 12 percent, and 15 percent. The firm with the 10 percent profit rate has assets of $2 billion whereas the other two firms have assets of $1 billion each. If a firm's assets are used to weight its profit rate, the weighted arithmetic mean of the profit rates of these three firms is

$$\bar{X}_w = \frac{2(10) + 1(12) + 1(15)}{2 + 1 + 1} = \frac{47}{4} = 11.75.$$

Thus, the weighted mean is 11.75 percent.

Comparing equation (2.3) with equation (2.4) shows that the arithmetic mean based on grouped data is a type of weighted arithmetic mean: It is a weighted mean of the midpoints of the class intervals, the weight attached to each particular midpoint being the number of measurements falling within that class interval. In business and economics many types of weighted means are encountered and used. The Consumer Price Index, a prominent measure of the rate of inflation, is a weighted arithmetic mean of the relative changes in the prices of various goods and services, as we shall see in detail in Chapter 14.

THE MEDIAN

Other than the mean, the most widely used measure of central tendency is the median, which is defined as the *middle value* of the relevant set of data. In other words, the median is the *value that divides the set of data in half, 50 percent of the measurements being above it and 50 percent being below it.* Let's consider the profit rates of the petroleum firms in Table 2.1 again. If we list these 23 firms in the order of their 1973 profit rates (from lowest to highest) the profit rate of the twelfth firm must be the middle value. In other words, as many firms have profit rates exceeding this value as have profit rates falling below it. According to this method of listing we find that the *median* profit rate of the oil firms is 13.3 percent.

If there is an even number of observations in a frequency distribution, *none* of the observations can be the middle value. For example, among the four numbers 2, 4, 6, and 8 there can be no "middle" number or median because the middle lies *between* two of the numbers—specifically, between 4 and 6. To resolve this difficulty, convention dictates that *if there is an even number of observations, the mean of the middle pair of observations is regarded as the median.* (In this case the median would be the mean of 4 and 6, or 5.)

Just as the mean frequently must be approximated from a frequency distribution, so the median also must often be approximated in this way. The first step in calculating the median from a frequency distribution is to find the class interval that contains the median. To do this, we start with the lowest class interval, cumulate the number of measurements in one, two, three, and subsequent class intervals, stopping with the interval where the cumulated number of measurements first exceeds or equals $n/2$ if the measurements are a sample or $N/2$ if they are the whole population. This particular class interval contains the median.

To calculate the median from the frequency distribution of the 15 chemical firms in Table 2.2, we cumulate the number of firms with profit rates less than the upper limit of each class interval. Thus, as shown in the third column of Table 2.3, the number of firms with profit rates under 11 percent is 3, the number with profit rates under 13 percent is 6, and so on. As specified in the previous paragraph, we must keep on cumulating until we reach the first class interval where the cumulated number of measurements exceeds $n/2$ (7.5 in this case, because $n = 15$). Since the cumulated number is 3 for the first class interval, 6 for the second, and 11 for the third, it is clear that the third class interval (13 and under 15 percent) is the first where the cumulated number exceeds 7.5. Thus, this is the class interval in which the median is located.

To estimate where in this class interval the median is situated, we assume that the interval's measurements are *spaced evenly* along its width of 2 percentage points. If so, the five measurements are 13.2; 13.6; 14.0; 14.4; and 14.8. The median must be the eighth lowest measurement, since there are 15 measurements in the sample. Because there are six measurements in lower

TABLE 2.3

Finding the Class Interval in Which the Median Profit Rate Is Situated, 15 Chemical Firms

Profit rate	Number of firms	Cumulated number of firms[a]
9 and under 11 percent	3	3
11 and under 13 percent	3	6
13 and under 15 percent	5	11
15 and under 17 percent	3	14
17 and under 19 percent	1	15

[a] This is the number of firms with profit rates *less than* the *upper limit* of the relevant class interval.

class intervals, the median must therefore be the second lowest value in this class interval. Thus, the median must be 13.6.

A general expression for finding the median in grouped data of this sort is

$$M = \left(\frac{n/2 - c}{f_m}\right) l + L_m, \tag{2.5}$$

where c = the number of measurements in class intervals below the one containing the median; f_m = the number of measurements in the class interval containing the median; l = the width of the class interval containing the median; and L_m = the lower limit of the class interval containing the median. Since $n = 15$, $c = 6$, $f_m = 5$, $l = 2$, and $L_m = 13$,

$$\begin{aligned} M &= \left(\frac{7.5 - 6}{5}\right) 2 + 13 \\ &= \left(\frac{1.5}{5}\right) 2 + 13 \\ &= 13.6. \end{aligned}$$

Thus, this formula results in precisely the same answer (13.6) as was obtained in the previous paragraph.[3] This will always be true.

USES OF THE MEAN AND THE MEDIAN

Both the mean and the median are important and useful measures of central tendency (that is, of the "average level" of a set of data). In some circumstances the mean is a better measure than the median, and in others the reverse is true. The following factors are among the most important determinants of whether the mean or the median should be used.

Sensitivity to Extreme Observations. The median is often preferred over the mean when the latter can be influenced strongly by extreme observations. For example, how do we go about computing the average income of the families in an apartment building containing 19 families, 8 of which earn \$10,000 per year, 10 of which earn \$12,000 per year, and 1 of which earns \$1 million per year? (The latter presumably has the penthouse.) The mean income of the 19 families equals

$$\frac{8(\$10{,}000) + 10(\$12{,}000) + 1(\$1{,}000{,}000)}{19} = \frac{\$1{,}200{,}000}{19}$$

$$= \$63{,}157.$$

However, this figure is not a very good description of the yearly income level

3. It is clear from Table 2.3 that the number of measurements in class intervals below the one containing the median is 6, since the "13 and under 15 percent" class interval includes the median. The width of this class interval equals 2 percent. The number of observations in this class interval is 5, and the lower limit of this class interval is 13. Thus, $c = 6$, $l = 2$, $f_m = 5$, and $L_m = 13$.

of the majority of the families in the building. A better measure might be the median, which in this case is $12,000 per year. The median is much less affected by the one extreme point (the millionaire), which raised the mean very considerably.

Open-ended Class Intervals. As you will recall from our discussion of class intervals in Chapter 1, it is not unusual for frequency distributions to have open-ended class intervals—that is, class intervals with no finite upper or lower limits. For example, in a frequency distribution of the annual income of American families, two class intervals might be "less than $1,000" and "$30,000 and more." Each of these class intervals is open-ended.[4] If one needs to calculate an average from a frequency distribution with one or more open-ended class intervals, there may be no alternative but to use the median, since calculation of the mean requires a knowledge of the sum of the measurements in the open-ended classes. Unless knowledge of this sort is available (and it frequently is not), the median is often preferable.

Mathematical Convenience. The mean rather than the median is often the preferred measure of central tendency because it possesses convenient mathematical properties that the median lacks. For example, the mean of two combined populations or samples is a weighted mean of the means of the individual populations or samples. On the other hand, given the medians of two populations or samples, there is no way to determine what the median of the two populations combined or two samples combined would be.

Extent of Sampling Variation. As pointed out earlier, sample statistics such as the sample mean or the sample median are often used to estimate the population mean. A major reason for preferring the mean to the median is that the sample mean tends to be more reliable than the sample median in estimating the population mean. In other words, the sample mean is less likely than the sample median to depart considerably from the population mean. This is a very important consideration which will be more fully appreciated in Chapter 7, where we shall cover this topic in greater detail.

THE MODE

Another often-used measure of central tendency is the *mode*, which is defined as the *most frequently observed value of the measurements in the relevant set of data.* For example, if a breakfast food manufacturer were to ask people to indicate which of several colors they preferred on a particular cereal carton, and if 100 people preferred red, 50 preferred green, and 20 preferred yellow, then the mode would be at the color red. Or if there were 19 families in your apartment building, 8 of which earned $10,000 per year, 10 of which earned $12,000 per year, and 1 of which earned $1 million per year, the mode would be at $12,000. Why? Because $12,000 is the most frequently observed value of family income in the building.

4. At first glance, you may think that the "less than $1,000" class interval has a finite lower limit—namely, zero. This is incorrect because some people have negative incomes, and there is no limit on how large losses of this sort can be.

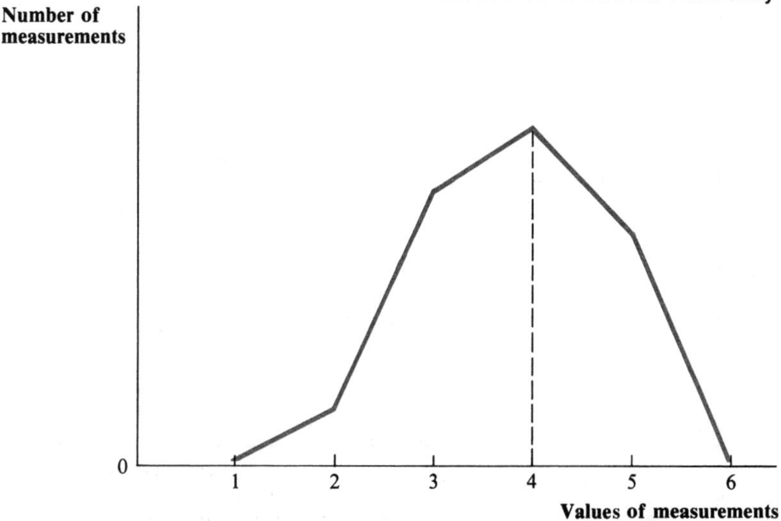

FIGURE 2.1 Mode of a Frequency Polygon

When data are presented in the form of a frequency distribution, the mode can be estimated as *the midpoint or class mark of the class interval containing the largest number of measurements*. The mode of the profit rates of the chemical firms in Table 2.2 is 14 percent. Why? Because the largest number of measurements is contained in the class interval "13 and under 15 percent," and the midpoint of this class interval is 14 percent. The class interval containing the largest number of measurements is called the ***modal class***. Thus, the modal class in Table 2.2 is "13 and and under 15 percent."

Based on a graphical representation of a frequency distribution such as a frequency polygon, it is easy to find the mode of a body of data. We need only find the value along the horizontal axis where the frequency polygon achieves its *maximum* vertical height. For example, the mode of the frequency distribution portrayed in Figure 2.1 equals 4. Some frequency distributions (like the one shown in Figure 2.2) have more than one mode. If a frequency distribution has more than one mode, it is called ***multimodal***; if it has two modes, it is called ***bimodal***. Frequency distributions with more than one mode often arise because two or more quite different types of measurements or observations are included. If we were to form a frequency distribution of the heights of American adults, we might find two modes, one at the modal height for men and one at the modal height for women. Great care must be exercised in constructing and interpreting measures of central tendency for multimodal distributions since *measures like the mean or the median may fall between the modes and be unrepresentative of the bulk of the measurements lying near the separate modes*. In cases where a multimodal frequency distribution arises because two or more quite different types of measurements or observations are included, it often is wise to construct a *separate* frequency distribution for each type rather than combine them. In the example above, one frequency distribution might be constructed for men's heights and another frequency distribution for the heights of women.

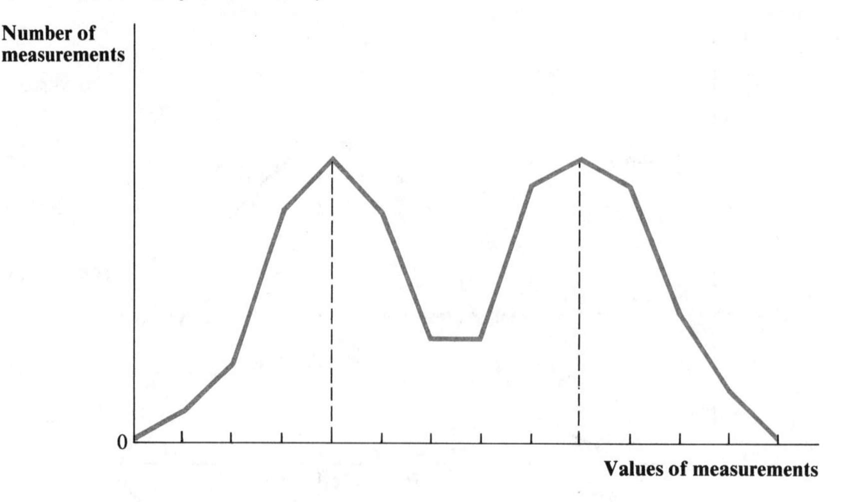

FIGURE 2.2 Frequency Polygon of a Bimodal Frequency Distribution

RELATIONSHIPS AMONG THE MEAN, MEDIAN, AND MODE

Having discussed the three principal measures of central tendency (the mean, median, and mode), we must describe how these three measures are related to one another. If the frequency distribution of a set of data has a single mode and is ***symmetrical***, as in panel A of Figure 2.3, the mean, median, and mode coincide. (Symmetrical means that if we were to "fold" the distribution at its mean, the part of the distribution to the left of the mean would be a perfect match for the part to the right of the mean.) Many frequency distributions are not symmetrical, but are ***skewed to the right*** (as in panel B of Figure 2.3) or ***skewed to the left*** (as in panel C of Figure 2.3). A frequency distribution that is skewed to the right has a long tail to the right, whereas one that is skewed to the left has a long tail to the left. As shown in Figure 2.3, if the frequency distribution is skewed to the right, the mean generally exceeds the median, which in turn exceeds the mode. If the frequency distribution is skewed to the left, the mode generally exceeds the median, which in turn exceeds the mean (also shown in Figure 2.3).

EXERCISES

2.1 According to data presented in the previous chapter, the profit after taxes (as a percent of sales) of each manufacturing industry in the United States in 1974 was as follows:

Durable goods (industry)	Profit (percent of sales)	Nondurable goods (industry)	Profit (percent of sales)
Transportation equipment	3.9	Food and kindred products	2.6
Electrical equipment	4.3	Tobacco	5.8

(*table continues on p. 38*)

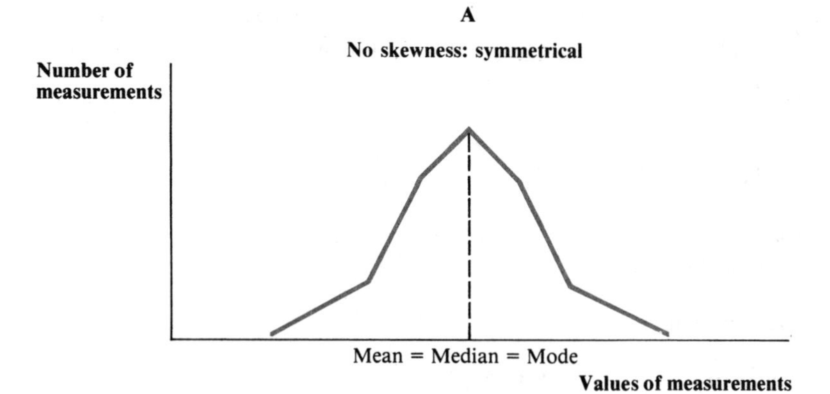

B

Skewed to the right

Number of measurements

Mode Mean
Median

Values of measurements

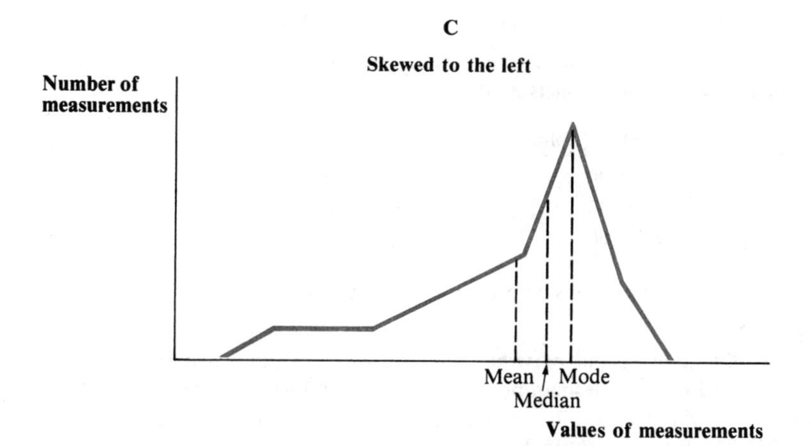

FIGURE 2.3 Relationship of Mean, Median, and Mode

Durable goods (industry)	Profit (percent of sales)	Nondurable goods (industry)	Profit (percent of sales)
Machinery (except electrical)	5.6	Textiles	2.8
Fabricated metal products	4.0	Paper	5.4
Primary iron and steel	4.1	Printing	4.8
Primary nonferrous metals	5.4	Chemicals	6.8
Stone, clay, and glass	4.8	Petroleum and coal	7.6
Instruments	8.4	Rubber	4.0
Other durable goods	5.0	Other nondurable products	2.1

(a) Calculate the mean profit rate of the durable manufacturing industries (left-hand column) in 1974.
(b) Calculate the mean profit rate of the nondurable manufacturing industries (right-hand column) in 1974.
(c) Interpret and compare the results of (a) and (b) of this question.

2.2 Based on the data in Exercise 2.1, calculate the median profit rate of nondurable manufacturing industries in 1974. Calculate the median profit rate of durable manufacturing industries in 1974. How big is the difference between these two medians? Interpret your results.

2.3 Given the mean profit rate for nondurable manufacturing industries and the mean profit rate for durable manufacturing industries, describe a simple procedure for obtaining the mean profit rate for all manufacturing industries (if the number of non-durable manufacturing industries is known to equal the number of durable manufacturing industries).

2.4 Given the median profit rate for nondurable manufacturing industries and the median profit rate for durable manufacturing industries, can you determine the median profit rate for all manufacturing industries from this information alone?

2.5 A salesman made 100 visits to customers. The frequency distribution of the amount of commission he earned per visit is as follows:

Amount of commission (dollars)	Number of visits
0 and under 20	60
20 and under 40	30
40 and under 60	10

Calculate the mean amount of his commission per visit. Estimate the total commission he earned for all 100 visits.

2.6 Based on the data in the previous exercise, calculate the median amount of commission earned by the salesman per visit. Based on information solely concerning this median, can you tell whether the total commission earned for all 100 visits exceeded $2,000? Why, or why not?

2.7 In a township in Pennsylvania, all lots are 1/4 acre, 1/2 acre, 1 acre, or 2 acres. The frequency distribution of lot sizes for all residential property in this township is:

Size of lot (acres)	Number of lots
1/4	100
1/2	500
1	50
2	20

What is the mode of this frequency distribution? Is the mode bigger than the mean size of lot? Is it bigger than the median size of lot?

2.8 Do you think that the median amount paid by Americans in income tax in 1978 was less than the mean amount paid? Explain.

2.9 For *any* set of measurements, what is the sum of the deviations of these measurements from their mean?

2.10 A set of measurements has a symmetrical frequency distribution, and the median is 3. If there are 1,000 measurements, can you calculate their sum?

2.5 Measures of Dispersion

IMPORTANCE OF DISPERSION

In the previous section we saw that measures of central tendency provide useful summary information concerning the general level or average value of a body of data. However, this obviously does not mean that such measures alone can provide a complete or adequate description of the data. A case that is close to home can illustrate the limitations of such measures. Suppose that you have taken an examination and that the instructor, after grading the exam, announces to the class that the mean grade is 75. Then he or she hands back the exams, and you find that your paper has received a grade of 80. Clearly, it is hard to interpret this grade on the basis of information concerning only the average grade. The variability about the average is very important, too. If the grades are highly variable (as in panel A of Figure 2.4), then a large number of your classmates may have received a grade higher than yours. On the other hand, if there is little variability in the grades (as in panel B of Figure 2.4), then you may have received close to the highest grade.

If there is enough variability about an average, the average may not mean much. If Mr. Rich, one of the world's wealthiest men, is driven to the airport by his chauffeur, what is the average income level of the two occupants of the car? Mr. Rich's income is $10 million a year and his chauffeur's income is $10 thousand a year; thus, the mean income of the two is $5,005,000 per year. This is a misleading figure, however, since it vastly overstates the chauffeur's income and vastly understates the income of Mr. Rich. In cases of this sort, then, we see that an average can be quite misleading. (More will be said about this later in the chapter.)

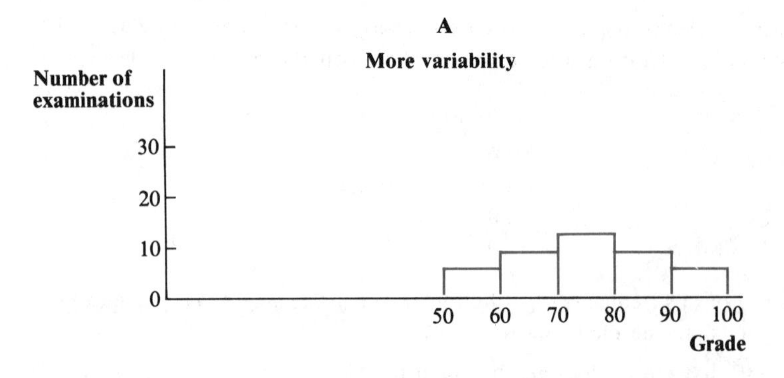

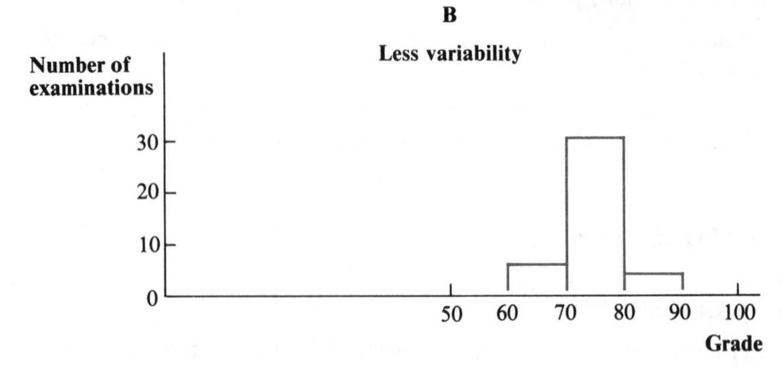

FIGURE 2.4 Histogram of Examination Grades

DISTANCE MEASURES OF DISPERSION

Both summary measures of dispersion and summary measures of central tendency are important in describing a body of data, but neither alone will suffice. There are two types of summary measures of dispersion: distance measures and measures of average deviation. ***Distance measures*** *describe the variation in the data in terms of the distance between selected measurements.* The most frequently used distance measures are the range and the interquartile range.

Range. Perhaps the simplest measure of variability is the *range*, which is *the difference between the highest and lowest values in the body of data.* In the example of the exam grades discussed above, if the lowest grade in the class is 50 and the highest is 99, then the range is 99 − 50 = 49. Although the range is a popular measure of variability, particularly because it is so easy to compute, it has the important disadvantage of being unaffected by the values of all observations other than the highest and the lowest. For example, the range of the exam grades would be the same (49) if the grades were distributed evenly from 50 to 99, or if all grades other than the highest and lowest fell between 70 and 80. Yet the variability certainly is less in the latter case.

Interquartile Range. Another common measure of variability is the *interquartile range*, which is defined as *the difference between the third quartile and the first quartile.* The *third quartile* is the value such that 75 percent of the observations lie below it; the *first quartile* is the value such that 25 percent of the observations lie below it. Thus, the interquartile range measures *the spread bounding the middle 50 percent of the values of the observations.* (Note that the *second quartile*, which is the value such that 50 percent of the observations lie below it, is another name for the median.) One advantage of the interquartile range is that it can be calculated from frequency distributions with open-ended class intervals, while the range cannot.

Measures Based on Percentiles. In addition to the range and the interquartile range, other distance measures of dispersion can be based on percentiles. The *Xth percentile* is defined as the value that exceeds *X* percent of the observations. Thus, the 90th percentile is the value such that 90 percent of the observations lie below it. The difference between the 90th percentile and the 10th percentile is a possible measure of dispersion; so is the difference between the 99th percentile and the 1st percentile. The former measures the spread bounding the middle 80 percent of the values of the observations. The latter measures the spread bounding the middle 98 percent of them.

THE VARIANCE AND STANDARD DEVIATION

Although the distance measures of dispersion are sometimes used, they are not as important as measures of average deviation, the most significant of which are the variance and the standard deviation. The *variance* of the measurements in a population, denoted by σ^2 (sigma squared), is defined as *the arithmetic mean of the squared deviations of the measurements from their mean.* Thus, if $X_1, X_2, \ldots, X_N$ are the measurements in the population,

$$\sigma^2 = \frac{(X_1 - \mu)^2 + (X_2 - \mu)^2 + \ldots + (X_N - \mu)^2}{N} = \frac{\sum_{i=1}^{N} (X_i - \mu)^2}{N}. \tag{2.6}$$

The variance is a measure of dispersion, but it is expressed in units of squared deviations or squares of the values of the measurements rather than in the same units as the measurements. The *standard deviation* is a measure of dispersion which is expressed in the same units as the measurements. The standard deviation is denoted by σ (sigma), which is the positive square root of the variance. In other words, the standard deviation is

$$\sigma = \sqrt{\frac{\sum_{i=1}^{N} (X_i - \mu)^2}{N}}. \tag{2.7}$$

Since the standard deviation is so important in statistics, its definition is worth discussing in greater detail. To begin with, note that if X_i is the *i*th measurement in the population, then the difference between this measurement and the population mean equals $(X_i - \mu)$. This is the deviation of the ith measurement from the population mean, and the square of this deviation obviously equals $(X_i - \mu)^2$. Next, let's find the mean of these squared deviations. Since there are N of these squared deviations, this mean equals

$$\frac{\sum_{i=1}^{N} (X_i - \mu)^2}{N}$$

Recall from our earlier discussion that Σ is the summation sign, which means that the numbers to its right—that is, $(X_i - \mu)^2$—should be summed from the lower limit on i (given below the summation sign) to the upper limit on i (given above the summation sign). Thus, in this case, $(X_i - \mu)^2$ is to be summed from $i = 1$ to $i = N$. (In other words, the squared deviations are to be summed for all observations.) Then, to obtain the mean of the squared deviations, the resulting sum is divided by N. Finally, we must find the square root of this mean, the result being the standard deviation. Equation (2.7) shows this complete procedure.

Intuitively, it seems clear that *the more dispersion there is in a body of data the bigger the standard deviation will be.* If there is no dispersion at all, every observation will equal the population mean, with the result that every one of the deviations will equal zero. (Thus, the standard deviation will equal zero.) As the dispersion in the data increases, the deviations of the observations from the population mean will tend to increase as well, and so will the mean of the squared deviations. (Thus, the standard deviation will also increase.) For this reason, if one knows that the standard deviation of the measurements in one population is higher than the standard deviation of the measurements in another population, this indicates that there is more dispersion in the former population than in the latter.[5]

If the body of data is a sample rather than a population, the formulas for the variance and the standard deviation are somewhat different from those used for the entire population. Specifically, the sample variance, denoted by s^2, is defined as

$$s^2 = \frac{\sum_{i=1}^{n} (X_i - \bar{X})^2}{n - 1}. \tag{2.8}$$

And the sample standard deviation, denoted by s, is defined as

$$s = \sqrt{\frac{\sum_{i=1}^{n} (X_i - \bar{X})^2}{n - 1}}. \tag{2.9}$$

5. However, if the means of the two populations differ, a better procedure for comparing their dispersion may be to calculate the coefficient of variation in each population. (See Exercises 2.15 and 2.16.)

There are three differences between these formulas and those given for the population variance and standard deviation. First, the *sample* mean $\bar{X}$ is substituted for the *population* mean μ. Second, the squared deviations from the mean are summed over all measurements in the *sample*, not the *population*. Third, the sum of the squared deviations from the mean is divided by $(n - 1)$, not by N.

Generations of statistics students have been puzzled over why the denominator of the sample variance is $(n - 1)$, while the denominator of the population variance is N. Basically, the reason is that the sample variance, if its denominator were n, would tend to underestimate the population variance. (More will be said about this in later chapters.)

CALCULATING THE STANDARD DEVIATION

To illustrate the computations involved in calculating the population standard deviation, let's take two cases in which we have all the data in the population and thus are able to compute the population standard deviation. The first case is the population of incomes in Mr. Rich's limousine, as discussed previously. Since there are only two numbers in this population, the income of the chauffeur ($10,000) and Mr. Rich's income ($10,000,000), the population mean is $5,005,000 and the deviations from the mean (in dollars) are

$$(X_1 - \mu) = 10{,}000 - 5{,}005{,}000 = -4{,}995{,}000,$$

$$(X_2 - \mu) = 10{,}000{,}000 - 5{,}005{,}000 = 4{,}995{,}000.$$

Thus, the mean of the squared deviations (the variance) equals

$$\frac{\sum_{i=1}^{2} (X_i - \mu)^2}{2} = \frac{(-4{,}995{,}000)^2 + (4{,}995{,}000)^2}{2}$$

$$= 24{,}950{,}025{,}000{,}000,$$

and the standard deviation is [6]

$$\sqrt{\frac{\sum_{i=1}^{2} (X_i - \mu)^2}{2}} = \sqrt{24{,}950{,}025{,}000{,}000} = 4{,}995{,}000.$$

Next, let's take a somewhat more complicated case—that of the 1973 profit rates of petroleum firms in Table 2.1. What is the standard deviation of these profit rates? The second column of Table 2.4 shows the deviation from the mean (that is, $X_i - \mu$) for each of the firms. (Recall from an earlier section that $\mu = 13.1$.) In the third column, the square of this deviation—that is, $(X_i - \mu)^2$—is calculated for each firm. The sum of these squared deviations is shown to equal 272.27:

6. In this very simple case, the standard deviation equals the deviation of each observation from the mean. For populations containing more than two observations, this generally will not be the case.

$$\sum_{i=1}^{23} (X_i - \mu)^2 = 272.27.$$

Thus,

$$\frac{\sum_{i=1}^{23} (X_i - \mu)^2}{N} = \frac{272.27}{23} = 11.84$$

and

$$\sigma = \sqrt{\frac{\sum_{i=1}^{23} (X_i - \mu)^2}{N}} = \sqrt{11.84} = 3.4.$$

Consequently, the standard deviation of the 1973 profit rates of the petroleum firms in Table 2.1 is 3.4 percentage points.

Having calculated the standard deviation for the population in two cases, let's turn now to the calculation of the sample standard deviation; at the

TABLE 2.4

Calculation of $\Sigma(X_i - \mu)^2$ for the Population of 1973 Profit Rates of 23 Major Petroleum Firms

X_i	$(X_i - \mu)$	$(X_i - \mu)^2$
17.8	4.7	22.09
16.2	3.1	9.61
14.9	1.8	3.24
14.4	1.3	1.69
14.5	1.4	1.96
12.4	−0.7	0.49
10.7	−2.4	5.76
13.4	0.3	0.09
8.7	−4.4	19.36
9.0	−4.1	16.81
11.7	−1.4	1.96
10.5	−2.6	6.76
11.9	−1.2	1.44
15.5	2.4	5.76
9.6	−3.5	12.25
9.0	−4.1	16.81
16.2	3.1	9.61
7.9	−5.2	27.04
13.3	0.2	0.04
11.2	−1.9	3.61
22.3	9.2	84.64
17.7	4.6	21.16
13.4	0.3	0.09
Total		272.27

same time, we will illustrate how the standard deviation can be calculated from a frequency distribution. Specifically, we will estimate the standard deviation of the profit rates of the sample of 15 chemical firms in Table 2.2. As in the calculation of the arithmetic mean from a frequency distribution in equation (2.3), we assume that *the midpoint of each class interval (that is, the class mark) can be used to represent each value of the measurements in that class interval.* This means that we assume that the profit rates of the firms in the "9 percent and under 11 percent" class interval can be approximated by 10 percent, that the profit rates of the firms in the "11 percent and under 13 percent" class interval can be approximated by 12 percent, and so on. Thus, the sample standard deviation can be approximated by

$$s = \sqrt{\frac{\sum_{j=1}^{k} f_j (X'_j - \bar{X})^2}{n - 1}}, \tag{2.10}$$

where f_j is the number of measurements in the jth class interval, X'_j is the midpoint of the jth class interval, and k is the number of class intervals.

Table 2.5 shows how the formula for the sample standard deviation in equation (2.10) can be applied to the frequency distribution of profit rates of chemical firms. The first column of Table 2.5 provides the class intervals of this frequency distribution. The second column gives the number of measurements in each class interval (the values of f_j). The third column shows the

TABLE 2.5
Calculation of the Standard Deviation of the Sample of 15 Profit Rates of Chemical Firms

Profit rate	Number (f_j) of firms	Midpoint (X'_j) of class interval	$X'_j - \bar{X}$	$(X'_j - \bar{X})^2$	$f_j(X'_j - \bar{X})^2$
9 and under 11 percent	3	10	−3.47	12.0409	36.1227
11 and under 13 percent	3	12	−1.47	2.1609	6.4827
13 and under 15 percent	5	14	0.53	0.2809	1.4045
15 and under 17 percent	3	16	2.53	6.4009	19.2027
17 and under 19 percent	1	18	4.53	20.5209	20.5209
					83.7335

Using equation (2.10), we find that

$$s = \sqrt{\frac{\sum_{j=1}^{k} f_j (X'_j - \bar{X})^2}{n - 1}} = \sqrt{\frac{83.7335}{14}} = 2.4.$$

(table continues)

Using the shortcut in equation (2.13),

$$s = \sqrt{\frac{\sum_{j=1}^{k} f_j X_j'^2 - \frac{1}{n}\left(\sum_{j=1}^{k} f_j X_j'\right)^2}{n - 1}}$$

Since

$$\sum_{j=1}^{k} f_j X_j'^2 = 3(10^2) + 3(12^2) + 5(14^2) + 3(16^2) + 1(18^2) = 2804,$$

and

$$\sum_{j=1}^{k} f_j X_j' = 3(10) + 3(12) + 5(14) + 3(16) + 1(18) = 202.$$

$$s = \sqrt{\frac{2804 - \frac{1}{15}(202)^2}{14}} = \sqrt{\frac{2804 - 2720.27}{14}} = 2.4.$$

Thus, the answer is the same as for equation (2.10). Except for the effects of rounding errors, these answers must always be identical.

midpoint of each class interval (the values of X_j'). Since we know from our earlier discussion (on page 30) that the sample mean is 13.47 percent, the values of $(X_j' - \bar{X})$ are as shown in the fourth column, and the values of $(X_j' - \bar{X})^2$ are as shown in the fifth column. Finally, the sixth column shows the product of f_j and $(X_j' - \bar{X})^2$. Summing up the figures in the sixth column, we have the sum of the squared deviations from the sample mean, which is 83.7335. Dividing this sum by 14, which is $(n - 1)$, we get 5.98, the sample variance. Taking the square root of 5.98, we get 2.4, the sample standard deviation.

SHORTCUTS IN CALCULATING THE VARIANCE AND STANDARD DEVIATION

Modern electronic computers are often used to compute the variances and standard deviations of populations and samples. The advent of computers has transferred a great deal of drudgery from human beings to these mechanical aids. However, since many computations of this sort are still done by hand, it is useful to note that modifications of equations (2.8) and (2.9) often simplify the calculations. Specifically, these modifications are

$$s^2 = \frac{\sum_{i=1}^{n} X_i^2 - \frac{1}{n}\left(\sum_{i=1}^{n} X_i\right)^2}{n - 1} \tag{2.11}$$

and

$$s = \sqrt{\frac{\sum_{i=1}^{n} X_i^2 - \frac{1}{n}\left(\sum_{i=1}^{n} X_i\right)^2}{n - 1}}. \tag{2.12}$$

Similarly, if you are calculating the sample standard deviation based on a frequency distribution, it is often quicker and easier to use the following modification of equation (2.10) to carry out the calculations:

$$s = \sqrt{\frac{\sum_{j=1}^{k} f_j X_j'^2 - \frac{1}{n}\left(\sum_{j=1}^{k} f_j X_j'\right)^2}{n - 1}} \tag{2.13}$$

These modifications do not alter the answers given by the formulas in equations (2.8), (2.9), and (2.10). They are merely different ways of obtaining the same result. To verify this, see the continuation of Table 2.5 on page 46, which shows how the formula in equation (2.13) is applied to the frequency distribution of profit rates of chemical firms. Clearly, the result is the same as when the formula in equation (2.10) was used, but the calculations are easier and less time-consuming.

INTERPRETATION OF THE STANDARD DEVIATION

The standard deviation is the most important summary measure of dispersion. If the frequency distribution of a population conforms to the so-called normal distribution (to be discussed in detail in Chapter 5), then we know the percentage of measurements in the population that fall within 1, 2, or 3 standard deviations of the population mean. Specifically, 68.3 percent of the measurements lie within ± 1 standard deviation of the mean, 95.4 percent of the measurements lie within ± 2 standard deviations of the mean, and 99.7 percent of the measurements lie within ± 3 standard deviations of the mean.

Thus, if we know that the population of diameters of pieces of pipe produced by a firm conforms to the normal distribution with a mean of 4 inches and a standard deviation of 0.05 inches, it follows that 68.3 percent of the pieces produced by this firm will have diameters of 3.95 to 4.05 inches, that 95.4 percent will have diameters of 3.90 to 4.10 inches, and that 99.7 percent will have diameters of 3.85 to 4.15 inches. This is useful information. For example, if the diameters of the pieces must be between 3.90 and 4.10 inches to meet specifications, it follows that 4.6 percent of the pieces will be unacceptable.

Of course, it is essential to realize that results such as these pertain only to populations that conform to the normal distribution. However, even if a population does not conform to the normal distribution, we can still make statements (based on Chebyshev's inequality, to be discussed in Chapter 4) about the percentage of any population that will be within a certain number of standard deviations of the population mean. Much more will be said on this subject in later chapters.

EXERCISES

2.11 In an Ohio township, the frequency distribution of the sizes of all residential lots is given below:

Size of lot (acres)	Number of lots
1/4	300
1/2	400
1	200
2	100

(a) What is the standard deviation of the lot sizes?
(b) What is the variance of the lot sizes?
(c) What is the range of the lot sizes?
(d) What is the interquartile range of the lot sizes?

2.12 Based on the data in Exercise 2.1 in this chapter, what is the range of the profit rates among nondurable manufacturing industries in 1974? What is the range among durable manufacturing industries? What is the range among all manufacturing industries? Can the range for nondurable or durable manufacturing industries be greater than that for all manufacturing industries?

2.13 Based on the data in Exercise 2.5, can you determine the range of the amount of commission earned per visit by the salesman? Can you obtain upper and/or lower bounds for this range? Explain.

2.14 Use the data in Exercise 2.1 to calculate the standard deviation of the industry profit rates (a) for nondurable manufacturing industries; (b) for durable manufacturing industries; and (c) for all manufacturing industries. (Assume that the entire population is given.)

2.15 The standard deviation is a measure of *absolute* dispersion or variability, and it is affected by the units of measurement. To illustrate this fact, calculate the standard deviation of the amount of commission per visit in Exercise 2.5. Then express the commissions in cents (not dollars) and calculate the standard deviation. What is the ratio of the latter standard deviation (for commissions expressed in cents) to the former standard deviation (for the commissions expressed in dollars)? Why?

2.16 To obtain a measure of dispersion or variability that is normalized for the units of measurement, statisticians often use a measure of *relative* variability such as the ***coefficient of variation***. The coefficient of variation equals the standard deviation as a percentage of the mean. In other words, for a population it equals

$$V = \frac{\sigma}{\mu} \cdot 100.$$

(a) Prove that the coefficient of variation of the commissions in Exercise 2.5 is the same whether the commissions are measured in cents or dollars.
(b) From the data in Exercise 2.1 calculate the coefficient of variation of the profit rates for nondurable manufacturing industries.
(c) From the data in Exercise 2.1 calculate the coefficient of variation of the profit rates for durable manufacturing industries.
(d) Based on the coefficient of variation, are the profit rates more variable among durable goods industries or nondurable goods industries?

2.17 Based on the data in Exercise 2.5, is the distribution of commissions earned by the salesman skewed to the left? To the right?

2.6 Misuses of Descriptive Statistics

The famous British prime minister Disraeli once said that there were lies, damned lies, and statistics. You are no doubt already on guard against many kinds of misuses of statistics. However, it is important for you to develop as much skill in rooting out statistical fallacies and chicanery as possible, since the business and economic world is full of pitfalls for the statistically unwary. This section covers five kinds of errors that frequently occur in descriptive statistics. These are errors that can result in costly mistakes, and even trained statisticians occasionally fall prey to them.

Inappropriate Comparisons. Suppose you read in the newspaper that the crime rate in your area is 2 percent lower than last year. Before accepting this conclusion, it would be wise to question whether the *definition* of crime on which these statistics were based has remained the same from last year to the present. In other words, might the apparent decrease in the crime rate be due to the fact that some types of behavior were defined as crimes last year which are no longer classified as such? Also, do the figures pertain to the same kind of population during the two periods, or could the apparent drop in the crime rate be due to the fact that the figures for the most recent period pertain to a different set of people than the earlier figures? For example, the later figures might represent all individuals, including young children, whereas the earlier ones might pertain only to individuals over 10 years old. Furthermore, are the later and earlier periods really comparable? Perhaps the later figures pertain to a portion of the year when crime is always relatively low, whereas the earlier figures pertain to a full year. Unless you can be reasonably certain that each of these possible discrepancies is nonexistent or relatively minor it is difficult, if not impossible, to interpret this kind of statistical item in the newspaper.

Very Small or Obviously Biased Samples. Suppose you receive a report stating that a new device will increase the number of miles an automobile can travel per gallon of gasoline by 5 percent. On the surface this certainly sounds good, but this figure of 5 percent might be based on the effect of the device on the mileage of only three cars. If so, the figure of 5 percent may not apply to your car, because so small a sample may not be at all representative of the entire population of automobiles. Therefore, it is generally wise to ask about the size of the sample on which a particular statistical result is based. This is a good question to ask even in areas like medicine and the natural sciences, since despite the scientific and precise nature of the work, results sometimes are based on very small samples.

Some organizations and people use statistics as a drunk uses a lamppost: for support, not illumination. Thus individuals or firms sometimes carry out studies on one small sample after another until eventually and by chance one of the samples indicates what they want to prove about the product they sell. They then publicize the results of this small sample, hoping thereby to influence their customers or potential customers. The makers of the new device to increase mileage per gallon may have tested the effect of their invention on one three-car sample after another. (They may also have discarded the results of many earlier small samples that showed the device had no effect!) Finally, when one sample indicated, by chance, a 5 percent increase in mileage per gallon, they may have publicized the results. Clearly, statistics based on such improperly selected samples can be worse than valueless.

Improper Choice of Average. For a given set of data, a person can use one measure of central tendency to indicate one thing and another measure of central tendency to indicate another. A real estate salesman trying to impress a potential customer with the high income level in a particular suburb quotes the *mean* family income in this suburb as $30,000 per year. His statement is perfectly true, but it is also misleading because the income *distribution* in the suburb happens to be highly skewed to the right. Perhaps 2 percent of the suburb's residents earn about $750,000 per year, and the rest earn between $10,000 and $20,000 per year. Thus, the mean of $30,000 is not a very representative figure.

On the other hand, the mayor of this particular suburb is trying to impress the state government with the *low* income level of the area (and hence, with its need for state aid). In contrast to the real estate salesman, the mayor chooses as a measure of central tendency the *mode* of the frequency distribution of families by income level. Because the income distribution is highly skewed to the right, the mode is likely to be considerably less than the mean and somewhat less than the median. The mode of the income distribution in the community may be only $14,000. Thus, as a measure of central tendency, the real estate salesman quotes $30,000, while the mayor quotes $14,000. Both are right; the difference is due to the fact that different measures of central tendency are being used. The moral, of course, is clear: Whenever someone quotes an "average" or a measure of central tendency, be sure to find out what kind of a measure it is and how representative it is likely to be.

Neglect of the Variation about an Average. Even if the proper kind of average is chosen, the average may be surrounded by so much variation that it alone may be misleading. It is sometimes argued that home builders, focusing their attention on the average number of people in the American family, build too many medium-sized homes and too few small and large ones. In other words, if the average family size in the United States is 3.6 persons, home builders may tend to neglect the considerable variation about this average, and build too many homes for families of 3 persons or 4 persons. (According to some observers, this has in fact occurred.)[7]

Another case of this sort is encountered in ads which announce that a certain toothpaste will result, on the average, in a such-and-such percent reduction in dental cavities. Besides the fact that some of these claims may be subject to some of the other problems discussed above, they may be misleading because there may be such great variation about the average. What is important to a particular consumer is whether the toothpaste will reduce his or her cavities. Because the variation among individuals in the effects of the toothpaste may be so great, the chance that it will have a beneficial effect on one particular person may be scarcely better than 50-50, even though on the average it may offer some protection against tooth decay.

Misinterpretation of Graphs and Charts. Sometimes graphs and charts are presented so as to give a misleading impression. For example, a real estate salesman wants to run an ad showing the increase over time in the average price of houses in the suburb where he works. He decides to use a *pictogram*, a

7. See Darrell Huff and Irving Geis, *How to Lie with Statistics* (New York: Norton, 1954), part of which is contained in E. Mansfield, *Statistics for Business and Economics: Readings and Cases* (New York: Norton, 1980). Also, see Oskar Morgenstern's article in the latter book.

graph in which the size of the object in the picture indicates the relative size of the thing the object represents. The salesman uses the pictogram in Figure 2.5 to show that the average price of a house (represented by the size of the deed) doubled between 1965 and 1979. This chart certainly makes it appear that real estate prices have risen, and indeed, it is misleading in this respect. Why? Because the salesman doubled both the width and the height of the deed in Figure 2.5, thus *quadrupling* the area of the deed. Thus, based on the *area* of the deed, the chart makes it appear that real estate prices have quadrupled, not doubled.

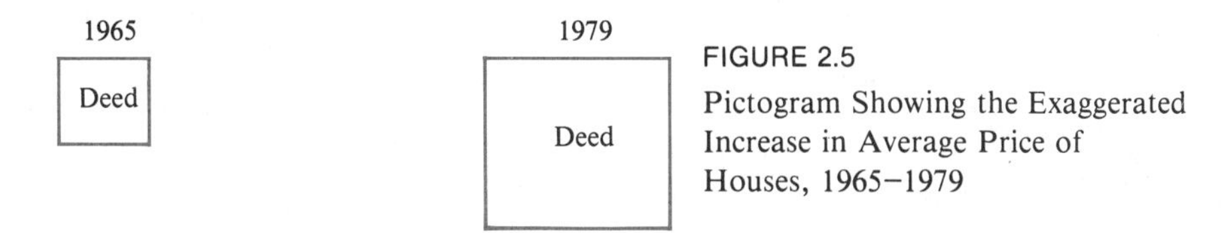

FIGURE 2.5

Pictogram Showing the Exaggerated Increase in Average Price of Houses, 1965–1979

Another frequently encountered error occurs in the construction of histograms where the class intervals are not of equal width. The frequency distribution of profit rates of the 23 major oil firms that we constructed in Chapter 1 is reproduced in Table 2.6. If we combine the first two class intervals in Table 2.6, we obtain the frequency distribution in Table 2.7. We then construct bars on the line segments of the horizontal axis corresponding to the classes of the frequency distribution, and we make the height of each bar equal to the number of cases in each class interval. The result is shown in panel A of Figure 2.6. Comparing panel A of Figure 2.6 with Figure 1.2, it is clear that panel A gives a distorted picture. The reason is simple: In the class interval from 6.95 to 10.95 percent, there are 7 firms, which means that on the average there are 3.5 firms for every two percentage points in this class interval. Thus, to be comparable with the other vertical bars, the height of the

TABLE 2.6

Frequency Distribution of Population of 1973 Profit Rates of 23 Major Petroleum Firms

Profit rate	Number of firms
6.95 and under 8.95 percent	2
8.95 and under 10.95 percent	5
10.95 and under 12.95 percent	4
12.95 and under 14.95 percent	6
14.95 and under 16.95 percent	3
16.95 and under 18.95 percent	2
18.95 and under 20.95 percent	0
20.95 and under 22.95 percent	1
Total	23

TABLE 2.7

Frequency Distribution of Population of Profit Rates of 23 Petroleum Firms, Unequal Widths of Class Intervals

Profit rate	Number of firms
6.95 and under 10.95 percent	7
10.95 and under 12.95 percent	4
12.95 and under 14.95 percent	6
14.95 and under 16.95 percent	3
16.95 and under 18.95 percent	2
18.95 and under 20.95 percent	0
20.95 and under 22.95 percent	1
Total	23

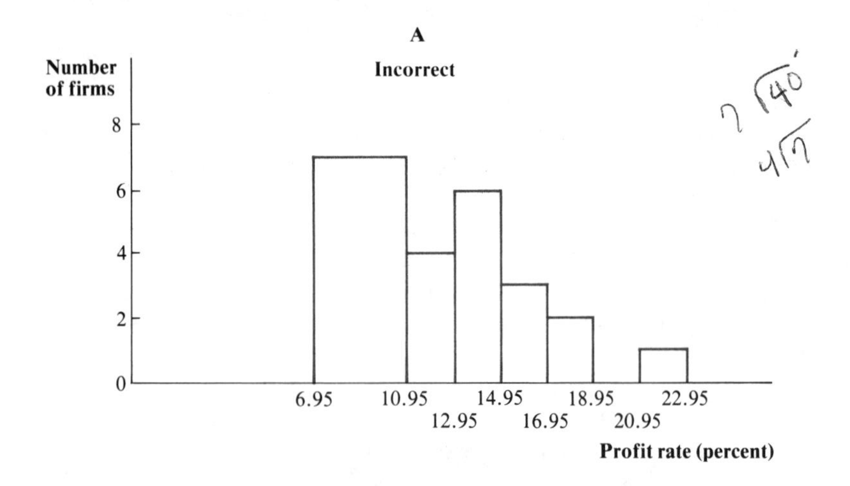

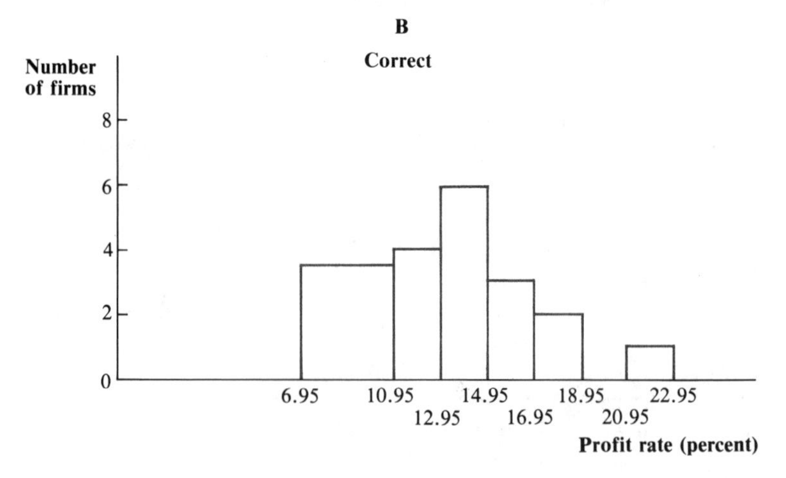

FIGURE 2.6 Incorrect and Correct Histograms of Profit Rates of Petroleum Firms, Unequal Widths of Class Intervals

bar for this class interval should be 3.5, not 7. Put differently, the *area* of each bar in a histogram should be proportional to the number of observations in the relevant class interval. Thus, since area equals width times height, the height of the bar in the 6.95–10.95 percent class interval should be 7 divided by 2 because this class interval is twice as wide as the others. The resulting histogram, corrected in this way, is shown in panel B of Figure 2.6. Put bluntly, panel A is wrong, panel B is right.

2.7 Snow Removal in New York City: A Case Study in Descriptive Statistics

An actual crisis that faced the city of New York illustrates how some of the statistical techniques described in this and the previous chapter have been used to help solve important real problems. On February 9, 1969 a major snowstorm hit New York and paralyzed the city for days. Mayor John Lindsay, faced with a consequent political crisis, asked a group of statisticians and systems analysts to carry out a thorough study to find out what, if anything, was wrong with the city's snow-removal procedures.

Clearly, the first question the study faced was: How unusual was this snowstorm? That is, was it much more severe that those typically experienced in New York? In order to answer this question, the statisticians obtained U.S. Weather Bureau records of the depth of each snowfall in New York City in recent years. Based on these records, the statisticians constructed the frequency distribution shown in Table 2.8. The results indicate that the 1969 storm, with a snowfall of over 15 inches, was unusual indeed. (From 1948 to 1967 only three storms resulted in more than 15 inches of snow.)

The next question was: How much work has to be done to cope with a snowstorm? The three basic activities that must be performed are spreading salt, plowing, and snow removal; and the amount of each of these activities depends on the number of street-miles to be serviced. The Department of

TABLE 2.8

Frequency Distribution of Depths of Snowstorms, 1948–67, New York City

Depth (inches)	Number of snowstorms during 20-year period (1948–67)
1 and under 3	67
3 and under 5	26
5 and under 7	13
7 and under 9	4
9 and under 11	3
11 and under 13	3
13 and under 15	2
15 and under 17	3

SOURCE: E. Savas, "The Political Properties of Crystalline H_2O: Planning for Snow Emergencies in New York," *Management Science*, October 1973.

Sanitation divides all streets into three priority classes: primary, secondary, and tertiary (the clearance of primary streets being the highest priority and the clearance of tertiary streets the lowest). The analysts further subdivided the primary streets into emergency streets (ones of the very highest priority) and other primary streets. Based on this classification, the statisticians determined the number of miles of New York's streets in each of these classes. The resulting frequency distribution is shown in Table 2.9.

TABLE 2.9
Miles of New York City Streets, by Snow-Clearance Priority

Priority class	Miles of streets
Primary	
Emergency	1600
Other	930
Secondary	1978
Tertiary	1331
Total	5839

SOURCE: See Table 2.8.

Next, the study had to investigate the city's existing capacity for spreading salt, plowing, and snow removal. The main types of equipment used to cope with snowstorms are spreaders and plows. Based on data regarding the frequency of downtime on such equipment, it was assumed that about 40 percent of the spreaders and plows would be out of commission at the onset of a storm. As a result, it was concluded that about 134 spreaders and 1,050 plows would be usable. Based on estimates of the time taken to reach various streets and the length of rest breaks, meal breaks, and refueling time, as well as the speed at which spreaders and salters could move along the snowy streets, it was possible to determine how long it would take the existing number of spreaders and plows to service the city's street network. The results indicated that "there is sufficient equipment available, in the aggregate, to plow *every* mile of *every* street in the city in only six hours, and to plow the high-priority streets in less than two hours."[8]

This finding, based on the simplest kind of statistical analysis, was startling, since it indicated that the city had enough plowing capability to do the required work in a very short period of time. Why, then, had it taken so long to clean up the snowstorm of February 9? The analysts reasoned that the distribution of snow plows might have been the problem and to find out they obtained data concerning the distribution of plows among the various boroughs of New York. The results, shown in Table 2.10, indicate clearly that relative to their number of miles of primary streets, Manhattan, Brooklyn, and

8. E. Savas, "The Political Properties of Crystalline H_2O: Planning for Snow Emergencies in New York," *Management Science*, October 1973, p. 142.

TABLE 2.10
Allocation among Boroughs of Plows and Primary Street Mileage,New York City

Borough	Borough's percentage of total mileage of primary streets	Borough's percentage of city's snow plows
Manhattan West	8.6	9.2
Manhattan East	6.5	10.7
Bronx West	6.6	8.9
Bronx East	9.9	8.4
Brooklyn West	9.6	11.0
Brooklyn North	7.5	11.5
Brooklyn East	6.8	9.8
Queens West	14.7	12.8
Queens East	20.8	13.5
Richmond	9.2	4.0
Total[a]	100.00	100.0

SOURCE: See Table 2.8.
[a] Due to rounding errors, the percentages do not sum to exactly 100.0.

the Bronx had a relatively large number of plows, whereas Queens and Richmond had relatively few. It was therefore no wonder that snow removal was relatively slow in Richmond, which with over 9 percent of the city's mileage of primary streets, had only about 4 percent of the city's plows.

Why were the plows so poorly allocated, and how could the allocation be improved? The reason for the misallocation was linked to the fact that snow plows are merely sanitation trucks fitted with plows. The city had allocated these vehicles among geographical areas in accord with their primary function, refuse collection. Thus, densely populated areas like Manhattan and Brooklyn, which produce much more refuse per street-mile than other areas, received a relatively large number of these vehicles. To improve the allocation of plows, the analysts recommended that plows be mounted on vehicles other than refuse trucks and that these other vehicles be distributed so as to make the plowing capability of each borough proportional to its mileage of primary streets.[9]

The results of the snow-crisis study were presented to Mayor John Lindsay in June 1969. Since this was a time when Lindsay was running for reelection, and since the snow-removal crisis of the previous February figured prominently in the campaign, it is not surprising that the mayor called for immediate adoption and implementation of the study's recommendations. During the following winter New York's new snow-emergency plan went into operation and, based on experience to date, seems to have been a success.

9. Other recommendations were made as well. This brief sketch cannot do justice to the study. For further details, see E. Savas, ibid.

During the next four years, only one big snowstorm occurred, and this time the snow was cleaned up within a few hours.

EXERCISES

2.18 "There were more civilian than military amputees during the war. During the period of the war, 120,000 civilians suffered amputations, but only 18,000 military personnel."[10] Does this prove that civilians were more likely to suffer amputations? If not, where does the fallacy lie?

2.19 A national magazine once published a story saying that farmers lead other groups in the consumption of alcohol. As evidence, it pointed to the fact that a rehabilitation center in rural Illinois treated more farmers than other occupational groups for alcoholism. Do you regard this evidence as unbiased? Why, or why not?

2.20 Data have been published which indicate that the more children a couple has, the less likely the couple is to get a divorce. Does this indicate that increases in the number of children are related causally to the likelihood of divorce? Why, or why not?

2.21 Based on the data in Exercise 2.7, a real estate agent says that the typical size of a lot in the relevant township is .54 acres. What sort of average is the agent using? How many lots are of this "typical" size?

2.22 In 1973, the total net income per farm was $19,685 in Iowa and $19,456 in Minnesota. Based on these data, a television commentator maintains that Iowa farmers were better off than Minnesota farmers in 1973. Do you consider this statement to be very meaningful? What sort of pitfall is present here? What would be a better way of interpreting the data concerning these two states?

2.23 "Patents are of little value since the Supreme Court invalidates most of the patents that come before it."[11] Do you agree with this statement? If not, in what way does it represent a misuse of statistics?

2.24 Based on the data in Table 2.8, a newspaper reporter writes that the average snowfall in New York is less than 2 inches and consequently the expense of preparing for snowfalls of more than 10 inches is far in excess of the potential benefits. Do you agree? If not, in what way is this statement a misuse of statistics? What sort of pitfall is present here?

CHAPTER REVIEW

1. There are several frequently used *measures of central tendency:* the *mean*, the *median*, and the *mode*. The *mean* is the sum of the numbers contained in the body of data divided by how many numbers there are. The *median* is a figure which is chosen so that one-half of the numbers in the body of data are below it and one-half are above it. The *mode* is the number that occurs most often. These three kinds of measures may differ substantially from one another. For example, if there are a few extremely high observations

10. W. A. Wallis and H. Roberts, *Statistics* (Glencoe: Free Press, 1956), p. 91.
11. Ibid., p. 98.

(as in most income distributions) the mean will be considerably higher than the median.

2. *Measures of variability* or *dispersion* tell us how much variation there is among the numbers in a body of data. Perhaps the simplest measure of variability is the *range*, which is defined as the difference between the highest and the lowest number in the body of data. However, the most important measure of dispersion is the *standard deviation*, which is defined as the square root of the mean of the squared deviations of the observations from their mean. The square of the standard deviation is called the *variance*.

3. There are many misuses of descriptive statistics. Figures are sometimes presented in such a way that they seem comparable when in fact they are based on different definitions, concepts, time periods, areas, and so forth. Sometimes figures are presented which are based on very small or obviously biased samples. Individuals and companies sometimes choose the type of average that supports their case best, even if this information is misleading. An average is sometimes presented which has so much variability about it that the average alone is misleading. In addition, graphs and charts (such as pictograms) are sometimes presented so as to give a misleading impression. Be on your guard against improper statistical procedures of this sort.

Getting Down to Cases:

A DESCRIPTIVE ANALYSIS OF DIE STAMPINGS[12]

A manufacturing firm (which allowed the actual data below to be published) wanted very much to control the distance between two holes stamped in a piece of metal. For a sample of 49 die stampings, it was found that the distances (in inches) between the two hole centers were as follows:

3.008	3.007	3.007	3.006	3.006
3.006	3.006	3.006	3.006	3.006
3.005	3.005	3.005	3.005	3.005
3.004	3.004	3.004	3.004	3.004
3.004	3.003	3.002	3.002	3.000
3.004	3.003	3.003	3.003	3.002
3.002	3.001	3.001	3.001	3.001
3.001	3.001	3.001	3.000	3.000
3.000	3.000	2.999	2.999	2.999
2.998	2.997	2.997	2.996	

1. The tolerances stated that the distance between the two hole centers should be 3,000 ± .004 inches (that is, not less than 2.996 or more than 3.004 inches).

(a) Construct a frequency distribution of the distances between the hole centers for the sample.

(b) Use the frequency distribution to determine the percent of the sample not meeting the tolerances.

12. This case is based on a section of Theodore Brown, "Quality Control," *Harvard Business Review*, November 1951, pp. 69–80.

(c) What is the mean and the standard deviation of the distances in the sample?

(d) Is the frequency distribution unimodal or multimodal?

(e) What possible reasons can you give for the unsatisfactory performance of this production process? Can you determine the actual reasons?

2. After some research by the firm's production engineers, it was found that the 25 distances shown in the first five horizontal rows of the array of data (in the paragraph before last) were stampings based on the use of die A, while the 24 distances in the last five horizontal rows were stampings based on the use of die B.

(a) Does this information shed any light on the unsatisfactory performance of this production process?

(b) Does this information help to explain your answer to part (d) of question 1 above?

(c) Based on this information, what advice would you give the firm?

APPENDIX 2.1

Rules of Summation

In this chapter, we encountered Σ, the mathematical summation sign. Since this sign will be used frequently in later chapters, it is worthwhile to summarize some of the rules of summation. We know from this chapter's discussion that

$$\sum_{i=1}^{n} X_i = X_1 + X_2 + \ldots + X_n .$$

From this fact, we can establish the validity of the following three rules.

The first rule is:

$$\boxed{\sum_{i=1}^{n} aX_i = a\sum_{i=1}^{n} X_i ,}$$

where a is a constant. To prove that this rule is correct, note that

$$\begin{aligned}\sum_{i=1}^{n} aX_i &= aX_1 + aX_2 + \ldots + aX_n \\ &= a(X_1 + X_2 + \ldots + X_n) \\ &= a\sum_{i=1}^{n} X_i ,\end{aligned}$$

which proves the rule.

The second rule is

$$\boxed{\sum_{i=1}^{n} a = na.}$$

To prove that this rule is correct, note that

$$\sum_{i=1}^{n} a = a \sum_{i=1}^{n} 1$$

$$= a(\underbrace{1 + 1 + \ldots + 1}_{n \text{ terms}})$$

$$= na,$$

which proves the rule.

The third rule is

$$\boxed{\sum_{i=1}^{n} (X_i + Y_i) = \sum_{i=1}^{n} X_i + \sum_{i=1}^{n} Y_i.}$$

To prove that this rule is correct, note that

$$\sum_{i=1}^{n} (X_i + Y_i) = X_1 + Y_1 + X_2 + Y_2 \ldots + X_n + Y_n$$

$$= (X_1 + X_2 + \ldots + X_n) + (Y_1 + Y_2 + \ldots + Y_n)$$

$$= \sum_{i=1}^{n} X_i + \sum_{i=1}^{n} Y_i,$$

which proves the rule.

Finally, let's also consider the concept of double summation. The expression

$$\sum_{i=1}^{n} \sum_{j=1}^{m} X_i Y_j$$

means the sum of the products of X and Y where X takes its first, second, . . ., nth values and Y takes its first, second, . . ., mth values. For example,

$$\sum_{i=1}^{3} \sum_{j=1}^{2} X_i Y_j = X_1Y_1 + X_1Y_2 + X_2Y_1 + X_2Y_2 + X_3Y_1 + X_3Y_2.$$

The following example illustrates the use of these rules of summation.

Example 2.1 If $X_1 = 4$, $X_2 = 6$, and $X_3 = -3$, evaluate the following sums:

(a) $\sum_{i=1}^{3} X_i$

(b) $\sum_{i=1}^{2} X_i^2$

(c) $\sum_{i=1}^{3} 3X_i$

Solution: (a) $\sum_{i=1}^{3} X_i = 4 + 6 - 3 = 7$

(b) $\sum_{i=1}^{2} X_i^2 = 4^2 + 6^2 = 52$

(c) $\sum_{i=1}^{3} 3X_i = 3 \sum_{i=1}^{3} X_i = 3(7) = 21$

3

Probability

3.1 Introduction

Much of statistical theory and practice rests on the concept of probability since, as stressed in Chapter 1, any conclusion concerning a population based only on a sample is subject to a certain amount of uncertainty. The concept of probability is by no means unfamiliar. For example, based on the past examination questions given by a professor, you make rough judgments as to the probability that he or she will include certain kinds of questions in a forthcoming test. Or based on the past performance of certain football teams, you make rough estimates of the probability that one team will defeat another. This chapter will provide an introduction to probability theory; in subsequent chapters, we shall build on this foundation and go deeper into probability theory.

3.2 Definition of Probability

EXPERIMENTS AND SAMPLE SPACES

Any probability pertains to the results of a situation which we call an experiment. An *experiment* is *any process by which data are obtained through the observation of uncontrolled events in nature or through controlled procedures in a laboratory*. If you roll a die, this is an experiment which may have one of six outcomes, depending on which number comes up. Or if you flip a coin, this is also an experiment, which has as possible outcomes either a head or a tail. Or if you take a statistics course, this too is an experiment (although you may not prefer to think of it that way), which has as possible outcomes your passing or your not passing.

Any experiment can result in various ***outcomes***. For example, one possible outcome of your experiment with the die is that a 4 may come up. And one possible (and hopefully very likely) outcome of your experiment with the statistics course is that you will pass it. The ***sample space*** *is the set of* ***all*** *possible outcomes that may occur as a result of a particular experiment. For example, the sample space for the experiment in which you rolled the die is*

$$S = \{1, 2, 3, 4, 5, 6\}. \tag{3.1}$$

In other words, S is a set composed of the numbers that can come up when a die is thrown; that is, it is a set composed of 1, 2, ..., 6. The symbol S is conventionally used to designate the sample space, and the outcomes in this set are called the *elements* of the sample space.

It is important to recognize that the outcomes in a sample space need not be numbers. In the case of your experiment with the statistics course, there are two possible outcomes: You pass, or you don't pass. In this case, the sample space is

$$S = \{\text{pass, not pass}\}.$$

Sometimes it is convenient in cases of this sort to designate the elements of the sample space as 0 and 1, where 0 stands for "pass" and 1 stands for "not pass." Thus, the sample space becomes

$$S = \{0, 1\}.$$

It frequently is useful to represent the sample space visually. One way to do this is to express the possible outcomes of the experiment as points on a graph. For example, if two dice are rolled simultaneously, then a graph showing the sample space can be drawn where the number coming up on the first die is located along the horizontal axis, and the number coming up on the second die is located along the vertical axis. This is shown in Figure 3.1. Each of the 36 points in the graph represents a possible outcome of the roll of the dice.

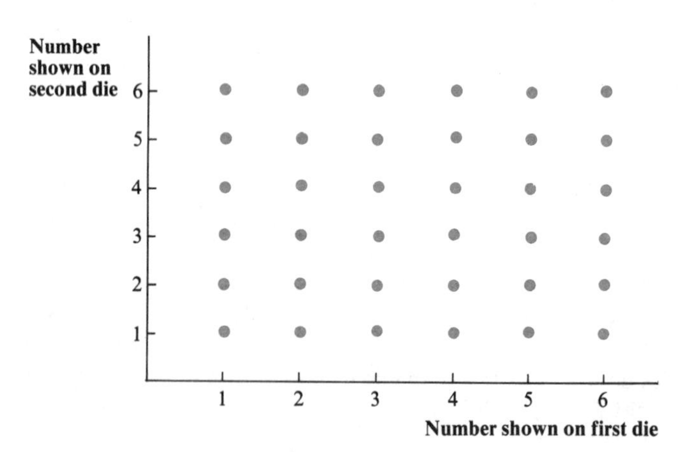

FIGURE 3.1 Sample Space for Simultaneous Roll of Two Dice

Another way to represent the sample space, particularly when an experiment is carried out in stages, is to use a *tree diagram*. Each fork in such a diagram shows the possible outcomes that may occur at a certain stage of the experiment. For example, if you take examinations two days in a row, on the first day there are two possibilities: You pass, or you don't pass. These possibilities are represented by the first fork in Figure 3.2. On the second day there are again two possibilities: Either you pass or you don't pass. These possibilities are represented by the two forks to the right of the first fork. The

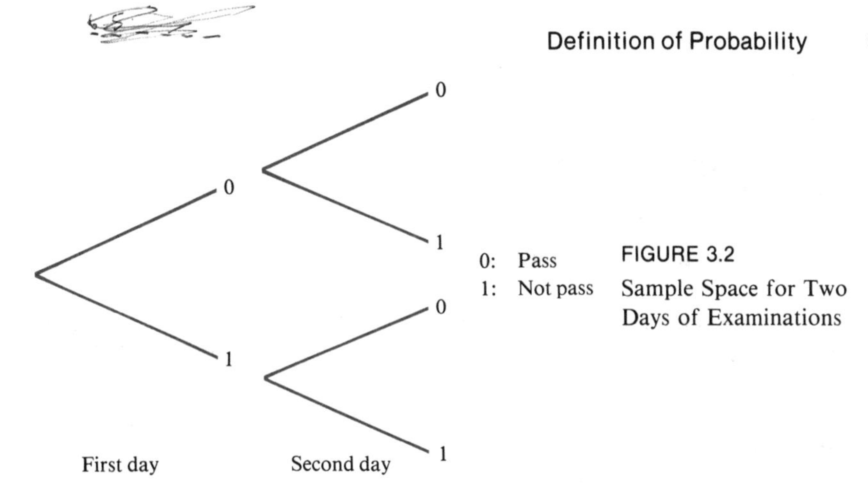

FIGURE 3.2
Sample Space for Two Days of Examinations

upper fork represents the outcomes if you passed on the first day, whereas the bottom fork represents the outcomes if you did not pass on the first day. At the right-hand end of the tree diagram we wind up with four points representing the four elements in the sample space, which are (1) passing on neither the first day nor the second; (2) passing on the first day but not on the second; (3) passing on the second day but not on the first; and (4) passing on both days.

The following two examples should help illustrate the important concept of sample space.

Example 3.1 A market-research firm is interested in the buying habits of families in Topeka, Kansas. The Martin family is among those the firm is studying. The firm is interested in which laundry presoak and detergent boosters the Martins will buy during August of this year. It wants to know whether the Martins buy Axion (a Colgate-Palmolive product), or Biz (a Procter and Gamble product), or both, or neither. Depict graphically the sample space generated by this experiment.

Solution: We can depict this sample space as four points on a graph, as shown in the figure below. In this graph, a 0 on the horizontal axis means that the Martins do not buy Axion in August, while a 1 means that they do

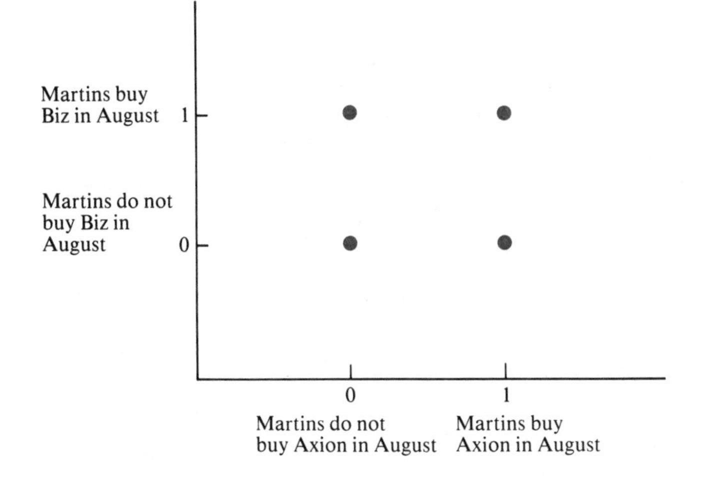

buy it. On the vertical axis, a 0 means that the Martins do not buy Biz in August, while a 1 means that they do buy it.

Example 3.2 Each year, Americans (and particularly New Yorkers) wonder whether there will be a massive electric power failure in New York City (such as occurred in July 1977). Next year, there may or may not be a failure; and in the year after next, there may or may not be one. (Only God and perhaps Consolidated Edison, the local power company, know for sure.) Use a tree diagram to depict the sample space for the next two years.

Solution: We can depict this sample space by constructing the graph below, which shows a fork for the first year leading to a 0 (if a failure occurs) or a 1 (if a failure does not occur). Whether or not there is a failure in the first year, another such fork is given for the second year. Thus, at the right-hand end of the graph, we arrive at the four points in the sample space.

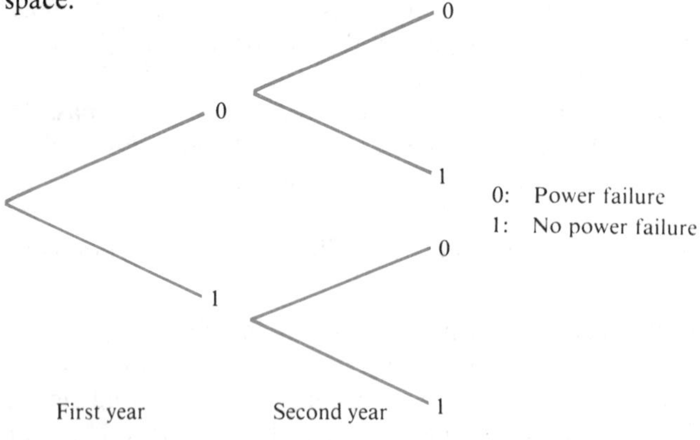

PROBABILITIES

Each element of the sample space is a *possible* outcome of the experiment under consideration. In equation (3.1), each element of the sample space is a possible outcome of throwing the die. In each occurrence of the experiment, *one and only one* outcome can take place. ***Probabilities*** *are numbers that are assigned to each of the elements of the sample space. The number assigned to a particular outcome (that is, to a particular element of the sample space) is the proportion of times that specific outcome occurs over the long run if the experiment is repeated many times under uniform conditions.* Thus, the probability that a particular die will come up a 1 is the proportion of times this will occur if the die is thrown many, many times; and the probability that the same die will come up a 2 is the proportion of times this will occur if the die is thrown many, many times. And so on.

In general, *if an experiment is repeated a very large number of times* M, *and if outcome* U *occurs* m *times, the probability of* U *is*

$$P(U) = \frac{m}{M}. \tag{3.2}$$

Thus, if a die is "true" (meaning that each of its sides is equally likely to come up when the die is rolled), the probability of its coming up a 1 is 1/6, because if it is rolled many, many times, this will occur one-sixth of the time. Moreover, even if the die is not true, this definition can be applied. Suppose that a local mobster injects some loaded dice into a crap game, and that one of the players (who is suspicious) asks to examine one of them. If he rolls this die, what is the probability that it will come up a 1? To answer the question, we must imagine the die in question being rolled again and again. After many thousands of rolls, if the proportion of times that it has come up a 1 is 0.195, then this is the probability of its coming up a 1.

Based on our definition of a probability, the following four fundamental propositions must be true:

1. *The Probability of an Impossible Outcome Must Be Zero.* This follows from the definition of a probability in equation (3.2), because if an outcome is impossible, the number of times the outcome occurs (that is, m) must equal zero.

2. *The Probability of an Outcome that Is Certain Must Equal 1.* This also follows from the definition of a probability in equation (3.2), because if an outcome is certain, the number of times the outcome occurs (m) must equal the number of times the experiment takes place (that is, M).

3. *The Probability of Any Outcome Must Be No Less than Zero and No Greater than 1.* This, too, follows from the definition of a probability in equation (3.2). Since the number of times any outcome occurs (m) cannot be negative, its probability cannot be less than zero. Since the number of times any outcome occurs cannot exceed the number of times the experiment takes place (M), its probability cannot exceed 1.

4. *The Sum of the Probabilities of All Possible Outcomes of the Experiment Must Equal 1.* This is true because one and only one of the outcomes will occur each time the experiment is performed. That is, if $P(U_i)$ is the probability that the ith outcome will occur, we can be sure that

$$\Sigma P(U_i) = 1,$$

where the summation is over all the outcomes in the sample space. Why? Because if we add up the total number of times that each outcome occurs (that is, if we add up the values of m for all possible outcomes), the result must equal the number of times the experiment takes place (that is, M). Consequently, it follows from the definition of a probability in equation (3.2) that the sum of the probabilities of all possible outcomes of the experiment must equal 1.

To illustrate the meaning of these four propositions, consider once again the roll of a single die. According to the first proposition, the probability of the die's coming up a 9 is zero. Why? Because it is impossible for a die to come up a 9. According to the second proposition, if a die is loaded so that it is certain to come up a 5, the probability of its coming up a 5 equals 1. Why? Because the probability of a certain outcome must equal 1. According to the third proposition, the probability of any number coming up must be no less than zero and no greater than 1. According to the fourth proposition, if $P(1)$ is the probability that a particular die comes up a 1, $P(2)$ is the probability that it comes up a 2, . . . , and $P(6)$ is the probability that it comes up a 6, we can be sure that

$$P(1) + P(2) + P(3) + P(4) + P(5) + P(6) = 1.$$

3.3 Probability of an Event

EVENTS

One of the most important functions of probability theory is to enable us to calculate the probability of an event. *An **event** is a subset of a sample space.* A *subset* is any part of a set (including the whole set and the empty set, which has no elements at all). Put less formally, an event can be defined as a group of zero, one, two, or more outcomes of an experiment. For example, if the experiment is rolling a particular die, one event is that the number that comes up is odd. Another event is that the number that comes up is a 3 or a 6. Still another event is that the number that comes up is a 5. There are many practical reasons (other than preparing for trips to Las Vegas or Atlantic City) why statisticians need to be able to calculate the probability of an event.

In a previous section, we said that probabilities are numbers that are attached to each of the elements of the sample space. In other words, they are numbers attached to each possible outcome of the experiment. These numbers can be used to calculate the probability of an event. To do so, the following proposition is used.

> ***Probability of an Event:*** *The probability that an event will take place is the sum of the probabilities of the outcomes that comprise the event.*

This proposition is true because (1) if any of the outcomes comprising the event occurs, the event itself occurs, and (2) more than one outcome cannot occur simultaneously.[1]

For example, if a true die is rolled, what is the probability that it comes up an odd number? The event in which the die comes up an odd number can be broken down into three outcomes: (1) the die comes up a 1; (2) the die comes up a 3; and (3) the die comes up a 5. In other words, these three outcomes comprise this event. Since the probability of each of these outcomes is 1/6, the probability of this event is 1/6 + 1/6 + 1/6, or 1/2.

A somewhat more complex illustration involves the buying habits of the

1. An *elementary event* is a single possible outcome of an experiment. Thus, it is an event that cannot be subdivided further into other events. For example, the event that a 2 appears on a die is an elementary event since it cannot be subdivided into other events, whereas the event that an odd number appears on a die is *not* an elementary event, since it can be subdivided into three elementary events: (1) a 1 appears; (2) a 3 appears; and (3) a 5 appears. The probability assigned to each elementary event, like any probability, must be no less than zero and no greater than 1. Let $E_1, E_2, \ldots, E_r$ be elementary events (where only one of them is certain to occur), and let $P(E_i)$ be the probability that the ith elementary event occurs. Then it must be true that $\Sigma P(E_i) = 1$ (where the summation is from $i = 1$ to $i = r$), and that the probability that either of two elementary events occurs equals the sum of the probabilities of these two elementary events.

Martin family with respect to liquid detergents and scouring powders. If the Martin family can buy up to 3 containers of liquid detergent and up to 3 cans of scouring powder in the next 3 months, the sample space showing the amount of each product purchased during this period is shown in Figure 3.3. For simplicity, let's assume that each of the 16 elements of this sample space is equally likely. This means that the probability of each point in Figure 3.3 is 1/16. Given these probabilities of the outcomes, we can readily calculate the probabilities of various other events.

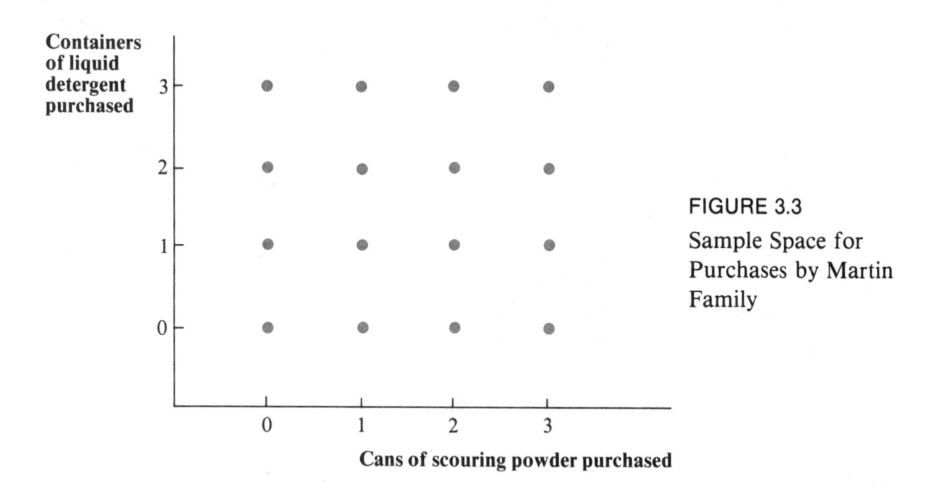

FIGURE 3.3
Sample Space for Purchases by Martin Family

To illustrate, what is the probability that the Martin family buys 3 cans of scouring powder and up to 3 containers of liquid detergent in the next 3 months? This event includes the 4 elements in the sample space designated as subset C in Figure 3.4. Thus, the probability of this event is the sum of the probabilities of these 4 outcomes, or $1/16 + 1/16 + 1/16 + 1/16 = 1/4$. What is the probability that the Martin family buys both 1 can of scouring powder and less than 2 containers of liquid detergent? This event includes the

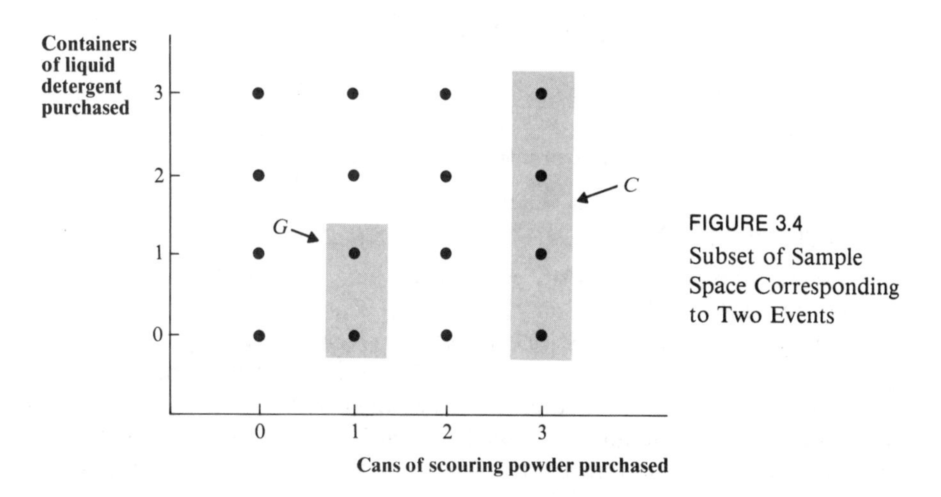

FIGURE 3.4
Subset of Sample Space Corresponding to Two Events

2 elements in the sample space designated as subset G in Figure 3.4. Thus, the probability of this event is the sum of the probabilities of these 2 outcomes, or $1/16 + 1/16 = 1/8$.

PROBABILITY OF EITHER EVENT *A* OR EVENT *B* OR BOTH

A ***composite event*** is an event that is defined by using combinations of other events. One type of composite event occurs if *either* or *both* of two other events occur. For example, if A is the event that you pass your economics course and B is the event that you pass your statistics course, your graduation may depend on *either* or *both* of these events occurring. If so, the composite event that you graduate will occur only if *either* event A or event B (*or both*) occur. Turning to a nonacademic illustration, consider the composite event that the Martin family buys during the next 3 months *either* (1) less than 2 cans of scouring powder; or (2) less than 2 containers of liquid detergent; or (3) *both* (that is, less than 2 cans of scouring powder *and* less than 2 containers of liquid detergent). Subset D of the sample space in Figure 3.5 is the set of outcomes where fewer than 2 cans of scouring powder are bought (event 1). Subset E is the set of outcomes where fewer than 2 containers of liquid detergent are bought (event 2). *The set of outcomes where either or both of these events occurs is composed of all outcomes in either subset* D *or* E. This set consists of the 12 points in the shaded area in Figure 3.5. Since the probability of each of these points is 1/16, the probability of this composite event is 12/16.

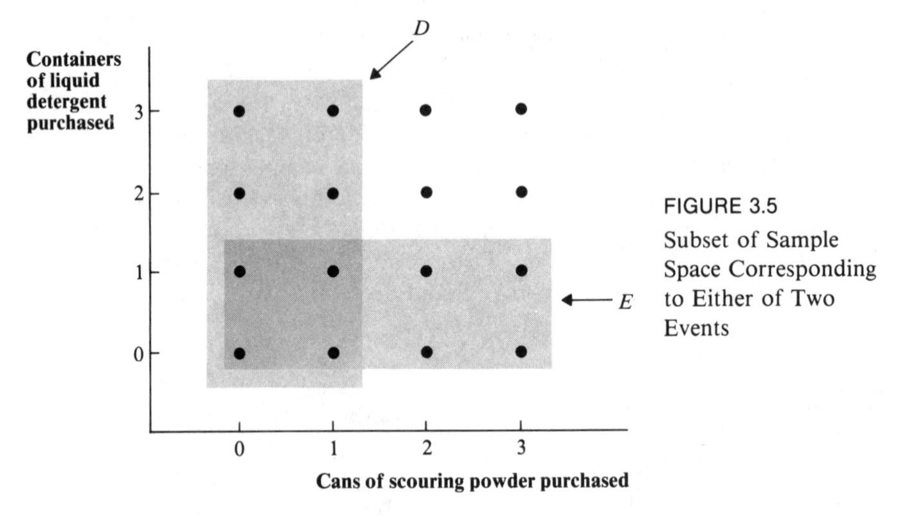

FIGURE 3.5
Subset of Sample Space Corresponding to Either of Two Events

PROBABILITY OF BOTH EVENT *A* AND EVENT *B*

A second type of composite event occurs *only* if *both* of two other events occur. Thus, if A is the event that you pass your economics course and B is the event that you pass your statistics course, your receiving a good job offer may be dependent on the occurrence of *both.* If so, the event that you receive a good job offer will occur only if *both* event A and event B occur. Consider the event that the Martin family buys *both* less than 2 cans of scouring powder *and* less than 2 containers of liquid detergent during the next 3 months. This event

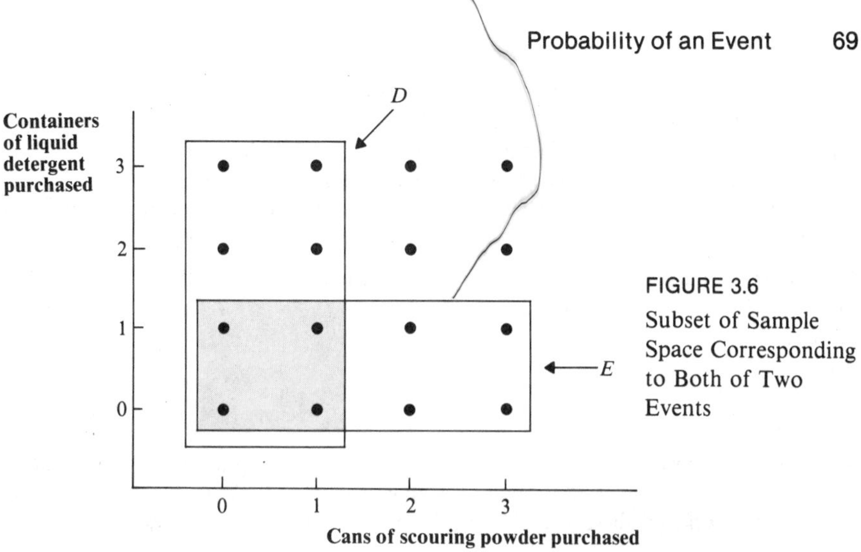

FIGURE 3.6
Subset of Sample Space Corresponding to Both of Two Events

occurs only if *both* of the following events occur: (1) the Martins buy less than 2 cans of scouring powder, and (2) the Martins buy less than 2 containers of liquid detergent. Since the set of outcomes comprising the first event is subset *D* of the sample space in Figure 3.6 and the set of outcomes comprising the second event is subset *E* of the sample space in Figure 3.6, *the set of outcomes comprising both of these events is composed of all outcomes in both subsets* D *and* E. In other words, it is the set of 4 points in the shaded area in Figure 3.6. Since the probability of each of these points is 1/16, the probability of this composite event is 4/16.

EXERCISES

3.1 The Bona Fide Washing Machine Company has decided to buy a firm which manufactures rivets, but has not decided which one to buy. There are three states in which rivet producers exist: Michigan, Illinois, and New York. In each state, a firm producing rivets may have one of three forms of legal organization; it may be a proprietorship, a partnership, or a corporation. Draw a diagram similar to Figure 3.1 showing the nine different ways in which the Bona Fide Washing Machine Company can choose a rivet producer to buy. Put the location of the company along the horizontal axis, and its form of legal organization along the vertical axis. On the horizontal axis, let 1 equal Michigan, let 2 equal Illinois, and let 3 equal New York. On the vertical axis, let 1 equal proprietorship, let 2 equal partnership, and let 3 equal corporation.

3.2 Using the diagram you drew in Exercise 3.1, designate the subset of the sample space corresponding to each of the following events:
(a) The Bona Fide Washing Machine Company buys a Michigan rivet producer.
(b) The Bona Fide Washing Machine Company buys a corporation.
(c) The Bona Fide Washing Machine Company buys a proprietorship in New York.
(d) The Bona Fide Washing Machine Company buys either a corporation or a proprietorship.
(e) The Bona Fide Washing Machine Company buys either a proprietorship or a New York firm (or both).

3.3 Using the results you obtained in Exercises 3.1 and 3.2, indicate the set of outcomes, at least one of which must occur if
(a) the Bona Fide Washing Machine Company buys a Michigan rivet producer;
(b) the Bona Fide Washing Machine Company buys a corporation;
(c) the Bona Fide Washing Machine Company buys either a corporation or a proprietorship.

3.4 A gambler makes a single roll with a pair of true dice. What is the probability that each of the following numbers comes up?
(a) 7
(b) 8
(c) 11
(d) either a 2, a 12, or both

3.5 Using your results from Exercises 3.1 and 3.2, indicate the set of events, all of which must occur if (a) the Bona Fide Washing Machine Company buys a proprietorship in New York; (b) the Bona Fide Washing Machine Company buys a corporation in Michigan.

3.6 In cases where a composite event occurs whenever all of a number of other events occur, the set of outcomes corresponding to the composite event is composed of all outcomes where all the other events occur. Show that this is true for both composite events listed in Exercise 3.5.

3.4 Addition Rule

CASE OF TWO EVENTS

As pointed out in the previous section, statisticians and decision makers frequently must calculate the probability that at least one of two events occurs. For instance, if a fair coin is flipped twice and we want to calculate the probability that it comes up heads at least once, we know that this will occur if (1) the coin comes up heads on the first flip; or (2) the coin comes up heads on the second flip. The probability of the first event is 1/2, and the probability of the second event is 1/2. The probability of both events (that is, that the coin will come up heads *both* times) is 1/4. Can we use these probabilities to determine the likelihood that the coin will come up heads at least once in the two flips?

To solve problems of this sort, statisticians have devised the so-called addition rule, which is as follows.

> ***Addition Rule:*** *If* A *and* B *are two events, and the probability of* A *is denoted by* P(A) *and the probability of* B *is denoted by* P(B), *then the probability of either* A *or* B *(or both), denoted by* P(A *or* B), *equals* P(A) + P(B) − P(A *and* B), *where* P(A *and* B) *is the probability that both* A *and* B *will occur.*

Let's define A as the event that the coin comes up heads on the first flip, and B as the event that the coin comes up heads on the second flip. Then $P(A)$ is the probability of heads on the first flip, which is 1/2. And $P(B)$ is the probability of heads on the second flip, which is 1/2. And $P(A \text{ and } B)$ is the

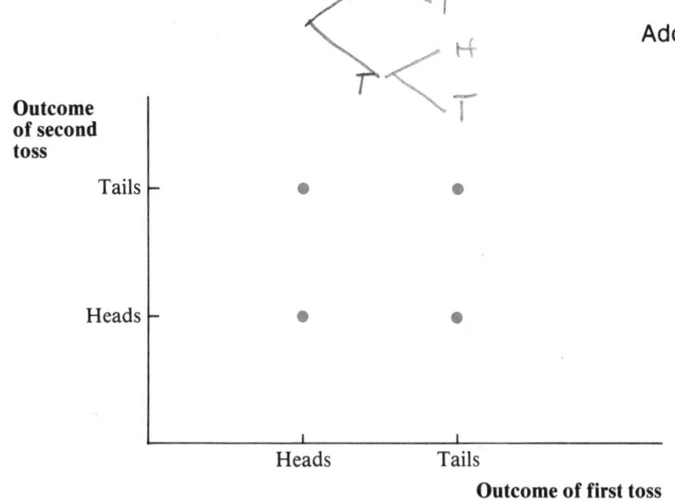

FIGURE 3.7
Sample Space for Two Flips of a Coin

probability of heads on both the first and second flips, which is 1/4. Thus, the probability of at least one heads is

$$P(A \text{ or } B) = P(A) + P(B) - P(A \text{ and } B) = 1/2 + 1/2 - 1/4 = 3/4.$$

This result accords with Figure 3.7, which shows the relevant sample space. There are four possible outcomes in this sample space. The outcomes correspond to

(1) heads/heads (heads first flip, heads second flip)
(2) heads/tails (heads first flip, tails second flip)
(3) tails/heads (tails first flip, heads second flip)
(4) tails/tails (tails first flip, tails second flip)

Each outcome has a probability of 1/4. Since at least one heads comes up in three of these four outcomes, the probability of at least one heads is 3/4, which accords with the result obtained from the addition rule.

This example indicates why $P(A$ and $B)$ must be subtracted from the sum of $P(A)$ and $P(B)$ to obtain $P(A$ or $B)$. If $P(A$ and $B)$ were not subtracted, the outcome where both events A and B occur (that is heads/heads) would be counted twice, since it is included in the outcomes where A (heads on the first flip) occurs, as well as in the outcomes where B (heads on the second flip) occurs. If it were not subtracted, the answer would be incorrectly given as 1, not 3/4.

MUTUALLY EXCLUSIVE EVENTS

In some cases, two events are ***mutually exclusive***; that is, they cannot occur together. For example, if you roll a die once, it is impossible for the die to come up with *both* a 4 *and* a 5, since only one number can appear at a given time. Thus, the event that "the die shows a 4" and the event that "the die shows a 5" are mutually exclusive. *If two events are mutually exclusive, the probability that they both occur is zero; consequently, if the events,* A *and* B, *are mutually exclusive,* P(A *and* B*)* = *0,* with the result that we have the following addition rule for mutually exclusive events.

Addition Rule for Mutually Exclusive Events: *If* A *and* B *are two mutually exclusive events, the probability of either* A *or* B, *denoted by* P(A *or* B), *equals* P(A) + P(B).

Clearly, this addition rule follows directly from the addition rule given earlier in this section since, if $P(A \text{ and } B) = 0$, the earlier addition rule implies that this rule will be true. To illustrate the present addition rule, let's define A as "a die shows a 4" and B as "a die shows a 5." As we have seen, these two events are mutually exclusive (if only one die is thrown only once), so $P(A \text{ or } B) = P(A) + P(B) = 1/6 + 1/6 = 1/3$. Thus, the probability of either a 4 or a 5 is 1/3.

CASE OF MORE THAN TWO EVENTS

Both the general addition rule and the special addition rule for mutually exclusive events can be extended to cases where more than two events are considered. For present purposes it is not necessary to give the extension of the more general addition rule, but it is important to give the extension of the addition rule for mutually exclusive events, which is as follows.

Addition Rule for Any Number of Mutually Exclusive Events: *If* E_1, $E_2 \ldots E_n$ *are* n *events, the probability of* E_1 *or* E_2 *or* $E_3 \ldots$ *or* E_n, *denoted by* $P(E_1 \text{ or } E_2 \text{ or } E_3 \ldots \text{ or } E_n)$, *equals* $P(E_1) + P(E_2) + P(E_3) + \ldots + P(E_n)$, *if these events are mutually exclusive.*

To illustrate the use of this rule, let's define E_1 as "a die shows a 1," E_2 as "a die shows a 2," E_3 as a "die shows a 3," and E_4 as "a die shows a 4." Then since these four events are mutually exclusive (so long as only one die is thrown once), $P(E_1 \text{ or } E_2 \text{ or } E_3 \text{ or } E_4) = P(E_1) + P(E_2) + P(E_3) + P(E_4) = 1/6 + 1/6 + 1/6 + 1/6 = 2/3$. Thus, the probability of a 1, a 2, a 3, or a 4 is 2/3.

COMPLEMENT OF AN EVENT

The ***complement*** of an event occurs when the event itself does not occur. Thus, since the event or its complement is sure to occur, and since the event and its complement are mutually exclusive, the addition rule implies that if $P(A)$ is the probability that event A will occur, and if P (not A) is the probability that event A will not occur, then

$$P(A) + P(\text{not } A) = 1,$$

or

$$P(A) = 1 - P(\text{not } A).$$

This result is useful because, if it is easier to determine P (not A) than $P(A)$, then $P(A)$ can be obtained by deducting P (not A) from 1. Suppose that we want to determine the probability that if two dice are thrown the number that comes up *will not equal* 2. The complement of this event is that the number *will equal* 2. Since the probability of this complementary event is 1/36, the probability that the number will not equal 2 must be $1 - 1/36$, or 35/36.

APPLICATIONS

The addition rule has a host of important applications. The following two examples illustrate ways in which it is used.

Example 3.3 A new chemical product is about to be introduced commercially, and a great many chemical firms are competing to be first to put this product on the market. An industrial analyst believes that the probability is 0.30 that DuPont will be first, 0.15 that Dow will be first, 0.15 that Monsanto will be first, and 0.15 that Union Carbide will be first. Based on the analyst's beliefs, what is the probability that any one of these four firms will be first, assuming that a tie does not occur?

Solution: Let E_1 be the event that DuPont is first, E_2 be the event that Dow is first, E_3 be the event that Monsanto is first, and E_4 be the event that Union Carbide is first. If a tie is not allowed, these events are mutually exclusive, since more than one firm cannot be first. Thus, $P(E_1 \text{ or } E_2 \text{ or } E_3 \text{ or } E_4) = P(E_1) + P(E_2) + P(E_3) + P(E_4) = 0.30 + 0.15 + 0.15 + 0.15 = 0.75$. Consequently, the probability that any one of these four firms will be first is 0.75.

Example 3.4 A bank gives summer jobs to two business school students, Mary Carp and John Minelli. The bank's personnel manager hopes that at least one of these students will decide to go to work for the bank upon graduation. If the probability that each student will decide to work for the bank is 0.3, and the probability that both will decide to work for the bank is 0.1, what is the probability that the personnel manager's hopes will be fulfilled?

Solution: Let E_1 be the event that Mary Carp will decide to go to work for the bank, and E_2 be the event that John Minelli will decide to do so. Since $P(E_1) = 0.3$, $P(E_2) = 0.3$, and $P(E_1 \text{ and } E_2) = 0.1$,

$$P(E_1 \text{ or } E_2) = 0.3 + 0.3 - 0.1 = 0.5.$$

Thus, the probability that the personnel manager's hopes will be fulfilled equals 0.5.

3.5 Multiplication Rule

The addition rule is very useful in cases where we are interested in the probability that *at least one* of several events will take place. Thus, in the previous section, we used it to determine the probability that *at least one* heads would come up in two flips of a true coin. In many situations, however, the statistician or decision maker is interested in the probability that *all* of several events will occur. We may want to determine the probability that *both* flips of a coin will be heads. Or we may want to determine the probability that *all four* of a firm's salesmen will be sick tomorrow. In cases of this sort, the multiplication rule, not the addition rule, is the one to apply. To understand the multiplication rule, it is essential that you be familiar with the concepts of joint probability, marginal probability, and conditional probability.

JOINT PROBABILITIES

In our discussion of the addition rule, we used $P(A \text{ and } B)$, the probability that *both* events A and B will occur. This is an example of a joint probability, which is defined as the *probability of the joint occurrence of two or more events.* As a concrete illustration of joint probabilities, consider Table 3.1, which shows the results of a hypothetical survey in which 10,000 individuals were asked whether or not they favored increased defense spending. As indicated in Table 3.1, 6,000 of those asked were Democrats and 4,000 were Republicans. Let A be the event that a person favored increased defense spending, let C be the event that he or she did not favor increased defense spending, let B be the event that the person was a Democrat, and let D be the event that he or she was a Republican.

TABLE 3.1

10,000 Persons Classified by Attitude toward Increased Defense Spending and by Political Party

Political party	Favors increased defense spending (A)	Does not favor increased defense spending (C)	Total
Democrat (B)	2,500	3,500	6,000
Republican (D)	2,500	1,500	4,000
Total	5,000	5,000	10,000

Joint probabilities can be illustrated as follows. If a person is chosen at random from this group of 10,000, the joint probability that he or she is a Democrat and favors increased defense spending is

$$P(A \text{ and } B) = \frac{2{,}500}{10{,}000} = 0.25.$$

Similarly, the probability that a randomly selected person is a Democrat and does not favor increased defense spending is

$$P(B \text{ and } C) = \frac{3{,}500}{10{,}000} = 0.35.$$

The probability that a randomly selected person is a Republican and favors increased defense spending is

$$P(A \text{ and } D) = \frac{2{,}500}{10{,}000} = 0.25.$$

The probability that a randomly selected person is a Republican and does not favor increased defense spending is

$$P(C \text{ and } D) = \frac{1{,}500}{10{,}000} = 0.15.$$

TABLE 3.2

Joint Probability Table for 10,000 Persons Classified by Attitude toward Increased Defense Spending and by Political Party

Political party	Favors increased defense spending (A)	Does not favor increased defense spending (C)	Marginal probabilities
Democrat (B)	0.25	0.35	0.60
Republican (D)	0.25	0.15	0.40
Marginal probabilities	0.50	0.50	1.00

These joint probabilities can be shown in a *joint probability table* like Table 3.2. The probabilities in Table 3.2 are obtained by dividing each of the numbers in Table 3.1 by the total number of persons in the survey (10,000).

MARGINAL PROBABILITIES

In addition to showing the joint probabilities just mentioned, Table 3.2 shows the probability that a randomly chosen person is a Democrat, is a Republican, is in favor of increased defense spending, or is not in favor of increased defense spending. These probabilities, contained in the margins of the joint probability table, are called *marginal probabilities* or *unconditional probabilities*. (The name *marginal* stems from the position of these probabilities in the margins of the table.) For example, the marginal probability that a randomly chosen person favors increased defense spending is 0.50. The marginal probability that a randomly chosen person in this group is a Democrat is 0.60.

Each marginal probability can be derived by summing the appropriate joint probabilities. For example, the marginal probability that a person favors increased defense spending is the sum of (1) the joint probability that a person is a Democrat and favors increased defense spending; and (2) the joint probability that a person is a Republican and favors increased defense spending. That this is true follows from the addition rule.[2] Consequently, if we did not know the marginal probability that a person favors increased defense spending, but knew the joint probabilities, we could find the marginal probability by adding .25 (the joint probability that a person is a Democrat and favors increased defense spending) and .25 (the joint probability that a person is a Republican and favors increased defense spending). The result of adding the two is of course .50.

2. A person favors increased military spending if at least one of the following events occurs: (1) the person is a Democrat and in favor of increased military spending; (2) the person is a Republican and in favor of increased military spending. Since a person cannot be both a Democrat and a Republican, these two events are mutually exclusive. Thus, according to the addition rule, the probability that a person favors increased military spending equals the sum of the probability of the first of these events and the probability of the second of these events.

CONDITIONAL PROBABILITIES

Statisticians frequently are interested in how the probability of one event is influenced by whether or not another event occurs. *The probability that one event will occur, given that another event is certain to occur, is a* ***conditional probability***. In Table 3.2 we may be interested in calculating the probability that a person favors increased defense spending, given that he or she is a Democrat. This conditional probability is denoted by $P(A|B)$ and is read "the probability of A, given B." (Recall that event A is a person's favoring increased defense spending and event B is his or her being a Democrat.) The vertical line in $P(A \mid B)$ is read "given," and the event following the line (B in this case) is the one that is certain to occur.

Regardless of which events A and B stand for, *the conditional probability of* A, *given that* B *must occur, is*

$$P(A \mid B) = \frac{P(A \text{ and } B)}{P(B)}. \tag{3.3}$$

In other words, the conditional probability of A, given the occurrence of B, is the joint probability of A and B divided by the marginal probability of B. (To rule out the possibility of dividing by zero it is assumed, of course, that the marginal probability of B is non-zero.)

To illustrate the use of the above definition, let's return to Table 3.2 and calculate the probability that a person favors increased defense spending, given that he or she is a Democrat. Since $P(A \text{ and } B) = .25$ and $P(B) = .6$,

$$P(A \mid B) = \frac{P(A \text{ and } B)}{P(B)} = \frac{.25}{.60} = .42.$$

This conditional probability is .42. To verify the correctness of this result we can go back to Table 3.1, which shows that the proportion of Democrats favoring increased defense spending is 2,500/6,000, or .42. Thus, the conditional probability of A given B is simply the proportion of times that A occurs out of the total number of times that B occurs.

STATEMENT OF RULE

Now that we have the definition of conditional probability in equation (3.3) at hand, the statement of the multiplication rule is quite simple. To obtain this rule we need only multiply both sides of equation (3.3) by $P(B)$, the result being as follows.

> ***Multiplication Rule:*** *If* A *and* B *are two events, the joint probability that both* A *and* B *will occur equals the conditional probability of* A, *given* B, *times the probability of* B. *In other words,*

$$P(A \text{ and } B) = P(A \mid B)P(B). \tag{3.4}$$

Using the multiplication rule, we can determine the probability that a randomly selected person is both a Democrat and not in favor of increased defense spending. If we let C be the event that the person is not in favor of increased defense spending, and if we let B be the event that the person is a Democrat, then what we want to determine is $P(C \text{ and } B)$. We know that the probability that a person is a Democrat equals .6. In other words, $P(B) = .6$. We also know that the probability that a person is not in favor of increased defense spending, given that he or she is a Democrat, equals .58. That is, $P(C|B) = .58$. From these facts alone we can determine $P(C \text{ and } B)$, because the multiplication rule implies that

$$P(C \text{ and } B) = P(C \mid B)P(B)$$
$$= (.58)(.6) = .35.$$

Thus, the desired probability equals .35.

Going a step further, the multiplication rule can be extended to include more than two events, the result being as follows.

Multiplication Rule for n Events: *If* $E_1, E_2, \ldots, E_n$ *are* n *events, the joint probability that all these events will occur equals the probability of* E_1 *times the conditional probability of* E_2*, given* E_1*, times the conditional probability of* E_3*, given* E_1 *and* E_2*, . . . times the conditional probability of* E_n*, given* $E_1, E_2, \ldots, E_{n-1}$*. In other words,*

$$P(E_1 \text{ and } E_2 \text{ and} \ldots \text{and } E_n) = P(E_1) \cdot P(E_2|E_1) \cdot P(E_3|E_1 \text{ and } E_2) \cdot \ldots \cdot P(E_n|E_1 \text{ and } E_2 \text{ and} \ldots \text{and } E_{n-1}). \tag{3.5}$$

Consider the following example: (1) the probability that an undergraduate at the University of Michigan will go to graduate school is 1/2; (2) the probability that a Michigan undergraduate will get a master's degree, given that he or she goes to graduate school, is 2/3; and (3) the probability that a Michigan undergraduate will get a Ph.D., given that he or she goes to graduate school and gets a master's degree, is 1/5. What is the probability that a Michigan undergraduate will go to graduate school, will get a master's degree, and will get a Ph.D.? Using equation (3.5), this probability equals (1/2)(2/3)(1/5), or 1/15.

STATISTICAL INDEPENDENCE

The probability of the occurrence of an event is sometimes dependent on whether or not another event occurs. In Table 3.2, the probability that a person favors increased defense spending depends on whether the person is a Democrat or a Republican.[3] As we saw in a previous section, the probability that a person favors increased defense spending, given that he or she is a Democrat equals .42. On the other hand, the probability that an individual

3. Note once again that the figures in Table 3.2 are hypothetical. Neither Democrats, Republicans, nor independents should assume that we regard the figures as accurate. They are used only for illustrative purposes.

favors increased defense spending, given that he or she is a Republican, equals .25/.40, or .62. Thus, these two events—a person's attitude toward increased defense spending and his or her political party—are *dependent* in the sense that the probability of one's occurring depends on whether or not the other occurs.

Not all events are dependent. For example, the probability that the Martin family purchases 2 cans of scouring powder, given that they purchase 2 containers of liquid detergent, is equal to the marginal (or unconditional) probability that they purchase 2 cans of scouring powder. (Both probabilities equal 1/4, as is evident from our discussion of Figure 3.3.) In other words, the probability that the Martins will purchase 2 cans of scouring powder is not influenced by whether or not they purchase 2 containers of liquid detergent. Similarly, if a fair coin is flipped a number of times the probability of its coming up heads does not depend on whether the coin came up heads on the last flip. Why? Because the coin's behavior on one flip is not influenced by its behavior on the last flip (or on any other flip, for that matter).

The definition of statistical independence is as follows.

Statistical Independence. *If events* A *and* B *are statistically independent, the probability of the occurrence of one event is not affected by the occurrence of the other. That is, each of the following equations is true:*

$$P(A \mid B) = P(A) \tag{3.6a}$$

$$P(B \mid A) = P(B). \tag{3.6b}$$

The example below illustrates how this definition can be used.

Example 3.5 Two machines are drawn at random from a population of 10 machines, 3 of which are defective and 7 of which are not defective. The sampling is *not* carried out with replacement. (In other words, the first machine chosen is *not* put back into the population before the second machine is chosen.) Is whether or not the second machine is defective statistically independent of whether or not the first is defective?

Solution: Given that the first machine selected is defective, what is the probability that the second machine selected will be defective? If the machine selected first is *not* put back into the population (and thus has no chance of being selected again), this probability equals 2/9, since only 2 of the 9 machines left in the population are defective. On the other hand, if the first selection were not defective, this probability would be 3/9. Thus, the probability of the second selection's being defective depends on whether the first selection is defective, which means that the results of each selection are *not* statistically independent. (However, *if the sampling were done with replacement, the results of each selection would be statistically independent, since the probability of getting a defective would be 3/10, regardless of the outcome of earlier selections.*)

THE MULTIPLICATION RULE WITH INDEPENDENT EVENTS

When two events are statistically independent, the multiplication rule can be simplified in the following way.

Multiplication Rule for Two Independent Events. *If* A *and* B *are statistically independent events, the joint probability that both* A *and* B *will occur equals the unconditional probability of* A *times the unconditional probability of* B. *In other words,*

$$P(A \text{ and } B) = P(A) \cdot P(B) \tag{3.7}$$

Moreover, the extension of the multiplication rule to situations where there are more than two events can be simplified in the following way when all of the events are statistically independent.

Multiplication Rule for** n **Independent Events. *If* E_1, E_2, . . , E_n *are events, the joint probability that all* n *events will occur equals the product of their unconditional probabilities of occurrence if all the events are statistically independent. In other words,*

$$P(E_1 \text{ and } E_2 \text{ and } \ldots \text{ and } E_n) = P(E_1) \cdot P(E_2) \cdot \ldots \cdot P(E_n). \tag{3.8}$$

APPLICATIONS

The multiplication rule is a powerful aid to the solution of a wide variety of practical problems in business, economics, and other fields. The following are three examples of how the multiplication rule and the concept of statistical independence can be used.

Example 3.6 Two cards are chosen at random and without replacement from an ordinary deck of playing cards. What is the probability that both are hearts?

Solution: Let A be the event that the first card is a heart and B be the event that the second card is a heart. Clearly, $P(A) = 13/52$, since 13 cards out of the 52 in the deck are hearts. And $P(B|A) = 12/51$, since the probability that the second card is a heart (*given that the first is a heart*) is 12/51 because 12 cards out of the remaining 51 in the deck are hearts. Applying the multiplication rule,

$$P(A \text{ and } B) = P(B|A) \cdot P(A)$$
$$= 12/51(13/52) = 1/17.$$

Note that the multiplication rule implies that $P(A \text{ and } B)$ equals $P(B|A) \cdot P(A)$, as well as $P(A|B) \cdot P(B)$, which is the expression in equation (3.4).[4]

Example 3.7 A salesman must call on all four of his customers in a certain area. The probability that he finds each of them in his or her office on a particular day is 1/2, and whether or not one customer is in is statistically independent of whether any of the others is in. What is the probability that the salesman will find all of the customers in their offices on this day?

4. Because the definition of a conditional probability in equation (3.3) implies that $P(B|A) = P(A \text{ and } B) \div P(A)$, it follows that $P(A \text{ and } B) = P(B|A)P(A)$.

Solution: Let E_1 be the event that the first customer is in his or her office, E_2 be the event that the second customer is in, E_3 be the event that the third customer is in, and E_4 be the event that the fourth customer is in. Since these events are statistically independent, the multiplication rule implies that

$$P(E_1 \text{ and } E_2 \text{ and } E_3 \text{ and } E_4) = P(E_1) \cdot P(E_2) \cdot P(E_3) \cdot P(E_4)$$
$$= (1/2)(1/2)(1/2)(1/2) = 1/16.$$

Example 3.8 An automobile company has 4,000 dealers, who are polled to find out their annual incomes and their ages. The number of dealers in each income and age category is as follows:

Incomes (dollars)	Dealers under 45	Dealers 45 and above	Total
Under 25,000	1,000	1,000	2,000
25,000 and over	500	1,500	2,000
Total	1,500	2,500	4,000

(a) What is the probability that a randomly chosen dealer (1) will have an income of under $25,000; (2) will be under 45; (3) will have an income of under $25,000, given that he or she is under 45?

(b) Are age and income statistically independent?

Solution: (a) The probability that a dealer will have an income of under $25,000 is 2,000/4,000, or 1/2. The probability that a dealer will be under 45 is 1,500/4,000, or 3/8. The probability that a dealer will have an income of under $25,000, given that he or she is under 45, is 1,000/1,500, or 2/3.

(b) Age and income are not statistically independent because the probability that a dealer has an income of under $25,000 is different if he or she is under 45 than if he or she is 45 or over. If a dealer is under 45 this probability is 1,000/1,500, or 2/3. If a dealer is 45 or over this probability is 1,000/2,500, or 2/5.

EXERCISES

3.7 The Jones family decides to buy a new car and narrows the choice down to a Ford, a Chevrolet, or a Toyota. The probability that they will buy a Ford is 0.3, and the probability that they will buy a Chevrolet is 0.4.
(a) What is the probability that they will purchase either a Ford or a Chevrolet?
(b) What is the probability that they will purchase a Toyota?
(c) If the Joneses were undecided as to whether or not they would purchase a car at all, would this influence your answer to (b)? If so, would you increase or decrease your answer to (b)?

3.8 The probability that the Jones family will purchase a Ford is 0.3 and the probability that they will purchase a station wagon is 0.2. If the probability that they will purchase a Ford station wagon is 0.05, what is the probability that they will purchase either a Ford or a station wagon (or both)?

3.9 The Jones family is undecided about whether or not to buy a new car. If the probability is 0.9 that they will buy one, and if the probability is 0.3 that they will buy a

Ford, and if the probability is 0.4 that they will purchase a car getting more than 20 miles per gallon, what is the probability
(a) that they will buy either a car getting more than 20 miles per gallon or a Ford, if there are no Ford cars that get more than 20 miles per gallon;
(b) that they will buy either a car getting more than 20 miles per gallon or a Ford (or both), if all Fords get more than 20 miles per gallon?

3.10 The Black Belt Sewing Machine Company has four plants. The number of motors shipped in 1978 by each supplier to each of Black Belt's plants is as follows:

	Supplier				
Plant	I	II	III	IV	V
1	200	100	300	100	400
2	300	200	400	200	300
3	200	300	300	300	100
4	100	300	200	400	200

(a) If the firm picks a motor from among those received by plant 2 (and if each motor has the same probability of being chosen) what is the probability that this motor came from supplier III?
(b) If the firm picks a motor from among those received at any of its plants, what is the probability that it came from supplier III?
(c) If event X is that a motor comes from supplier III, is this event independent of the plant receiving the motor?

3.11 Based on the data in Exercise 3.10, calculate the probability
(a) that a motor goes to plant 1, given that it comes from supplier II;
(b) that a motor comes from supplier II, given that it goes to plant 1;
(c) that a motor goes to plant 1, given that it comes from either supplier II or supplier III;
(d) that a motor comes from supplier II, given that it goes to either plant 1 or plant 2.

3.12 The Black Belt Sewing Machine Company finds that the probability that a motor both comes from supplier I and is defective is 0.1. Given the data in Exercise 3.10, what is the probability that a motor purchased from supplier I is defective?

3.13 If 20 percent of all the motors received by the Black Belt Sewing Machine Company are defective, what is the probability that a motor comes from supplier I, given that it is found to be defective? (You will need to use some of the data from Exercise 3.12.)

3.14 An article in a New York newspaper by two well-known columnists stated that if the probability of downing an attacking airplane were 0.15 at each of five defense stations, and if a plane had to pass all five stations before arriving at the target, the probability that the plane would be downed before reaching the target was 0.75.[5]
(a) Do you agree with this reasoning?
(b) If not, what is the correct answer?

3.15 Suppose that $P(A) = 0.6$, $P(B) = 0.3$, and $P(A \text{ and } B) = 0.1$.
(a) Are A and B mutually exclusive events? Why, or why not?
(b) Are A and B statistically independent events? Why, or why not?

3.16 Grace Jones is selling her house. She believes there is a 0.2 chance that each

5. W. A. Wallis and H. Roberts, *Statistics* (Glencoe: Free Press, 1956), p. 96.

person who inspects the house will purchase it. What is the probability that more than two people will have to inspect the house before Grace Jones finds a buyer? (Assume that the decisions of the people inspecting the house are independent.)

3.17 A true die is rolled twice. What is the probability of getting (a) a total of 6; (b) a total of less than 6; (c) a total of 7 or more?

3.18 Three cards are to be drawn from a deck with the five of hearts missing.
(a) What is the probability that all three cards will be clubs?
(b) What is the probability that all three cards will be of the same suit?

3.6 Bayes' Theorem

STATEMENT OF THE THEOREM

The Reverend Thomas Bayes, an eighteenth-century British scholar, put forth a theorem concerned with the calculation of the probability that a particular hypothesis is true, given that particular events occur. Or, put more crudely, the theorem is concerned with the probability of a particular cause, given the observation of a particular effect. For example, suppose that a firm receives goods in lots from three trucking companies. If this firm receives a lot with more than 2 percent defectives, Bayes' theorem might be used to calculate the probability that each trucking company delivered it. Because it reasons backward from effects to causes, Bayes' theorem has been a controversial part of statistics. Although no one denies that it is formally valid, many great statisticians have felt that it could be applied correctly in very few cases. One reason why it has been viewed with skepticism is that in order to apply this theorem, one must know the probabilities that various hypotheses or causes are true, which often is far from easy. Nonetheless, in recent years more and more statisticians have come to use Bayes' theorem, as we shall see in subsequent chapters.

Bayes' Theorem: *Suppose that there are* m *hypotheses—*$H_1, H_2, \ldots, H_m$. *Only one of these hypotheses can be true, and one of them must be true. The probability that the* i*th hypothesis is true is* $P(H_i)$. *Then if an event* E *occurs, the conditional probability that the* i*th hypothesis is true equals*

$$P(H_i|E) = \frac{P(E|H_i)P(H_i)}{\sum_{i=1}^{m} P(E|H_i)P(H_i)}. \qquad (3.9)$$

The result in equation (3.9) can be derived from our previous findings. From our definition of a conditional probability in equation (3.3), we know that

$$P(H_i|E) = \frac{P(H_i \text{ and } E)}{P(E)}. \qquad (3.10)$$

Moreover, since the unconditional probability that E occurs is the sum of the joint probability that E occurs and H_1 is true, that E occurs and H_2 is true, that E occurs and H_3 is true, and so on, it follows that

$$P(E) = \sum_{i=1}^{m} P(H_i \text{ and } E).$$

Substituting the right-hand side of this expression for $P(E)$ in equation (3.10), we have

$$P(H_i|E) = \frac{P(H_i \text{ and } E)}{\sum_{i=1}^{m} P(H_i \text{ and } E)}. \tag{3.11}$$

Using the multiplication rule in equation (3.4), it is evident that

$$P(H_i \text{ and } E) = P(E|H_i)P(H_i).$$

Thus, substituting the right-hand side of this expression for $P(H_i \text{ and } E)$ in equation (3.11), we have

$$P(H_i|E) = \frac{P(E|H_i)P(H_i)}{\sum_{i=1}^{m} P(E|H_i)P(H_i)},$$

which is what we set out to derive.

Two things should be noted concerning the conditions underlying Bayes' theorem. First, it is assumed that the m hypotheses are *mutually exclusive* and *exhaustive*. In other words, *only one* of them can be true, but one of them *must* be true. Second, it is assumed that the probability that each of these hypotheses is true is *known*. These probabilities are called ***prior probabilities***. Much more will be said about prior probabilities in Chapters 15 and 16.

AN APPLICATION

To illustrate the application of Bayes' theorem, we return to the case where a firm receives goods in lots from three trucking companies. The probability that each trucking company (Keepon, Superior, or Never-Fail) delivers a lot to this firm is shown in Table 3.3. And the probability that a lot

TABLE 3.3

Probability that Lot Was Delivered by Each of Three Trucking Companies, and Contains More than 2 Percent Defectives, Given that Each Company Delivered It

A. Probability that lot is delivered by	
Keepon Trucking Company	0.20
Superior Trucking Company	0.40
Never-Fail Trucking Company	0.40
B. Probability that lot contains more than 2 percent defectives if delivered by	
Keepon Trucking Company	0.010
Superior Trucking Company	0.020
Never-Fail Trucking Company	0.025

will contain more than 2 percent defectives, given that each trucking company delivers the lot, is also shown in Table 3.3. If this firm receives a lot with more than 2 percent defectives, what is the probability that each trucking company delivered it?

At first glance, the application of Bayes' theorem to this problem may not be obvious. To see how it is applicable, it is important to recognize what the hypotheses are and what the event is. Clearly, there are three hypotheses: (1) Keepon delivered the lot (H_1); (2) Superior delivered the lot (H_2); and (3) Never-Fail delivered the lot (H_3). These hypotheses are mutually exclusive and exhaustive, as the theorem requires. (Hypotheses of this sort are often called *states of nature*. For example, a particular state of nature—H_1—is that Keepon delivered the lot.)

What is the event? It is the fact that a lot with more than 2 percent defectives was delivered. If we call this event E, it follows from equation (3.9) that

$$P(H_1|E) = \frac{P(E|H_1)P(H_1)}{P(E|H_1)P(H_1) + P(E|H_2)P(H_2) + P(E|H_3)P(H_3)},$$

$$P(H_2|E) = \frac{P(E|H_2)P(H_2)}{P(E|H_1)P(H_1) + P(E|H_2)P(H_2) + P(E|H_3)P(H_3)},$$

$$P(H_3|E) = \frac{P(E|H_3)P(H_3)}{P(E|H_1)P(H_1) + P(E|H_2)P(H_2) + P(E|H_3)P(H_3)}.$$

And if we insert the numbers in Table 3.3 into these formulas, we have

$$P(H_1|E) = \frac{(.01)(.20)}{(.01)(.20) + (.02)(.40) + (.025)(.40)} = \frac{.002}{.02} = 0.10,$$

$$P(H_2|E) = \frac{(.02)(.40)}{(.01)(.20) + (.02)(.40) + (.025)(.40)} = \frac{.008}{.02} = 0.40,$$

$$P(H_3|E) = \frac{(.025)(.40)}{(.01)(.20) + (.02)(.40) + (.025)(.40)} = \frac{.010}{.02} = 0.50.$$

In other words, the probability that Keepon delivered the lot is 0.10, the probability that Superior delivered the lot is 0.40, and the probability that Never-Fail delivered the lot is 0.50.

3.7 Subjective or Personal Probability

The probabilities used in the application of Bayes' theorem often are based on a different definition of probability than that presented in Section 3.2. There the probability of an event was defined as the proportion of times that the event will occur in the long run if the relevant experiment is repeated over and over. This is the so-called *frequency definition of probability*. For example, as pointed out in Section 3.2, the probability that a particular die will come up a 1 can be viewed as the proportion of times that this will occur if the die is thrown innumerable times.

Some experiments are not easy to interpret in these terms because they cannot be repeated over and over. A new commercial product may have a different probability of succeeding if it is put on the market this month rather than next month. This is an "experiment" that cannot be performed over and over because market and other conditions vary from month to month. If the new product is *not* introduced this month but next, the experiment will be performed under different conditions and thus will be a different experiment. There are many events of this type in business and economics.

In dealing with events and experiments of this sort statisticians and decision makers sometimes use a *subjective* or *personal definition of probability*. According to this definition, *the probability of an event is the degree of confidence or belief on the part of the statistician or decision maker that the event will occur*. For example, if the decision maker believes that event A is more likely to occur than event B, the probability of A is higher than the probability of B. If the decision maker believes that the odds are 50-50 that a particular event will occur, the probability attached to this event equals 0.50. The important factor in this concept of probability is what the decision maker believes.

To illustrate subjective probability, suppose that a marketing manager must decide whether to substitute a new product label for an old one. The manager must estimate as accurately as possible the probability that the new label is superior to the old. The relevant experiment cannot be conducted over and over; that is, the new label cannot be substituted for the old one again and again to determine the proportion of cases in which the substitution improves the product's profitability. Instead, the manager must use his or her knowledge, experience, and intuition, together with whatever objective information can be obtained, to estimate the chances that the new label is superior to the old. If the conclusion is that the probability of its being superior is 0.50, then this is the manager's subjective probability of this event.

The same rules of probability apply, regardless of whether probabilities are given a frequency definition or a subjective definition. Therefore, all the principles presented in this chapter apply to both types of probabilities. Since subjective probabilities must conform to the same mathematical rules as probabilities based on the frequency definition, and since they can be manipulated in essentially the same way to solve problems, we shall not distinguish subjective probabilities from frequency-based probabilities in most of the succeeding parts of this book.[6]

3.8 A Case Study in Marketing[7]

The following is a realistic illustration of how both Bayes' theorem and subjective probabilities are used in business situations.

6. For some classic papers concerning alternative definitions of probability, see R. von Mises, "Probability: An Objectivist View," and L. J. Savage, "Probability: A Subjectivist View," in E. Mansfield, *Statistics for Business and Economics: Readings and Cases* (New York: Norton, 1980).
7. This example is based to some extent on P. Green and R. Frank, "Bayesian Statistics and Marketing Research," *Applied Statistics*, 1966, pp. 173–90. The numbers have been changed.

A firm is concerned about the sales appeal of the label on one of its products. A designer is hired to create some new labels, and when the company's executives see the various new labels, one is consistently regarded as the best. However, the firm's marketing manager is still uncertain as to whether or not the new label will substantially increase the product's sales.

To obtain more information as to whether or not the new label is superior to the old one, the marketing manager sets in motion a survey of consumers. This survey is designed in such a way that if the new label is really superior, the probability that the survey will indicate that it is superior is 0.8, the probability that it will yield ambiguous results is 0.1, and the probability that the survey will indicate that the label is not superior is 0.1. Thus,

$$P(b|B) = 0.8; P(a|B) = 0.1; P(n|B) = 0.1,$$

where $P(b|B)$ is the probability that the survey will indicate that the label is superior, given that in fact it is superior; $P(a|B)$ is the probability that the survey will yield ambiguous results if the new label is in fact superior; and $P(n|B)$ is the probability that the survey will indicate that the label is not superior, given that it is in fact superior.

The survey is also designed so that if the new label is *not* superior, the probability that the survey will indicate that it is not superior is 0.7, the probability that the survey will be ambiguous is 0.1, and the probability that the survey will indicate that it is superior is 0.2. Thus,

$$P(n|N) = 0.7; P(a|N) = 0.1; P(b|N) = 0.2,$$

where $P(n|N)$ is the probability that the survey will indicate that the label is not superior given that in fact it is not superior; $P(a|N)$ is the probability that the survey will yield ambiguous results if in fact the new label is not superior; and $P(b|N)$ is the probability that the survey will indicate that the label is superior, given that in fact it is not superior.

Despite the enthusiasm of the firm's executives, the marketing manager believes that the probability that the new label is in fact superior is only 0.5. As we saw in the previous section, this is a subjective probability, based on the manager's experience and judgment. This probability is formulated prior to knowing the results of the survey. When the results of the survey come in, they indicate that the new label is superior. Given these survey results, what is the probability that the new label is *in fact* superior? Can you use Bayes' theorem to answer this question?

To apply Bayes' theorem to this problem, one must begin by sorting out the hypotheses and the event. The two hypotheses are (1) the new label is in fact superior (hypothesis B) and (2) the new label is in fact not superior (hypothesis N). The event is the fact that the survey indicates that the new label is superior. Clearly, on the basis of equation (3.9),

$$P(B|b) = \frac{P(b|B)P(B)}{P(b|B)P(B) + P(b|N)P(N)} \tag{3.12}$$

and

$$P(N|b) = \frac{P(b|N)P(N)}{P(b|B)P(B) + P(b|N)P(N)}, \tag{3.13}$$

where $P(B|b)$ is the probability that the new label is really superior, given that the survey indicates that this is so, and where $P(N|b)$ is the probability that the new label is really not superior, given that the survey indicates that it is superior.

Substituting the figures presented in the several previous paragraphs for the symbols in equations (3.12) and (3.13), we have

$$P(B|b) = \frac{(0.8)(0.5)}{(0.8)(0.5) + (0.2)(0.5)} = 0.8$$

and

$$P(N|b) = \frac{(0.2)(0.5)}{(0.8)(0.5) + (0.2)(0.5)} = 0.2.$$

Thus, given the marketing manager's prior probabilities and the fact that the survey indicates that the new label is superior, the probability that it is in fact superior is 0.8, while the probability that it is not in fact superior is 0.2.

EXERCISES

3.19 The Bona Fide Washing Machine Company loses a shipment of its goods due to a mistake by one of its executives. The firm has three executives (Tom, Dick, and Jane) who could have made the mistake, and it can't tell with certainty which one was responsible. However, on the basis of past performance, it is known that the probability that Tom would make such a mistake (if he were responsible for a shipment) is .001, the probability that Dick would do so is .002, and the probability that Jane would do so is .001. If Tom is responsible for 50 percent of the shipments, while the rest are split evenly between Dick and Jane, what is the probability that each executive was responsible for the shipment that was lost?

3.20 The Bona Fide Washing Machine Company is considering John Jones for a key position in marketing. After interviewing Jones and reviewing his credentials, Bona Fide's president feels there is a 75 percent chance that Jones would do well in the job and a 25 percent chance that he would not. The president decides to have a management consulting firm evaluate Jones. According to the experience of other companies, the consulting firm is reasonably accurate in evaluating marketing managers, as indicated by the following table:

	Conditional probability (given actual experience) that consulting firm predicts:	
Actual experience	Manager will do well	Manager will not do well
Manager does well	0.9	0.1
Manager does not do well	0.2	0.8

(a) If the consulting firm predicts Jones will do well, what is the probability that this will be true?
(b) If the consulting firm predicts Jones will not do well, what is the probability that this will be true?

3.21 There are four roads (A, B, C, and D) from Avon to Burgundy and three roads (a, b, and c) from Burgundy to Coventry. The Bona Fide Washing Machine Company sends a shipment from Avon to Coventry via Burgundy. This shipment gets lost after reaching Coventry, and the traffic agent is asked what route it took from Avon to Coventry. He doesn't know, and says that each possible route was equally likely. Eventually, he is forced to guess, and he guesses that the shipment went via road A to Burgundy and via road b to Coventry. What is the probability that he is correct?

3.22 Distinguish between a subjective probability and one based on the frequency definition. Why are subjective probabilities ever used?

3.23 If a reporter feels that the odds are 2 to 1 that he can get an interview with the governor of his state, what is his personal probability that this will happen?

3.24 If you believe the odds are 3 to 1 that you will get at least a B in your statistics course, what is the probability (your subjective probability) that this is the case?

3.25 According to R. A. Fisher, advocates of Bayes' theorem "seem forced to regard mathematical probability, not as an objective quantity measured by observed frequencies, but as measuring merely psychological tendencies, theorems respecting which are useless for scientific purposes."[8] Do you agree? Do you think Bayes' theorem and subjective probability are useless for business decision making? Why, or why not?

CHAPTER REVIEW

1. Statisticians use the term *experiment* to describe any process by which data are obtained through the observation of uncontrolled events in nature or through controlled laboratory procedures. The *sample space* is defined as the set of all possible outcomes that may occur as a result of a particular experiment. In each occurrence of the experiment, one and only one outcome takes place. Sample spaces are often depicted as points on a graph or in tree diagrams. *Probabilities* may be regarded as numbers that are assigned to each of the elements of the sample space. It is essential that each of these numbers be greater than or equal to zero, and that the sum of the numbers assigned all of the elements in the sample space be equal to 1.

2. The *addition rule* states that, if $P(A)$ is the probability that event A occurs and if $P(B)$ is the probability that event B occurs, then $P(A \text{ or } B) = P(A) + P(B) - P(A \text{ and } B)$. Of course, if A and B are mutually exclusive, it follows that $P(A \text{ or } B) = P(A) + P(B)$. If $E_1, E_2, \ldots, E_n$ are mutually exclusive events, then $P(E_1 \text{ or } E_2 \text{ or} \ldots \text{or } E_n) = P(E_1) + P(E_2) + \ldots + P(E_n)$.

3. The *conditional probability* of event A given the occurrence of event B—that is, $P(A \mid B)$—equals $P(A \text{ and } B) \div P(B)$. If $P(A \mid B) = P(A)$, event A is said to be statistically independent of event B. The *multiplication rule* states that if A and B are two events, the probability that both occur—$P(A \text{ and } B)$—equals $P(A \mid B)P(B)$. Of course, if A and B are independent, it follows that $P(A \text{ and } B)$ equals $P(A)P(B)$. If $E_1, E_2, \ldots, E_n$ are statistically independent events, $P(E_1 \text{ and } E_2 \text{ and} \ldots \text{and } E_n) = P(E_1) \cdot P(E_2) \cdot \ldots \cdot P(E_n)$.

4. *Bayes' theorem* is concerned with the calculation of the probabilities

8. R. A. Fisher, "Statistical Inference," in E. Mansfield, op. cit.

that particular hypotheses are true, given that particular events occur. This theorem assumes that there are a number of hypotheses—$H_1, H_2, \ldots, H_m$—that are mutually exclusive and exhaustive. It is assumed that we know the probability that each of these hypotheses is true. These probabilities, called *prior probabilities*, are $P(H_1), P(H_2), \ldots, P(H_m)$. If E is an event and if $P(E|H_i)$ is the probability of the event occurring given that H_i is true, Bayes' theorem states that the probability that H_i is true (given that the event occurs) is

$$P(H_i|E) = \frac{P(E|H_i)P(H_i)}{\sum_{i=1}^{m} P(E|H_i)P(H_i)}.$$

Getting Down to Cases:

THE RELIABILITY OF THE APOLLO SPACE MISSION[9]

The Apollo mission with its objective of landing men on the moon involved severe problems of reliability. For the mission to be successful, all (or at least a great many) components of the Apollo system had to operate properly. A malfunction of any one of many components could have resulted in mission failure. Experts used statistical theory to develop sophisticated methods for estimating the probability that the undertaking would be successful. We are able here to present only simplified versions of what was done. A schematic version of the basic Apollo module is shown in Figure 3.8. We assume that the module works if, and only if, all five components function properly.

(a) Suppose that the probability is 0.99 that each of the five Apollo components functions properly. If the components are independent, what is the probability that a mission will succeed?

One way of increasing the reliability of the Apollo system was through the use of a parallel configuration of components. For example, the second stage of the Saturn rocket used in the Apollo program had five rocket motors, and if any one of the motors failed, the others could be used for a satisfactory earth orbit. To see how the use of a parallel configuration of components increases reliability, suppose that the main engine component

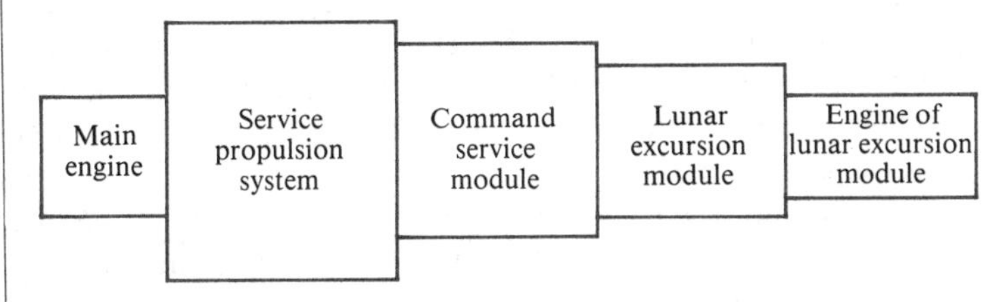

FIGURE 3.8 Simplified Representation of Apollo Module

9. This case is based to some extent on a section from G. Lieberman, "Striving for Reliability," in J. Tanur, F. Mosteller, W. Kruskal, R. Link, R. Pieters, and G. Rising. *Statistics: A Guide to the Unknown* (San Francisco: Holden-Day, 1972).

in Figure 3.8 is replaced by two engines, either of which is able to perform the tasks required by the mission. If either of these engines malfunctions, the other can take over.

(b) If the probability is 0.99 that each of the engines functions properly, what is the probability that at least one of the engines functions properly?

(c) If the main engine component in Figure 3.8 is replaced by two such engines, what is the probability that a mission will succeed?

(d) Compare your result in (c) with that in (a) and interpret the difference between the two results.

(e) What are the disadvantages of a parallel configuration of components? Why not add as many redundant components as possible in order to increase reliability?

APPENDIX 3.1
Counting Techniques

In many practical situations, it is important to be able to calculate the number of ways in which an event can occur. Such calculations must often be carried out in order to compute the probability of an event.

SELECTION OF ITEMS

Frequently, a selection or a choice must be made in several steps, and at each step there are a number of alternatives. For example, suppose that a top student at a leading business school is dressing for an interview with a potential employer, and that she has a choice of a beige, blue, or green dress. Given her choice of dress color, she has a choice of beige, black, or navy shoes. In situations of this sort, one often must be able to calculate the total number of different outcomes that are possible. For example, in how many different ways can the applicant dress for the interview?

A tree diagram is a useful graphic device for solving such problems because it shows at each step the choices or alternatives that can be made and the possible outcomes at the end of the selection process. (Recall that at the beginning of this chapter tree diagrams were also used to represent sample spaces.) Figure 3.9 shows a tree diagram for the choices open to the applicant getting ready for the interview. As you can see, three branches are open in the first step, each branch corresponding to the color of dress. Once the young woman has picked one of these initial branches, she goes to another fork where there again are three branches, each corresponding to a different color of shoes. Once the second step has been completed, there clearly are nine different outcomes corresponding to the points on the right-hand side of the diagram where she can wind up. That is, the young woman can choose a beige dress and beige shoes, a beige dress and black shoes, and so on.

In this case, there are three items of one kind (dress colors) and three items of a second kind (shoe colors), and we find that the number of ways we can pair an item of the first type with an item of the second type is

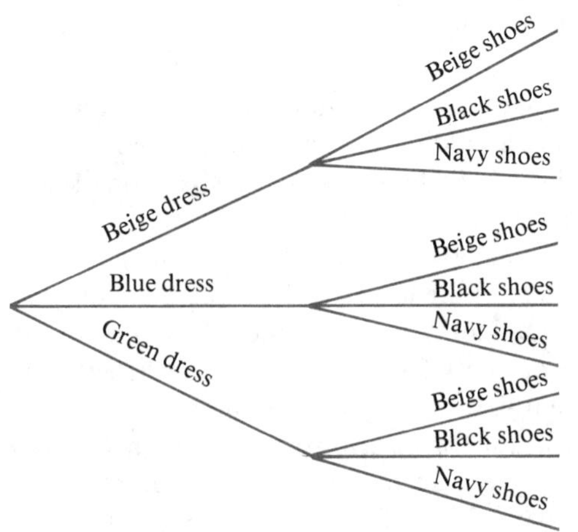

FIGURE 3.9
Tree Diagram of Student's Choices of Dress Color and Shoe Color

(3)(3), or 9. Let's generalize this result. If there are m kinds of items, and *if there are* n_1 *items of the first kind,* n_2 *items of the second kind, . . . , and* n_m *items of the* m*th kind, the number of ways we can select one of each of the* m *kinds of items is* $n_1 \times n_2 \times \ldots \times n_m$. To illustrate, if subsequent to the interview the business student is offered a job by each of five companies, and if she is asked to list her first and second choices, how many different pairs could she list? There are five firms that she could list as her first choice. But once she makes her first choice, there are only four firms left that could be her second choice. Thus, in this case, $m = 2$, $n_1 = 5$, and $n_2 = 4$, so the answer is (5)(4), or 20. Figure 3.10 shows the relevant tree diagram.

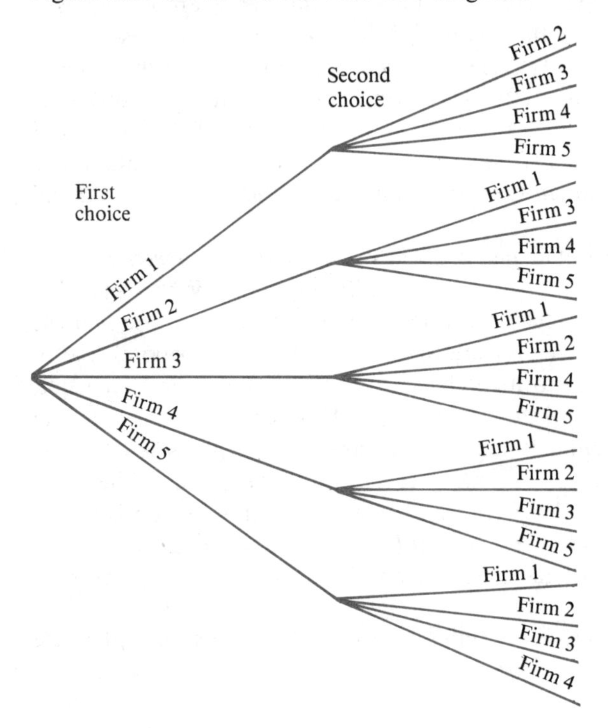

FIGURE 3.10
Tree Diagram of Student's Choices with Five Job Offers

The following example provides another illustration of how this result can be used.

Example 3.9 Thirty firms bid on a particular job, and each firm submits a different bid. The names of the lowest bidder, the second lowest bidder, and the third lowest bidder are published. In how many ways can one select three firms ordered in this way?

Solution: Clearly, any one of the 30 firms may be the lowest bidder. But once one selects any one of them as the lowest bidder, there are only 29 left that can be the second lowest bidder. And once one selects the lowest and second lowest bidders, there are only 28 firms that can be selected as the third lowest bidder. Consequently, one can pick the lowest bidder in 30 different ways, the pair of lowest and second lowest bidders in (30)(29) different ways, and the triad of lowest, second lowest, and third lowest bidders in (30)(29)(28) different ways. Thus, the answer is (30)(29)(28), or 24,360. In this case, $m = 3$, $n_1 = 30$, $n_2 = 29$, and $n_3 = 28$.

PERMUTATIONS

At this point, we can define a permutation. *If* x *items are selected (without replacement) from a set of* n *items, any particular sequence of these* x *items is called a* ***permutation***. For example, Yale Harvard Brown is a permutation of three of the Ivy League colleges. That is, it is a possible sequence in which three of the eight names of the Ivy League colleges (Harvard, Yale, Princeton, Columbia, Pennsylvania, Dartmouth, Cornell, and Brown) can be listed. In this case, $x = 3$ and $n = 8$, since three items (college names) are being selected from a set of eight items. Turning to another example, *ab* is a permutation of two of the first three letters of the alphabet. In this case, $x = 2$ and $n = 3$ since two items (letters) are being selected from a set of three items. Another such permutation is *bc*. In other words, it is another pair of letters that can be made up out of *a*, *b*, and *c*. Note that the order of the items is of importance in a permutation. Thus, *ab* and *ba* are different permutations (from one another) even though they are composed of the same letters, *a* and *b*.

In general, *the number of permutations of* x *items that one can select from a set of* n *items is* $n(n-1)(n-2)\ldots(n-x+1)$. Based on our previous results, this is not difficult to prove. There are x items to be chosen. Suppose that the first item that is chosen is put into the first slot along the horizontal line in Figure 3.11. Suppose the second item chosen is put into the second such slot, the third into the third slot, and so on. For our first choice, there are n possible items. Once the first selection is made, there are $(n - 1)$ items left for the second slot. Once the first and second selections have been made, there are $(n - 2)$ items left for the third slot and once the $(x - 1)$th selection is made, there are $(n - x + 1)$ items left for the last (that is, the xth) slot. So the total number of sequences of x items that can be chosen from n items must be $n(n - 1)(n - 2)\ldots(n - x + 1)$.

To check this result, let's see whether it gives the right answer in the case

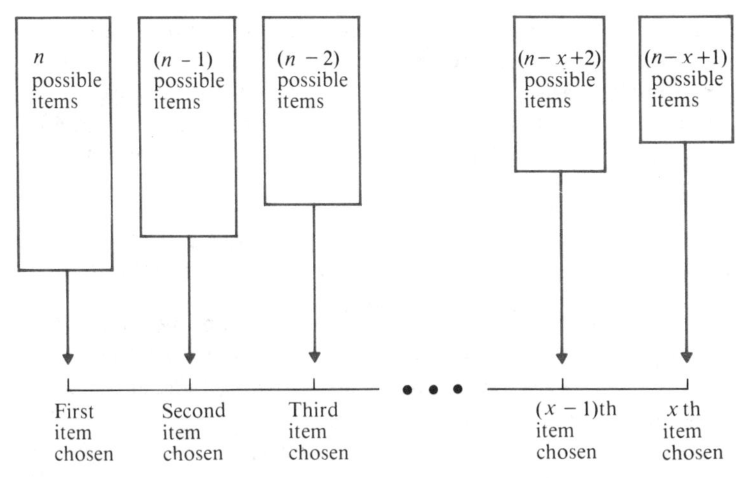

FIGURE 3.11 Number of Permutations of x Items from a Set of n Items

of the business student with the job offers and in Example 3.9. With regard to the student offered a job by the five firms, we want to know the number of permutations of two items (firms) that one can select from a set of five items. According to the previous paragraph, this number should be (5)(4), since $n = 5$ and $x = 2$. This result is just what we got before. In Example 3.9, the case of the 30 firms bidding on the job, we want the number of permutations of three items (firms) that one can select from a set of 30 items. Since $n = 30$ and $x = 3$, the answer according to the previous paragraph is (30)(29)(28), which jibes with our previous result.

COMBINATIONS

In contrast to a permutation, where the order of the items matters, *a* ***combination*** *is a selection of items where the order does not matter.* For example, in the case of the Ivy League colleges, each permutation of three of the colleges shows a possible ranking of the top three schools in football. Clearly, the order here is important, since the first name in the sequence indicates first place and the last name indicates third place. But if instead we were interested in the number of three-way crew races that could occur in the Ivy League the order wouldn't matter, since Harvard Cornell Brown and Cornell Brown Harvard would mean the same thing. Thus, in this latter case, one would want to calculate the number of combinations of three items that can be selected from a set of eight items.

Using the results we obtained concerning the number of permutations, it is easy to find out how many combinations there are. After all, the crucial point is that, from the point of view of a combination, many permutations are indistinguishable. For example, as we just observed, Harvard Cornell Brown and Cornell Brown Harvard are the same thing. Taking each particular combination, how many permutations correspond to this single combination?

Clearly, there are (3)(2)(1), since this is the number of permutations that can be created from three items. (This, of course, is an application of our formula for the number of permutations, both x and n being 3 in this case). To illustrate this, you can see that there are the following six permutations corresponding to the single combination consisting of Harvard, Cornell, and Brown:

Harvard Cornell Brown	Cornell Brown Harvard	Brown Cornell Harvard
Harvard Brown Cornell	Cornell Harvard Brown	Brown Harvard Cornell

Thus, since there are six permutations for each combination and since there are (8)(7)(6) = 336 permutations (because $n = 8$ and $x = 3$), it follows that there must be 336 ÷ 6 = 56 combinations. In other words, there must be 56 different three-way crew races that can be arranged in the Ivy League.

In general, *the number of combinations of* x *items that one can select from a set of* n *items is*

$$\frac{n(n-1)\dots(n-x+1)}{x(x-1)\dots(2)(1)}. \qquad (3.14)$$

To prove this, recall that the numerator of this ratio is the number of permutations of x items that can be selected from a set of n items. Since there are $x(x-1)\dots(2)(1)$ permutations that correspond to each combination of x items, the number of combinations must equal the numerator of (3.14) divided by this amount. Frequently, $x(x-1)\dots(2)(1)$ is referred to as $x!$, which is read as "x factorial." For example, $4! = (4)(3)(2)(1) = 24$, and $6! = (6)(5)(4)(3)(2)(1) = 720$. (Note that by definition $0! = 1$.) Thus, the number of combinations of x items that can be chosen from a set of n items can also be written as

$$\frac{n(n-1)\dots(n-x+1)}{x!} = \frac{n(n-1)\dots(2)(1)}{(n-x)(n-x-1)\dots(2)(1)x!} = \frac{n!}{(n-x)!x!}.$$

To switch from crew to basketball, if we want to know the number of different basketball teams of five members each that can be formed from a group of eight players, this is the number of combinations of five items that can be selected from a set of eight items. Thus, the answer must be

$$\frac{(8)(7)(6)(5)(4)}{(5)(4)(3)(2)(1)} = \frac{6720}{120} = 56.$$

Expressed in terms of factorials, the answer is

$$\frac{8!}{3!5!} = \frac{40{,}320}{(6)(120)} = 56.$$

GALILEO ON GAMBLING: A CASE STUDY[10]

To illustrate the application of combinations and permutations, let's go back several hundred years to a case involving the famous mathematician and astronomer Galileo. In 1613, Galileo accepted a position as First and Extraordinary Mathematician of the University of Pisa and Mathematician to his Serene Highness Cosimo II of Tuscany. Besides receiving a title that betrayed no false modesty, Galileo received a large salary and no duties (a good combination in any age), except that he was asked by Cosimo to work on certain problems in which his Serene Highness was interested. One problem that Cosimo apparently asked Galileo to solve was the following. Three dice are thrown. Although there are six ways of getting a 9—621 (that is, one die showing a 6, another a 2, and another a 1), 531, 522, 441, 432, and 333—and six ways of getting a 10—631, 622, 541, 532, 442, and 433—the probability of throwing a 9 seems in fact to be less than that of throwing a 10. Why?

Before considering Galileo's solution, it is important to recognize that this was not a purely academic exercise. At that time, there was a popular dice game in which three ordinary six-sided dice were thrown. Cosimo apparently had gambled sufficiently often at this game to find that the probability of getting a 9—that is, the probability that the sum of the numbers on the three dice would be 9—was lower than the probability of getting a 10. He wanted to know why this was the case.

Galileo began his reply by pointing out that if each side of each die is equally likely to come up, there are (6)(6)(6) = 216 configurations of the dice that can arise. To see this, note that the first die can have six outcomes (1 to 6). And since the second die can also have six outcomes, the number of pairs of outcomes is (6)(6). Moreover, since the third die can also have six outcomes, the number of triads of outcomes is (6)(6)(6). Further, since each side of each die is equally likely to come up, the probability of each of these triads is 1/216. This is, of course, a simple illustration of what we discussed in previous sections of this chapter.

Next, Galileo noted that although it was true that both 9 and 10 could be rolled in six ways, the probability of each of these ways was not the same. For example, the probability of 333 was lower than that of 432. To see why, let's consider each of the three dice, and let's denote the outcome of a throw by three numbers in brackets, the first being the number on the first die, the second being the number on the second die, and the third being the number on the third die. Clearly, 432 can occur on the basis of six of these triads—[432], [423], [342], [324], [243], [234]—whereas 333 can occur on the basis of only one triad, [333]. Why six triads in the case of 432? Because this is the number of permutations of three items selected from a set of three items. Why 1 triad in the case of 333? Because there is only one triad in which all dice show a 3.

Galileo constructed a table showing the number of triads resulting in

10. The discussion in this section is based on F. N. David, *Games, Gods, and Gambling* (New York: Hafner, 1962).

TABLE 3.4
Number of Triads Resulting in Dice Showing a 9

Way of rolling the number	Total number of triads corresponding to this way of rolling the number	(Individual triads) First die	Second die	Third die
621	6	6	2	1
		6	1	2
		2	6	1
		2	1	6
		1	6	2
		1	2	6
531	6	5	3	1
		5	1	3
		3	5	1
		3	1	5
		1	5	3
		1	3	5
522	3	5	2	2
		2	5	2
		2	2	5
441	3	4	4	1
		4	1	4
		1	4	4
432	6	4	3	2
		4	2	3
		3	4	2
		3	2	4
		2	4	3
		2	3	4
333	1	3	3	3
Total	25			

each of the six ways that 9 could be formed. As indicated in Table 3.4, there are 25 such triads. Similarly, he showed that there are 27 triads resulting in the six ways that 10 can be formed (as shown in Table 3.5). Since each triad has a probability of 1/216, it follows that the probability of rolling a 9 is 25/216, whereas the probability of rolling a 10 is 27/216. This was how Galileo solved the problem.

EXERCISES

3.26 The Black Belt Sewing Machine Company relies on five suppliers (I, II, III, IV, and V) to provide it with motors. Each of these suppliers can ship motors to each of Black Belt's four plants (1, 2, 3, and 4). Black Belt stamps a letter on each motor to indicate which supplier it came from and which plant used it. For example, it stamps an

TABLE 3.5
Number of Triads Resulting in Dice Showing a 10

Way of rolling the number	Total number of triads corresponding to this way of rolling the number	(Individual triads) First die	Second die	Third die
631	6	6	3	1
		6	1	3
		3	6	1
		3	1	6
		1	6	3
		1	3	6
622	3	6	2	2
		2	6	2
		2	2	6
541	6	5	4	1
		5	1	4
		4	5	1
		4	1	5
		1	5	4
		1	4	5
532	6	5	3	2
		5	2	3
		3	5	2
		3	2	5
		2	5	3
		2	3	5
442	3	4	4	2
		4	2	4
		2	4	4
433	3	4	3	3
		3	4	3
		3	3	4
Total	27			

A on a motor from supplier I that is put into a sewing machine coming out of plant 1, a *B* on a motor from supplier II that is put into a sewing machine coming out of plant 2, a *C* on a motor from supplier II that is put into a sewing machine coming out of plant 1, and so on. How many letters will it take to represent all possible combinations of motors and plants?

3.27 If the Black Belt Sewing Machine Company bought the same number of motors in 1979 from each supplier, and if each supplier shipped the same number of motors to each of Black Belt's plants, what proportion of all Black Belt's motors had an *A* stamped on them in 1979? (See Exercise 3.26.)

3.28 Every week the Black Belt Sewing Machine Company tests one motor shipped by each of its five suppliers. The motors tested are ranked according to performance, from highest to lowest. The firm then sends this ranking to its plants. After repeating these tests hundreds of times, all possible orderings of the five suppliers have occurred. How many different rankings has the firm sent out to its plants?

3.29 There are seven persons on the board of directors of the Bona Fide Washing Machine Corporation. The chairman of the board wants to designate three members of the board as a committee to investigate personnel problems. In how many ways can the chairman choose such a committee?

3.30 If there are five men and two women on the board of directors, in how many ways can the chairman of the board choose a committee of three if he wants two men and one woman on the committee?

3.31 The number of combinations of x items that one can select from a set of n items is often represented as $\binom{n}{x}$. Prove that $\binom{n}{x} = \binom{n}{n-x}$.

3.32 How many different poker hands can be drawn in which there is a straight (or straight flush)?

3.33 A shipment of goods contains 50 items, 5 of which are defective. To determine whether the shipment should be accepted, the buyer picks one item at random and then (without replacing the first item) picks another at random. If neither is defective, the shipment will be considered acceptable. What is the probability that the buyer will accept the shipment?

4

Probability Distributions, Expected Values, and the Binomial Distribution

4.1 Introduction

In the previous chapter, we discussed many basic concepts of probability. To understand statistical techniques, it is necessary to go more deeply into probability theory. In this chapter we will discuss the nature and characteristics of random variables and probability distributions. Then we will describe expected values and take up the most important discrete probability distribution, the binomial distribution. We shall also indicate some of the ways in which the binomial distribution is applied in business and government.

4.2 Random Variables

One of the basic concepts in probability theory is that of the random variable, which is defined as follows.

> ***Random Variable:*** *A random variable is a numerical quantity the value of which is determined by an experiment. In other words, its value is determined by chance. More formally, it is a numerically valued function defined on a sample space.*

In succeeding paragraphs we shall go into considerable detail to clarify what is meant by a random variable, but it is sufficient here to note two points concerning this definition. First, the definition assumes that some sort of an experiment is conducted. (As stressed in Chapter 3, the term *experiment* should be interpreted very broadly and covers more than just laboratory procedures.) The outcomes of this experiment constitute a sample space, as pointed out in Chapter 3. Second, our definition implies that a random variable is one whose value is defined for each element of the sample space. This means that its value is defined for each outcome of the experiment, which is equivalent to saying that its value is determined by chance (because the outcome of the experiment is determined by chance).

To illustrate, if we take the sum of the numbers showing on two dice, this sum is clearly a random variable since its value is determined by the "experiment" of throwing the pair of dice. As pointed out in the previous chapter, the sample space generated by this experiment can be represented by the 36 points in Figure 3.1, which are reproduced in Figure 4.1. The value of this sum corresponding to each element of the sample space is shown above the point representing this element in Figure 4.1. Obviously, this sum fulfills the condition of the definition that a random variable must assume a value for each element of the sample space.

As another illustration, suppose that a gambler proposes the following bet on the outcome of the throw of a single die. He will give you \$10 if you throw a 4, a 5, or a 6, and you will give *him* \$10 if you throw a 1, a 2, or a 3. The amount you win (or lose)—+\$10 (or −\$10)—is a random variable since its value is determined by the outcome of an experiment. (That is, its value is determined by chance.) This experiment can result in six possible outcomes, corresponding to the die's coming up 1, 2, 3, 4, 5, or 6. For each of these outcomes (or points in the sample space), the amount you win is defined, which agrees with our definition of a random variable.

At this point, the various characteristics of a random variable should be coming into focus. First, *a random variable must take on numerical values.* In the experiment with the statistics course discussed in the previous chapter, the outcomes "pass" and "not pass" are not the values of a random variable

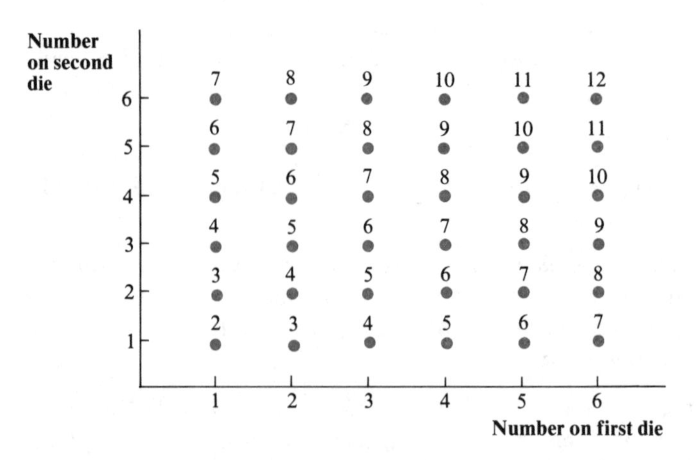

FIGURE 4.1 Sample Space for Roll of Two Dice

because they are not in numerical form. However, if we arbitrarily let "pass" equal 0 and "not pass" equal 1, the result is a random variable. Second, *the value of a random variable must be defined for all possible outcomes of the experiment in question (that is, for all elements of the sample space).* For example, in the $10 bet based on the throw of the die, suppose that you win $10 if you throw a 4, a 5, or a 6, and that you lose $10 if you throw a 1 or a 2. But suppose also that *the amount you win or lose is undefined if you throw a 3.* Under these new rules the amount you win (or lose) is no longer a random variable. Why? Because how much you win (or lose) is no longer defined for *all* possible outcomes of the experiment. Specifically, if the outcome of the experiment is a 3, the value of the amount you win (or lose) is undefined.

It is important to recognize that the value of a random variable is unknown *before* the experiment in question is carried out. *After* the experiment is carried out, the value of the random variable is always known. For example, if you roll a pair of dice, the sum of the numbers on them is unknown before the roll; but after the roll, the value of the sum is known. Similarly, if you accept the gamble involving the roll of a single die, the amount you win or lose is unknown before you roll the die; after the roll, this amount is known.

DISCRETE AND CONTINUOUS RANDOM VARIABLES

Statisticians distinguish between two types of random variables: discrete and continuous. A *discrete random variable* can assume *only a finite or countable*[1] *number of distinct values.* For example, the sum of the numbers showing on two dice is a discrete random variable since the sum can only assume one of 11 possible values: 2, 3, 4, . . . , 11, or 12. The amount at stake in the gamble with the single die is also a discrete random variable since it can assume only one of two possible values: +$10 and −$10. Many—but by no means all—important random variables are discrete. Some random variables assume *any numerical value on a continuous scale.* Such random variables are called *continuous random variables.*

To illustrate the concept of a continuous random variable, let's return to the case study at the end of Chapter 2, which dealt with a manufacturing plant that produces pieces of metal in which two holes are stamped. The specifications require that the distance between the hole centers should be 3.000 ± .004 inches. In fact, however, the distance between the hole centers varies from piece to piece, because of differences in machines, dies, workers, and other factors. The productive process of turning out each piece of metal can be regarded as an experiment, and the outcome of this experiment is the distance between the hole centers. This distance is a random variable; but because it can vary continuously it is not a discrete random variable. This means that the distance between the hole diameters is not confined to certain rounded values like 3.001 inches, 3.002 inches, and so on, but can also assume *any* value in between them. Of course, in practice we might only measure the

1. Some discrete random variables, like the Poisson variable described in Chapter 5, assume a countably infinite number of values.

distance to the nearest .001 inch, in which case the rounded distances are discrete. But the true distance is continuous.

In this chapter we shall be concerned entirely with discrete random variables, and in the next we shall take up continuous random variables. Meanwhile, the following examples should be useful in illustrating the meaning and characteristics of a random variable.

Example 4.1 A fair coin is flipped three times. If it comes up heads three times, you receive $100; if it comes up heads twice, you receive $50; if it comes up heads once, you lose $75; and if it never comes up heads, you lose $100.

(a) Is the number of heads that comes up a random variable? If so, is it discrete or continuous?

(b) Is the amount you win (or lose) a random variable? If so, is it discrete or continuous?

Solution: (a) The number of heads that comes up is a random variable, since there are eight possible outcomes (listed below) of this experiment and to each outcome there corresponds a certain number of heads:

heads/heads/heads	tails/tails/heads
heads/heads/tails	tails/heads/tails
heads/tails/heads	heads/tails/tails
tails/heads/heads	tails/tails/tails

This random variable is discrete because it can assume only four possible values: 0, 1, 2, and 3.

(b) The amount you win is a random variable, because for each of the outcomes of this experiment you win (or lose) a corresponding amount. The amount you win (or lose) is a discrete random variable because it can assume only four possible values: +$100, +$50, −$75, and −$100.

Example 4.2 A shipment contains 20 machines, 4 of which are defective. The firm receiving the shipment chooses a random sample of 3 machines (without replacement). If any of the machines in the sample is defective, the firm rejects the shipment.

(a) Is the number of defective machines in the sample a random variable? If so, is it discrete or continuous?

(b) Is whether or not the shipment is rejected a random variable? If so, is it discrete or continuous?

Solution: (a) The number of defective machines in the sample is a random variable since there are eight possible outcomes (listed below) of this experiment:

NNN	NND	NDN	DNN
NDD	DND	DDN	DDD

The first letter stands for the first machine in the sample and is D if it is defective or N if it is not defective. The second letter stands for the second machine in the sample and is D or N. The third letter, which is also D or N, stands for the third machine. To each of these outcomes there corresponds a number of defective machines in the sample. This random variable is discrete since it can assume only four possible values: 0, 1, 2, and 3.

(b) Whether or not the shipment is rejected is not a random variable because it is not in numerical form. However, it can be turned into a random variable by letting zero stand for rejection of the shipment and 1 stand for acceptance of it. (Obviously, any other pair of arbitrarily chosen numbers could also be used for this purpose.)

4.3 Probability Distributions

Based on the previous section, we know that the value of a random variable is determined by the outcome of a corresponding experiment. For example, the value of one particular sum of the numbers shown on a pair of dice is determined by one particular roll of the dice. And the amount you win or lose in the gamble involving the single die is determined by each roll of the die. Thus, since there is a certain probability of each outcome of the experiment, there must also be a certain probability of each value of the random variable. For example, since there is a 1/36 chance that each of the points in Figure 4.1 will occur (if the dice are true), it must be possible to deduce the probability that the sum of the numbers showing on the dice will equal 2, 3, 4, and so on. Similarly, since there is a 1/6 chance that a true die will show a 1, 2, 3, 4, 5, or 6, it must be possible to deduce the probability that the amount you will win in your gamble involving the single die will be +\$10 or − \$10. Such probabilities are provided by the probability distribution of the random variable, which is defined as follows.

> ***Probability Distribution:*** *The probability distribution of a random variable X provides the probability of each possible value of the random variable. If P(x) is the probability that x is the value of the random variable, we can be sure that $\Sigma P(x) = 1$, where the summation is over all values that X takes on. This is because these values of X are mutually exclusive and one of them must occur.*

To illustrate a probability distribution, consider once again the sum of the numbers showing on two true dice. What is the probability distribution of this sum? Clearly, this sum can assume only the following values: 2, 3, 4, . . . , 11, and 12. *What is the probability of a 2?* This value of the sum can arise only if each die shows a 1, and the probability of this is 1/36. *What is the probability of a 3?* This value of the sum can arise if the first die shows a 1 and the second shows a 2, or if the first shows a 2 and the second shows a 1. Since the probability of each is 1/36, the probability of either one is 2/36. (Remember the addition rule.) *What is the probability of a 4?* This value of the sum can arise in three ways (a 1 on the first die and a 3 on the second, a 2 on the first die and a 2 on the second, or a 3 on the first die and a 1 on the second), each of which has a probability of 1/36. Thus, the probability of a 4 is 3/36. Based on similar reasoning, it is easy to figure out the probability of a 5, 6, . . . , 12. (One way to do this is to count the number of points in Figure 4.1 that result in a particular sum and multiply this number by 1/36, which is the probability of each point.) The results are shown in Table 4.1. As you can see, the sum of the probabilities in the probability distribution equals 1, which agrees with the definition above.[2]

Although it sometimes is convenient to represent a probability distribution by a *table* like Table 4.1, it is sometimes even more convenient to use a *graph*. We can present the probability distribution of the sum of two dice in a *line chart*, as shown in Figure 4.2. Along the horizontal axis are ranged the various possible values of the random variable. At each point on the horizontal axis corresponding to such a value a vertical line is erected, the length of which is equal to the probability of this value's occurrence. Thus, in Figure 4.2 the vertical line at 3 is twice as long as that at 2 since the probability of a 3 is twice that of a 2. As you can see from Figure 4.2, the highest vertical line is at 7, and the line chart is symmetrical about this line in this case. (That is, the probability of a 6 equals the probability of an 8, the probability of a 5 equals the probability of a 9, and so on.)

Still another way to represent a probability distribution is by a mathematical function. As pointed out in our definition of a probability distribution, the probability that a random variable equals x is denoted by $P(x)$. In some cases we can express $P(x)$ as a relatively simple mathematical function of x. For example, if you flip a fair coin twice the number of heads that comes up is a random variable whose probability distribution can be represented by the following equation:

TABLE 4.1
Probability Distribution of Sum of Numbers Showing on Two True Dice

Sum of numbers	Probability
2	1/36
3	2/36
4	3/36
5	4/36
6	5/36
7	6/36
8	5/36
9	4/36
10	3/36
11	2/36
12	1/36
Total	1

2. Why must the sum of $P(x)$ equal 1? Since the various values of x are mutually exclusive, the probability that *some* value of the random variable occurs is the sum of $P(x)$, as we know from the addition rule. Since some value of the random variable *must* occur, the probability that some value of the random variable occurs must equal 1. Thus, it follows that the sum of $P(x)$ must equal 1.

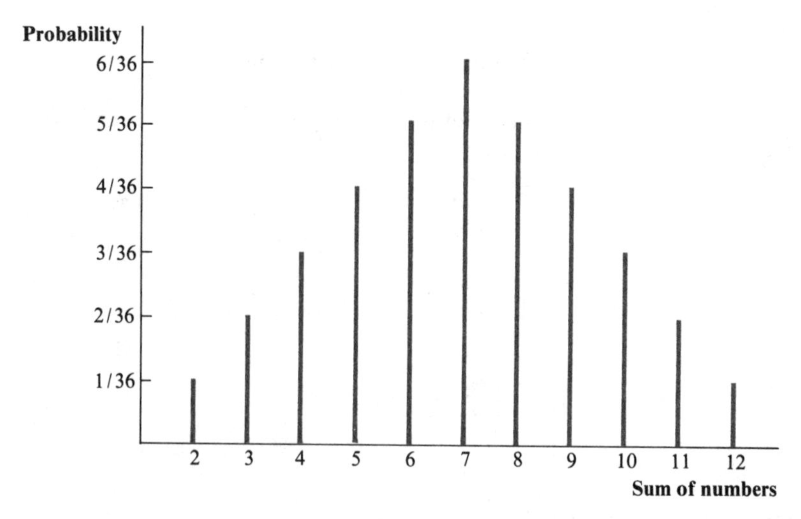

FIGURE 4.2 Probability Distribution of Sum of Numbers Showing on Two True Dice

$$P(x) = \frac{1/2}{x!(2-x)!}, \text{ for } x = 0, 1, 2, \tag{4.1}$$

where $x!$ is defined as $x(x-1)(x-2)\ldots(2)(1)$. Thus, $2! = (2)(1) = 2$, and $1! = 1$. (Also, note that $0! = 1$.)[3] The derivation of equation (4.1) is discussed in detail in Section 4.7. For now, it is essential only that we make sure that this equation really does represent the probability distribution of the number of heads. To see that this is true, note that the equation says that $P(0) = 1/2 \div [0!2!] = 1/4$; $P(1) = 1/2 \div [1!1!] = 1/2$; $P(2) = 1/2 \div [2!0!] = 1/4$. Clearly, each of these probabilities is right. Note, too, that $P(0) + P(1) + P(2) = 1$, as it should according to our definition of a probability distribution.

PROBABILITY DISTRIBUTIONS AND RELATIVE FREQUENCY DISTRIBUTIONS

It is important to note the similarity between a probability distribution and a frequency distribution (discussed in Chapter 1). To see how similar they are, consider the probability distribution of the number coming up on a single die. If the die is true, this probability distribution is

$$P(x) = 1/6, \text{ for } x = 1, 2, 3, 4, 5, 6. \tag{4.2}$$

Suppose that this die is thrown 1,000 times and that the number of times each number comes up is as shown in Table 4.2. If we divide the number of times each number comes up by the total number of throws (1,000), we get the proportion of times each number comes up (as shown in the third column of Table 4.2). Each of these proportions is the empirical counterpart of the

3. Readers of Appendix 3.1 will already have encountered $x!$, which is read as "x factorial."

TABLE 4.2
Number of Times Each Number Shows on a Die Cast 1,000 Times

Number on die	Number of times number comes up	Proportion of times number comes up
1	170	.170
2	159	.159
3	172	.172
4	158	.158
5	160	.160
6	181	.181
Total	1,000	1.000

corresponding probability. That is, the proportion of cases in which a 3 comes up is the empirical counterpart of the probability of a 3. If the frequency concept of probability applies, each of these proportions will tend to get closer and closer to the corresponding probabilities as the number of cases included in the frequency distribution becomes larger and larger.

Another way of stating this is to say that a probability distribution of a certain type is often regarded as the *relative frequency distribution* of the population. *A relative frequency distribution shows the proportion (not the number) of cases falling within each class interval.* For example, if we were to roll a true die again and again until finally data were collected concerning millions of rolls, what would be the relative frequency distribution of this population? After dividing the number of throws resulting in a 1, 2, 3, 4, 5, or a 6 by the total number of throws, we would obtain the proportion of throws coming up a 1, 2, 3, 4, 5, or a 6. This is the relative frequency distribution in this situation. In essence, this relative frequency distribution would be like the probability distribution in equation (4.2). Thus, *probability distributions are often used to represent or approximate population relative frequency distributions*, as we shall see in later chapters.

Since the concept of a probability distribution is so important in statistics, it is essential that it be well understood. The following two examples should help to clarify and illustrate its meaning.

Example 4.3 As in Example 4.1, a fair coin is flipped three times. If it comes up heads three times you receive \$100; if it comes up heads twice you receive \$50; if it comes up heads once you lose \$75; if it never comes up heads you lose \$100.

(a) Construct a line chart of the probability distribution of the number of heads.

(b) Construct a line chart of the probability distribution of the amount you win.

Solution: (a) There are the following eight possible outcomes of this experiment, each with a probability of 1/8:

heads/heads/heads
heads/heads/tails
heads/tails/heads
tails/heads/heads
tails/tails/heads
tails/heads/tails
heads/tails/tails
tails/tails/tails

Thus, the probability of no heads is 1/8; the probability of heads once is 3/8; the probability of heads twice is 3/8; and the probability of heads three times is 1/8. The line chart is as follows.

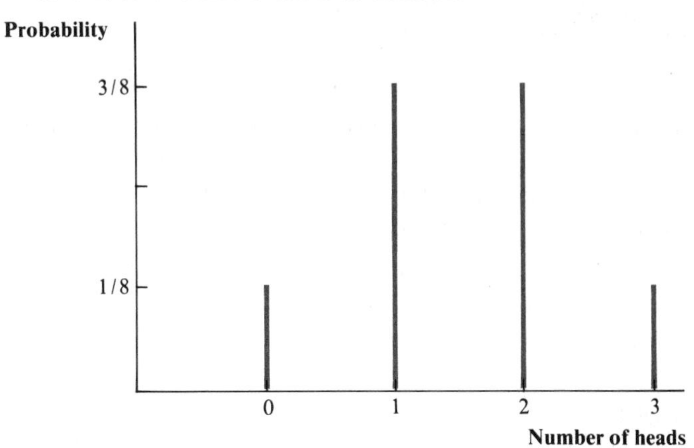

(b) Since the probability of no heads is 1/8, the probability that you will lose \$100 is 1/8. Since the probability of heads once is 3/8, the probability that you will lose \$75 is 3/8. Since the probability of heads twice is 3/8, the probability that you will win \$50 is 3/8. Since the probability of heads three times is 1/8, the probability that you will win \$100 is 1/8. Thus, the line chart is as follows.

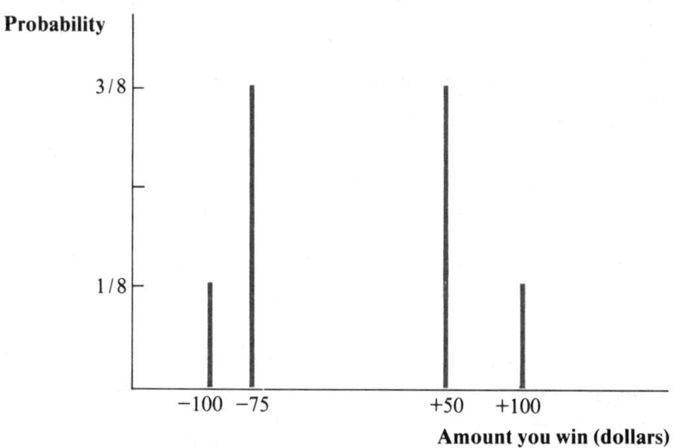

Example 4.4 John Gillicuddy buys two tickets at \$1 each on a car that is being raffled off. The organizers of the raffle sell a total of 5,000 tickets, and one ticket is picked at random to determine the winner. The car is worth \$5,000.

(a) Construct a table showing the probability distribution of the outcome of the raffle, if 0 represents Gillicuddy's not winning the car and 1 represents his winning it.

(b) Construct a table showing the probability distribution of the amount that Gillicuddy wins or loses in the raffle.

Solution: (a) Since John Gillicuddy has two tickets out of a total of 5,000, his probability of winning the car is 2/5,000 = 1/2,500 and his probability of not winning the car is 2,499/2,500. Thus, if zero stands for his not winning and 1 stands for his winning, the probability of each value of this random variable is

Value of random variable	Probability
0	2,499/2,500
1	1/2,500

(b) If Gillicuddy wins the car, he makes $4,998 (the value of the car less the amount spent on the tickets); if he does not win the car, he loses $2 (the amount spent on the tickets). Thus, the amount he wins or loses can assume two possible values, with the following probabilities:

Amount he wins or loses	Probability
− $2	$\frac{2,499}{2,500}$
+ $4,998	$\frac{1}{2,500}$

EXERCISES

4.1 The Alpha Corporation sells bicycles. Based on past experience, it feels that in the summer months it is equally likely that it will sell 0, 1, 2, 3, or 4 bicycles in a day. (The firm has never sold more than 4 bicycles per day.)
(a) Is the number of bicycles sold in a day a random variable?
(b) If so, what values can this random variable assume?
(c) If the number of bicycles sold in a day is a random variable, construct a table showing its probability distribution.

4.2 (a) Based on the information in Exercise 4.1, plot the probability distribution of the number of bicycles sold in a day in a line chart. (b) What mathematical function can represent this probability distribution?

4.3 Based on the information in Exercise 4.1, what is the probability that the number of bicycles sold on a given day will be less than 1? Less than 3? Less than 4? Less than 6? Less than 7?

4.4 The Alpha Corporation has only one salesman, whose income depends on the number of bicycles he sells per day. Specifically, he receives no commission on the first bicycle sold per day, a $20 commission on the second bicycle sold in a day, a $30 commission on the third, and a $40 commission on the fourth. (Thus, if he sells 3 bicycles in a given day, his commissions for this day total $50.) His income consists entirely of these commissions.
(a) Is his income on a particular day a random variable?
(b) If so, what values can this random variable assume?
(c) If his income is a random variable, construct a table showing its probability distribution.

4.5 Based on the information in Exercise 4.4, plot the probability distribution of the salesman's income on a particular day in a line chart.

4.6 Based on the information in Exercise 4.4, what is the probability that his income in a particular day will exceed $20? $30? $40?

4.7 Suppose that the number of bicycles sold one day is independent of the number sold the next day. (a) Based on the data in Exercise 4.1, construct a table showing the probability distribution of the total number of bicycles sold in a two-day period. (b) What is the probability that this number will exceed 2? 4? 6? 8?

4.8 Unfortunately, the bicycle salesman has a weakness for gambling and is in debt to the Mob for $100. The Mob's local enforcer tells the salesman that he has two days to pay up, or he'll need a hospital bed. If the number of bicycles sold one day is independent of the number sold the next day, what is the probability that the salesman will earn enough during this two-day period to stay out of the hospital? (Assume that commissions are net of taxes and other deductions.)

4.9 At the end of the two-day period described in Exercise 4.8 the salesman has earned only $70. The Mob's enforcer says that he will give the salesman a chance to win the extra $30. He will allow the salesman to flip a coin twice. If it comes up heads both times, he will give the salesman the $30 he needs; if it doesn't come up heads both times, he will take the salesman's $70 and begin the hospital-filling mayhem. Plot in a line chart the probability distribution for the salesman's monetary gains or losses in this game.

4.4 Expected Value of a Random Variable

As pointed out in the previous section, the probability distribution of a random variable is analogous to a frequency distribution. It is therefore not surprising that just as we found it useful to calculate the mean of a frequency distribution, we now find it useful to calculate the mean of a probability distribution. Another term often used to denote the mean of a probability distribution is the *expected value of the random variable*, which is defined as follows.

> ***Expected Value of a Random Variable:*** *The expected value of a discrete random variable* X, *denoted by* E(X), *is the weighted mean of the possible values that the random variable can assume, where the weight attached to each value is the probability that the random variable will assume this value. In other words,*
>
> $$E(X) = \sum_{i=1}^{m} x_i P(x_i), \tag{4.3}$$
>
> *where the random variable* X *can assume* m *possible values,* $x_1, x_2, \ldots, x_m$, *and the probability of its equaling* x_i *is* $P(x_i)$.

Clearly, the expected value of a random variable is analogous to the mean of a frequency distribution. If we compare the definition in equation (4.3) with our definition of the mean of a set of data calculated from a frequency distribution in equation (2.3), we see that the definitions are essentially the same.[4]

For the reason just stated, the expected value of any random variable X is often called the *mean* of X. If, as described in the previous section, the probability distribution in question is viewed as the population relative frequency distribution (which shows the proportion of cases with each value), the expected value of the random variable is really the population mean. To illustrate: We roll a true die a million times, and we regard the resulting relative frequency distribution of outcomes as the population. Clearly, the expected value of the number showing on a true die is

$$E(X) = \sum_{i=1}^{6} x_i P(x_i) = 1(1/6) + 2(1/6) + 3(1/6) + 4(1/6) +$$

$$5(1/6) + 6(1/6) = 3\frac{1}{2}.$$

Why? Because the random variable in question (the number coming up) can assume 6 possible values (thus, $m = 6$). These possible values are 1 to 6: thus $x_1 = 1, x_2 = 2, \ldots, x_6 = 6$. And the probability of each value is 1/6: thus, $P(x_1) = P(x_2) = \ldots = P(x_6) = 1/6$. The mean of the population of 1 million outcomes of rolling a true die is also 3 1/2, since the number of rolls in this population is so large that the proportion of cases when each number comes up is equal, for all practical purposes, to the probability of the number's coming up.

ROLE OF EXPECTED VALUES IN DECISION MAKING

The concept of a random variable's expected value (or mean, or *mathematical expectation*, as it is sometimes called) is extremely important in statistics. For example, consider the following gamble, which has already been described in Examples 4.1 and 4.3. A fair coin is flipped three times. If it comes up heads three times, you receive \$100; if it comes up heads twice, you receive \$50; if it comes up heads once, you lose \$75; if it never comes up heads, you lose \$100. Should you accept this gamble? Or put differently, if you were to repeat this gamble again and again, would you tend to come out ahead or would you tend to lose? The answer depends on the expected value of the amount you win, an amount which (as we saw in Example 4.1) is a random variable. Why does it depend on this expected value? Because this expected value is the mean amount you will win (or lose) if you accept the gamble again and again for an indefinite number of times. In other words, the expected value is the mean of the infinite population of amounts you would win (or lose) if you accepted this gamble an infinite number of times.

4. To see how similar these definitions are, suppose that in each class interval the observations are all equal. (In other words, assume that every observation equals its class mark.) Then, since $f_j + n$ in equation (2.3) is the relative frequency with which x_j occurs in the set of data, it can be regarded as the probability that x_j occurs in this set of data. And if $P(x_j)$ is substituted for $f_j + n$ in equation (2.3), the result is equivalent to the right side of equation (4.3).

What is the expected value of the amount you would win (or lose) from this gamble? Once again, let X stand for the relevant random variable, which in this case is the amount you would win (or lose). This random variable can assume only four values, $x_1 = 100$, $x_2 = 50$, $x_3 = -75$, and $x_4 = -100$, with the result that

$$E(X) = \sum_{i=1}^{4} x_i P(x_i) = 100(1/8) + 50(3/8) - 75(3/8) - 100(1/8)$$

$$= -9\frac{3}{8},$$

since (as we know from Example 4.3) $P(x_1) = P(x_4) = 1/8$ and $P(x_2) = P(x_3) = 3/8$. Because the expected value of the amount you would win from this gamble is negative, it is clear that you would lose money if you were to repeat the gamble over and over. More specifically, *if you averaged the results of repeating this gamble over and over, you would lose $9.375 per gamble.* Thus, the gamble is not a "fair bet."

Whether the expected value of one's winnings from a particular wager is positive or negative is obviously of importance in determining whether or not one should accept the wager. For example, if you have the choice of accepting a gamble where the expected value of your winnings is +$10 (as opposed to −$9.375 for the gamble described above), you may decide to accept the former rather than the latter. Under some circumstances, a decision maker may choose the action or gamble that has the largest expected value of winnings or profits. But in general, although the expected value of one's winnings or profits is of importance, this alone is not the sole indicator of what action a decision maker should take. Much more will be said on this subject in Chapter 15.

The two examples below illustrate the calculation and interpretation of expected values.

Example 4.5 At a certain gambling casino the following game is played: The dealer, who works for the house, allows you to pick two cards from a full deck. If both are hearts, you win $15; otherwise, you lose $1. What is the expected value of the amount you win (or lose) each time you play this game?

Solution: The amount you would win is a random variable that can assume two possible values, +$15 or −$1. The probability that it equals $15 is the probability of your getting two hearts, which is 12/51(13/52) = 1/17, or .059. (Recall Example 3.6.) The probability that the amount will be −$1 must therefore be 1 − .059, or .941. Thus, the expected value of this random variable is

$$(\$15)(.059) + (-\$1)(.941) = -\$.056.$$

Example 4.6 A marketing executive must decide whether or not to use a new label on a product. The firm will gain $800,000 if he adopts the new label and it turns out to be superior to the old label. The firm will lose $500,000 if the executive adopts the new label and it proves to be not superior to the old one. The firm will neither gain nor lose money if the executive sticks with the old label. The executive feels there is a 50-50 chance that the new label is superior to the old and a 50-50 chance that it is not. If he wants to take the action with the higher expected gain to the firm, should he decide to use the new label or not?

Solution: If the executive decides to adopt the new label, the expected value of the firm's gain is

$$(\$800{,}000)(1/2) + (-\$500{,}000)(1/2) = \$150{,}000,$$

because the firm's gain is a random variable that can assume two possible values, +\$800,000 and −\$500,000, and the probability of each value is 1/2. If the executive decides not to adopt the new label, the expected value of the firm's gain is zero, since zero is the only possible value that it can assume. Since \$150,000 exceeds zero, the expected gain if he adopts the new label is higher than if he does not. Thus, if he wants to take the action with the higher expected gain, he should adopt the new label.

4.5 Variance and Standard Deviation of a Random Variable

Just as the expected value of a random variable is analogous to the mean of a frequency distribution, the variance of a random variable is analogous to the variance of a frequency distribution. In particular, the variance of a random variable is defined as follows.

Variance of a Random Variable: *The variance of a random variable* X, *denoted by* $\sigma^2(X)$, *is the expected value of the squared deviations of the random variable from its expected value. In other words,*

$$\sigma^2(X) = E([X - E(X)]^2) = \sum_{i=1}^{m} [(x_i - E(X)]^2 P(x_i), \tag{4.4}$$

where the random variable can assume m *possible values,* $x_1, x_2, \ldots, x_m$, *and* $P(x_i)$ *is the probability of its equaling* x_i.

To see the similarity between the variance of a random variable and the variance of a body of data, recall from Chapter 2 that the latter was defined as the mean of the squared deviations of the observations from their mean. Substitute "expected value" for "mean" in the previous sentence, and you have the definition of the variance of a random variable.

Similarly, the standard deviation of a random variable is analogous to the standard deviation of a body of data. Its definition is as follows.

Standard Deviation of a Random Variable: *The standard deviation of a random variable is the positive square root of the random variable's variance. In other words, the standard deviation of a random variable* X, *denoted by* $\sigma(X)$, *is*

$$\sigma(X) = \sqrt{E([X - E(X)]^2)} = \sqrt{\sum_{i=1}^{m} [x_i - E(X)]^2 P(x_i)}. \tag{4.5}$$

Recall from Chapter 2 that the standard deviation of a set of data is the square root of the variance of the data. Thus, the definition here is analogous to the one given there.

Just as the standard deviation of a set of data indicates the extent of the dispersion or variability among the individual measurements within the set of data, *the standard deviation of a random variable indicates the extent of the dispersion or variability among the values that the random variable may assume.* For example, if it is *certain* that a random variable X will equal a certain number, then the difference between X and its expected value—that is, $X - E(X)$—will always be zero. Thus, the expected value of $[X - E(X)]^2$ will also be zero. Consequently, the variance and standard deviation of X will be zero. On the other hand, if it is likely that X will assume values *far removed* from its expected value, then the difference between X and its expected value—that is $X - E(X)$—will have a large probability of being big. Therefore, the expected value of $[X - E(X)]^2$ will also be big; and consequently, the variance and standard deviation of X will both be big.

To illustrate how one can calculate the variance and standard deviation of a random variable, let's calculate the variance and standard deviation of the number coming up on a true die. If this random variable is designated as X, it will be recalled from the previous section of this chapter that the expected value of this random variable, $E(X)$, equals $3\frac{1}{2}$. Thus, the variance of this random variable is

$$\begin{aligned}\sigma^2(X) &= \sum_{i=1}^{6}\left[x_i - 3\frac{1}{2}\right]^2 P(x_i)\\ &= \left[1 - 3\frac{1}{2}\right]^2\left(\frac{1}{6}\right) + \left[2 - 3\frac{1}{2}\right]^2\left(\frac{1}{6}\right) + \left[3 - 3\frac{1}{2}\right]^2\left(\frac{1}{6}\right)\\ &\quad + \left[4 - 3\frac{1}{2}\right]^2\left(\frac{1}{6}\right) + \left[5 - 3\frac{1}{2}\right]^2\left(\frac{1}{6}\right)\\ &\quad + \left[6 - 3\frac{1}{2}\right]^2\left(\frac{1}{6}\right)\\ &= \left(6\frac{1}{4}\right)\left(\frac{1}{6}\right) + \left(2\frac{1}{4}\right)\left(\frac{1}{6}\right) + \left(\frac{1}{4}\right)\left(\frac{1}{6}\right) + \left(\frac{1}{4}\right)\left(\frac{1}{6}\right)\\ &\quad + \left(2\frac{1}{4}\right)\left(\frac{1}{6}\right) + \left(6\frac{1}{4}\right)\left(\frac{1}{6}\right)\\ &= \frac{35}{12}.\end{aligned}$$

And since the standard deviation of a random variable is the square root of its variance,

$$\sigma(X) = \sqrt{35/12}.$$

The following example provides additional practice in calculating the variance and standard deviation of a random variable.

Example 4.7 A fair coin is flipped three times. What is the variance and standard deviation of the number of times it comes up heads?

Solution: If X denotes the number of heads that come up, we know from Example 4.3 that the probability that $X = 0$ is 1/8, the probability that $X = 1$ is 3/8, the probability that $X = 2$ is 3/8, and the probability that $X = 3$ is 1/8. The expected value of X is

$$E(X) = (0)(1/8) + (1)(3/8) + (2)(3/8) + (3)(1/8)$$
$$= 3/2.$$

Thus, the variance of X is

$$\sigma^2(X) = [0 - 3/2]^2(1/8) + [1 - 3/2]^2(3/8) + [2 - 3/2]^2(3/8) + [3 - 3/2]^2(1/8)$$
$$= (9/4)(1/8) + (1/4)(3/8) + (1/4)(3/8) + (9/4)(1/8)$$
$$= 3/4.$$

And the standard deviation of X is

$$\sigma(X) = \sqrt{3/4}.$$

4.6 Chebyshev's Inequality

Once we know the standard deviation of a random variable, we can make some interesting statements about the extent of the dispersion or variability among the values that the random variable can assume. In particular, we can apply the following theorem developed by the nineteenth-century Russian mathematician P. Chebyshev.

> ***Chebyshev's Inequality:*** *For any random variable, the probability that the random variable will assume a value within* k *standard deviations of the random variable's expected value is at least* $1 - 1/k^2$.

Thus, the probability that a random variable will assume a value within *two* standard deviations of its expected value (that is, its mean) is ***at least*** $1-1/2^2$, or 3/4. And the probability that a random variable will assume a value within *three* standard deviations of its expected value is *at least* $1 - 1/3^2$, or 8/9.

In other words, this theorem tells us that the probability that a random variable will assume a value *more than k* standard deviations from the random variable's expected value is *less than* $1/k^2$. Consider the following example. You are given the expected value and standard deviation of the profits to be made from a particular business venture. The expected value is $400,000 and the standard deviation is $100,000. What is the probability that the profits from this venture will be below zero or above $800,000? If you know the probability distribution of the profits you can figure out this probability exactly, but suppose that *you are not given this probability distribution.* Using Chebyshev's inequality, you can still determine the *maximum* amount this probability can possibly be: $1/4^2$, or $1/16$. Since the probability that the profits will be below zero or above $800,000 is the same as the probability that the profits will assume a value more than four standard deviations from the profits' expected value, the maximum amount this probability can be is $1/4^2$.

A key point to remember is that Chebyshev's inequality is true for *all* probability distributions.[5] The following is another illustration of its usefulness.

Example 4.8 A manufacturer (cited in the case study at the end of Chapter 2) produces pieces of metal in which two holes are stamped. The specifications call for the distance between the hole centers to be $3.000 \pm .004$ inches. It is believed that the plant is turning out parts such that the expected value of this distance is 3.000 inches and the standard deviation of this distance is .0005. What is the maximum probability that a piece of metal produced by this plant will fail to meet the specifications?

Solution: If the distance between hole centers in a piece of metal is less than 2.996 inches or greater than 3.004 inches, the piece does not meet specifications. Since the expected value of this distance is 3.000 inches and the standard deviation is .0005 inches, a piece will fail to meet the specifications if the distance between its hole centers assumes a value greater than eight standard deviations from the expected value. (Why? Because 2.996 inches is eight standard deviations below the expected value and 3.004 is eight standard deviations above the expected value.) According to Chebyshev's inequality, the probability of this occurring is less than $1/8^2$, or $1/64$.

EXERCISES

4.10 A pair of true dice are thrown. If the total number rolled is less than 6, you win $1; if it is greater than 8, you lose $1; otherwise you neither gain nor lose any amount. What is the expected value of the amount gained or lost?

4.11 Recall from Exercise 4.1 that the Alpha Corporation is equally likely to sell 0, 1, 2, 3, or 4 bicycles in a day. (a) What is the expected value of the number of bicycles sold by the Alpha Corporation in a particular day? (b) Is this a value that the random variable in question can assume?

4.12 (a) Under the circumstances described in Exercise 4.4, what is the expected value of the income of the Alpha Corporation's salesman in a particular day? (b) Is this a value that the random variable in question can assume?

4.13 (a) Under the circumstances described in Exercise 4.7, what is the expected value of the income of the Alpha Corporation's salesman in a particular two-day period? (For his rate of compensation, see Exercise 4.4.) (b) Is this answer twice the answer to Exercise 4.12? Why, or why not?

5. It is also worth emphasizing that Chebyshev's inequality can be applied to any body of data. Thus, if you know the mean and standard deviation of a particular set of data Chebyshev's inequality says that the proportion of observations (in the set of data) within k standard deviations of the mean must be at least $1 - 1/k^2$.

4.14 (a) Considering only the monetary gains and losses of the gamble described in Exercise 4.9, is this a fair gamble? (b) What is the expected value of the salesman's winnings (or losses) in this game? (b) Omitting the nonmonetary considerations (namely, the threat of injury), would the salesman agree to this gamble if he is interested in maximizing the expected value of his monetary gains?

4.15 The Alpha Corporation is considering the purchase of a large number of bicycles being sold by another firm which is going out of business. If these bicycles are free of any major defects the Alpha Corporation would make $5,000 by buying and reselling them. On the other hand, if these bicycles contain major defects the Alpha Corporation would lose $10,000 by having to repair the defects before selling them. Alpha's president believes the probability is 0.9 that these bicycles are free of major defects (and hence the probability is 0.1 that they do contain major defects). If he is interested in maximizing the expected value of his firm's gains, should he purchase the bicycles? Why, or why not?

4.16 If the Alpha Corporation is equally likely to sell 0, 1, 2, 3, or 4 bicycles in a day, what is (a) the variance and (b) the standard deviation of the number of bicycles sold in a day?

4.17 Under the circumstances described in Exercise 4.4, what is (a) the variance and (b) the standard deviation of the income of the Alpha Corporation's salesman in a particular day?

4.18 If $Y = 2X$, what is the expected value of Y if the expected value of X is (a) 14; (b) -3? (For further discussion of problems of this kind, see Appendix 4.1 of this chapter.)

4.19 A firm manufactures metal rods which must be rejected if they are not between 8.250 and 8.500 inches in diameter. The expected value of the diameters of the rods is 8.375 inches, and the standard deviation of these diameters is 1/40 inch. What is the maximum probability that a diameter will be rejected?

4.7 The Binomial Distribution

Earlier in this chapter we discussed the nature and characteristics of random variables and probability distributions in general. We turn now to the description and analysis of the most important discrete probability distribution in statistics: the binomial distribution. Specifically, we indicate the conditions that generate the binomial distribution, the formula for the distribution (as well as its mean and standard deviation), and some of the ways in which this distribution has been used to help solve important problems in business and economics.

BERNOULLI TRIALS

To understand the circumstances under which the binomial distribution arises, it is convenient to begin by describing a *Bernoulli process* or, what is the same thing, a series of *Bernoulli trials*. (James Bernoulli was a seventeenth-century Swiss mathematician who performed some of the early work on the binomial distribution.) Each Bernoulli trial takes place under the following circumstances. First, *each trial results in one of two possible outcomes, which is termed either "success" or "failure."* Second, *the probability of a success remains the same from one trial to the next.* Third, *the outcomes of the trials are independent of one another.*

An example of a Bernoulli trial is a game in which a true die is thrown, and if a 4, a 5, or a 6 comes up, you win. If a 1, a 2, or a 3 comes up, you lose. To make sure that this is a Bernoulli trial, let's see whether it meets the three conditions specified in the previous paragraph. Certainly, the first condition is met, since there are only two possible outcomes—win (success) or lose (failure). And the second condition is met, since the probability of winning (that is, of a success) remains constant (at 1/2 from one trial to the next). Moreover, there is no reason to believe that your chances of winning are influenced in any way by the outcome of previous trials, since the die has no memory. Thus, all three conditions are met.

THE BINOMIAL DISTRIBUTION

Suppose that n Bernoulli trials occur and that the probability of a success on each trial equals P. Under these circumstances[6] the number of successes occurring in these n trials has a binomial probability distribution, which means

$$P(x) = \frac{n!}{x!(n-x)!} P^x(1-P)^{n-x}, \text{ for } x = 0, 1, 2, \ldots, n \tag{4.6}$$

where $P(x)$ is the probability that the number of successes equals x. The number of successes, X, is a random variable, whereas n and P are constants. It is customary to refer to X as a binomial random variable.

To illustrate the calculation of binomial probabilities, consider a case where a fair coin is flipped four times. What is the probability distribution of the number of heads? Since there are two mutually exclusive outcomes each time the coin is flipped, since the probability of heads is 1/2 each time, and since the outcomes of the tosses are statistically independent, it follows that this is a situation where there are four Bernoulli trials and where the probability of a success (heads) equals 1/2. Substituting 4 for n and 1/2 for P in equation (4.6), we can calculate the probability distribution of the number of heads, with the following result:

$$P(0) = \frac{4!}{0!4!}\left(\frac{1}{2}\right)^0\left(\frac{1}{2}\right)^4 = \frac{1}{16},$$

$$P(1) = \frac{4!}{1!3!}\left(\frac{1}{2}\right)^1\left(\frac{1}{2}\right)^3 = \frac{1}{4},$$

$$P(2) = \frac{4!}{2!2!}\left(\frac{1}{2}\right)^2\left(\frac{1}{2}\right)^2 = \frac{3}{8},$$

$$P(3) = \frac{4!}{3!1!}\left(\frac{1}{2}\right)^3\left(\frac{1}{2}\right)^1 = \frac{1}{4},$$

$$P(4) = \frac{4!}{4!0!}\left(\frac{1}{2}\right)^4\left(\frac{1}{2}\right)^0 = \frac{1}{16}.$$

6. Of course, the probability of failure on each trial equals $1 - P$, since success and failure on each trial are complementary events (as defined in Chapter 3).

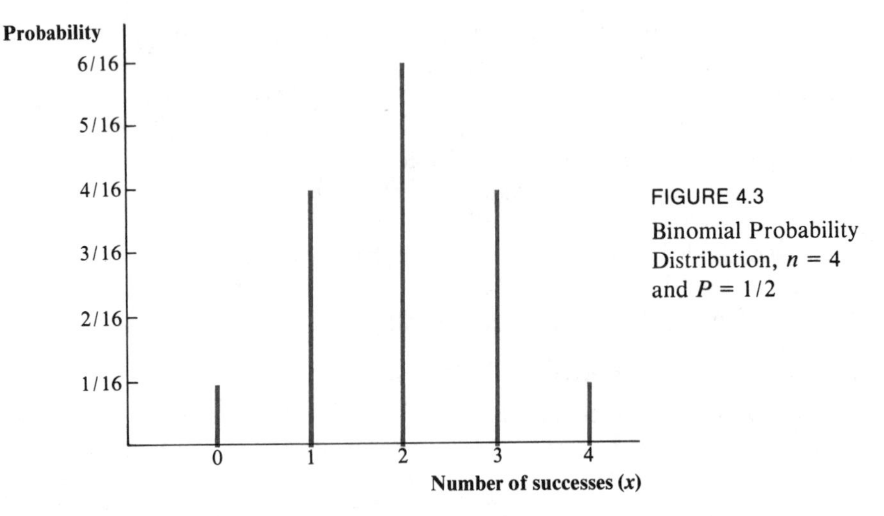

FIGURE 4.3
Binomial Probability Distribution, $n = 4$ and $P = 1/2$

Thus, the probability of no heads is 1/16, of heads once is 1/4, of heads twice is 3/8, of heads three times is 1/4, and of heads four times is 1/16. This probability distribution is plotted in Figure 4.3.

It is important to recognize that many random variables of considerable practical importance in business and economics have a binomial distribution. For example, later in this chapter we show how this probability distribution has played a central role in quality control in industry and government.

TESTING THE FORMULA FOR $P(x)$

Before undertaking some practical applications of the binomial distribution, we want to verify the correctness of the formula for the binomial probability distribution in equation (4.6). To do this, let's show that the probabilities of 0, 1, or 2 heads (in 4 tosses) based on equation (4.6) are correct.

Probability of No Heads. Clearly, there is only one way for this outcome to occur: The coin must come up tails *four times in a row*. Recalling the multiplication rule, the probability of tails four successive times must be equal to $(1/2)^4$ or 1/16, since the probability of tails each time is 1/2 and the outcomes of various tosses are statistically independent of one another. This result agrees with our calculation of $P(0)$, as computed from equation (4.6).

Probability of Heads Once. There are four mutually exclusive ways that one heads can occur in four tosses; namely, heads can come up on the first, second, third, or fourth toss. The probability of each of these ways equals 1/16. Why? Because the probability both of heads on a *particular* toss and tails on the *other three* tosses equals $(1/2)^4$, or 1/16, since the probability of heads and the probability of tails equal 1/2, and the outcomes of various tosses are statistically independent. (Again, recall the multiplication rule.) Since the probability of the occurrence of *each* of these ways is 1/16 (and since they are mutually exclusive), the addition rule dictates that the probability that *any one* of these ways occurs must be 4(1/16), or 1/4. This result is in accord with our calculation of $P(1)$ as computed from equation (4.6).

Probability of Heads Twice. If we designate heads by H and tails by T, there are the following six ways that two successes can be distributed among the four tosses of the coin: (1) $HHTT$; (2) $HTHT$; (3) $HTTH$; (4) $THHT$; (5) $THTH$; and (6) $TTHH$. The first way occurs when the first and second tosses come up heads and the third and fourth tosses come up tails; the second way occurs when the first and third tosses come up heads and the second and fourth tosses come up tails; and so on. Each of these six ways has a probability of occurrence of $(1/2)^4$ or 1/16, because of the multiplication rule. Thus, because of the addition rule, the probability that any one of these (mutually exclusive) ways occurs must be 6(1/16), or 3/8. This result is in accord with our calculation of $P(2)$, as computed from equation (4.6).

DERIVATION[7] OF THE FORMULA FOR $P(x)$

Although the foregoing discussion demonstrates that the formula for $P(x)$ results in the correct probabilities of 0, 1, and 2 heads out of four tosses of a coin, it does not demonstrate that this formula is always valid. To show that this is the case, it is necessary to derive the formula in equation (4.6). In the following four paragraphs, we provide such a derivation. Since the derivation is somewhat more technical than other parts of this chapter, some readers may want to skip these paragraphs. An understanding of subsequent sections does not depend on reading them.

To derive this formula, the first step is to note that there are a variety of ways that n Bernoulli trials can give rise to exactly x successes and $n - x$ failures. One way is that successes occur on each of the first x trials, after which all the failures occur. If S is a success and F is a failure, this sequence can be represented by

$$\overbrace{SSS \ldots S}^{x} \; \overbrace{FFF \ldots F}^{n-x}$$

Another possible sequence is the following, where the first $x - 1$ trials result in successes, the next $n - x$ trials result in failures, and the last trial results in a success:

$$\overbrace{SS \ldots S}^{x-1} \; \overbrace{FFF \ldots F}^{n-x}S$$

Because the trials are independent, the probability that the first of these two sequences will occur is

$$\overbrace{PPP \ldots P}^{x} \; \overbrace{(1-P)(1-P)(1-P) \ldots (1-P)}^{n-x} = P^x(1-P)^{n-x}.$$

7. Some instructors may prefer to omit the next five paragraphs, which can be skipped without loss of continuity.

And the probability of occurrence of the second sequence is:

$$\overbrace{PP \ldots P}^{x-1} \; \overbrace{(1-P)(1-P)(1-P) \ldots (1-P)}^{n-x} P = P^x(1-P)^{n-x},$$

which equals the probability of the first sequence. Clearly, the probability of obtaining *any* particular sequence of x successes and $(n - x)$ failures must be the same as the probability of each of these two sequences.

How many ways can n Bernoulli trials give rise to exactly x successes and $n - x$ failures? In other words, how many different sequences of this type can occur? There are n trials, which we can represent by n different cards (with a 1 on the first card, a 2 on the second card, . . . , and an n on the nth card). The problem is to determine how many different ways x of these cards can be chosen, since any set of x cards can be used to represent a sequence where successes occur on the trials corresponding to these cards (and where failures occur on the rest of the trials). Expressed in the language of Appendix 3.1, this problem amounts to asking how many combinations of x cards can be drawn from a set of n cards, since the order in which the cards are drawn does not matter. From Appendix 3.1 we know that the answer is $n! \div [(n - x)!x!]$. Consequently, this is the number of sequences in which exactly x successes occur.

Because each of these sequences constitutes one of the mutually exclusive ways in which x successes can occur in n trials, and because each such sequence has a probability of occurrence of $P^x(1 - P)^{n-x}$, the probability of x successes is obtained by adding $P^x(1 - P)^{n-x}$ as many times as there are sequences. Since there are $n! \div [(n - x)!x!]$ such sequences, it follows that this sum equals $P^x(1 - P)^{n-x}$ multiplied by $n! \div [(n - x)!x!]$. Thus, the probability that x successes will occur in n trials must equal

$$\frac{n!}{(n - x)!x!} P^x(1 - P)^{n-x},$$

which is the expression in equation (4.6). We have therefore proved that this formula is correct.

THE BINOMIAL DISTRIBUTION: A FAMILY OF DISTRIBUTIONS

It is important to recognize that the binomial probability distribution is not a single distribution, but a family of distributions. Depending on the values of n and P, one can obtain a wide variety of probability distributions, all of which are binomial. For example, Figure 4.4 shows the binomial probability distribution when $n = 6$ and $P = 0.2, 0.3, 0.7$, and 0.8. As you can see, it is skewed to the right when P is less than 1/2 and skewed to the left when P is greater than 1/2. Figure 4.5 shows the binomial probability distribution when $P = 0.4$ and $n = 5, 10$, and 20. As you can see, the binomial probability distribution becomes increasingly bell-shaped as n increases in value. More will be said about this in the next chapter.

If n is fairly large, it is laborious to carry out the calculations involved in

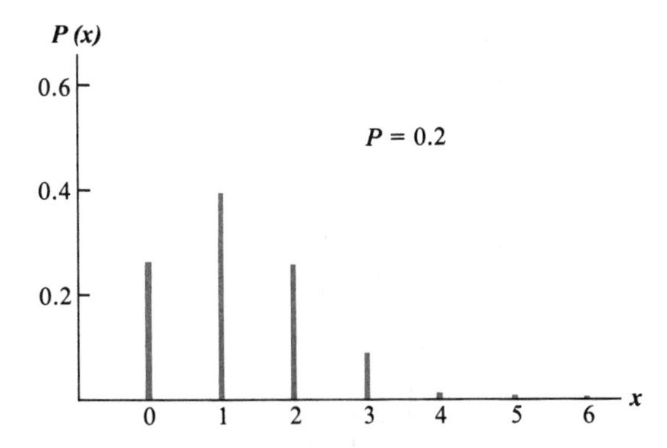

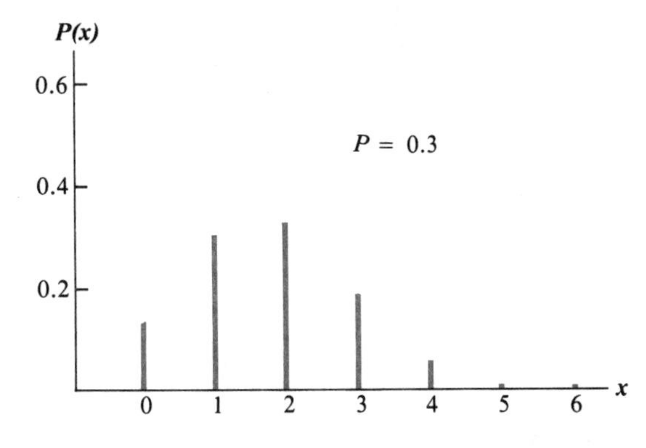

FIGURE 4.4
Four Binomial Distributions, All with $n = 6$

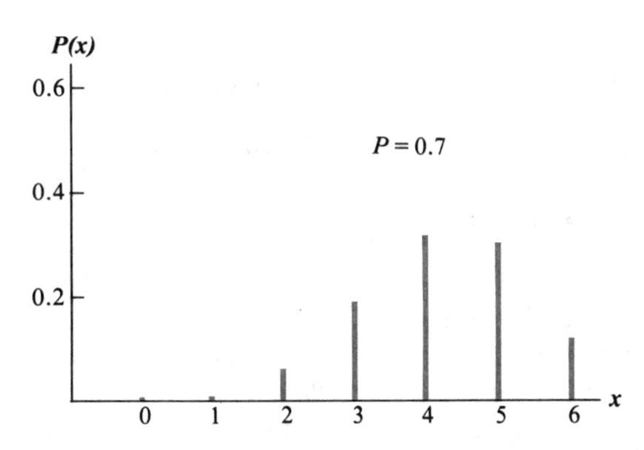

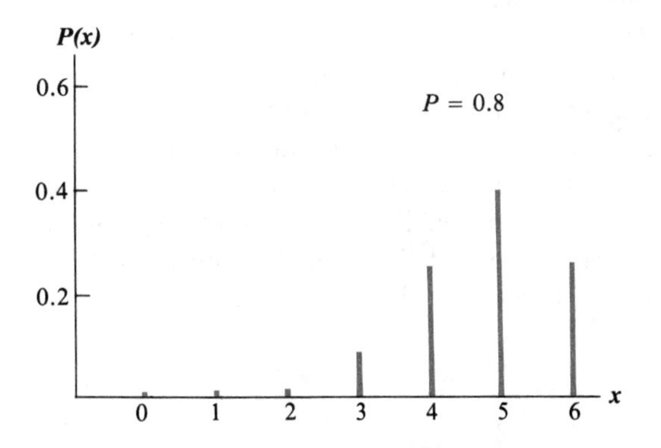

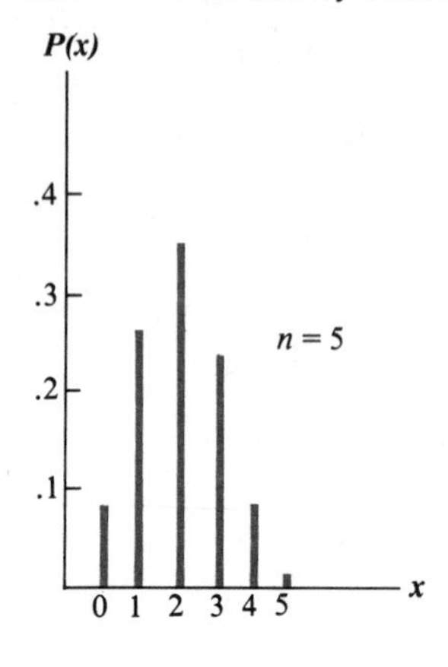

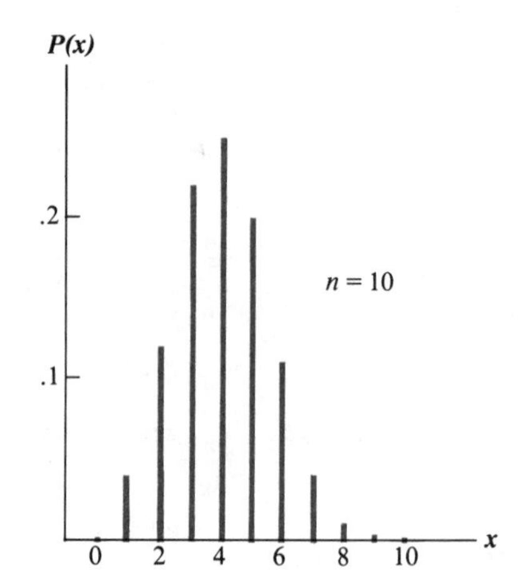

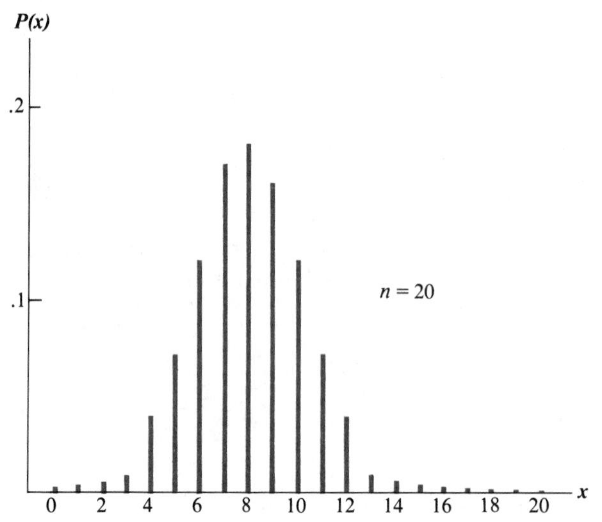

FIGURE 4.5
Three Binomial Probability Distributions, All with $P = 0.4$

using equation (4.6) to compute $P(x)$. Fortunately, there are tables which give the value of $P(x)$ corresponding to various values of n and P. Appendix Table 1 supplies the values of $P(x)$ for values of n from 1 to 20, and for values of P from 0.05 to 0.50. For large values of n, approximations of the binomial distribution can be found, as we shall see in the following chapter.

If the value of P exceeds 0.5, Appendix Table 1 can still be used. We need only switch the definitions of success and failure. For example, if the probability of a success is 0.8, what is the probability of four successes in 10 trials? To use Appendix Table 1, let's reverse the definitions of success and failure; that is, let's call a success what we formerly called a failure and vice versa. If we do this, what we want is the probability of six successes (formerly regarded as failures) in 10 trials. Since the probability of a success (formerly a failure) is 0.2, Appendix Table 1 shows that this probability equals .0055. (Look it up and see.)

MEAN AND STANDARD DEVIATION OF A BINOMIAL RANDOM VARIABLE

In Sections 4.4 and 4.5, we defined the expected value and standard deviation of a random variable. What are the expected value and standard deviation of a binomial random variable? First, let's consider the expected value, which (you will recall) is defined as the weighted mean of the possible values that a random variable can assume, each value being weighted by its probability of occurrence. Thus, the mean of a binomial random variable must equal

$$E(X) = \sum_{x=0}^{n} x P(x), \tag{4.7}$$

since the possible values of X are from 0 to n.

To illustrate the calculation of the expected value of a binomial random variable, suppose once again that $n = 4$ and $P = 1/2$. Then it follows from equation (4.7) that

$$E(X) = (0)P(0) + (1)P(1) + (2)P(2) + (3)P(3) + (4)P(4).$$

Substituting into this formula the values of $P(0), \ldots, P(4)$ obtained on page 117, we find that

$$E(X) = 0(1/16) + 1(1/4) + 2(3/8) + 3(1/4) + 4(1/16) = 2.$$

Thus, the expected value (or mean) of X, if $n = 4$ and $P = 1/2$, is 2; or put somewhat differently, the mean number of successes under these conditions is 2.

Calculating the expected value of a binomial random variable in this way shows clearly what the basic computations are. In practical work, however, this is a highly inefficient procedure because it can be shown that for a binomial random variable

$$E(X) = nP. \tag{4.8}$$

Thus, all one has to do is multiply the number of trials by the probability of success on each trial. Regardless of the values of n or P, this simple computation will give the expected value of a binomial random variable. To make sure that this shortcut is correct, let's apply it to the case where $n = 4$ and $P = 1/2$. Using equation (4.8), the expected value is 4(1/2), or 2, which is precisely the result we obtained by the more laborious procedure carried out in the previous paragraph. Of course, equation (4.8) certainly appeals to common sense. If n trials are carried out, and if the probability of a success on each trial is P, it stands to reason that the mean number of successes will be nP.

Next we turn to the standard deviation of a binomial random variable. From Section 4.5 we know that the standard deviation of any random variable is defined as

$$\sqrt{\Sigma(x_i - E(X))^2 P(x_i)},$$

where the summation is over all possible values of X. Since the possible values of a binomial random variable are 0, 1, . . . , n, its standard deviation must therefore equal

$$\sigma(X) = \sqrt{\sum_{x=0}^{n} (x - E(X))^2 P(x)}. \tag{4.9}$$

To illustrate the calculation of the standard deviation of a binomial random variable, suppose once again that $n = 4$ and $P = 1/2$. Then it follows from equation (4.9) that

$$\begin{aligned}\sigma(X) &= [(0 - 2)^2(1/16) + (1 - 2)^2(1/4) + (2 - 2)^2(3/8) \\ &\quad + (3 - 2)^2(1/4) + (4 - 2)^2(1/16)]^{1/2} \\ &= \sqrt{1} = 1.\end{aligned}$$

As in the case of the expected value, this rather laborious computation is useful because it shows clearly what the definition means. But in practical work there is no need to carry out these extensive calculations, since it can be shown that for a binomial random variable

$$\sigma(X) = \sqrt{nP(1 - P)}. \tag{4.10}$$

Thus, *all one has to do is multiply the number of trials by the product of the probability of success and the probability of failure, and find the square root of the result.* Regardless of the values of n and P, this will yield the standard deviation of a binomial random variable. For example, if $n = 4$ and $P = 1/2$, $\sigma(X) = \sqrt{4(1/2)(1/2)} = 1$, according to equation (4.10), which is precisely the answer we got by the more laborious process in the previous paragraph.

The following two examples are designed to illustrate how binomial probabilities are calculated and interpreted, as well as to show how the expected value and standard deviation of a binomial random variable are computed.

Example 4.9 An oil exploration firm plans to drill six holes. It is believed that the probability that each hole will yield oil is 0.1. Since the holes are in quite different locations, the outcome of drilling one hole is statistically independent of that of drilling any of the other holes.

(a) If the firm will be able to stay in business only if two or more holes produce oil, what is the probability of its staying in business?

(b) Give the expected value and the standard deviation of the number of holes that result in oil.

Solution: (a) If the firm can stay in business only if two or more holes produce oil, it follows that the probability that it will stay in business equals 1 minus the probability that the number of holes resulting in oil is zero or 1. Each hole drilled can be viewed as a Bernoulli trial where the probability of success is 0.1. Thus, the probability that the number of successes is zero or 1 equals:

$$P(0 \text{ or } 1) = P(0) + P(1) = \frac{6!}{6!0!}(.9^6) + \frac{6!}{5!1!}(.1)(.9^5)$$

$$= .531 + .354 = .885$$

Consequently, the probability that the firm will be able to stay in business is $1 - .885 = .115$.

(b) The expected value of the number of holes yielding oil is 6(.1), or 0.6, since $n = 6$ and $P = .1$. The standard deviation of the number of holes yielding oil is $\sqrt{6(.1)(.9)} = \sqrt{.54}$, or .73, since $n = 6$, $P = .1$, and $(1 - P) = .9$.

Example 4.10 A drug firm administers a new drug to 20 people with a certain disease. If the probability is 0.15 that the drug will cure each person of the disease, and if the result for one person is independent of that for another person, what is the probability that 3 or more of the 20 people will be cured? Use Appendix Table 1 to find the answer.

Solution: In Appendix Table 1, look at the section where $n = 20$, and find the column corresponding to $P = .15$. Let X be the number of persons cured by the drug. The probability that X is greater than or equal to 3 equals 1 minus the probability that $X = 0$, 1, or 2. According to the relevant column of Appendix Table 1, $P(0) = .0388$, $P(1) = .1368$, and $P(2) = .2293$. Thus, the probability that X is greater than or equal to 3 is $1 - (.0388 + .1368 + .2293) = 1 - .4049$, or .5951. In other words, the probability that 3 or more people will be cured is approximately .60.

4.8 Acceptance Sampling

We have already emphasized that the binomial distribution is of great practical importance in business and economics. A major example of such application is acceptance sampling, which is one important field of quality control. As an illustration of acceptance sampling, suppose that a firm purchases a large number of metal fixtures from a particular supplier. To make sure that only a reasonably small percentage of the fixtures in any particular shipment is defective, *the firm takes a sample of 30 fixtures from the shipment and tests each to determine whether it is defective. If none is defective, the firm accepts the shipment; otherwise, it rejects it*. This is a very simple type of acceptance-sampling scheme, but it is a convenient starting point.

To see why the binomial distribution is important in acceptance sampling, we must note that if the sample is chosen at random,[8] the probability that each fixture in the sample will be defective will equal the proportion of defective fixtures in the shipment. (This assumes that the sample is only a small proportion of the shipment, which is generally true, or that

8. By *random sample* we mean one where each fixture in the sample has the same probability of being chosen. A more complete definition is given in Chapter 6.

sampling is with replacement.[9]) For example, if 10 percent of the fixtures in a particular shipment are defective, the probability that a randomly chosen fixture in the shipment will be defective is .10. Thus, each choice of a fixture by the inspector can be regarded as a Bernoulli trial, where a success occurs if the fixture is found to be defective. On each such trial, the probability of a success equals P, the proportion of defective fixtures in the shipment. Consequently, if the inspector chooses a sample of 30 fixtures from a shipment, the number of defective fixtures in the sample will be a binomial random variable. Specifically, the probability that x of these fixtures will be defective is

$$P(x) = \frac{30!}{(30 - x)!x!} P^x(1 - P)^{30-x},$$

and thus the probability that none is defective is

$$P(0) = (1 - P)^{30}.$$

This result is important because it enables the firm to calculate the probability that it will accept a shipment with various percentages of fixtures that are defective. For example, if 20 percent of the fixtures in a shipment are defective, what is the probability that the firm will accept such a shipment? Substituting .20 for P, we find that this probability equals $(1 - .20)^{30}$, or .001. (Why? Because the firm will accept the shipment only if none are defective.) On the other hand, suppose that 10 percent of the fixtures in a shipment are defective. Under these circumstances, what is the probability that the firm will accept the shipment? Substituting .10 for P, we find that this probability equals $(1 - .10)^{30}$, or .04. Finally, if 5 percent of the fixtures in a shipment are defective what is the probability that the firm will accept the shipment? Substituting .05 for P, we find that this probability equals $(1 - .05)^{30}$, or .21.

4.9 Acceptance Sampling in the Department of Defense: A Case Study

In the previous section we used the binomial distribution to calculate the probability that a shipment of goods would be accepted, given that it contained a certain percentage of defective items. Calculations of this sort are used by firms and government agencies to determine the sample size and the type of sampling plan they should use. The sampling plan in the previous section may be quite appropriate if the firm wants to accept shipments with 5 percent defectives about 21 percent of the time, but the firm may not want to do this. It may, for example, want to accept such shipments only about 8 percent of the time. If so, this objective can be achieved by taking a sample of 50 rather than 30 fixtures from each shipment, and by accepting the shipment

9. If the sampling is without replacement and if the sample is not a small proportion of the shipment, the hypergeometric distribution, not the binomial distribution, should be used. For a discussion of the hypergeometric distribution, see Appendix 4.2 of this chapter.

only if none are defective. (As an exercise, prove that this is true. If you have difficulty, consult the note at the bottom of this page.[10])

A very important set of sampling plans has been established by the Department of Defense, which requires that they be used where applicable by the army, navy, and air force. Since the Department of Defense purchases goods and services so extensively, these sampling plans have had a very wide and significant influence throughout American industry. The purpose of these plans is to determine whether the Department of Defense or some component thereof should accept a shipment or lot of goods received from a supplier. Each plan is characterized by (1) a sample size; (2) an acceptance number; and (3) a rejection number. If the number of defective items in the sample is equal to or less than the *acceptance number*, the lot or batch is accepted. If the number of defective items in the sample is equal to or greater than the *rejection number*, the lot or batch is rejected. Of course, the acceptance and rejection numbers and the sample size are under the control of the statistician. It is not necessary to set the acceptance number equal to zero, as was done in the previous paragraph and in the preceding section.

Suppose that the Department of Defense receives a shipment of 1,000 motors. According to its sampling plans, it would draw a random sample of 80 motors, and the acceptance and rejection numbers would depend on the *acceptable quality level* (AQL), which is defined as the maximum percentage of items in a shipment that can be defective and yet have the shipment acceptable. For example, if the acceptable quality level is 1 percent, the acceptance number is 2 and the rejection number is 3. On the other hand, if the acceptable quality level is 4 percent, the acceptance number is 7 and the rejection number is 8. Table 4.3 shows the acceptance and rejection numbers corresponding to various acceptable quality levels for the shipment of motors.

TABLE 4.3

Acceptance and Rejection Numbers for Various Acceptable Quality Levels[a]

Acceptable quality level (percentage of defective items in shipment)	Acceptance number	Rejection number
0.1	0	1
1.0	2	3
2.5	5	6
4.0	7	8
6.5	10	11
10.0	14	15

SOURCE: Department of Defense, *Military Standard 105D*.

[a] This table assumes that there are 501 to 1200 items in the shipment and that the sample size is 80.

10. If $P = .05$, $(1 - P)^{50} = (1 - .05)^{50} = .078$. Thus, the probability of accepting a shipment with 5 percent defectives with this acceptance-sampling plan is about 8 percent.

TABLE 4.4
Probability of Accepting Shipment of Items (Based on the Defense Department's Acceptance-Sampling Plan when the AQL = 4 Percent), Given that the Shipment Contains Various Percentages of Defectives[a]

Percentage of defective items in shipment	Probability of accepting shipment
5	.95
6	.90
7.5	.75
11.9	.25
14.2	.10
15.8	.05

SOURCE: Table 4.3.
[a]See Table 4.3.

Once the sample size and the acceptance and rejection numbers have been chosen, the binomial distribution can be used to determine the probability that a shipment of goods will be accepted (given that it contains a certain percentage of defectives). The Department of Defense statisticians have calculated these probabilities and have put them into tabular form for use in characterizing and quantifying the risks associated with the department's sampling plan. For example, if the acceptable quality level is 4 percent (and thus the acceptance number is 7 and the rejection number is 8) the probability of accepting the shipment of motors is as shown in Table 4.4, given that the shipment contains various percentages of defectives. Thus, if the shipment contains 5 percent defectives this probability equals .95; if the shipment contains 14.2 percent defectives this probability equals .10. These and other acceptance-sampling plans of the Department of Defense are a major application of the binomial distribution.[11]

EXERCISES

4.20 The Uphill Manufacturing Company, a maker of bicycle pedals, has seven suppliers that provide it with screws. The screw producers are faced with work stoppages because of labor problems, and Uphill's management feels that in the next six months there is a 10 percent probability that each supplier will be unable to provide Uphill with screws. Because the prospective labor problems are quite different from one supplier to another and the employees of the various suppliers do not act together, Uphill's management also thinks that whether any one supplier is unable to provide screws is independent of whether any other supplier can do so.
(a) What is the probability that none of the suppliers will be able to provide Uphill with screws?

11. For further discussion of the Defense Department's acceptance-sampling plans, see E. Mansfield, "Acceptance Sampling by the Defense Department," in E. Mansfield, *Statistics for Business and Economics: Readings and Cases* (New York: Norton, 1980).

(b) What is the probability that more than half of the suppliers will be unable to provide Uphill with screws?
(c) What is the expected number of suppliers that will be unable to provide Uphill with screws?
(d) What is the standard deviation of the number of suppliers that will be unable to provide Uphill with screws?
(e) If the employees of the various suppliers band together to negotiate with all their employers, what effect do you think this will have on your answers to (a), (b), (c), and (d)?

4.21 The Uphill Manufacturing Company wants to know whether its own employees are satisfied with their working conditions. Since it would be impossible for Uphill's management to talk in depth with each of the company's 20,000 employees, the president decides to pick 20 employees and see how well each is satisfied with existing conditions. Suppose that the probability that each employee in this sample will express dissatisfaction is 0.20, and that whether or not one employee expresses dissatisfaction is statistically independent of whether or not another employee does so.
(a) What is the probability that a majority of the sample will express dissatisfaction with existing working conditions?
(b) What is the probability that 2 or less of the sample will express dissatisfaction with existing working conditions?
(c) Uphill's personnel director proposes the following gamble to the company president: If the number of employees expressing dissatisfaction is more than 20 percent of the sample, he will pay the president $100. If this is not the case, the president will pay him $100. Is this a fair bet? Why, or why not?
(d) What is the expected number of employees who will express dissatisfaction? What is the standard deviation?
(e) Using Chebyshev's inequality, determine an upper bound for the probability that the number of employees will be more than three standard deviations from the expected number you determined in (d). By how much does this upper bound exceed the true probability?

4.22 It is important that the Uphill Manufacturing Company maintain careful control over the quality of the screws it uses. Each shipment (which contains 10,000 screws) is subjected to the following acceptance-sampling procedure: A sample of 15 screws is taken from the shipment and each is tested for defects. If more than one of the screws are found to be defective, the shipment is rejected.
(a) If a shipment contains 20 percent defectives, what is the probability that it will be accepted?
(b) If a shipment contains 10 percent defectives, what is the probability that it will be accepted?
(c) If a shipment contains 5 percent defectives, what is the probability that it will be accepted?
(d) Uphill's suppliers protest that this inspection procedure is inappropriate because they guarantee only that no more than 10 percent of a shipment is defective. Do you agree that it is inappropriate? Why, or why not?

4.23 In Example 4.10, what is the expected value of the number of people in the sample cured by the drug? What is the standard deviation of this number?

4.24 In Example 4.10, what is the expected value of the number of people not cured by the drug? What is the standard deviation of this number?

4.25 The ratio of the variance of a binomial random variable to its expected value is 1/2. Can you determine n? Can you determine P?

4.26 In how many ways can eight successes be distributed among 10 trials? In how many ways can two successes be distributed among 10 trials? Are your answers to these two questions the same?

4.27 A shipment contains 20 machines, 3 of which are defective. The firm receiving the shipment chooses a random sample of 3 machines *without replacement*. Prove that the probability distribution of the number of defective machines in the sample is

$$P(x) = \frac{\dfrac{3!}{x!(3-x)!}\,\dfrac{17!}{(3-x)!(14+x)!}}{\dfrac{20!}{3!17!}}, \text{ for } x = 0, 1, 2, 3,$$

where x is the number of defective machines in the sample. This is the so-called *hypergeometric distribution*. For some discussion of this distribution, see Appendix 4.2 of this chapter.[12]

CHAPTER REVIEW

1. A *random variable* is a quantity the value of which is determined by an experiment; in other words, its value is determined by chance. A random variable must assume numerical values, and its value must be defined for all possible outcomes of the experiment in question; that is, for all elements of the sample space. Random variables are of two types: discrete and continuous. *Discrete random variables* can assume only a finite or countable number of numerical values, whereas *continuous random variables* can assume any numerical value on a continuous scale.

2. The *probability distribution of a discrete random variable* provides the probability of each possible value of the random variable. If $P(x)$ is the probability that x is the value of the random variable, the sum of $P(x)$ for all values of x must be 1. Probability distributions are represented by tables, graphs, or equations. A probability distribution can often be viewed as the relative frequency distribution (which shows the proportion of cases with each value) of a population.

3. If $x_1, x_2, \ldots, x_m$ are the possible values of a random variable X and if $P(x_1)$ is the probability that x_1 occurs, $P(x_2)$ is the probability that x_2 occurs, and so on, then the expected value of this random variable $E(X)$ is $\Sigma x_i P(x_i)$. In other words, the *expected value* is the weighted mean of the values that the random variable can assume, each value being weighted by the probability that it occurs. To shed light on decision problems, statisticians frequently calculate and compare the expected value of monetary gain if various courses of action are followed. However, except under special circumstances, such a comparison alone cannot provide a complete solution to these problems.

4. The *variance of a random variable X*, denoted by $\sigma^2(X)$, is the expected value of the squared deviation of the random variable from its expected value.

12. Exercises 4.26 and 4.27 assume that the reader is familiar with Appendix 3.1.

The *standard deviation of a random variable* is the positive square root of the random variable's variance. The standard deviation of a random variable is the most frequently used measure of the extent of dispersion or variability of the values assumed by the random variable. If it is certain that a random variable will equal a certain value, its standard deviation (and variance) is zero.

5. According to *Chebyshev's inequality*, the probability that any random variable will assume a value within k standard deviations of the random variable's expected value is at least $1 - 1/k^2$. Stated differently, the probability that any random variable will assume a value more than k standard deviations from the random variable's expected value is less than $1/k^2$.

6. The most important discrete probability distribution is the *binomial distribution*, which represents the number of successes in n Bernoulli trials. That is, if each trial can result in a "success" or a "failure," and if the probability of a success equals P on each trial (and if the outcomes of the trials are independent), the number of successes, x, that occur in n trials has a binomial distribution. Specifically,

$$P(x) = \frac{n!}{x!(n-x)!} P^x(1 - P)^{n-x}, \text{ for } x = 0, 1, 2, \ldots n.$$

The expected value of x is nP, and its standard deviation is $\sqrt{nP(1-P)}$. A very important application of the binomial distribution is *acceptance sampling*, a technique used extensively in business and government.

Getting Down to Cases:

QUALITY CONTROL IN THE MANUFACTURE OF RAILWAY-CAR SIDE FRAMES[13]

A producer of railway-car side frames wanted to establish a system to control the proportion of defective frames produced. Based on past data, the firm knew that 20 percent of the frames it produced were defective. It wanted to establish a procedure which would signal when the fraction of defectives jumped above 20 percent. After considerable discussion the firm decided to sample 10 frames from each day's output and find out the number of defectives. If this number exceeded 2 out of 10, the firm would stop its productive process to attempt to find out why such a relatively high percent were defective.

(a) Assuming that the 10 frames sampled were a very small proportion of the day's output, what was the probability that the firm's productive process would be stopped when in fact 20 percent of the day's output was defective?

(b) If defectives were to increase to 40 percent of a particular day's

13. The first half of this case is based in part on a section from A. J. Duncan, *Quality Control and Industrial Statistics* (Homewood, Ill.: Irwin, 1959). The numbers have been changed for pedagogical reasons. The reader may be surprised that so large a percentage (20 percent) of the frames were defective, but this was actually the case.

output, what is the probability that this inspection procedure would result in a stoppage of the firm's productive process?

(c) Based on your results in (a) and (b), write a one-paragraph report concerning the adequacy of the firm's sampling plan.

The firm sends a customer a shipment of five frames chosen at random from the day's output under the conditions described in (b). According to the terms of the firm's agreement with the customer, the firm must pay the customer $100 for every defective frame shipped as compensation for expenses incurred in receiving and handling defective materials.

(d) Graph the probability distribution of the amount the firm will have to pay the customer with regard to this shipment of five frames.

(e) What is the expected value of the amount the firm will have to pay the customer with regard to this shipment?

(f) What is the standard deviation of the amount the firm will have to pay the customer with regard to this shipment?

APPENDIX 4.1

Expected Value of a Linear Function of a Random Variable

Frequently, one is interested in the expected value of a linear function of a random variable. (If X is a random variable and $Y = a + bX$, where a and b are constants, then Y is a *linear function* of X.) For example, suppose that a firm's total annual costs are a linear function of the number of units of output it sells per year, and that the price per unit of output is constant. Then the firm's annual profit is a linear function of the number of units of output it sells per year, as shown in the illustration below. If the number of units of output sold per year is a random variable with a known expected value, what is the expected value of the firm's annual profit? To solve this sort of problem, we use the following result:

> ***Expected Value of a Linear Function of a Random Variable.*** *If* Y *is a linear function of a random variable* X*—that is, if* Y = a + bX*—the expected value of* Y, E(Y)*, equals* a + bE(X)*, where* E(X) *is the expected value of* X.

This result is not difficult to prove. Given the definition of an expected value in equation (4.3), it follows that

$$E(Y) = \sum_{i=1}^{m} y_i P(y_i), \tag{4.11}$$

where the summation is over all possible values of Y, and $P(y_i)$ is the probability that Y equals y_i. Since y_i equals $a + bx_i$, the probability that $Y = y_i$ must equal the probability that $X = x_i$. (Why? Because Y can equal y_i if and only if X equals x_i.) Thus, $P(y_i)$ must equal $P(x_i)$. Substituting $P(x_i)$

for $P(y_i)$ and $(a + bx_i)$ for y_i in equation (4.11), we have

$$E(Y) = \sum_{i=1}^{m} (a + bx_i)P(x_i),$$

and since

$$(a + bx_i)P(x_i) = aP(x_i) + bx_iP(x_i),$$

$$E(Y) = \sum_{i=1}^{m} [aP(x_i) + bx_iP(x_i)] = \sum_{i=1}^{m} aP(x_i) + \sum_{i=1}^{m} bx_iP(x_i).$$

Since a and b are constants, they can be moved in front of the summation signs, so

$$E(Y) = a\sum_{i=1}^{m} P(x_i) + b\sum_{i=1}^{m} x_iP(x_i).$$

Finally, since

$$\sum_{i=1}^{m} P(x_i) = 1 \text{ and } \sum_{i=1}^{m} x_iP(x_i) = E(X),$$

$$E(Y) = a + bE(X),$$

which is the result we set out to prove.

To illustrate the application of this result, consider the following. A certain plant manufacturing TV sets has a fixed cost of $1 million per year. The gross profit from each TV set sold—that is, the price less the unit variable cost—is $20. The number of sets the plant sells per year is a random variable with an expected value of 100,000. What is the expected value of this plant's annual profit? Let π equal the plant's annual profit. Since this profit equals its gross profit less its fixed costs,

$$\pi = -1{,}000{,}000 + 20X,$$

where X is the number of TV sets sold per year. Thus,

$$E(\pi) = -1{,}000{,}000 + 20E(X) = -1{,}000{,}000 + 20(100{,}000)$$
$$= 1{,}000{,}000.$$

Consequently, the expected value of the plant's annual profit is $1 million.

In some cases, one also is interested in the standard deviation of a linear function of a random variable. To solve such a problem, we use the following result.

Standard Deviation of a Linear Function of a Random Variable. *If* Y *is a linear function of a random variable* X*—that is, if* Y = a + bX*—the standard deviation of* Y*, σ(*Y*), equals* bσ(X)*, where σ(*X*) is the standard deviation of* X.

To illustrate the above, in the case of the TV manufacturer, what is the standard deviation of the annual profit if the standard deviation of the number of TV sets sold per year is 10,000? Since $\pi = -1{,}000{,}000 + 20X$, the standard deviation of π equals $20\sigma(X)$, or \$200,000, since $\sigma(X)$ equals 10,000.

APPENDIX 4.2

The Hypergeometric Distribution[14]

In acceptance sampling and in many other areas, the hypergeometric distribution is an important probability distribution. To understand the conditions under which this distribution arises, suppose that a finite population consists of two types of items which we call successes and failures. If there are A successes and $(N - A)$ failures in the population, and if we draw a random sample (without replacement) of n items from the population, what is the probability distribution of the number of successes in the sample? Letting x equal the number of successes in the sample, the answer is

$$P(x) = \frac{\dfrac{A!}{x!(A - x)!} \dfrac{(N - A)!}{(n - x)!(N - A - n + x)!}}{\dfrac{N!}{(N - n)!n!}}, \text{ for } x = 0, 1, 2, \ldots, n. \tag{4.12}$$

This is called the hypergeometric distribution.[15]

It is not too difficult to derive the result in equation (4.12). How many different combinations of x successes can be made from the A successes in the population? As we know from Appendix 3.1, the answer is $A! \div [x!(A - x)!]$. How many different combinations of $(n - x)$ failures can be made from the $(N - A)$ failures in the population? As we know from Appendix 3.1, the answer is $(N - A)! \div [(n - x)!(N - A - n + x)!]$. Thus, the total number of different ways that we can combine the x successes with $(n - x)$ failures is

$$\left(\frac{A!}{x!(A - x)!}\right) \left(\frac{(N - A)!}{(n - x)!(N - A - n + x)!}\right)$$

As we also know from Appendix 3.1, the total number of different combinations of n items that can be made from the N items in the population is $N! \div [(N - n)!n!]$. Thus, since the probability of x successes in a sample of n is the number of equally likely samples with x successes and $(n - x)$ failures divided by the total number of equally likely samples of size n, this probability is

$$\frac{\dfrac{A!}{x!(A - x)!} \dfrac{(N - A)!}{(n - x)!(N - A - n + x)!}}{\dfrac{N!}{(N - n)!n!}},$$

which is what we set out to prove.

14. We assume here that the reader has read Appendix 3.1.
15. Note that x cannot exceed A, and that $n - x$ cannot exceed $N - A$.

To illustrate the use of the hypergeometric distribution, suppose that a shipment of motors contains 10 motors, 2 of which are defective and 8 of which are not. What is the probability that a sample of three motors chosen at random from this shipment *without replacement* will contain one defective? Since $A = 2$, $N = 10$, $n = 3$, and $x = 1$, equation (4.12) implies that this probability equals

$$\frac{\left(\frac{2!}{1!1!}\right)\left(\frac{8!}{2!6!}\right)}{\frac{10!}{7!3!}} = \frac{2\left[\frac{8(7)}{2}\right]}{\frac{10(9)(8)}{3(2)(1)}} = \frac{56}{120} = \frac{7}{15}.$$

If one is sampling without replacement from a finite population, the hypergeometric distribution, not the binomial distribution, is the correct one to use. *(If the sampling is with replacement, the binomial is the correct one.) However, if the sample is a small percentage of the population, the binomial distribution provides a very good approximation to the hypergeometric distribution.* For example, suppose that a shipment of goods contains 1,000 items, 20 of which are defective and 980 of which are not defective. A random sample of 20 items is drawn from the shipment without replacement, and we want to know the probability of x defectives in this sample. Even though the sample is without replacement, the sample is so small a proportion of the shipment that for an adequate approximation we should use the binomial probability distribution based on $n = 20$ and $P = .02$ (that is, 20/1000). In general, if the sample is 5 percent or less of the population, this approximation is adequate.

If sampling is without replacement, it can be shown that the mean and standard deviation of the number of successes is

$$E(X) = nP$$

$$\sigma(X) = \sqrt{nP(1 - P)\left(\frac{N - n}{N - 1}\right)},$$

where $P = A/N$, the proportion of successes in the population. In other words, these are the mean and standard deviation of the hypergeometric distribution. If the sample is a small percent of the population, $(N - n)/(N - 1)$ will be approximately equal to 1, since n/N is very small. Thus, under these conditions, the mean and standard deviation of the number of successes will be approximately equal to those given for the binomial distribution in equations (4.8) and (4.10).

APPENDIX 4.3
Joint Probability Distributions and Sums of Random Variables

JOINT PROBABILITY DISTRIBUTIONS

In Chapter 3 we discussed experiments involving two events, and characterized the probability of each outcome in terms of a joint probability table. (Recall the joint probability table for persons classified by attitude

TABLE 4.5
Joint Probability Distribution for X and Y

Value of Y	Value of X 0 (no tails)	1 (one tail)	Total (marginal probabilities)
0 (no tails)	1/4	1/4	1/2
1 (one tail)	1/4	1/4	1/2
Total (marginal probabilities)	1/2	1/2	1

toward increased defense spending and by political party, shown in Table 3.2.) Let's now consider cases where each event is represented by a numerical value. Experiments of this sort involve two random variables. For example, suppose that two coins (coin A and coin B) are each flipped once. Let X be the random variable corresponding to the number of times that coin A comes up tails, and let Y be the random variable corresponding to the number of times that coin B comes up tails. Table 4.5 shows the ***joint probability distribution*** for X and Y. The joint probabilities in Table 4.5 show the probabilities that X and Y assume particular values. For example, the probability that $X = 1$ and $Y = 0$, denoted by $P(X = 1 \text{ and } Y = 0)$, equals 1/4.

A joint probability distribution can be represented graphically, as illustrated in Figure 4.6. The value of X is measured along one axis, the value of Y is measured along another axis, and the joint probability is measured

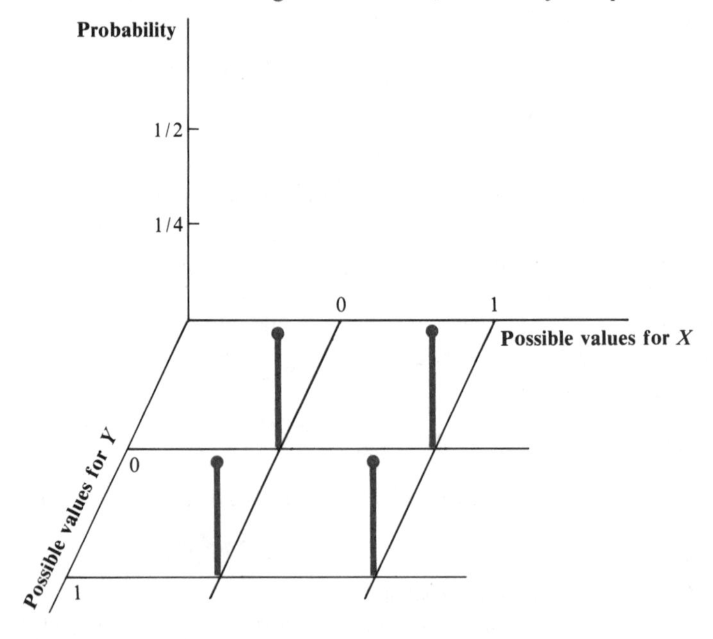

FIGURE 4.6 Joint Probability Distribution for X and Y

along the third (vertical) axis. Each possible outcome is represented by a spike from the floor of the graph (at the point corresponding to this outcome's value of X and Y); the height of the spike measures the probability of this outcome. (Of course, the sum of these heights for all possible outcomes equals 1.) In the particular case in Figure 4.6, the height of each of the four spikes equals 1/4, in accord with Table 4.5.

In Chapter 3 we referred to the row and column totals in a joint probability table as marginal probabilities. The column totals in Table 4.5 correspond to $P(x)$ and constitute the *marginal probability distribution of* X. The row totals in Table 4.5 correspond to $P(y)$ and constitute the *marginal probability distribution of* Y. In other words, the marginal probability distributions are as shown in Table 4.6.

TABLE 4.6
Marginal Probability Distributions of X and of Y

Marginal probability distribution of X		Marginal probability distribution of Y	
Value of X	Probability	Value of Y	Probability
0	1/2	0	1/2
1	1/2	1	1/2
Total	1	Total	1

It is important to distinguish between a marginal probability distribution and a conditional probability distribution. To obtain the conditional probability that Y assumes a particular value (say, 1) given that X assumes a particular value (say, 0), we divide the joint probability that $Y = 1$ and $X = 0$ by the marginal probability that $X = 0$. In other words,

$$P(Y = 1|X = 0) = \frac{P(Y = 1 \text{ and } X = 0)}{P(X = 0)} = \frac{.25}{.50} = .50.$$

To obtain the conditional probability that $Y = 0$ given that $X = 0$, we divide the joint probability that $Y = 0$ and $X = 0$ by the marginal probability that $X = 0$:

$$P(Y = 0|X = 0) = \frac{P(Y = 0 \text{ and } X = 0)}{P(X = 0)} = \frac{.25}{.50} = .50.$$

Thus, the *conditional probability distribution of* Y *given that* X $= 0$ is as shown in Table 4.7. (Prove, as an exercise, that in this case the conditional probability distribution of Y, given that $X = 1$, is the same as the conditional probability distribution of Y, given that $X = 0$.)

The conditional probability distribution of X can be derived in a similar way. To obtain the conditional probability that X assumes a particular value (say, 1), given that Y assumes a particular value (say, 1), we divide the joint

TABLE 4.7
Conditional Probability Distributions of Y

Conditional probability distribution of Y, given that $X = 0$		Conditional probability distribution of Y, given that $X = 1$	
Value of Y	Probability	Value of Y	Probability
0	1/2	0	1/2
1	1/2	1	1/2
Total	1	Total	1

probability that $X = 1$ and $Y = 1$ by the marginal probability that $Y = 1$. In other words,

$$P(X = 1 | Y = 1) = \frac{P(X = 1 \text{ and } Y = 1)}{P(Y = 1)} = \frac{.25}{.50} = .50.$$

To obtain the conditional probability that $X = 0$, given that $Y = 1$, we divide the joint probability that $X = 0$ and $Y = 1$ by the marginal probability that $Y = 1$:

$$P(X = 0 | Y = 1) = \frac{P(X = 0 \text{ and } Y = 1)}{P(Y = 1)} = \frac{.25}{.50} = .50.$$

Thus the *conditional probability distribution of* X *given that* Y $= 1$ is as shown in Table 4.8. (Prove, as an exercise, that in this case the conditional probability distribution of X, given that $Y = 0$, is the same as the conditional probability distribution of X, given that $Y = 1$.)

In Chapter 3 we took up the definition of statistical independence of events. Now we define statistical independence of random variables.

> ***Statistical Independence of Random Variables:*** *Two random variables* X *and* Y *are statistically independent if the conditional probability distribution of* X, *given any value of* Y, *is identical to the marginal probability distribution of* X, *and if the conditional probability distribution of* Y, *given any value of* X, *is identical to the marginal probability distribution of* Y.

TABLE 4.8
Conditional Probability Distributions of X

Conditional probability distribution of X, given that $Y = 0$		Conditional probability distribution of X, given that $Y = 1$	
Value of X	Probability	Value of X	Probability
0	1/2	0	1/2
1	1/2	1	1/2
Total	1	Total	1

In the case of the two coins, a comparison of Table 4.6 with Tables 4.7 and 4.8 shows that X and Y are independent. Why? Because both conditional probability distributions of Y in Table 4.7 are identical to the marginal probability distribution of Y in Table 4.6, and both conditional probability distributions of X in Table 4.8 are identical to the marginal probability distribution of X in Table 4.6.

Example 4.11 Let X equal 0 if a particular flight from Seattle to Chicago is on time, and 1 if it is not. Let Y equal 1 if the flight encounters severe turbulence, and 0 if it does not. The joint probability distribution for X and Y is as follows:

	Value of Y	
Value of X	0 (No turbulence)	1 (Turbulence)
0 (on time)	.75	.05
1 (not on time)	.15	.05

What is the marginal probability distribution of X? What is the conditional probability distribution of X, given that $Y = 0$? Are X and Y statistically independent?

Solution: The marginal probability distribution of X is given by the horizontal row totals; that is, the marginal probability that $X = 0$ is $.75 + .05 = .80$, and the marginal probability that $X = 1$ is $.15 + .05 = .20$. The conditional probability that $X = 0$, given that $Y = 0$, is

$$P(X = 0 \mid Y = 0) = \frac{P(X = 0 \text{ and } Y = 0)}{P(Y = 0)} = \frac{.75}{.90} = .83,$$

and the conditional probability that $X = 1$, given that $Y = 0$, is

$$P(X = 1 \mid Y = 0) = \frac{P(X = 1 \text{ and } Y = 0)}{P(Y = 0)} = \frac{.15}{.90} = .17.$$

Since the marginal probability distribution of X is not identical to the conditional probability distribution of X, given that $Y = 0$, X and Y are not statistically independent.

SUMS OF RANDOM VARIABLES

To solve problems in business and economics, statisticians often find it necessary to determine the expected value and variance of the sum of a number of random variables. The following propositions are very helpful in solving such problems.

Expected Value of a Sum of Random Variables: *If* $X_1, X_2, \ldots, X_m$ *are* m *random variables,*

$$E(X_1 + X_2 + \ldots + X_m) = E(X_1) + E(X_2) + \ldots + E(X_m).$$

That is, the expected value of the sum of random variables is equal to the sum of the expected values of the random variables. This proposition is true whether or not the random variables are statistically independent.

Variance of a Sum of Random Variables: *If* $X_1, X_2, \ldots, X_m$ *are* m *statistically independent random variables,*

$$\sigma^2(X_1 + X_2 + \ldots + X_m) = \sigma^2(X_1) + \sigma^2(X_2) + \ldots + \sigma^2(X_m).$$

That is, the variance of the sum of statistically independent random variables is equal to the sum of the variances of the random variables.

The example below illustrates how these propositions can be applied.

Example 4.12 In order to commercialize a new product, the Ozone Chemical Company must carry out research, do pilot-plant work, and build a production facility. The cost of each of these three steps is a random variable. The expected value and standard deviation of each of these random variables are as follows:

Random variable	Expected value (dollars)	Standard deviation (dollars)
Cost of carrying out research	25,000	20,000
Cost of pilot-plant work	50,000	20,000
Cost of production facility	200,000	30,000

What is the expected value of the total cost incurred in these three steps? If the costs incurred in various steps are statistically independent, what is the standard deviation of the total cost incurred in these three steps?

Solution: Let X_1 be the cost of carrying out research, X_2 be the cost of pilot-plant work, and X_3 be the cost of building a production facility. Based on the above proposition,

$$E(X_1 + X_2 + X_3) = E(X_1) + E(X_2) + E(X_3).$$

Since $E(X_1) = \$25{,}000$, $E(X_2) = \$50{,}000$, and $E(X_3) = \$200{,}000$,

$$E(X_1 + X_2 + X_3) = \$25{,}000 + \$50{,}000 + \$200{,}000 = \$275{,}000.$$

That is, the expected value of the total cost incurred in these three steps is \$275,000. If X_1, X_2, and X_3 are statistically independent,

$$\sigma^2(X_1 + X_2 + X_3) = \sigma^2(X_1) + \sigma^2(X_2) + \sigma^2(X_3).$$

Since $\sigma^2(X_1) = (\$20{,}000)^2$, $\sigma^2(X_2) = (\$20{,}000)^2$, and $\sigma^2(X_3) = (\$30{,}000)^2$,

$$\sigma^2(X_1 + X_2 + X_3) = (\$20{,}000)^2 + (\$20{,}000)^2 + (\$30{,}000)^2,$$

and

$$\sigma(X_1 + X_2 + X_3) = \sqrt{(\$20{,}000)^2 + (\$20{,}000)^2 + (\$30{,}000)^2}$$
$$= \$41{,}231.$$

That is, the standard deviation of the total cost incurred in these three steps is \$41,231.

5

The Normal and Poisson Distributions

5.1 Introduction

In the previous chapter we studied the nature and characteristics of the binomial distribution, the most important discrete probability distribution. We now turn to the normal distribution, a continuous probability distribution that plays a central role in statistics. We will take up the nature and characteristics of the normal distribution and some of the ways in which this distribution is used in business and industry. We will also describe the Poisson distribution, another important probability distribution. In the appendix to this chapter we show how the Poisson distribution and the exponential distribution are used to analyze queueing problems in business and government.

5.2 Continuous Distributions

First we will describe how the probability distribution of a continuous random variable can be characterized graphically. To see how this is done, let's return to the actual case of a manufacturing plant that produces pieces of metal in which two holes are stamped. (See p. 57.) The distance between the hole centers varies from one piece of metal to another and is a continuous random variable since it can assume any numerical value on a continuous scale. If we were to examine 1,000 pieces of metal produced by this plant, we could find the proportion of these pieces where the distance between the hole centers is 2.994 inches and under 2.995 inches, the proportion where the distance is 2.995 inches and under 2.996 inches, and so on. Then, as shown in panel A of Figure 5.1, we could construct a histogram in which the area of the bar representing each class interval is equal to the proportion of all pieces within this class interval. (In other words, the height of the bar equals the proportion of all pieces within this class interval divided by the width of the class interval.) The area under this histogram must equal 1 because the sum of the proportions in all class intervals must equal 1. Such a histogram is said to use a *density* scale.

To obtain more complete information about the relative frequency

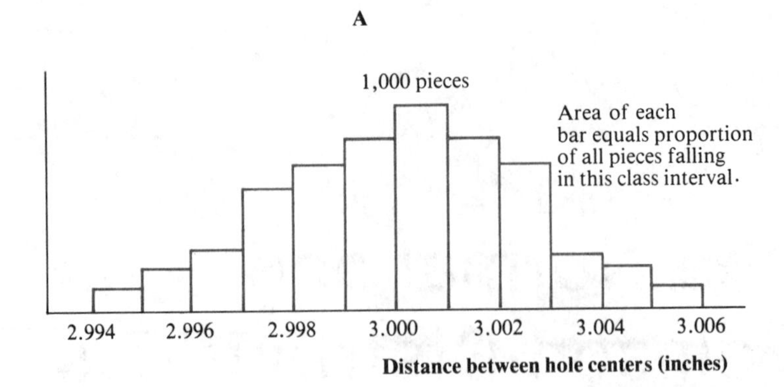

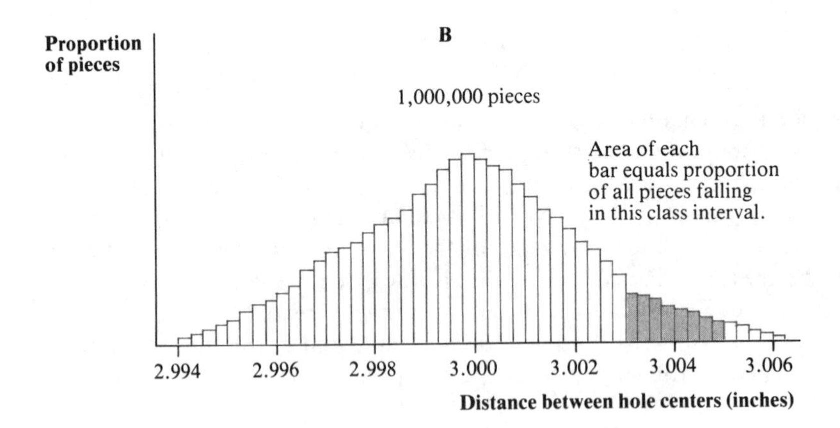

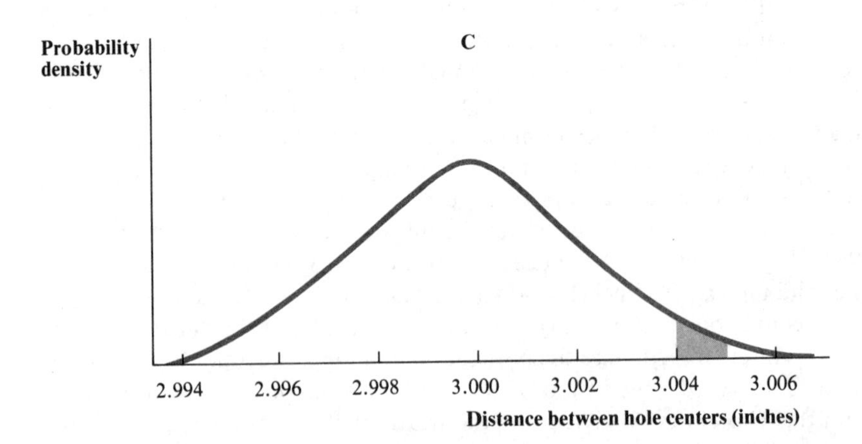

FIGURE 5.1 Histograms of Distances between Hole Centers of Pieces of Metal, Based on Data for 1,000 Pieces and 1,000,000 Pieces; and Probability Density Function

distribution of these distances, we could examine 1,000,000 rather than 1,000 pieces of metal and construct another histogram like the one in panel A of Figure 5.1. Because of the greatly increased number of observations, we would now be able to divide the class intervals *more finely*. As a result, we would have *more* bars, each of which is *narrower* than in the histogram in panel A. This second histogram is shown in panel B of Figure 5.1. The proportion of pieces of metal in which the distance between hole centers lies in a particular range can be read from this histogram by measuring the area of the bars lying within this range. For example, the proportion of pieces of metal where the distance between the hole centers is 3.003 inches and under 3.005 inches is measured by the shaded area of the histogram in panel B of Figure 5.1.

Finally, in order to obtain even more complete information, we could examine countless numbers of these pieces of metal, with the result that the class intervals could (and would) be made ever finer and more numerous. In the limit, the histogram would become a *smooth curve*, as shown in panel C of Figure 5.1. As in panels A and B, the total area under this smooth curve would equal 1, since the proportion of observations in all class intervals must total 1. Also, in panel C, as in the other panels of Figure 5.1, the proportion of pieces of metal in which the distance between the hole centers lies in a particular range can be found by measuring the area under the smooth curve in this range. Thus, if we wanted to know the proportion of pieces of metal in which the distance between the hole centers is between 3.004 inches and 3.005 inches we would measure the shaded area under the smooth curve in panel C.

The smooth curve in panel C of Figure 5.1 is important because we can use it to determine the probability that the distance between hole centers lies within a particular range, such as between 3.004 inches and 3.005 inches. As we have just seen, the proportion of cases in the long run where the distance lies in this range is equal in value to the area under the smooth curve in this range. Thus, *the probability that the distance between hole centers lies within a particular range is equal in value to the area under the smooth curve in this range.* In Figure 5.1 the probability that the distance between hole centers is between 3.004 inches and 3.005 inches equals the shaded area under the smooth curve in panel C.

5.3 Probability Density Function of a Continuous Random Variable

The smooth curve in panel C of Figure 5.1 is called a probability density function. We turn now from the specific case in Figure 5.1 to some generalizations concerning the probability density function of a continuous random variable.

> ***Probability Density Function of a Continuous Random Variable.*** *In the limit, as more and more observations are gathered concerning a continuous variable, and as class intervals become narrower and more numerous, the histogram (using a density scale) of the variable becomes a smooth curve (as in Figure 5.1) called a* probability density function. *The total area under*

any probability density function must equal 1. *The probability that a random variable will assume a value between any two points,* a *and* b, *equals the area under the random variable's probability density function over the interval from* a *to* b.[1]

To illustrate the use and interpretation of a random variable's probability density function, consider the diameters of tires produced by a particular manufacturer. The diameter of a randomly chosen tire can be viewed as a random variable. Suppose the tire's probability density function is as shown in Figure 5.2. This curve is of basic importance because it enables us to calculate the probability that a tire's diameter will be in any particular range in which we are interested. If we want to know the probability that a tire's diameter will be between C and D in Figure 5.2, all we have to do is determine the area between C and D under the probability density function. This is the shaded area in Figure 5.2

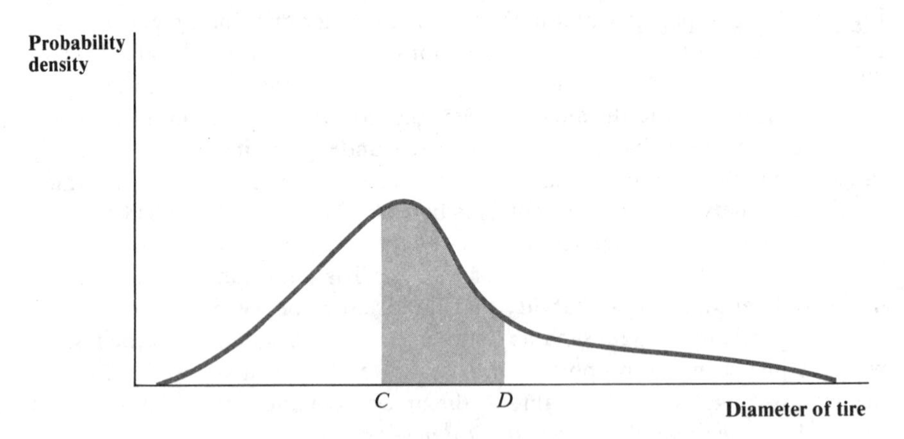

FIGURE 5.2 Probability Density Function of Diameters of Tires

5.4 The Normal Distribution

The most important continuous probability distribution is the normal distribution. The formula for the probability density function of a normal random variable (that is, a random variable with a normal distribution) is

$$f(x) = \frac{1}{\sqrt{2\pi}\,\sigma} e^{-\frac{1}{2}[(x-\mu)/\sigma]^2}, \quad (5.1)$$

where μ is the random variable's expected value (or mean), σ is its standard deviation, e is approximately 2.718 and is the base of the natural logarithms, and π is approximately 3.1416. Like the binomial distribution, the normal

1. The probability that a continuous random variable is precisely equal to a particular value is zero since the area under the probability density function at this particular value is a line of zero width.

distribution is really a family of distributions. Depending on its mean and standard deviation, the location and shape of the normal probability density function—or ***normal curve***, as we shall call it for short—can vary considerably.

To show how much the normal curve can vary, Figure 5.3 presents three normal curves, one with a mean of 15 and a standard deviation of 2.50, one with a mean of 40 and a standard deviation of 5, and one with a mean of 60 and a standard deviation of 1. As you can see, all three are bell-shaped and symmetrical, but the curves are located at quite different points along the horizontal axis because they have different means, and they exhibit quite different amounts of spread or dispersion because they have different standard deviations. Because of the differences in their means and standard deviations, some normal curves (like the middle one in Figure 5.3) are short and squat whereas others (like the one on the right in Figure 5.3) are tall and skinny. But in accord with the definition of a probability density function in the previous section, the total area under any normal curve must equal 1.

Although normal curves vary in shape because of differences in mean and standard deviation, all normal curves have the following characteristics in common:

1. *Symmetrical and Bell-shaped.* All normal curves are symmetrical about the mean. In other words, the height of the normal curve at a value that is a certain amount *below* the mean is equal to the height of the normal curve at a value that is the same amount *above* the mean. Because of this symmetry, the mean of a normal random variable equals both its median and its mode. (Recall the discussion of the relative position of the mean, median, and mode in Chapter 2.) Besides being symmetrical, the normal curve is bell-shaped, as in Figure 5.3. And a normal random variable can assume values ranging from $-\infty$ to $+\infty$.

2. *Probability that a Value Will Lie within* k *Standard Deviations of the Mean.* Regardless of its mean or standard deviation, the probability that the

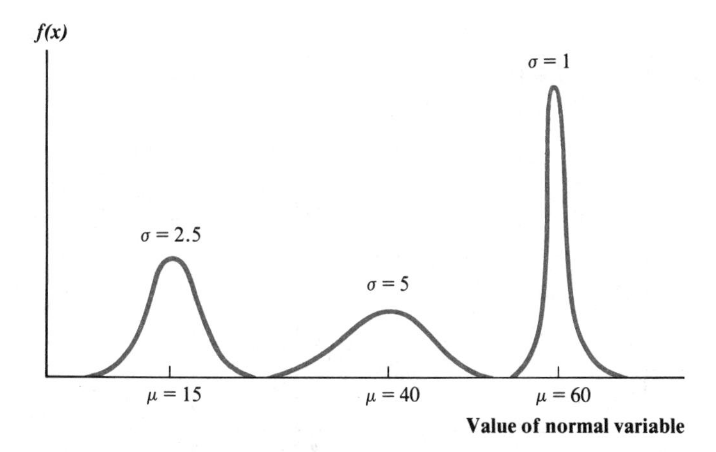

FIGURE 5.3 Three Normal Curves, with $\mu = 15$ and $\sigma = 2.5$; $\mu = 40$ and $\sigma = 5$; and $\mu = 60$ and $\sigma = 1$

value of a normal random variable will lie within *one* standard deviation of its mean is 68.3 percent, the probability that it will lie within *two* standard deviations of its mean is 95.4 percent, and the probability that it will lie within *three* standard deviations of its mean is 99.7 percent. Panel B of Figure 5.4 shows the distance from the mean, μ, in units of the standard deviation, σ. Clearly, almost all the area under a normal curve lies within 3 standard deviations of the mean.

3. *Location and Shape Determined Entirely by μ and σ.* The location of a normal curve along the horizontal axis is determined *entirely* by its mean μ. For example, if the mean of a normal curve equals 4, it is centered at 4; if its mean equals 400, it is centered at 400. The amount of spread in a normal curve is determined *entirely* by its standard deviation σ. If σ increases, the curve's spread widens; if σ decreases, the curve's spread narrows.

Why is the normal distribution so important in statistics? Basically, for

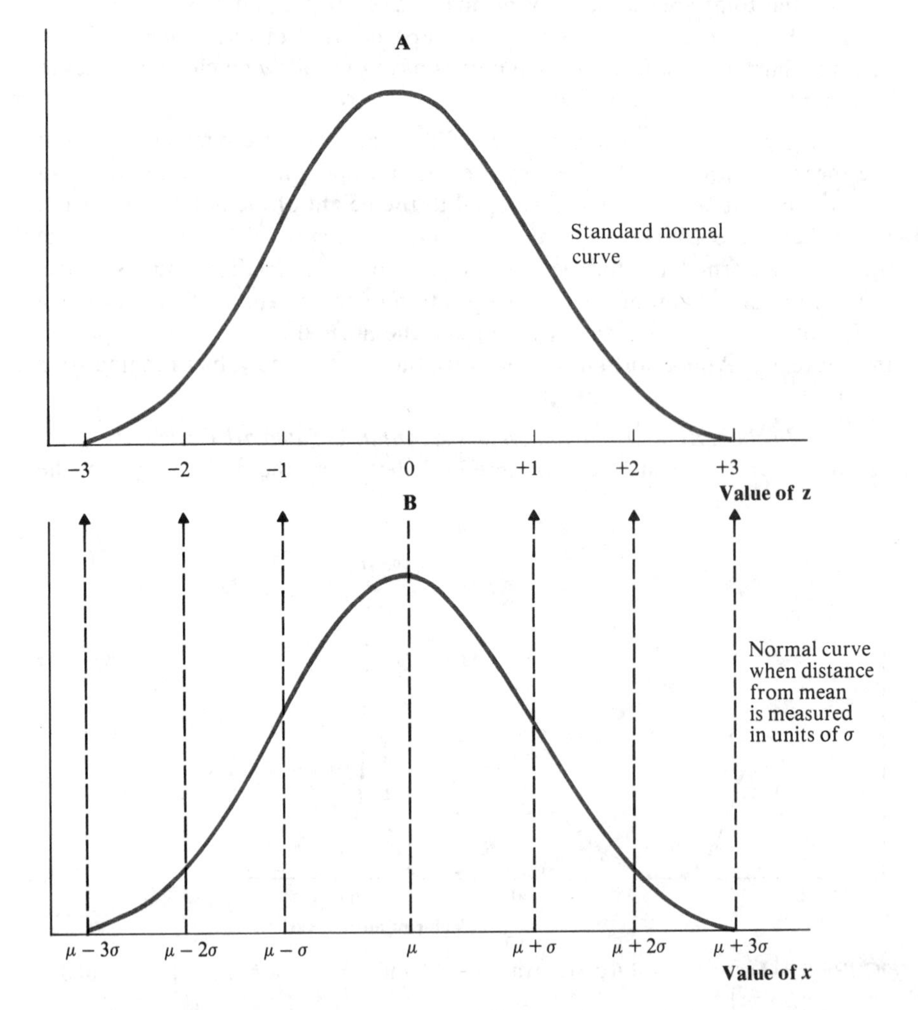

FIGURE 5.4 Comparison between Standard Normal Curve and Any Normal Curve when Distance from Mean Is Measured in Units of σ

three reasons. First, *the normal distribution is a reasonably good approximation to many populations.* Experience has shown that many (but by no means all) population histograms (using a density scale) are approximated quite well by a normal curve. For example, the histogram (using a density scale) of heights, weights, or IQs is likely to be reasonably close to a normal curve. Second, it can be shown that under circumstances described in the following chapter, *the probability distribution of the sample mean should be close to the normal distribution.* This is one of the most fundamental results in statistics. We must postpone discussing it until the next chapter, but it is an important reason for the key role played by the normal distribution in statistics.[2] Third, and related to the previous point, *the normal distribution can be used in many instances to approximate the binomial distribution.* In Section 5.8, we shall show how this approximation can be employed.

5.5 The Standard Normal Curve

As stressed in the previous section, normal curves vary greatly in shape because of differences in the mean μ, and in the standard deviation σ. However, if one expresses any normal random variable as a deviation from its mean, and measures these deviations in units of its standard deviation, the resulting random variable, called a ***standard normal variable***, has the probability distribution shown in panel A of Figure 5.4. This probability distribution is called the ***standard normal curve***.

If the weights of adult males are normally distributed, with a mean of 170 pounds and a standard deviation of 20 pounds, it is possible to express the weight of each adult male in *standard units* by finding the deviation of his weight from the mean and expressing this deviation in units of the standard deviation. For example, if William Morris's weight is 190 pounds, it is +1.0 in standard units. Why? Because his weight is 20 pounds above the mean, and since the standard deviation is 20 pounds, this amounts to a positive (+) deviation from the mean of 1 standard deviation. On the other hand, if John Jarvis's weight is 160 pounds, it is −0.5 in standard units because his weight is 10 pounds below the mean, which amounts to a negative (−) deviation from the mean of 0.5 standard deviations.

The important point to note is that if any normal variable is expressed in standard units, its probability distribution is given by the standard normal curve. Thus, if the weights of adult males are normally distributed and if we express them in standard units, their probability distribution is given by the standard normal curve. Put more formally, if X is a normally distributed random variable, then

$$\boxed{Z = \frac{X - \mu}{\sigma}} \tag{5.2}$$

has the standard normal distribution regardless of the values of μ and σ. Thus,

2. Under some circumstances, statistics other than the sample mean also have a normal distribution. We single out the sample mean only because of its great importance in statistical applications.

if X is the weight of an adult male, $(X - 170) \div 20$ has the standard normal distribution.

As a further illustration, given the fact that the heights of adult females are normally distributed with a mean of 66 inches and a standard deviation of 2 inches, if X is the height of an adult female, what is Z? Z is X expressed in standard units. That is,

$$Z = \frac{X - 66}{2}.$$

What is the value of Z corresponding to a height of 67 inches? It is $(67 - 66) \div 2$, or 0.5. What is the height corresponding to a Z value of -2.0? Since $(X - 66) \div 2 = -2.0$, X must equal 62 inches.

Figure 5.4 shows what happens when we express a normal variable, X, in standard units. Panel B shows the probability distribution of X. Note that in this panel the value of X is measured in units of the standard deviation (σ) from the mean (μ). When we express X in standard units, the value of Z corresponding to each value of X is shown by the arrows. Thus, if X equals $\mu - 3\sigma$, the corresponding Z value is -3; if X equals $\mu - 2\sigma$, the corresponding Z value is -2; and so on. Clearly, *the mean of the standard normal distribution is zero*, since zero in panel A corresponds to μ in panel B. Also, *the standard deviation of the standard normal distribution is* 1, since a distance of σ along the horizontal axis in panel B corresponds to a distance of 1 in panel A.[3]

5.6 Calculating Normal Probabilities

It frequently is necessary to calculate the probability that the value of a normal random variable lies between two points. To calculate this probability, two steps must be carried out:

1. *Find the Points on the Standard Normal Distribution Corresponding to These Two Points.* For example, if the heights of adult women are normally distributed with a mean of 66 inches and a standard deviation of 2 inches and if we want to know the probability that the height of an adult woman lies between 65 and 68 inches, our first step is to find the standard values corresponding to 65 and 68 inches. Since $\mu = 66$ and $\sigma = 2$, these standard values are $(65 - 66) \div 2$ and $(68 - 66) \div 2$, respectively. Simplifying terms, they are -0.5 and $+1.0$.

3. Using the results of Appendix 4.1, it is easy to prove that the mean of the standard normal variable is zero. Since $Z = (X - \mu) \div \sigma$, it follows that

$$Z = -\frac{\mu}{\sigma} + \frac{1}{\sigma}X.$$

Thus, in accord with Appendix 4.1,

$$E(Z) = -\frac{\mu}{\sigma} + \frac{1}{\sigma}E(X) = -\frac{\mu}{\sigma} + \frac{\mu}{\sigma} = 0.$$

Also, the results of Appendix 4.1 imply that the standard deviation of Z must equal the standard deviation of X (that is, σ) multiplied by $1/\sigma$, or 1.

2. *Determine the Area under the Standard Normal Curve between the Two Points We Have Found.* If we want to know the probability that the height of an adult woman lies between 65 and 68 inches, we determine the area under the standard normal curve between the points on the curve corresponding to 65 and 68 inches. Since these standard values are −0.5 and 1.0 (as we know from the preceding paragraph), we must determine the area under the standard normal curve between −0.5 and 1.0.

Why does this procedure give the correct answer? *Because the area under any normal curve between two points is equal to the area under the standard normal curve between the corresponding two points.* A comparison between panels A and B in Figure 5.5 demonstrates that this is true. In panel B the normal curve shows the distribution of adult female heights. The probability that the height of an adult woman lies between 65 and 68 inches equals the area under the curve between 65 and 68 inches (that is, the shaded area in panel B). In panel A we show the standard normal distribution. The points on

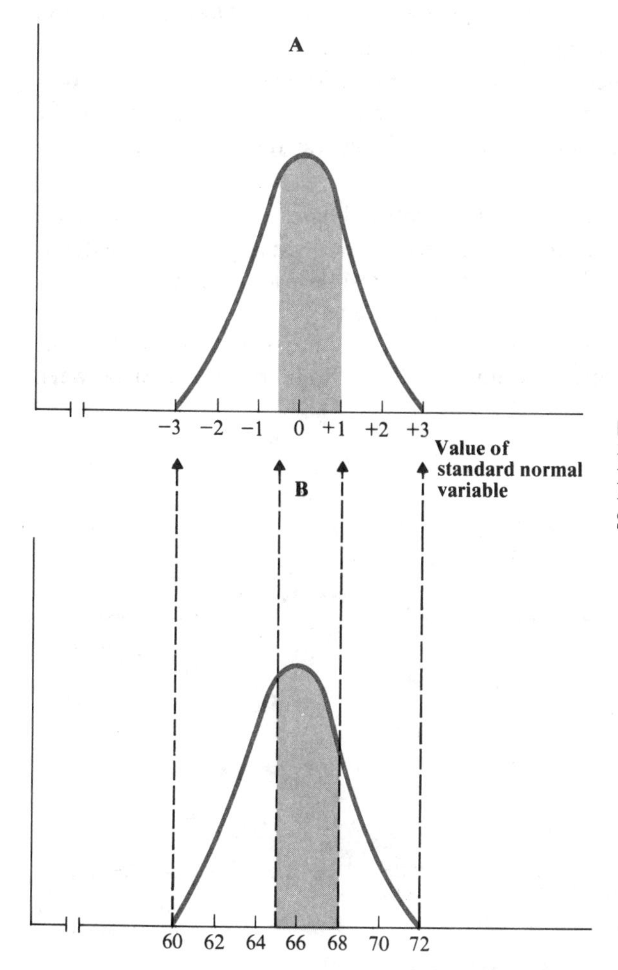

FIGURE 5.5
Normal Distribution of Female Heights and Standard Normal Curve

the standard normal distribution corresponding to 65 and 68 inches are −0.5 and 1.0, as we know from previous paragraphs. Thus, the shaded area under the standard normal curve equals the probability that the standard normal variable lies between −0.5 and 1.0. As you can see, the two shaded areas are equal. Thus, the probability that a height lies between 65 and 68 inches equals the area under the standard normal curve between −0.5 and 1.0.

5.7 Using the Table of the Standard Normal Distribution

The area under the standard normal distribution between various points is tabled. To carry out the second step in the procedure described above, one must be able to use this table, which is contained in Appendix Table 2. In the following paragraphs, we indicate how this table is used in various situations.

1. *Area between Zero and Some Positive Value.* Each number in the body of Appendix Table 2 shows the area between zero (the mean of the standard normal distribution) and the positive number (z) given in the left-hand column (and top) of the table. For example, to determine the area between zero and 1.10, look at the row labeled 1.1 and the column labeled .00; the area is .3643. This is the shaded area in panel A of Figure 5.6. Similarly, to determine the area between zero and 1.63, look at the row labeled 1.6 and the column labeled .03; the area is .4484.

2. *Area between Zero and Some Negative Value.* Because the standard normal curve is symmetrical, the area between zero and any negative value is equal to the area between zero and the same positive value. Hence Appendix Table 2 can readily be used to evaluate the desired area. For example, the area between zero and −1.10 equals that between zero and +1.10; thus, this area (shaded in panel B of Figure 5.6) must be .3643. Similarly, the area between

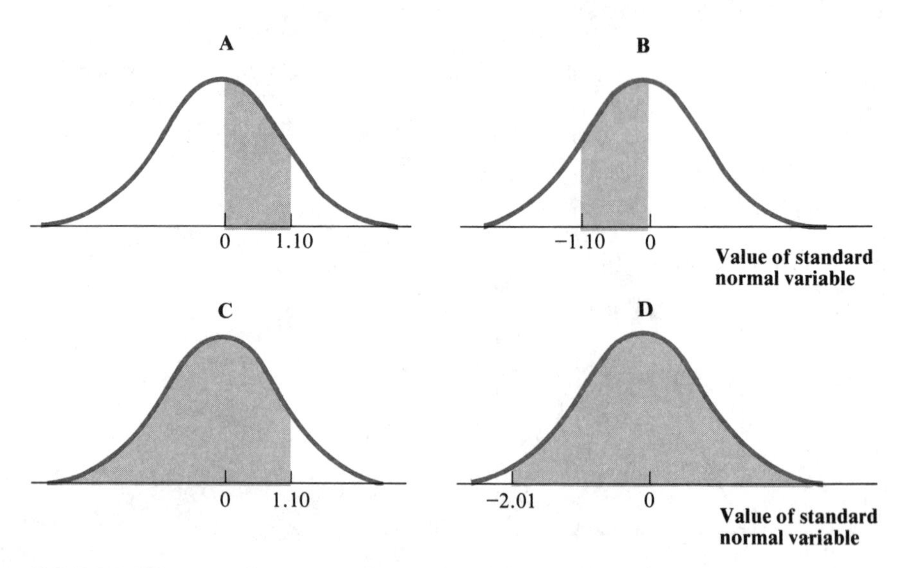

FIGURE 5.6 Areas under the Standard Normal Curve

zero and −1.63 equals that between zero and +1.63, which we know to be .4484.

3. *Area to the Left of Some Positive Value.* Suppose that we want to determine the area to the left of 1.10. This area (shaded in panel C of Figure 5.6) is composed of two parts: the area to the left of zero, and the area between zero and 1.10. The area to the left of zero is .5 because the standard normal curve is symmetrical about zero and because the area under the entire curve equals 1. The area between zero and 1.10 can be determined from Appendix Table 2, as we already know. Since it is .3643, the area we want is .5000 + .3643 = .8643.

4. *Area to the Right of Some Negative Number.* What is the area to the right of −2.01? This area (shaded in panel D of Figure 5.6) is composed of two parts: the area to the right of zero, and the area between zero and −2.01. The area to the right of zero is .5 because the standard-normal curve is symmetrical about zero and because the area under the entire curve equals 1. The area between zero and −2.01 can be determined from Appendix Table 2, since it equals the area between zero and 2.01, which is .4778. Thus, the area we want equals .5000 + .4778 = .9778.

5. *Area to the Right of Some Positive Value.* What is the area to the right of 1.65? This area (shaded in panel A of Figure 5.7) plus the area between zero and 1.65 must equal .5000, because the total area to the right of zero equals .5000. Thus, the area we want equals .5000 minus the area between zero and 1.65. Since Appendix Table 2 shows that the area between zero and 1.65 is .4505, the area we want equals .5000 − .4505 = .0495.

6. *Area to the Left of Some Negative Value.* What is the area to the left of

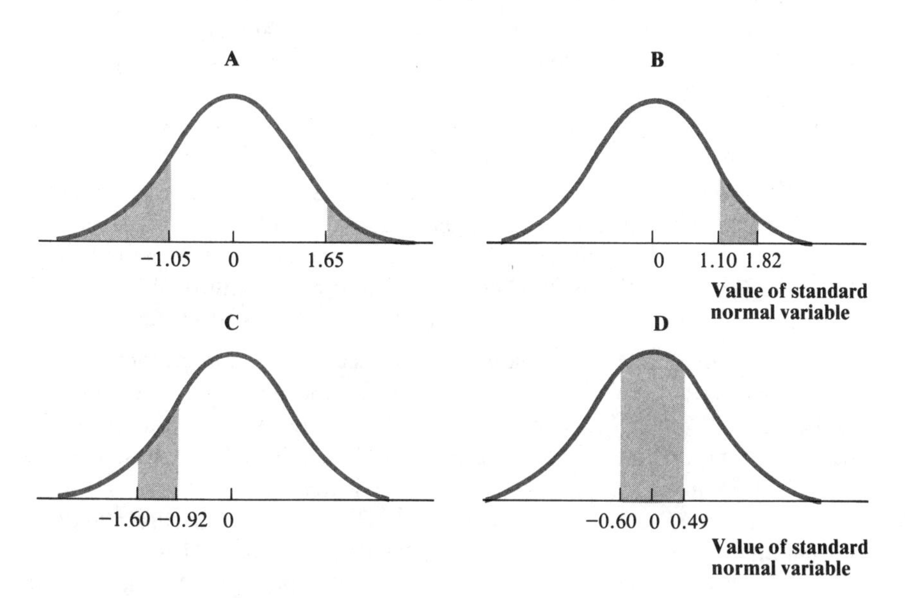

FIGURE 5.7 Areas under the Standard Normal Curve

-1.05? This area (shaded in panel A of Figure 5.7) plus the area between zero and -1.05 must equal .5000, because the total area to the left of zero is .5000. Thus, the area we want equals .5000 minus the area between zero and -1.05. The area between zero and -1.05 equals the area between zero and 1.05, which is .3531, according to Appendix Table 2. Hence the area we want equals $.5000 - .3531 = .1469$.

7. *Area between Two Positive Values.* What is the area between 1.10 and 1.82? This area (shaded in panel B of Figure 5.7) equals the difference between (*a*) the area between zero and 1.82, and (*b*) the area between zero and 1.10. Appendix Table 2 shows that the former area is .4656 and that the latter area is .3643. Thus, the area we want is $.4656 - .3643 = .1013$.

8. *Area between Two Negative Values.* What is the area between -1.60 and -0.92? This area (shaded in panel C of Figure 5.7) equals the difference between (*a*) the area between zero and -1.60, and (*b*) the area between zero and -0.92. The area between zero and -1.60 equals the area between zero and 1.60, which is .4452, according to Appendix Table 2. The area between zero and -0.92 equals the area between zero and 0.92, which is .3212, according to Appendix Table 2. Thus, the area we want is $.4452 - .3212 = .1240$.

9. *Area between a Negative and a Positive Value.* Finally, suppose we want to determine the area between -0.60 and 0.49. This area (shaded in panel D of Figure 5.7) equals the sum of (*a*) the area between zero and -0.60, and (*b*) the area between zero and 0.49. The area between zero and -0.60 equals the area between zero and 0.60, which is .2257, according to Appendix Table 2. The area between zero and 0.49 is .1879, according to Appendix Table 2. Thus, the area we want is $.2257 + .1879 = .4136$.

Because of the central importance of the normal distribution in statistics, it is essential that you be able to calculate the probability that a normal random variable lies in a given range. The following three examples are designed to illustrate how this is done.

Example 5.1 Find the probability that the value of the standard normal variable will lie between -1.23 and $+1.14$.

Solution: Appendix Table 2 shows that the area under the standard normal curve between 0 and 1.23 is .3907, so the area between 0 and -1.23 must also be .3907. Appendix Table 2 shows that the area between 0 and 1.14 is .3729. Thus, the area between -1.23 and $+1.14$ equals $.3907 + .3729 = .7636$, which means that the probability we want equals .7636.

Example 5.2 The diameters of the tires produced by a tire manufacturer are normally distributed with a mean of 36 inches and a standard deviation of .001 inches. What is the probability that a tire produced by this firm will have a diameter that is (a) between 35.9990 and 36.0005 inches; (b) less than 35.9985 inches; (c) greater than 36.0004 inches?

Solution: (a) The first step is to find the points on the standard normal distribution corresponding to 35.9990 and 36.0005. These points are $(35.9990 - 36) \div .001$ and $(36.0005 - 36) \div .001$, or -1.0 and $+0.5$, respectively. According to Appendix Table 2, the area under the standard normal curve between zero and 1.0 is .3413, which means that the area between zero and -1.0 also is .3413. According to Appendix Table 2, the area between zero and 0.5 is .1915. Thus, the area between -1.0 and 0.5 is $.3413 + .1915 = .5328$. This is the probability that a tire's diameter is between 35.9990 and 36.0005 inches.

(b) The point on the standard normal distribution corresponding to 35.9985 inches is $(35.9985 - 36) \div .001$, or -1.5. According to Appendix Table 2, the area under the standard normal curve between 0 and 1.5 is .4332, so the area between 0 and -1.5 also is .4332. Thus, the area to the left of -1.5 equals $.5000 - .4332 = .0668$. This is the probability that a tire's diameter is less than 35.8885 inches.

(c) The point on the standard normal distribution corresponding to 36.0004 inches is $(36.0004 - 36) \div .001$, or $+0.4$. According to Appendix Table 2, the area under the standard normal curve between zero and 0.4 is .1554, so the area to the right of 0.4 must equal $.5000 - .1554 = .3446$. This is the probability that a tire's diameter is more than 36.0004 inches.

Example 5.3 The president of the tire manufacturing firm in Example 5.2 makes a statement to reporters that 90 percent of the tires produced by the firm have diameters of 36.0020 inches or less, while 10 percent of the tires have diameters of more than 36.0020 inches. He is incorrect. What figure should be substituted for 36.0020?

Solution: The first step is to find the number which the value of the standard normal variable will exceed with a probability of .10. Since the probability must be $.50 - .10$ (or .40) that the value of the standard normal variable lies between zero and this number, we must look in Appendix Table 2 for the value of Z corresponding to a probability of .40. This value is 1.28. The next step is to find the value of the tire diameter that corresponds to this value of the standard normal variable. In other words, we must find the value of X in equation (5.2) that corresponds to $Z = 1.28$. Clearly, this value is $\mu + 1.28\sigma$, which equals $36 + 1.28(.001)$, or 36.00128 inches. Thus, the probability is 0.10 that a tire's diameter will exceed 36.00128 inches. The figure of 36.00128 inches, not the president's, is correct.

5.8 The Normal Distribution as an Approximation to the Binomial Distribution

As pointed out in an earlier section of this chapter, one reason why the normal distribution is so important is that it can be used as an approximation to the binomial distribution under certain circumstances. These are described below.

Normal Approximation to the Binomial Distribution. *If* n *(the number of trials) is large and* P *(the probability of success) is not too close to zero or* 1, *the probability distribution of the number of successes occurring in* n *Bernoulli trials can be approximated by a normal distribution. Experience indicates that the approximation is fairly accurate as long as* $nP \geq 5$ *when* $P \leqslant 1/2$ *and* $n(1 - P) > 5$ *when* $P > 1/2$.

The fact that the normal distribution can approximate the binomial distribution under the circumstances described above is useful because, as noted in Chapter 4, it is tedious to calculate the binomial probabilities when n is large.

The following illustration shows how the normal distribution is used to estimate binomial probabilities. If a true coin is flipped 1,600 times, what is the probability distribution of the number of times that the coin comes up heads? Since there are 1,600 Bernoulli trials, the number of times heads comes up is clearly a binomial random variable. Moreover, since $n = 1{,}600$ and $P = 1/2$, its mean is 800 $(= nP)$ and its standard deviation is 20 $(= \sqrt{nP(1 - P)})$. The probability distribution of the number of times heads comes up is shown in panel A of Figure 5.8. Since it is difficult to evaluate each of the binomial probabilities (in equation 4.6) when n is as large as 1,600, we would like to approximate this probability distribution with another that is easier to calculate. Fortunately, as noted above, the normal distribution—with the same

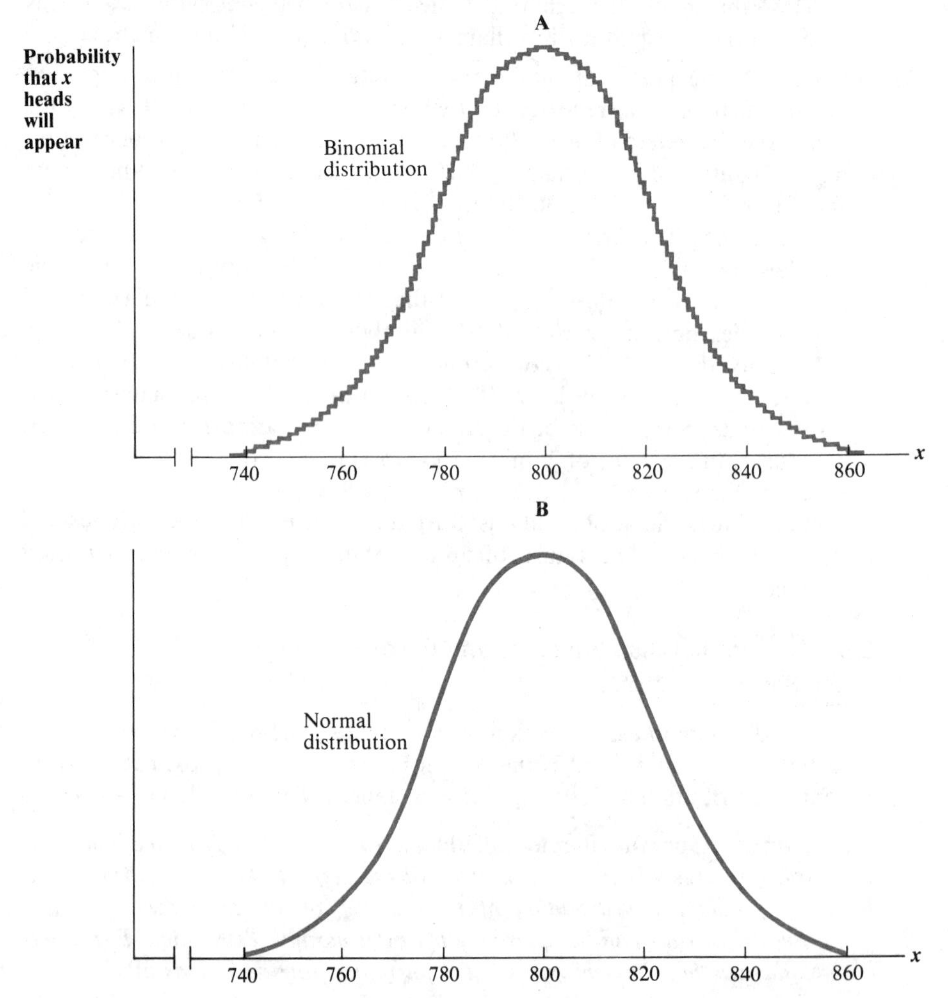

FIGURE 5.8 Comparison of Binomial Distribution and Normal Distribution, Both with Mean of 800 and Standard Deviation of 20

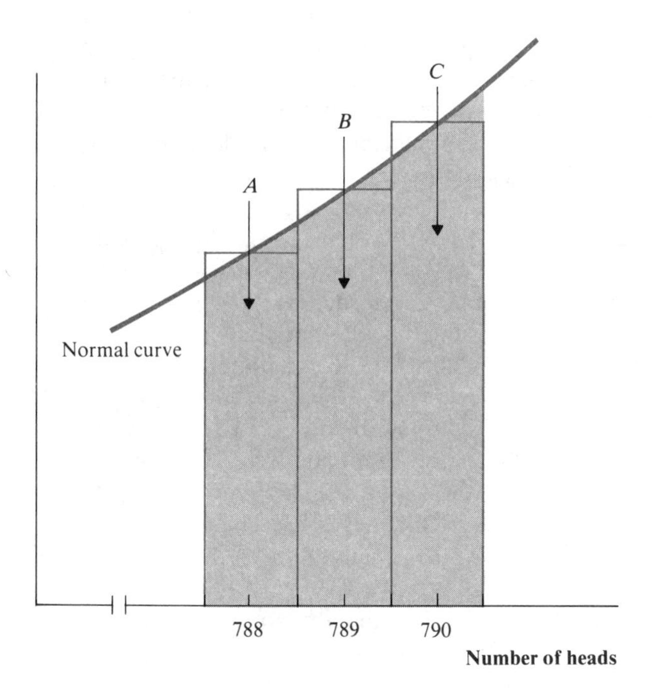

FIGURE 5.9 Normal Approximation to the Binomial Distribution in the Range between 787.50 and 790.50

mean (800) and standard deviation (20) as the binomial distribution—is a good approximation.

A visual comparison of the normal distribution in panel B of Figure 5.8 with the binomial distribution in panel A (each of which has a mean of 800 and a standard deviation of 20) certainly indicates that the former is shaped much like the latter. But to make sure that this is a good approximation, we must investigate in greater detail. Let's look carefully at the segment of both probability distributions between $x = 788$ and $x = 790$ (where x is the number of heads that comes up). Figure 5.9 shows a "blow up" of each of the probability distributions over the relevant range. If the approximation is accurate, the area under the normal curve between 787.50 and 790.50 must be approximately equal to the area under the binomial distribution between 787.50 and 790.50. In other words, the shaded area under the continuous curve should be approximately equal to the sum of the areas of the three rectangles (A, B, and C) shown in Figure 5.9. The sum of the areas of these three rectangles equals the true probability that the number of heads is 788, 789, or 790, whereas the shaded area is the approximation to this probability. Based on Figure 5.9, it certainly appears that the approximation is good.

To find the probability that the number of heads is 788, 789, or 790, we do *not* find the area under the normal curve between 788 and 790; instead, we find the area *between 787.50 and 790.50.* As shown in Figure 5.9, in order to approximate the three rectangles (*A, B,* and *C*) corresponding to the probabilities of 788, 789, and 790 heads, we must include the area under the continuous curve from 787.50 to 790.50. This is often called a *continuity correction*, a correction due to the fact that a discrete probability distribution is being approximated by a continuous one. In general, to find the probability that a binomial variable equals at least c but no more than d (where $c < d$), we find the probability that a normal variable (with mean nP and standard deviation $\sqrt{nP(1 - P)}$) lies between $(c - 1/2)$ and $(d + 1/2)$.

Based on the normal approximation, what is the probability that the number of heads is 788, 789, or 790? As we have seen, this probability is approximately equal to the probability that the value of a normal random variable with mean equal to 800 and standard deviation equal to 20 lies between 787.50 and 790.50. The value of the standard normal variable corresponding to 787.50 is $-.625$, and that corresponding to 790.50 is $-.475$. Using Appendix Table 2, we find that the area under the standard normal curve between zero and 0.625 is approximately .234, which means that the area between zero and -0.625 also is approximately .234. Similarly, the area between zero and 0.475 is approximately .183, which means that the area between zero and -0.475 also is approximately .183. Thus, the probability that the number of heads is 788, 789, or 790 equals (approximately) $.234 - .183 = .051$.

The following is another illustration of how the normal distribution can be used to approximate the binomial distribution.

Example 5.4 The probability that a machine will be down for repairs next week is 1/2. A firm has 100 such machines, and whether one is down is statistically independent of whether another is down. What is the probability that at least 60 machines will be down?

Solution: The number of machines down for repairs has a binomial distribution with mean equal to 100(1/2), or 50, and standard deviation equal to $\sqrt{100(1/2)(1/2)}$, or 5. Because of the continuity correction, the probability that the number down for repairs is 60 or more can be approximated by the probability that the value of a normal variable with mean equal to 50 and standard deviation equal to 5 exceeds 59.50. The value of the standard normal variable corresponding to 59.50 is $(59.50 - 50) \div 5$, or 1.9. Appendix Table 2 shows that the area under the standard normal curve between zero and 1.9 is .4713, so the area to the right of 1.9 must equal $.5000 - .4713 = .0287$. This is the (approximate) probability that at least 60 machines will be down for repair.

5.9 Wind Changes for Hurricanes: A Case Study[4]

Since hurricanes annually cause property damage totaling billions of dollars, government agencies and private organizations have a strong interest in

4. This case study is based on R. Howard, J. Matheson, and D. North, "The Decision to Seed Hurricanes," *Science*, June 1972. An abridged version of this article is contained in E. Mansfield, *Statistics for Business and Economics: Readings and Cases* (New York: Norton, 1980). The present discussion is simplified, and some numbers have been changed. For a much more complete and accurate (but also more technical) account, see the above article in *Science*.

It is assumed that *if no change occurs in maximum sustained wind speed*, a hurricane of the sort considered here will cause $100 million in property damage.

studying hurricanes in order to reduce their impact. In 1970, statisticians at the Stanford Research Institute (now SRI International) began a study for the U.S. Department of Commerce of the behavior and impact of hurricanes. The normal distribution was used to compute the probability that wind changes of various magnitudes would occur for an unseeded hurricane during a 12-hour period before landfall. Based on various meteorological studies, it appears that *the percentage change in maximum sustained wind speed during this period is normally distributed with a mean of zero and a standard deviation of 15.6 percent.* In other words, the distribution of the percentage change in an unseeded hurricane's wind speed during such a period is as shown in Figure 5.10.

The amount of property damage caused by a hurricane is related to the extent of the change in its maximum sustained wind speed. Thus, if a change of +32 percent or more occurs, the Stanford statisticians estimate that property damage of well over $300 million might be expected. On the other hand, if a change of −34 percent or less occurs, the estimated property damage is less than $20 million. From this we see that it is important to know the probability that the change in maximum sustained wind speed will lie in various ranges. What is the probability that the change in wind speed will exceed +32 percent, thus resulting in well over $300 million in property damage? Obviously, this is an important question for many firms and government agencies.

If X is the percentage change in maximum sustained wind speed, we want to evaluate the probability that $X > 32$. This probability is denoted by

$$Pr\{X > 32\}.$$

Using the procedures described in earlier sections of this chapter, it is relatively simple to evaluate this probability. To do so, we must find the value of the standard normal variable corresponding to 32 percent. This value is (32 − 0) ÷ 15.6, or 2.05. (Why? Because the mean of X is zero and its standard deviation is 15.6, as pointed out above.) Appendix Table 2 shows that the area under the standard normal curve between zero and 2.05 is .4798, so the area to the right of 2.05 equals .5000 − .4798 = .0202. Thus, the probability is about

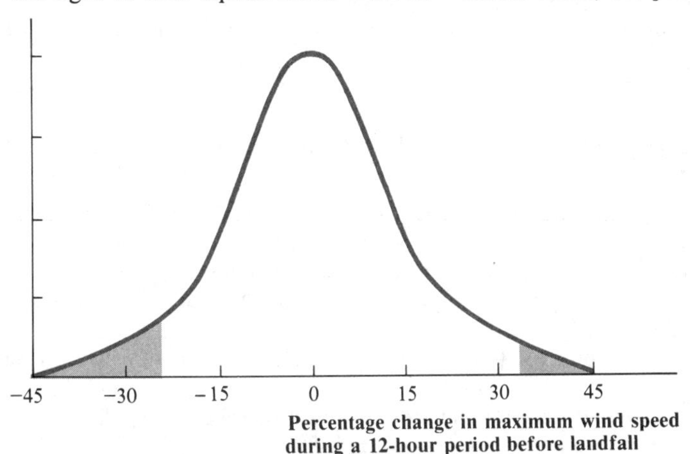

FIGURE 5.10 Distribution of Extent of Wind Change for an Unseeded Hurricane

.02 that an increase of more than 32 percent in wind speed will occur. (The shaded area to the right of zero in Figure 5.10 is equal to this probability.)

It is also useful to be able to answer a somewhat different type of question: What is the percentage change in wind speed that will be exceeded with a specified probability? For example, what is the percentage change in wind speed that will be exceeded with a probability of .95? To answer this question, we must find the value of x_0 such that

$$Pr\{X > x_0\} = .95.$$

As a first step, note that Appendix Table 2 shows that the area under the standard normal curve between zero and 1.64 is .45. Thus, the area between zero and -1.64 also is .45, which means that the area to the right of -1.64 is .95. Having determined that the probability is .95 that a standard normal variable will exceed -1.64, we must now find the value of the normal variable corresponding to this value of the standard normal variable. In other words, we must find the value of X in equation (5.2) when $Z = -1.64$. Clearly, the desired value of the normal variable is $\mu - 1.64\sigma$, which here equals $0 - 1.64(15.6) = -25.6$ percentage points. Thus, a -25.6 percentage point change in wind speed is the value that will be exceeded with a probability of .95. (The shaded area to the left of zero in Figure 5.10 equals .05.)

EXERCISES

5.1 Find the probability that the standard normal variable lies (a) above 2.3; (b) below -3.0; (c) above 0.7; (d) between 1 and 2; (e) between -1 and 2.

5.2 If X is a normal variable with $\mu = 2$ and $\sigma = 3$, show how it can be converted into the standard normal variable.

5.3 "If you know that the probability that a normal variable exceeds a certain number, Q, is .10, you can be sure that the probability that this variable is less than $-Q$ is also .10." Do you agree? Why, or why not?

5.4 From past experience, the Uphill Manufacturing Corporation knows that the deviations of the width of its bicycle pedals from their mean width are normally distributed with a standard deviation of .02 inches.

(a) What is the probability that a pedal's width is more than .03 inches above the mean?

(b) What is the probability that a pedal's width is more than .05 inches below the mean?

(c) What is the probability that a pedal's width differs (either positively or negatively) from the mean by less than .015 inches?

5.5 During 1978, the mean width of all bicycle pedals produced by Uphill was exactly equal to what the design called for. However, during 1979, the mean width of all bicycle pedals produced by the firm was .01 inches greater than the design called for. In both years, the standard deviation of the pedal widths was .02 inches, and the pedal widths were normally distributed.

(a) In 1978, what was the probability that a pedal would be wider than called for by the design?

(b) In 1979, what was the probability that a pedal would be wider than called for by the design?

(c) In 1978, what was the probability that a pedal would be more than .04 inches wider than called for by the design?
(d) In 1979, what was the probability that a pedal would be more than .04 inches wider than called for by the design?

5.6 There is a 1/3 probability that a pedal produced by the Uphill Manufacturing Company will not survive more than 10 years of normal wear. The firm sells these pedals in cartons of 300. What is the probability that in any such carton less than 85 or more than 115 pedals will not survive more than 10 years of normal wear? (Use the normal approximation to the binomial distribution.)

5.7 Suppose X is a normal random variable with mean μ and standard deviation σ.
(a) Under what circumstances is X/σ the standard normal variable?
(b) Under what circumstances is $(X - \mu)$ the standard normal variable?

5.8 If the weights of adult males are normally distributed with mean equal to 170 pounds and standard deviation equal to 20 pounds, the probability that a certain weight will be exceeded is .05. What is this weight?

5.9 In the previous exercise, the probability that an adult male's weight will be less than a certain amount is .10. What is this amount?

5.10 The probability that a marksman will hit the target is 1/3. If he takes 50 shots, what is the probability that he will hit the target less than 10 times? (Use the normal distribution as an approximation to the binomial distribution.)

5.10 The Poisson Distribution[5]

Another important probability distribution is the Poisson distribution, which is named after a nineteenth-century Swiss mathematician. The *Poisson distribution* is a discrete probability distribution which has the following formula:

$$P(x) = \frac{\mu^x e^{-\mu}}{x!}, \text{ for } x = 0, 1, 2, \ldots \tag{5.3}$$

where $P(x)$ is the probability that a variable with a Poisson distribution equals x, μ is the mean or expected value of the Poisson distribution and e is approximately 2.718 and is the base of the natural logarithms.[6] Like the binomial and normal distributions, the Poisson distribution is really a family of distributions. Depending on the value of μ, the shape of the probability distribution will vary considerably.

One reason why the Poisson distribution is so important in statistics is

5. This material and the remainder of this chapter (including the Appendix) can be omitted without loss of continuity.
6. A Poisson random variable, like a binomial random variable, is discrete. Unlike the binomial, it does not assume a finite number of possible values; instead, it assumes a countably infinite number of possible values.

that it can be used as an approximation to the binomial distribution under the circumstances described below.

> ***Poisson Approximation to the Binomial Distribution.*** *If* n *(the number of trials) is large and* P *(the probability of success) is small, the probability of* x *successes occurring in* n *Bernoulli trials can be approximated by the Poisson distribution where* $nP = \mu$. *Experience indicates that this approximation is adequate for most practical purposes if* n *is at least 20 and* P *is no greater than .05.*

Whereas the normal distribution approximates the binomial distribution when P is *not* very small, the Poisson distribution approximates it when P *is* very small; thus, the two approximations complement one another.

To illustrate how the Poisson distribution can be used in this way, let's consider the following situation. You drive to work 15,000 times in a 30-year period, and the probability of your having an accident each time you drive to work is .0001. In this case, each trip can be considered a "trial" and each accident can be considered a "success" (although only for your garage mechanic and/or mortician). Thus, $n = 15{,}000$ and $P = .0001$. Since n is very large and P is very small, the Poisson distribution should be a good approximation to the binomial distribution. Since $\mu = nP = 15{,}000(.0001)$, or 1.5, equation (5.3) can be used to obtain the probability of 0, 1, 2, . . . accidents during this thirty-year period, the results being:

$$P(0) = \frac{1.5^0 e^{-1.5}}{0!} = e^{-1.5} = 0.22,$$

$$P(1) = \frac{1.5^1 e^{-1.5}}{1!} = 1.5e^{-1.5} = 0.33,$$

$$P(2) = \frac{(1.5)^2 e^{-1.5}}{2!} = \frac{2.25e^{-1.5}}{2} = 0.25,$$

$$P(3) = \frac{(1.5)^3 e^{-1.5}}{3!} = \frac{3.375e^{-1.5}}{6} = 0.13,$$

$$P(4) = \frac{(1.5)^4 e^{-1.5}}{4!} = \frac{5.062e^{-1.5}}{24} = 0.05,$$

$$P(5) = \frac{(1.5)^5 e^{-1.5}}{5!} = \frac{7.594e^{-1.5}}{120} = 0.01.$$

We could, of course, compute the probability of 6, 7, 8, . . . accidents, but these probabilities are less than .005. Figure 5.11 shows this Poisson distribution graphically. As you can see, the distribution is not symmetrical, but skewed to the right. (Recall our discussion of skewness in Chapter 2.)

As pointed out earlier in this section, *the expected value of any Poisson*

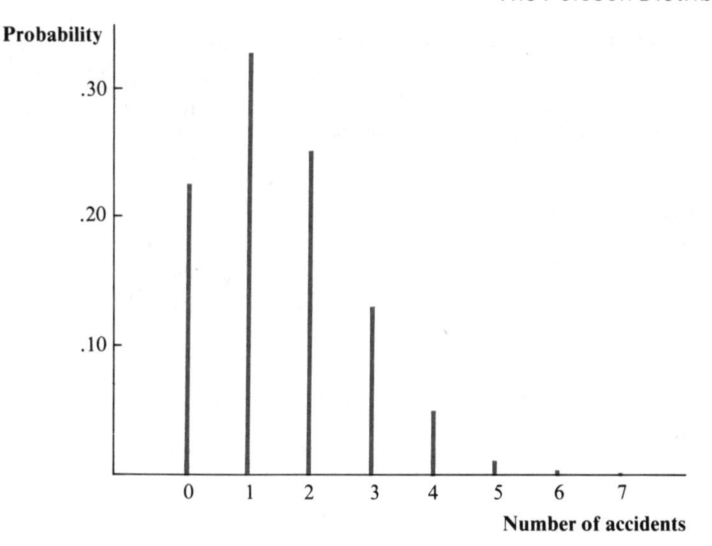

FIGURE 5.11 Poisson Distribution, with $\mu = 1.5$

random variable equals μ. To demonstrate this, in the present illustration the expected number of accidents equals

$$E(X) = (0)(0.22) + (1)(0.33) + (2)(0.25) + (3)(0.13) + (4)(.05) + (5)(.01) + (6)(.004) + (7)(.001) + \ldots = 1.50.$$

Also, *the standard deviation of any Poisson random variable equals* $\sqrt{\mu}$. (As an exercise, prove that this is true in this illustration; that is, prove that the standard deviation of the number of accidents equals $\sqrt{1.50}$. If you have difficulty, consult the footnote on this page.)[7]

7. To determine the standard deviation, first determine the variance by inserting the values of $P(x)$ into the following expression:

$$\sum (x - \mu)^2 P(x).$$

The result is

$$\sigma^2 = (0 - 1.5)^2(0.22) + (1 - 1.5)^2(0.33) + (2 - 1.5)^2(0.25) + (3 - 1.5)^2(0.13) + (4 - 1.5)^2(.05) + (5 - 1.5)^2(.01) + (6 - 1.5)^2(.004) + (7 - 1.5)^2(.001) + \ldots.$$

Thus,

$$\sigma^2 = 2.25(0.22) + 0.25(0.33) + 0.25(0.25) + 2.25(0.13) + 6.25(.05) + 12.25(.01) + 20.25(.004) + 30.25(.001) + \ldots$$

The computations involved in evaluating $P(x)$ can be onerous. To reduce the computational burden, Appendix Table 3 can be used. This table shows $P(x)$ for selected values of μ. The following example will provide additional practice in using the Poisson distribution.

Example 5.5 A machine turns out engine parts, 2 percent of which are defective. These parts are packaged in boxes of 100 and are shipped to the plant where they are used.

(a) What is the probability of 0, 1, 2, or 3 defectives in a box?

(b) What is the expected number of defectives in a box?

(c) What is the standard deviation of the number of defectives in a box?

Solution: (a) Since $\mu = nP = 100(.02)$ or 2.0, equation (5.3) yields the following probabilities:

$$P(0) = 2^0 e^{-2} \div 0! = .1353,$$

$$P(1) = 2^1 e^{-2} \div 1! = .2707,$$

$$P(2) = 2^2 e^{-2} \div 2! = .2707,$$

$$P(3) = 2^3 e^{-2} \div 3! = .1804.$$

These probabilities can be found in the column of Appendix Table 3 where $\mu = 2.0$.

(b) The expected number of defectives in a box equals $\mu = nP = 100(.02)$, or 2.

(c) The standard deviation of the number of defectives in a box equals $\sqrt{\mu} = \sqrt{nP} = \sqrt{2}$.

5.11 Additional Uses for the Poisson Distribution

Besides being a useful approximation to the binomial distribution, the Poisson distribution is very important in its own right. Assume that events of a particular kind occur at random during a particular time span. To make things more concrete, suppose that the events in question are demands by a firm's customers for a particular type of spare part. If the following four conditions are met, the probability distribution of the number of such events (that is, the number of demands for this type of spare part) in a fixed period of time will be a Poisson distribution:

1. *The Probability that Each Event Occurs in a Very Short Time Interval Must Be Proportional to the Length of This Time Interval.* Thus, the probability that a spare part of this type is demanded in a two-minute interval must be double the probability that a spare part of this type is demanded in a

---

$$= 0.495 + .082 + .063 + .292 + .313 + .122 + .081 + .030 + \ldots$$

$$= 1.50.$$

The standard deviation is the square root of the variance, or $\sqrt{1.50}$.

one-minute interval. Why? Because the former time interval is twice as long as the latter time interval.

2. *The Probability that Two or More Events of the Relevant Kind Occur in a Very Short Time Interval Must Be So Small that It Can Be Regarded as Zero.* Thus, the probability that more than one order will occur for this type of spare part in a one-second time interval must be essentially zero. This assumption seems reasonable in this case. If the time interval is only one second long, it would be difficult indeed for two different orders for this type of spare part to be received by the company.

3. *The Probability that a Particular Number of These Events Occurs in a Particular Time Interval Must Not Depend on When This Time Interval Begins.* Thus, the probability that an order is received by the company in a one-minute time interval beginning at noon tomorrow must be the same as the probability that an order will be received in a one-minute time interval beginning at 2 P.M. today. This is because it is assumed that this probability depends only on the length of the time interval, not on when the time interval begins. The fact that the one time interval begins at noon tomorrow and the other time interval begins at 2 P.M. today must not influence this probability at all.

4. *The Probability that a Particular Number of These Events (Demands) Occurs in a Particular Time Interval Must Not Depend on the Number of These Events that Occurred Prior to the Beginning of This Time Interval* (or in some shorter time interval prior to the beginning of this time interval). For example, suppose that five orders for this type of spare part were received prior to 2 P.M. today. The fact that five orders (rather than four, six, or some other number) were received prior to that time should not influence the probability of receiving an order in the one-minute time interval beginning at 2 P.M. today. Of course, this assumption may be violated if there is a tendency for these events to bunch together in time. For example, if orders tend to bunch together, the probability of receiving an order in the one-minute time interval beginning at 2 P.M. today may be dependent on whether an order was received just before 2 P.M. If so, the Poisson distribution is not appropriate.

If these four conditions are met, it can be shown that *the probability that* x *such events will occur in a time interval of length* Δ *(delta) is*

$$P(x) = \frac{(\lambda\Delta)^x e^{-\lambda\Delta}}{x!}, \tag{5.4}$$

where λ *(lambda) is the mean number of such events per unit of time.* This, of course, is the same probability distribution as in equation (5.3), the only difference being that $\lambda\Delta$ is used here in place of μ. However, since $\lambda\Delta$ is the expected value of x, it is the same as what we formerly called μ.[8]

8. Sometimes the Poisson distribution is used to characterize events distributed at random in space rather than in time. For example, the Poisson might be used to find the probability of a submarine's being located in a particular area.

5.12 Replacement of Parts on Polaris Submarines: A Case Study[9]

The United States Department of Defense faced a problem in the operation of Polaris submarines that illustrates the practical utility of the Poisson distribution. Each submarine goes out on a mission of relatively fixed length (about 60 days), after which it is resupplied by a tender. During each mission a submarine must rely upon its own supply of spare parts. At the end of the mission, the tender replenishes the items that have been taken out of the submarine's supplies. How many spare parts should each tender carry in order to replace the spare parts that are used up in the preceding mission? Analysts have made substantial use of the Poisson distribution in solving this problem.

Obviously, the number of spare parts of a particular type that a tender must replenish for a particular submarine equals the number of such parts that failed during the preceding mission. Thus, the answer to the Defense Department's problem depends in considerable part (but not wholly)[10] on the probability distribution of the number of parts of a particular type which will fail during a mission. To estimate this probability distribution, analysts have found it useful to assume that failures correspond to the four conditions discussed in the previous section. In other words, the probability that a particular kind of part will fail during a short time interval is proportional to the length of the time interval and is independent of when the interval occurs; it is also independent of how many such parts have failed prior to that time interval. Also, the probability that more than one part of a particular type will fail in a very short interval is so small that it can be regarded as zero.

Given that failures of a particular type of part can be represented in this way, we know (from the previous section) that the probability that x failures of a particular type of part will occur during a mission is

$$P(x) = \frac{(\lambda\Delta)^x e^{-\lambda\Delta}}{x!},$$

where Δ is the length of the mission and λ is the average number of failures per unit of time for the particular type of part. According to studies based on the first 61 patrols of Polaris submarines, the value of $\lambda\Delta$ varies from very close to zero to as high as 5.0, depending on the particular part.

If the Defense Department is specifically interested in a part where the value of $\lambda\Delta$ is estimated to be 1.0, what is the probability distribution of the number of such parts that will fail during a mission? Based on the column of Appendix Table 3 where $\mu = 1.0$, the answer is

$$P(0) = \frac{1^0 e^{-1}}{0!} = .3679,$$

$$P(1) = \frac{1^1 e^{-1}}{1!} = .3679,$$

9. This section is based in part on S. Haber and R. Sitgreaves, "An Optimal Inventory Model for the Intermediate Echelon when Repair Is Possible," *Management Science*, February 1975.
10. Ibid.

$$P(2) = \frac{1^2 e^{-1}}{2!} = .1839,$$

$$P(3) = \frac{1^3 e^{-1}}{3!} = .0613,$$

$$P(4) = \frac{1^4 e^{-1}}{4!} = .0153.$$

We can ignore $P(5)$, $P(6)$, . . . since each of these probabilities is less than .005.

Obviously, decision makers in the Defense Department have found it very helpful to know that the chances are about 37 out of 100 that *no* spare parts of this type will need to be replaced after a mission, that the chances are about 37 out of 100 that *one* will have to be replaced, that the chances are about 18 out of 100 that *two* will have to be replaced, and so on. Information of this sort, if properly applied, can promote a much more effective and economical inventory policy for submarine tenders.

EXERCISES

5.11 If $\mu = 2$, what is the probability that a Poisson random variable, X, equals (a) 1; (b) 2; (c) 3?

5.12 If $P = 1/3$ and $n = 100$, should you use the normal distribution or the Poisson distribution as an approximation to the binomial distribution?

5.13 Given $P = .01$ and $n = 300$, should you use the normal distribution or the Poisson distribution as an approximation to the binomial distribution?

5.14 If a Poisson random variable has an expected value of 3.0, what is its variance?

5.15 If a Poisson random variable's coefficient of variation—that is, its standard deviation divided by its mean—equals 2, what is its mean?

5.16 Given $n = 20$ and $P = .05$, what is the probability that $X = 0$, based on (a) the binomial distribution; and (b) the Poisson approximation to the binomial distribution?

5.17 The Uphill Manufacturing Company wants no more than 1/10 of 1 percent of the bicycle pedals it produces to be defective. Production quality is checked by examining a certain number of pedals chosen at random from each day's output, and the manufacturing process is stopped if any pedal is defective. If the firm wants the probability to be about .05 that the process will be stopped when it is producing 1/10 of 1 percent defectives, how many pedals must be examined from each day's output?

5.18 The Uphill Manufacturing Company's switchboard receives an average of four incoming calls per minute.
(a) What conditions must be satisfied if the number of incoming calls in any given minute is to be represented by a Poisson distribution?
(b) If these conditions are met, what is the probability that there will be exactly five incoming calls in a given minute? Exactly six incoming calls in a given minute? Exactly seven incoming calls in a given minute?
(c) Describe how the probability distribution of the number of incoming calls in a given minute might be useful in deciding how much capacity the firm should have to handle calls.

(d) What is the standard deviation of the number of incoming calls in a given minute?
(e) Using Chebyshev's inequality, determine an upper bound for the probability that the number of incoming calls in a given minute is more than 2 standard deviations above or below its mean. By how much does this upper bound exceed the true probability?

5.19 R. D. Clarke reported the following data concerning the number of hits by buzz bombs during World War II in south London. (Each area covers 1/4 square kilometer.)

Number of hits	Number of areas
0	229
1	211
2	93
3	35
4	7
5 or more	1
Total	576

Are the results what might be expected on the basis of the Poisson distribution, where the mean number of hits per area was 1.0?

5.20 The number of accidents occurring in a given month at one of the Uphill Manufacturing Company's plants is known to conform to the Poisson distribution. The standard deviation of this distribution is 1.732 accidents per month. What is the probability that no accidents will occur in this plant during this month?

CHAPTER REVIEW

1. In the limit, as more and more observations are gathered concerning a continuous random variable, and as class intervals become narrower and more numerous, the histogram (using a density scale) of the variable becomes a smooth curve called a *probability density function*. The total area under any probability density function must equal 1. The probability that a random variable will assume a value between any two points is equal in value to the area under the random variable's probability density function between these two points.

2. The most important continuous probability distribution is the *normal distribution*, whose probability density function is called the *normal curve*. The location and spread of a normal curve depend on its mean and standard deviation, but all normal curves are symmetrical and bell-shaped. If one expresses any normal variable as a deviation from its mean and measures these deviations in units of its standard deviation, the result is called the *standard normal variable*. The standard normal variable is a normal variable with a mean of zero and a standard deviation of 1. The areas under the standard normal curve are tabled.

3. To calculate the probability that the value of any normal variable (with mean μ and standard deviation σ) lies between two points, a and b, find the points on the standard normal distribution corresponding to a and b. These points are $(a - \mu) \div \sigma$ and $(b - \mu) \div \sigma$. Then use Appendix Table 2 to determine the area under the standard normal curve between $(a - \mu) \div \sigma$ and $(b - \mu) \div \sigma$. This area equals the probability that the value of the normal variable is between a and b.

4. If n (the number of trials) is large and P (the probability of success) is not too close to zero or 1, the probability distribution of the number of successes in n Bernoulli trials can be approximated by a normal distribution. Specifically, to approximate the probability that the number of successes is from c to d, find the probability that the value of a normal variable (with mean nP and standard deviation $\sqrt{nP(1 - P)}$) lies between $(c - 1/2)$ and $(d + 1/2)$. (Of course, c is presumed to be less than d.)

5. An important discrete probability distribution is the Poisson distribution, which is $P(x) = \mu^x e^{-\mu} \div x!$. The mean of the Poisson distribution is μ, and its standard deviation is $\sqrt{\mu}$. If n (the number of trials) is large and P (the probability of success) is small, the probability distribution of the number of successes in n trials can be approximated by the Poisson distribution where $nP = \mu$. Experience indicates that this approximation is adequate for most practical purposes when n is at least 20 and P is no greater than .05.

6. Besides being a useful approximation to the binomial distribution, the Poisson distribution is of importance in its own right. For example, the Poisson distribution is the probability distribution of the number of events that occur in a time interval under the following circumstances: (a) The probability that an event occurs in a very short time interval is proportional to the length of the time interval, and does not depend on when the interval occurs or on how many events occurred before the beginning of the interval. (b) The probability of more than one event occurring in a very short time interval is negligible. For instance, the Poisson distribution has been used to represent the probability distribution of the number of parts of a particular type which fail during one mission of a Polaris submarine.

Getting Down to Cases:

A TRUNKING PROBLEM IN THE TELEPHONE INDUSTRY[11]

The Bell Telephone System has been a pioneer in using probability theory to solve many kinds of industrial problems. The following case (with some simplifications) is derived from actual practice.

A telephone exchange at A was to serve 2,000 telephones in a nearby exchange at B. Since it would have been too expensive to install 2,000 trunklines from A to B, it was decided to install enough trunklines so that only 1 out of every 100 calls would fail to find an unutilized trunkline immediately at its disposal.

During the busiest hour of the day, each of these 2,000 telephone subscribers requires a trunkline to B for an average of two minutes. Thus, at a fixed moment during the busiest hour, there are 2,000 telephone subscribers each of which has a probability of 1/30 that it will require a trunkline to B. Under normal conditions, whether or not one subscriber requires a trunkline to B is independent of whether another subscriber does so. (Under abnormal conditions, such as a flood or an earthquake, this assumption of independence is unlikely to hold, since many people are

11. This case was described by W. Feller in *An Introduction to Probability Theory and Its Applications* (New York: Wiley, 1950). Feller based his discussion on work published by E. C. Molina in a Bell Telephone System Publications monograph.

likely to want to make calls; however, the telephone company was interested in solving the problem under typical conditions.)

As stated above, the telephone company wanted to determine how many trunklines it should install so that when 1 out of the 2,000 subscribers put through a call requiring a trunkline to B during the busiest hour of the day, he or she would find an unutilized trunkline to B immediately at his or her disposal in 99 out of 100 cases.

(a) Solve this problem using the normal distribution.

(b) Write a one-paragraph report to the telephone company describing the results you have obtained.

APPENDIX 5.1
Waiting Lines and the Exponential Distribution

Many business situations are concerned with waiting lines or *queues*. People wait in line at railroad stations to buy tickets and at airports to have themselves and their baggage inspected for weapons. Cars must wait in line at toll booths, and airplanes must wait in line to take off and land. Statisticians and management scientists are interested in such waiting lines because they want to be sure that the capacity to service the people, cars, or airplanes is in the proper relation to the rate at which they arrive. If this service capacity is too small, the queue will be disproportionately large and the people, cars, or airplanes will waste an inordinate amount of time waiting for service. On the other hand, if the service capacity is too large, the waiting time will tend to be very short; but much of the service capacity will be underutilized or perhaps not utilized at all.

In this appendix, we shall be concerned with queueing theory, a branch of statistics that analyzes the factors determining how long a waiting line will be, and how much time a person, car, or airplane must spend in order to get through the line. We shall present some simple analytical results that apply under special but not unusual circumstances.

POISSON ARRIVALS

To begin with, let's use the term ***customers*** to designate the persons, cars, airplanes, or whatever else that is arriving. Further, let's regard the facility that services and eventually releases the customers as a set of ***service counters***. Thus, the general situation that we are analyzing is as shown in Figure 5.12. Clearly, *the average length of time that a customer waits in line depends, among other things, on how frequently customers arrive at the service counters and on how rapidly the service counters can perform the service that is required.* This section will focus on the arrivals; the service times will be taken up in a later section.

For many types of situations, it is realistic to assume that customers arrive *at random*. Put somewhat crudely, this means that the probability that a customer will arrive in one small interval of time is no different from the probability that a customer will arrive in any other small interval of time of equal duration. This means also that the probability that a certain number of

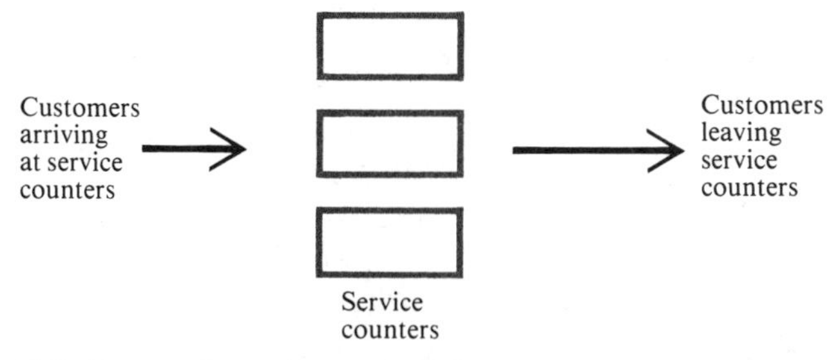

FIGURE 5.12 General Queueing Situation

customers will arrive in a certain time interval does not depend on the number of customers that arrived prior to the beginning of this time interval. Thus, the probability distribution of arrivals conforms to the four assumptions described in Section 5.11, so that the probability that x customers will arrive in a time interval of length Δ is

$$P(x) = \frac{(\lambda\Delta)^x e^{-\lambda\Delta}}{x!}, \tag{5.5}$$

where λ is the mean number of arrivals per unit of time. In other words, *the number of arrivals in a given time interval has a Poisson distribution.*

THE EXPONENTIAL DISTRIBUTION

If the number of customers arriving within a given time interval has a Poisson distribution, it can be shown that *the interval of time between consecutive arrivals* has an exponential probability distribution, which is defined below.

Exponential Distribution. *If* Y *is the time interval between consecutive arrivals,* Y *is a continuous random variable with the following probability density function:*

$$P(y) = \lambda e^{-\lambda y}, \tag{5.6}$$

where λ is the mean number of arrivals per unit of time. The probability that Y *lies between any two numbers,* c *and* d, *equals*

$$e^{-\lambda c} - e^{-\lambda d}. \tag{5.7}$$

Figure 5.13 shows the exponential probability density function, assuming that $\lambda = 1.0$. The shaded area under the curve equals the probability that the value of the exponential random variable will lie between c and d; as indicated in equation (5.7), this probability equals $e^{-\lambda c} - e^{-\lambda d}$. The mean of an exponential random variable equals $1/\lambda$.

To illustrate the use of the exponential distribution, suppose that customers arrive randomly at a shoe store, and that the mean number of arrivals per hour is 6. What is the probability that the time interval between two consecutive arrivals is between 1/4 and 1/2 hour? Since the mean number

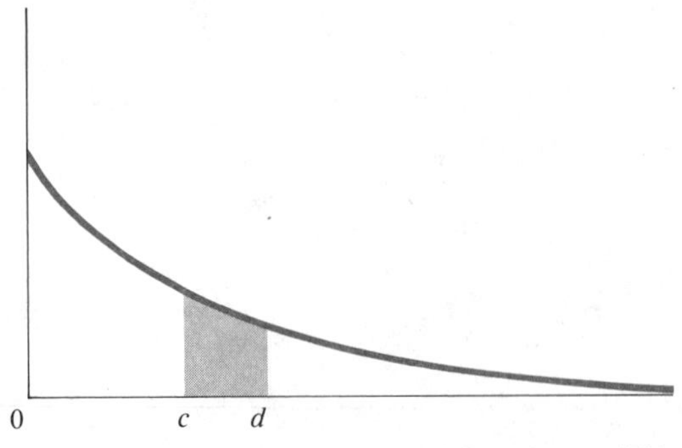

FIGURE 5.13
Probability Density Function for Exponential Random Variable, with $\lambda = 1$

of arrivals per hour is 6, λ equals 6. Consequently, using equation (5.7) the probability that the time interval between two consecutive arrivals is between 1/4 and 1/2 hour is

$$e^{-6(1/4)} - e^{-6(1/2)} = e^{-1.5} - e^{-3.0} = .223 - .050 = .173,$$

since $\lambda = 6$, $c = 1/4$, and $d = 1/2$. (Appendix Table 4 shows the value of e^{-x} for various values of x.) Thus, this probability equals .173.

EXPONENTIAL SERVICE TIMES

Now let's turn to the probability distribution of service time. Once service begins, a certain length of time elapses before a customer has received service and leaves the service counter. This length of time, known as the ***service time***, has a probability distribution *which is often assumed to be exponential.* In other words, the probability that the service time is between u and v is assumed to equal

$$e^{-mu} - e^{-mv}, \tag{5.8}$$

where m is the mean number of customers that can be serviced per unit of time. This assumption is adopted frequently because it is analytically convenient and because it often conforms quite well to reality.

To keep things simple, we also assume that there is only one service counter, and that the ***line discipline***—that is, the rules governing the behavior of the customers in line—is that a single line is formed, that customers are served on a first come, first served basis, and that no customer leaves the line before being served. Obviously, the line discipline as well as the probability distribution of arrivals, the probability distribution of service times, and the number of service counters, influences the probability that a customer will have to wait as well as the expected length of the waiting line. The rules that are adopted are clearly of importance, and the results presented below must be modified if the line discipline departs significantly from these assumptions.

Based on these assumptions, a number of interesting and important conclusions can be derived concerning the waiting lines that will occur under various circumstances. First, it can be shown that the *probability that a newly arrived customer will have to wait is*

$$P_w = \lambda/m. \tag{5.9}$$

Thus, if the mean number of customers arriving per hour is 3, and if the mean number of customers that can be serviced per hour is 4, the probability is 3/4, or 0.75, that a newly arrived customer will have to wait. The ratio, λ/m, is often called the *utilization factor*. We assume that λ is less than m; if not, the waiting line will grow beyond any finite bounds.

Second, *the mean time that a customer spends waiting in line is*

$$\frac{\lambda/m}{m - \lambda} = \frac{P_w}{m - \lambda}. \tag{5.10}$$

Thus, under the circumstances described in the previous paragraph, the mean time that a customer spends waiting is 0.75 ÷ (4 − 3) = 0.75 hours. *The mean time that a customer spends waiting in line and being served is*

$$\frac{1}{m - \lambda} \tag{5.11}$$

So, under the circumstances described in the previous paragraph, the mean time that a customer spends waiting and being served is 1 ÷ (4 − 3) = 1 hour.

Third, *the mean length of the waiting line is*

$$E_L = \frac{(\lambda/m)^2}{1 - \lambda/m} = \frac{P_w^2}{1 - P_w}. \tag{5.12}$$

Thus, under the circumstances described above, the mean length of the waiting line is 0.75^2 ÷ (1 − .75) = 2.25 customers. Note that *the total time lost (due to waiting) by all customers during a unit of time is also given by* E_L since the total time spent waiting by all customers is equal to the average number of customers waiting at any point in time multiplied by the unit of time used in the analysis. So, under the circumstances described above, 2.25 customer-hours are lost (spent on line) every hour.

DOCKING SHIPS: AN ILLUSTRATION

Queueing theory can be applied to a wide variety of practical situations, as in the case of a shipping company that must dock and service its ships at a facility it owns. The ships arrive randomly at this facility; five ships, on the average, arrive per month. The probability distribution of the length of time that a ship must stay at the facility for docking and servicing is known to be an exponential distribution. The mean number of ships that the facility can dock and service per month is 10. Given these circumstances, it follows from equation (5.9) that the probability that a ship will have to wait before being docked and serviced is 5/10. It also follows from equation (5.10) that the mean time a ship will spend waiting before being docked and serviced is

$5/10 \div (10 - 5) = 1/10$ months, or about three days. From equation (5.12) it follows that the mean length of the waiting line is $(5/10)^2 \div (1 - 5/10) = 1/2$ ships and that about 1/2 ship-months are lost (in waiting) each month.

Faced with this situation, the firm must decide whether or not to increase the capacity of its docking and servicing facility. In particular, if it is willing to spend an extra $10,000 per month, it can increase the mean number of ships that the facility can dock and service per month from 10 to 12. What effect will this have on the amount of time lost? Using equation (5.12) again, it is clear that the number of ship-months which will be lost each month if this extra expenditure is made equals $(5/12)^2 \div (1 - 5/12) = 25/84$. In other words, if the extra expenditure is made, there will be a monthly reduction of lost time of $(1/2 - 25/84) = 17/84$ ship-months (about six ship-days). Assume that $3,000 is the cost of losing each ship-day since this is the amount the firm could earn if a ship were working for a day rather than waiting before being docked and serviced. If so, the firm should spend the extra $10,000 on the facility, since this will reduce the expected monthly costs due to lost time by about $18,000. (Reducing lost time by about six days is worth about $18,000.)

EXERCISES

5.21 The Alpha Corporation, which sells bicycles, has only one salesman who waits on customers on a first come, first served basis. Customers arrive at the firm's sales office at random, and the mean number arriving in an hour is 3.
(a) Alpha's salesman (who, as we know from Exercise 4.8, has a weakness for gambling) is considering leaving the sales office untended for an hour to visit a local gambling parlor. If a customer arrives while the salesman is out, there is a good chance that he will be fired. What is the probability that a customer will arrive in his absence?
(b) A customer arrives at 11 A.M. What is the probability that the next customer will arrive between 11:30 A.M. and noon?

5.22 The length of time the Alpha Corporation's salesman spends with a customer conforms to the exponential distribution, and the mean number of customers which can be serviced per hour is 6.
(a) What is the probability that the amount of time he spends with a particular customer exceeds 20 minutes?
(b) What is the probability that the amount of time he spends with a particular customer is between 10 and 20 minutes?

5.23 Given the circumstances described in Exercises 5.21 and 5.22, what is the probability that a newly arrived customer will have to wait before the salesman can see him or her?

5.24 Given the circumstances described in Exercises 5.21 and 5.22, what is the mean time a customer spends waiting? What is the mean time a customer spends waiting and being served?

5.25 Given the circumstances described in Exercises 5.21 and 5.22, what is the mean length of the waiting line? What is the expected number of customer-hours lost (spent waiting) each hour?

5.26 The president of the Alpha Corporation, after receiving complaints about the poor service at its sales office, reprimands the salesman, who admits that he has been slow in servicing customers. The salesman promises to double the mean number of customers that can be serviced per hour (that is, increase it from 6 to 12).

(a) If the salesman keeps his promise and if the service times continue to conform to the exponential distribution, how much of a reduction will occur in the probability that a newly arrived customer will have to wait?

(b) If he keeps his promise and if the service times continue to conform to the exponential distribution, how much of a reduction will occur in the mean length of the waiting line?

6

Sample Designs and Sampling Distributions

6.1 Introduction

As emphasized throughout the previous chapters, the field of statistics is concerned with the nature and effectiveness of sampling techniques. To comprehend how statistical methods are used in business and economics, it is essential that you be familiar with the major kinds of sample designs and that you understand the concept of a sampling distribution. In this chapter, we begin by describing the various kinds of commonly used sample designs. Then we discuss the concept of a sampling distribution and present some fundamental results concerning the sampling distributions of the sample mean and proportion. Our treatment of these topics makes extensive use of the probability theory contained in Chapters 3 to 5.

6.2 Probability Samples and Judgment Samples

At the outset of any sampling investigation, one must determine whether the sample is to be a probability sample or a judgment sample. These are the two broad classes of sample designs that can be used, and each is defined as follows.

> ***A probability sample*** *is one where the probability that each element (that is, each member) of the population is included in the sample is known. In a* ***judgment sample****, personal judgment plays a major role in determining which elements of the population are selected, and this probability is not known.*

For example, in constructing a sample of steel firms in the United States, if we pick randomly 5 of the 10 biggest steel firms and if we pick randomly 10 of the other steel firms, the resulting sample of 15 steel firms is a probability sample because we know the probability that each steel firm in the population is included. On the other hand, if we go down the list of steel firms and choose 15 firms that we consider "typical" or "average," this is a judgment sample because it is not based on random methods of selection and we do not know the probability that each steel plant in the population will be included.

The most important disadvantage of a judgment sample is that there is no way to tell how "far off" a sample result is likely to be. That is, one cannot estimate the difference between the sample result and the population parameter one is trying to measure. In contrast, *if a probability sample is taken, we can estimate how large the sampling error is likely to be,* as we shall see in subsequent sections of this chapter. Nonetheless, there are situations where judgment samples are used. In some cases, a probability sample is too expensive or impractical. For example, if our sample of steel firms must be confined to a single city because of budget limitations, we might well decide to choose the city on the basis of expert judgment rather than on the basis of chance. In a case of this sort, the geographical coverage of the sample will be so narrow that what can be inferred concerning the population as a whole will be largely a matter of judgment in any event, regardless of whether a probability or a judgment sample is used.

One type of judgment sample encountered frequently is a quota sample. *In a quota sample, the population as a whole is split into a number of groups or strata, and whoever is charged with drawing the sample is instructed to include a certain number of members of each group.* For example, if it is known that 20 percent of the individuals in a particular town are black men, 25 percent are black women, 30 percent are white men, and 25 percent are white women, a quota sample of 100 people in the town might specify that the interviewers pick 20 black men, 25 black women, 30 white men, and 25 white women. In this way an attempt is made to construct a sample that seems representative of the population as a whole. *The reason why quota sampling is a form of judgment sampling is that the choice of members from each group or stratum to be included in the sample is not determined by random selection, but by the interviewers.* Since interviewers tend to choose members of each group that can be contacted most readily, a variety of unknown biases may result. And like any form of judgment sample, there is no way to determine how large the sampling errors are likely to be.[1] Nonetheless, quota sampling is often used because it is convenient and inexpensive.

6.3 Types of Probability Samples

Four types of probability samples are frequently used: simple random samples, systematic samples, stratified random samples, and cluster samples. Following are brief descriptions of each type.

SIMPLE RANDOM SAMPLE

Simple random sampling is the method that serves as the best introduction to probability sampling.

1. For further comparison of probability and judgment samples, see M. Hansen and W. Hurwitz, "Dependable Samples for Market Surveys," in E. Mansfield, *Statistics for Business and Economics: Readings and Cases* (New York: Norton, 1980).

> *If the population contains* N *elements, a* ***simple random sample*** *of* n *elements is a sample chosen so that every combination of* n *elements has an equal chance of selection. Assuming that the sampling is without replacement, this means that each element in the population has a probability of* 1/N *of being the first chosen, that each of the* (N − 1) *elements not chosen on the first draw has a probability of* 1/(N − 1) *of being the second chosen, . . . , and that each of the* (N − n + 1) *elements not chosen on the* (n − 1)*th draw has a probability of* 1/(N − n + 1) *of being the last chosen.*

An *alternative and equally useful definition is as follows.*

> *A* **simple random sample** *is a sample chosen so that the probability of selecting each element in the population is the same for each and every element, and the chance of selecting one element is independent of whether some other element is chosen.*

To illustrate the application of a simple random sample, suppose that a firm has four manufacturing plants (A, B, C, and D) and that it wants to select a simple random sample of two of these plants. There are six different samples of size 2 that can be drawn from this population of four plants.[2] (As shown in Appendix 3.1, the number of different samples of size n that can be drawn without replacement from a population of N elements equals $N! \div [(N - n)!n!]$.) Specifically, these six samples are (AB); (AC); (AD); (BC); (BD); and (CD). For this sample to be a simple random sample, each of these six samples must have a probability of 1/6 of being selected. If sampling is without replacement, this can be achieved by choosing the first plant in such a way that each of the four plants has a probability of 1/4 of being chosen, and then by choosing the second plant in such a way that each of the remaining three plants has a probability of 1/3 of being chosen.

If the population is infinite, the number of possible samples of size n is also infinite. Thus, the alternative definition (given above) of a simple random sample must be used: A simple random sample is one where the probability of selecting each element in the population is the same, and the chance of selecting one element is independent of whether some other element is chosen. For example, if a coin were tossed repeatedly for an indefinite period of time, an infinite population would result and each element contained in this population would be heads or tails. If the coin were tossed six times, this would be a simple random sample of size 6 if the probability of heads (or tails) remained the same from one toss to the next and if the result of one toss was independent of the result of another toss.

How do statisticians actually pick a simple random sample? In cases where the number of elements in the population is small, we can (1) number each of the elements in the population; (2) record each number on a slip of paper: (3) place all the slips of paper in a hat or bowl, where they are mixed

2. More accurately, in keeping with definitions in Chapter 1, the population consists of *measurements* concerning these four plants, and the sample consists of *measurements* concerning two of them. Also, we assume here that sampling is without replacement.

well; and (4) draw n slips of paper from the hat or bowl. The numbers on the slips of paper will indicate which elements in the population are included in the sample. Although this procedure is straightforward, we can run into problems if the number of elements in the population is large or if the slips of paper are not well mixed. To avoid such problems statisticians generally use a table of random numbers in choosing a simple random sample. We shall describe how such a table is used in Section 6.5.

SYSTEMATIC SAMPLE

To illustrate how a systematic sample is chosen, suppose that a statistician has a list of 1,000 tool and die firms from which to pick a sample of 50. Since there are 1,000 firms, this can be accomplished by taking every 20th firm on the list. Once a choice has been made among the first 20 firms on the list, the entire sample has been chosen since there are only 20 samples that can be drawn. The first possible sample consists of the 1st, 21st, 41st, . . . firms on the list. The second possible sample consists of the 2nd, 22nd, 42nd, . . . firms on the list. And so on. To determine which of these possible samples will be drawn, the statistician draws at random a number between 1 and 20. The chosen number identifies which of the first 20 firms he or she will begin with, which in turn identifies which sample will be drawn. In general, a systematic sample is defined as follows.

> *A* ***systematic sample*** *is obtained by taking every* k*th element on a list of all elements in the population. To determine which of the first* k *elements is chosen, a number from* 1 *to* k *is chosen at random.*

A systematic sample is often viewed as being essentially the same as a simple random sample. It is important to recognize that this is true only *if the elements of the population are in random order on the list.* One cannot always be sure that the phenomenon being measured does not have a periodicity or other type of pattern on the list. For example, W. A. Wallis and H. Roberts have pointed out that on census record sheets "the first names on the sheets tend to be predominantly male, gainfully employed, and above average in income. The reason is that the enumerators are instructed to start in a certain block at the corner house (which tends to have a higher rental value than houses in the middle of the block), and in the household to start with the head (usually male and the breadwinner)."[3] If there is a periodicity of this sort, a systematic sample may be far from random and may be less precise than a simple random sample.

On the other hand, circumstances exist where because the elements of the population are not listed in random order, a systematic sample may be *more* precise than a simple random sample. Basically, the reason is that a systematic sample insures that the elements in the sample are distributed evenly throughout the list. If there is a strong tendency for the characteristic being measured to increase or decrease steadily as one progresses from the

3. W. A. Wallis and H. Roberts, *Statistics* (Glencoe: Free Press, 1956), pp. 488–489.

beginning to the end of the list, this "even distribution" increases the precision of the sample estimate.

STRATIFIED RANDOM SAMPLE

In designing a sample, it frequently is useful to recognize that the population can be divided into various groups or *strata*. For example, if a sample of students is to be drawn in order to estimate the mean height of freshmen at your college or university, it would probably be wise to stratify the population (in this case, the freshman class) into two groups: males and females.

> *In general,* ***stratified random sampling*** *is sampling in which the population is divided into strata and a random sample is taken from the elements in each stratum.*

For reasons discussed below, a more precise estimate can often be obtained from a sample of a given size if stratified random sampling is used rather than simple random sampling.

To illustrate the application of stratified random sampling, consider once again the problem of sampling the tool and die firms. A sample survey is being made in order to estimate the proportion of all tool and die firms which have introduced numerically controlled machine tools. If the statistician believes that the proportion of tool and die firms that have introduced numerically controlled machine tools is higher among larger firms than smaller ones, he or she can divide the population into two *strata*, one containing firms with more than 20 employees, and one containing firms with 20 or fewer employees. Then, if 50 percent of all tool and die companies have more than 20 employees, and 50 percent have 20 employees or fewer, the proportion of all firms in the population with numerically controlled machine tools is

$$P = 0.5P_1 + 0.5P_2,$$

where P_1 is the proportion of *all* firms with more than 20 employees having introduced numerically controlled machine tools, and P_2 is the proportion of *all* firms with 20 or fewer employees having introduced numerically controlled machine tools. Consequently, *if a simple random sample is taken from each stratum,* an estimate of the proportion of all tool and die firms that have introduced numerically controlled machine tools can be obtained by computing

$$p = 0.5p_1 + 0.5p_2, \tag{6.1}$$

where p_1 is the proportion of the *sample* of firms taken from the stratum with more than 20 employees having introduced numerically controlled machine tools, and p_2 is the proportion of the *sample* of firms taken from the stratum with 20 or fewer employees having introduced such machine tools.

The basic idea in formulating strata is to subdivide the population so that these subdivisions differ greatly with regard to the characteristic being measured, and so that there is as little variation as possible within each stratum (or subdivision) with regard to the characteristic under measurement. For example, in the problem discussed above, the statistician should stratify firms so that the differences *among* the strata in the proportion using numerically controlled machine tools is *large,* while the variation *within* each stratum in this regard is *small.*

If properly constructed, a stratified sample can generally result in more precise results than those obtained by using a simple random sample. This is because the error in the estimate for the population as a whole is due only to errors in the estimates for each stratum, whereas in simple random sampling there are also errors due to weighting the strata incorrectly.[4]

Once the strata have been defined, there remains the problem of determining how the total sample is to be divided among the strata. In other words, how many elements are to be chosen from each stratum? Two possible answers are proportional allocation and optimal allocation.

Proportional Allocation. This method of allocating the sample *makes the sample size in each stratum proportional to the total number of elements in the stratum.* For example, in the case of the tool and die firms, this would mean that 1/2 of the sample would be chosen from firms with more than 20 employees while 1/2 would be chosen from firms with 20 or fewer employees. At first glance, it certainly seems reasonable to sample the same proportion of elements in each stratum; and this allocation method is frequently used in business and economic surveys. But if the statistician has some knowledge of the population standard deviation in each stratum, it may be preferable to use optimum allocation, described below.

Optimum Allocation. This method prescribes that *the sample size in each stratum be proportional to the product of the number of elements (in the population) in the stratum and the standard deviation of the characteristic being measured in the stratum.* For example, suppose that we want to estimate the average assets of banks in a given state, and that we subdivide banks into two strata: national banks and state banks. If there are 300 national banks and 500 state banks, and if the standard deviation of the assets of the national banks is $100 million and that of the assets of the state banks is $20 million, then the number of national banks in the sample should be proportional to 300 × $100

4. For example, in the case discussed above, the proportion of firms that have introduced numerically controlled machine tools *in a simple random sample* would be

$$\frac{n_1}{n}p_1 + \frac{n_2}{n}p_2,$$

where n_1 is the number of firms in the *sample* with more than 20 employees, p_1 is the proportion of these larger firms that have introduced numerically controlled machine tools, n_2 is the number of firms in the *sample* with 20 or less employees, p_2 is the proportion of these smaller firms having introduced numerically controlled machine tools, and $n = n_1 + n_2$. This expression is the same as equation (6.1) except that n_1/n will differ (due to chance variation) from the true proportion of firms with more than 20 employees, whereas the true proportion, 0.5, is used in equation (6.1). Similarly, n_2/n in this expression will differ (due to chance variation) from the true proportion of firms with 20 or less employees, whereas the true proportion, 0.5, is used in equation (6.1).

million, while the number of state banks in the sample should be proportional to 500 × \$20 million. In other words, 3/4 of the banks in the sample should be national banks and 1/4 should be state banks. The reason why optimum allocation is "optimum" is that it minimizes the expected sampling errors in the estimate of the population mean (or proportion). Despite its advantages, it may be impractical to use optimum allocation if little or nothing is known about the population standard deviation in each stratum.

CLUSTER SAMPLE

Still another important sampling technique is cluster sampling.

> *In a* **cluster sample**, *one divides the elements in the population into a number of clusters or groups. One then begins by choosing at random a sample of these clusters, after which a simple random sample of the elements in each chosen cluster is selected.*

The problem of sampling tool and die firms can be used to illustrate the application of a cluster sample. In this case, clusters can be formed by geographical location. All tool and die firms can be classified by the cities or areas in which they are located, and a simple random sample of these cities or areas can be chosen. Then, within each of the chosen cities or areas a simple random sample of the firms themselves can be picked.

The major advantage of cluster sampling is that *it is cheaper to sample elements that are physically or geographically close to one another*. Thus, it is cheaper to sample 50 tool and die firms, all of which are concentrated in five cities, than to sample 50 tool and die firms that are scattered all over the country. In general, the results of a cluster sample are less precise than those of a simple random sample (assuming that sample size is constant) because the elements in a particular cluster tend to be relatively similar. However, *per dollar spent on the survey* a cluster sample may be more effective than simple random sampling. Why? Because, for the same total cost, cluster sampling provides a much larger sample than simple random sampling.

6.4 Inventory Valuation: A Case Study

To illustrate how the techniques described in previous sections have been applied successfully to actual business problems, consider the case of a major manufacturing firm that had a large inventory of materials. The firm wanted to sample the items in this inventory (rather than attempting to include them all) in order to estimate the change during the previous year in the total monetary value of the inventory. Information of this sort was important in estimating the firm's earnings and its taxes. After studying the problem, the firm's statisticians and consultants recommended a sampling plan that combined stratified sampling (with optimum allocation) and systematic sampling.

First, this sampling plan stipulated that all items be divided into four strata: (1) Items worth \$10,000 or more; (2) items worth \$1,000 to \$9,999;

(3) items worth $100 to $999; and (4) items worth $99 or less. The reason for this stratification was that the increase in monetary value during the year of a high-priced item (in current prices) was liable to be greater than that of a low-priced item. Given the four strata, the sampling plan called for the selection of all 395 items in the first stratum, 1350 of the 3800 items in the second stratum, 260 of the 7280 items in the third, and 110 of the 7700 items in the fourth.[5] In all, 2115 out of the approximately 19,000 items in the inventory were to be included in the sample.

How did the firm's statisticians determine this allocation of the sample among the strata? Basically, by using the principles of optimum allocation discussed previously in this chapter. To see this, consider the three strata from which samples were taken. (In the first stratum, all items were selected.) Table 6.1 shows the number of items in each of these three strata, as well as the standard deviation of the change in the monetary values of items in each. In the last column of Table 6.1 the product of the number of items and the standard deviation is shown for each stratum. Optimum allocation requires that the sample size in each stratum be proportional to the figures in the last column of Table 6.1. The firm's statisticians picked the sample size in each stratum to conform to this requirement. That is, since $1,900,000 ÷ $364,000 = 5.2, the sample size in the second stratum is about 5.2 times as large as the sample size in the third stratum. And since $1,900,000 ÷ $154,000 = 12.3, the sample size in the second stratum is about 12.3 times as large as the sample size in the fourth stratum.

TABLE 6.1
Selected Characteristics of Three Strata, Case Study of Inventory Valuation

Stratum (dollar value of item)	(1) Number of items in stratum	(2) Standard deviation of change in monetary value of items in stratum (dollars)	(1) × (2) (dollars)
1,000 to 9,999	3,800	500	1,900,000
100 to 999	7,280	50	364,000
99 or less	7,700	20	154,000

To carry out the sampling in each stratum, all items were listed and numbered. Then 10 systematic subsamples were chosen in each stratum. For example, in the third stratum a number between 1 and 280 was chosen at random. If this number turned out to be 119, then the 119th, 399th, 679th, . . . items in this stratum were chosen, the result being a subsample consisting of 26 items. Then another such random number was chosen. If the number was 108, then the 108th, 388th, 668th, . . . items were chosen, the result being another

5. The numbers have been changed slightly to simplify the exposition. This case is from W. Edward Deming, *Sample Design in Business Research* (New York: Wiley, 1960). It has been simplified in various respects.

subsample composed of 26 items. After 10 such subsamples had been chosen, the entire sample of 260 items had been drawn from the third stratum. Since there was every reason to believe that the items were listed at random, the result could be considered a simple random sample. Once the sample in each stratum had been gathered,[6] the firm could compute the mean change in monetary value of the sample items in each stratum. Then, to estimate the total change in the monetary value of the inventory, it could compute

$$395\bar{x}_1 + 3800\bar{x}_2 + 7280\bar{x}_3 + 7700\bar{x}_4,$$

where $\bar{x}_1$ is the mean change in monetary value of the sample items in the first stratum, $\bar{x}_2$ is the mean change in monetary value of the sample items in the second stratum, and so on. This, of course, was the estimate that the firm wanted, the "pay dirt" of the analysis.

6.5 Using a Table of Random Numbers

Here and in subsequent chapters we shall concentrate on simple random sampling because it provides the simplest and best introduction to probability sampling. In order to select a simple random sample, statisticians generally use a table of random numbers.

A *table of random numbers* is a table of numbers generated by a random process. For example, suppose that we want to construct a table of five-digit random numbers. To do so, we could write a 0 on one slip of paper, a 1 on a second slip, a 2 on a third, . . . , and a 9 on a tenth slip. We could then put the 10 slips of paper into a hat and draw out one at random. The number on this slip of paper would be the first digit of a random number. After replacing the slip of paper in the hat, we could make another random draw. The number on the new slip of paper would be the second digit of the random number. This procedure could then be repeated three more times, yielding the third, fourth, and fifth digits of the random number. The result is the first five-digit random number. Given enough time and persistence, we could formulate as many such random numbers as needed. Fortunately, it is not necessary to go through such a time-consuming process because tables of random numbers have already been formulated. Although these tables generally have been calculated on a computer rather than by drawing slips of paper from a hat, the principles of their construction are essentially the same as if slips of paper had been used.

A table of random numbers is given in Appendix Table 5. As you can see, it contains column after column of five-digit random numbers. (The left-hand column shows the line number and should not be confused with the random numbers.) To illustrate how the table can be used to pick a simple random sample, suppose that we want to draw a simple random sample of 50 tool and die firms from the population of 1,000 tool and die firms. Our first step would be to number each of the tool and die firms from 0 to 999. We would then turn to any page of the table and proceed in any systematic way to

6. Actually, of course, the sample in the first stratum includes all the items, not just some of them.

pick our sample. For example, we might begin at the top, left-hand column of page A14 and work down. Since we need three-digit, not five-digit, random numbers, we would pay attention to only the first three digits of each five-digit random number. The topmost five-digit number in the left-hand column of page A14 is 53535; its first three digits are 535. Thus, the first firm to be picked is number 535. Reading down, the next five-digit number in the left-hand column of page A14 is 41292; its first three digits are 412. Thus, the second firm to be picked is number 412. This procedure should be repeated until the entire sample of 50 has been selected.

Several points should be noted concerning this procedure. (1) If any previously chosen number comes up again, it should be ignored. For example, if 412 were to come up again before the 50 firms were chosen, it should be ignored. (2) If a number comes up which is not admissible because it does not correspond to any item in the population, it also should be ignored. For example, if there were 500, not 1,000, tool and die firms, and if 680 came up, it should be ignored. (3) It makes no difference whether you read the table down, up, to the left, to the right, or diagonally. As long as you proceed from number to number in a systematic fashion the table will work properly.

Finally, it is worth noting that tables of random numbers are used in statistics in so-called Monte Carlo studies, as well as in the selection of random samples. (In the appendix to this chapter, we present a brief discussion of Monte Carlo studies.)

EXERCISES

6.1 If a finite population consists of the elements *A, B, C, D,* and *E*, list all possible samples of size 2 that can be drawn (without replacement) from this population.

6.2 List all possible samples of size 3 that can be drawn (without replacement) from the population in Exercise 6.1. If each of these possible samples has the same probability of being selected, what is the probability that *A* will be included in the sample that is selected? What is the probability that both *A* and *B* will be included in this sample?

6.3 How many different samples of size 4 can be selected (without replacement) from a finite population consisting of the five boroughs of New York City (Manhattan; the Bronx; Brooklyn; Queens; Staten Island)? How many different samples of size 3 can be selected (without replacement) from this population?

6.4 Use a table of random numbers to select a sample of two drugstores from among those listed in the yellow pages of your telephone directory.

6.5 The tickets issued by a movie theater are numbered serially. On October 28, 1979 the theater sold tickets with numbers beginning at 20860 and ending at 23102. Use a table of random numbers to select a sample of 10 of these tickets.

6.6 The Uphill Manufacturing Company is interested in determining how consumers in a particular town rate the performance of various makes of bicycles. It hires a polling organization to call every hundredth number listed in the telephone directory and ask whoever answers to rate the comparative performance of various makes.
(a) Is this a random sample? Why, or why not?
(b) What pitfalls and disadvantages do you see in the design of this sample?

6.7 The polling organization carries out the survey described in Exercise 6.6. The report of its findings indicates that the telephone calls were made between 10 A.M. and 2 P.M. on Mondays, Tuesdays, and Wednesdays; if there was no answer at a particular telephone number, that number was dropped from the sample. Give several reasons why this survey might yield distorted results.

6.8 An agricultural research firm wants to estimate the total acres of corn planted in a certain county of 3,000 farms where corn is the major crop. If the firm can obtain an alphabetical list of all the names of farm owners in the county, describe how the firm can draw a random sample of 100 farms. (Assume each farm has a single owner.)

6.9 The same research firm decides to pick a number from 1 to 30 at random, and the number turns out to be 16. The firm then picks the 16th, 46th, 76th, 106th, . . . farms on the list in Exercise 6.8. (a) Is this a random sample? (b) If the number of acres of a farm planted with corn is not related in any way to the name of its owner, will a sample of this sort be essentially equivalent to a simple random sample?

6.10 A statistician working for a local university criticizes the agricultural research firm's survey design (given in the previous exercise). The statistician says that the firm could obtain more precise results by stratifying the sample according to last year's farm size. Do you agree? Why, or why not?

6.11 Continuing the discussion in Exercise 6.10, available data indicate that the number of farms, classified by last year's size, is as shown in the middle column below:

Farm size last year (acres)	Number of farms	Standard deviation of number of acres of corn
0–50	1,000	10
51–100	500	10
101–150	500	10
151–200	400	15
201–250	400	15
Over 250	200	50
Total	3,000	

If proportional allocation is used, and if the total sample size is 100, what will be the number of farms chosen at random from each of the six strata?

6.12 The right-hand column in Exercise 6.11 shows the estimated standard deviation of the number of acres of corn among the farms in each stratum. (For example, among farms with a total acreage last year of 50 or less, the standard deviation of the number of acres planted with corn this year is estimated to be 10.) If optimum allocation is used, and if the total sample size is 100, what will be the number of farms chosen at random in each of the six strata?

6.13 Still another statistician takes issue with the stratified sample proposed by the statistician in Exercise 6.10. The second statistician suggests that the list of farms in alphabetical order by owners' names be used to construct a cluster sample. All farms owned by people with names beginning with *A* would make up one cluster, all farms owned by people with names beginning with *B* would constitute a second cluster, and so on. The statistician suggests that a random sample of five clusters be chosen and that 20 farms be chosen at random in each cluster. If the principal cost of the survey is the expense of driving from one farm to another to obtain information concerning the number of acres planted with corn, is this suggestion likely to reduce the survey's costs? Why, or why not?

6.6 Concept of a Sampling Distribution

Sampling distributions are of central importance in statistics. The idea of a sampling distribution stems directly from the fact that there generally are a large number of possible samples that can be selected from a population. Thus, the value of any statistic computed from a sample will vary from sample to sample. For example, suppose that a population consists of the eight numbers in Table 6.2, that we are going to take a simple random sample of two of these numbers, and that the sample statistic in which we are interested is the sample mean. Since the sample is randomly chosen, the probability that each possible sample will be selected is known and it is possible to deduce the probability distribution of the value of the sample statistic. This is an example of a sampling distribution, which is defined as follows.

*A **sampling distribution** is the probability distribution of the value of a statistic.*

What is the sampling distribution of the mean of a simple random sample of size 2 from the population in Table 6.2? There are 28 different samples of size 2 that can be drawn without replacement from this population (since there are only eight numbers in the population). Table 6.3 lists each of these samples, as well as its mean. Since the probability of selecting each of these samples is 1/28, the probability that the sample mean assumes each of the 28 values (not all of which are different from one another) is 1/28. Thus, the probability distribution of the value of the sample mean in this case is as shown in Figure 6.1. This probability distribution is a sampling distribution.

TABLE 6.2
Population of Eight Numbers

Measurements included in the population
2
4
5
7
8
0
1
3

Sampling distributions can be constructed for any sample statistic, not just the sample mean. For example, Table 6.3 also shows the sample range for each of the 28 samples of size 2 from the population in Table 6.2. Since the probability of each of the 28 values (not all of which are different from one another) is 1/28, the probability distribution of the value of the sample range in this case is as shown in Figure 6.2. This is another example of a sampling distribution.

TABLE 6.3

Mean and Range of 28 Possible Samples of Size 2 from the Population in Table 6.2

Sample	Numbers included in sample	Sample mean	Sample range
1	2,4	3.0	2
2	2,5	3.5	3
3	2,7	4.5	5
4	2,8	5.0	6
5	2,0	1.0	2
6	2,1	1.5	1
7	2,3	2.5	1
8	4,5	4.5	1
9	4,7	5.5	3
10	4,8	6.0	4
11	4,0	2.0	4
12	4,1	2.5	3
13	4,3	3.5	1
14	5,7	6.0	2
15	5,8	6.5	3
16	5,0	2.5	5
17	5,1	3.0	4
18	5,3	4.0	2
19	7,8	7.5	1
20	7,0	3.5	7
21	7,1	4.0	6
22	7,3	5.0	4
23	8,0	4.0	8
24	8,1	4.5	7
25	8,3	5.5	5
26	0,1	0.5	1
27	0,3	1.5	3
28	1,3	2.0	2

Any sampling distribution is based on the assumption that the sample size and the sample design remain fixed. Thus, the sampling distributions in Figures 6.1 and 6.2 are based on the assumption that samples of size 2 are taken, and that the samples are simple random samples. In general, it is not possible to derive sampling distributions in the way employed in Table 6.3 because most populations contain so many elements that it is essentially impossible to list all possible samples. Even if there are only 1,000 elements in a population, there are 1,000!/(998! 2!) or 499,500 different samples of size 2 that can be drawn from this population. Obviously, the listing of all such samples would be a formidable task; and fortunately (as we shall see in later sections) there are other ways of finding the sampling distribution of a sample statistic. The reason why we presented this laborious procedure is not that it is used to obtain sampling distributions in actual cases, but that it is useful in explaining the concept of a sampling distribution.

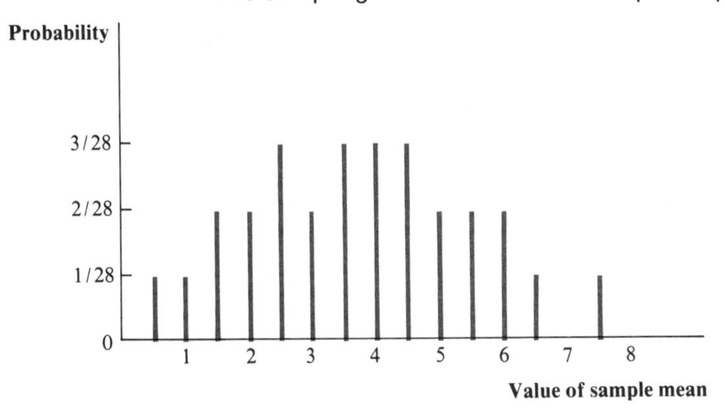

FIGURE 6.1 Probability Distribution of the Mean of a Simple Random Sample of Size 2 from Population in Table 6.2

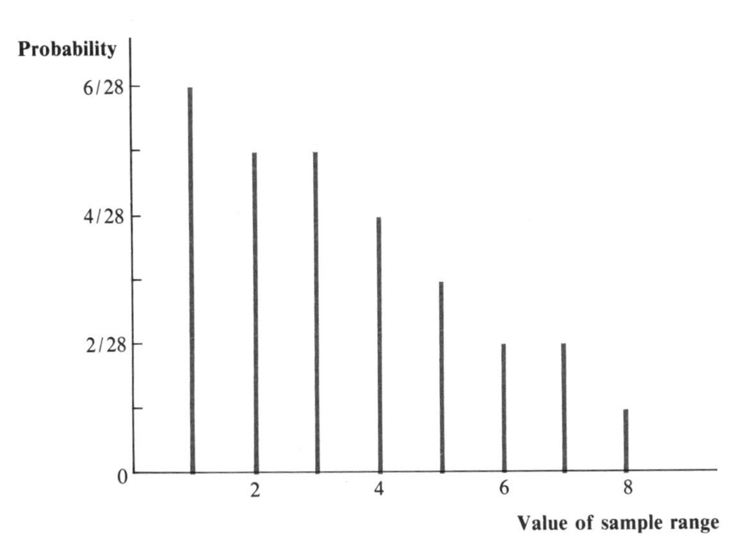

FIGURE 6.2 Probability Distribution of the Range of a Simple Random Sample of Size 2 from Population in Table 6.2

6.7 The Sampling Distribution of the Sample Proportion

Many statistical investigations are aimed at estimating the proportion of the members of some population who have a specified characteristic. For example, we may be interested in the proportion of St. Louis women who prefer Brand A beer to Brand B, or in the proportion of automobiles of a particular make that have a certain defect. In investigations of this sort, the sample proportion is generally used as an estimate of the population proportion. If the sample is a small percentage of the whole population (say, 5 percent or less), we can determine the sampling distribution of the sample proportion without much difficulty, based on our results in Chapter 4 concerning the binomial distribution.

If the sample is a simple random sample, and if it is a small percentage of the population, each observation has a probability P of having the specified characteristic, where P is the proportion of the population with this characteristic. If the sample contains n observations, each observation selected can be viewed as a Bernoulli trial where there is a probability P of success, where success is defined as the observation's having the specified characteristic. Thus, as we know from Chapter 4, the number of observations in the sample that have the specified characteristic must be a binomial random variable. In other words,

$$P(x) = \frac{n!}{(n-x)!x!} P^x(1-P)^{n-x}, \text{ for } x = 0,1,2,\ldots n,$$

where $P(x)$ is the probability that x observations in the sample will have the specified characteristic.

The sample statistic whose probability distribution we want to derive is the sample proportion. It is important to note that the value of the sample proportion is determined entirely by the number of observations in the sample with the specified characteristic. For example, if the sample size is 5 the sample proportion can be 1/5 if and only if exactly one observation in the sample has the desired characteristic; it can be 2/5 if and only if exactly two observations in the sample have the desired characteristic; and so on. Thus, *the probability that the value of the sample proportion is x/n must equal the probability that the number of observations in the sample with the specified characteristic is x.* Consequently, the sampling distribution of the sample proportion must be as follows:

$$\boxed{Pr\left(p = \frac{x}{n}\right) = \frac{n!}{(n-x)!x!} P^x(1-P)^{n-x},} \tag{6.2}$$

where $Pr(p = x/n)$ denotes the probability that the sample proportion p equals x/n.

To illustrate how this formula can be used, suppose that a market-research firm is about to ask a sample of beer drinkers whether they prefer Budweiser over Miller (presenting both in unmarked cans). If the firm selects a simple random sample of 20 beer drinkers, and if the proportion of the population preferring Budweiser over Miller is 0.50, what is the sampling distribution of the sample proportion? The formula in equation (6.2) tells us that the probability that the sample proportion equals zero can be determined by finding (in Appendix Table 1) the probability that a binomial variable (with $n = 20$ and $P = 0.5$) equals zero, that the probability that the sample proportion equals 1/20 can be determined by finding the probability that this binomial variable equals 1, that the probability that the sample proportion equals 2/20 can be determined by finding the probability that this binomial variable equals 2, and so on. The resulting sampling distribution is shown in Figure 6.3.

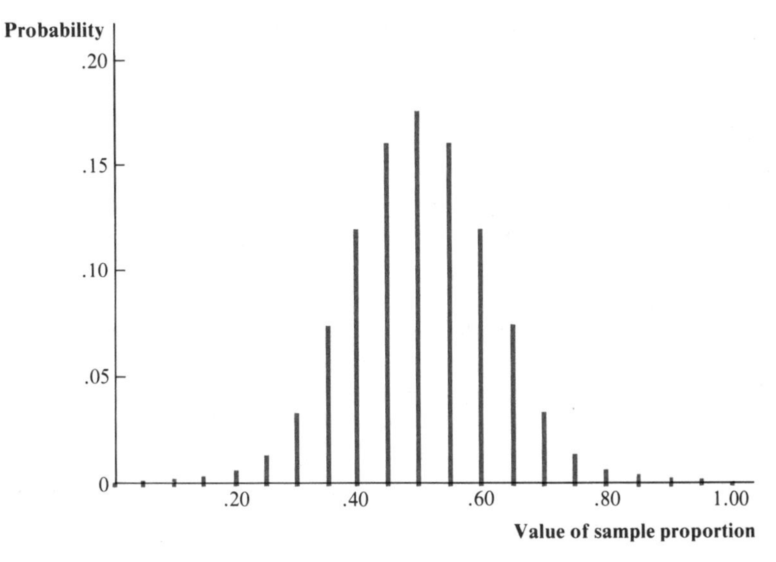

FIGURE 6.3 Sampling Distribution of the Sample Proportion, $n = 20$ and $P = 0.50$

The sampling distribution of a sample statistic is useful in indicating the amount of error the statistic is likely to contain. The following example shows how the sampling distribution of the sample proportion can be used in this way.

Example 6.1 A shipment of 1,000 engines arrives at a firm, which draws a random sample of 20 engines and calculates the proportion of defectives in the sample. If the proportion of defective engines in the entire shipment equals .15, what is the probability that the sample proportion will be in error by more than .10? (That is, what is the probability that the sample proportion will differ by more than .10 from the population proportion?)

Solution: Since the population proportion is .15, the sample proportion will *not* differ by more than .10 from the population proportion if the sample proportion equals .05, .10, .15, .20, or .25. Using Appendix Table 1, we can obtain the probability of the sample proportion's equaling each of these values. The probability that it equals .05 is .1368, since this is the probability of one success (in this case, one defective) in 20 trials, given that $P = .15$. The probability that it equals .10 is .2293, since this is the probability of two successes in 20 trials, given that $P = .15$. The probability that it equals .15 is .2428, since this is the probability of three successes in 20 trials, given that $P = .15$. The probability that it equals .20 is .1821, since this is the probability of four successes in 20 trials, given that $P = .15$. The probability that it equals .25 is .1028, since this is the probability of five successes in 20 trials, given that $P = .15$. Thus, the probability that the sample proportion differs from the population proportion by more than .10 is $1 - (.1368 + .2293 + .2428 + .1821 + .1028)$, or .1062.

6.8 The Sampling Distribution of the Sample Mean

Perhaps the most frequent objective of statistical investigations is to estimate the mean of some population. For example, a government agency may want to estimate the mean profit rate of firms in a certain industry, or a firm may want to estimate the mean longevity of a particular type of material or piece of equipment. In investigations of this sort the sample mean is generally used as an estimate of the population mean. To measure the extent of the sampling errors that may be present in the sample mean, it is essential to know the sampling distribution of the sample mean, which is the subject of this section. We will begin with the case where the population is known to be normal, then take up the case where it is not normal, and conclude with the case where the population contains relatively few observations.

WHERE THE POPULATION IS NORMAL

Suppose that simple random samples of size 10 are selected repeatedly from a normal population and that the mean of each such sample is calculated. If a very large number of such sample means are calculated, what does the probability distribution of their values look like? As shown in Figure 6.4, this probability distribution has the same mean as the population, namely μ. Thus, if a large number of sample means were calculated, *on the average* the sample mean would equal the population mean. Also, the probability distribution of the sample mean is bell-shaped and symmetrical about the population mean, which means that it is equally likely that a sample mean will fall below or above the population mean. Further, because of the averaging the dispersion of the sampling distribution of the sample mean is less than the dispersion of the population. In other words, the standard deviation of the sampling distribution of the sample mean, denoted by $\sigma_{\bar{x}}$, is less than the stan-

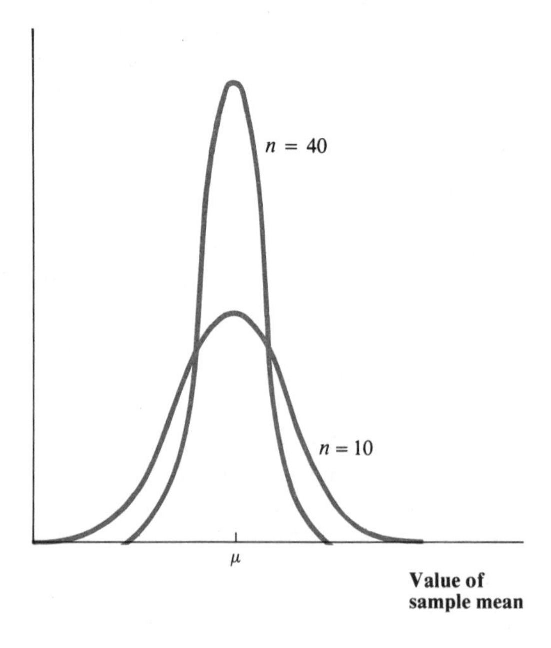

FIGURE 6.4
Sampling Distribution of Sample Mean, $n = 10$ and $n = 40$

dard deviation of the population, denoted by σ.

Now suppose that simple random samples of size 40 are selected repeatedly from the same normal population, and that the mean of each such sample is calculated. If a very large number of such sample means are calculated, what does the probability distribution of their values look like? As shown in Figure 6.4, this probability distribution, like the one for samples of size 10, has the same mean as the population and is bell-shaped and symmetrical. The most obvious difference between this sampling distribution and the one for samples of size 10 is this distribution's much smaller standard deviation. Because the sample size is larger, it is likely that the mean of a sample of size 40 will be closer to the population mean than the mean of a sample of size 10. This, of course, accounts for the smaller dispersion in the sampling distribution of the sample mean for samples of size 40.

A close inspection of both sampling distributions in Figure 6.4 suggests that they are normal distributions. At least this is how they look, and mathematicians have proved that this is indeed the case. A remarkable feature of a normal population is that the sampling distribution of means of simple random samples from such a population is also normal. This is true for a sample of any size. Given this result as well as the others noted above, it is possible to provide the following complete description of the sampling distribution of the sample mean from a normal population.

Sampling Distribution of the Sample Mean (Normal Population): *The sample mean is normally distributed, the mean of its sampling distribution equals the mean of the population (μ), and the standard deviation of its sampling distribution ($\sigma_{\bar{x}}$) equals the standard deviation of the population divided by the square root of the sample size. That is,*

$$\sigma_{\bar{x}} = \frac{\sigma}{\sqrt{n}}, \tag{6.3}$$

where σ is the population standard deviation and n *is the sample size.*

Equation (6.3) is a very important result. *The standard deviation of the sampling distribution of the sample mean, $\sigma_{\bar{x}}$, is a measure of "how far off" the sample mean is likely to be from the population mean.* If $\sigma_{\bar{x}}$ is *large*, the distribution of the sample mean contains *a great deal of dispersion,* which means there is a relatively *high* probability that the sample mean will depart considerably from the population mean. If $\sigma_{\bar{x}}$ is *small*, the distribution of the sample mean contains *relatively little dispersion,* which means there is a relatively *low* probability that the sample mean will depart considerably from the population mean. For example, panel A of Figure 6.5 shows the distribution of the sample mean in a case where its standard deviation—often called the ***standard error of the sample mean***—is relatively large. Panel B shows the distribution of the sample mean in a case where its standard deviation is relatively small. As shown in Figure 6.5, the probability that the sample mean differs from the population mean by more than an arbitrary amount (say 3, which is shown in Figure 6.5) is much higher in the case shown in panel A than

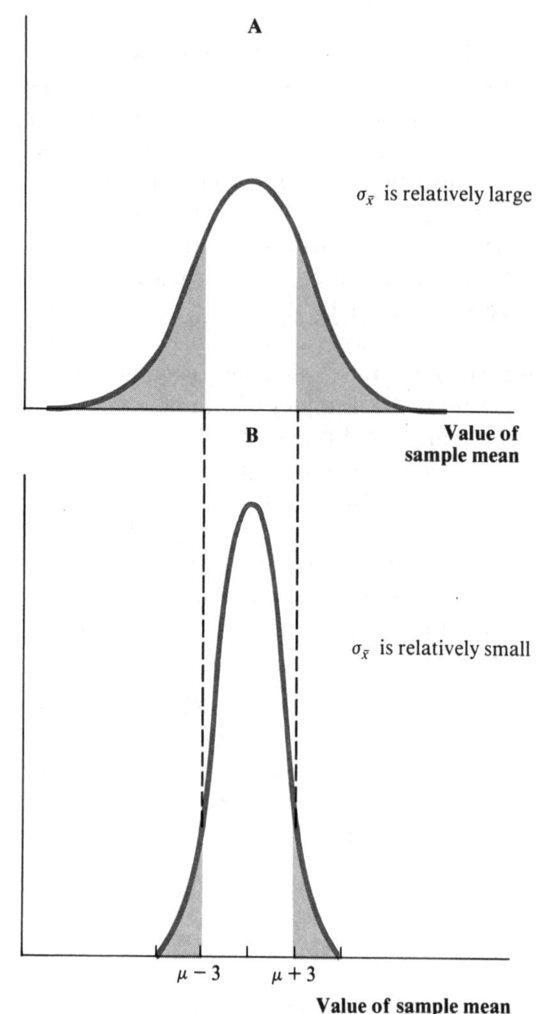

FIGURE 6.5

Sampling Distribution of the Sample Mean, when $\sigma_{\bar{x}}$ Is Relatively Large and Relatively Small

in that shown in panel B. (In each case this probability is equal in value to the shaded area in Figure 6.5.)

Given that $\sigma_{\bar{x}}$ is a measure of how "far off" the sample mean is likely to be, it is obviously important to specify what determines $\sigma_{\bar{x}}$. According to equation (6.3), $\sigma_{\bar{x}}$ is determined by two things: the standard deviation of the population, σ, and the sample size, n. *Holding sample size constant, $\sigma_{\bar{x}}$ is proportional to the standard deviation of the population.* This is reasonable because one would expect that sample means drawn from more variable populations would themselves tend to be more variable. *Holding constant the standard deviation of the population, $\sigma_{\bar{x}}$ is inversely proportional to the square root of the sample size.* This means that, in a sense, diminishing returns set in as the sample size is increased. If the sample size is quadrupled, $\sigma_{\bar{x}}$ is cut only in half; if the sample size is multiplied by 25, $\sigma_{\bar{x}}$ is cut only to one-fifth of its previous amount.

To prevent confusion, it is extremely important that you understand the distinction between $\sigma_{\bar{x}}$ (the standard deviation of the sampling distribution of the sample mean) and σ (the standard deviation of the population). The former is the standard deviation of the distribution of the sample mean. (Such distributions have been presented in Figure 6.4 and Figure 6.5) The latter is the standard deviation of the population from which the samples are drawn. Since the distribution of the sample mean is obviously quite different from the population being sampled, the standard deviations of the two are quite different, although not unrelated. (As pointed out in the previous paragraph, $\sigma_{\bar{x}}$ is proportional to σ, if sample size is held constant.)

Our results concerning the sampling distribution of the sample mean have important applications, some of which are illustrated by the following example.

Example 6.2 The diameters of tires produced by a firm are known to be normally distributed with a standard deviation of .01 inches. A simple random sample of 100 tires is selected from the plant's output, and the mean of the diameters is computed. What is the probability that this sample mean will exceed the population mean by more than .0015 inches?

Solution: The distribution of the sample mean is normal with mean equal to the population mean μ and standard deviation equal to $\sigma/\sqrt{n}$, which in this case is $.01 \div \sqrt{100}$, or .001 inches. Figure 6.6 shows the distribution of the sample mean. The probability we want equals the area under this curve to the right of $\mu + .0015$. The value of the standard normal variable corresponding to $\mu + .0015$ is $[(\mu + .0015) - \mu] \div \sigma_{\bar{x}} = .0015 \div .001 = 1.5$. Thus, we must evaluate the area under the standard normal curve to the right of 1.5. Appendix Table 2 shows that the area between zero and 1.5 is .4332, so the area to the right of 1.5 must equal $.5000 - .4332 = .0668$. This is the probability that the sample mean will exceed the population mean by more than .0015 inches.

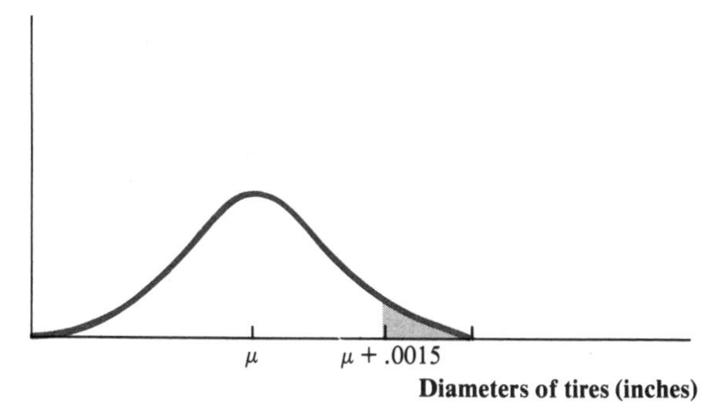

FIGURE 6.6
Probability that the Sample Mean Exceeds the Population Mean by More than .0015 Inches

WHERE THE POPULATION IS NOT NORMAL

As we know from the previous paragraphs, if the population is normal the sampling distribution of the sample mean is normal. But what if, as frequently occurs in business and economics, the population is *not* normal? A remarkable mathematical theorem, known as the central limit theorem, has shown that if the sample size is moderately large, the distribution of the sample mean can be approximated by the normal distribution, even if the population is not normal. This very important theorem can be stated as follows.

> ***Central Limit Theorem.*** *As the sample size* n *becomes large, the sampling distribution of the sample mean can be approximated by a normal distribution with a mean of* μ *and a standard deviation of* $\sigma/\sqrt{n}$, *where* μ *is the mean of the population and* σ *is its standard deviation.*

What is particularly impressive about this result is that the normal approximation seems to be quite good so long as the sample size is larger than about 30, regardless of the nature of the population. And if the population is reasonably close to normal, sample sizes of much less than 25 are likely to result in the normal approximation's being serviceable.[7]

To illustrate the working of the central limit theorem, suppose that the population frequency distribution is as shown in panel A of Figure 6.7. This population is such that it is equally likely that an observation in the population falls anywhere between the lower limit A and the upper limit B. (A population of this sort is said to have a *uniform distribution*.) Clearly, this population is far from normally distributed. Panels B, C, and D of Figure 6.7 show the sampling distribution of the sample mean for samples of size 2, 4, and 25 from this population. Even for samples of size 4, this sampling distribution is bell-shaped and close to normal. For samples of size 25, it is very close to normal.

The central limit theorem applies to discrete populations as well as to continuous ones, a fact which has important implications for the sampling distribution of the sample proportion. *The sample proportion is really a sample mean from a population where a success is denoted by a* 1 *and a failure is denoted by a zero.* The mean of such a population is P, the proportion of successes in the population; and the mean of the sample is p, the proportion of successes in the sample. Since the sample proportion is really a sample mean from a population consisting of zeros and 1's, its sampling distribution must tend toward normality as the sample size increases, according to the central limit theorem. This fact will be used frequently in subsequent chapters.

The following example illustrates some of the practical implications of the central limit theorem. In subsequent chapters this theorem will have repeated application.

Example 6.3 A manufacturer of light bulbs tests a random sample of 64 of the

7. Using the results of Appendix 4.3 and 4.1, we can prove that the expected value of $\bar{x}$ equals μ. From Appendix 4.3 we know that $E(\Sigma x_i) = E(x_1) + E(x_2) + \ldots + E(x_n)$. Since the expected value of each observation in a random sample equals μ, it follows that $E(\Sigma x_i) = n\mu$. And using the results of Appendix 4.1, $E(\Sigma x_i/n) = \mu$. Note that this proof does not depend on an assumption that the population is normal.

Based on Appendix 4.3 and 4.1, it is also possible to prove that $\sigma_{\bar{x}} = \sigma/\sqrt{n}$. Since the observations in a random sample are statistically independent, we know from Appendix 4.3 that $\sigma^2(\Sigma x_i) = \sigma^2(x_1) + \sigma^2(x_2) + \ldots + \sigma^2(x_n)$. Since the variance of each observation equals σ^2, it follows that $\sigma^2(\Sigma x_i) = n\sigma^2$. And using the results in Appendix 4.1, $\sigma^2(\Sigma x_i/n) = n\sigma^2/n^2 = \sigma^2/n$. This proof does not assume that the population is normal.

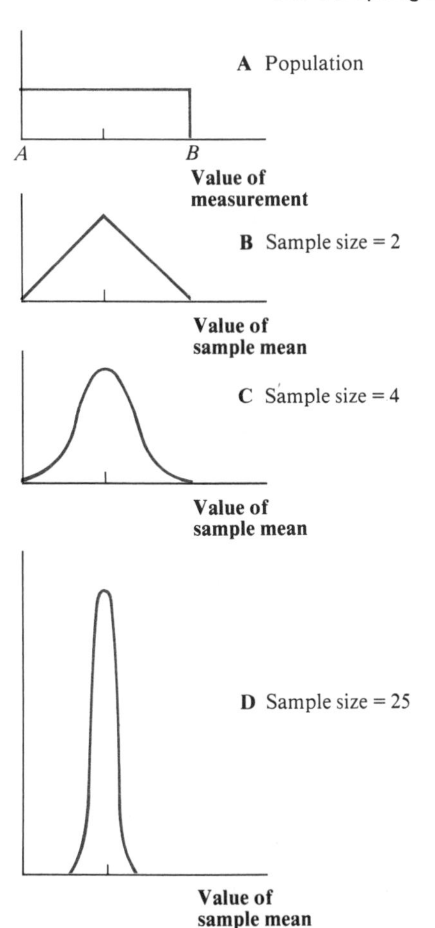

FIGURE 6.7

Sampling Distribution of Means of Samples of Sizes 2, 4, and 25 from a Population with a Uniform Distribution

light bulbs produced today to determine the longevity of each. From past experience, the firm knows that the standard deviation of the longevity of its light bulbs is 160 hours. What is the probability that the sample mean will differ from the population mean by more than 40 hours?

Solution: Because the sample size is well above 30, we can be reasonably certain that the sampling distribution of the sample mean is very close to normal, regardless of the nature of the population. More specifically, the distribution of the sample mean should be essentially normal with a mean equal to the population mean μ and a standard deviation equal to $\sigma/\sqrt{n}$, where σ is the population standard deviation. Since $\sigma = 160$ and $n = 64$, $\sigma_{\bar{x}} = 160 \div \sqrt{64} = 20$ hours. What we want is the probability that the sample mean lies above $\mu + 40$ or below $\mu - 40$. Figure 6.8 shows the distribution of the sample mean. The probability we want equals the sum of the values of the two shaded areas. The value of the standard normal variable corresponding to $\mu + 40$ is $[(\mu + 40) - \mu] \div \sigma_{\bar{x}} = 40 \div 20 = 2$; the value corresponding to $\mu - 40$ is $[(\mu - 40) - \mu] \div \sigma_{\bar{x}} = -40 \div 20 = -2$. Thus, we must evaluate the areas under the standard normal curve to the right of 2 and to the left of

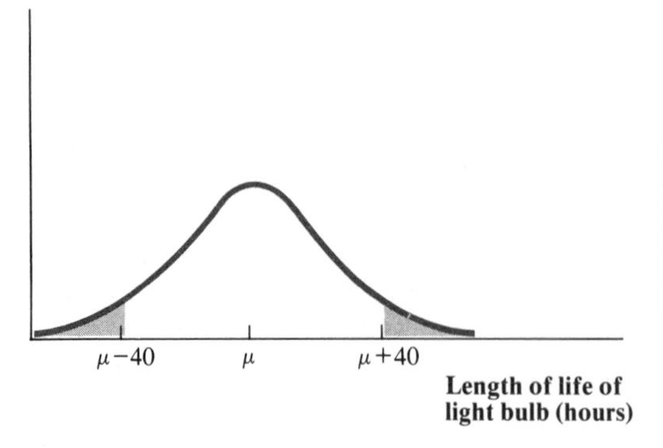

FIGURE 6.8
Probability that the Sample Mean Differs from the Population Mean by More than 40 Hours

−2. Since Appendix Table 2 shows that the area between zero and 2 is .4772, the area to the right of 2 must be .5000 − .4772 = .0228. Because of the symmetry of the standard normal curve, the area to the left of −2 must also be .0228. Thus, the total area to the right of 2 and to the left of −2 equals .0228 + .0228 = .0456. This is the probability that the sample mean will differ from the population mean by more than 40 hours.

WHERE THE POPULATION IS SMALL

In this section we have assumed that the population is infinite, while in fact many populations in business and economics are not. If the population is large relative to the size of the sample, our previous results can be applied without modification even though the population is finite. However, if the population is less than 20 times the sample size, the following simple correction should be applied to the standard deviation of the sample mean.

Standard Deviation of the Sample Mean (Finite Populations). *Whereas* $\sigma_{\bar{x}}$ *equals* $\sigma/\sqrt{n}$ *for infinite populations, in the case of finite populations the following formula applies:*

$$\sigma_{\bar{x}} = \frac{\sigma}{\sqrt{n}} \sqrt{\frac{N - n}{N - 1}}, \tag{6.4}$$

where σ *is the standard deviation of the population,* N *is the number of observations in the population, and* n *is the sample size.*

In other words, the value of $\sigma_{\bar{x}}$ for an infinite population must be multiplied by $\sqrt{(N - n)/(N - 1)}$ to make it appropriate for a finite population. The multiplier $\sqrt{(N - n)/(N - 1)}$ is often called the *finite population correction factor*. When n is small relative to N, this correction factor is so close to 1 that it can be ignored.[8]

If sampling is with replacement (that is, if each element that is included

8. In Appendix 4.2 it was stated that the standard deviation of the hypergeometric distribution equals the standard deviation of the binomial distribution times the finite population correction factor.

in the sample is put back in the population so that it can be chosen more than once), there is no need to use the finite population correction factor, even if the number of elements in the population is small relative to the sample size. In other words, equation (6.3) is valid under these circumstances. However, in most practical situations in business and economics, sampling is not carried out with replacement.

The following illustrates the application of equation (6.4). A taxi company has 17 cabs and chooses a simple random sample of 13 of them. On the basis of the sample, the firm computes the mean number of miles that a taxi has been driven since its tires were inspected. If the population standard deviation is 5,000 miles, the standard deviation of the sample mean equals

$$\sigma_{\bar{x}} = \frac{\sigma}{\sqrt{n}}\sqrt{\frac{N-n}{N-1}} = \frac{5{,}000}{\sqrt{13}}\sqrt{\frac{17-13}{17-1}}$$
$$= \left(\frac{5{,}000}{3.606}\right)\left(\frac{1}{2}\right) = 693.$$

Thus, the standard deviation of the sample mean is 693 miles, which is one-half of what it would have been had the population been infinite.

Since the standard deviation of the sample mean is a measure of how far off the sample mean is likely to be, an important implication of equation (6.4) is that *if the sample is small relative to the population, the accuracy of the sample mean depends entirely on the sample size and not on the fraction of the population included in the sample.* To test this, we can draw a simple random sample of size 100 from two populations, both of which have a standard deviation of 1,000. One population contains 10,000 observations, so the sample is 1 percent of the population; the other contains 100,000 observations, so the sample is 1/10 of 1 percent of the population. In the first population, the standard deviation of the sample mean is

$$\sigma_{\bar{x}} = \frac{1{,}000}{\sqrt{100}}\sqrt{\frac{10{,}000-100}{10{,}000-1}} = 100\sqrt{\frac{9{,}900}{9{,}999}} \doteq 100.$$

In the second population, it is

$$\sigma_{\bar{x}} = \frac{1{,}000}{\sqrt{100}}\sqrt{\frac{100{,}000-100}{100{,}000-1}} = 100\sqrt{\frac{99{,}900}{99{,}999}} \doteq 100.$$

Thus, the standard deviation of the sample mean is essentially the same, although the fraction of the population included in the sample is 10 times bigger in the first population than in the second. This is an important point, and one some people find hard to believe. To repeat, the sample size, not the fraction of the population included in the sample, determines the accuracy of the sample mean if the sample is small relative to the population.

Once the modification of $\sigma_{\bar{x}}$ in equation (6.4) has been made, the results for infinite populations can be applied to finite populations where the sample size is large relative to the population. The following example shows how this modification can be made and applied.

Example 6.4 A shipment of 100 chairs is received by a furniture store. A simple random sample of 36 chairs is chosen, and each is tested for its closeness to specifications. Based on this test, each chair is given a rating from zero to 10 points. If the store knows from past experience that the standard deviation of the ratings in a shipment is 2 points, what is the probability that the mean rating in the sample will be more than 1 point below the mean rating for all chairs in the shipment?

Solution: Since the sample is considerably more than 5 percent of the population, equation (6.4) must be used to calculate $\sigma_{\bar{x}}$, which is $(2/\sqrt{36})\sqrt{(100-36)/99} = (2/6)\sqrt{64/99} = .27$ points. Because the sample size is well above 25, the sampling distribution of the sample mean should be close to normal. The probability that the sample mean is more than 1 point below the population mean equals the probability that the sample mean is more than $1/.27 = 3.70$ standard deviations below the population mean. To evaluate this probability we find the area under the standard normal curve to the left of -3.70. Appendix Table 2 shows that the area between zero and 3.70 exceeds .499, so the area to the right of 3.70 must be less than .001. Because of the symmetry of the standard normal curve, the area to the left of -3.70 must also be less than .001. Thus, the probability that the mean rating for the sample will be more than 1 point below the mean rating for all chairs in the shipment is less than .001.

6.9 Accounts Receivable in a Department Store: An Experiment

To illustrate the fact that the central limit theorem (which played an important role in the previous section) can really be counted on to work, consider the following experiment that was carried out by Robert Trueblood, a partner in the accounting firm of Touche, Ross and Company, and Richard Cyert, now president of Carnegie-Mellon University. Trueblood and Cyert obtained data from a department store concerning the total dollar-amounts of 59 accounts-receivable ledgers. (Each ledger contained the accounts of 200 to 300 individual customers of the department store.) The amount for each ledger is shown in Table 6.4. Based on these data, the mean dollar-amount for a ledger is $10,263, and the standard deviation is $1,670.

Trueblood and Cyert were interested in what would happen if the department store, rather than going through all the ledgers, drew a random sample of 10 and used the sample mean to estimate the population mean.[9] To find out, 100 random samples (each containing 10 ledgers) were drawn, and the sample mean for each was computed. From these 100 samples was

9. Actually, they were more concerned with the use of the sample in estimating the population *total.* But since the population total is simply 59 times the population mean, the population total can be estimated by multiplying the sample mean by 59. Thus, the reliability of the estimate of the population total is directly related to the reliability of the sample mean. Note, too, that the samples were drawn *with replacement.* Thus, the finite population correction factor should not be used. See R. Trueblood and R. Cyert, *Sampling Techniques in Accounting* (New York: Prentice Hall, 1957).

TABLE 6.4
Total Dollar-Amounts of 59 Accounts-Receivable Ledgers

Ledger number	Amount (dollars)	Ledger number	Amount (dollars)
1	10,811	31	9,534
2	13,977	32	11,453
3	10,167	33	10,432
4	10,956	34	11,211
5	9,983	35	9,474
6	16,026	36	8,774
7	9,468	37	11,338
8	11,362	38	10,644
9	10,034	39	11,779
10	9,281	40	10,438
11	9,395	41	10,853
12	10,221	42	8,211
13	10,873	43	11,244
14	8,611	44	13,698
15	12,795	45	10,841
16	11,436	46	8,753
17	8,342	47	10,067
18	8,882	48	12,325
19	10,731	49	12,039
20	8,192	50	12,147
21	12,541	51	9,013
22	8,844	52	9,441
23	12,420	53	10,479
24	11,098	54	8,476
25	8,050	55	9,029
26	8,100	56	8,871
27	8,754	57	7,125
28	9,922	58	9,561
29	9,877	59	9,541
30	7,593		
Total dollar amount of accounts receivable			605,533

constructed the relative frequency distribution of the sample mean, which showed the proportion of these samples where the sample mean fell into various class intervals. The results are plotted in the histogram (using a density scale) in Figure 6.9.

According to our discussion in the previous section, the sample mean should be distributed approximately normally, the mean of its sampling distribution should equal the population mean, and the standard deviation of its sampling distribution should equal the population standard deviation divided by the square root of the sample size (because the sample was drawn with replacement). Thus, since we know that the mean of the population is $10,263 and that the population standard deviation is $1,670, it follows that

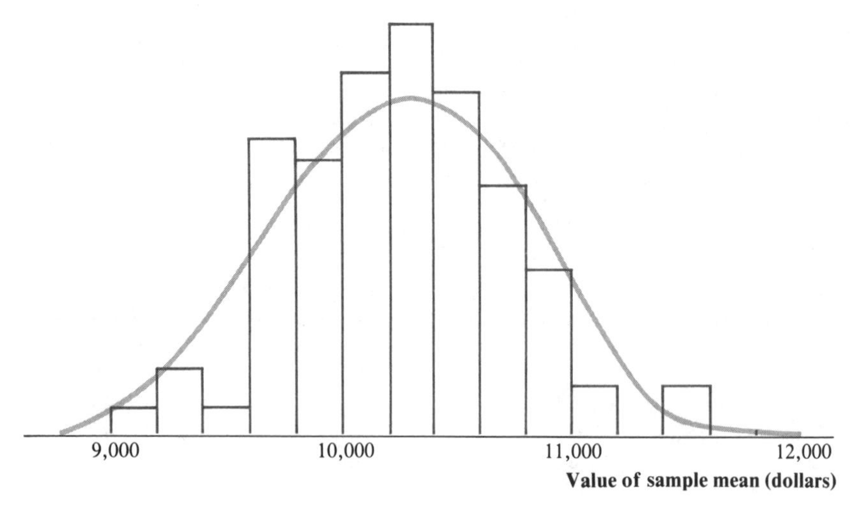

FIGURE 6.9 Histogram of 100 Sample Means, and Normal Approximation[a]

[a]The height of each bar equals the number of sample means falling within the relevant class interval divided by the width of the class interval ($200). The area of each bar equals the proportion of sample means falling within this class interval.

the mean of the sampling distribution of the sample mean in this case should be $10,263 and that its standard deviation should be $\$1{,}670 \div \sqrt{10} = \$528$.

The point of Trueblood and Cyert's experiment was to demonstrate that the distribution of the sample means in Figure 6.9 would be quite close to theoretical expectations. To test this, compare the actual distribution of sample means (shown in the histogram in Figure 6.9) with the normal curve with mean of $10,263 and standard deviation of $528 (also shown in Figure 6.9). The two are quite close, as you can see. They are not exact replicas of one another, but no statistician would expect exact agreement, since the distribution in Figure 6.9 is based only on 100 random samples. If one had the patience to draw 1,000 such samples, the approximation would be closer than in Figure 6.9; and if one were to draw 10,000 such samples, the approximation would be even better.

As pointed out in the previous section, statisticians frequently use the sampling distribution of the sample mean to compute the probability that a sample mean will differ from the population mean by at least a certain amount. The results in Figure 6.9 provide some idea of the validity of these computations. For example, as we shall see in greater detail in the next chapter, statisticians are fond of stating that the probability that the sample mean will differ from the population mean by more than $1.96\sigma_{\bar{x}}$ is approximately .05. This follows from the fact that the probability is approximately .95 that the value of any normal variable will lie within 1.96 standard deviations of its mean.

Using the results in Figure 6.9, we can see how close this statement is to being true for the 100 samples drawn by Trueblood and Cyert. Since

$\sigma_{\bar{x}}$ = \$528, and since (1.96)(\$528) = \$1,035, this statement amounts to saying that the probability is .05 that the sample mean will differ by more than \$1,035 from the population mean. And since the population mean is \$10,263, this amounts to saying that the probability is about .05 that the sample mean will fall below \$9,228 (that is, \$10,263 − \$1,035) or will exceed \$11,298 (that is, \$10,263 + \$1,035). In fact, the sample mean did fall below \$9,228 or exceed \$11,298 in 5 of the 100 samples (or 5 percent of the time). Of course, the fact that this experimental result is so close to what theory predicts is to some extent a coincidence, but it does illustrate the usefulness of our propositions in the previous section concerning the sampling distribution of the sample mean.

EXERCISES

6.14 The Energetic Corporation is a producer of electric light bulbs. The company's production process does not result in precisely similar bulbs. Instead, bulb length is normally distributed, with a mean of 3.00 inches and a standard deviation of .10 inches. What is the sampling distribution of the mean length of a simple random sample of four bulbs? That is, what is the mean of this sampling distribution? What is its standard deviation? What is its shape?

6.15 If the population is normal, the distribution of the sample mean is normal, regardless of how small the sample may be. Using this fact, determine the probability that the sample mean in Exercise 6.14 will exceed 3.01 inches.

6.16 In Exercise 6.14, how large a sample must the firm take to make the standard deviation of the sample mean equal .01 inches?

6.17 The Energetic Corporation wants to estimate the mean amount spent on light bulbs in 1979 by the nation's 100 largest firms. A random sample of 25 of these firms is drawn (without replacement). If the actual mean expenditure by these 100 firms was \$50,000 and the standard deviation was \$5,000, what is the expected value of the sample mean? What is the standard deviation of the sample mean?

6.18 Using Chebyshev's inequality, determine an upper-bound for the probability that the sample mean in Exercise 6.17 will *not* be between \$48,000 and \$52,000. If the population in Exercise 6.17 is reasonably close to normal, what is a good approximation to this probability?

6.19 If we are sampling from an infinite population, how much reduction occurs in the standard deviation of the sample mean if (a) the sample size is increased from 2 to 4; (b) the sample size is increased from 4 to 6; (c) the sample size is increased from 100 to 102?

6.20 The Energetic Corporation wants to estimate the mean life of the light bulbs produced on a particular day. It is known that the standard deviation of the longevity of the light bulbs produced on any given day is about 100 hours. A random sample of 100 bulbs produced on the day in question is drawn and the mean length of life of this sample is determined. What is the probability that this sample mean will differ from the population mean by more than 15 hours?

6.21 If the population is roughly normal, the *standard deviation of the sample median* equals $\sqrt{\pi/2} \cdot \sigma/\sqrt{n}$. If a sample median is to have a standard deviation equal to that of a sample mean (from the same population), how much bigger must the sample size be?

6.22 The Energetic Corporation draws a simple random sample of its bills to customers to determine what proportion of bills contain numerical errors. In the population as a whole, 10 percent contain such errors. If there are 10,000 such bills, and if the sample contains 100 bills, what is the probability that the proportion of bills in the sample containing numerical errors exceeds 16 percent?

6.23 In Exercise 6.22, how large must the sample be if the standard deviation of the sample proportion equals .06?

6.24 A simple random sample of size 5 is taken (without replacement) from a population of 10 items. If the standard deviation of the population is 10, what is the standard error of the sample mean?

6.25 Show that for a finite population the standard error of the sample mean equals approximately

$$\frac{\sigma}{\sqrt{n}}\sqrt{1-\frac{n}{N}}$$

if N is of reasonable size. The quantity, n/N, is often called the *sampling fraction.*

6.26 Using the formula in the previous exercise, find the standard error of the sample mean if $\sigma = 2$, $n = 16$, and (a) $N = 100$; (b) $N = 10{,}000$.

CHAPTER REVIEW

1. Sample designs can be divided into two broad classes: probability samples and judgment samples. A *probability sample* is one where the probability that each element of the population will be chosen in the sample is known. A *judgment sample* is chosen in such a way that this probability is not known. One of the most important disadvantages of a judgment sample is that one cannot calculate the sampling distribution of a sample statistic and thus there is no way of knowing how big the sampling errors in the sample results are likely to be.

2. If a population contains N elements, a *simple random sample* is one chosen so that each of the different samples of size n that could be chosen has an equal chance of selection. Alternatively, a simple random sample can be defined as one in which the probability of selecting each element in the population is the same and in which the chance that any one element is chosen is independent of the choice of any other element. A table of random numbers is used in choosing a simple random sample. A *systematic sample* is obtained by taking every kth element of the population on a list. If the elements of the population are in random order, this is tantamount to a simple random sample.

3. One of the most important ways that expert judgment can be used to design a sample is in the construction of strata, or subdivisions of the population. In *stratified random sampling* (where the population is stratified first, and a simple random sample is chosen from each stratum), the resulting estimate is often more precise for a sample of given size than if simple random sampling is used. In formulating strata, one should subdivide the population so that the strata differ greatly with regard to the characteristic being measured, and so that there is as little variation as possible *within* each stratum with regard to this characteristic. To maximize the precision of the sample results, the sample size in each stratum should be proportional to the product

of the number of elements in the stratum and the standard deviation of the characteristic being measured in the stratum. In *cluster sampling*, the population is divided into a number of clusters and a sample of these clusters is chosen, after which a simple random sample of the elements in each chosen cluster is selected. The major advantage of cluster sampling is that it is relatively cheap to sample elements that are physically or geographically close together.

4. A *sampling distribution* is the probability distribution of the value of a particular sample statistic. If the sample is a small proportion of the population (or if the sampling is with replacement), the binomial distribution can be used to derive the sampling distribution of the sample proportion. If the population is normal, the sampling distribution of the sample mean is normal with a mean equal to the population mean and with a standard deviation equal to the population standard deviation divided by the square root of the sample size. Even if the population is not normal, the sampling distribution of the sample mean will have approximately these properties as long as the sample size is larger than about 30. (If the population is at all close to normal, a sample size of much less than 30 often will suffice.)

5. The *standard deviation of the sampling distribution of the sample mean* $\sigma_{\bar{x}}$ is a measure of the extent of the sampling error that is likely to be contained by the sample mean. This is often called the *standard error of the sample mean*. If the population is finite, it equals the value of $\sigma_{\bar{x}}$ for an infinite population multiplied by $\sqrt{(N - n)/(N - 1)}$; this multiplier is called the *finite population correction factor*. If the sample is a small percentage of the population, this multiplier is so close to 1 that it can be ignored. Thus, if the sample is small relative to the population, the accuracy of the sample mean depends entirely on the sample size, not on the fraction of the population included in the sample.

Getting Down to Cases:

SAMPLING AT THE EXXON RESEARCH AND ENGINEERING COMPANY[10]

The Exxon Corporation, a huge U.S.-based multinational oil firm, has a scientific and engineering affiliate, the Exxon Research and Engineering Company. This affiliate maintains a storehouse containing supplies necessary for its operations. The characteristics of the items stored there are shown in Table 6.5. Originally, a 100 percent annual inventory was taken in

TABLE 6.5

Characteristics of Materials-and-Supplies Inventory

Number of catalogue items	3,900
Number of units of issue	260,000
Value of inventory (dollars)	185,000
Minimum unit cost of item (dollars)	0.005
Maximum unit cost of item (dollars)	120

10. This case is based on Raymond Obrock, "A Case Study of Statistical Sampling," *The Journal of Accountancy*, March 1958, and is also contained in E. Mansfield, *Statistics for Business and Economics: Readings and Cases* (New York: Norton, 1980). Table 6.7 has been simplified somewhat for pedagogical purposes.

which every catalogue item was counted and the amount compared with what the accounting records said it should be. Adjustments were then made in the inventory account to bring the two into equality. The adjustments made during a five-year period are shown in Table 6.6. At the end of this period Exxon decided to sample the items in its storehouse. About 3,900 types of items are stored there, and the adjustment to the inventory account is the sum of the adjustments for each of the individual types of items.

TABLE 6.6
History of Inventory Adjustments

Year	Excess of book value over actual value (dollars)
1	260
2	406
3	1,026
4	1,339
5	566

(a) What parameter should Exxon try to estimate?

(b) What is the population from which Exxon is sampling?

(c) What are the advantages of a probability sample rather than a judgment sample in this situation?

(d) Would stratification be useful in this case? If so, how?

The types of items in Exxon's storehouse vary considerably in their unit costs. For example, a roll of tape costs less than a pair of scissors. When types of items were classified by their unit costs, Exxon found that the number of types of items in each class and the standard deviation of the adjustments (in dollars) for the individual types of items in each class were as shown in Table 6.7.

TABLE 6.7
Number of Types of Items and Standard Deviation of Adjustments of Types of Items by Unit Cost

Unit cost of type of item (dollars)	Number of types of items	Standard deviation of adjustments (dollars)
Under 1	2,200	4
1 to 19.99	1,600	5
20 and over	100	18

(e) What advice would you give Exxon as to the way in which a stratified sample should be allocated among classes?

APPENDIX 6.1

Monte Carlo Methods: Another Application of a Table of Random Numbers[11]

Here and in previous chapters we have shown cases where mathematical analysis has been used to determine probability distributions. (For example, the central limit theorem uses mathematical analysis to determine the sampling distribution of the sample mean.) What techniques can be applied when the situation is too complicated for mathematical analysis? In such a case the statistician can use Monte Carlo methods, which are based on experimentation or simulation of the relevant situation. Although Monte Carlo methods do not result in an *analytical* solution to a *general* type of problem, they can provide an *approximate numerical answer* to a *particular concrete problem*—and in many practical cases this is all that is needed.

To illustrate how Monte Carlo methods work, suppose that a firm receives a lot of 100 items, 30 of which fail within a year. The firm would like to determine the probability that 30 or more out of 100 items will fail, given that the supplier is correct in its claim that the probability is 1/6 that each item will fail within a year. One way of solving this problem is to take 100 dice and throw them simultaneously. The probability that 30 or more dice will come up with a 1 is the same as the probability the firm requires, since the probability that each (true) die will come up with a 1 is 1/6. If the firm throws the 100 dice 1000 times and records the number of throws on which 30 or more dice come up with a 1, then this number divided by 1000 should be a fairly accurate approximation of the desired probability.

The essence of Monte Carlo methods is that *an analogue of the relevant situation is created, and one simulates the relevant process to generate a body of data from which the desired information can be estimated.* Let's take the following situation as an example. The shipping firm discussed in Appendix 5.1 is contemplating a change in its facilities such that it will take exactly 0.1 months to dock and service each ship. Given that ships arrive at random and that the mean number of arrivals per month is 5, the firm wants to know the probability that a ship will wait more than 0.2 months under this proposed setup. How can Monte Carlo methods be used to answer this question?

To simulate this situation, the firm's statisticians must generate a hypothetical "history" (record) of arrivals which conforms to the fact that ships arrive at random and that the mean number of arrivals per month is 5. Under these circumstances, as we know from Appendix 5.1, the probability that the interval between arrivals is between c and d months equals

$$e^{-5c} - e^{-5d}.$$

Letting $c = 0$, we obtain the probability that this time interval, Y, is less than d months:

$$Pr\ \{Y < d\} = 1 - e^{-5d}. \tag{6.5}$$

11. Some of the material in this Appendix assumes that the reader has covered the latter part of Chapter 5 and Appendix 5.1. Readers who have not covered these topics will probably want to skip this appendix as well.

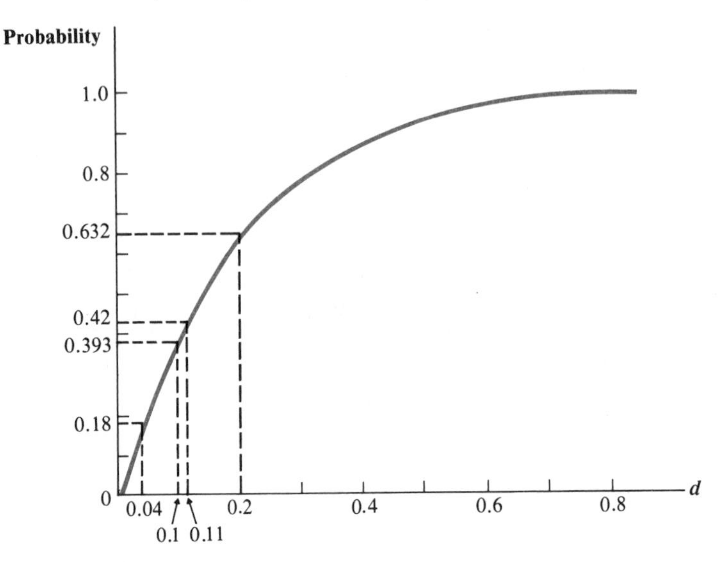

FIGURE 6.10 Probability that Time Interval between Successive Arrivals Is Less than *d* Months[a]

[a]To derive the function graphed here, we insert $d = 0, 0.1, 0.2, \ldots$ into $1 - e^{-5d}$, the result being

d	$1 - e^{-5d}$	d	$1 - e^{-5d}$
0	.000	0.5	.918
0.1	.393	0.6	.950
0.2	.632	0.7	.970
0.3	.777	0.8	.982
0.4	.865	0.9	.989

Figure 6.10 shows this probability for various values of d.

In generating a history of arrivals, we want the probability distribution of the values we assign to the intervals between successive arrivals to agree with equation (6.5). This can be done by picking a point between zero and 1 at random on the vertical axis of the graph in Figure 6.10, and finding the horizontal coordinate of the point on the curve corresponding to this randomly chosen point. The resulting interval will have the desired probability distribution.[12] To pick a point at random along the vertical axis of Figure 6.10,

12. To see that an interval chosen in this way has the desired probability distribution, let's begin by considering two probabilities: (1) the probability that an interval chosen in this way will be less than .1 months; and (2) the probability that an interval chosen in this way will be less than .2 months. The probability that the time interval between successive arrivals will be less than 0.1 months should be .393, according to equation (6.5). What is the probability that an interval chosen in this way will be less than 0.1 months? It equals .393, because if we pick a random number between zero and 1, there is a .393 chance that it will be less than .393. If it is less than .393, Figure 6.10 shows that this procedure will result in an interval that is less than .1 months. The probability that the time interval between successive intervals is less than 0.2 months should be .632, according to equation (6.5). What is the probability that an interval chosen in this way will be less than 0.2 months? It equals .632, because if we pick a random number between zero and 1, there is a .632 chance that it will be less than .632. If it is less than .632, Figure 6.10 shows that this procedure will result in an interval that is less than .2 months. It should now be clear that the probability of an interval chosen in this way being less than *any* length of time equals what is dictated by equation (6.5). This means that an interval chosen in this way has the desired probability distribution.

we draw a number from a table of two-digit random numbers and divide the number by 100. Thus, to generate a history of arrivals, we must draw a large number of two-digit random numbers. Suppose that they are as follows:

18, 42, 00, 02, 24, 43, 64, 33, 42, 65, 11, 11, 32, 40,

62, 28, 35, 20, 86, 51, 49, 62, 01, and so forth.

Dividing each of these numbers by 100, we can convert them into probabilities; and, as we have indicated, each such probability can be used to generate a time interval between ship arrivals. The first random number is 18, so the first probability is 0.18. Corresponding to 0.18 on the vertical axis in Figure 6.10, the time interval on the horizontal axis is 0.04 months.[13] Thus, assuming that the first ship arrives at time zero, the next arrives at 0.04 months. Since the next random number is 42, the next probability is 0.42. Figure 6.10 shows that the time interval on the horizontal axis corresponding to 0.42 on the vertical axis is 0.11 months, so the third ship arrives 0.11 months after the second, or at time 0.15 months. Repeating this procedure, we can generate as long a history of arrivals as we like. Table 6.8 shows the first 23 such results.

TABLE 6.8
Monte Carlo Study of Waiting Line in Shipping Example

Random number	Time between arrivals	Time of arrival	Time wait ends	Time of servicing	Waiting time
18	.04	.04	.10	.20	.06
42	.11	.15	.20	.30	.05
00	.00	.15	.30	.40	.15
02	.00	.15	.40	.50	.25
24	.05	.20	.50	.60	.30
43	.11	.31	.60	.70	.29
64	.20	.51	.70	.80	.19
33	.08	.59	.80	.90	.21
42	.11	.70	.90	1.00	.20
65	.21	.91	1.00	1.10	.09
11	.02	.93	1.10	1.20	.17
11	.02	.95	1.20	1.30	.25
32	.08	1.03	1.30	1.40	.27
40	.10	1.13	1.40	1.50	.27
62	.19	1.32	1.50	1.60	.18
28	.07	1.39	1.60	1.70	.21
35	.09	1.48	1.70	1.80	.22
20	.04	1.52	1.80	1.90	.28
86	.39	1.91	1.91	2.01	.00
51	.15	2.06	2.06	2.16	.00
49	.14	2.20	2.20	2.30	.00
62	.19	2.39	2.39	2.49	.00
1	.00	2.39	2.49	2.59	.10

13. To keep things as simple as possible we use only two decimal places here for the time intervals.

Once the firm's statisticians have established this history of arrivals, they can determine how long each ship must wait before it can be docked and serviced. As shown in Table 6.8, the second ship to arrive appears at time 0.15 and has to wait until time 0.20, when the first ship has been serviced. Then the third ship arrives at approximately the same time as the second ship, and must wait until time 0.30 when the second ship has been serviced. Table 6.8 shows the history of arrivals and waiting times for the first 23 ships (after the first ship, which is assumed to appear at time 0.00). The firm's statisticians can determine the percentage of ships that have to wait more than 0.2 months to be serviced, and this proportion can be used to estimate the desired probability. In a real Monte Carlo study, a history of this sort might be run for several hundred or even several thousand ships, depending on how precise the results needed to be.

It is important to emphasize that the results in Table 6.8 are intended only to demonstrate the methods used, not to provide the solution to a real problem. As stated above, many more than 23 ships would have to be included to obtain the solution (and a computer might be used to cut down on the drudgery). Monte Carlo methods are used in many aspects of business and economics, not just in the analysis of waiting lines, and large-scale electronic computers make it possible to carry out Monte Carlo studies in situations where they would otherwise be too complex or too expensive.

EXERCISES

6.27 Use one-digit random numbers to simulate 100 flips of a fair coin.

6.28 Use one-digit random numbers (excluding 7, 8, 9, and zero) to simulate 100 rolls of a single true die.

7

Statistical Estimation

7.1 Introduction

Having covered the necessary aspects of probability theory and sampling techniques in previous chapters, we can now go on to statistical inference, the branch of statistics that shows how rational decisions can be made on the basis of sample information. Statistical inference deals with two types of problems: *estimation* and *hypothesis testing*. This chapter covers estimation, and the next two chapters deal with hypothesis testing.

Since statistical estimation is of enormous importance in business and economics, we will provide a rather detailed discussion of various estimation techniques and how they can be used.

7.2 Point Estimates and Interval Estimates

Many statistical investigations are carried out in order to estimate the parameter of some population. For example, at the end of Chapter 6 we considered a case where the object was to estimate the mean dollar-amount of a number of accounts-receivable ledgers; earlier in that chapter we examined a case where a firm wanted to estimate the total change in the value of its inventory. Decision makers are interested in estimating a particular parameter because their proper course of action depends on the value of this parameter. For example, a breakfast food firm may want to estimate the proportion of consumers who prefer brand C to brand D because this information will influence whether or not the firm tries to develop a competing breakfast food that imitates brand C.

In estimating a particular parameter, the decision maker uses a statistic calculated from a sample. For example, in the investigation of the accounts-receivable ledgers, the sample mean was used as an estimate of the population mean; and in the study to estimate the proportion of people preferring brand C breakfast food to brand D the sample proportion might well be the statistic used to estimate the population proportion. A statistic which is used to estimate a parameter is an ***estimator.*** An **estimate** is the ***numerical value*** of the estimator that is used. It is important to distinguish between an *estimator* and an *estimate*. For example, if a sample mean is used to estimate a population mean, and if the sample mean equals 10, the estimator used is the sample mean, whereas the estimate is 10.

Statisticians differentiate between two broad classes of estimates: point estimates and interval estimates. A ***point estimate*** *is a single number.* For example, if a firm estimates that 9 percent of the items in an incoming shipment are defective, 9 percent is a point estimate. Or if the proportion of individuals preferring brand C breakfast food to brand D is estimated to be .81 this, too, is a point estimate. An ***interval estimate****, on the other hand, is a range of values within which the parameter is thought to lie.* Thus, if a firm estimates that between 8 and 10 percent of the items in an incoming shipment are defective, 8 to 10 percent is an interval estimate. If the proportion of people preferring brand C breakfast food to brand D is estimated at between .75 and .90 this also is an interval estimate.

In some circumstances, only a point estimate is required. The estimate may be used in a complex series of computations, and the users may want only a single number. More often, however, an interval estimate is preferable to a point estimate because the former indicates how much error is likely to be in the estimate. For example, consider the point estimate that 9 percent of a certain shipment of items are defective. Such an estimate provides no idea of how much error it is likely to contain. That is, it is impossible to tell whether the data indicate that the proportion defective is likely to be very close to 9 percent, or whether it may well depart considerably from 9 percent. The advantage of interval estimates is that they provide such information. The decision maker can construct an interval estimate so that he or she has a specified amount of confidence that it will include the desired parameter. For example, in constructing its interval estimate of the percent of defective items, the firm in question can establish a .95 probability that such an interval will include the population percent defective.

At the outset of an attempt to estimate a parameter of any given population, the statistician must make basic decisions about the nature of the sample design and the type of estimator to be used. (For reasons pointed out in Chapter 6, we assume that simple random sampling is used here.) Once the sample design has been chosen, the statistician must determine *how large a sample to take.* The sample size will depend on the costliness of errors in the estimate and on the costliness of sampling. If substantial errors in the estimate will result in large penalties, the optimal sample size will tend to be large because the cost of increased sample size is likely to be outweighed by the resulting reduction in the sampling errors contained in the estimate. Also, if it is relatively inexpensive to increase the sample size, the optimal sample size will tend to be larger than if it is relatively expensive to do so.

The choice of *which estimator to use* will depend on the nature of the sampling distributions of various estimators. As stressed in the previous chapter, an estimator's sampling distribution shows the extent to which the estimator will vary from sample to sample and how far we can expect it to depart from the parameter it is intended to measure. An estimator with a sampling distribution showing a low probability of its departing greatly from this parameter is preferable to one whose sampling distribution indicates a high probability of its departing greatly. Why? Because for the same sample size and cost one can obtain greater *accuracy* with the former rather than the latter estimator. More will be said about this in the following section.

7.3 Point Estimation

Generally, there are a variety of estimators that can be used to make a point estimate of a particular parameter of a given population. For example, if we want to estimate the mean of a certain population, we can use either the mean or the median of a sample drawn from this population. Which should we use? Obviously, we want to choose the one that will be closer to the population mean, but there is no way of knowing which is closer because we do not know the value of the population mean. All we can do is compare the sampling distribution of the sample mean with that of the sample median. Such a comparison will show which of these estimators is more likely to depart considerably from the population mean. In choosing among estimators, statisticians consider the following three criteria: unbiasedness, efficiency, and consistency. Each of these is described below.

UNBIASEDNESS

As pointed out in previous chapters, the expected value of a statistic's sampling distribution is a measure of central tendency which shows the statistic's long-term mean value. If a statistic's sampling distribution is as shown in Figure 7.1, and if this statistic is used as an estimator of θ, this statistic clearly is not a very reliable estimator. Why? Because the expected value of this estimator is 3θ, which means that if such an estimator were used repeatedly, the average estimate would be about three times the parameter we

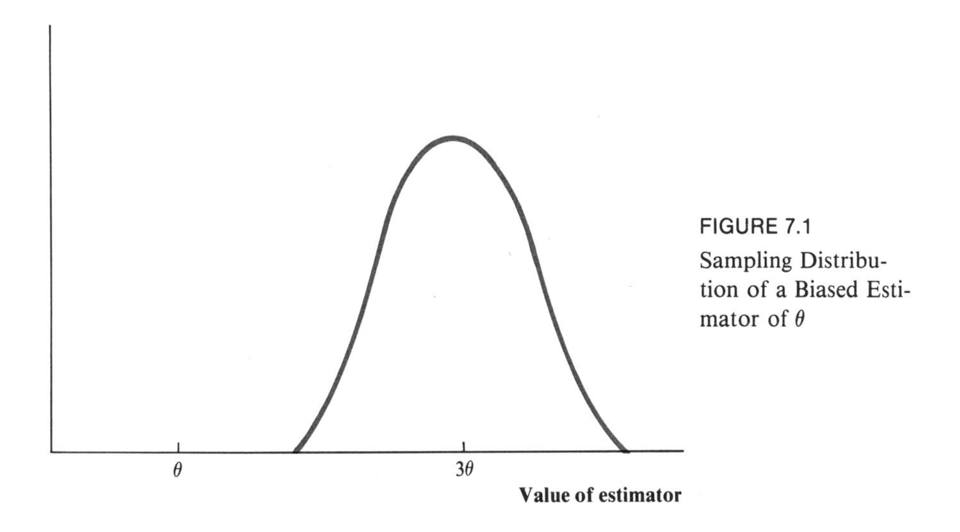

FIGURE 7.1
Sampling Distribution of a Biased Estimator of θ

wish to estimate. To avoid estimators of this sort, statisticians use unbiasedness as one criterion.

> ***Unbiasedness.*** *An unbiased estimator is a statistic the expected value of which equals the parameter being estimated.*

To illustrate how this definition can be applied, the sample mean is an unbiased estimator of the population mean because, as we saw in the previous chapter, the mean of its sampling distribution equals the population mean.

The concept of bias used here is somewhat different from that discussed in Chapter 1 where bias was described as a systematic, persistent sort of error due to faulty selection of a sample. Even if a sample is a properly chosen random sample, a bias can result if the estimator does not, on the average, equal the parameter being estimated. For example, had we defined the sample variance as

$$\frac{\sum_{i=1}^{n}(x_i - \bar{x})^2}{n},$$

it would have been a biased estimator of the population variance, σ^2. As shown in panel A of Figure 7.2, its expected value would have been $[(n - 1)/n]\sigma^2$, not σ^2. It is for this reason (as we pointed out in Chapter 2) that we define the sample variance with $(n - 1)$, not n, in the denominator. If $(n - 1)$ is the denominator, the sample variance is an unbiased estimate of σ^2, as shown in panel B of Figure 7.2.

EFFICIENCY

The unbiasedness of an estimator does not necessarily mean that the estimator is likely to be close to the parameter we want to estimate. For example, the estimator whose sampling distribution is shown in panel A of Figure 7.3 has a high probability of differing considerably from its mean, which (since it is unbiased) is the parameter we want to estimate. This is because its sampling distribution contains a great deal of dispersion or variability. On the other hand, the estimator whose sampling distribution is shown in panel B of Figure 7.3 has a low probability of differing considerably from its mean because its sampling distribution exhibits little dispersion or variability. Statisticians would say that the estimator in panel B is "more efficient" than the one in panel A, since its sampling distribution is concentrated more tightly about the parameter we want to estimate. Efficiency, defined as follows, is an important criterion used by statisticians to choose among estimators.

> ***Efficiency.*** *If two estimators are unbiased, one is more efficient than the other if its variance is less than the variance of the other.*[1]

To illustrate, let's go back to the question of whether the sample mean or

1. For biased estimators, one estimator is more efficient than another if its mean-square error is less than that of the other. If X is an estimator of θ, its mean-square error is the expected value of its squared deviation from θ. That is, it equals

$$E[(X - \theta)^2].$$

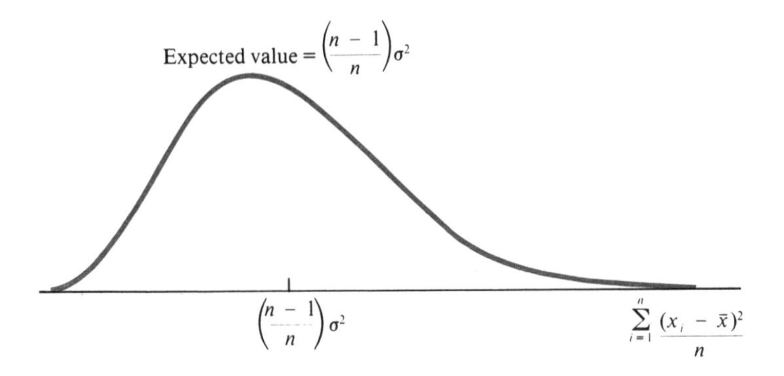

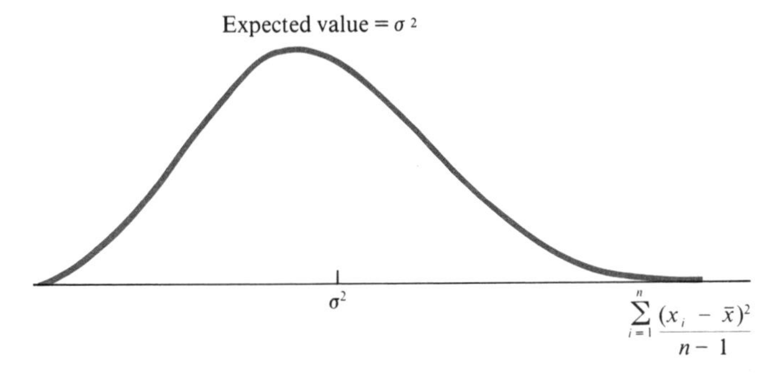

FIGURE 7.2 Sampling Distributions of $\sum_{i=1}^{n} \frac{(x_i - \bar{x})^2}{n}$ and $\sum_{i=1}^{n} \frac{(x_i - \bar{x})^2}{n-1}$

the sample median should be used to estimate the population mean. As we know from Chapter 6, the variance of the sample mean is $\sigma^2 \div n$; and it can be shown that if the population is normal, the variance of the sample median is approximately $1.57\sigma^2 \div n$ if the sample is large. (See Exercise 6.21.) Thus, the sample mean is *more efficient* than the sample median because the sample median's variance is 57 percent greater than that of the sample mean. This implies that the sample median is more likely than the sample mean to differ considerably from the population mean. (In other words, holding constant the sample size, the sample mean is more likely than the sample mean to be close to the population mean.) Because it is more efficient, the sample mean is

A

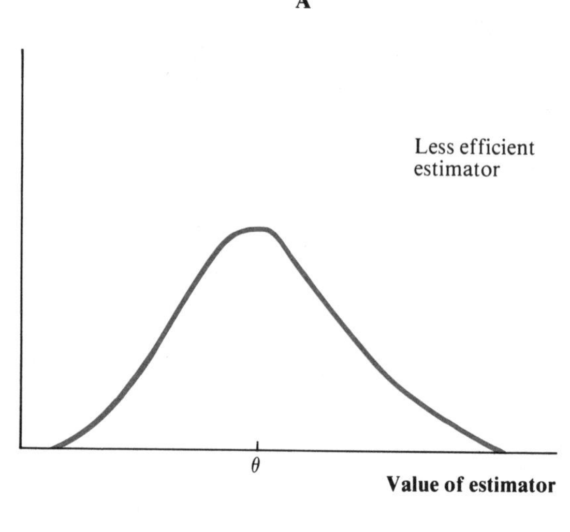

B

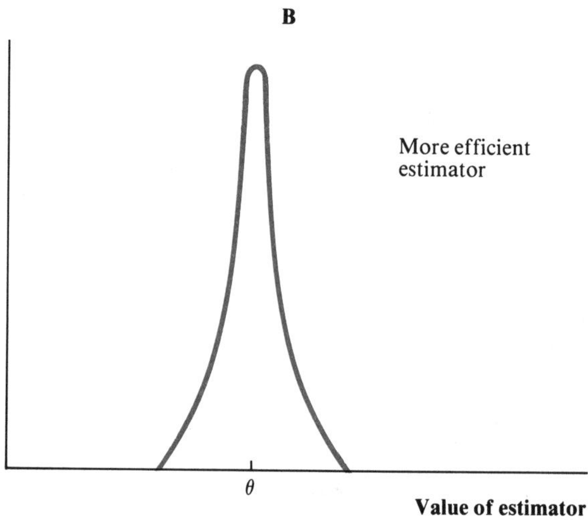

FIGURE 7.3
Sampling Distributions of Two Unbiased Estimators of θ

generally preferred over the sample median as an estimator of the population mean.

CONSISTENCY

Still another criterion used for choosing among estimators is consistency. Some estimators are consistent, while others are not. The statistical definition of consistency is as follows.

> ***Consistency.*** *A statistic is a consistent estimator of a parameter if the probability that the statistic's value is very near the parameter's value increasingly approaches* 1 *as the sample size increases.*

In other words, if a statistic is a consistent estimator of a particular parameter, the statistic's probability distribution becomes increasingly concentrated on this parameter as the sample size increases.

It is desirable for an estimator to be consistent because this means that the estimator becomes more reliable as the sample size increases. For example, consider the sample mean, which is a consistent estimator of the population mean. Since the standard deviation of the sample mean's sampling distribution equals $\sigma/\sqrt{n}$, it clearly tends to zero as n increases in value. This means that the probability distribution of the sample mean becomes concentrated ever more tightly about the population mean. If the population is normal, the sample median is also a consistent estimator; thus, both the sample mean and the sample median satisfy this criterion.

7.4 Point Estimates for μ, σ, and P

We have just described the criteria that statisticians frequently use to choose among estimators. Based on these criteria, certain statistics are generally preferred over others as estimators of the population mean (μ), population variance (σ^2), population standard deviation (σ), or population proportion (P). This does not mean that other estimators are not preferred under special circumstances or that it is incorrect to use other estimators. It does mean that based on the criteria given in the previous section, the following estimators are the standard ones used by statisticians to estimate these parameters.

Sample Mean. This is the most common estimator of the population mean. As we know, it is unbiased and consistent. Moreover, it can be shown that if the population is normal the sample mean is the most efficient unbiased estimator available. For these reasons the sample mean is generally the preferred estimator of the population mean.

Sample Variance and Standard Deviation. The sample variance is an unbiased and consistent estimator of the population variance. It is relatively efficient as compared with other estimators. Its square root, the sample standard deviation, is generally used as an estimator of the population standard deviation even though it is not unbiased. The sample standard deviation is also relatively efficient.

Sample Proportion. This is an unbiased, consistent, and relatively efficient estimator of the population proportion. For these reasons, it is generally the preferred estimator of the population proportion.

EXERCISES

7.1 The Bona Fide Washing Machine Company chooses a random sample of 25 motors from those it receives from one of its suppliers (supplier I). It determines the length of life of each of the motors. The results (expressed in thousands of hours) are as follows:

4.1	4.6	4.6	4.6	5.1
4.3	4.7	4.6	4.8	4.8

(table continues)

4.5	4.2	5.0	4.4	4.7
4.7	4.1	3.8	4.2	4.6
3.9	4.0	4.4	4.0	4.5

The firm's management is interested in estimating the mean length of life of the motors received from supplier I. Provide a point estimate of this population parameter.

7.2 A public-interest law firm picks a random sample of 60 hi-fi repair stores in a particular area, and asks each of them to repair a hi-fi set. In each case the law firm determines whether the store makes unnecessary repairs in order to inflate its bill. The law firm finds that eight of the stores are guilty of this practice. Provide a point estimate of the proportion of all such stores in the area that inflate bills in this way.

7.3 After the law firm in Exercise 7.2 has presented the results of its study, an attorney representing the hi-fi repair stores objects that a sample of this size is quite unreliable. The attorney also maintains that the sample percentage of hi-fi repair stores engaging in such shady practices is a biased estimate of the percentage of all such stores engaging in such practices. Evaluate the attorney's objections.

7.4 A statistic's mean-square error can be used as a measure of its reliability as an estimator. If X is a statistic that is used as an estimator of θ, X's mean-square error is

$$E[(X - \theta)^2].$$

If X is an unbiased estimator of θ, what is another name for its mean-square error?

7.5 It can be shown that an estimator's mean-square error equals

$$(\mu - \theta)^2 + \sigma^2,$$

where μ is its mean, θ is the parameter to be estimated, and σ is its standard deviation. Explain why $(\mu - \theta)$ is often called its bias.

7.6 Based on the formula in Exercise 7.5, explain why a biased estimator may be preferred to an unbiased one if the former has a much smaller variance than the latter.

7.7 Is the sample proportion a consistent estimator of the population proportion? Why, or why not?

7.5 Confidence Intervals for the Population Mean

Interval estimates are generally preferred over point estimates because the latter provide no information concerning how much error they are likely to contain. Interval estimates, on the other hand, do provide such information. To illustrate an interval estimate and how it is constructed, in this section we show how such an estimate is made of the population mean. We begin with the case where the population standard deviation is known and the sample size is large ($n > 30$). Then we take up the more realistic case where the standard deviation is unknown, both when the sample size is large and when it is small.

WHERE σ IS KNOWN: LARGE SAMPLE

In Chapter 5 we defined $Pr\{X > 32\}$ as the probability that X is greater than 32. Now we define $Pr\{a < X < b\}$ as the probability that X lies between a and b. Thus, the probability that the value of the sample mean lies between $\mu - 1.96\sigma/\sqrt{n}$ and $\mu + 1.96\sigma/\sqrt{n}$ is denoted by $Pr\{\mu - 1.96\sigma/\sqrt{n}$

$< \bar{X} < \mu + 1.96\sigma/\sqrt{n}\}$. To construct an interval estimate of the population mean, we begin by noting that our results from the previous chapter concerning the sampling distribution of the sample mean imply that

$$Pr\left\{\mu - 1.96\frac{\sigma}{\sqrt{n}} < \bar{X} < \mu + 1.96\frac{\sigma}{\sqrt{n}}\right\} = 0.95, \tag{7.1}$$

where μ is the population mean, σ is the population standard deviation, n is the sample size, and $\bar{X}$ is the sample mean. What equation (7.1) says is that the probability that the sample mean will lie within 1.96 standard errors of the population mean equals 0.95. (Recall that the standard error of the sample mean equals $\sigma/\sqrt{n}$.) Since we know from the previous chapter that if $n > 30$ (and if the population is large relative to the sample size), the sample mean is normally distributed with a mean of μ and a standard deviation of $\sigma/\sqrt{n}$, and since we know (from Appendix Table 2) that the probability that any normal random variable will lie within 1.96 standard deviations of its mean is 0.95, it follows that equation (7.1) is true.[2]

To construct an interval estimate for the population mean, we rearrange the terms inside the brackets on the left side of equation (7.1). If we subtract μ from $\mu - 1.96\sigma/\sqrt{n}$, $\bar{X}$, and $\mu + 1.96\sigma/\sqrt{n}$, we get

$$Pr\left\{-1.96\frac{\sigma}{\sqrt{n}} < \bar{X} - \mu < 1.96\frac{\sigma}{\sqrt{n}}\right\} = .95.$$

And if we subtract $\bar{X}$ from all three terms of the inequality in brackets and multiply each of the resulting terms by -1, we get

$$Pr\left\{\bar{X} - 1.96\frac{\sigma}{\sqrt{n}} < \mu < \bar{X} + 1.96\frac{\sigma}{\sqrt{n}}\right\} = .95. \tag{7.2}$$

(See footnote 3 if you have difficulty deriving equation (7.2).)[3]

2. How do we know (from Appendix Table 2) that the probability that any normal variable will be within 1.96 standard deviations of its mean is 0.95? The reasoning is as follows. We want to determine the probability that a normal variable lies between $\mu - 1.96\sigma$ and $\mu + 1.96\sigma$. The points on the standard normal curve corresponding to these points are $[(\mu - 1.96\sigma) - \mu] \div \sigma$ and $[(\mu + 1.96\sigma) - \mu] \div \sigma$. That is, they are -1.96 and $+1.96$. Appendix Table 2 shows that the area under the standard normal curve between zero and 1.96 equals .4750. Thus, the area between -1.96 and $+1.96$ equals $2(.4750) = .95$.

3. If we subtract $\bar{X}$ from all three terms, we get

$$Pr\left\{-\bar{X} - 1.96\frac{\sigma}{\sqrt{n}} < -\mu < -\bar{X} + 1.96\frac{\sigma}{\sqrt{n}}\right\} = .95.$$

The important fact to note at this point is that if all three terms inside the brackets are multiplied by -1 the inequalities must be reversed. (To see that this is true, consider the simple inequality $y \geqslant 1$. If this is true, it follows that $-y \leqslant -1$. As we indicated, the inequality is reversed after multiplication by -1.) Applying this fact, it follows that

$$Pr\left\{\bar{X} - 1.96\frac{\sigma}{\sqrt{n}} < \mu < \bar{X} + 1.96\frac{\sigma}{\sqrt{n}}\right\} = .95,$$

which is what we set out to derive.

Since equation (7.2) is of basic importance in the construction of an interval estimate for a population mean, it is essential that you know exactly what it means. Does it mean that the probability is .95 that the population mean lies between $\bar{X} - 1.96\sigma/\sqrt{n}$ and $\bar{X} + 1.96\sigma/\sqrt{n}$? In one sense yes, in another no. Before the sample has been drawn and while the value of the sample mean is still unknown, there is a .95 probability that the interval between $\bar{X} - 1.96\sigma/\sqrt{n}$ and $\bar{X} + 1.96\sigma/\sqrt{n}$ will include the population mean. Thus, if we were to draw one sample after another, in 95 percent of the samples (over the long run) this interval—which varies, of course, from sample to sample because $\bar{X}$ varies—would include the population mean.

To see that this is true, look at Figure 7.4, which shows what would occur if we were to construct a large number of interval estimates for a certain population mean μ. Panel A of Figure 7.4 shows the sampling distribution of the sample mean, $\bar{X}$. If the sample mean lies between $\mu - 1.96\sigma/\sqrt{n}$ and $\mu + 1.96\sigma/\sqrt{n}$, the interval estimate will contain μ. Since the area under the sampling distribution between $\mu - 1.96\sigma/\sqrt{n}$ and $\mu + 1.96\sigma/\sqrt{n}$ equals .95,

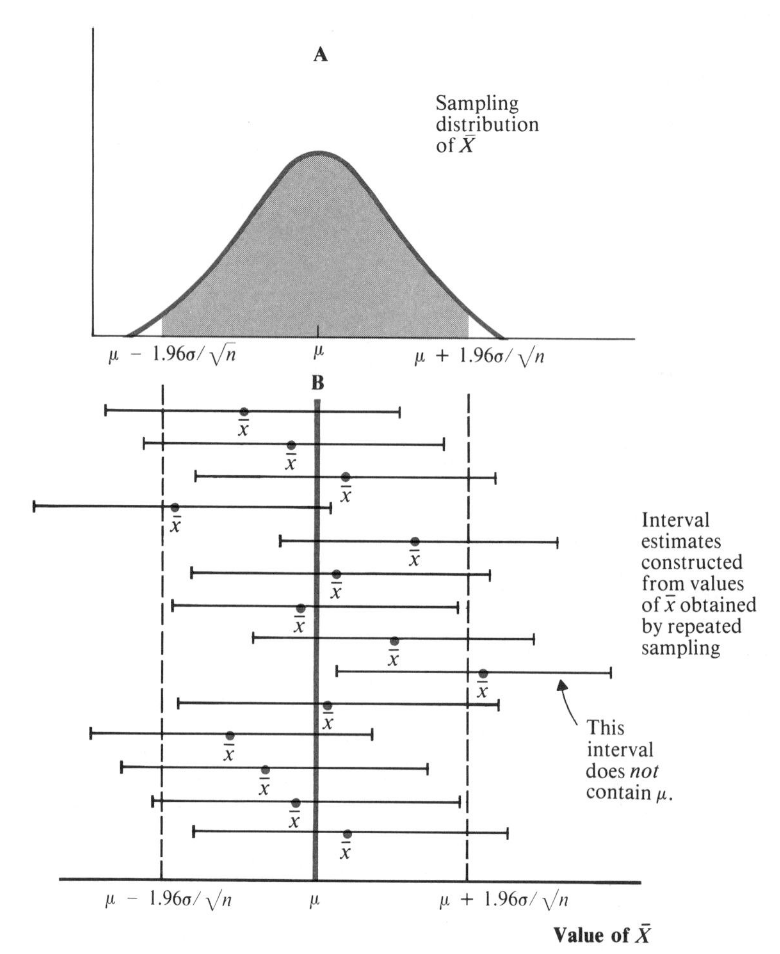

FIGURE 7.4 Display of Interval Estimates Constructed from Values of $\bar{X}$ Obtained from Repeated Sampling

the probability that $\bar{X}$ will lie in this range equals .95. Panel B of Figure 7.4 shows the interval estimates that would result from repeated sampling. Over the long run, 95 percent of these intervals would contain μ.

Note, however, that the previous discussion has pertained only to the situation before the sample has been drawn. Once a particular sample has been drawn and the sample mean has been calculated, it no longer is correct to say that the probability is 0.95 that the interval between $\bar{x} - 1.96\sigma/\sqrt{n}$ and $\bar{x} + 1.96\sigma/\sqrt{n}$ includes the population mean. To see why, suppose that $\sigma = 10$, $n = 100$, and $\bar{x} = 2.00$. Under these circumstances, equation (7.2) may be incorrectly interpreted to say that

$$Pr\left\{2.00 - 1.96\frac{10}{\sqrt{100}} < \mu < 2.00 + 1.96\frac{10}{\sqrt{100}}\right\} = .95,$$

or

$$Pr\{.04 < \mu < 3.96\} = .95.$$

But since the population mean μ is a constant, it makes no sense to say that the probability that it is between .04 and 3.96 is .95. Either its value lies in this interval or it doesn't. *All that one can say is that if intervals of this sort are calculated repeatedly, they will include the population mean in about 95 percent of the cases.*

The interval within the brackets on the left side of equation (7.2) is called a ***confidence interval.*** As you can see, this is an interval which has a certain probability of including the population mean, this probability being called the ***confidence coefficient.*** Thus, the confidence coefficient in equation (7.2) is .95. Although .95 is frequently used as a confidence coefficient, there is no reason why other confidence coefficients should not be chosen. In general, *if the confidence coefficient is set equal to $(1 - \alpha)$, the confidence interval for the population mean is*

$$\boxed{\bar{x} - z_{\alpha/2}\frac{\sigma}{\sqrt{n}} < \mu < \bar{x} + z_{\alpha/2}\frac{\sigma}{\sqrt{n}},} \tag{7.3}$$

where $z_{\alpha/2}$ is the value of the standard normal variable that is exceeded with a probability of $\alpha/2$.

To illustrate the use of expression (7.3), suppose that we want to construct a 90 percent confidence interval rather than a 95 percent confidence interval in the case presented in the paragraph before last. (Note that confidence coefficients are often expressed as percentages.) Since $1 - \alpha = .90$, it follows that $\alpha/2 = .05$. From Appendix Table 2, we find that $z_{.05} = 1.64$. Thus, inserting $z_{.05}$, $\bar{x}$, σ, and n into expression (7.3), we get

$$2.00 - 1.64\frac{10}{\sqrt{100}} < \mu < 2.00 + 1.64\frac{10}{\sqrt{100}},$$

which implies that the 90 percent confidence interval for the population mean is .36 to 3.64.

Two important points should be noted here. First, *holding constant the size of the sample, the width of the confidence interval tends to increase as the confidence coefficient increases.* For example, if the confidence coefficient is 90 percent, the width of the confidence interval is $2(1.64)\sigma/\sqrt{n}$, whereas if the confidence interval is 95 percent, the width[4] of the confidence interval is $2(1.96)\sigma/\sqrt{n}$. This makes sense. If you want to be more and more confident that the interval estimate includes the population mean, you must widen the interval if the sample size is fixed. Second, *for a fixed confidence coefficient (and a fixed population standard deviation), the only way to reduce the width of the confidence interval is to increase the sample size.* For example, if the confidence coefficient is 95 percent, the width of the confidence interval is $2(1.96)\sigma/\sqrt{n}$, as we saw above. If σ is fixed, the only way to reduce this width is to increase n.

WHERE σ IS UNKNOWN: LARGE SAMPLE

In most actual cases, the standard deviation of the population is unknown. If the sample size exceeds 30 it is a relatively simple matter to adapt the results presented previously to the situation where σ is unknown. As indicated earlier in this chapter, the sample standard deviation s is generally used as an estimator of the population standard deviation. Mathematicians have shown that if the sample is large, we can simply substitute the sample standard deviation for the population standard deviation in the results obtained in the previous part of this section. Thus, if we want to construct a 95 percent confidence interval—that is, a confidence interval with a confidence coefficient of 95 percent—we can substitute s for σ in equation (7.2), the result being

$$Pr\left\{\bar{X} - 1.96\frac{s}{\sqrt{n}} < \mu < \bar{X} + 1.96\frac{s}{\sqrt{n}}\right\} = .95.$$

Consequently, the interval estimate is from $\bar{x} - 1.96s/\sqrt{n}$ to $\bar{x} + 1.96s/\sqrt{n}$.

In general, *if the confidence coefficient is set equal to $(1 - \alpha)$, the confidence interval for the population mean is*

$$\bar{x} - z_{\alpha/2}\frac{s}{\sqrt{n}} < \mu < \bar{x} + z_{\alpha/2}\frac{s}{\sqrt{n}}, \tag{7.4}$$

where s is the sample standard deviation, and $z_{\alpha/2}$ is the value of the standard normal variable that is exceeded with a probability of $\alpha/2$. Like expression (7.3), expression (7.4) is applicable only if the population is large relative to the sample or if sampling is with replacement.[5] As we saw in the previous chapter,

4. The 95 percent confidence interval is $\bar{x} - 1.96\sigma/\sqrt{n}$ to $\bar{x} + 1.96\sigma/\sqrt{n}$; thus, the difference between its upper and lower limits (that is, its width) is $2[1.96\sigma/\sqrt{n}]$. Similarly, since the 90 percent confidence interval is $\bar{x} - 1.64\sigma/\sqrt{n}$ to $\bar{x} + 1.64\sigma/\sqrt{n}$, the difference between its upper and lower limits (that is, its width) is $2[1.64\sigma/\sqrt{n}]$.

5. If the population is not large relative to the sample and if sampling is without replacement, $\sigma/\sqrt{n}$ in (7.3) should be multiplied by the finite population correction factor $\sqrt{(N-n)/(N-1)}$.

if these two conditions are both not true the finite population correction factor must be included. Thus, *if sampling is without replacement, and if the population is not large*, the confidence interval for the population mean is

$$\bar{x} - z_{\alpha/2}\frac{s}{\sqrt{n}}\sqrt{\frac{N-n}{N-1}} < \mu < \bar{x} + z_{\alpha/2}\frac{s}{\sqrt{n}}\sqrt{\frac{N-n}{N-1}}, \tag{7.5}$$

where N is the number of items in the population.

These results concerning the confidence interval for a population mean are of enormous importance in business and economics, as well as in many other fields. The following example based on a real case[6] (although the numbers have been changed) should help illustrate how these results are used and interpreted.

EXAMPLE 7.1 A chemical firm wants to estimate the mean strength of a new synthetic fiber. To measure this fiber's strength the firm determines the number of pounds that can be supported by one strand before breaking. A random sample of 36 strands of the fiber is taken, with the following results:

Strand	Breaking load (pounds)	Strand	Breaking load (pounds)	Strand	Breaking load (pounds)
1	2.2	13	2.2	25	2.3
2	2.2	14	2.3	26	2.4
3	2.2	15	2.3	27	2.3
4	2.3	16	2.3	28	2.4
5	2.3	17	2.2	29	2.4
6	2.3	18	2.2	30	2.3
7	2.3	19	2.2	31	2.3
8	2.3	20	2.2	32	2.3
9	2.4	21	2.2	33	2.4
10	2.4	22	2.4	34	2.3
11	2.4	23	2.3	35	2.4
12	2.2	24	2.4	36	2.3

Construct a 95 percent confidence interval for the mean breaking load of a strand of this new fiber.

Solution: If x_i is the breaking load (in pounds) of the ith strand in the sample, we find that

$$\sum_{i=1}^{36} x_i = 82.8$$

$$\bar{x} = 82.8/36 = 2.3$$

$$\sum_{i=1}^{36}(x_i - \bar{x})^2 = \sum_{i=1}^{36} x_i^2 - \frac{1}{36}\left(\sum_{i=1}^{36} x_i\right)^2 = 0.20$$

$$\sum_{i=1}^{36}(x_i - \bar{x})^2/(n-1) = 0.20/35 = .00571$$

$$s = \sqrt{.00571} = .0756.$$

6. Owen Davies, *The Design and Analysis of Industrial Experiments* (London: Oliver and Boyd, 1956), p. 72. The actual analysis was more extensive and complicated than our discussion in this section, which is simplified to emphasize the basic points considered here.

Since the population is very large (because a very large number of such strands of fiber can be produced), expression (7.4) is appropriate. Because a 95 percent confidence interval is wanted, $z_{\alpha/2} = z_{.025} = 1.96$. Thus, the desired confidence interval is

$$2.30 - 1.96\left(\frac{.0756}{\sqrt{36}}\right) < \mu < 2.30 + 1.96\left(\frac{.0756}{\sqrt{36}}\right).$$

Simplifying terms, this confidence interval is 2.275 to 2.325 pounds.

As stressed before, the proper interpretation of this result is *not* that the probability is 95 percent that the population mean lies between 2.275 and 2.325 pounds. Instead, this result means that if confidence intervals of this sort were constructed in a great number of cases they would include the population mean 95 percent of the time.

WHERE σ IS UNKNOWN: SMALL SAMPLE

In many cases, a confidence interval must be constructed on the basis of a sample where $n \leqslant 30$. In such cases, the expressions given in the previous part of this section are not appropriate. No longer can we simply substitute the sample standard deviation for the population standard deviation as we did in expression (7.4). However, if the population is normal it is possible to construct a confidence interval for the population mean even if the sample size is 30 or less. Such a confidence interval is based on the t distribution described below.

> ***The t distribution.*** *If the population sampled is normally distributed,* $(\bar{X} - \mu) \div s/\sqrt{n}$ *has the* t *distribution. The* t *distribution is symmetrical, bell-shaped, and has zero as its mean.*

The t distribution is a sampling distribution: Specifically, it is the distribution of the statistic $(\bar{X} - \mu) \div s/\sqrt{n}$. Suppose that simple random samples of size n are taken repeatedly from a normal population with expected value of μ and standard deviation of σ. If we calculate the value of $(\bar{X} - \mu) \div s/\sqrt{n}$ for each sample, we can construct a sampling distribution for this statistic. If a sufficient number of samples are chosen this sampling distribution will conform to the t distribution. The t distribution is really a family of distributions, each of which corresponds to a particular *number of degrees of freedom*. In this context the number of degrees of freedom equals $(n - 1)$; in other contexts it will equal other amounts, as we shall see in later chapters.

It is not easy to give an adequate intuitive, nonmathematical interpretation of the number of degrees of freedom. From a mathematical point of view, the number of degrees of freedom is simply a parameter in the formula for the t distribution. However, one way to interpret the number of degrees of freedom (that is, $[n - 1]$) in the present context is to say that it equals the number of independent deviations from the sample mean (that is, $[x_i - \bar{x}]$) in the computation of the sample standard deviation s. Since the sum of these deviations—that is, $\Sigma(x_i - \bar{x})$—equals zero, it follows that if we know $(n - 1)$ of these deviations we can determine the value of the remaining deviation by

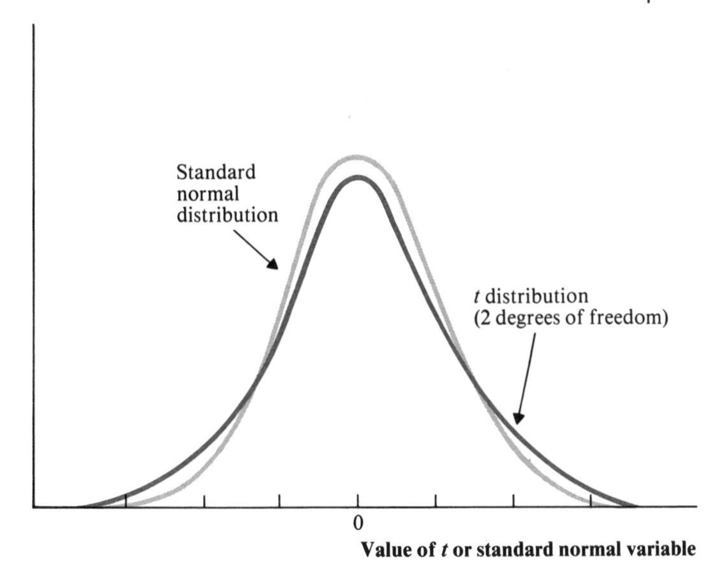

FIGURE 7.5 The t Distribution (with 2 Degrees of Freedom) Compared with the Standard Normal Distribution

using the fact that their sum equals zero. Thus, only $(n - 1)$ of the deviations are independent. In other words, as the statistician might put it, there are $(n - 1)$ degrees of freedom.

The shape of the t distribution is rather like that of the standard normal distribution. Figure 7.5 compares the t distribution (with 2 degrees of freedom) to the standard normal distribution. As you can see, both are symmetrical, bell-shaped, and have a mean of zero. The t distribution is somewhat flatter at the mean and somewhat higher in the tails than the standard normal distribution. As the number of degrees of freedom becomes larger and larger, the t distribution tends to become exactly the same as the standard normal distribution. The t distribution is often called Student's t distribution because the statistician W. S. Gosset, who first derived this distribution, published his findings under the pseudonym Student.[7]

To find the probability that the value of t exceeds a certain number, we can use Appendix Table 6. As you can see, each row of this table corresponds to a particular number of degrees of freedom. The numbers in each row are the numbers that are exceeded with the indicated probability by a t variable. For example, the first row indicates that if a t variable has one degree of freedom, there is a .40 probability that its value will exceed .325, a .25 probability that its value will exceed 1.000, a .05 probability that its value will exceed 6.314, a .01 probability that its value will exceed 31.821, and so on. Since the t distribution is symmetrical, it follows that if a t variable has one degree of freedom there is a .40 probability that its value will lie below −.325, a .25 probability that its value will lie below −1.000, and so on.

7. Gosset used this pseudonym because his employer, Guinness Brewery, forbade publication of scientific research of this sort by employees under their own names.

If the sample size is 30 or less and the population is normal (and large relative to the sample), a confidence interval for the population mean can be constructed by using the t distribution in place of the standard normal distribution in expression (7.4). In other words, *if the confidence coefficient is set equal to* $(1 - \alpha)$, *the confidence interval for the population mean is:*

$$\bar{x} - t_{\alpha/2}\frac{s}{\sqrt{n}} < \mu < \bar{x} + t_{\alpha/2}\frac{s}{\sqrt{n}}, \tag{7.6}$$

where $t_{\alpha/2}$ *is the value of a* t *variable (with* $n - 1$ *degrees of freedom) that is exceeded with a probability of* $\alpha/2$. Thus, if a sample of 16 observations is chosen, and if the sample mean is 20 and the sample standard deviation is 4, the 95 percent confidence interval for the population mean is

$$20 - 2.131\left(\frac{4}{\sqrt{16}}\right) < \mu < 20 + 2.131\left(\frac{4}{\sqrt{16}}\right),$$

since Appendix Table 6 shows that for 15 degrees of freedom the value of t that will be exceeded with a probability of .025 is 2.131. Simplifying terms, it follows that the confidence interval in this case is 17.869 to 22.131. Statisticians frequently construct confidence intervals in this way. The following example illustrates how it is done.

Example 7.2 A manufacturer of light bulbs wants to estimate the mean length of life of a new type of bulb which is designed to be extremely durable. The firm's engineers test nine of these bulbs and find that the length of life (in hours) of each is as follows:

5000	5100	5400
5200	5400	5000
5300	5200	5200

Previous experience indicates that the lengths of life of individual bulbs of a particular type are normally distributed. Construct a 90 percent confidence interval for the mean length of life of all bulbs of this new type.

Solution: If x_i is the length of life of the ith light bulb in the sample, we find that

$$\sum_{i=1}^{9} x_i = 46{,}800$$

$$\bar{x} = 5{,}200$$

$$\sum_{i=1}^{9} (x_i - \bar{x})^2 = 180{,}000$$

$$\sum_{i=1}^{9} (x_i - \bar{x})^2/(n - 1) = 22{,}500$$

$$s = \sqrt{22{,}500} = 150.$$

Since $n = 9$, expression (7.6) is appropriate. Because a 90 percent confidence interval is wanted, $t_{\alpha/2} = t_{.05}$; and the number of degrees of freedom $(n - 1)$ is 8. Appendix Table 6 shows that if there are 8 degrees of freedom, $t_{.05} = 1.86$. Thus, the desired confidence interval is

$$5200 - 1.86\left(\frac{150}{\sqrt{9}}\right) < \mu < 5200 + 1.86\left(\frac{150}{\sqrt{9}}\right)$$

Simplifying terms, the confidence interval is 5107 to 5293 hours.

EXERCISES

7.8 What is the probability that the value of a random variable with the t distribution with 4 degrees of freedom will lie
(a) above 3.747;
(b) below -4.604;
(c) between 2.132 and 2.776?

7.9 The Bona Fide Washing Machine Company knows that the standard deviation of the lengths of life of motors received from supplier I is 400 hours. Calculate the 95 percent confidence interval for the mean length of life of the motors received from this supplier, based on a sample of 40 motors where the mean length of life is 4,500 hours.

7.10 The Bona Fide Washing Machine Company's statistician says that the 90 percent confidence interval for the mean length of life of motors received from supplier II is 4,500 to 4,800 hours, based on a sample of 36 motors. The statistician also says that the standard deviation of the lengths of life of motors received from supplier II is 500 hours. Is there any contradiction between these statements? If so, what is the contradiction?

7.11 The statistician in Exercise 7.10 says that if the standard deviation of lengths of life of motors received from supplier II is known, the width of the 95 percent confidence interval for the mean is always about 20 percent greater than the width of the 90 percent confidence interval (if the sample size is held constant). Is this true? Why, or why not?

7.12 Bona Fide does not know the standard deviation of the lengths of life of motors received from supplier III because it has had very little experience with this supplier. Bona Fide therefore chooses a random sample of 36 motors received from supplier III and obtains the following (in thousands of hours):

4.2	5.0	4.6	4.9	5.0	5.1
4.3	4.9	4.5	4.8	4.9	4.6
4.4	5.1	4.7	4.4	4.8	4.6
4.8	4.7	4.4	4.5	4.8	4.8
4.9	4.8	4.3	4.6	4.7	4.5
5.1	4.8	4.6	4.6	4.7	5.0

(a) Compute a 90 percent confidence interval for the mean length of life of motors received from supplier III.
(b) Compute a 95 percent confidence interval for the mean length of life of motors received from supplier III.

7.13 Since Bona Fide also does not know the standard deviation of the lengths of life of motors received from supplier IV, it chooses a random sample of 9 motors from supplier IV and determines the life of each. The results (in thousands of hours) are as follows:

4.3	4.6	3.8
4.2	4.3	3.9
4.1	3.9	4.0

(a) Compute a 90 percent confidence interval for the mean length of life of motors received from supplier IV.
(b) Compute a 95 percent confidence interval for the mean length of life of motors received from supplier IV.
(c) What major assumption underlies your calculations in (a) and (b)?

7.14 A firm with 50 overseas plants chooses (without replacement) a random sample of 40 plants. For each plant in this sample the firm determines the number of days the plant was shut down in 1979 by labor disputes. The sample mean turns out to be 9.8 days. If the standard deviation of the number of days the firm's overseas plants were shut down by labor disputes in 1979 was 2, calculate a 90 percent confidence interval for the mean number of days that all the firm's overseas plants were shut down for this reason in 1979.

7.15 In Exercise 7.14, if the firm had 100 rather than 50 overseas plants, calculate a 90 percent confidence interval for the mean number of days that all the firm's overseas plants were shut by labor disputes in 1979. Explain the difference between your answer here and in Exercise 7.14.

7.16 Compare the t distribution with an infinite number of degrees of freedom in Appendix Table 6 to the standard normal distribution in Appendix Table 2. In particular, show that the probability is the same that each will exceed (a) 1.645; (b) 1.960; (c) 0.674. Do you find this surprising? Why, or why not?

7.17 For sufficiently large samples it can be shown that the sample standard deviation s is approximately normally distributed, with a mean equal to the population standard deviation σ, and with a standard deviation equal to $\sigma \div \sqrt{2n}$. Use these results to show that if n is sufficiently large *a confidence interval for* σ *is*

$$\frac{s}{1 + \frac{z_{\alpha/2}}{\sqrt{2n}}} < \sigma < \frac{s}{1 - \frac{z_{\alpha/2}}{\sqrt{2n}}}$$

where the confidence coefficient equals $(1 - \alpha)$. (In Chapter 9 we take up the χ^2 distribution, which can be used to construct a confidence interval for σ when n is small.)

7.6 Confidence Intervals for the Population Proportion

Statisticians find it important to estimate the population proportion as well as the population mean. We have seen that the purpose of a statistical investigation may be to estimate the proportion of people in a particular city who prefer brand C breakfast food to brand D. Or a political pollster may want to estimate the percentage of voters who say they will vote for a particular candidate in the next election. In this section we describe how confidence intervals are constructed for population proportions. We will begin with the case where the normal distribution can be used and then take up the case where special graphs are required.

USE OF THE NORMAL DISTRIBUTION

The sample proportion equals X/n, where X is the number of successes in the sample and n is the sample size. A success occurs when a person prefers brand C breakfast food to brand D, or when a voter says he or she will vote for the candidate in the next election. As we know from Chapter 4, the expected value of X, the number of successes, is nP, where P is the population proportion. Since the sample proportion is X divided by n, its expected value must equal the expected value of X divided by n. (For a proof, see Appendix 4.1.)[8] Consequently,

$$E\left(\frac{X}{n}\right) = P.$$

In other words, the expected value of the sample proportion equals the population proportion. For example, if 70 percent of the population prefers brand C breakfast food to brand D, the expected value of the sample proportion equals 0.70.

The standard deviation of the sampling distribution of the sample proportion equals

$$\sigma\left(\frac{X}{n}\right) = \sqrt{\frac{P(1 - P)}{n}}.$$

To see why, recall from Chapter 4 that the standard deviation of X equals $\sqrt{nP(1 - P)}$. The standard deviation of X/n equals the standard deviation of X divided by n—that is, $\sqrt{nP(1 - P)}$ divided by n or $\sqrt{P(1 - P)/n}$. (This follows from Appendix 4.1.) For example, if a random sample of 10 people is taken (with replacement) and if $P = .70$, the standard deviation of the sample proportion preferring brand C breakfast food to brand D equals $\sqrt{.7(.3)/10} = .145$.

Thus, the sampling distribution of the sample proportion has a mean equal to P and a standard deviation equal to $\sqrt{P(1 - P)/n}$. We know from previous chapters that if the sample size is sufficiently large (and if P is not very close to zero or 1) the sampling distribution can be approximated by the normal distribution. Thus, under these conditions, the sample proportion is approximately normally distributed with a mean of P and a standard deviation of $\sqrt{P(1 - P)/n}$, which means that

$$Pr\left\{P - z_{\alpha/2}\sqrt{\frac{P(1 - P)}{n}} < \frac{X}{n} < P + z_{\alpha/2}\sqrt{\frac{P(1 - P)}{n}}\right\} = 1 - \alpha,$$

where $z_{\alpha/2}$ is the value of the standard normal variable that is exceeded with

8. In Appendix 4.1 we showed that if a random variable is a linear function of another random variable, its expected value is a linear function of the expected value of the other random variable. The sample proportion is a linear function of X; specifically, it equals $(1/n)X$. Thus, its expected value equals $1/nE(X) = 1/n(nP) = P$.

the probability of $\alpha/2$. If we rearrange terms within the brackets on the left side of this equation, it follows that[9]

$$Pr\left\{\frac{X}{n} - z_{\alpha/2}\sqrt{\frac{P(1-P)}{n}} < P < \frac{X}{n} + z_{\alpha/2}\sqrt{\frac{P(1-P)}{n}}\right\} = 1 - \alpha.$$

As it stands, this equation cannot be used to construct a confidence interval for P because without a knowledge of P we cannot compute $\sqrt{P(1-P)/n}$. However, if the sample is sufficiently large it is permissible to substitute the sample proportion for P in this expression, the result being

$$Pr\left\{\frac{X}{n} - z_{\alpha/2}\sqrt{\frac{\left(\frac{X}{n}\right)\left[1-\frac{X}{n}\right]}{n}} < P < \frac{X}{n} + z_{\alpha/2}\sqrt{\frac{\left(\frac{X}{n}\right)\left[1-\left(\frac{X}{n}\right)\right]}{n}}\right\}$$

$$= 1 - \alpha. \tag{7.7}$$

Based on equation (7.7), it is clear that *if the sample is sufficiently large and if the confidence coefficient is set equal to* $(1 - \alpha)$ *the confidence interval for the population proportion is*

$$\boxed{p - z_{\alpha/2}\sqrt{\frac{p(1-p)}{n}} < P < p + z_{\alpha/2}\sqrt{\frac{p(1-p)}{n}},} \tag{7.8}$$

where p *is* x/n, *the sample proportion.* Of course, this expression assumes that the population is large relative to the sample or that sampling is carried out with replacement; otherwise $\sqrt{p(1-p)/n}$ must be multiplied by $\sqrt{(N-n)/(N-1)}$. This result is of widespread usefulness in business and economic statistics. The following example illustrates how it is applied.

Example 7.3 A polling organization selects a random sample of 400 residents of Dallas, Texas and asks each person whether he or she would prefer gasoline rationing to higher gasoline prices as a means of cutting gasoline consumption. Sixty percent of the people in the sample prefer gasoline

9. The reasoning here is like that leading up to equation (7.2). Specifically, if we subtract P from all three terms inside the brackets,

$$Pr\left\{-z_{\alpha/2}\sqrt{\frac{P(1-P)}{n}} < \frac{X}{n} - P < z_{\alpha/2}\sqrt{\frac{P(1-P)}{n}}\right\} = 1 - \alpha.$$

Subtracting X/n from all three terms inside the brackets, we get

$$Pr\left\{-\frac{X}{n} - z_{\alpha/2}\sqrt{\frac{P(1-P)}{n}} < -P < -\frac{X}{n} + z_{\alpha/2}\sqrt{\frac{P(1-P)}{n}}\right\} = 1 - \alpha.$$

Multiplying all three terms inside the brackets by -1, and reversing the inequalities, we get

$$Pr\left\{\frac{X}{n} - z_{\alpha/2}\sqrt{\frac{P(1-P)}{n}} < P < \frac{X}{n} + z_{\alpha/2}\sqrt{\frac{P(1-P)}{n}}\right\} = 1 - \alpha,$$

which is what we set out to derive.

rationing. Calculate a 95 percent confidence interval for the percentage of Dallas residents in favor of rationing.

Solution: Since $(1 - \alpha) = .95$, $z_{\alpha/2} = z_{.025} = 1.96$. Inserting .60 for p and 400 for n in expression (7.8), we have

$$.60 - 1.96\sqrt{\frac{.60(.40)}{400}} < P < .60 + 1.96\sqrt{\frac{.60(.40)}{400}}.$$

Simplifying terms, this confidence interval is

$$.60 - .048 < P < .60 + .048,$$

which means that our interval estimate of the percentage of Dallas residents favoring rationing is 55.2 percent to 64.8 percent.

USE OF SPECIAL GRAPHS

Expression (7.8) provides a confidence interval for the population proportion when the sample size is sufficiently large. But how large is "sufficiently large"? According to William Cochran,[10] the minimum sample size needed to insure the validity of this expression is about 30 if the population proportion is about .50; about 50 if the population proportion is about .40 or .60; about 80 if it is about .30 or .70; about 200 if it is about .20 or .80; and about 600 if it is about .10 or .90. If the sample size does not meet these standards one must then use special graphs that have been computed for this purpose. Appendix Tables 7a and 7b respectively contain 95 percent confidence intervals and 99 percent confidence intervals for the population proportion. These graphs are used in the following way.

If the sample proportion is less than 50 percent, one finds the point on the bottom horizontal axis that equals the sample proportion. For example, if the sample proportion is 45 percent, the appropriate point on the horizontal axis is .45. Then, using the two curves that pertain to the relevant sample size, one finds the two points on the vertical axis that correspond to this point on the horizontal axis. For example, if $n = 100$, the two curves are as shown in Figure 7.6 for a 95 percent confidence interval. If the sample proportion is 0.45, the two points on the vertical axis corresponding to 0.45 on the horizontal axis are .35 and .55, as shown in Figure 7.6. Thus, the 95 percent confidence interval is 35 percent to 55 percent.

If the sample proportion is greater than 50 percent, one uses the top (rather than the bottom) horizontal scale and the right-hand (rather than the left-hand) vertical scale in Appendix Tables 7a and 7b. For example, suppose that the sample proportion is 0.65 and $n = 100$. As shown in Figure 7.7, we begin by finding .65 on the *top* horizontal scale, after which we find the two points on the *right-hand* vertical scale corresponding to this location on the

10. See W. G. Cochran, *Sampling Techniques* (New York: Wiley, 1953), p. 41. As Cochran points out, smaller sample sizes than those given above may be acceptable if the statistician is willing to accept somewhat larger risks of error than is Cochran. As pointed out in Chapter 5, many statisticians regard the normal approximation as acceptable so long as $nP > 5$ when $P \leqslant 1/2$ and $n(1 - P) > 5$ when $P > 1/2$.

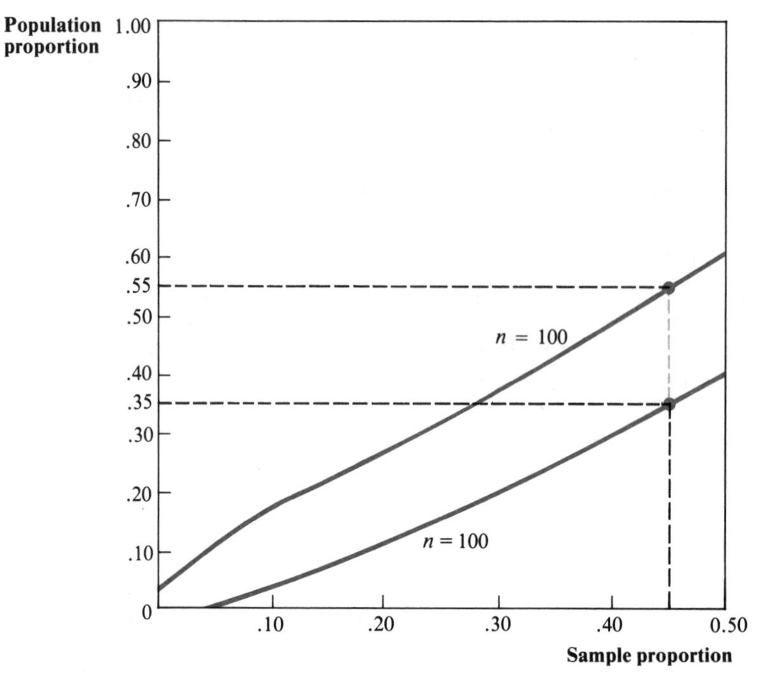

FIGURE 7.6 Derivation of a Confidence Interval for the Population Proportion, when the Sample Proportion Is 0.45

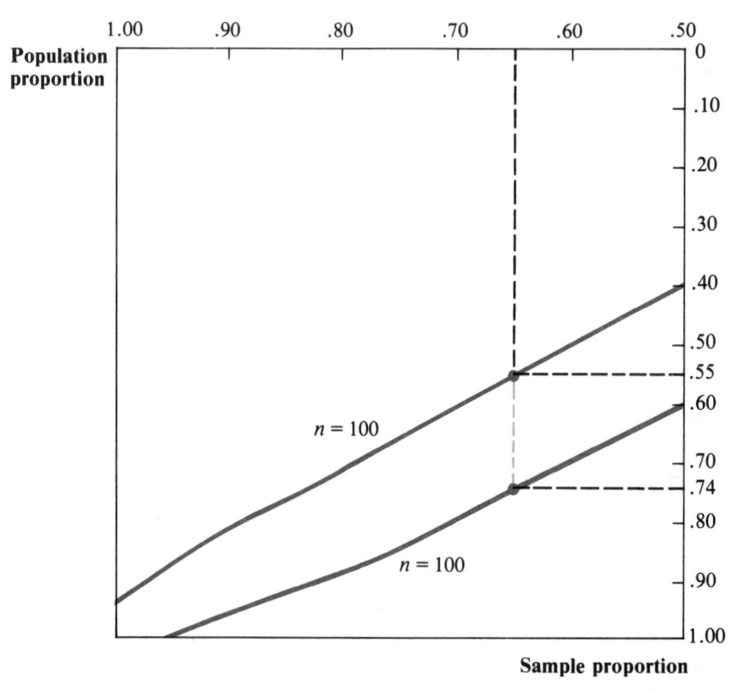

FIGURE 7.7 Derivation of a Confidence Interval for the Population Proportion, when the Sample Proportion Is 0.65

two curves. Since these two points are 55 and 74 percent, the 95 percent confidence interval is 55 percent to 74 percent.

These graphs are of considerable use to statisticians. The following is a further illustration of how they are employed.

Example 7.4 An advertising agency shows two television commercials to a sample of 60 people. Seventy percent of these individuals prefer the first commercial over the second. Obtain a 95 percent confidence interval for the proportion of all members of the relevant population who feel this way.

Solution: Since the sample proportion exceeds 50 percent, we must use the top horizontal and right-hand vertical scales in Appendix Table 7a. First, we find .70 on the top horizontal scale. Then, we must find the points on the two curves (for $n = 60$) that correspond to this horizontal location. Since these two points correspond to 57 percent and 81 percent on the right-hand vertical axis, the desired confidence interval is 57 percent to 81 percent.

7.7 Confidence Interval for the Difference between Two Means or Two Proportions[11]

Frequently, the purpose of a statistical investigation is to estimate the difference between two population means or two population proportions. For example, suppose that a firm buys thread from two suppliers and wants to estimate the difference between the mean strength of the thread of one supplier and that of the other. In this section we will show how a confidence interval can be constructed for the difference between two population means. We will also show how a confidence interval can be constructed for the difference between two population proportions.

DIFFERENCE BETWEEN TWO MEANS: INDEPENDENT SAMPLES

Let's take two populations, one with a mean of μ_1 and a standard deviation of σ_1, the other with a mean of μ_2 and a standard deviation of σ_2. A simple random sample of n_1 observations is chosen from the first population and a simple random sample of n_2 observations is chosen from the second. These two random samples are entirely *independent*; in particular, the observations in one sample are not paired in any way with those in the other sample.[12] *If both samples are large, and if the confidence coefficient is set equal to* $(1 - \alpha)$, *the confidence interval for the difference between the population means is*

$$\bar{x}_1 - \bar{x}_2 - z_{\alpha/2}\sqrt{\frac{s_1^2}{n_1} + \frac{s_2^2}{n_2}} < \mu_1 - \mu_2 < \bar{x}_1 - \bar{x}_2 + z_{\alpha/2}\sqrt{\frac{s_1^2}{n_1} + \frac{s_2^2}{n_2}}, \quad (7.9)$$

11. Some instructors may want to skip this section. If so, this can be done without loss of continuity. Note that the methods discussed in this section are by no means the only ones that can be applied to obtain a confidence interval for the difference between two means or two proportions.
12. If they are paired, the results of this paragraph are not applicable. For a description of how a confidence interval can be obtained from paired comparisons, see Section 7.9. Also, it is assumed here and in expression (7.10) that the population is large relative to the sample or that sampling is with replacement.

where s_1^2 is the variance of the sample taken from the first population and s_2^2 is the variance of the sample taken from the second population.

The following example shows how this result can be used under practical circumstances.

Example 7.5 A pea-canning company has two plants and wants to estimate the difference between the mean drained weight of the contents of the cans filled at the two plants. It is suspected that the mean drained weight is higher at plant 1 than at plant 2 because of the gradual deterioration of some of the equipment at plant 1. A random sample of 100 cans is drawn at each plant, and the sample mean is found to be 23.02 ounces at plant 1 and 22.83 ounces at plant 2. The sample variance is .64 (ounces)^2 at plant 1 and .36 (ounces)^2 at plant 2. Construct a 90 percent confidence interval for the difference between the mean drained weight of cans filled at plant 1 and the mean drained weight of cans filled at plant 2.

Solution: Since $(1 - \alpha) = .90$, $z_{\alpha/2} = z_{.05} = 1.64$. Inserting 23.02 for $\bar{x}_1$, 22.83 for $\bar{x}_2$, .64 for s_1^2, .36 for s_2^2, and 100 for n_1 and n_2, expression (7.9) becomes the following:

$$.19 - 1.64(.1) < \mu_1 - \mu_2 < .19 + 1.64(.1),$$

or

$$.026 < \mu_1 - \mu_2 < .354.$$

Thus, the desired confidence interval for the difference between the mean drained weight of cans filled at plant 1 and the mean drained weight of cans filled at plant 2 is .026 to .354 ounces.

DIFFERENCE BETWEEN TWO PROPORTIONS: INDEPENDENT SAMPLES

Suppose that there are two populations, one where the proportion having a certain characteristic is P_1, the other where the proportion with this characteristic is P_2. A simple random sample of n_1 observations is chosen from the first population and a simple random sample of n_2 observations is chosen from the second. These two random samples are entirely *independent*; in particular, the observations in one sample are not paired in any way with those in the other. *If both samples are sufficiently large, and if the confidence coefficient is set at $(1 - \alpha)$, the confidence interval for the difference between the population proportions is*

$$p_1 - p_2 - z_{\alpha/2}s_{p_1-p_2} < P_1 - P_2 < p_1 - p_2 + z_{\alpha/2}s_{p_1-p_2}, \tag{7.10}$$

where p_1 is the sample proportion in the first population, p_2 is the sample proportion in the second population, and $s_{p_1-p_2} = \sqrt{\dfrac{p_1(1-p_1)}{n_1} + \dfrac{p_2(1-p_2)}{n_2}}$.

The following example shows how this result can be used.

Example 7.6 A polling organization wants to estimate the difference between the proportion of Republicans favoring the abolition of double taxation of dividends and the proportion of Democrats favoring this measure. A simple random sample of 400 Republicans is drawn, and it is found that 80 percent favor the measure. A random sample of 400 Democrats is also drawn, 40 percent of whom favor this measure. Construct a 95 percent confidence interval for the difference between the proportion of Republicans and the proportion of Democrats favoring the tax-abolition proposal.

Solution: Since $(1 - \alpha) = .95$, $z_{\alpha/2} = z_{.025} = 1.96$. If we substitute .80 for p_1, .40 for p_2, and 400 for n_1 and n_2 in expression (7.10), we get

$$.40 - \frac{1.96}{20}\sqrt{.40} < P_1 - P_2 < .40 + \frac{1.96}{20}\sqrt{.40},$$

or

$$.338 < P_1 - P_2 < .462.$$

Thus, the desired confidence interval for the difference between the percentage of Republicans and percentage of Democrats favoring the tax-abolition measure is 33.8 to 46.2 percentage points.

EXERCISES

7.18 A certain firm has a large inventory of spare parts. In order to estimate the proportion of these parts that have deteriorated to the point of being no longer usable, the firm draws a random sample of 24 of the parts and finds that 25 percent are no longer usable. Construct a 95 percent confidence interval for the nonusable proportion of the entire inventory.

7.19 A market-research firm plans to estimate the difference between the mean rating of two television programs. It asks 100 viewers to rate the first program on a scale of zero to 10; the sample mean is 5.3 and the sample variance is 1.6. Another sample of 400 viewers are asked to rate the second program on the same scale; the sample mean is 5.8 and the sample variance is 1.8. Calculate a 90 percent confidence interval for the difference between the mean ratings of the two programs in the population as a whole.

7.20 There are two methods of re-roofing a house, method A and method B. A research organization chooses a random sample of 200 houses re-roofed according to method A and finds that 18 percent experienced leaks within four years. Another random sample of 200 houses re-roofed according to method B is chosen. It is found that 29 percent of these houses experienced leaks within four years. Compute a 95 percent confidence interval for the difference between method B and method A in the percentage of houses experiencing leaks within four years.

7.21 A public-interest law firm picks a random sample of 100 hi-fi repair stores and asks them to fix a hi-fi set. In 34 of the cases the customer is undercharged. (That is, the bill is lower than the law firm considers appropriate.) Use the normal approximation to obtain a 90 percent confidence interval for the proportion of cases where such undercharging occurs.

7.22 If a sample is large, the sample standard deviation is approximately normally distributed, with a mean equal to the population standard deviation σ and with a standard deviation equal to $\sigma \div \sqrt{2n}$. In the circumstances described in Exercise 7.21, the sample standard deviation of the amounts charged by the 100 hi-fi repair stores to fix the hi-fi set was $8.70. Provide a 90 percent confidence interval for the population standard deviation. (See Exercise 7.17.)

7.23 An airline wants to determine the proportion of passengers on its New York-Chicago flights who carry only hand luggage. The airline picks a random sample of 40 passengers traveling on these flights and finds that 14 percent carry only hand luggage. Calculate a 95 percent confidence interval for the proportion of hand-luggage passengers on these flights.

7.24 A bank wants to determine the proportion of its depositors who also have deposits in any of the local savings and loan associations. A random sample is constructed of 100 depositors, and it is determined that 46 percent have deposits in a local savings and loan association.
(a) Use the normal approximation to determine a 95 percent confidence interval for this proportion.
(b) Use Appendix Table 7a to determine this confidence interval, and compare the result with your finding in (a).

7.25 The bank in Exercise 7.24 finds that the sample standard deviation of the size of its depositors' deposits in local savings and loan associations is $8,200 (in the sample described in Exercise 7.24). Calculate a 95 percent confidence interval for the population standard deviation. (Use the information provided in Exercise 7.22, and see Exercise 7.17.)

7.8 Determining the Size of the Sample

Earlier in this chapter we learned that in setting out to estimate a particular parameter a statistician must choose the kind of estimator to be used and the sample size. We have already discussed the various kinds of estimators, and now we take up the determination of the size of the sample. First, we will indicate how this decision can be made in estimating a population mean; then we will indicate how to pick a sample size when estimating a population proportion.

WHERE THE POPULATION MEAN IS ESTIMATED

In many cases, the decision maker authorizing the experiment or survey wants the resulting estimate to have a specified degree of precision.[13] For example, consider the manufacturer of light bulbs (Example 7.2) which needed to estimate the mean longevity of a new type of bulb. This firm might want the probability to be .90 that the sample mean will differ from the population mean by no more than 30 hours. Given this specified degree of precision, we can determine how large the sample must be if we know the standard deviation of the life of bulbs of the new type.

To see how the sample size can be determined, recall once again that for

13. Ideally, the degree of precision specified should reflect the basic factors described in Section 7.2: the costliness of errors in the estimate and the costliness of sampling. However, it frequently is difficult to quantify the cost of errors of a particular size; and in practice decision makers often specify degrees of precision based on intuitive and informal judgments concerning these more basic factors.

reasonably large samples the sample mean $\bar{X}$ is distributed normally with mean equal to the population mean μ and standard deviation equal to $\sigma/\sqrt{n}$ (where σ is the population standard deviation and n is the sample size). Thus,

$$Pr\left\{\mu - 1.64\frac{\sigma}{\sqrt{n}} < \bar{X} < \mu + 1.64\frac{\sigma}{\sqrt{n}}\right\} = .90, \tag{7.11}$$

which says that the probability that the sample mean will lie within 1.64 standard deviations of the population mean equals .90. As we know from previous discussions, this is true of any normal variable. Another way of stating the same thing is

$$Pr\left\{-1.64\frac{\sigma}{\sqrt{n}} < \bar{X} - \mu < 1.64\frac{\sigma}{\sqrt{n}}\right\} = .90.$$

In other words, the probability that the sample mean will differ from the population mean by less than $1.64\sigma/\sqrt{n}$ is .90.

If the desired precision is to be obtained, the probability that the sample mean will differ from the population mean by less than 30 hours must equal .90. This means that

$$1.64\sigma/\sqrt{n} = 30,$$

since we know from the previous paragraph that the probability that the sample mean will differ from the population mean by less than $1.64\sigma/\sqrt{n}$ equals .90. Suppose (contrary to Example 7.2) that the firm knows that σ equals 160 hours. Then

$$\frac{1.64(160)}{\sqrt{n}} = 30$$

or

$$n = \left[\frac{1.64(160)}{30}\right]^2,$$

which means that n must equal 77.

In general, *if it is desired that the probability be* $(1 - \alpha)$ *that the sample mean differ from the population mean by no more than some number* δ *(delta), the sample size must equal*

$$\boxed{n = \left(\frac{z_{\alpha/2}\sigma}{\delta}\right)^2,} \tag{7.12}$$

where σ *is the population standard deviation and* $z_{\alpha/2}$ *is the value of the standard normal variable which has a probability* $\alpha/2$ *of being exceeded.*[14]

Equation (7.12) is of great practical importance. Even if you do not know the value of σ, rough estimates of its value can be inserted into (7.12) to get

14. This assumes that the sample is large and that the population is large relative to the sample (or that sampling is with replacement).

some idea of how large the sample must be. The following example shows how one can determine the sample size using this formula.

Example 7.7 A bank wants to estimate the mean balance in the checking accounts of its depositors 65 years old or over. There are a very large number of such accounts, and the bank manager believes that the standard deviation of the balances held by such individuals is about \$160. If the bank wants the probability to be .95 that the sample mean will differ from the population mean by no more than \$20, how big a sample must be taken?

Solution: Since $(1 - \alpha) = .95$, $z_{\alpha/2} = z_{.025} = 1.96$. Substituting 160 for σ and 20 for δ in (7.12), we have

$$n = \left[\frac{(1.96)(160)}{20}\right]^2 = 246.$$

Thus, the sample size should be 246.

WHERE THE POPULATION PROPORTION IS ESTIMATED

In investigations aimed at estimating a population proportion, a degree of precision is generally specified. For example, consider the polling organization (Example 7.3) that needs to estimate the proportion of Dallas residents who prefer gasoline rationing over higher gasoline prices as a means of reducing gasoline consumption. The sponsors of this survey may decide that for the results to be useful to them the probability must be .90 that the sample percentage differs from the population percentage by no more than five percentage points. To see how this statement of desired precision can be used to determine the sample size, recall that for sufficiently large samples the sample proportion p is approximately normally distributed with mean equal to the population proportion P and standard deviation equal to $\sqrt{P(1-P)/n}$. Thus,

$$Pr\left\{-1.64\sqrt{\frac{P(1-P)}{n}} < p - P < 1.64\sqrt{\frac{P(1-P)}{n}}\right\} = .90, \tag{7.13}$$

which says that the probability that the sample proportion will differ from the population proportion by no more than 1.64 standard deviations equals .90. This, of course, is true of all normal variables.

If the desired precision is to be obtained, the probability that the sample proportion will differ from the population proportion by .05 (that is, five percentage points) must be .90. This means that

$$1.64\sqrt{\frac{P(1-P)}{n}} = .05,$$

since we know from the previous paragraph that the probability that the sample proportion will differ from the population proportion by less than $1.64\sqrt{P(1-P)/n}$ equals .90. Although the polling organization does not know the value of P, it is likely to have some idea of its approximate value. For example, suppose that P is believed to be in the neighborhood of 0.5.

Then

$$1.64 \sqrt{\frac{(.5)(.5)}{n}} = .05,$$

or

$$n = \frac{(1.64)^2}{(.05)^2}(.5)(.5),$$

which means that n must equal about 269.

In general, *if it is desired that the probability be $(1 - \alpha)$ that the sample proportion differs from the population proportion by no more than some number δ, and if the population proportion is believed to be approximately equal to $\hat{P}$, the sample size must equal*

$$n = \left(\frac{z_{\alpha/2}}{\delta}\right)^2 \hat{P}(1 - \hat{P}). \tag{7.14}$$

This result assumes that the sample is large enough for the normal approximation to be used and that the population is large relative to the sample (or that sampling is with replacement). It is worth noting that if one wants a conservative estimate of n (that is, an estimate that tends to err on the high side), it is best to shade one's estimate of P in the direction of 0.5. Why? Because the right side of equation (7.14) gets larger as $\hat{P}$ approaches 0.5. Thus, if $\hat{P}$ is shaded toward 0.5, the estimate of n will tend to err on the high side.

Equation (7.14), like equation (7.12), is of great practical importance. The following example illustrates how this formula is used.

Example 7.8 A government agency plans to estimate the percentage of welfare recipients in a particular area who are over 60 years of age. A reasonable estimate is that this percentage is about 30. The agency wants the probability to be .99 that the sample percentage differs from the population percentage by less than 5 percentage points. How large a sample should the agency take?

Solution: Since $(1 - \alpha) = .99$, $z_{\alpha/2} = z_{.005} = 2.58$. Substituting .30 for P and .05 for δ in equation (7.14), we have

$$\begin{aligned} n &= \left(\frac{2.58}{.05}\right)^2 \cdot (.30)(.70) \\ &= 2662.56(.21) \\ &= 559. \end{aligned}$$

Thus, the sample size should be about 559.

7.9 Statistical Estimation in the Chemical Industry: A Case Study

In Chapter 1 we pointed out that the huge British chemical firm Imperial Chemical Industries (ICI) carried out the following experiment to estimate the effect of a chlorinating agent on the abrasion resistance of a certain type of rubber.[15] Ten pieces of this type of rubber were cut in half, and one half-piece was treated with the chlorinating agent, while the other half-piece was untreated. Then the abrasion resistance of each half was evaluated on a machine, and the difference between the abrasion resistance of the treated half-piece and the untreated half-piece was computed. Table 7.1 shows the 10 differences (one corresponding to each of the pieces of rubber in the sample). Based on this experiment, ICI was interested in estimating the mean difference between the abrasion resistance of a treated and untreated half-piece of this type of rubber. In other words, if this experiment were performed again and again, an infinite population of such differences would result. ICI was interested in estimating the mean of this population, since the mean is a good measure of the effect of the chlorinating agent on this type of rubber's abrasion resistance.

If you were a statistical consultant for ICI, how would you analyze these data? Recalling the material in Section 7.4, you would recognize that a good point estimate of the mean of this population is the sample mean, which Table 7.1 shows to be 1.27. Thus, your first step would be to advise ICI that if it wants a single number as an estimate, 1.27 is a good number to use. Next, mindful of one of the central points of this chapter, you would point out that such a point estimate contains no indication of how much error it may

TABLE 7.1

Differences in Abrasion Resistance (Treated Material Minus Untreated Material), 10 Pieces of Rubber

Piece	Difference
1	2.6
2	3.1
3	−0.2
4	1.7
5	0.6
6	1.2
7	2.2
8	1.1
9	−0.2
10	0.6

$$\sum_{i=1}^{10} x_i = 12.7$$

$$\bar{x} = 1.27$$

$$s = 1.1265$$

SOURCE: Owen L. Davies, *The Design and Analysis of Industrial Experiments* (London: Oliver and Boyd, 1956), p. 13.

15. Owen Davies, op. cit., p. 13.

contain, whereas a confidence interval does contain such information. Since the population standard deviation is unknown and the sample is small, expression (7.6) should be used in this case to calculate a confidence interval. Assuming that the firm wants a confidence coefficient of 95 percent, the confidence interval is 0.464 to 2.076, because $t_{.025} = 2.262$, $s = 1.1265$, and $n = 10$. The chances are 95 out of 100 that such a confidence interval would include the population mean.[16]

The above analysis is, in fact, exactly how ICI's statisticians proceeded. Despite the fact that the sample consisted of only 10 observations, the evidence was very strong that the chlorinating agent had a positive effect on abrasion resistance. After all, the 95 percent confidence interval was that the mean difference between abrasion resistance of rubber with and without treatment was an increase of between 0.464 and 2.076. (For that matter, the ICI statisticians found that the 98 percent confidence interval was that the mean difference was an increase of between 0.265 and 2.275.) The best estimate was that the chlorinating agent resulted in an increase of about 1.27 in abrasion resistance.

In conclusion, note that it would have been incorrect to have viewed the abrasion resistance of the 10 treated half-pieces as one sample and the abrasion resistance of the 10 untreated half-pieces as another, and to have used expression (7.9) to obtain a confidence interval for the difference between the mean abrasion resistance of treated half-pieces and the mean abrasion resistance of untreated half-pieces. This would have been incorrect because the two samples are *paired* or *matched*; that is, each half-piece in one sample has a mate (the other half of the piece it comes from) in the other sample. Thus, the observations in one sample are not independent of those in the other sample, because if one half-piece is relatively resistant to abrasion due to chance variation in its production or other factors, its mate in the other sample is likely to be resistant as well. Because the samples are not independent, expression (7.9) is not appropriate, since it assumes independence.[17] Instead, the proper technique is to consider the difference between each observation in one sample and its mate in the other sample as a single observation, as we did in Table 7.1.

EXERCISES

7.26 A magazine for business executives plans to determine the mean annual income of its subscribers. For the results to be useful, it is believed that the probability that the sample mean will differ by less than \$5,000 from the population mean should equal .95. A rough estimate of the standard deviation of the annual income of the magazine's subscribers is \$15,000.

(a) How large a sample should be drawn?

16. Note that this analysis assumes that the population is normally or approximately normally distributed.

17. Also, the formula in expression (7.9) is not appropriate because it assumes that the sample sizes are large.

(b) What sorts of biases may be present in such a survey?
(c) How can bias be present when the sample mean is an unbiased estimator?

7.27 A local government plans to estimate the percentage of vacant buildings in a particular area of several square miles. A reasonable guess is that about 20 percent of the buildings in this area are vacant. The government wants the probability to be .90 that the sample percentage differs from the population percentage by no more than 2 percentage points. How large should the sample of buildings be?

7.28 The president of the Bona Fide Washing Machine Company tells the firm's statistician that it is important that he be able to estimate the mean longevity of motors received from supplier I with a .95 probability of an error of no more than 20 hours. Based on the information in Exercise 7.9, how large a sample must the firm take of the motors received from supplier I to be certain that this is true?

7.29 A spokesman for a hi-fi repair shop claims that in about 40 percent of the cases where a hi-fi set is repaired the customer is undercharged due to clerical errors. A public-interest law firm decides to estimate the proportion of cases where under-charging occurs. The law firm plans to construct a random sample so that the probability of the sample proportion's being in error by more than .01 is .05. How large should the sample be?

7.30 Suppose that in the experiment described in Section 7.9 the chlorinated half-pieces of rubber were evaluated on a different machine than the untreated half-pieces. Would this make the results of the experiment more difficult to interpret? If so, how?

CHAPTER REVIEW

1. Estimates are of two types: *point estimates* and *interval estimates.* A statistic used to estimate a population parameter is called an *estimator.* In choosing among estimators, statisticians consider the following three criteria: lack of bias, efficiency, and consistency. An *unbiased estimator* is one whose expected value equals the parameter being estimated. If two estimators are unbiased, one is more efficient than the other if its variance is less than the variance of the other. A statistic is a consistent estimator of a parameter if the probability that the statistic's value is very near that of the parameter approaches 1 as the sample size increases. Based on these criteria, the sample mean, sample proportion, and sample standard deviation are judged to be very good estimators of the population mean, population proportion, and population standard deviation, respectively.

2. If the population standard deviation σ is known, the large-sample confidence interval for the population mean μ is

$$\bar{x} - z_{\alpha/2}\frac{\sigma}{\sqrt{n}} < \mu < \bar{x} + z_{\alpha/2}\frac{\sigma}{\sqrt{n}},$$

where $z_{\alpha/2}$ is the value of the standard normal variable that is exceeded with a probability of $\alpha/2$. Before the sample is drawn, there is a probability of $(1 - \alpha)$ that this interval will include μ. Thus, the confidence coefficient is said to be $(1 - \alpha)$. If σ is unknown, the sample standard deviation s can be substituted for σ in this expression if sample size n is at least 30.

3. When the population standard deviation is unknown and the sample size is less than 30, the t distribution can be used to formulate a confidence

interval for the population mean, providing the population is at least approximately normal. The *t distribution* is a family of distributions, each of which corresponds to a certain number of degrees of freedom. As the number of degrees of freedom increases, the t distribution moves increasingly closer to the standard normal distribution. In this context, the number of degrees of freedom equals $(n - 1)$. If the confidence coefficient is set equal to $(1 - \alpha)$, the confidence interval for the population mean is

$$\bar{x} - t_{\alpha/2}\frac{s}{\sqrt{n}} < \mu < \bar{x} + t_{\alpha/2}\frac{s}{\sqrt{n}},$$

where $t_{\alpha/2}$ is the value of a t variable (with $[n - 1]$ degrees of freedom) that is exceeded with a probability of $\alpha/2$.

4. If the sample size is sufficiently large, and if the confidence coefficient is set equal to $(1 - \alpha)$, the confidence interval for the population proportion P is

$$p - z_{\alpha/2}\sqrt{\frac{p(1 - p)}{n}} < P < p + z_{\alpha/2}\sqrt{\frac{p(1 - p)}{n}},$$

where p is the sample proportion. Otherwise, one must use special graphs (in Appendix Tables 7a and 7b) to obtain a confidence interval for the population proportion.

5. If independent samples of sizes n_1 and n_2 are chosen from two populations, we can calculate a confidence interval for the difference between the population means $\mu_1 - \mu_2$. The following formula is applicable if both n_1 and n_2 are large:

$$\bar{x}_1 - \bar{x}_2 - z_{\alpha/2}\sqrt{\frac{s_1^2}{n_1} + \frac{s_2^2}{n_2}} < \mu_1 - \mu_2 < \bar{x}_1 - \bar{x}_2 + z_{\alpha/2}\sqrt{\frac{s_1^2}{n_1} + \frac{s_2^2}{n_2}},$$

where the confidence coefficient equals $(1 - \alpha)$. Also, we can calculate a confidence interval for the difference between the population proportions, $P_1 - P_2$. The following formula is applicable if both n_1 and n_2 are sufficiently large:

$$p_1 - p_2 - z_{\alpha/2}s_{p_1 - p_2} < P_1 - P_2 < p_1 - p_2 + z_{\alpha/2}s_{p_1 - p_2},$$

where $s_{p_1 - p_2} = \sqrt{\frac{p_1(1 - p_1)}{n_1} + \frac{p_2(1 - p_2)}{n_2}}$.

Neither of these formulas is correct if the two samples are matched, since this violates the assumption that the samples are independent.

6. If it is desired that the probability be $(1 - \alpha)$ that the sample mean differs from the population mean by no more than a given number δ the sample size must equal

$$n = \left(\frac{z_{\alpha/2}\sigma}{\delta}\right)^2.$$

If it is desired that the probability be $(1 - \alpha)$ that the sample proportion differs from the population proportion by no more than some number δ, the sample size must equal

$$n = \left(\frac{z_{\alpha/2}}{\delta}\right)^2 \hat{P}(1 - \hat{P}),$$

where $\hat{P}$ is an estimate of P. These results assume that simple random sampling is used, that the sample is large enough so that the normal distribution can be used, and that the population is large relative to the sample (or that sampling is with replacement). The last assumption is made throughout this Chapter Review.

Getting Down to Cases:

THE EFFECT OF A NEW ENZYME ON A PHARMACEUTICAL MANUFACTURING PROCESS[18]

A drug firm was attempting to estimate the extent to which a new enzyme altered the yield of a certain manufacturing process. For a given batch of product, the yield was measured by the ratio of the actual output to the theoretical output as calculated from formulas based on past experience. Thus, a yield of 1.03 meant that 3 percent more output was gotten from the batch than the formula indicated; 0.98 meant that 2 percent less output was gotten from the batch than the formula indicated.

The drug firm tested the new enzyme on 36 batches and obtained the following yields:

1.28	1.31	1.48	1.10	0.99	1.22
1.65	1.40	0.95	1.25	1.32	1.23
1.43	1.24	1.73	1.35	1.31	0.92
1.10	1.05	1.39	1.16	1.19	1.41
0.98	0.82	1.22	0.91	1.26	1.32
1.71	1.29	1.17	1.74	1.51	1.25

(a) Calculate a 90 percent confidence interval for the true mean yield gotten from the enzyme.

(b) Calculate a 95 percent confidence interval for the true mean yield gotten from the enzyme.

(c) Calculate a 99 percent confidence interval for the true mean yield gotten from the enzyme.

(d) Specify the assumptions underlying your results in (a), (b), and (c).

(e) Provide an unbiased point estimate of the true mean yield from the enzyme.

(f) If the yields are normally distributed, is the estimator you used in (e) at least as efficient as any other unbiased estimator?

(g) If the assumptions specified in (d) are correct, would you conclude

18. This case is based on a section from W. Allen Wallis and Harry V. Roberts, *Statistics: A New Approach* (New York: Free Press, 1956). Some of the numbers have been changed for expository purposes.

that the true mean yield gotten from the enzyme almost certainly exceeds 1.00? Explain.

(h) What is the mistake (if any) in the following argument, put forth by one member of the firm's research department: "There is no good theoretical reason for believing that the mean yield should be higher than 1.00 with the enzyme. Moreover, in the 36 batches studied, the standard deviation of individual yields was .228. In my opinion, .228 represents a large fraction of the difference between [the sample mean] 1.268 and 1.00. There is no real evidence that the enzyme increases yield . . . ."[19]

APPENDIX 7.1
Bayesian Estimation

The basic difference between *Bayesian estimation* and the estimation procedures discussed in previous sections of this chapter is that Bayesian techniques view the parameter to be estimated as a random variable, not a constant. In the case of the new fiber in Example 7.1, suppose that the firm's statisticians feel strongly (based on previous experience with the new fiber and on *a priori* reasoning) that the mean breaking load of a strand of this fiber should be about 2.5 pounds. Specifically, suppose that the statisticians' subjective probability distribution of the population mean is as shown in Figure 7.8. In other words, their prior probabilities regarding the mean can be represented by this distribution. (Recall our discussion of prior probabilities in Chapter 3). If this is the case, then Bayesian estimation techniques allow the firm's statisticians to take these prior probabilities into account, whereas the techniques discussed previously in this chapter make no allowance for such prior probabilities.

If the prior distribution of the population mean is normal, as in Figure 7.8, then the Bayesian estimate of the population mean is

$$\tilde{\mu} = \frac{n\sigma_p^2\bar{x} + \sigma^2\mu_p}{n\sigma_p^2 + \sigma^2}, \tag{7.15}$$

where μ_p is the mean of the prior (subjective) probability distribution of the population mean, σ_p is the standard deviation of the prior (subjective) probability distribution of the population mean, σ is the population standard deviation, $\bar{x}$ is the sample mean, and n is the sample size.

At first glance, the reasonableness of this estimate is by no means obvious. But another way of writing this estimate is

$$\tilde{\mu} = \left(\frac{n\sigma_p^2}{n\sigma_p^2 + \sigma^2}\right)\bar{x} + \left(\frac{\sigma^2}{n\sigma_p^2 + \sigma^2}\right)\mu_p.$$

Written this way, it is clear that this estimate of the mean is a weighted average of (1) the sample mean and (2) the mean of the prior distribution of the population mean. In other words, this estimate is a compromise between μ_p

19. Ibid. Some of the numbers have been changed.

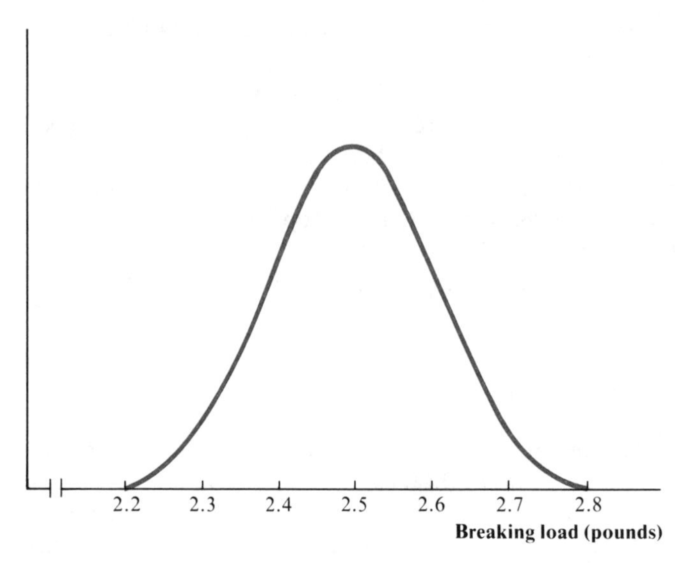

FIGURE 7.8 Prior (Subjective) Distribution of Population Mean

(based on the decision maker's prior feelings) and $\bar{x}$ (what the sample actually turns up). The larger the sample size and the greater the decision maker's uncertainty concerning the population mean (that is, the greater n and σ_p^2), the more weight is given to $\bar{x}$ and the less weight is given to μ_p. This, of course, is sensible.

To show how Bayesian estimation works, let's return once more to the case of the new synthetic fiber. If the firm's research department's prior distribution concerning the mean breaking load of a strand of this fiber is as shown in Figure 7.8, then $\mu_p = 2.5$ and $\sigma_p = 0.1$. (Why? Because the mean and standard deviation of the distribution in Figure 7.8 are 2.5 and 0.1, respectively.) Suppose that the firm's statisticians know that the standard deviation of the population equals .075. Then if the firm draws a sample of 36 strands and if the sample mean is 2.3, the Bayesian estimate is

$$\begin{aligned}\tilde{\mu} &= \left[\frac{36(.01)}{36(.01) + .0056}\right](2.3) + \left[\frac{.0056}{36(.01) + .0056}\right](2.5)\\ &= (.98)(2.3) + (.02)(2.5) = 2.30.\end{aligned}$$

Thus, in this case the Bayesian estimate is very close to $\bar{x}$. This is because $n\sigma_p^2$ is large relative to σ^2.

8

Hypothesis Testing

8.1 Introduction

As pointed out in the previous chapter, statistical inference, the part of statistics that indicates how probability theory can be used to help make decisions based on sample data, deals with two types of problems: estimation and hypothesis testing. Having discussed estimation problems in the previous chapter, we turn now to hypothesis testing. Since statisticians are continually engaged in testing hypotheses, and since the methods they use and the results they obtain play a major role in business and economics, it is important to understand the concepts involved in hypothesis testing and how these concepts can be applied to practical problems.

8.2 Hypothesis Testing: An Illustration

Hypothesis testing deals with decision making—that is, with the rules for choosing among alternatives. Since statistical decisions must be made under conditions of uncertainty, there is a non-zero probability of error; and the object of the statistician's theory of hypothesis testing is to develop decision rules that will control and minimize the probability of error.

To introduce some of the essential features of the statistical theory of hypothesis testing, let's look at a problem that confronted the American Stove Company. Some time ago this firm was engaged in manufacturing a metal piece whose height was of central importance in determining whether it met the relevant specifications. Based on previous experience, the firm's statisticians knew that the mean height of the metal pieces produced was .8312 inches. A test was needed for detecting any change in this mean height so that whatever factors were responsible for such a change could be corrected.[1]

The company's statisticians picked a random sample from each day's

1. Lester Kauffman, "Statistical Quality Control at the St. Louis Division of American Stove Company," as quoted and described in Acheson Duncan, *Quality Control and Industrial Statistics* (Homewood, Ill.: Irwin, 1959).

output. Based on this sample, they wanted to test the hypothesis that a change had *not* occurred in the mean height of all the metal pieces produced in a day. In other words, the statisticians wanted to test the hypothesis that the mean height still was .8312 inches. Let's denote this hypothesis by H_0 and call it the null hypothesis. In statistics the *null hypothesis* is the basic hypothesis that is being tested for possible rejection. The firm had to choose between this hypothesis and the *alternative hypothesis*, denoted by H_1, which maintained that a change *had* occurred in the mean height of all the metal pieces produced in a day. In other words, the alternative hypothesis maintained that the mean height no longer was .8312 inches. (Much more will be said in the following section about the ways in which the null and alternative hypotheses are defined.)

To understand the nature of the decision that faced the American Stove Company it is essential to recognize that, on any given day, four possible situations could arise:

1. No change occurs in the mean height of the metal pieces produced (H_0 is true), and the firm concludes that this is the case. Thus, the firm accepts H_0, which is the correct decision under these circumstances.

2. No change occurs in the mean height of the metal pieces produced (H_0 is true), but the firm concludes that such a change has occurred. Thus, the firm rejects H_0 (and accepts H_1), which is the wrong decision under the circumstances. (One cost of this wrong decision is the output that is lost while the firm's productive process is shut down needlessly so that the nonexistent deterioration in quality can be corrected.)

3. A change occurs in the mean height of the metal pieces produced (H_0 is not true), and the firm concludes that this is the case. Thus, the firm rejects H_0 (and accepts H_1), which is the correct decision under the circumstances.

4. A change occurs in the mean height of the metal pieces produced (H_0 is not true), but the firm concludes that no change has occurred. Thus, the firm accepts H_0 (and rejects H_1), which is the wrong decision under the circumstances. (One cost of this wrong decision is that the deterioration in quality is allowed to go uncorrected, with the result that an unusually high percentage of the firm's output must be scrapped because it fails to meet specifications.)

Based on the listing of the possibilities above, it is clear that two kinds of error can be committed. First, one can reject the null hypothesis when it is true—which is possibility 2 above. Second, one can accept the null hypothesis when it is false—which is possibility 4 above. These two kinds of error are called Type I and Type II errors and are defined as follows:

> *A* ***Type I error*** *occurs if the null hypothesis is rejected when it is true. A* ***Type II error*** *occurs if the null hypothesis is accepted when it is not true.*

Thus, if the American Stove Company concludes that a change occurs in the mean height of the metal pieces produced when in fact no such change takes place, it commits a Type I error. If it concludes that no change occurs when there is a change, it commits a Type II error.[2]

Table 8.1 provides a condensed description of the situation facing the American Stove Company. The two possible *states of nature* shown in the table are that hypothesis H_0 is true or that it is not true. It is not known which of these states of nature—or states of the world—is valid, but one of the two alternative *courses of action* (the acceptance or rejection of the null hypothesis) must be chosen. The *outcome*—that is, the result of each course of action when each state of nature is true—is shown in Table 8.1. As you can see, a correct decision is the result of the two cases corresponding to possibilities 1 and 3 above; a Type I error is the result of the case corresponding to possibility 2 above; and a Type II error is the result of the case corresponding to possibility 4 above.

TABLE 8.1
Possible Outcomes of the American Stove Company's Decision Problem

Alternative courses of action	State of nature: H_0 is true	H_0 is not true[a]
Accept H_0	Correct decision	Type II error
Reject H_0	Type I error	Correct decision

[a] In other words, H_1 is true.

8.3 Basic Concepts of Hypothesis Testing

To carry out the desired test, the American Stove Company needs some criterion or decision rule for choosing between the null hypothesis (that the mean height is still .8312 inches) and the alternative hypothesis (that the mean height is no longer .8312 inches). The classical statistical theory of hypothesis testing presented in this chapter provides such decision rules based on the results of a random sample. These rules are constructed so that the probability of a Type I error and that of a Type II error can each be measured (if possible) and, to some extent at least, be reduced. Before describing these rules, we must discuss how one specifies the null hypothesis and the alternative hypothesis, and we must distinguish between one-tailed and two-tailed tests.

2. If the hypothesis that a change occurs in the mean height of the metal pieces were regarded as the null hypothesis rather than as the alternative hypothesis, and if the hypothesis that no such change occurs were regarded as the alternative hypothesis (rather than as the null hypothesis), what is now regarded as a Type I error would become a Type II error, and what is now regarded as a Type II error would become a Type I error. To see this, note that a Type I error is now to reject the hypothesis that no change occurs when in fact this is true. But with the redefinition of hypotheses, this would be a Type II error (since it would mean accepting the null hypothesis when the alternative hypothesis is true). The point is that what is a Type I or Type II error depends on how the null and alternative hypotheses are defined.

NULL AND ALTERNATIVE HYPOTHESES

As mentioned above, the null hypothesis is the basic hypothesis that is being tested for possible rejection. Typically, the null hypothesis corresponds to the *absence* of the effect that is being investigated. For example, the American Stove Company wants to detect a change in the mean height of the metal pieces produced. If such a change is *absent*, the mean height is still .8312 inches. Thus, the null hypothesis is that the mean height is .8312 inches. In testing any null hypothesis, there is an alternative hypothesis which the decision maker accepts if the null hypothesis is rejected. For example, the alternative hypothesis in the previous section is that the mean height differs from .8312 inches. There are two quite different kinds of alternative hypotheses, which correspond to one-tailed tests and two-tailed tests.

TWO-TAILED TESTS

In many statistical investigations, the purpose is to see whether a certain population parameter has changed or whether it differs from a particular value. This is true in the case of the American Stove Company, which wants to detect changes *in either direction* in the mean height of the metal pieces produced. Because the proportion of its output not meeting specifications will increase if the mean height is *either* too large *or* too small, the firm wants to detect changes in either direction. Thus, the null hypothesis and alternative hypothesis in this case are

$$H_0: \mu = .8312 \qquad H_1: \mu \neq .8312,$$

where μ is the population mean.

How can the firm carry out such a test? The firm can calculate the sample mean of the heights, $\bar{x}$. If the null hypothesis is true (that is, if the population mean is .8312 inches), the sampling distribution of the sample mean is as shown in Figure 8.1. Specifically, as we know from Chapter 6, the sample mean is approximately normally distributed and has a mean of .8312 inches and a standard deviation of $\sigma/\sqrt{n}$ (where σ is the population standard deviation and n is the sample size). Thus, if σ is known, the firm can establish certain values of the sample mean that are very unlikely to occur, given that the null hypothesis is true. In particular, as pointed out in Figure 8.1, the probability is only .05 that the value of the sample mean will differ from .8312 inches by more than $1.96\sigma/\sqrt{n}$ under these circumstances. Consequently, if the value of the sample mean does in fact differ by that much from .8312 inches, it is evidence that the null hypothesis is not true.

Several things should be noted about this hypothesis-testing procedure. First, *the decision maker or statistician bases his or her decision on the value of a* ***test statistic****, which is a statistic computed from the sample.* In this case, the test statistic is the sample mean since this is the statistic used by the firm to test the null hypothesis. Second, *the set of all possible values of the test statistic is referred to as the* **sample space**. For example, in this case the set of all possible values of the sample mean is the sample space. Third, *the testing procedure divides the sample space into two mutually exclusive parts: the* **acceptance region**

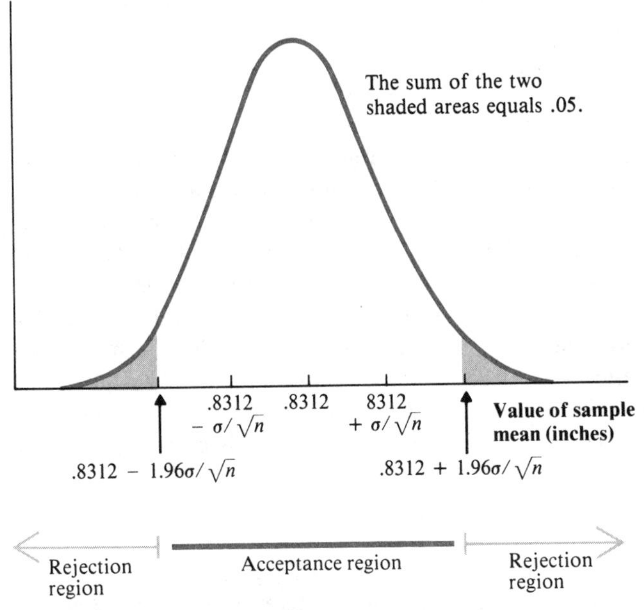

FIGURE 8.1 Sampling Distribution of the Mean Height of Metal Pieces in Sample (if Null Hypothesis Is True)

and the **rejection region.** *If the test statistic's value falls in the acceptance region, the null hypothesis is accepted; if its value falls in the rejection region, the null hypothesis is rejected.* For example, in this case the acceptance region consists of values of the sample mean that differ from .8312 inches by less than $1.96\sigma/\sqrt{n}$.

Why is this hypothesis-testing procedure called a two-tailed test? The answer is indicated in Figure 8.1, which shows that the rejection region contains values of the test statistic (that is, the sample mean) that lie under *both* tails of its (that is, the sample mean's) sampling distribution. Basically, the reason why the rejection region is of this sort is that the firm wants to reject the null hypothesis (that is, $\mu = .8312$ inches) *either if μ is less than .8312 inches or if μ is greater than .8312 inches.* Regardless of whether the mean height is too small or too large, the firm wants to detect departures from the specified value of .8312 inches.

ONE-TAILED TESTS

In many statistical investigations, the decision maker or statistician is concerned about detecting departures from the null hypothesis *in only one direction.* For example, suppose that an automobile firm wants to test whether the proportion of defective tires in a particular shipment is .06, as claimed by the manufacturer. In this case, the firm does not want to reject the null hypothesis (that the proportion defective is .06) if the proportion defective is in fact less than .06 since, regardless of whether this proportion is .06 or less than .06, the firm wants to keep the shipment. *Only if the proportion defective* exceeds *.06 does the firm want to reject the shipment.* Thus, in this case the null and alternative hypotheses are

$$H_0: P = .06 \qquad H_1: P > .06,$$

where P is the proportion of the shipment that is defective.[3]

Whereas the automobile firm wants to reject the null hypothesis only if the parameter in question is too high, other situations call for a rejection of the null hypothesis only if the parameter in question is too low. For example, a company wants to test whether the mean length of life of the light bulbs in a particular incoming shipment is less than 2,000 hours. In this case, the firm does not want to reject the null hypothesis (that the mean length of life is 2,000 hours) if the population mean is in fact greater than 2,000 hours, since, regardless of whether the mean is 2,000 hours or more than 2,000 hours, the firm wants to keep the shipment. *Only if the population mean is* less *than 2,000 hours does the firm want to reject the shipment.* Thus, in this case the null and alternative hypotheses are

$$H_0: \mu = 2{,}000 \qquad H_1: \mu < 2{,}000,$$

where μ is the population mean.

When the decision maker or statistician is interested only in detecting departures from the null hypothesis in one direction, the hypothesis-testing procedure is a one-tailed test. To see why, consider the case of the company that wants to test whether the mean length of life of the light bulbs in the shipment equals 2,000 hours. If the population mean equals 2,000 hours, the sampling distribution of the sample mean is as shown in Figure 8.2. That is, if a random sample of n light bulbs is selected from the shipment, the mean length of life of the bulbs in the sample will be distributed approximately normally with a mean of 2,000 hours and a standard deviation of $\sigma/\sqrt{n}$ (where σ is the population standard deviation). Thus, if the null hypothesis is true (that is, if $\mu = 2{,}000$), the probability is only .05 that the value of the sample mean will fall below 2,000 hours by an amount exceeding $1.64\sigma/\sqrt{n}$. Consequently, if the value of the sample mean does in fact fall below 2,000 hours by an amount exceeding $1.64\sigma/\sqrt{n}$, this is evidence that the null hypothesis is not true. In other words, the sample mean is the test statistic and the rejection region contains all values of the sample mean falling below 2,000 hours by an amount exceeding $1.64\sigma/\sqrt{n}$. This rejection region (shown in Figure 8.2) is under only *one* tail of the test statistic's (that is, the sample mean's) sampling distribution, which accounts for the term ***one-tailed test.***

DECISION RULES AND THE NULL HYPOTHESIS

It is important to recognize that *a test procedure (whether a one-tailed or two-tailed test) specifies a* **decision rule** *indicating whether the null hypothesis*

3. In this case, it is possible to define the null hypothesis as $P \leqslant .06$ rather than $P = .06$. (And in the case in the next paragraph, one can define the null hypothesis as $\mu \geqslant 2{,}000$ rather than $\mu = 2{,}000$.) However, it simplifies the exposition to define the null hypothesis in the manner shown above. More is said about this in footnotes 4 and 8.

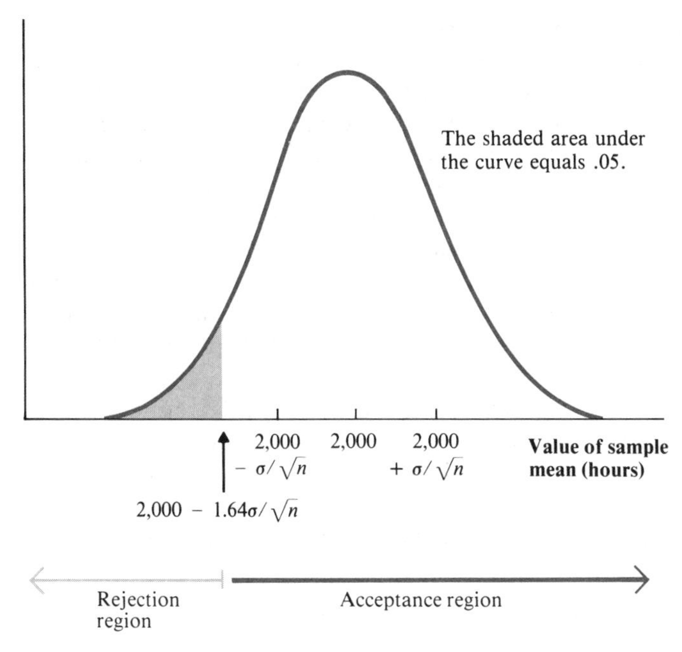

FIGURE 8.2 Sampling Distribution of Mean Length of Life of Bulbs in Sample (if Null Hypothesis Is True)

should be accepted or rejected. For example, in the case of the American Stove Company, the decision rule is (1) reject the null hypothesis if the sample mean differs by more than $1.96\sigma/\sqrt{n}$ from .8312 inches; (2) otherwise, accept the null hypothesis. In the case of the firm receiving the shipment of light bulbs, the decision rule is (1) reject the null hypothesis if the sample mean falls below 2,000 hours by more than $1.64\sigma/\sqrt{n}$; (2) otherwise, accept the null hypothesis. Each such decision rule can be expected to result in a wrong decision for a certain percentage of the time. In other words, each such decision rule contains a certain probability of a Type I error and a certain probability of a Type II error. In later sections, we shall indicate how these probabilities can be calculated.

It is also important to note that *the null hypothesis always states that a population parameter is equal to some value, not that it is unequal to some value.*[4] For example, a beer manufacturer needs to determine whether beer drinkers prefer a new brand of its beer over the old brand. To find out, a sample of beer drinkers is given some of the new brand, and these beer drinkers are asked to rate the new brand on a scale from 1 (very poor) to 90 (perfect). Then the

4. In footnote 3 we pointed out that in the one-tailed tests discussed there, the null hypothesis may be defined as $P \leqslant .06$ or $\mu \geqslant 2{,}000$. If such a definition is used, the point in the text still stands: The null hypothesis cannot be that a population parameter is unequal to some value. For example, neither the hypothesis that $P \leqslant .06$ nor that $\mu \geqslant 2{,}000$ is a statement that a population parameter is *unequal* to some value; they are both statements that a population parameter is *at most* or *at least* some value.

mean rating in the sample is compared with 60, which is known to be the population mean rating for the old brand. In a situation of this sort the preferred procedure, as noted earlier, is to regard the existence of *no difference* between the mean rating of the new brand and that of the old brand as the null hypothesis. One reason why statisticians prefer this procedure is that, if we regard the null hypothesis as being that the population mean for the new brand *equals* 60, we can calculate the sampling distribution of the sample mean. Thus, we can control the probability of a Type I error (as will be indicated in greater detail in the next section). But if we regard the null hypothesis as being that the mean rating for the new brand *exceeds* that of the old brand, we cannot calculate the probability of a Type I error unless we know how large the difference between the means is. Since statisticians find it desirable to specify and control the probability of a Type I error, they have a preference for making the null hypothesis specific in this way. Thus, if one is testing whether a coin is true, the null hypothesis is that the probability of heads is 1/2. And if one is testing whether the mean profit rate of oil firms exceeds that of chemical firms, the null hypothesis is that the means are equal.[5]

8.4 One-Sample Test of a Mean: Large Samples

Having described some of the basic concepts in the statistical theory of hypothesis testing, we are ready now for a detailed consideration of the most important statistical tests. In this and the following sections, we are concerned with the case where data are available concerning a single sample. This section covers the test of a mean; the following section discusses the test of a proportion. In both sections,[6] we assume that the sample is large ($n > 30$).

SETTING UP THE TEST

Although there is no cut-and-dried procedure to be followed, certain steps are essential in testing any hypothesis. The first step is to formulate the null hypothesis (the hypothesis that is being tested) and the alternative hypothesis. Exactly what parameter is relevant for the problem at hand, and what value of this parameter constitutes the null hypothesis? Is the alternative hypothesis a *one-sided alternative*, which means that the decision maker cares about departures from the null hypothesis in one direction? Or is the alternative hypothesis a *two-sided alternative*, which means that the decision maker wants to detect departures from the null hypothesis in both directions?[7]

To be specific, suppose that we take the case of the firm that receives the

5. For a classic discussion of these and related matters, see R. A. Fisher, "Statistical Inference," reprinted in E. Mansfield, *Statistics for Business and Economics: Readings and Cases* (New York: Norton, 1980). As pointed out in footnote 4, the null hypothesis may sometimes be defined as a population parameter's being at most or at least some value. More is said about this in footnote 8.

6. In the next section, it is sometimes assumed that n is considerably in excess of 30. See footnote 13.

7. Of course, one-sided alternative hypotheses lead to one-tailed tests, and two-sided alternative hypotheses lead to two-tailed tests.

shipment of light bulbs. In this case, as we know from the previous section, the null hypothesis is that the mean length of life of the bulbs in the shipment is 2,000 hours; and the alternative hypothesis is one-sided (namely, that the mean length of life is less than 2,000 hours). Suppose that the firm knows from past experience that the standard deviation of the length of life of light bulbs in a shipment of this sort is 200 hours, and that it tests a random sample of 100 bulbs. What sort of test procedure or decision rule should the firm use to determine whether, once it obtains the results of the sample, it should accept the null hypothesis (and accept the shipment) or reject the null hypothesis (and reject the shipment)? Clearly, the firm should be more inclined to reject the null hypothesis if the sample mean is relatively low than if it is relatively high, but how low must the sample mean be to justify rejecting the null hypothesis?

To answer this question, the decision maker or statistician must first recognize that any such test procedure or decision rule can make either a Type I error or a Type II error. In this case, a Type I error will arise if the firm rejects a shipment in which the mean length of life of the bulbs is 2,000 hours; a Type II error will arise if the firm accepts a shipment where the mean length of life of the bulbs is less than 2,000 hours. Table 8.2 shows the possibilities in this case, much as Table 8.1 did in the case of the American Stove Company. As stressed in an earlier section, any test procedure or decision rule is characterized by a certain probability of a Type I error and a certain probability of a Type II error, denoted as follows:

> *The* ***probability of a Type I error*** *is designated as α (alpha), and the* ***probability of a Type II error*** *is designated as β (beta).*

Whether a particular test procedure or decision rule is appropriate is dependent on whether α and β are set at the right levels from the standpoint of the decision maker. In this case, whether a particular test procedure or decision rule is appropriate depends on whether α and β are set at levels that are satisfactory to the firm receiving the shipment of light bulbs.

TABLE 8.2

Possible Outcomes of the Decision Problem concerning the Shipment of Light Bulbs

Alternative courses of action	State of nature[a] $\mu = 2{,}000$	$\mu < 2{,}000$
Accept shipment	Correct decision	Type II error
Reject shipment	Type I error	Correct decision

[a]If $\mu > 2{,}000$, the correct decision is to accept the shipment. See footnotes 3, 4, and 8.

The value assigned to α is called the ***significance level*** of the test. Typically, the significance level is set at .05 or .01, which means that the probability of a Type I error is .05 or .01. The decision maker can set the significance level at any amount that he or she chooses. However, in setting this level it generally is wise to keep in mind the relative costs of committing various types of errors. If a Type I error is very costly, a very low level for α should be chosen; on the other hand, if a Type I error is not very costly, a higher value for α is acceptable. At the same time it should be recognized that for a fixed sample size the probability of a Type I error cannot be reduced without increasing the probability of a Type II error. (For example, if the firm establishes a decision rule whereby the shipment is rejected if the sample mean is less than 1,600 hours, this rule will have a lower value of α than if it rejects only those where the sample mean is less than 1,800 hours; but the value of β will be higher for the former rule.) Thus, in choosing α, the decision maker should keep β in mind as well, picking a value of α such that α and β bear a sensible relation to the relative costs of Type I and Type II errors. (In a subsequent part of this section, we shall indicate how the value of β can be calculated.)

DECISION RULES

Once the value of α has been specified, it is relatively simple to determine the decision rule that should be applied to test whether a population mean equals a specified value. For example, in the case of the firm that is testing whether the mean length of life of the light bulbs is 2,000 hours, we know that if the null hypothesis is true, the sampling distribution of the sample mean is as shown in Figure 8.3. Since the sample size is large, the central limit theorem assures us that the sample mean will be approximately normally distributed. And since $\sigma = 200$ and $n = 100$, $\sigma/\sqrt{n} = 20$. Thus, if the null hypothesis is true (that is, if the population mean is 2,000) the sample mean must be

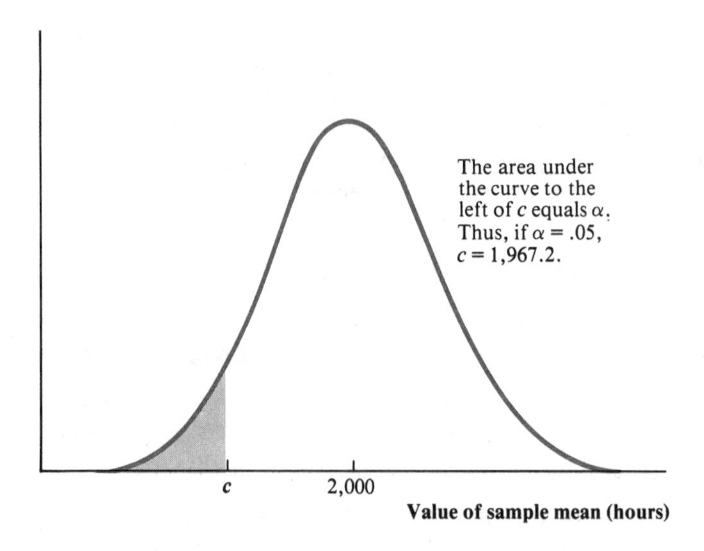

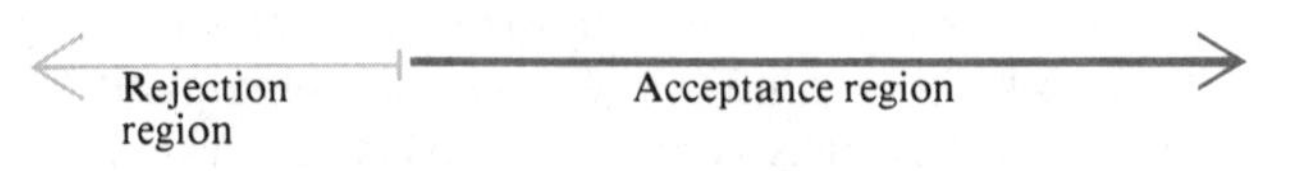

FIGURE 8.3 Sampling Distribution of Mean Length of Life of Bulbs in Sample (if Null Hypothesis Is True)

(approximately) normally distributed, with a mean of 2,000 and a standard deviation of 20, as shown in Figure 8.3.

Since (for reasons given in the previous section) this is a one-tailed test, the rejection region is entirely in the lower tail of the sampling distribution in Figure 8.3. In other words, our test consists of determining whether the value of the sample mean is below a certain number (labeled c in Figure 8.3). If the sample mean is below this number, the null hypothesis is rejected; if it is not below this number, the null hypothesis is accepted. The value of c that should be used depends on the choice of α, because α equals the probability that the sample mean's value will fall below c when the null hypothesis is true. In other words, the value of α equals the area under the sampling distribution to the left of c in Figure 8.3. Thus, since this sampling distribution is (approximately) normal with a mean of 2,000 and a standard deviation of 20,

$$c = 2{,}000 - z_\alpha \cdot 20,$$

where z_α is the value of the standard normal variable that is exceeded with a probability of α. Why is this equation correct? Because the area under the standard normal curve to the left of $-z_\alpha$ equals α. Thus, the area in Figure 8.3 under the normal curve to the left of the value corresponding to $-z_\alpha$ must also equal α. What is the value of $\bar{X}$ corresponding to a Z value of $-z_\alpha$? Based on equation (5.2), the answer is $\mu - z_\alpha\sigma_{\bar{x}}$, or $2{,}000 - z_\alpha \cdot 20$, which is what is shown above.

To illustrate the use of this equation, suppose that the firm testing the light bulbs decides to set α at .05. In other words, it wants the probability of a Type I error to equal .05. In this case, since $z_{.05} = 1.64$, the null hypothesis should be rejected if the value of the sample mean is less than 1,967.2 hours (because $c = 2{,}000 - 1.64(20)$, or 1,967.2). If the sample mean is not less than 1,967.2 hours, the null hypothesis should be accepted. Or suppose that the firm decides to set α at .01, which means that it wants the probability of a Type I error to equal .01. In this case, since $z_{.01} = 2.33$, the null hypothesis should be rejected if the value of the sample mean is less than 1,953.4 hours (because $c = 2{,}000 - 2.33(20)$, or 1,953.4). If the sample mean is not less than 1,953.4 hours, the null hypothesis should be accepted.

In general, the decision rule for testing the null hypothesis that a population mean equals a certain amount μ_0 against the alternative hypothesis that it is less than μ_0 is as follows.

> ***Decision Rule when Alternative Hypothesis is*** $\mu < \mu_0$: *Accept the null hypothesis if* $\bar{x} \geq \mu_0 - z_\alpha\sigma/\sqrt{n}$; *reject the null hypothesis if* $\bar{x} < \mu_0 - z_\alpha\sigma/\sqrt{n}$.

This is no more than a restatement of what we have already said, since $c = \mu_0 - z_\alpha\sigma/\sqrt{n}$, according to our previous results.[8]

8. This decision rule often is stated in terms of $z = \sqrt{n}(\bar{x} - \mu_0)/\sigma$, rather than $\bar{x}$. In these terms, the rule is: Accept the null hypothesis if $z \geq -z_\alpha$; reject the null hypothesis if $z < -z_\alpha$.
In footnotes 3 and 4 we noted that the null hypothesis can be defined as $\mu \geq 2{,}000$ rather than $\mu = 2{,}000$. If this definition is used, α is the *maximum* probability of a Type I error. That is, if the population mean exceeds 2,000 hours, the probability of a Type I error will be less than α.

In addition, we should present the decision rule for testing the null hypothesis that a population mean equals a certain amount μ_0 against the alternative hypothesis that it is *more than* μ_0.

> **Decision Rule when Alternative Hypothesis is** $\mu > \mu_0$: *Accept the null hypothesis if* $\bar{x} \leqslant \mu_0 + z_\alpha \sigma / \sqrt{n}$; *reject the null hypothesis if* $\bar{x} > \mu_0 + z_\alpha \sigma / \sqrt{n}$.

The reasoning leading up to this decision rule is, of course, precisely the same as that leading up to the decision rule in the case where the alternative hypothesis is $\mu < \mu_0$. The only difference is that the rejection region is in the upper tail, not the lower tail, of the sampling distribution of the sample mean.[9]

Finally, if the test is two-tailed—that is, if the alternative hypothesis is that the population mean *differs from* μ_0—the decision rule is as follows.

> **Decision Rule when Alternative Hypothesis is** $\mu \neq \mu_0$: *Accept the null hypothesis if* $\mu_0 - z_{\alpha/2}\sigma/\sqrt{n} \leqslant \bar{x} \leqslant \mu_0 + z_{\alpha/2}\sigma/\sqrt{n}$; *reject the null hypothesis if* $\bar{x} < \mu_0 - z_{\alpha/2}\sigma/\sqrt{n}$ *or if* $\bar{x} > \mu_0 + z_{\alpha/2}\sigma/\sqrt{n}$.

Since the probability of a Type I error must equal α, the probability that the sample mean will fall in the rejection region in *each* tail of its sampling distribution, given that the null hypothesis is true, is set equal to $\alpha/2$.[10]

OPERATING-CHARACTERISTIC CURVE

Up to now the decision rules presented in this section have been constructed so that the probability of a Type I error—that is, α—is fixed at a predetermined level. But how large is β, the probability of a Type II error? For example, if the firm receiving the shipment of light bulbs sets α equal to .05, how large is the probability that the firm will accept the shipment, given that the mean length of life of the bulbs is less than 2,000 hours? Clearly, the answer depends upon the extent to which the mean length of life of the bulbs falls below 2,000 hours. To see this, recall from the previous part of this section that if $\alpha = .05$ the test procedure is the following: Reject the null hypothesis (and thus the shipment) if $\bar{x}$ is less than 1.967.2 hours; otherwise, accept the null hypothesis. Thus, the probability that the firm will accept the shipment, given that the mean length of life of the light bulbs is less than 2,000 hours, equals the probability that the sample mean will be 1,967.2 hours or more

9. As pointed out in footnote 8, decision rules of this sort can be expressed in a variety of equivalent ways. For example, when stated in terms of $z = \sqrt{n}(\bar{x} - \mu_0)/\sigma$, this rule is: Accept the null hypothesis if $z \leqslant z_\alpha$; reject the null hypothesis if $z > z_\alpha$. Of course, the other decision rules presented below can also be stated in a variety of different ways.

10. Note that there is a close connection between the test procedure described above and the confidence interval described in Chapter 7. *If the null hypothesis is that the population mean equals* μ_0, *this hypothesis will be rejected if and only if a confidence interval (with confidence coefficient equal to* $1 - \alpha$*) for* μ *does not include* μ_0. For example, if one computes a 95 percent confidence interval for the population mean, and it turns out to be 15 to 20, then it follows that a two-tailed test (with $\alpha = .05$) will reject the null hypothesis that the population mean is any value below 15 or above 20. This intimate connection between hypothesis testing and confidence intervals is present for other tests too.

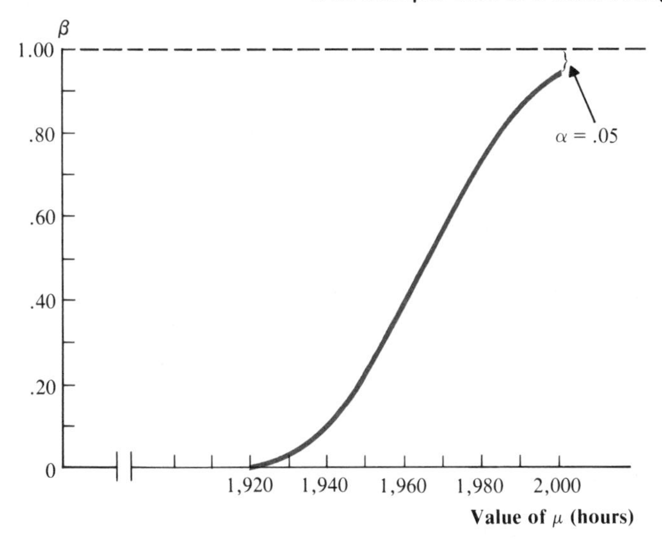

FIGURE 8.4 Operating-Characteristic Curve of Test that Mean Length of Life of Bulbs Equals 2,000 Hours

under these circumstances. It is obvious that this probability will decrease as the population mean falls increasingly farther below 2,000 hours.

Figure 8.4 shows the value of β at each value of the population mean. In accord with our argument in the previous paragraph, the value of β decreases as the population mean falls farther and farther below 2,000 hours. To show how the value of β is calculated, suppose that the mean length of life of the bulbs in the shipment is 1,960 hours. The probability that the shipment will be accepted equals the probability that the sample mean will be 1,967.2 hours or more. (See panel B of Figure 8.5.) Since $\bar{X} \geqslant 1{,}967.2$ if and only if $(\bar{X} - 1{,}960) \div 20 \geqslant (1{,}967.2 - 1{,}960) \div 20$, the probability that $\bar{X} \geqslant 1{,}967.2$ equals the probability that $(\bar{X} - 1{,}960) \div 20 \geqslant 0.36$, because $(1{,}967.2 - 1{,}960) \div 20 = 0.36$. Since $\bar{X}$ is distributed approximately normally with a mean of 1,960 hours and a standard deviation of 20 hours, $(\bar{X} - 1{,}960) \div 20$ has the standard normal distribution. Thus, the probability that $(\bar{X} - 1{,}960) \div 20 \geqslant 0.36$ equals the area under the standard normal curve to the right of .36, which equals .36. (See Appendix Table 2.)

What if the population mean is 1,940 hours, not 1,960 hours? Under these conditions, what is β? The probability that the shipment will be accepted equals the probability that the sample mean will be 1,967.2 hours or more. (See panel C of Figure 8.5.) Since $\bar{X} \geqslant 1{,}967.2$ if and only if $(\bar{X} - 1{,}940) \div 20 \geqslant (1{,}967.2 - 1{,}940) \div 20$, the probability that $\bar{X} \geqslant 1{,}967.2$ equals the probability that $(\bar{X} - 1{,}940) \div 20 \geqslant 1.36$, because $(1{,}967.2 - 1{,}940) \div 20 = 1.36$. Since $\bar{X}$ is distributed approximately normally with a mean of 1,940 hours and a standard deviation of 20 hours, $(\bar{X} - 1{,}940) \div 20$ has the standard normal distribution. Thus, the probability that $(\bar{X} - 1{,}940) \div 20 \geqslant 1.36$ equals the area under the standard normal curve to the right of 1.36, which equals .09. (See Appendix Table 2.)

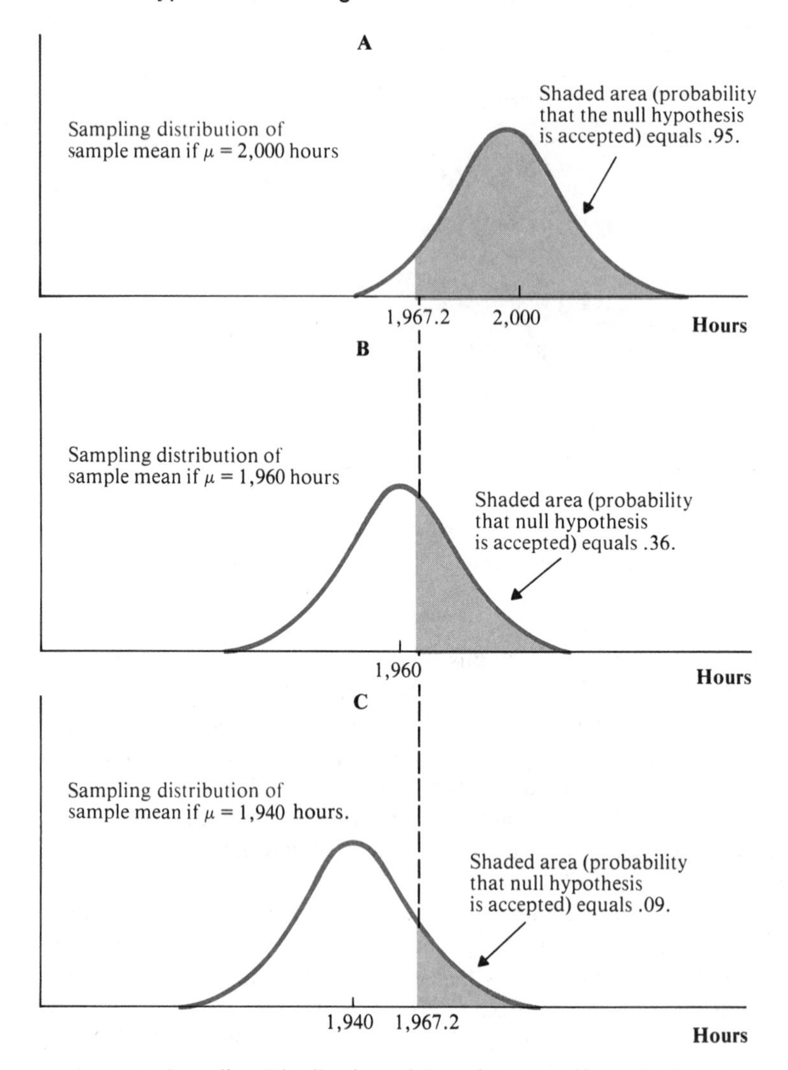

FIGURE 8.5 Sampling Distribution of Sample Mean, if μ = 2,000, 1,960, and 1,940 Hours

Using the methods described in the previous two paragraphs, we can calculate the value of β at a large number of values of the population mean. (For example, panel A of Figure 8.5 corresponds to the case where the population mean is 2,000 hours.) If we plot each value of β against the corresponding value of the population mean, we get the curve in Figure 8.4, which is called the operating-characteristic curve of this test. The definition of an operating-characteristic curve (or *OC curve*) is as follows.

Operating-Characteristic Curve: *A test's operating-characteristic curve shows the value of β (that is, the probability of a Type II error) if the*

population parameter assumes various values other than that specified by the null hypothesis.

Sometimes statisticians use the *power curve* rather than the operating-characteristic curve. The power curve shows the value of $(1 - \beta)$—whereas the operating-characteristic curve shows the value of β—at each value of the parameter under test. Clearly, one can easily deduce the power curve from the operating-characteristic curve, and vice versa, since they provide the same information.[11]

A test's operating-characteristic curve can be used to determine the value of α as well as the value of β. At the point where the parameter equals the value specified by the null hypothesis, the OC curve shows the probability that the null hypothesis will be accepted. Thus 1 minus this probability must equal the probability that the null hypothesis will be rejected, given that it is true (which is α, the probability of a Type I error). Consequently, at the point where the parameter equals the value specified by the null hypothesis, the distance between the OC curve and 1 equals α. For example, Figure 8.4 shows that at the point where the mean length of life of the bulbs is 2,000 hours, the distance between the OC curve and 1 equals .05, which we know equals α in this case.[12]

Not all tests have operating-characteristic curves shaped like that in Figure 8.4. If a test is designed to detect positive rather than negative departures of the population mean from the value specified by the null hypothesis, its operating-characteristic curve will be shaped like that in panel A of Figure 8.6. If a test is designed to detect departures of the population mean from the value specified by the null hypothesis in either direction (both positive and negative) the operating-characteristic curve will be shaped like that in panel B of Figure 8.6.

Once a test's OC curve is available, one is in a much better position to determine whether the test is appropriate. For example, if the executives of the firm receiving the light bulbs are informed of the OC curve for this test, they may feel that it results in too high a probability of both Type I error and Type II error. Figure 8.7 shows the OC curve for another test, one whereby the firm would choose a sample of 400 bulbs and reject the shipment if the sample mean were less than 1,980.4 hours. As you can see, this test results in a much smaller probability of Type I error (.025 rather than .05) and of Type II error

11. We have encountered an operating-characteristic curve before, although we did not call it that. Table 4.4 is an operating-characteristic curve. As you can see from Table 4.4, the probability of a Type II error—that is, of accepting the shipment when the percentage defective exceeds 4 percent—falls as the percentage defective increases.

12. Recall that we pointed out in an earlier part of this section that if the sample size is fixed, a reduction in the probability of a Type I error can be achieved only at the expense of an increase in the probability of a Type II error. The reason for this can be explained more fully now. Basically, the reason is that, if α is reduced, the OC curve must be shifted upwards, which means an increase in β.

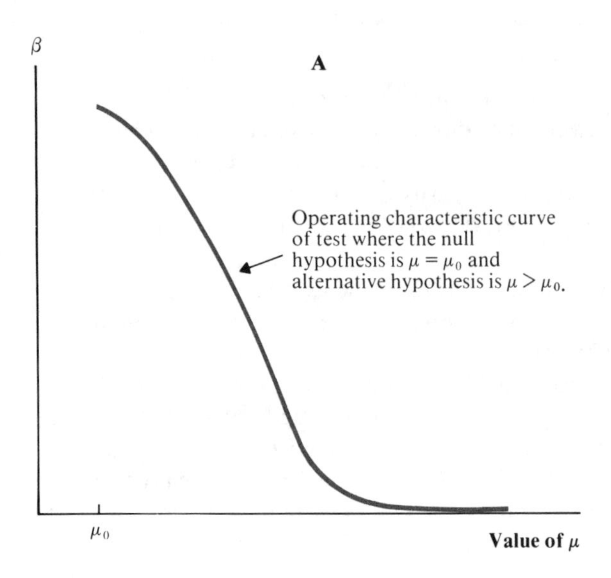

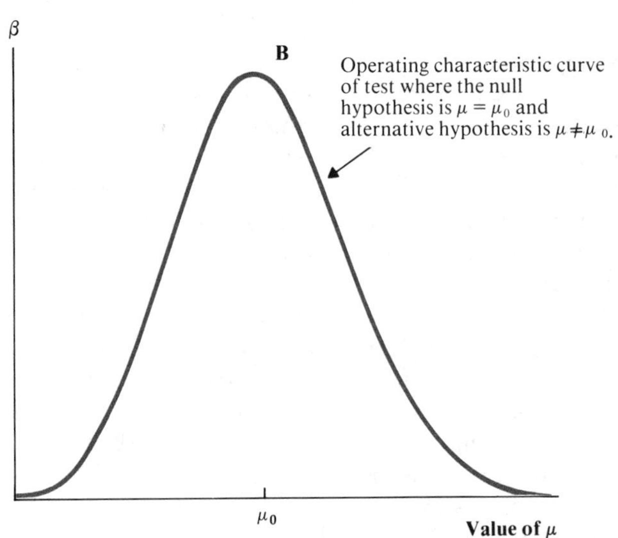

FIGURE 8.6 Shape of Operating-Characteristic Curve if Alternative Hypothesis Is $\mu > \mu_0$ or $\mu \neq \mu_0$

(for practically all values of the population mean) than the earlier test. Why does this test have smaller probabilities of both types of error? Because it is based on a much larger sample. Whether or not the reduction in risk is worth the extra cost of inspecting the larger sample must be determined by the firm's management. After weighing the relevant costs, the firm can specify the sort of

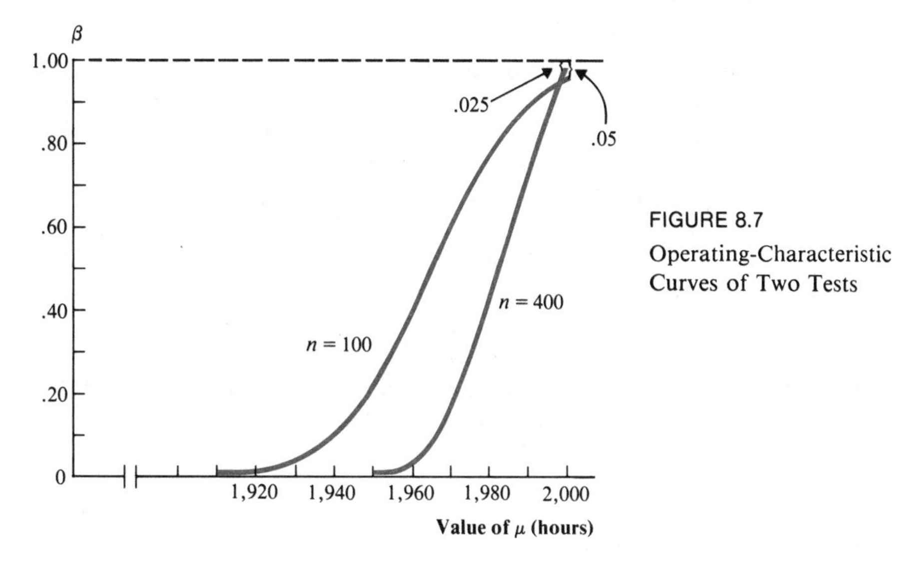

FIGURE 8.7
Operating-Characteristic Curves of Two Tests

OC curve it wants. Then, by a proper choice of sample size and significance level a test can be chosen that is best suited to its requirements.

EXTENSIONS OF THE TEST PROCEDURES

At this point we must indicate how the test procedures described in this section can be modified to suit circumstances somewhat different from those specified above. In particular, we will describe how a test of this sort can be constructed when the standard deviation is unknown and when the population is not large relative to the sample.

Where the Population Standard Deviation Is Unknown. If σ is unknown and if the sample size is large ($n > 30$), the sample standard deviation s can be substituted for the population standard deviation σ in the decision rules given above. For example, if the alternative hypothesis is $\mu < \mu_0$, the decision rule says accept the null hypothesis if $\bar{x} \geqslant \mu_0 - z_\alpha s/\sqrt{n}$; and reject the null hypothesis if $\bar{x} < \mu_0 - z_\alpha s/\sqrt{n}$. In Section 8.8 we shall indicate how a test can be constructed if σ is unknown and the sample size is small.

Where the Population Is Small. If the population is less than 20 times the sample (and if sampling is without replacement), the decision rules given above must be modified. In particular, $\sigma/\sqrt{n}$ (or $s/\sqrt{n}$ if σ is unknown) must be multiplied by the finite population correction factor $\sqrt{(N - n)/(N - 1)}$. For example, if the alternative hypothesis is $\mu < \mu_0$ (and σ is known), the decision rule says accept the null hypothesis if $\bar{x} \geqslant \mu_0 - z_\alpha(\sigma/\sqrt{n})\sqrt{(N - n)/(N - 1)}$; and reject the null hypothesis if $\bar{x} < \mu_0 - z_\alpha(\sigma/\sqrt{n})\sqrt{(N - n)/(N - 1)}$.

INTERPRETATION OF TEST RESULTS

When we test the hypothesis that the mean of a particular population equals a specified value μ_0, we base our test on the value of the sample mean. Because of sampling variation, we cannot expect the mean of the sample to be exactly equal to the population mean. Thus, if the sample mean is different from μ_0 we cannot conclude that the population mean does not equal μ_0. What matters is the probability that the size of the difference between the sample mean and μ_0 could have arisen by chance. If this probability is so great that the null hypothesis is not rejected, we say that the difference between the sample mean and μ_0 (the value of the population mean specified by the null hypothesis) is *not statistically significant*. That is, this difference could be attributable to chance. On the other hand, if this probability is so low that the null hypothesis is rejected this difference between the sample mean and μ_0 is termed *statistically significant*.

The concept of *statistical* significance should not be confused with *practical* significance. Even if there is a statistically significant difference between the sample mean and the postulated value of the population mean, this difference may not really matter. For example, a market researcher wants to determine whether the mean income of families in a certain town is $14,000. Based on a very large sample, the researcher finds that the sample mean is $14,152, which differs significantly (in a statistical sense) from the postulated $14,000. Whether this difference is of any practical significance depends on whether the market researcher (or his client) really cares about a difference of $152. For many purposes, such a difference may not be important. For instance, the researcher may only want to determine whether the mean falls between $13,500 and $14,500, or whether it falls outside this range. If so, the observed difference of $152 between the sample mean and the postulated value of the population mean is of no practical importance even though it is statistically significant.

It is also worth noting that the fact that the null hypothesis is rejected by any of the decision rules given above does not *prove* that the null hypothesis is false. Moreover, the fact that the null hypothesis is accepted by any one of these decision rules does not *prove* that the null hypothesis is true. These test procedures are designed to detect cases where it is unlikely that the sample evidence could have occurred if the null hypothesis was true: In such cases the null hypothesis is rejected. But as long as there is *some chance* that the sample evidence could have occurred if the null hypothesis was true, there is no way to *guarantee* that the results of the test will be absolutely correct. As stressed throughout this section, statistical test procedures are generally designed so that there is a non-zero probability of a Type I or a Type II error. To eliminate such errors entirely would generally be impossible or foolishly expensive.

8.5 One-Sample Test of a Proportion: Large Samples

Many statistical investigations are aimed at testing whether a population proportion equals a specified value. For example, in acceptance sampling the object is to test whether the proportion of defective items in a shipment equals a specified value. (Recall our discussion of acceptance sampling in Chapter 4.) In this section we will describe the statistical test procedures that can be used in such investigations.

SETTING UP THE TEST

As emphasized in the previous section, the first step in any test procedure (whether of a mean, a proportion, or whatever) is to formulate the hypothesis that is being tested. This is a crucial aspect of the work, since, if the null hypothesis and alternative hypothesis are formulated incorrectly, the test procedure will be designed to answer the wrong question. Suppose that an oil company receives a report from a Washington lobbyist, stating that 70 percent of American voters prefer rationing to higher gasoline prices as a means of cutting gasoline consumption. The oil company is skeptical of this report since it believes that the percentage favoring rationing is lower than this figure. The company's statisticians draw a simple random sample of 400 voters and ask these voters' preferences in this regard. What is the null hypothesis in this case? It is that the proportion P of all voters favoring rationing equals .70. What is the alternative hypothesis in this case? It is that P is less than .70 since the company is interested only in determining whether P is below .70. It makes no difference to the oil company whether P equals .70 or is greater than .70.

Having formulated the null and alternative hypotheses, the next step is to determine the proper values of α (the probability of a Type I error) and of β (the probability of a Type II error). As stressed in the previous section, these probabilities should be chosen on the basis of the relative costs of a Type I and a Type II error. In the present case, a Type I error occurs if the oil company rejects the hypothesis that $P = .70$ when in fact this hypothesis is true. A Type II error occurs if the oil company accepts the hypothesis that $P = .70$ when in fact P is less than .70. If a Type I error is much more costly than a Type II error, the value of α should be low relative to the value of β. If a Type II error is much more costly than a Type I error, the value of β should be low relative to the value of α. For each possible value of α one can calculate the test's operating-characteristic curve which shows β. (The method for calculating the operating-characteristic curve is demonstrated later in this section.) The value of α should be chosen so that the OC curve is in line with the relative costs in the case at hand. For the sake of concreteness, suppose that in this case α is set equal to .05.

DECISION RULES

Once the value of α has been specified, we can determine the decision rule that should be applied to test whether a population proportion equals a specified value P_0. As one might guess, the test statistic is the sample proportion p. If the null hypothesis is true (that is, if $P = P_0$), the sample proportion is approximately normally distributed with a mean of P_0 and a standard deviation of $\sqrt{P_0(1 - P_0)/n}$, if the sample size n is sufficiently large. (Recall Chapter

7.) Thus, if the null hypothesis is true, the sampling distribution of the sample proportion is as shown in Figure 8.8. In the case of the oil company the rejection region is entirely in the lower tail of this sampling distribution. (It is a one-tailed test.) Thus, our test consists of determining whether the value of the sample proportion is below a certain number, labeled b in Figure 8.8. If the value of the sample proportion is below this number the null hypothesis is rejected; if it is not below this number the null hypothesis is accepted. The value of b depends on the choice of α, because α equals the probability that the sample proportion's value will fall below b when the null hypothesis is true. In other words, as shown in Figure 8.8, α equals the value of the area under the sampling distribution to the left of b. Thus,

$$b = P_0 - z_\alpha \sqrt{\frac{P_0(1 - P_0)}{n}},$$

where z_α is the value of the standard normal variable that is exceeded with a probability of α.

In the case of the oil company the null hypothesis is that $P = .70$, which means that $P_0 = .70$. Since $n = 400$, it follows that if the null hypothesis is true, the sample proportion will be approximately normally distributed with a mean of 0.70 and a standard deviation of $\sqrt{(.70)(.30)/400}$ or .023. Given that α is set equal to .05, what must be the value of b? Applying the formula in the previous paragraph, b must equal $.70 - 1.64(.023)$, or .662, since $z_{.05} = 1.64$. Thus, the decision rule that the oil company should use is the following: Reject the null hypothesis if the sample proportion is less than .662; accept the null hypothesis if the sample proportion is greater than or equal to .662.

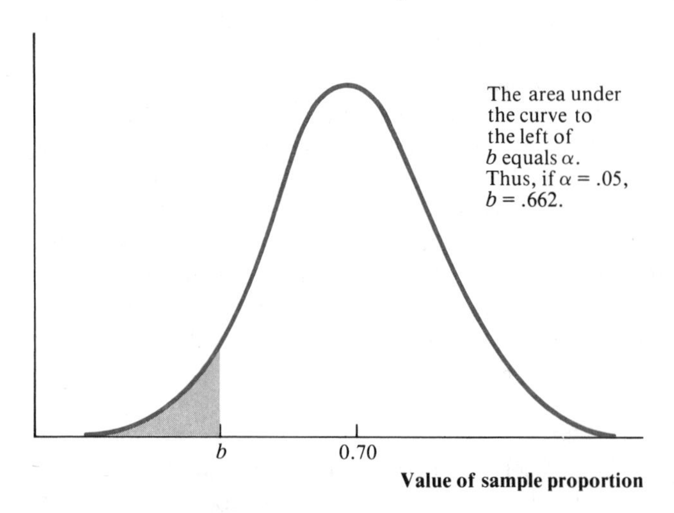

FIGURE 8.8

Sampling Distribution of Proportion of Voters in Sample Preferring Rationing, if Null Hypothesis Is True

In general, when n is sufficiently large,[13] the decision rule for testing the null hypothesis that a population proportion equals a certain amount P_0 against the alternative hypothesis that the population proportion is less than P_0 is as follows.

***Decision Rule when Alternative Hypothesis is* $\mathbf{P < P_0}$:** *Accept the null hypothesis if* $p \geqslant P_0 - z_\alpha\sqrt{P_0(1 - P_0)/n}$; *reject the null hypothesis if* $p < P_0 - z_\alpha\sqrt{P_0(1 - P_0)/n}$.

This restates what we have already said, since $b = P_0 - z_\alpha\sqrt{P_0(1 - P_0)/n}$.

The decision rule for testing the null hypothesis that a population proportion equals a certain amount P_0 against the alternative hypothesis that it is *greater* than P_0 is as follows.

***Decision Rule when Alternative Hypothesis is* $\mathbf{P > P_0}$:** *Accept the null hypothesis if* $p \leqslant P_0 + z_\alpha\sqrt{P_0(1 - P_0)/n}$; *reject the null hypothesis if* $p > P_0 + z_\alpha\sqrt{P_0(1 - P_0)/n}$.

The reasoning behind this decision rule is the same as that for the decision rule in the previous paragraph. The only difference is that the rejection region is in the upper tail, not the lower tail, of the sampling distribution of the sample proportion.

Finally, if the test is two-tailed—that is, if the alternative hypothesis is that the population proportion *differs* from P_0—the decision rule is the following:

***Decision Rule when Alternative Hypothesis is* $\mathbf{P \neq P_0}$:** *Accept the null hypothesis if* $P_0 - z_{\alpha/2}\sqrt{P_0(1 - P_0)/n} \leqslant p \leqslant P_0 + z_{\alpha/2}\sqrt{P_0(1 - P_0)/n}$; *reject the null hypothesis if* $p < P_0 - z_{\alpha/2}\sqrt{P_0(1 - P_0)/n}$ *or if* $p > P_0 + z_{\alpha/2}\sqrt{P_0(1 - P_0)/n}$.

Since the probability of a Type I error must equal α, the probability that the sample proportion will fall in the rejection region in *each* tail of its sampling distribution, given that the null hypothesis is true, is set equal to $\alpha/2$.[14]

OPERATING-CHARACTERISTIC CURVE

The decision rules presented in the previous part of this section are constructed so that the probability of a Type I error equals α. For example, in the case of the oil company α was set equal to .05. But how large is β, the probability of a Type II error? The answer is given by the test's operating-characteristic curve. Figure 8.9 shows the operating-characteristic curve of the oil company's test. This curve indicates that the probability that the company will accept the hypothesis that $P = .70$ when in fact $P = 2/3$ is .58, and the

13. In Chapters 5 and 7, we discussed how large the sample size must be for the normal approximation to be useful.

14. Of course, if the population is small relative to the sample, $\sqrt{P_0(1 - P_0)/n}$ should be multiplied by $\sqrt{(N - n)/(N - 1)}$ in each of these decision rules.

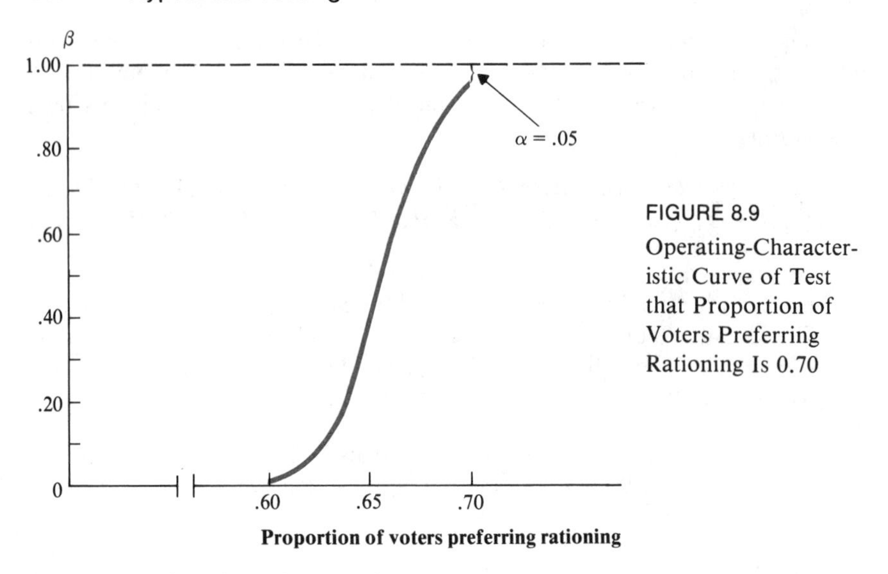

FIGURE 8.9
Operating-Characteristic Curve of Test that Proportion of Voters Preferring Rationing Is 0.70

probability that the company will accept the hypothesis that $P = .70$ when in fact $P = .60$ is less than .01. As we pointed out at the beginning of this section, the decision maker should set the value of α (and n) so that the operating-characteristic curve of the test is suitable for the problem at hand.

To illustrate how the OC curve in Figure 8.9 was calculated, consider the point on this curve corresponding to $P = .64$. To determine the value on the OC curve that corresponds to $P = .64$, recall that the null hypothesis ($P = .70$) will be accepted by the oil company if the sample proportion p is .662 or greater. Thus, the probability that the oil company will accept the null hypothesis equals the probability that the sample proportion will be .662 or greater. If $P = .64$, the sample proportion will be approximately normally distributed with mean equal to .64 and standard deviation equal to $\sqrt{(.64)(.36)/400}$, or .024. Consequently, the probability that the sample proportion will be .662 or more is equal to the probability that a normal variable with a mean of .64 and a standard deviation of .024 will be .662 or more. Applying the techniques we learned in Chapter 5, this probability is .18.[15] Thus, .18 is the value of the OC curve when $P = .64$.

The following example further illustrates the application of these test procedures.

Example 8.1 A ketchup producer samples a large shipment of tomatoes received from a supplier. A random sample of 100 tomatoes is inspected, and each tomato is rated either "superior" or "less-than-superior." The ketchup producer wants to test the hypothesis that 40 percent of the

15. The probability that the value of a normal variable with a mean of .64 and a standard deviation of .024 will be at least .662 equals the probability that the value of the standard normal variable will be at least $(.662 - .64) \div .024$, or .92. Using Appendix Table 2, we find that this probability is .18.

tomatoes in the shipment are "superior." Whether the population proportion is greater or less than 40 percent, the firm wants the test to detect this fact. If 34 of the tomatoes in the sample are rated "superior," should the ketchup producer reject the hypothesis that 40 percent of the tomatoes in the shipment fall in this category? (Assume that $\alpha = .05$.)

Solution: Let P be the proportion of "superior" tomatoes in the shipment. Then $H_0: P = .40$ and $H_1: P \neq .40$. (That is, the null hypothesis is that $P = .40$ and the alternative hypothesis is that $P \neq .40$.) Since $z_{.025} = 1.96$,

$$z_{\alpha/2}\sqrt{P_0(1 - P_0)/n} = 1.96\sqrt{\frac{(.40)(.60)}{100}} = .096,$$

$$P_0 + z_{\alpha/2}\sqrt{P_0(1 - P_0)/n} = .40 + .096 = .496,$$

$$P_0 - z_{\alpha/2}\sqrt{P_0(1 - P_0)/n} = .40 - .096 = .304.$$

Thus, the null hypothesis should be rejected if the sample proportion is greater than .496 or less than .304. Since the sample proportion is .34, the null hypothesis should not be rejected.

EXERCISES

8.1 Suppose that the Crooked Arrow National Bank is testing the hypothesis that the proportion of deposit slips filled out incorrectly is 1 percent.
(a) In this case, what is H_0?
(b) Under what circumstances will the bank incur a Type I error?
(c) Under what circumstances will the bank incur a Type II error?
(d) In this case, what considerations will determine the proper values of α and β?

8.2 The Crooked Arrow National Bank is interested in testing the hypothesis that a particular applicant is qualified for a teller's position.
(a) If this is the null hypothesis, what is the alternative hypothesis?
(b) What is the consequence of a Type I error and of a Type II error?
(c) What factors should guide the choice of α and β?

8.3 In Exercise 8.2 suppose that the bank hypothesizes that the applicant is not qualified.
(a) If this is the null hypothesis, what is the alternative hypothesis?
(b) What is the consequence of a Type I error and of a Type II error?
(c) What factors should guide the choice of α and β?

8.4 It is necessary for an automobile producer to test the hypothesis that the mean number of miles per gallon achieved by its cars is 28 against the alternative hypothesis that it is not 28. If the standard deviation of the number of miles per gallon achieved by the company's cars is 6, and if the company decides to base its test on a random sample of 100 of its cars, provide a suitable test procedure if α is set to equal .05.

8.5 In Exercise 8.4 suppose that the mean number of miles per gallon for the sample of 100 cars is 26.2. On the basis of this result, should the company reject the hypothesis that the population mean is 28? Why, or why not?

8.6 Suppose that the automobile producer in Exercise 8.4 is interested in rejecting the null hypothesis only if the mean number of miles per gallon achieved by its cars is less than 28. Provide the firm with a suitable test procedure for this set of circumstances.

8.7 If the automobile producer is interested in rejecting the null hypothesis only if the mean number of miles per gallon achieved by its cars is less than 28, should it reject the null hypothesis on the basis of the result in Exercise 8.5? Why, or why not?

8.8 The president of the Crooked Arrow National Bank wants to test whether 60 percent of the bank's loans are made to persons who reside in the city where the bank is located. The bank's statistician chooses a random sample of 200 of the people to whom the bank has made loans and finds that 52 percent reside in this city.
(a) If a 5 percent significance level is used, should this hypothesis be accepted or rejected? (Use a two-tailed test.)
(b) If a 1 percent significance level is used, should this hypothesis be accepted or rejected? (Use a two-tailed test.)

8.9 A public-interest law firm picks a random sample of 100 hi-fi repair stores and asks them to repair a hi-fi set. In 34 of the cases the customer is undercharged. Use these results to test (at the 10 percent significance level) the hypothesis that in one-half of all such cases customers are undercharged. (Use a two-tailed test.)

8.10 Use a one-tailed test in Exercise 8.9, where the alternative hypothesis is that $P < .50$.

8.11 To test whether its tellers are performing adequately, the Crooked Arrow National Bank takes a random sample of 10 of the transactions performed by each teller each day. If 1 or more of a certain teller's transactions contains an error, the teller is reprimanded.
(a) What is the decision rule in this case?
(b) If the hypothesis being tested is that 1 percent of the transactions contain errors, what is α in this case? (*Hint:* The normal distribution should *not* be used here.)

8.12 (a) In the situation in Exercise 8.11, what is the probability of a Type II error if in fact 2 percent of the transactions contain errors?
(b) What is this probability if 5 percent of the transactions contain errors?

8.13 Draw the operating-characteristic curve of the test used by the Crooked Arrow National Bank in Exercise 8.11.

8.14 The president of the Crooked Arrow National Bank makes a speech to the bank employees in which he hints that the existing system of sampling transactions is too likely to overlook tellers with a high incidence of errors. The bank's statistician, in response to this speech, suggests that a random sample of 20 of the transactions performed by each teller should be inspected each day and that the teller should be reprimanded if any transaction contains an error.
(a) What is the decision rule if this suggestion is accepted?
(b) What is the value of α if this suggestion is accepted (and if the hypothesis being tested is that 1 percent of the transactions contain errors)?
(c) If this suggestion is adopted, what is the value of β if 5 percent of the transactions contain errors?

8.6 Two-Sample Test of Means: Large Samples

In previous sections, we have described how a statistical procedure can be formulated to test whether a population mean or a population proportion equals a specified value. We turn now to a case where a random sample is drawn from each of two populations. In this section we describe how one can test the hypothesis that the means of two populations are equal. In the next section, we will describe how one can test the hypothesis that two population proportions are equal. In both sections,[16] we assume that the sample size in each population is large (that is, $n > 30$).

SETTING UP THE TEST

Suppose that a market research firm wants to determine whether the mean rating consumers give to their favorite beer differs from that which is given to brand X, when both beers are unidentified. Specifically, the firm picks a random sample of n_1 people, gives each person his or her favorite beer in an unmarked can, and asks the individual to rate it on a scale from 1 (very poor) to 90 (perfect). The firm then picks another random sample of n_2 people, gives each person brand X (also in an unmarked can) and asks the person to rate it on the same scale. The market research firm would like to test whether the mean rating of the favorite beer differs from the mean rating of brand X.

In contrast to the previous two sections, we are concerned here with two populations, not one. The first population is the population of ratings that beer drinkers will give to their favorite beer if it is served to them in an unmarked can. The mean of this population is μ_1 and the standard deviation is σ_1. The second population is the population of ratings that beer drinkers will give to brand X when it is given to them in an unmarked can. The mean of this population is μ_2 and the standard deviation is σ_2. The market research firm draws a random sample of n_1 ratings from the first population. In other words, n_1 individuals are asked to rate their favorite beer in an unmarked can. Also, the market-research firm draws a random sample of n_2 ratings from the second population. This means that n_2 people are asked to rate brand X beer in an unmarked can. There is no relation between the individuals chosen in the two samples, which are completely independent. How can the market research firm use these two samples to test whether μ_1 equals μ_2?

As we have stressed before, the first step in constructing a test is to formulate the null hypothesis and the alternative hypothesis. The null hypothesis here is that $\mu_1 = \mu_2$—or that $\mu_1 - \mu_2 = 0$. The alternative hypothesis is that $\mu_1 - \mu_2 \neq 0$, since the firm is interested in rejecting the null hypothesis both when $\mu_1 - \mu_2 < 0$ and when $\mu_1 - \mu_2 > 0$. In other words, the firm wants to reject the null hypothesis both when brand X's mean rating is higher than the mean rating of the favorite beer and when the reverse is true. Stated differently, what the firm wants is a two-tailed test.

The next step is to specify the value of α, the significance level of the test. As we have emphasized in previous sections, the value of α should be determined on the basis of the relative costs of both Type I and Type II errors. An effort should be made to strike a proper balance between α and β, even if (as is sometimes the case) one has only a vague idea of the relative costs. One important point is that α should always be specified *before the data are*

16. In the next section, n may have to be substantially greater than 30 for the normal approximation to apply. As pointed out in footnote 13, this depends on the value of P. In the rest of this chapter, we assume that the population is large relative to the sample (or that sampling is with replacement).

examined. (This is true for any type of test, not just those described in this section.) Otherwise, it would be possible for the investigator to choose a significance level small enough so that the null hypothesis is accepted (if the investigator wants to accept it) or a significance level large enough so that the null hypothesis is rejected (if the investigator wants to reject it). Such a procedure would be a mockery of correct statistical practice. For the sake of concreteness, suppose that α in this case is set equal to .01.

DECISION RULES

Once the value of α has been specified, we can determine the decision rule that should be applied to test whether the difference between the two population means (that is, $\mu_1 - \mu_2$) is zero. If both n_1 and n_2 are large, it can be shown that the sampling distribution of the difference between the sample means (that is, $\bar{X}_1 - \bar{X}_2$) is approximately normal, with a mean of $(\mu_1 - \mu_2)$ and a standard deviation of $\sqrt{\sigma_1^2/n_1 + \sigma_2^2/n_2}$. Thus, if the null hypothesis is true (which means that $\mu_1 - \mu_2 = 0$), the difference between the sample means is approximately normally distributed, with a mean of zero and a standard deviation of $\sqrt{\sigma_1^2/n_1 + \sigma_2^2/n_2}$. Consequently, if the market research firm knows that σ_1 (the standard deviation of the ratings of the favorite beer) is 12 and σ_2 (the standard deviation of the ratings of brand X) is 10, and if $n_1 = 100$ and $n_2 = 200$, the difference between the sample means is approximately normally distributed, with a mean of zero and a standard deviation of $\sqrt{12^2/100 + 10^2/200}$ or 1.39. In other words, if the null hypothesis is true, the sampling distribution of $(\bar{X}_1 - \bar{X}_2)$ is as shown in Figure 8.10.

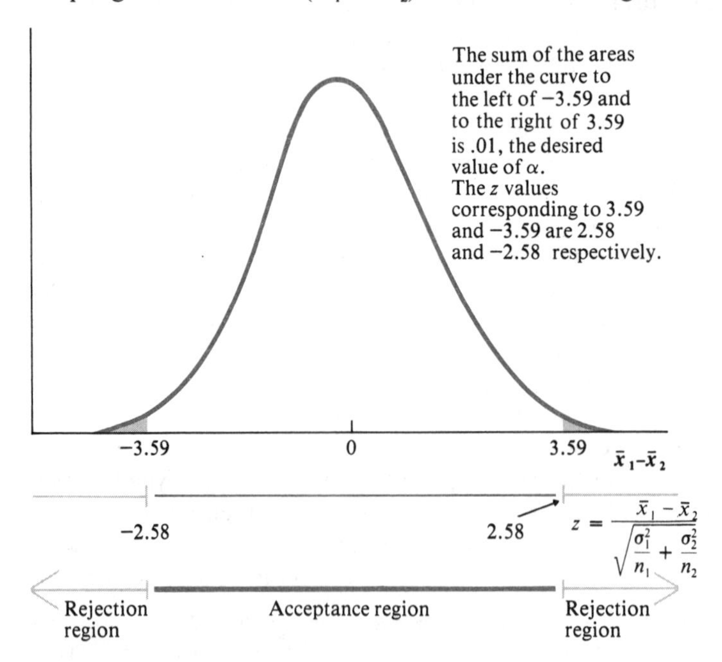

FIGURE 8.10 Sampling Distribution of Difference between Sample Means, if Null Hypothesis Is True

Since the market research firm wants to reject the null hypothesis either when $\mu_1 - \mu_2 < 0$ or when $\mu_1 - \mu_2 > 0$, it must set up a rejection region under each tail of the sampling distribution of $(\bar{X}_1 - \bar{X}_2)$, as shown in Figure 8.10. The rejection region in *each* tail must be constructed so that the probability is $\alpha/2$ that the difference between the sample means will fall in this region if the null hypothesis is true. (Since there are two tails, the total probability of a Type 1 error will be α.) If the rejection region in the lower tail is

$$\bar{x}_1 - \bar{x}_2 < -z_{\alpha/2}\sqrt{\frac{\sigma_1^2}{n_1} + \frac{\sigma_2^2}{n_2}} = -2.58(1.39) = -3.59,$$

and if the rejection region in the upper tail is

$$\bar{x}_1 - \bar{x}_2 > z_{\alpha/2}\sqrt{\frac{\sigma_1^2}{n_1} + \frac{\sigma_2^2}{n_2}} = 2.58(1.39) = 3.59,$$

this condition is met. That is, the probability is .005 that $(\bar{X}_1 - \bar{X}_2)$ will be less than -3.59 if the two population means are equal, and the probability is .005 that $(\bar{X}_1 - \bar{X}_2)$ will be greater than 3.59 if the two population means are equal. Thus, the market research firm should use the following decision rule: Accept the null hypothesis if $-3.59 \leqslant (\bar{x}_1 - \bar{x}_2) \leqslant 3.59$; reject the null hypothesis if $(\bar{x}_1 - \bar{x}_2) < -3.59$ or if $(\bar{x}_1 - \bar{x}_2) > 3.59$. The decision rule has the desired probability of a Type I error, namely .01.

Like any random variable, the difference between the sample means can be expressed in standard units (that is, in z values). Since its mean is zero (if the null hypothesis is true) and its standard deviation is $\sqrt{\sigma_1^2/n_1 + \sigma_2^2/n_2}$, its observed value in standard units is

$$\boxed{z = (\bar{x}_1 - \bar{x}_2) \div \sqrt{\frac{\sigma_1^2}{n_1} + \frac{\sigma_2^2}{n_2}}.}$$

Stated in terms of z, the rejection region in the lower tail is

$$z < -z_{\alpha/2} = -2.58$$

and the rejection region in the upper tail is

$$z > z_{\alpha/2} = 2.58.$$

Why? Because z equals $(\bar{x}_1 - \bar{x}_2) \div 1.39$, since $\sqrt{\sigma_1^2/n_1 + \sigma_2^2/n_2} = 1.39$. Hence, if$(\bar{x}_1 - \bar{x}_2) < -3.59$, it follows that $z < -3.59 \div 1.39$, or -2.58. Similarly, if $(\bar{x}_1 - \bar{x}_2) > 3.59$, it follows that $z > 3.59 \div 1.39$, or 2.58. Thus, the market-research firm should accept the null hypothesis if $-2.58 \leqslant z \leqslant 2.58$, and reject it otherwise.

In general, the decision rules for testing the null hypothesis that $\mu_1 - \mu_2 = 0$ are as follows.

Decision Rules:[17] *When the alternative hypothesis is $\mu_1 - \mu_2 \neq 0$, reject the*

17. If the population standard deviations are unknown, the sample standard deviations can be substituted in these decision rules if both samples are large.

null hypothesis if z *exceeds* $z_{\alpha/2}$ *or is less than* $-z_{\alpha/2}$. *When the alternative hypothesis is* $\mu_1 - \mu_2 < 0$, *reject the null hypothesis if* $z < -z_\alpha$. *When the alternative hypothesis is* $\mu_1 - \mu_2 > 0$, *reject the null hypothesis if* $z > z_\alpha$.

The following example is provided to further illustrate the application of these tests.

Example 8.2 A firm has two plants, each of which produces screws. It is necessary to test whether the mean diameter of the screws produced at one plant equals the mean diameter of those produced at the other plant. If the mean diameter of screws at either plant is larger than that at the other plant, the hypothesis that the mean diameters at the two plants are equal will have to be rejected. At both plants the standard deviation of the diameters of the screws is .01 inches. A random sample of 100 screws is taken from each plant's output, and the sample mean is found to be .41 inches at the one plant and .45 inches at the other. If the significance level is .05, should the firm reject the hypothesis that the means are equal?

Solution: Let μ_1 be the mean of the diameters of the screws produced at the one plant and μ_2 be the mean of the diameters of the screws produced at the other plant. The null hypothesis is that $\mu_1 - \mu_2 = 0$, and the alternative hypothesis is that $\mu_1 - \mu_2 \neq 0$. According to the decision rule given above, the null hypothesis should be rejected if $z > 1.96$ or if $z < -1.96$, since $z_{.025} = 1.96$. Because $z = (.41 - .45) \div \sqrt{.01^2/100 + .01^2/100} = -.04/.0014 = -28.6$, the firm should reject the null hypothesis that the mean diameters are equal.

8.7 Two-Sample Test of Proportions: Large Samples

In this section we will describe a statistical procedure to test the hypothesis that two population proportions are equal. This test procedure is based on two independent samples, one from each population. It is assumed that each of these samples is large.

SETTING UP THE TEST

Suppose that a chair manufacturer needs to find out whether a new technique should be substituted for an old one. If the proportion of chairs meeting specifications is greater using the new technique, it is economical for the firm to make the switch; otherwise, it is not economical to do so. The firm uses the new technique to produce 200 chairs and the old technique to produce 400 chairs. It then determines the proportion of chairs in each sample which meets specifications. How can the firm test whether the proportion of chairs meeting specifications is the same for both techniques?

As we have stressed, the first step in setting up a test is to formulate the null hypothesis and the alternative hypothesis. Clearly, the null hypothesis is that P_1, the proportion of chairs meeting specifications when the new technique is used, equals P_2, the proportion meeting specifications when the old technique is used. The alternative hypothesis is that $P_1 > P_2$. Since the firm wants to switch to the new technique only if $P_1 > P_2$, this is the kind of departure from the null hypothesis that it wants to detect. It is of no concern whether $P_2 < P_1$ since, if this is the case, the implications are no different than if the null hypothesis is true: In either case the firm will not switch to the new technique.

The next step, as we know, is to set a significance level α. The considerations influencing this choice are no different here than in earlier sections. Suppose that the chair producer decides to set α equal to .05.

DECISION RULES

Given that the value of α has been chosen, we can determine the decision rule that should be applied to determine whether the difference between the two population proportions (that is, $P_1 - P_2$) is zero. If the null hypothesis is true and if both n_1 and n_2 are large, the difference between the sample proportions ($p_1 - p_2$) is approximately normally distributed with a mean of zero and a standard deviation of $\sqrt{P(1 - P)(1/n_1 + 1/n_2)}$, where P is the value of the population proportion in both populations. (In other words, if the null hypothesis is true, the sampling distribution of the difference between the sample proportions is as shown in Figure 8.11.)

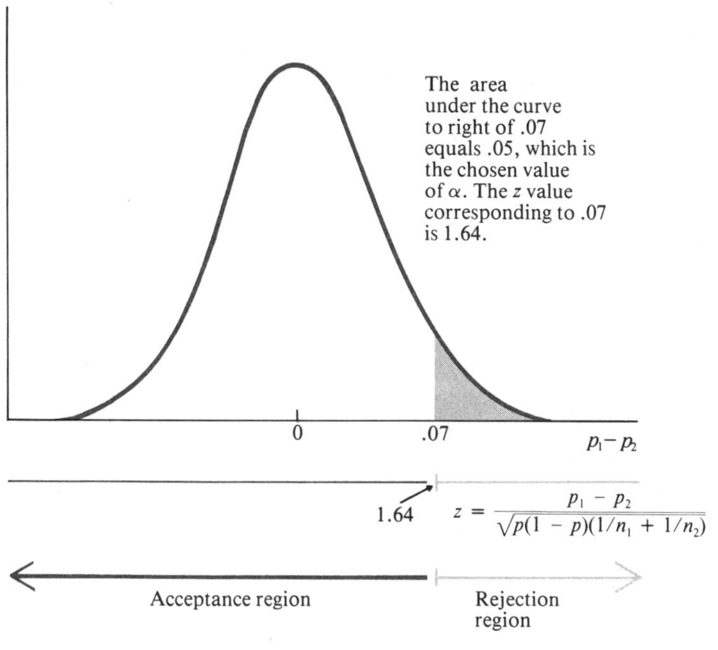

FIGURE 8.11 Sampling Distribution of Difference between Sample Proportions, if Null Hypothesis Is True

If the null hypothesis is true, a good estimate of P is

$$p = \frac{n_1 p_1 + n_2 p_2}{n_1 + n_2}.$$

Clearly, p is the weighted mean of the sample proportions in the two samples, each being weighted by the sample size. If n_1 and n_2 are large we can substitute the above estimate for P in the expression in the previous paragraph. Thus, if the null hypothesis is true, the difference between the sample proportions is approximately normally distributed, with a mean of zero and a standard deviation of $\sqrt{p(1-p)(1/n_1 + 1/n_2)}$.

Since the chair manufacturer wants to reject the null hypothesis only if $P_1 > P_2$, the rejection region must be set up under the upper tail of the sampling distribution of $(p_1 - p_2)$, as shown in Figure 8.11. This rejection region must be constructed so that the probability is α that the difference between the sample proportions will fall in this region, if the null hypothesis is true. If this rejection region is

$$p_1 - p_2 > z_\alpha \sqrt{p(1-p)\left(\frac{1}{n_1} + \frac{1}{n_2}\right)},$$

this condition is met. Thus, since $\alpha = .05$, the chair company should use the following decision rule: Accept the null hypothesis if $p_1 - p_2 \leqslant 1.64\sqrt{p(1-p)(3/400)}$; but if $p_1 - p_2 > 1.64\sqrt{p(1-p)(3/400)}$, reject the null hypothesis.

Like any random variable, the difference between the sample proportions can be expressed in standard units. Since its mean is zero (if the null hypothesis is true) and its standard deviation is $\sqrt{p(1-p)(1/n_1 + 1/n_2)}$, its observed value in standard units is

$$\boxed{z = (p_1 - p_2) \div \sqrt{p(1-p)(1/n_1 + 1/n_2)}.}$$

Stated in terms of z, the chair manufacturer should reject the null hypothesis if $z > 1.64$, and accept it otherwise. Why? Because the z value exceeds 1.64 if $p_1 - p_2 > 1.64\sqrt{p(1-p)(1/n_1 + 1/n_2)}$ which, according to the previous paragraph, is the condition for rejecting the null hypothesis.

To illustrate the application of this decision rule, suppose that the chair company finds that the proportion of chairs meeting specifications in the sample where the new technique is used is .60 and that the proportion in the sample where the old technique is used is .50. Then,

$$p = \frac{(200)(.60) + (400)(.50)}{600} = \frac{320}{600} = .53,$$

and

$$z = (.60 - .50) \div \sqrt{.53(.47)(3/400)}$$
$$= .10 \div .043 = 2.33.$$

Consequently, the chair manufacturer should reject the null hypothesis (that $P_1 = P_2$) because $z > 1.64$.

In general, when n_1 and n_2 are sufficiently large, the decision rules for testing the null hypothesis that $P_1 - P_2 = 0$ are as follows.

> **Decision Rules:** *When the alternative hypothesis is* $P_1 - P_2 \neq 0$, *reject the null hypothesis if* z *exceeds* $z_{\alpha/2}$ *or is less than* $-z_{\alpha/2}$. *When the alternative hypothesis is* $P_1 - P_2 < 0$, *reject the null hypothesis if* $z < -z_\alpha$. *When the alternative hypothesis is* $P_1 - P_2 > 0$, *reject the null hypothesis if* $z > z_\alpha$.

A further illustration of the application of these tests follows.

Example 8.3 A polling organization is interested in determining whether candidate Jones will run better in urban or in rural areas. A random sample is drawn of 400 urban voters and 400 rural voters, and it is determined that 55 percent of the urban voters and 49 percent of the rural voters prefer candidate Jones over the others. Should the polling organization conclude that the popularity of this candidate differs between urban and rural areas? (Assume that $\alpha = .05$.)

Solution: Let P_1 be the proportion of urban voters who prefer candidate Jones, and let P_2 be the proportion of rural voters who prefer Jones. Since the polling organization wants to reject the null hypothesis (that $P_1 = P_2$) when either $P_1 > P_2$ or $P_2 > P_1$, the correct decision rule is to reject the null hypothesis if z is greater than 1.96 or less than -1.96 (because $z_{\alpha/2} = 1.96$). Since $n_1 = n_2 = 400$,

$$p = \frac{(400)(.55) + (400)(.49)}{800} = .52,$$

and

$$z = (.55 - .49) \div \sqrt{.52(.48)(.005)}$$

$$= .06 \div .035 = 1.71.$$

Thus, the polling organization should not reject the null hypothesis (that $P_1 = P_2$), since z is not greater than 1.96 or less than -1.96.

EXERCISES

8.15 A personnel agency decides to test whether the mean aptitude test score of engineering graduates differs from that of business graduates. The agency knows that the standard deviation of the scores among engineering graduates is 10 points and that the same is true for business graduates. A random sample is taken of 100 engineering graduates and of 100 business graduates. The mean score of the engineering graduates is 80 and the mean score of the business graduates is 78.
(a) What is the null hypothesis? What is the alternative hypothesis?
(b) What is the appropriate decision rule?
(c) If $\alpha = .10$, should the agency accept or reject the null hypothesis?
(d) If $\alpha = .05$, should the agency accept or reject the null hypothesis?
(e) If $\alpha = .01$, should the agency accept or reject the null hypothesis?

8.16 The personnel agency in the previous exercise wants to reject the hypothesis that the mean scores of engineering and business graduates are the same only if the mean engineering score exceeds the mean business score.

(a) What is the null hypothesis? What is the alternative hypothesis?
(b) What is the appropriate decision rule?
(c) If $\alpha = .10$, should the agency accept or reject the null hypothesis?
(d) If $\alpha = .05$, should the agency accept or reject the null hypothesis?
(e) If $\alpha = .01$, should the agency accept or reject the null hypothesis?

8.17 The personnel agency in Exercise 8.15 wants to reject the hypothesis that the mean scores of engineering and business graduates are the same only if the mean business score exceeds the mean engineering score.
(a) What is the null hypothesis? What is the alternative hypothesis?
(b) What is the appropriate decision rule?
(c) If $\alpha = .10$, should the agency accept or reject the null hypothesis?
(d) If $\alpha = .05$, should the agency accept or reject the null hypothesis?
(e) If $\alpha = .01$, should the agency accept or reject the null hypothesis?

8.18 An economist wants to determine whether the proportion of tool and die firms now using numerically controlled machine tools is different in Canada than in the United States. The economist draws a random sample of 81 tool and die firms in Canada and 100 tool and die firms in the United States, and finds that 20 of the Canadian firms and 30 of the American firms have introduced numerically controlled machine tools.
(a) What is the null hypothesis? What is the alternative hypothesis?
(b) What is the appropriate decision rule?
(c) If $\alpha = .10$, should the null hypothesis be accepted or rejected?
(d) If $\alpha = .05$, should the null hypothesis be accepted or rejected?
(e) If $\alpha = .01$, should the null hypothesis be accepted or rejected?

8.19 Suppose that the economist in the previous exercise wants to reject the hypothesis that the proportion of tool and die firms now using numerically controlled machine tools is the same in Canada as in the United States only if the proportion is higher in Canada.
(a) What is the null hypothesis? What is the alternative hypothesis?
(b) What is the appropriate decision rule?
(c) If $\alpha = .10$, should the null hypothesis be accepted or rejected?
(d) If $\alpha = .05$, should the null hypothesis be accepted or rejected?
(e) If $\alpha = .01$, should the null hypothesis be accepted or rejected?

8.20 Suppose that the economist in Exercise 8.18 wants to reject the hypothesis that the proportion of tool and die firms now using numerically controlled machine tools is the same in Canada as in the United States only if the proportion of users is higher in the United States.
(a) What is the null hypothesis? What is the alternative hypothesis?
(b) What is the appropriate decision rule?
(c) If $\alpha = .10$, should the null hypothesis be accepted or rejected?
(d) If $\alpha = .05$, should the null hypothesis be accepted or rejected?
(e) If $\alpha = .01$, should the null hypothesis be accepted or rejected?

8.8 One-Sample Test of a Mean: Small Samples

In previous sections of this chapter we have assumed that the sample was large. We now turn to the case where the sample is small ($n \leqslant 30$). In this section, we indicate how one can test the hypothesis that the population mean equals a specified value. In the next section, we will indicate how one can test the hypothesis that two population means are equal. In both sections we assume that the populations are normal or approximately normal. In neither section do we assume that any population standard deviation is known.

SETTING UP THE TEST

Suppose that the beer manufacturer (discussed briefly in Section 8.3) wants to test whether the mean rating for a new brand of its beer equals 60 (which is known to be the population mean rating for its old brand). If the mean rating for the new brand is higher than 60, the beer manufacturer will substitute the new brand for the old. If the mean rating for the new brand is less than 60, the implications are the same as if it equals 60. That is, in either of these two latter cases the new brand will be dropped. Suppose that the beer manufacturer samples 16 beer drinkers and obtains ratings on a scale of 1 to 90 for the new brand of beer, the results being those in Table 8.3. Should the manufacturer accept or reject the hypothesis that the mean rating for the new brand is 60?

As usual, we begin by formulating the null hypothesis and the alternative hypothesis. In this case, the null hypothesis is that the population mean rating for the new brand μ equals 60. The alternative hypothesis is that this mean exceeds 60. (The beer manufacturer will not reject the null hypothesis if $\mu < 60$, since if $\mu < 60$ the implications are the same as if μ equals 60.) To sum up, H_0 (the null hypothesis) is that $\mu = 60$, and H_1 (the alternative hypothesis) is that $\mu > 60$. We also need to know the significance level: Assume that the beer manufacturer wants α to equal .05.

TABLE 8.3

Ratings of a New Brand of Beer by 16 Randomly Selected Consumers

Person	Rating	Person	Rating	Person	Rating
1	46	7	71	12	54
2	64	8	77	13	64
3	62	9	69	14	63
4	58	10	67	15	57
5	54	11	59	16	68
6	65				

DECISION RULES

In contrast to our earlier discussion (in Section 8.4) of tests that μ equals a specified value, we do not know the population standard deviation, and the sample size is small. The decision rules in Section 8.4 are therefore not appropriate here. Nonetheless, if the population is normal we can obtain the decision rule we want by using the t distribution. As pointed out in Chapter 7, $(\bar{X} - \mu) \div s/\sqrt{n}$ has the t distribution (with $n - 1$ degrees of freedom) if the population is normal. (Of course, s is the sample standard deviation, and n is the sample size.) Thus, if $\mu = \mu_0$, the value specified by the null hypothesis, $(\bar{X} - \mu_0) \div s/\sqrt{n}$ has the t distribution. To test the null hypothesis, we compute

$$t = (\bar{x} - \mu_0) \div s/\sqrt{n},$$

which is used as follows.

Decision Rules: *When the alternative hypothesis is* $\mu \neq \mu_0$, *reject the null hypothesis if* t *exceeds* $t_{\alpha/2}$ *or is less than* $-t_{\alpha/2}$. *When the alternative hypothesis is* $\mu < \mu_0$, *reject the null hypothesis if* $t < -t_\alpha$. *When the alternative hypothesis is* $\mu > \mu_0$, *reject the null hypothesis if* $t > t_\alpha$.

Since the alternative hypothesis in the beer manufacturer's case is that $\mu > 60$, the third decision rule is the applicable one. Using the data in Table 8.3, we find that the sample standard deviation s is 7.60. Thus, the decision rule is: Reject the null hypothesis if t is greater than 1.753, since, as shown in Appendix Table 6, $t_{.05} = 1.753$. (Note that since $n = 16$, there are 16 − 1, or 15, degrees of freedom.) Because $\bar{x} = 62.38$, $t = (62.38 - 60) \div 7.6/\sqrt{16}$, or 1.25. Since t is not greater than 1.753, there is no reason to reject the null hypothesis that the mean rating of the new brand of beer is 60.

The example below provides a further application of these tests.

Example 8.4 A construction firm samples 9 pieces of pipe in a shipment and finds that their diameters (in inches) are as follows.

7.7	8.0	7.4
7.8	8.4	7.2
7.9	8.2	7.6

The firm would like to test (at the .05 significance level) the hypothesis that the mean diameter in the shipment is 8.0 inches. The firm wants to reject this hypothesis if the mean is less than 8.0 inches, but not if it is greater than 8.0 inches. The diameters are believed to be normally distributed. Should the firm accept or reject this hypothesis?

Solution: The null hypothesis is that the mean diameter μ equals 8.0 inches and the alternative hypothesis is that $\mu < 8.0$ inches. The sample standard deviation s is .38 inches. The null hypothesis should be rejected if $t < -1.86$, since Appendix Table 6 shows that $t_{.05}$ equals 1.86 when there are 8 degrees of freedom. Because $\mu_0 = 8.0$ and $\bar{x} = 7.8$,

$$t = (7.8 - 8.0) \div .38/\sqrt{9}$$
$$= -1.57.$$

Since the observed value of t is not less than −1.86, the null hypothesis should not be rejected. That is, the firm should not reject the hypothesis that the mean pipe diameter is 8.0 inches.[18]

18. For a case study in which an economic theory was subjected to this sort of test, see L. Fouraker and S. Siegel, "Tests of Hypotheses Concerning Bilateral Monopoly," reprinted in E. Mansfield, op. cit.

8.9 Two-Sample Test of Means: Small Samples

We will now describe how one can test the hypothesis that the means of two populations are equal, if the samples are small and the population standard deviations are unknown. (The tests assume that the population standard deviations in the two populations are equal.)

SETTING UP THE TEST

Suppose that the market-research firm in Section 8.6 wants to test whether the mean rating given by consumers to their favorite beer differs from that given to brand Y. The firm samples nine people and asks them to rate a beer (in reality their favorite brand) in an unmarked can. Suppose that the ratings are as shown in Table 8.4. Also, the firm samples another nine people and asks them to rate a beer (in reality brand Y) in an unmarked can. These ratings are also shown in Table 8.4. What should the firm conclude from these results?

Again, the first step is to specify the null and alternative hypotheses. The null hypothesis is that μ_1, the population mean of the ratings for the favorite beer, equals μ_2, the population mean of the ratings for brand Y. The alternative hypothesis is $\mu_1 - \mu_2 \neq 0$. The next step is to set the significance level, α. Suppose that the firm sets α at .05.

TABLE 8.4
Results of Beer Manufacturer's Study

Person	Rating of favorite brand in unmarked can	Person	Rating of brand Y in unmarked can
1	58	10	62
2	60	11	60
3	62	12	60
4	60	13	62
5	60	14	63
6	62	15	59
7	58	16	58
8	59	17	64
9	61	18	61

$\bar{x}_1 = 60$

$$s_1 = \sqrt{\frac{18}{8}} = 1.5$$

$\bar{x}_2 = 61$

$$s_2 = \sqrt{\frac{30}{8}} = 1.94$$

DECISION RULES

In contrast to our earlier discussion in Section 8.6 of tests of whether two population means are equal, we do not know the standard deviation of each population, and the sample size is small. Nonetheless, if both populations are normal, and if their standard deviations are equal, we can obtain the decision rules we need. First, an estimate of the variance in each population is

$$s^2 = \frac{(n_1 - 1)s_1^2 + (n_2 - 1)s_2^2}{n_1 + n_2 - 2}.$$

If the null hypothesis is true, it can be shown that $(\bar{X}_1 - \bar{X}_2) \div \sqrt{s^2(1/n_1 + 1/n_2)}$ has the t distribution with $n_1 + n_2 - 2$ degrees of freedom. Thus, to test the null hypothesis, we compute

$$\boxed{t = (\bar{x}_1 - \bar{x}_2) \div \sqrt{s^2(1/n_1 + 1/n_2)},}$$

which is used as follows.

> ***Decision rules:*** *When the alternative hypothesis is $\mu_1 - \mu_2 \neq 0$, reject the null hypothesis if t exceeds $t_{\alpha/2}$ or is less than $-t_{\alpha/2}$. When the alternative hypothesis is $\mu_1 - \mu_2 < 0$, reject the null hypothesis if $t < -t_\alpha$. When the alternative hypothesis is $\mu_1 - \mu_2 > 0$, reject the null hypothesis if $t > t_\alpha$.*

Because the alternative hypothesis is $\mu_1 - \mu_2 \neq 0$, the first decision rule is the appropriate one for the market-research firm. Using the data in Table 8.4, we must calculate s^2, which is

$$\frac{(n_1 - 1)s_1^2 + (n_2 - 1)s_2^2}{n_1 + n_2 - 2} = \frac{8(1.5)^2 + 8(1.94)^2}{16} = 3.$$

Thus,

$$t = (60 - 61) \div \sqrt{3(1/9 + 1/9)}$$
$$= -1 \div .82 = -1.22.$$

Since $t_{.025} = 2.12$ when there are 16 degrees of freedom (see Appendix Table 6), the observed value of t does not exceed $t_{.025}$ or fall below $-t_{.025}$. Consequently, the null hypothesis should not be rejected.

The above illustration is based on an actual study carried out by the director of research of Carling Brewing Company, one of the nation's largest breweries.[19] The actual study was more complicated and involved larger samples than those considered here, but the principles are essentially the same. Also, it is interesting to note that the results were much the same as the hypothetical results analyzed here. In particular, the actual study concluded that "participants, in general, did not appear to be able to discern the taste differences among the various beer brands. . . ."[20]

19. R. Allison and K. Uhl, "Influence of Beer Brand Identification on Taste Perception," *Journal of Marketing Research,* August 1964, pp. 36–39.
20. Ibid., p. 39.

The example below again illustrates the application of these tests.

Example 8.5 The Department of Defense tests two types of aircraft, type A and type B, to determine whether their mean speed under a particular set of conditions is the same. In four tests of type A aircraft the mean speed under the prescribed conditions was 590 miles per hour, and the sample standard deviation was 100 miles per hour. In four tests of type B aircraft the mean speed under these conditions was 750 miles per hour, and the standard deviation was 80 miles per hour. The Defense Department wants to detect differences in mean speed, regardless of which type of aircraft may be faster. There is good reason to believe that the speeds of each type of aircraft are normally distributed with the same standard deviation. If $\alpha = .05$, should the Department of Defense accept or reject the hypothesis that the mean speeds are equal?

Solution: Let μ_1 be the population mean speed of type A aircraft, and let μ_2 be the population mean speed of type B aircraft. Since the alternative hypothesis is $\mu_1 - \mu_2 \neq 0$, the first decision rule given above is the appropriate one. Firstly,

$$s^2 = \frac{(n_1 - 1)s_1^2 + (n_2 - 1)s_2^2}{n_1 + n_2 - 2} = \frac{3(100^2) + 3(80^2)}{6} = 8{,}200,$$

and

$$t = (\bar{x}_1 - \bar{x}_2) \div \sqrt{s^2\left(\frac{1}{n_1} + \frac{1}{n_2}\right)} = (590 - 750) \div \sqrt{8200\left(\frac{1}{4} + \frac{1}{4}\right)}$$

$$= -160 \div 64.03 = -2.50.$$

As shown in Appendix Table 6, $t_{.025} = 2.447$ (because there are $4 + 4 - 2$, or 6, degrees of freedom). Thus, the Department of Defense should reject the null hypothesis if the observed value of t exceeds 2.447 or is less than -2.447. Hence, since $t = -2.50$, it follows that the null hypothesis (that $\mu_1 = \mu_2$) should be rejected.

8.10 Limitations of Classical Hypothesis Testing

Our purpose in this chapter has been to present the basic principles of classical testing procedures and to describe and apply some of the most commonly used tests. Although these test procedures are of great importance, they are subject to a number of limitations described below. Many of these limitations will be remedied in later chapters.

Dependence upon Sample Information. The test procedures described in this chapter are totally dependent upon sample information. While sample information is often available, this is not always the case. For example, such information is not available concerning the effects of a merger of two firms because two companies cannot merge with one another a number of times in order to see what happens. Other kinds of statistical procedures, which will be covered in Chapters 15 and 16, allow us to come to decisions based on evidence other than purely sample information.

Incomplete Attention to Costs of Error and of Sampling. As we have seen, the test procedures described here are based on a choice by the decision maker of α and β, the probabilities of Type I and Type II error. However, the classical techniques do not indicate in any detail how α and β should be chosen. If one can specify the costs of various types and degrees of errors, it is possible to extend the analysis so that this choice of α and β is in effect subsumed within a more inclusive analysis. This kind of extension of classical theory falls under the heading of decision theory and is discussed in the last two chapters of this book.

Deciding Whether or Not to Sample. The test procedures described in this chapter are designed to indicate whether, on the basis of a given sample, a certain hypothesis should be accepted or rejected. These procedures do not tell us whether a sample should be taken in the first place. In some cases, the cost of collecting and analyzing a sample outweighs the value of the information the sample is likely to provide. If this is the case, the decision maker is better off not sampling at all. This kind of decision, which is not covered by the test procedures discussed here, is analyzed in detail in Chapter 16.

Inclusion of Prior Probabilities. The test procedures described here make no use of whatever personal or subjective judgments the decision maker may be willing to make. If reasonably dependable judgments of this sort can be made, it is obviously useful to be able to include them in one's analysis. In our discussion of decision theory in Chapters 15 and 16, we shall see how the analysis can be extended to include such judgments.

8.11 Statistical Process Control at the American Stove Company: A Case Study

One of the most important industrial applications of the statistical theory of hypothesis testing is in *process control.* In many kinds of manufacturing plants, the production process is repetitive. A bottle manufacturing plant turns out batch after batch of bottles, and a light bulb manufacturing plant turns out batch after batch of light bulbs. Obviously, it is extremely important that the quality of these bottles or light bulbs be maintained, but because of gradual changes in the underlying process, the shape or thickness of the bottles or light bulbs may change somewhat. The manufacturer wants to maintain a close surveillance over product quality so that whenever there is evidence that the plant is turning out an inferior product corrections can be made, thus avoiding the financial loss associated with the production of substandard items.

In carrying out this objective firms commonly use statistical process control. To illustrate this technique let's return to the actual case of the American Stove Company, which some time ago was producing a metal piece whose height, according to the specifications, was supposed to be about .83 inches. The firm studied the performance of its production process in order to establish a statistical process control procedure. The firm's statisticians calculated the mean value of the heights of the metal pieces produced during a long period when the production process was known to be in control. During this period the mean height of the metal pieces being produced was found to be .8312 inches. The standard deviation of the heights of the metal pieces produced during this period also was measured and was found to be .006169 inches.

The firm wanted a test procedure for detecting changes in the mean height of the metal pieces produced. For example, if (because of machine failure or mistakes by workers) the mean height increased to .841 inches or decreased to .828 inches, the firm wanted to be able to ascertain this change. Thus, what was really required was a test procedure for determining whether the mean height equaled .8312 inches against the two-sided alternative that the mean was greater than or less than .8312. But in contrast to our earlier discussion, the firm did not want α (the probability of a Type I error) to equal .05. Instead, it wanted α to be very low—about .003. The reason for setting such a low value of α was that the cost of a Type I error was very high. The cost was high because a Type I error in this context would have necessitated halting the production process (because the test indicates that the mean differs from .8312 inches) when in fact there was no reason to stop production. Such false alarms can be very costly.

The firm decided to pick a simple random sample of five metal pieces from each day's output and compute the mean height of the metal pieces in the sample. The following decision rule was chosen: Accept the null hypothesis (that the mean equals .8312 inches) if the sample mean $\bar{x}$ lies between $.8312 - 3(.006169/\sqrt{5})$ and $.8312 + 3(.006169/\sqrt{5})$—in other words, if the sample mean lies between .8229 and .8395 inches. Reject the null hypothesis if the sample mean lies below .8229 inches or above .8395 inches. This is a straightforward application of the results in Section 8.4,[21] which indicate that if the alternative hypothesis is $\mu \neq \mu_0$, the null hypothesis should be accepted if $\mu_0 - z_{\alpha/2}\sigma/\sqrt{n} \leqslant \bar{x} \leqslant \mu_0 + z_{\alpha/2}\sigma/\sqrt{n}$, and rejected otherwise. Since $\mu_0 = .8312$, $\sigma = .006169$, $n = 5$, and $z_{\alpha/2} \doteqdot 3$, this amounts to precisely the same thing as the firm's decision rule. (Since $z_{.0015} \doteqdot 3$, 3 is the value of $z_{\alpha/2}$ that is customarily used in quality control.)

Quality control procedures of this kind are generally embodied in a ***control chart***. For example, the American Stove Company's inspectors plotted each day's sample mean on the control chart in Figure 8.12. This chart contains one horizontal line which shows the mean height when the production process is in control (.8312 inches) and two other horizontal lines, one at $\mu_0 + 3\sigma/\sqrt{n}$ (that is, .8395 inches) and one at $\mu_0 - 3\sigma/\sqrt{n}$ (that is, .8229 inches). These latter two lines are called the upper and lower control limits. When a sample mean falls above the ***upper control limit*** (.8395 inches) or below the ***lower control limit*** (.8229 inches), the decision rule in the previous paragraph says we should reject the null hypothesis.

To illustrate the actual performance of this control chart, let's look at the

21. Since the sample size is only 5, the sampling distribution of the sample mean may not be exactly normal. (See Chapter 6.) But even if the approximation is rough, there is bound to be a very small probability of a Type I error, given the manner in which the control limits are constructed.

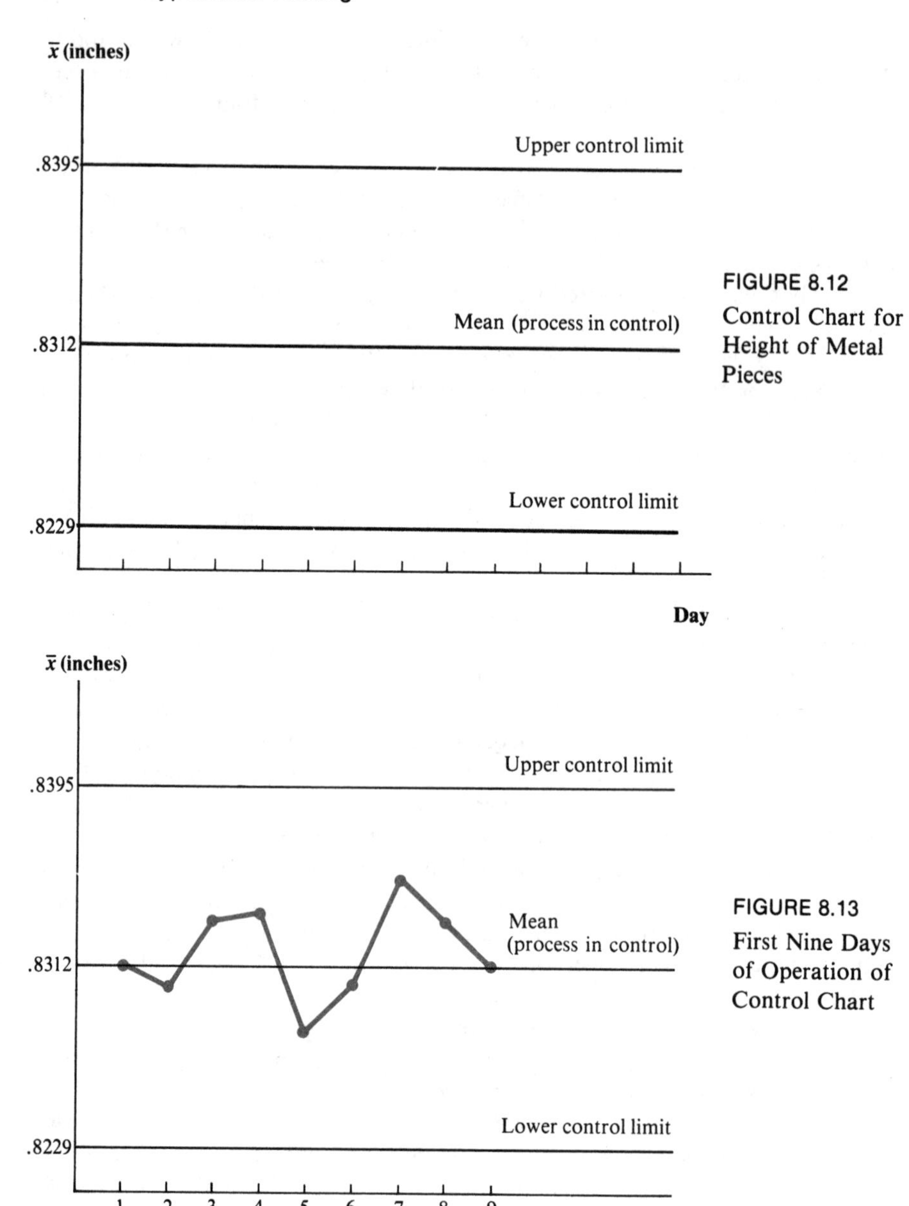

FIGURE 8.12
Control Chart for Height of Metal Pieces

FIGURE 8.13
First Nine Days of Operation of Control Chart

first nine days of its use. During the first day the sample mean height was .8310 inches; thus, .8310 inches was plotted for day 1 on the chart. During the second day the sample mean was .8304; thus, .8304 inches was plotted for day 2 on the chart. A similar procedure was carried out during the next seven days, as shown in Figure 8.13. On none of these days did the sample mean fall outside the control limits. Thus, there were no grounds for rejecting the null hypothesis (that $\mu = .8312$ inches) on any of these days. Of course, this brief period only illustrates the practical application of the control chart, which in fact was used over a much longer period of time.

When a sample mean falls outside the control limits, the firm must look for the reasons why the population mean seems to have shifted. In some cases the cause may be machine wear or a breakdown of equipment. (Such a degradation in machine performance can occur gradually without being apparent to the operatives.) In other instances the reason for a shift in the mean may be poor workmanship by the labor force. The control chart by itself cannot indicate the reason foı the shift in the mean. It can only signal that such a shift seems to have occurred, and it is then up to the firm to find and remedy the cause.[22]

Besides establishing a control chart for the sample mean (a so-called $\bar{x}$ *chart*), firms often establish a control chart to signal changes in the variability of the quality of the production output. For example, the American Stove Company set up a control chart to detect whether, for each day's output, the variation among the heights of the metal pieces differed from what it was when the process was within control. Such charts can be of considerable importance, since an increase in the amount of variation among the items produced can be a serious problem. For example, if the heights of the metal pieces become more variable, a larger proportion of the output will fall outside the tolerances. Generally, the sample range rather than the sample standard deviation is used in these charts because it is easier to compute and because in small samples it is approximately as reliable as the sample standard deviation. Control charts based on the range are often called *R charts*.

EXERCISES

8.21 The Bona Fide Washing Machine Company chooses a random sample of 25 motors received from supplier I. It determines the length of life of each of the motors, the results (in thousands of hours) being as follows:

4.1	4.6	4.6	4.6	5.1
4.3	4.7	4.6	4.8	4.8
4.5	4.2	5.0	4.4	4.7
4.7	4.1	3.8	4.2	4.6
3.9	4.0	4.4	4.0	4.5

Suppose that the firm wants to test the hypothesis that the mean length of life of motors

22. In some cases a shift in the mean may be favorable, not undesirable. For example, consider a manufacturer of metal parts that establishes a control chart based on the average amount of waste per part. Suppose that the control chart signals a downward shift in the mean. In other words, the mean amount of waste per part seems to be less than was formerly the case. In instances of this sort, the firm will stop the production process in an attempt to find the reasons for this shift in the mean so that this improvement in performance can be continued. Thus, even if a shift in the mean is favorable, the firm wants to detect it.

For further discussion and case studies of statistical quality control, see W. Edwards Deming, "Making Things Right," reprinted in E. Mansfield, op. cit.

received from supplier I equals 4,900 hours. If the Bona Fide Washing Machine Company does not know the standard deviation of the length of life of motors received from supplier I, should it accept this hypothesis (if it sets α equal to .05)? (Use a one-tailed test, the alternative hypothesis being that the mean is less than 4,900 hours.)

8.22 The Crooked Arrow National Bank wants to test whether the mean income of the individuals holding deposits at its southern branch equals \$15,000, the alternative hypothesis being that the mean income does not equal \$15,000. The bank takes a random sample of 36 of the depositors at this branch and finds that their incomes in 1979 (in dollars) were as follows:

20,100	8,200	13,200	5,200	11,100	51,100
19,400	8,900	14,300	10,100	10,800	9,600
10,100	10,300	15,800	12,300	9,100	10,900
23,000	26,000	16,100	14,000	7,200	12,000
24,200	11,400	17,200	15,100	4,300	13,200
25,100	12,900	18,900	16,000	38,000	15,100

(a) If the bank sets a 5 percent significance level, should it accept or reject this hypothesis? (Use a two-tailed test.)
(b) If it sets a 1 percent significance level, should it accept or reject this hypothesis? (Use a two-tailed test.)

8.23 The manager of the northern branch of the Crooked Arrow National Bank believes that the mean income of the depositors at his branch is \$20,000. He wants to test this hypothesis against the alternative that this mean is less than \$20,000. A random sample of 9 of the depositors at his branch is chosen, and their incomes (in dollars) turn out to be

24,000	13,400	18,400
22,900	13,800	8,200
11,100	9,300	14,600

(a) If the bank sets a 5 percent significance level, should it accept or reject the manager's hypothesis?
(b) If the bank sets a 1 percent significance level, should it accept or reject the manager's hypothesis?

8.24 The president of the Crooked Arrow National Bank asks the firm's statistician to test whether the mean income of the depositors at the southern branch is equal to the mean income of the depositors at the northern branch.
(a) If the significance level is set at 5 percent, should the statistician accept or reject this hypothesis, based on the data in Exercises 8.22 and 8.23? (Use a two-tailed test.)
(b) If the significance level is set at 1 percent should the statistician accept or reject this hypothesis, based on the data in Exercises 8.22 and 8.23? (Use a two-tailed test.)

8.25 Recall from Exercise 7.13 that the Bona Fide Washing Machine Company has no idea of the standard deviation of the life of the motors it receives from supplier IV, and that it chose a random sample of 9 motors received from supplier IV and determined the length of life of each. The results (in thousands of hours) were as follows:

4.3	4.6	3.8
4.2	4.3	3.9
4.1	3.9	4.0

Use these results to test the hypothesis that the mean life of the motors received from supplier IV equals 4,200 hours (against the alternative that it does not equal 4,200 hours) at the 5 percent level of significance. Do the same at the 10 percent level of significance. (Use two-tailed tests.)

8.26 Compare your results in Exercise 8.25 with your results in Exercise 7.13. Can you see how the results you obtained in Exercise 7.13 could have been used to provide the answer to Exercise 8.25?

8.27 A manufacturing plant is interested in establishing a control chart for a particular metal part that it produces. The part averages .1020 inches in length when the plant is operating normally, its standard deviation being .0004 inches. The plant decides to pick a random sample of four parts from each day's output and determine their mean length. Plot the control chart the firm should use under these circumstances. (Let $\alpha = .003$.)

8.28 Suppose that the firm in Exercise 8.27 adopts your control chart and that the mean length of parts (in inches) on eight successive days is as follows: .10203; .10204; .10198; .10186; .10188; .10181; .10180; and .10139. Plot these results on your control chart, and indicate whether (and if so, when) the firm should reject the null hypothesis.

CHAPTER REVIEW

1. The basic hypothesis that is being tested for possible rejection is called the *null hypothesis*, H_0. In testing any null hypothesis, there is an *alternative hypothesis*, H_1, that the decision maker accepts if the null hypothesis is rejected. A Type I error occurs when the null hypothesis is rejected when it is true. A Type II error occurs when the null hypothesis is accepted when it is not true. A test procedure specifies a *decision rule* indicating the conditions under which the null hypothesis should be accepted or rejected. This decision rule is based on the value of a test statistic, which is a statistic computed from a sample. All possible values of the test statistic are divided into two mutually exclusive regions: the *acceptance region* and the *rejection region*. If the test statistic's value falls within the acceptance region, the null hypothesis is accepted; if it falls within the rejection region, the null hypothesis is rejected.

2. The probability of a Type I error is designated as α, and the probability of a Type II error is designated as β. Whether or not a particular test procedure is appropriate depends on whether α and β are set at the right levels from the point of view of the decision maker. The value of α that is chosen is called the *significance level* of the test. A test's *operating-characteristic curve* shows the value of β if the population parameter assumes various values other than that specified by the null hypothesis. The shape of a test's operating-characteristic curve is influenced by the sample size and by the chosen level of α. With increases in sample size, it is possible to reduce the probability of both types of error (that is, both α and β).

3. If the alternative hypothesis is that the value of the parameter differs in either direction from the value specified by the null hypothesis, the test is a *two-tailed test*. If the alternative hypothesis is that the value of the parameter differs in only one direction from the value specified by the null hypothesis, the test is a *one-tailed test*.

4. For a two-tailed test of the null hypothesis that a population mean equals a specified value μ_0, the decision rule is the following: Accept the null hypothesis if $\mu_0 - z_{\alpha/2}\sigma/\sqrt{n} \leqslant \bar{x} \leqslant \mu_0 + z_{\alpha/2}\sigma/\sqrt{n}$; reject the null hypothesis if $\bar{x} < \mu_0 - z_{\alpha/2}\sigma/\sqrt{n}$ or if $\bar{x} > \mu_0 + z_{\alpha/2}\sigma/\sqrt{n}$. If the population standard deviation σ is unknown, it can be replaced with s, the sample standard deviation. This decision rule applies to large samples (that is, when $n > 30$). For small samples, the t distribution (with $n - 1$ degrees of freedom) must be used in place of the standard normal distribution. For small samples the decision rule is: Accept the null hypothesis if $-t_{\alpha/2} \leqslant t \leqslant t_{\alpha/2}$; reject it otherwise. For purposes of this test, $t = (\bar{x} - \mu_0) \div s/\sqrt{n}$. This small-sample test assumes that the population is approximately normal.

5. For a two-tailed test of the null hypothesis that a population proportion equals a specified value P_0, the decision rule is: Accept the null hypothesis if $P_0 - z_{\alpha/2}\sqrt{P_0(1 - P_0)/n} \leqslant p \leqslant P_0 + z_{\alpha/2}\sqrt{P_0(1 - P_0)/n}$; reject the null hypothesis if $p < P_0 - z_{\alpha/2}\sqrt{P_0(1 - P_0)/n}$ or if $p > P_0 + z_{\alpha/2}\sqrt{P_0(1 - P_0)/n}$. For a two-tailed test of the null hypothesis that two population proportions are equal, the decision rule is: Accept the null hypothesis if $-z_{\alpha/2} \leqslant z \leqslant z_{\alpha/2}$; reject it otherwise. For purposes of this test, $z = (p_1 - p_2) \div \sqrt{p(1 - p)(1/n_1 + 1/n_2)}$. Both tests assume that the samples are large.

6. For a two-tailed test of the null hypothesis that the mean of one population μ_1 equals the mean of another population μ_2, the decision rule is: Accept the null hypothesis if $-z_{\alpha/2} \leqslant z \leqslant z_{\alpha/2}$; reject it otherwise. For purposes of this test $z = (\bar{x}_1 - \bar{x}_2) \div \sqrt{\sigma_1^2/n_1 + \sigma_2^2/n_2}$. This rule assumes that the samples are large and that the standard deviation of each population is known. If neither of these assumptions is true, the decision rule is: Accept the null hypothesis if $-t_{\alpha/2} \leqslant t \leqslant t_{\alpha/2}$; reject it otherwise. For purposes of this test, $t = (\bar{x}_1 - \bar{x}_2) \div \sqrt{s^2(1/n_1 + 1/n_2)}$, where

$$s^2 = \frac{(n_1 - 1)s_1^2 + (n_2 - 1)s_2^2}{n_1 + n_2 - 2}.$$

The latter test assumes that both populations are normal and that their standard deviations are equal. (The number of degrees of freedom of the t distribution is $n_1 + n_2 - 2$.)

7. An important industrial application of statistical testing procedures is *statistical process control.* A control chart is established with a central horizontal line at μ_0, the value of the mean when the process is in control, and with two other lines at the upper and lower control limits. The upper control limit is at $\mu_0 + 3\sigma/\sqrt{n}$; the lower control limit is at $\mu_0 - 3\sigma/\sqrt{n}$. Sample means are plotted on the chart, and the process is stopped and examined when a mean falls outside these limits. There is approximately a .003 probability of the process being stopped if the population mean remains equal to μ_0. Besides control charts (so-called $\bar{x}$ charts) for the sample mean, firms often use so-called R charts, which plot the sample range and which are used to detect changes in the variability of the output.

Getting Down to Cases:

HOW DOW CHEMICAL CHECKS THE ACCURACY OF ITS CLERICAL WORK[23]

The Dow Chemical Company, like any large company, must audit invoices to determine whether they contain clerical errors. After some study, the company found that with regard to invoices covering purchases of bulk mailing and duplicating items, about 325 invoices are filled out each week, and each invoice contains 10 places where an error can be made. Thus, there are about 3,250 opportunities for invoice errors per week. Dow decided to sample these error opportunities in order to determine the number of cases where errors actually existed. The results were then used to determine whether or not the invoices for a given week should be accepted.

More specifically, Dow used the Department of Defense's standard procedures for acceptance sampling. Dow's statisticians wanted to test whether the proportion of error opportunities where there in fact was an error was equal to 0.25 percent; and they further wanted the probability of rejecting a week's batch of invoices with a .25 percentage of error to equal 5 percent. Also, they wanted the probability to be 90 percent that a batch of invoices would be rejected if 2.5 percent of its error opportunities contained errors.

(a) What was the null hypothesis?
(b) What was the alternative hypothesis?
(c) What value of α should be set?
(d) What value of β should be set?

The Defense Department's standard acceptance-sampling tables indicated that what Dow wanted could be provided by taking a random sample of 225 error opportunities each week and rejecting the batch of

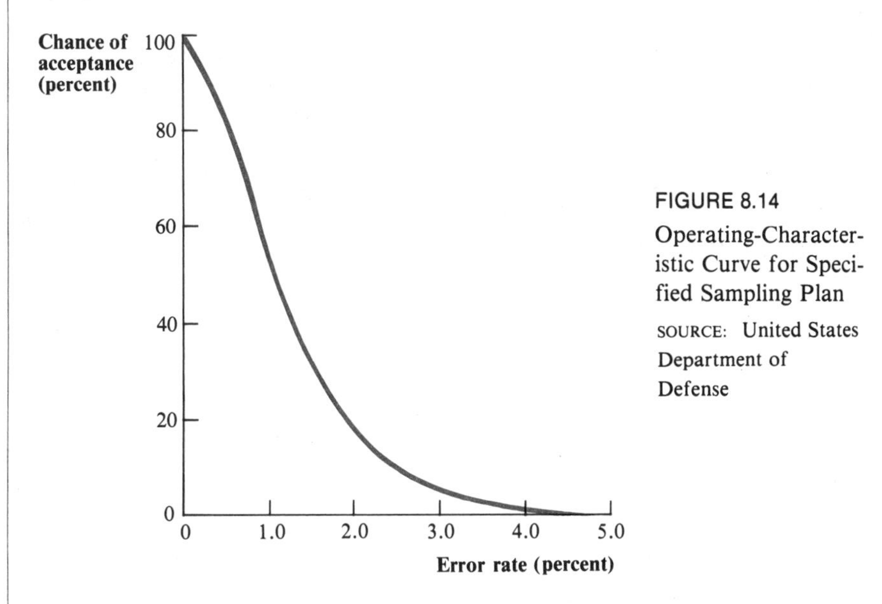

FIGURE 8.14
Operating-Characteristic Curve for Specified Sampling Plan
SOURCE: United States Department of Defense

23. This case is based on E. Yehle, "Accuracy in Clerical Work," *Systems and Procedures*, May 1960.

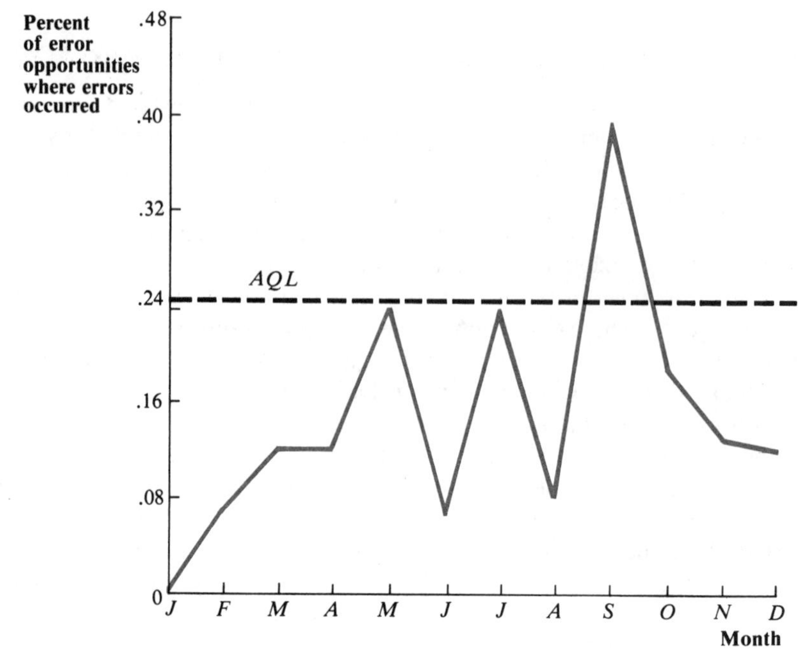

FIGURE 8.15 Error Rate in Invoices Sampled

invoices if more than two contained errors. Figure 8.14 contains the operating-characteristic curve for this test. For a year after adopting this plan, Dow continued to carry out a 100 percent audit of all invoices in order to check the accuracy of this plan. The results are shown in Figure 8.15.

(e) If 2 percent of the error opportunities in a particular batch of invoices contained errors, what was the probability that Dow's statistical test would have led to acceptance of this batch?

(f) Figure 8.15 shows that the actual error rate for September exceeded 0.25 percent, yet Dow's test procedure resulted in acceptance of all weekly invoice batches for September. Does this indicate that this statistical test doesn't really work? Why, or why not?

9

χ^2 (Chi-Square) Tests and Nonparametric Techniques*

9.1 Introduction

The procedures described in the previous chapter are only a small (albeit very important) sample of the many tests available for coping with the wide variety of problems that arise in business and economics. In this chapter we will present some additional tests which have a wide application. These tests are based on the principles described in Chapter 8, but they deal with different hypotheses and assumptions than those encountered there.

The χ^2 (chi-square) tests taken up in this chapter are based on a very important sampling distribution that we have not encountered before—the χ^2 or chi-square distribution. Besides describing these tests, we will also discuss the χ^2 distribution itself. Then we will take up nonparametric techniques, which make fewer assumptions about the nature of the population being sampled than the tests described in the previous chapter. Like the χ^2 tests, nonparametric techniques are very widely used in practical work.

9.2 Target Practice and Pearson's Test Statistic

To begin our discussion of χ^2 tests, let's consider the following sporting situation. Suppose that a marksman fires 200 shots at a target, and that each shot can result in three outcomes: (1) a bull's-eye; (2) a hit on the target, but not a bull's eye; and (3) a miss of the target. If the probability that a shot results in each of these outcomes is P_1, P_2, and P_3 respectively, and if the outcomes of successive shots are statistically independent, what is the expected number of bull's-eyes? Clearly, the answer is $200P_1$. What is the expected

* Some instructors may want to skip the latter part of this chapter, which deals with nonparametric techniques. Others may want to skip this chapter entirely. Subsequent chapters have been written so that this may be done without loss of continuity.

number of hits (but not bull's eyes)? Clearly, $200P_2$. What is the expected number of misses? Clearly, $200P_3$.

Given the outcomes of the 200 shots, we want to test the null hypothesis that P_1, P_2, and P_3 equal specified values (denoted by P_1^0, P_2^0, and P_3^0). To test this hypothesis, it seems reasonable to compare the actual number of shots having each outcome with the number that would be expected, given that these specified values of P_1, P_2, and P_3 are true. For example, if P_1^0 (the hypothesized value of P_1) is .2, we would expect about .2(200), or 40, shots to result in bull's-eyes. If the actual number of bull's-eyes is only 14, there is a seemingly large difference (14–40) between the actual and expected number of bull's-eyes. To tell whether we should view this difference as evidence that the hypothesis that $P_1 = .2$ is untrue, we must know the probability of a difference of this size, given that $P_1 = .2$.

In 1900 Karl Pearson, an English statistician who was one of the fathers of modern statistics, proposed that in a situation of this sort the following procedure be used to test the null hypothesis that P_1, P_2, and P_3 equal the specified values P_1^0, P_2^0, and P_3^0. First, we determine the expected number of shots that would be bull's-eyes if $P_1 = P_1^0$. This expected number (which equals $200P_1^0$) is denoted by e_1. Second, we determine the expected number of shots that would be hits (but not bull's-eyes) if $P_2 = P_2^0$. This expected number (which equals $200P_2^0$) is denoted by e_2. Third, we determine the expected number of shots that would be misses if $P_3 = P_3^0$. This expected number (which equals $200P_3^0$) is denoted by e_3. Then we calculate the following test statistic:

$$\sum_{i=1}^{3} \frac{(f_i - e_i)^2}{e_i}, \tag{9.1}$$

where f_1 is the actual number of shots that are bull's-eyes, f_2 is the actual number that are hits (but not bull's-eyes), and f_3 is the actual number of misses. Pearson showed that *if the null hypothesis is true (that is, if* $P_1 = P_1^0$, $P_2 = P_2^0$ *and* $P_3 = P_3^0$*), and if the sample size is large enough so that the smallest value of* e_i *is at least* 5, *this test statistic will have a sampling distribution which can be approximated adequately by the* χ^2 *distribution*, a very important probability distribution to which we now turn.

9.3 The χ^2 Distribution

The χ^2 distribution, which is the probability distribution of a χ^2 random variable, is defined as follows.

> ***χ^2 Distribution:*** *The χ^2 distribution, with ν (nu) degrees of freedom, is the probability distribution of the sum of squares of ν (nu) independent standard normal variables.*

To clarify this definition, let's begin by taking a single standard normal variable. (Recall from Chapter 5 that the standard normal variable is $(X - \mu)/\sigma$, where X is a normal variable with mean equal to μ and standard deviation equal to σ.) Instead of considering the probability distribution of its value, let's consider the probability distribution of the *square* of its value. If one knows that this variable has a standard normal distribution, it should be possible to figure out the distribution of the square of its value. This distribution, derived many years ago, is shown in panel A of Figure 9.1; it is a χ^2 distribution with 1 degree of freedom. Next, suppose that we consider two independent standard normal variables. Let's square the value of each and

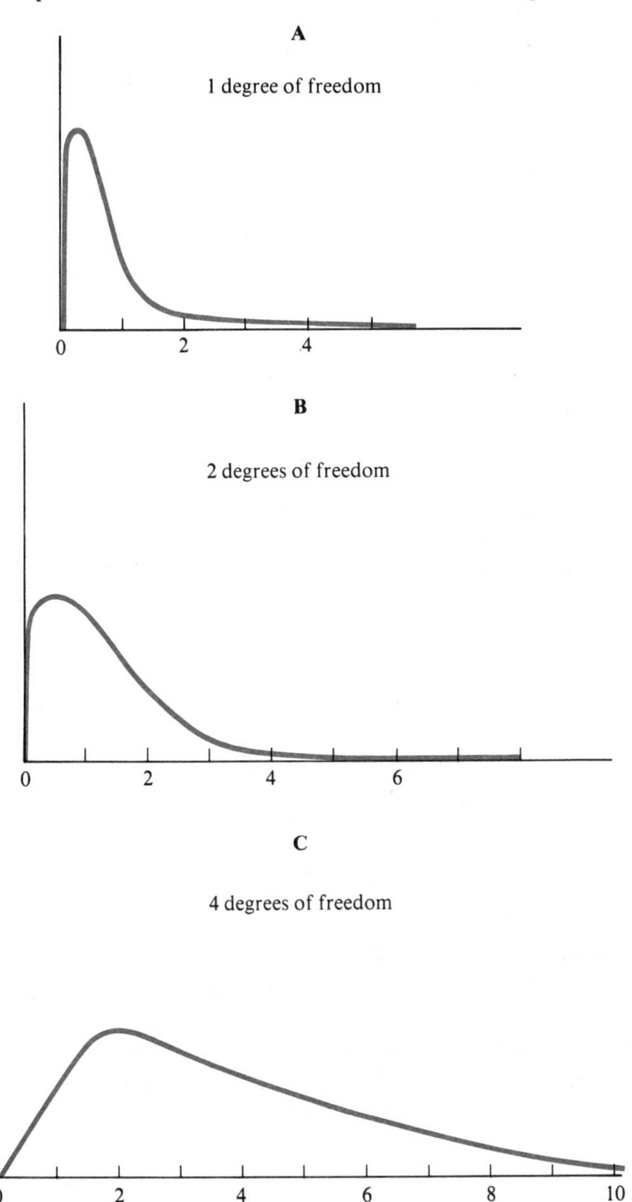

FIGURE 9.1 χ^2 Probability Density Functions with 1, 2, and 4 Degrees of Freedom

add the squares together. The distribution of this sum is shown in panel B of Figure 9.1; it is a χ^2 distribution with 2 degrees of freedom. Finally, suppose that we consider four independent standard normal variables. Let's square the value of each and add the squares together. The distribution of this sum, shown in panel C of Figure 9.1, is a χ^2 distribution (with 4 degrees of freedom).

Like the t distribution, the χ^2 distribution is a family of distributions, each of which is characterized by a certain number of *degrees of freedom.*[1] The number of degrees of freedom is the number of squares of standard normal variables that are summed up. Thus, in Figure 9.1, panel A shows a χ^2 distribution with 1 degree of freedom; panel B shows a χ^2 distribution with 2 degrees of freedom; and panel C shows a χ^2 distribution with 4 degrees of freedom.

As is evident from Figure 9.1, the χ^2 distribution generally is skewed to the right. Since the χ^2 random variable is a sum of squares, the probability that it is negative is zero. If there are ν degrees of freedom, it can be shown that

$$E(\chi^2) = \nu, \tag{9.2}$$

$$\sigma(\chi^2) = \sqrt{2\nu}. \tag{9.3}$$

In other words, the mean of a χ^2 random variable equals its number of degrees of freedom, and its standard deviation equals the square root of twice its number of degrees of freedom.

In applying the test procedures described in the following sections, it is essential that we be able to find the value of χ^2 that is exceeded with a probability of .05, .10, or some other such amount. That is, we need to be able to calculate χ^2_α, which is the value of χ^2 that is exceeded with a probability of α. Appendix Table 8 shows the value of χ^2_α for various numbers of degrees of freedom and for various values of α. Each row in this table corresponds to a certain number of degrees of freedom. For example, the tenth row shows that, if there are 10 degrees of freedom, the probability is .10 that the value of χ^2 will exceed 15.987, that the probability is .05 that it will exceed 18.307, and that the probability is .01 that it will exceed 23.2093. As the number of degrees of freedom becomes very large, the χ^2 distribution can be approximated by the normal distribution.

9.4 Test of Difference among Proportions

One very important application of the χ^2 distribution is to problems where the decision maker wants to determine whether a number of proportions are equal. In the previous chapter, we showed how one can test whether two population proportions are equal. Now we show how one can test whether any number of population proportions are equal. To illustrate the sort of situation to which this test is applicable, we will describe an actual case study involving

1. There is a close relationship between the t and χ^2 distributions. It can be shown that t equals the ratio of a standard normal variable to the square root of a χ^2 variable divided by its number of degrees of freedom. For further explanation one should consult a more advanced statistics text.

TABLE 9.1
Number of Defective Railway-Car Side Frames, 28 Days

Day	Number defective	Number accepted	Sample size
April 27	4	46	50
28	9	41	50
29	10	40	50
30	11	39	50
May 1	13	37	50
2	30	20	50
3	26	24	50
4	13	37	50
5	8	42	50
6	23	27	50
7	34	16	50
8	25	25	50
9	18	32	50
10	12	38	50
11	4	46	50
12	3	47	50
14	11	39	50
15	8	42	50
16	14	36	50
17	21	29	50
18	25	25	50
19	18	32	50
21	10	40	50
22	8	42	50
23	18	32	50
24	19	31	50
25	4	46	50
26	8	42	50

the production of side frames of railway cars.[2] Each day, a foundry producing these frames sampled 50 at random and, after inspecting them, determined how many were defective. Table 9.1 shows the number determined defective for each day during a 28-day period. The question that the executives of the foundry wanted answered was: Is the proportion defective constant from day to day, or does it vary? If the proportion was found to vary from day to day, this meant the production process was not in control; and the firm would then try to determine the reasons for the variation.

In a problem of this sort, *the null hypothesis is that the population proportions are all equal.* Thus, in this case study, the null hypothesis is that $P_1 = P_2 = P_3 = \ldots = P_{28}$, and *the alternative hypothesis is that these proportions are not all equal.* (Of course, P_1 is the population proportion defective on the

2. This case is reported in A. Duncan, *Quality Control and Industrial Statistics* (Homewood, Ill.: Irwin, 1959).

first day, P_2 is the population proportion defective on the second day, and so on.) If the null hypothesis is true, the proportion defective on all days can be estimated by pooling the data for all days. Clearly, the common proportion defective equals

$$\frac{\sum_{i=1}^{28} x_i}{\sum_{i=1}^{28} n_i},$$

where x_i is the number of defectives on the ith day, and n_i is the size of the total sample on the ith day. Thus, in this case the expected proportion defective on any day, if the null hypothesis is correct, is

$$\frac{4 + 9 + 10 + 11 + 13 + \ldots + 19 + 4 + 8}{50 + 50 + 50 + 50 + 50 + \ldots + 50 + 50 + 50} = .29.$$

To test whether the null hypothesis is correct, we calculate the expected number of defectives and the expected number of nondefectives on each day. Since the common proportion defective is .29, we would expect that 50(.29) = 14.5 frames would be defective each day, and that 50(.71) = 35.5 frames would not be defective each day. To test the null hypothesis, we compare these theoretical, or expected, frequencies with the actual ones. Clearly, the greater the difference between the theoretical frequencies and the actual ones, the less likely it is that the null hypothesis is true.

Specifically, the test procedure is as follows. Having calculated each expected frequency (shown in Table 9.2), we must compute the following test statistic,

$$\Sigma \frac{(f - e)^2}{e}, \tag{9.4}$$

where f is the actual frequency, e is the corresponding expected frequency, and the summation is over all items in Table 9.2.[3] If the null hypothesis is true, this test statistic has a sampling distribution that can be approximated by the χ^2 distribution with degrees of freedom equal to $(r - 1)$, where r is the number of population proportions that are being compared.

Why is $(r - 1)$ the appropriate number of degrees of freedom? The general rule is that *the appropriate number of degrees of freedom equals the number of comparisons between actual and expected frequencies, less the number of independent linear restrictions placed upon the frequencies.* Because the number of such restrictions varies from one case to another in this chapter, the number of degrees of freedom will be given by different formulas from one case to another.[4] Here the appropriate formula is $(r - 1)$, since the sum of the frequencies (50) in each day's population is given and the total number of

3. Note that this test statistic is the same as that in expression (9.1).

4. For example, in Section 9.5 the number of degrees of freedom is $(r - 1)(c - 1)$, as we shall see.

TABLE 9.2
Expected Frequencies, and Calculation of $\Sigma \frac{(f - e)^2}{e}$

	Number defective			Number not defective		
Day	Actual (f)	Expected (e)	$\frac{(f-e)^2}{e}$	Actual (f)	Expected (e)	$\frac{(f-e)^2}{e}$
April 27	4	14.5	$(4-14.5)^2/14.5$	46	35.5	$(46-35.5)^2/35.5$
28	9	14.5	$(9-14.5)^2/14.5$	41	35.5	$(41-35.5)^2/35.5$
29	10	14.5	$(10-14.5)^2/14.5$	40	35.5	$(40-35.5)^2/35.5$
30	11	14.5	$(11-14.5)^2/14.5$	39	35.5	$(39-35.5)^2/35.5$
May 1	13	14.5	$(13-14.5)^2/14.5$	37	35.5	$(37-35.5)^2/35.5$
2	30	14.5	$(30-14.5)^2/14.5$	20	35.5	$(20-35.5)^2/35.5$
3	26	14.5	$(26-14.5)^2/14.5$	24	35.5	$(24-35.5)^2/35.5$
4	13	14.5	$(13-14.5)^2/14.5$	37	35.5	$(37-35.5)^2/35.5$
5	8	14.5	$(8-14.5)^2/14.5$	42	35.5	$(42-35.5)^2/35.5$
6	23	14.5	$(23-14.5)^2/14.5$	27	35.5	$(27-35.5)^2/35.5$
7	34	14.5	$(34-14.5)^2/14.5$	16	35.5	$(16-35.5)^2/35.5$
8	25	14.5	$(25-14.5)^2/14.5$	25	35.5	$(25-35.5)^2/35.5$
9	18	14.5	$(18-14.5)^2/14.5$	32	35.5	$(32-35.5)^2/35.5$
10	12	14.5	$(12-14.5)^2/14.5$	38	35.5	$(38-35.5)^2/35.5$
11	4	14.5	$(4-14.5)^2/14.5$	46	35.5	$(46-35.5)^2/35.5$
12	3	14.5	$(3-14.5)^2/14.5$	47	35.5	$(47-35.5)^2/35.5$
14	11	14.5	$(11-14.5)^2/14.5$	39	35.5	$(39-35.5)^2/35.5$
15	8	14.5	$(8-14.5)^2/14.5$	42	35.5	$(42-35.5)^2/35.5$
16	14	14.5	$(14-14.5)^2/14.5$	36	35.5	$(36-35.5)^2/35.5$
17	21	14.5	$(21-14.5)^2/14.5$	29	35.5	$(29-35.5)^2/35.5$
18	25	14.5	$(25-14.5)^2/14.5$	25	35.5	$(25-35.5)^2/35.5$
19	18	14.5	$(18-14.5)^2/14.5$	32	35.5	$(32-35.5)^2/35.5$
21	10	14.5	$(10-14.5)^2/14.5$	40	35.5	$(40-35.5)^2/35.5$
22	8	14.5	$(8-14.5)^2/14.5$	42	35.5	$(42-35.5)^2/35.5$
23	18	14.5	$(18-14.5)^2/14.5$	32	35.5	$(32-35.5)^2/35.5$
24	19	14.5	$(19-14.5)^2/14.5$	31	35.5	$(31-35.5)^2/35.5$
25	4	14.5	$(4-14.5)^2/14.5$	46	35.5	$(46-35.5)^2/35.5$
26	8	14.5	$(8-14.5)^2/14.5$	42	35.5	$(42-35.5)^2/35.5$

$$\Sigma \frac{(f - e)^2}{e} \doteq 182$$

defectives in the sample is also given. Thus, there are $(r + 1)$ restrictions on the frequencies. Subtracting $(r + 1)$ from $2r$ (the number of comparisons between actual and expected frequencies), we get $(r - 1)$, the number of degrees of freedom.

As we just noted, if the null hypothesis is true, the distribution of $\Sigma(f - e)^2/e$ can be approximated by the χ^2 distribution with $(r - 1)$ degrees of freedom, which is shown in Figure 9.2. The rejection region is under the upper tail of this distribution, since large values of this test statistic indicate large discrepancies between the actual and expected frequencies. In general, to

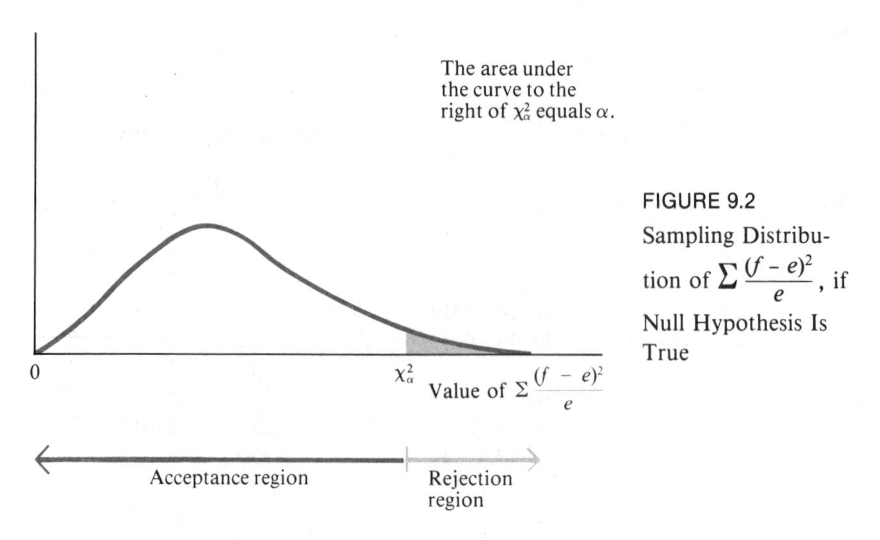

FIGURE 9.2
Sampling Distribution of $\Sigma \frac{(f - e)^2}{e}$, if Null Hypothesis Is True

test whether r population proportions are equal, the appropriate decision rule is as follows.

> ***Decision Rule:*** *Reject the null hypothesis that* $P_1 = P_2 = \ldots = P_r$ *if* $\Sigma(f - e)^2/e > \chi^2_\alpha$, *where* α *is the desired significance level of the test (and the number of degrees of freedom is* $r - 1$*). Accept the null hypothesis if* $\Sigma(f - e)^2/e \leqslant \chi^2_\alpha$.

Table 9.2 shows that the value of $\Sigma(f - e)^2/e$ is approximately 182. Appendix Table 8 shows that $\chi^2_{.05} = 40.1$ if the number of degrees of freedom equals 27. (Why are there 27 degrees of freedom? Because the number of degrees of freedom equals the number of population proportions to be compared—28 in this case—minus 1.) Since 182 exceeds 40.1, it is clear that the null hypothesis should be rejected. Even if the decision maker sets the significance level at .01, the null hypothesis should be rejected since $\chi^2_{.01} = 46.96$ if the number of degrees of freedom equals 27. Thus, there is every indication that the proportion defective varies from day to day, which in turn indicates that the foundry's performance was not in control. Based on this evidence, the firm was led to study the reasons for the observed variation in performance since it clearly was more than merely chance variation.

The test of differences among proportions that we have just described is commonly employed in business and economics. The following is another illustration of how it is used.

Example 9.1 An economist wants to test the hypothesis that the proportion of firms intending to raise their prices next year is the same in three industries, A, B, and C. The data for a sample of firms are as follows:

	Number of firms		
Decision	Industry A	Industry B	Industry C
To raise price	40	50	60
Not to raise price	60	50	40

Should the economist accept or reject this hypothesis? (Assume that the significance level equals .05.)

Solution: Summing up the data for all industries, 150 of the 300 firms intend to raise price. Thus, the common or overall proportion is 150/300, or .50. The expected number of firms intending to raise price, if the null hypothesis is true, is

	Number of firms		
Decision	Industry A	Industry B	Industry C
To raise price	50	50	50
Not to raise price	50	50	50

Thus,

$$\Sigma \frac{(f - e)^2}{e} = \frac{(40 - 50)^2}{50} + \frac{(60 - 50)^2}{50} + \frac{(50 - 50)^2}{50} + \frac{(50 - 50)^2}{50}$$

$$+ \frac{(60 - 50)^2}{50} + \frac{(40 - 50)^2}{50}$$

$$= 2 + 2 + 0 + 0 + 2 + 2 = 8.$$

Since there are three population proportions being compared, the number of degrees of freedom is 3 − 1, or 2. According to Appendix Table 8, $\chi^2_{.05} = 5.991$ when there are 2 degrees of freedom. Thus, since the observed value of $\Sigma(f - e)^2/e$ exceeds 5.991, the null hypothesis should be rejected. In other words, the economist should reject the hypothesis that the proportion of firms intending to raise their prices next year is the same in each of the three industries.

9.5 Contingency Tables

Decision makers are often concerned with problems involving contingency tables. A *contingency table* indicates whether two characteristics or variables are dependent on one another. In other words, a contingency table contains two variables of classification, and the point is to determine whether these two variables are related. For example, Table 9.3 shows a simple contingency table where the vertical columns represent Republicans and Democrats and the horizontal rows show the number of persons earning more or no more than $20,000 per year. Based on a random sample of 300 individuals, the results are

TABLE 9.3

A 2 × 2 Contingency Table

Income	Republicans	Democrats	Total
More than $20,000	30(20)	30(40)	60
No more than $20,000	70(80)	170(160)	240
Total	100	200	300

as shown in Table 9.3. This is a 2 × 2 contingency table since there are two rows and two columns. *A table of this sort can be used to test whether a person's income is independent of his or her political affiliation.*

The null hypothesis here is that the probability that a Republican will have an income above \$20,000 is the same as the probability that a Democrat will have an income above \$20,000. That is, the null hypothesis is that income is *independent* of political party. To test this hypothesis, we compute the expected frequencies in Table 9.3, assuming that the hypothesis is true. Since the probability of a person making over \$20,000 is 60 ÷ 300 or .20 in the sample as a whole, we would expect that 20 percent of the Republicans—that is, .20(100), or 20 Republicans—would earn over \$20,000 per year, and that 20 percent of the Democrats—that is, .20(200), or 40 Democrats—would earn over \$20,000 per year, if the null hypothesis is true. By the same token, we would expect that 80 percent of the Republicans—that is, .80(100), or 80 Republicans—would make no more than \$20,000 per year, and that 80 percent of the Democrats—that is, .80(200), or 160 Democrats—would make no more than \$20,000 per year, if the null hypothesis is true.

Given these expected frequencies, which are shown in parentheses in Table 9.3, we test the null hypothesis by computing $\Sigma(f - e)^2/e$. In other words, $\Sigma(f - e)^2/e$ is the test statistic. If α is the desired significance level, the decision rule is as follows.

> ***Decision Rule:*** *Reject the null hypothesis (of independence) if* $\Sigma(f - e)^2/e > \chi^2_\alpha$, *where there are* $(r - 1)(c - 1)$ *degrees of freedom* (r *being the number of rows and* c *being the number of columns). Accept the null hypothesis if* $\Sigma(f - e)^2/e \leqslant \chi^2_\alpha$.

Why is the appropriate number of degrees of freedom equal to $(r - 1)(c - 1)$? As noted in the previous section, the appropriate number of degrees of freedom equals the number of comparisons between actual and expected frequencies less the number of restrictions placed upon these frequencies. Since the number of entries in a contingency table equals cr, there are cr actual frequencies to be compared with the corresponding expected frequencies. Because the sum of the frequencies in each row and each column is a given quantity, there are $r + c - 1$ such restrictions.[5] Thus, the appropriate number of degrees of freedom is $cr - (r + c - 1)$, or $(r - 1)(c - 1)$.

In this case the value of the test statistic is

$$\Sigma\frac{(f - e)^2}{e} = \frac{(30 - 20)^2}{20} + \frac{(30 - 40)^2}{40} + \frac{(70 - 80)^2}{80} + \frac{(170 - 160)^2}{160}$$

$$= \frac{100}{20} + \frac{100}{40} + \frac{100}{80} + \frac{100}{160} = 9.375.$$

Suppose that the significance level is set at .05. If so, the null hypothesis should

5. There is a restriction for each row and column, since the sum of the frequencies in each row or column is given. However, one of these restrictions must hold if all the others are met, so the number of independent restrictions is $r + c - 1$.

be rejected, since $\Sigma(f - e)^2/e$ exceeds $\chi^2_{.05}$, which is 3.84. (Note that there is only 1 degree of freedom here, since $(r - 1) \cdot (c - 1) = 1$ in a 2×2 contingency table.)[6] In other words, based on this evidence, the probability that a person's income is above $20,000 does not seem to be independent of his or her political affiliation.

9.6 Contingency Tables and Metal Castings: A Case Study

Contingency tables are of great practical importance, and in this section we will describe their application to an actual industrial problem.[7] Table 9.4 provides actual data for a three-week period concerning the causes of rejects of metal castings in a certain manufacturing plant. The managers of this plant wanted to know whether the probability distribution of causes of rejects was the same from one week to the next. Clearly, this question is of importance, since increases in the probability of a particular cause are likely to indicate that the production process is becoming increasingly deficient in certain ways. A χ^2 test was used to answer this question. The null hypothesis is that the probability distribution of causes is the same for each week. Based on this assumption, the firm's statisticians calculated the expected frequencies for each item in the table, these frequencies being shown in parentheses to the right of the corresponding actual frequencies in Table 9.4.

TABLE 9.4
Causes of Rejects of Metal Castings, Three Weeks

	Week			
Cause of rejection	First	Second	Third	Total
Sand	97 (93.9)	120 (111.1)	82 (93.9)	299
Misrun	8 (8.5)	15 (10.0)	4 (8.5)	27
Shift	18 (9.4)	12 (11.2)	0 (9.4)	30
Drop	8 (10.4)	13 (12.3)	12 (10.4)	33
Corebreak	23 (25.8)	21 (30.5)	38 (25.7)	82
Broken	21 (19.8)	17 (23.4)	25 (19.8)	63
Other	5 (12.3)	15 (14.5)	19 (12.3)	39
Total	180	213	180	573

6. When there is only 1 degree of freedom, one should use a slightly different formula:

$$\Sigma\left(\frac{|f - e| - 1/2}{e}\right)^2.$$

In this case, the result would be

$$\frac{(9.5)^2}{20} + \frac{(9.5)^2}{40} + \frac{(9.5)^2}{80} + \frac{(9.5)^2}{160} = 90.25\left(\frac{3}{32}\right) = 8.46.$$

Thus, the results are essentially the same as in the text. Note that this so-called *continuity correction factor* also applies to the χ^2 tests in Sections 9.4 and 9.7, as well as to contingency tables, when there is only 1 degree of freedom.

7. This case is from A. Duncan, op. cit.

To obtain each of the expected frequencies, the (horizontal) row total for the item is multiplied by the (vertical) column total for the item, and the result is then divided by the overall total. For example, to get the expected frequency in the first row in the first column, we multiply the total for (horizontal) row 1 (299) by the total for (vertical) column 1 (180) and divide by the overall total (573). To see why this is correct, note that the total for row 1 divided by the overall total is an estimate of the probability of being in row 1, and the total for column 1 divided by the overall total is an estimate of the probability of being in column 1. Thus, to estimate the probability of being in *both* row 1 and column 1, assuming that these events are statistically independent, we obtain the product of these two estimated probabilities. This product times the overall total equals the expected frequency in the first row in the first column if the null hypothesis is true. Clearly, the same result can be obtained by multiplying the total for row 1 by the total for column 1 and dividing by the overall total.

Once the firm had obtained the expected frequencies in Table 9.4, it was a simple matter to test the null hypothesis. All that the firm had to do was compute $\Sigma(f - e)^2/e$, which equals

$$\frac{(97-93.9)^2}{93.9} + \frac{(8-8.5)^2}{8.5} + \frac{(18-9.4)^2}{9.4} + \frac{(8-10.4)^2}{10.4} + \frac{(23-25.8)^2}{25.8} +$$

$$\frac{(21-19.8)^2}{19.8} + \frac{(5-12.3)^2}{12.3} + \frac{(120-111.1)^2}{111.1} + \frac{(15-10.0)^2}{10.0} +$$

$$\frac{(12-11.2)^2}{11.2} + \frac{(13-12.3)^2}{12.3} + \frac{(21-30.5)^2}{30.5} + \frac{(17-23.4)^2}{23.4} +$$

$$\frac{(15-14.5)^2}{14.5} + \frac{(82-93.9)^2}{93.9} + \frac{(4-8.5)^2}{8.5} + \frac{(0-9.4)^2}{9.4} + \frac{(12-10.4)^2}{10.4} +$$

$$\frac{(38-25.7)^2}{25.7} + \frac{(25-19.8)^2}{19.8} + \frac{(19-12.3)^2}{12.3} = 45.2.$$

Since $(r - 1)(c - 1) = (7 - 1)(3 - 1) = 12$, the null hypothesis should be rejected at the .05 significance level if $\Sigma(f - e)^2/e$ exceeds $\chi^2_{.05}$, where χ^2 has 12 degrees of freedom. Since Appendix Table 8 shows that $\chi^2_{.05}$ is 21.03 when there are 12 degrees of freedom, it follows that the firm should reject the null hypothesis—which is what it did. Thus, contrary to the null hypothesis, it appears that the probability distribution of causes of rejects did vary significantly from week to week.

9.7 Tests of Goodness of Fit

Still another important application of the χ^2 distribution is where the decision maker wants to determine whether an observed frequency distribution conforms to a theoretical distribution. For example, suppose that a firm believes that the probability is .50 that a particular machine will need oiling in an eight-hour shift. Since the firm operates four such machines, it can use the binomial distribution to calculate the theoretical probabilities that zero, 1, 2,

3, or 4 machines will need oiling in a particular shift. (Recall Chapter 4.) These probabilities are shown in the first column of Table 9.5.

Let's assume that the firm, after collecting data concerning the actual frequency distribution of the number of machines needing oiling in 160 shifts, wants to determine whether the binomial distribution is an accurate representation of the actual distribution. The theoretical frequency distribution is provided in the third column of Table 9.5. To obtain this distribution, all that one has to do is multiply each of the numbers in the second column of Table 9.5 by 160. The actual frequency distribution is provided in the fourth column of Table 9.5. Of course, the actual and theoretical distributions do not coincide exactly, but this does not prove that the theoretical distribution is inappropriate, since some discrepancy between the two distributions would be expected due to chance. The question facing the firm is: Are the discrepancies between the actual and theoretical distributions so large that they cannot reasonably be attributed to chance?

TABLE 9.5
Goodness-of-Fit Test

Number of machines needing oiling	Theoretical probability	Theoretical frequency	Actual frequency
0	1/16	10	12
1	1/4	40	35
2	3/8	60	60
3	1/4	40	45
4	1/16	10	8

To answer this kind of question, statisticians use a procedure quite similar to those described in the previous two sections. *The null hypothesis is that the actual distribution can in fact be represented by the theoretical distribution, and that the discrepancies between them are due to chance.* To test this hypothesis, we calculate $\Sigma(f - e)^2/e$, where f is the observed frequency in a particular class interval of the frequency distribution and e is the theoretical, or expected, frequency in the same class interval of the frequency distribution. If the null hypothesis is true, it can be shown that $\Sigma(f - e)^2/e$ has approximately a χ^2 distribution, the number of degrees of freedom being 1 less than the number of values of $(f - e)^2/e$ that are summed up. (If some of the parameters of the theoretical distribution are estimated from the sample, the number of degrees of freedom is less than this amount by the number of parameters that are estimated.[8])

8. In this case there are two parameters (n and P) of the theoretical probability distribution, and their values (4 and 0.5) are given. However, often we want to test whether the data conform to a binomial distribution without specifying the value of P, since we have no *a priori* information concerning its value. More will be said about this below.

Once this test statistic has been calculated, the decision rule is as follows.

Decision Rule: *Reject the null hypothesis that the discrepancy between the actual and theoretical frequency distributions is due to chance if* $\Sigma(f - e)^2/e > \chi^2_\alpha$ *where the number of degrees of freedom equals the number of class intervals of the frequency distribution minus* $(h + 1)$*, where* h *is the number of parameters estimated from the sample. Accept the null hypothesis if* $\Sigma(f - e)^2/e \leqslant \chi^2_\alpha$.

Given the actual and theoretical frequency distributions in Table 9.5, the value of the test statistic is

$$\Sigma\frac{(f - e)^2}{e} = \frac{(12 - 10)^2}{10} + \frac{(35 - 40)^2}{40} + \frac{(60 - 60)^2}{60} + \frac{(45 - 40)^2}{40} + \frac{(8 - 10)^2}{10}$$

$$= 0.4 + 0.625 + 0 + 0.625 + 0.4 = 2.05$$

Since no parameter of the theoretical distribution is estimated from the data, the number of degrees of freedom is 1 less than the number of values of $(f - e)^2/e$ that are summed up; that is, $5 - 1 = 4$. Thus, if the .05 significance level is chosen, the null hypothesis should be rejected if $\Sigma(f - e)^2/e$ exceeds 9.488, since this is the value of $\chi^2_{.05}$ when there are 4 degrees of freedom. (See Appendix Table 8.) Since $\Sigma(f - e)^2/e$ does not exceed 9.488, it follows that there is no reason to reject the null hypothesis. In other words, the probability is greater than .05 that the observed discrepancies between the actual distribution and the binomial distribution could be due to chance.

In carrying out goodness-of-fit tests, statisticians sometimes force the mean of the theoretical distribution to equal the mean of the actual distribution. Frequently, the mean of the theoretical distribution is not stipulated on a priori grounds, and therefore this seems the sensible thing to do. For example, if the firm wanted to test whether the actual distribution in Table 9.5 was binomial, but was not willing to specify that the mean of the binomial distribution was 2.0, then the mean of the theoretical distribution would be set equal to the mean of the actual distribution. Since this would enable us to specify the entire theoretical distribution, we could calculate $\Sigma(f - e)^2/e$ and see whether it exceeds $\chi^2_{.05}$. But it is important to note that χ^2 would have one less degree of freedom, because we estimated an additional parameter of the theoretical distribution (namely, the mean) from the actual distribution.[9]

Finally, in carrying out all such goodness-of-fit tests, one should define the class intervals of the frequency distribution so that the *theoretical, or*

9. In testing whether a frequency distribution conforms to a normal population, statisticians frequently force the population mean to equal the sample mean and the population standard deviation to equal the sample standard deviation. Under these circumstances, the number of degrees of freedom is the number of class intervals in the frequency distribution minus 3 since two parameters are estimated.

expected, frequency in each and every class interval is at least 5. The reason is that the χ^2 distribution is not a good approximation of the distribution of $\Sigma(f - e)^2/e$ when the null hypothesis is correct, if the theoretical frequencies in particular class intervals are very small. To make sure that this rule of thumb is met, all that one has to do is to combine adjacent class intervals when any of them has a theoretical frequency of less than 5. (Example 9.2 illustrates this procedure.)

This rule of thumb should also be observed in the other applications of the χ^2 distribution described in the preceding sections. In the case of tests of differences among proportions, each expected or theoretical frequency must equal 5 or more; in contingency tables, the theoretical frequency in each cell must equal 5 or more. Recall that, when we discussed Karl Pearson's findings in Section 9.2, we pointed out that his results assumed that the smallest expected or theoretical frequency (that is, the smallest value of e_i) is at least 5. This rule of thumb is important and should be observed.

Goodness-of-fit tests are of considerable importance in business and economics. The following example illustrates how they can be used by government agencies also.

Example 9.2 Let's return to the replacement of parts on Polaris submarines, an actual case study presented in Chapter 5. Suppose that the Department of Defense believes that the probability distribution of the number of submarine parts of a certain type that will fail during a mission is as follows:[10]

Number of failures per mission	Theoretical probability
0	.368
1	.368
2	.184
3	.061
4 or more	.019

Data for 500 missions indicate that the number of these missions in which each number of failures occurred was as follows:

Number of failures per mission	Number of missions
0	190
1	180
2	90
3	30
4 or more	10

Test the hypothesis that the discrepancies between the actual and theoretical frequency distributions are due to chance. (Let the significance level equal .05.)

10. Readers who have covered the latter part of Chapter 5 should note that this probability distribution is a Poisson distribution in which the mean number of failures per mission is 1.0.

Solution: The theoretical frequency distribution can be obtained by multiplying the theoretical probabilities by 500, the results being as follows:

Number of failures per mission	Expected number of missions
0	.368(500) = 184.0
1	.368(500) = 184.0
2	.184(500) = 92.0
3	.061(500) = 30.5
4 or more	.019(500) = 9.5

Note that we have lumped together all numbers of failures exceeding 3 in one class interval to hold the theoretical frequency in the class interval to a level of five or more.

The value of the test statistic is

$$\Sigma \frac{(f - e)^2}{e} = \frac{(190 - 184)^2}{184} + \frac{(180 - 184)^2}{184} + \frac{(90 - 92)^2}{92} + \frac{(30 - 30.5)^2}{30.5} + \frac{(10 - 9.5)^2}{9.5}$$

$$= .196 + .087 + .043 + .008 + .026 = .360.$$

Since no parameters of the theoretical frequency distribution are estimated from the sample, the number of degrees of freedom is 5 − 1, or 4. Appendix Table 8 shows that if there are 4 degrees of freedom, $\chi^2_{.05} = 9.488$. Since the observed value of $\Sigma(f - e)^2/e$ is less than 9.488, the null hypothesis should not be rejected. The probability is much higher than .05 that the observed value of the test statistic could have arisen, given that the theoretical distribution was valid. In other words, the Defense Department should not reject the hypothesis that the data conform to the theoretical distribution.

9.8 Tests and Confidence Intervals Concerning the Variance

Before turning to nonparametric techniques, one final application of the χ^2 distribution should be discussed. If the population is normal, the χ^2 distribution can be used to test hypotheses concerning the variance or to construct confidence intervals for the variance. For example, a firm that produces screws wants to test the hypothesis that the variance of the diameters of the screws produced on a given day equals .0001. A random sample of 11 screws is drawn, and it is found that the sample variance s^2 equals .0003. If the significance level is set at .05, should the firm accept or reject the hypothesis that the population variance equals .0001?

The first step is to formulate the null hypothesis and the alternative hypothesis. In this case, the null hypothesis is that the population variance equals .0001, and the alternative hypothesis is that the population variance is either less than or greater than .0001. If the population is normal, and if its variance equals σ_0^2, it can be shown that $(n - 1)s^2 \div \sigma_0^2$ has the χ^2 distribution with $(n - 1)$ degrees of freedom, where n is the sample size. Thus, to test the null hypothesis that the variance equals σ_0^2, we compute

$$\frac{(n - 1)s^2}{\sigma_0^2},$$

which is used as follows.

Decision Rule: *When the alternative hypothesis is $\sigma^2 \neq \sigma_0^2$, reject the null hypothesis if $(n - 1)s^2/\sigma_0^2$ exceeds $\chi^2_{\alpha/2}$ or is less than $\chi^2_{1-\alpha/2}$. When the alternative hypothesis is $\sigma^2 > \sigma_0^2$, reject the null hypothesis if $(n - 1)s^2/\sigma_0^2$ exceeds χ^2_{α}. When the alternative hypothesis is $\sigma^2 < \sigma_0^2$, reject the null hypothesis if $(n - 1)s^2/\sigma_0^2$ is less than $\chi^2_{1-\alpha}$.*

The alternative hypothesis in this case is that $\sigma^2 \neq \sigma_0^2$. Appendix Table 8 shows that $\chi^2_{.025} = 20.48$ and $\chi^2_{.975} = 3.25$ because the number of degrees of freedom equals $(n - 1) = 11 - 1 = 10$. Thus, the decision rule is: Reject the null hypothesis if $(n - 1)s^2/\sigma_0^2$ is greater than 20.48 or less than 3.25. In fact,

$$\frac{(n - 1)s^2}{\sigma_0^2} = \frac{10(.0003)}{.0001} = 30,$$

because $n = 11$, $s^2 = .0003$, and $\sigma_0^2 = .0001$. Thus, the firm should reject the null hypothesis that the variance of the screw diameters equals .0001.

In many circumstances, statisticians must estimate the variance of a normal population based on the results of a random sample of n observations. *If the confidence coefficient is $(1 - \alpha)$, the confidence interval for the population variance is*

$$\boxed{\frac{(n - 1)s^2}{\chi^2_{\alpha/2}} < \sigma^2 < \frac{(n - 1)s^2}{\chi^2_{1-\alpha/2}},} \tag{9.5}$$

where $\chi^2_{\alpha/2}$ is the value of a χ^2 variable (with $n - 1$ degrees of freedom) that is exceeded with a probability of $\alpha/2$, and $\chi^2_{1-\alpha/2}$ is the value that is exceeded with a probability of $1 - \alpha/2$. Thus, if a sample of 20 observations is chosen, and the sample variance equals 5, the 95 percent confidence interval for the population variance is

$$\frac{19(5)}{32.85} < \sigma^2 < \frac{19(5)}{8.91},$$

since Appendix Table 8 shows that for 19 degrees of freedom, $\chi^2_{.025} = 32.85$ and $\chi^2_{.975} = 8.91$. Simplifying terms, it follows that the confidence interval in this case is 2.89 to 10.66.

EXERCISES

9.1 Find the value of $\chi^2_{.05}$ when there are (a) 5 degrees of freedom; (b) 10 degrees of freedom; (c) 20 degrees of freedom.

9.2 Find the value of $\chi^2_{.01}$ when there are (a) 8 degrees of freedom; (b) 14 degrees of freedom; (c) 26 degrees of freedom.

9.3 If the coefficient of variation of a χ^2 random variable equals 1, how many degrees of freedom does it have?

9.4 If $\chi^2_{.01} = 29.1413$, what is the number of degrees of freedom? If $\chi^2_{.95} = 10.8508$, what is the number of degrees of freedom?

9.5 We add up the squared values of 15 independent standard normal variables. (a) What sort of probability distribution does this sum have? (b) What is the expected value of this sum? (c) What is the variance of this sum?

9.6 The Crooked Arrow National Bank obtains data concerning a random sample of 100 deposit slips each day. The number of these slips containing errors is as follows for a period of 10 consecutive days: 9; 10; 8; 6; 12; 15; 12; 9; 8; 6. Test whether the proportion of deposit slips containing an error is the same each day. (Let $\alpha = .05$.)

9.7 The Crooked Arrow National Bank obtains data concerning a sample of 200 loan applications at each of its four branches in 1978. The number of these loan applications containing falsified information at each branch is 6; 8; 9; and 12. Test whether the proportion of loan applications containing falsified information is the same at each branch. (Let $\alpha = .05$.)

9.8 The Alpha Corporation, a bicycle maker, obtains tires from two suppliers, firm A and firm B. During 1978, it received the following number of acceptable and defective tires from each supplier:

Supplier	Number of acceptable tires	Number of defective tires	Total number of tires received
A	940	89	1,029
B	780	32	812

Is the probability that the Alpha Corporation would receive a defective tire the same, regardless of the supplier? Test this hypothesis, based on a significance level of .05.

9.9 The Alpha Corporation hires a market-research firm to interview 100 people between 20 and 30 years of age in each of four regions of the country who bought one of Alpha's competitors' bicycles in the past year. Each interviewee is asked whether he or she had ever heard of Alpha's bicycle, and if so, whether he or she felt it was overpriced. The results are as follows:

Region	Never heard of Alpha's bicycle	Thought Alpha's bicycle was		Total
		Overpriced	Not overpriced	
South	30	10	60	100
North	41	21	38	100
East	28	7	65	100
West	32	14	54	100
Total	131	52	217	400

Test whether there are regional differences in the probability that an interviewee who bought one of Alpha's competitors' bicycles in the past year (1) never heard of Alpha's bicycle; (2) thought it was overpriced; (3) did not think it was overpriced. (Let $\alpha = .01$.)

9.10 A gambler wants to determine whether a die is true. He throws the die 100 times, with the following results:

Number on face of die	Number of outcomes
1	15
2	18
3	20
4	17
5	13
6	17
Total	100

Is there any evidence that the die is not true? (Let $\alpha = .05$.)

9.11 The gambler in the previous exercise wants to determine whether a pair of dice is "loaded." He rolls the dice 200 times, with the following results:

Number on face of die	Number of outcomes
2	0
3	2
4	2
5	16
6	30
7	100
8	30
9	14
10	2
11	4
12	0
Total	200

Is there any evidence that these dice are not true? (Let $\alpha = .01$.)

9.12 A random sample of 30 observations is taken from a normal population. The sample variance equals 10. Test the hypothesis that the population variance equals 20, the alternative hypothesis being that $\sigma^2 \neq 20$. Use the .05 significance level.

9.13 Based on the results in the previous exercise, calculate a 98 percent confidence interval for the population variance.

9.9 Nonparametric Techniques

Many of the tests discussed here and in the previous chapter are based on the assumption that the population or populations under consideration are normal. Although this assumption frequently is close enough to the truth so that these tests are good approximations, cases often arise where statisticians prefer instead to use various kinds of nonparametric, or distribution-free, tests. The hallmark of these tests is that they avoid the assumption of normality. For practically every test discussed in this and the previous chapter, there is a nonparametric analogue. Besides having the advantage that normality is not assumed, nonparametric tests often are also easier to carry out because the computations are simpler. In addition, some experiments yield responses that can only be ranked, not measured along a cardinal scale. Nonparametric techniques are designed to analyze data of this sort, which is another important reason for their use.

In the following sections we shall describe three of the most commonly used nonparametric techniques: the sign test, the Mann-Whitney test, and the runs test. Each of these has important applications in business and economics.

9.10 The Sign Test

The sign test is used to test the null hypothesis that the population median equals a certain amount. For example, suppose that the Federal Trade Commission has 36 tires from a manufacturer tested to determine their length of life. The purpose is to find out whether the manufacturer's advertising claim that the tires average 25,000 miles of use is correct. If the median length of life is 25,000 miles, the probability that a tire chosen at random will last for less than 25,000 miles is 1/2, and the probability that it will last for more than 25,000 miles is also 1/2. If the results for the 36 tires are as shown in Table 9.6, we can place a minus sign to the right of the figure for each tire that lasts less than 25,000 miles. By the same token, we place a plus sign to the right of the figure for each tire that lasts more than 25,000 miles. If the null hypothesis holds (that is, if the median life is 25,000 miles), the number of pluses will have a binomial distribution:

$$P(x) = \frac{36!}{x!(36 - x)!}\left(\frac{1}{2}\right)^x\left(\frac{1}{2}\right)^{36-x},$$

where x is the number of pluses. (Why? Because there is a probability of 1/2 that each tire in the sample will last more than 25,000 miles and thus will have a plus beside it.) Since the sample size is relatively large ($n = 36$), we know from Chapter 5 that this binomial distribution can be approximated by a normal distribution with a mean of 18 and a standard deviation of 3. (Because $n = 36$ and $P = 1/2$, $nP = 36(1/2)$, or 18, and $\sqrt{nP(1 - P)} = \sqrt{36(1/2)(1/2)}$, or 3.) Thus, the sampling distribution of the number of pluses is as shown in Figure 9.3.

The null hypothesis, as noted above, is that the median length of life is

TABLE 9.6

Length of Life of Sample of 36 Tires

27,100 (+)	24,500 (−)	24,800 (−)	25,100 (+)
24,200 (−)	25,600 (+)	24,600 (−)	24,800 (−)
23,300 (−)	24,600 (−)	25,500 (+)	24,700 (−)
23,800 (−)	24,800 (−)	26,100 (+)	25,900 (+)
22,600 (−)	24,100 (−)	23,900 (−)	23,900 (−)
24,800 (−)	23,900 (−)	24,100 (−)	24,800 (−)
24,100 (−)	24,900 (−)	24,300 (−)	24,700 (−)
24,800 (−)	24,600 (−)	24,700 (−)	24,900 (−)
25,500 (+)	26,100 (+)	24,900 (−)	25,100 (+)

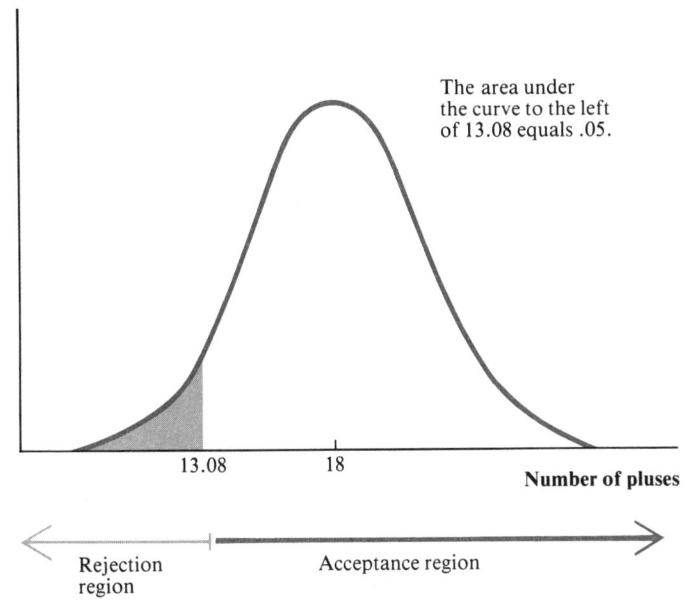

FIGURE 9.3 Sampling Distribution of Number of Pluses, if Null Hypothesis Is True

25,000 miles. The alternative hypothesis is that the median length of life is less than 25,000 miles. (Why is the alternative hypothesis one-sided? Because the FTC is interested in taking action against the advertising claims only if the median is less than 25,000 miles.) Thus, the FTC's statisticians will reject the null hypothesis only if the number of pluses—that is, x—is so small that it is unlikely that the sampling distribution in Figure 9.3 is valid. If the significance level is set at .05, it is clear that the null hypothesis should be rejected if the number of pluses is less than

$$18 - 1.64(3) = 13.08,$$

since there is a .05 probability that the number of pluses will be less than 13.08, given that the null hypothesis is true.[11] (See Figure 9.3.) Based on the data in Table 9.6, it follows that the FTC should reject the hypothesis that the median length of life is 25,000 miles, because the number of pluses is 9, which is considerably less than 13.08.

The sign test is based on the fact that if the population median (M) is equal to the value specified by the null hypothesis (say, M_0), x, the number of observations in the sample above M_0, has a binomial distribution with P equal to 1/2 and n equal to the sample size. Thus, if the sample size is large, the following decision rules can be used.

> ***Decision Rule if the Alternative Hypothesis is that*** $\mathbf{M < M_0}$: *Accept the null hypothesis if* $x \geqslant n/2 - z_\alpha\sqrt{n/4}$; *reject the null hypothesis if* $x < n/2 - z_\alpha\sqrt{n/4}$.

11. We have ignored the continuity correction discussed in Chapter 5 because it has no material effect on our results.

Decision Rule if the Alternative Hypothesis is that $\mathbf{M > M_0}$: *Accept the null hypothesis if* $x \leqslant n/2 + z_\alpha\sqrt{n/4}$; *reject the null hypothesis if* $x > n/2 + z_\alpha\sqrt{n/4}$.

Decision Rule if the Alternative Hypothesis is that $\mathbf{M \neq M_0}$: *Accept the null hypothesis if* $n/2 - z_{\alpha/2}\sqrt{n/4} \leqslant x \leqslant n/2 + z_{\alpha/2}\sqrt{n/4}$; *reject the null hypothesis if* $x < n/2 - z_{\alpha/2}\sqrt{n/4}$ *or if* $x > n/2 + z_{\alpha/2}\sqrt{n/4}$.

If the sample size is small, the binomial distribution, not the normal distribution, must be used; thus, these decision rules must be altered.

Note that the sign test does not depend on any assumptions about the nature or shape of the population. It rests only on the fact that the probability that a randomly chosen observation falls above the median is 1/2; and this is true for any population, normal or otherwise. (Basically, this is why the sign test is nonparametric.) Also, note that the computations involved in this test are comparatively simple. As pointed out previously, nonparametric tests are often used for this reason.

The sign test is useful in a wide variety of circumstances. The following example illustrates how it is employed in a sample consisting of matched pairs.

Example 9.3 A paint company conducts an experiment in which one half of a wood plank is painted with its paint and the other half is painted with its competitor's paint. The sample consists of 100 planks treated in this way. Each plank is placed outdoors in order to determine the difference between (1) the time it takes for the half covered with the company's paint to crack and (2) the time it takes for the half covered with the competitor's paint to crack. The paint firm wants to test the null hypothesis that the median difference equals zero. For 62 boards tested, it takes longer for the half covered with the company's paint to crack than for the half covered with the competitor's paint; for 38 boards the reverse is true. Use the sign test to test the null hypothesis. (Let the significance level equal .05.)

Solution: If the null hypothesis is true (that is, if the median difference is zero), the probability that the half of each board covered with the company's paint will last longer than the other half is 1/2. Thus, if we put a plus next to each board where the half covered by the company's paint lasts longer than the other half, the probability distribution of the number of pluses should be binomial with $n = 100$ and $P = 1/2$, if the null hypothesis is true. Since n is large, the binomial distribution can be approximated by the normal distribution, and the null hypothesis should be rejected if the number of pluses x is less than $n/2 - z_{.025}\sqrt{n/4}$ or greater than $n/2 + z_{.025}\sqrt{n/4}$. Since $n = 100$, the null hypothesis should be rejected if $x < 50 - 1.96(5)$, or 40.2, or if $x > 50 + 1.96(5)$, or 59.8. Since $x = 62$, the null hypothesis should be rejected. In other words, the evidence seems to indicate that the company's paint lasts longer than its competitor's.

Note that if the cracking times for both halves of any boards were exactly equal, they should be omitted from the analysis.

9.11 The Mann-Whitney Test

Another important nonparametric technique is the Mann-Whitney test, also known as the Wilcoxon test or the U test. This technique is used to find out whether two populations are identical. For example, the Federal Trade Commission might need to determine whether there is a difference between the population of lifetimes of firm C's tires and the population of lifetimes of firm D's tires. A random sample of 14 of firm C's tires and a random sample of 14 of firm D's tires are taken, the results being shown in Table 9.7. In this case, *the null hypothesis is that there is no difference between the two populations. The alternative hypothesis is that they are different.* Note that in contrast to Example 9.3, the two samples are entirely independent; *they are not matched pairs.*

The first step in applying the Mann-Whitney test is to rank all observations in the two combined samples. Table 9.7 shows the 28 observations ranked from lowest to highest, with the letter under each observation (a C or a D) indicating the firm from which it came. The members of firm C's sample are ranked 1, 2, 3, 4, 5, 7, . . . , and the members of firm D's sample are

TABLE 9.7

Lengths of Life of Sample of Tires from Two Firms

Firm C		Firm D	
19,800	19,700	19,850	20,850
19,900	19,600	31,250	20,650
20,100	19,500	20,380	21,050
20,200	20,400	19,950	19,820
31,300	20,300	20,260	20,350
18,700	21,000	19,750	31,150
18,900	21,100	20,450	20,750

Observations Ranked from Lowest to Highest

Observation	18,700,	18,900,	19,500,	19,600,	19,700,	19,750,	19,800,	19,820,
Firm	C	C	C	C	C	D	C	D
Rank	1	2	3	4	5	6	7	8
Observation	19,850,	19,900,	19,950,	20,100,	20,200,	20,260,	20,300,	20,350,
Firm	D	C	D	C	C	D	C	D
Rank	9	10	11	12	13	14	15	16
Observation	20,380,	20,400,	20,450,	20,650,	20,750,	20,850,	21,000,	21,050,
Firm	D	C	D	D	D	D	C	D
Rank	17	18	19	20	21	22	23	24
Observation	21,100,	31,150,	31,250,	31,300				
Firm	C	D	D	C				
Rank	25	26	27	28				

ranked 6, 8, 9, 11, 14, 16, 17 . . . .[12] If the null hypothesis is true, we would expect that the average rank for the items from one sample would be approximately equal to the average rank for the items from the other sample. On the other hand, if the alternative hypothesis is true, we would expect a marked difference between the average ranks in the two samples.

To carry out the Mann-Whitney test, it is more convenient to use the *rank-sum* instead of the average rank in a sample. Let R_1 be the sum of the ranks for one of the samples. (It doesn't matter which sample one chooses.) Then we compute

$$U = n_1 n_2 + \frac{n_1(n_1 + 1)}{2} - R_1, \tag{9.6}$$

where n_1 is the number of observations in the sample on which R_1 is based and n_2 is the number of observations in the other sample. If the null hypothesis is true, the sampling distribution of U has a mean equal to

$$E_u = \frac{n_1 n_2}{2} \tag{9.7}$$

and a standard deviation equal to

$$\sigma_U = \sqrt{\frac{n_1 n_2 (n_1 + n_2 + 1)}{12}}. \tag{9.8}$$

And if n_1 and n_2 are at least 10, the sampling distribution of U can be approximated quite well by the normal distribution.

Based on these facts concerning the sampling distribution of U (if the null hypothesis is true), it is easy to formulate an appropriate decision rule.

> **Decision Rule:** *Accept the null hypothesis that the two populations are identical if* $-z_{\alpha/2} \leqslant (U - E_u) \div \sigma_u \leqslant z_{\alpha/2}$; *reject the null hypothesis if* $(U - E_u) \div \sigma_u < -z_{\alpha/2}$ *or if* $(U - E_u) \div \sigma_u > z_{\alpha/2}$.

In applying this decision rule, note that

$$\frac{U - E_u}{\sigma_u} = \frac{n_1 n_2 + \dfrac{n_1(n_1 + 1)}{2} - R_1 - \dfrac{n_1 n_2}{2}}{\sqrt{n_1 n_2 (n_1 + n_2 + 1)/12}}. \tag{9.9}$$

Also, it should be recognized that this decision rule applies only to a two-tailed test. (The extension to a one-tailed test is straightforward.) Further, this decision rule assumes that both n_1 and n_2 are at least 10; if not, the test must be based on special tables, and the procedure described here is not appropriate.

To illustrate the use of the Mann-Whitney test, let's return to the data in Table 9.7. To compute R_1, let's use the sample of firm C's tires. Since the observations in this sample have ranks of 1, 2, 3, 4, 5, 7, 10, 12, 13, 15, 18, 23,

12. If there are ties, we must assign the tied items the mean of the ranks they have in common. For example, if two items are tied for 3d and 4th place, each would receive the rank of 3.5.

25, and 28, it is clear that $R_1 = 166$. Thus, since $n_1 = n_2 = 14$,

$$U = 14(14) + \frac{14(15)}{2} - 166 = 135,$$

and

$$E_u = \frac{14(14)}{2} = 98$$

$$\sigma_u = \sqrt{\frac{14(14)(29)}{12}} = 21.8.$$

Consequently,

$$\frac{U - E_u}{\sigma_u} = \frac{135 - 98}{21.8} = 1.70.$$

If the .05 significance level is chosen, it appears that the null hypothesis should not be rejected, since $z_{.025} = 1.96$, which is more than $(U - E_u) \div \sigma_u$. Thus, the Federal Trade Commission should conclude that the available evidence does not indicate that there is a difference between the population of lifetimes of firm C's tires and the population of lifetimes of firm D's tires.

Finally, it is worth noting that this test does not assume that the populations in question are normal. That is, there is no need to assume that the population of lifetimes of firm C's tires is normal, or that the population of lifetimes of firm D's tires is normal. Thus, this test avoids the assumption of normality underlying the t test, which was used in the previous chapter to test a similar kind of hypothesis. Moreover, this test requires far less computation than the t test, an important advantage under some circumstances.

The Mann-Whitney test is useful as a possible substitute for the t test. The following example further illustrates how it is used.

Example 9.4 A personnel agency gives an aptitude test to 12 men and 12 women, the scores being as follows:

Men	80, 79, 92, 65, 83, 84, 95, 78, 81, 85, 73, 52
Women	82, 87, 89, 91, 93, 76, 74, 70, 88, 99, 61, 94

Use these data to test the null hypothesis that the distribution of scores on this test is the same for men as for women. (Let the significance level equal .05.)

Solution: Rank the scores from lowest to highest, and put an M under each score if it is a man's and a W under each score if it is a woman's:

52	61	65	70	73	74	76	78	79	80	81	82	83	84	85	87	88
M	W	M	W	M	W	W	M	M	M	M	W	M	M	M	W	W

89	91	92	93	94	95	99
W	W	M	W	W	M	W

The sum of the ranks for men is $1 + 3 + 5 + 8 + 9 + 10 + 11 + 13 + 14 + 15 + 20 + 23 = 132$. Thus $U = 12(12) + [12(13)/2] - 132 = 90$. Also, $E_u = 12(12)/2 = 72$, and $\sigma_u = \sqrt{12(12)(25)/12} = \sqrt{300} = 17.3$. Thus, since $z_{\alpha/2} = 1.96$, the null hypothesis should be rejected if $(U - E_u) \div \sigma_u$ is less than -1.96 or greater than 1.96. In fact, $(U - E_u) \div \sigma_u = (90 - 72) \div 17.3 = 1.0$. Thus, the null hypothesis should not be rejected. The probability is greater than .05 that these results could have occurred, given that the distribution of scores on this test was the same for men as for women.

9.12 The Runs Test

Another important nonparametric technique is the runs test, which is designed to test the hypothesis that a sequence of numbers, symbols, or objects is in random order. A tennis fan interested in the outcomes of the matches between Jimmy Connors and Bjorn Borg finds that these outcomes are as follows (where C stands for a Connors victory and B stands for a Borg victory):

B C B C B C B C B C B C B C.

Does this sequence appear to be in a random order? No, since the winner seems to alternate back and forth. (If this were the case, one might suspect that tennis players, like some phoney wrestlers, decide in advance who will win and alternate the winner.) On the other hand, suppose that the sequence is

C C C C C C C C B B B B B B.

Does this sequence appear to be in a random order? No, since the winner of all the early matches is Connors, and the winner of all the later matches is Borg. (If this were the case, we might suspect that Borg had greatly improved—or that Connors's play had deteriorated—at about the middle of the sequence of matches.)

To understand the runs tests, you must know the statistician's definition of a run, which is the following:

> ***A run** is a sequence of identical numbers, symbols, objects, or events preceded and followed by different numbers, symbols, objects, or events (or by nothing at all).*

In the first version of the Connors-Borg matches given above, there are 14 runs; in the second version there are only 2 runs. Suppose that a third version is as follows:

$\underline{C}\ \underline{B}\ \underline{CCCC}\ \underline{BBB}\ \underline{C}\ \underline{B}\ \underline{C}\ \underline{BB}$.

How many runs are there? The answer is 8. (The lines under the letters designate each run.)

The runs test is based on the idea that if the probability of the occurrence of one number, symbol, object, or event is constant throughout the sequence, then there should be neither a very large number of runs (as in our first "history" of the Connors-Borg matches) nor a very small number of runs (as in our second such "history"). If there are two possible outcomes (such as B and C), and if the probability of each outcome remains constant throughout the sequence (and if successive outcomes are independent), it can be shown that the expected number of runs is:

$$E_r = \frac{2n_1n_2}{n_1 + n_2} + 1, \tag{9.10}$$

and the standard deviation of the number of runs is

$$\sigma_r = \sqrt{\frac{2n_1n_2(2n_1n_2 - n_1 - n_2)}{(n_1 + n_2)^2(n_1 + n_2 - 1)}}, \tag{9.11}$$

where n_1 is the number of outcomes of one type (such as B) and n_2 is the number of outcomes of the other type (such as C). If either n_1 or n_2 is larger than 20, it can be shown that the number of runs r is approximately normally distributed. When both n_1 and n_2 exceed 10, this approximation is good.

The runs test is designed to test the null hypothesis that a given sequence of outcomes is in random order—that is, that the probability of each outcome remains constant throughout the sequence and that successive outcomes are independent. If the significance level is set equal to α, it follows from the previous paragraph that the following decision rule can be applied.

> ***Decision Rule:*** *Accept the null hypothesis (of randomness) if* $E_r - z_{\alpha/2}\sigma_r \leqslant r \leqslant E_r + z_{\alpha/2}\sigma_r$; *reject the null hypothesis if* $r < E_r - z_{\alpha/2}\sigma_r$ *or if* $r > E_r + z_{\alpha/2}\sigma_r$.

This assumes that n_1 and n_2 meet the requirements in the previous paragraph.

To illustrate the application of this test, suppose that Connors and Borg play 37 times, with the following results:

C B C C B C C B C C B B B C B B C C C B C B C C B C C B B C C C B B C C B.

If n_1 is the number of Connors victories and n_2 is the number of Borg victories, $n_1 = 21$ and $n_2 = 16$. Thus,

$$E_r = \frac{2(21)(16)}{37} + 1 = 19.16$$

and

$$\sigma_r = \sqrt{\frac{2(21)(16)[2(21)(16) - 21 - 16]}{37^2(36)}} = \sqrt{\frac{672(635)}{49{,}284}} = \sqrt{8.7} = 2.94.$$

Thus, if α is set equal to .05,

$$E_r - z_{\alpha/2}\sigma_r = 19.16 - 1.96(2.94) = 13.4$$

$$E_r + z_{\alpha/2}\sigma_r = 19.16 + 1.96(2.94) = 24.9.$$

Since the actual number of runs r equals 22, the null hypothesis should not be rejected (r being neither less than 13.4 nor more than 24.9).

The runs test has many applications in business and economics. The example below shows how it can be used with quantitative rather than qualitative measurements.

Example 9.5 A winery measures daily the acidity of the wine it produces. During a 32-day period, it is found that the mean acidity equals 8. Each day during this period is classified as A (above average in acidity) or B (below average in acidity), the results being as follows:

B A A A A B B A B B A A A A B B B B A B A A A A A A A A B A A A A.

Use the runs test to determine whether there is a departure from randomness in the sequence. (Let the significance level equal .05.)

Solution: Since $n_1 = 21$ and $n_2 = 11$,

$$E_r = \frac{2(21)(11)}{32} + 1, \text{ or } 15.4,$$

and

$$\sigma_r = \sqrt{\frac{2(21)(11)[2(21)(11) - 21 - 11]}{32^2(31)}}$$

$$= \sqrt{\frac{198{,}660}{31{,}744}} = \sqrt{6.26}, \text{ or } 2.50.$$

Thus, the null hypothesis should be rejected if the number of runs is less than

$$E_r - z_{\alpha/2}\sigma_r = 15.4 - 1.96(2.50) = 10.5,$$

or greater than

$$E_r + z_{\alpha/2}\sigma_r = 15.4 + 1.96(2.50) = 20.3.$$

Since the number of runs equals 12, the winery should not reject the null hypothesis. In other words, it should not reject the hypothesis that the sequence is random.

9.13 Nonparametric Techniques: Pros and Cons

Based on the discussion in the previous sections, nonparametric techniques are clearly a very important statistical tool. Frequently, as we have seen, the statistician has the choice of using a nonparametric technique or a parametric procedure of the sort discussed in the previous chapter. For example, the sign test is to some extent a substitute for the test (in the previous chapter) of a specified value of the mean, and the Mann-Whitney test is to some extent a substitute for the parametric test (also in the previous chapter) of equality between two population means. The following factors should guide the choice between nonparametric and parametric techniques in situations where both can be used.

Simplicity of Calculations. Under some circumstances, it is very important that the computations in a statistical test be simple and rapid. For example, it may be necessary for a person with a relatively limited mathematical background to carry out the computations; or there may be a premium on speed. In such cases, nonparametric techniques like the sign test may be preferred over parametric tests.

Realism of Assumptions. Many parametric tests assume that the relevant populations are normal. Two such tests are the small-sample test that a population mean equals a specified value and the small-sample test for equality between two population means. Although these procedures are dependable in the face of moderate departures from normality, they can be quite misleading in the face of gross departures of this sort. Thus, if the populations are very far from normality, statisticians often prefer nonparametric techniques like the sign test and Mann-Whitney test.

Power of the Test. If the assumptions underlying the parametric tests in the previous chapter are reasonably close to reality, these tests generally are more powerful than nonparametric techniques. In other words, the parametric techniques are less likely to produce a Type II error, given that the probability of a Type I error is the same for both tests. This is an important advantage of parametric tests, as long as their assumptions are reasonably valid. Nonparametric techniques, as we have seen, often use rankings or orderings as distinct from the actual values of the observations, thus losing a certain amount of the information in the sample.

EXERCISES

9.14 A researcher claims that the median increase in output due to a new technique is 10 units of output per hour. A firm tries the new technique 100 times, and finds that in 41 of these trials the new technique results in an increase of more than 10 units of output per hour. In 59 of these trials there is an increase of less than 10 units of output per hour. Use the sign test to indicate whether the researcher's claim is correct, against a two-sided alternative. (Let $\alpha = .05$.)

9.15 Use the sign test to indicate whether the researcher's claim (in the previous exercise) is correct, against the alternative hypothesis that the median increase is less than 10 units of output per hour. (Let $\alpha = .05$.)

9.16 Recall from Exercise 8.22 that the Crooked Arrow National Bank chose a random sample of 36 depositors at its southern branch and found that the 1979 income of each was as follows:

Dollars					
20,100	8,200	13,200	5,200	11,100	51,100
19,400	8,900	14,300	10,100	10,800	9,600
10,100	10,300	15,800	12,300	9,100	10,900
23,000	26,000	16,100	14,000	7,200	12,000
24,200	11,400	17,200	15,100	4,300	13,200
25,100	12,900	18,900	16,000	38,000	15,100

Use the sign test to test the hypothesis that the median income of the depositors at the bank's southern branch equals $15,000, the alternative hypothesis being that it does not equal $15,000.

(a) Carry out this test at the 5 percent significance level.

(b) Carry out this test at the 1 percent significance level.

9.17 Compare the results you obtained in Exercise 9.16 with those obtained in Exercise 8.22.
(a) Are the results consistent?
(b) Are the two tests aimed at testing the same hypothesis?
(c) Are the two tests based on the same assumptions? If not, what are the differences in the assumptions?

9.18 Having installed a new procedure to reduce clerical errors, the Crooked Arrow National Bank wants to determine whether this new procedure has resulted in a difference in the distribution of the number of clerical errors per day in one of the bank's departments. The bank chooses a random sample of 11 days prior to the installation of the new procedure and finds that the number of errors per day was as follows: 7, 8, 10, 6, 5, 8, 9, 11, 10, 7, 9. The bank also chooses a random sample of 11 days following the installation of the new procedure and finds that the number of errors per day was as follows: 6, 5, 8, 7, 9, 6, 5, 4, 9, 7, 6. Use the Mann-Whitney test to determine whether the population distribution of the number of errors per day was different after vs. before the installation of the new procedure. (Let $\alpha = .05$.)

9.19 (a) Use a t test to test the hypothesis in Exercise 9.18, based on the data given there.
(b) Are the results consistent?
(c) Are the two tests aimed at testing the same hypothesis?
(d) Are the two tests based on the same assumptions? If not, what are the differences in the assumptions?

9.20 The Crooked Arrow National Bank decides that a better way to test error reduction under its new procedure is to choose 10 employees randomly and determine how many errors each made in a week before the introduction of the new procedure and how many each made in a week after its introduction. The results are as follows:

Type of procedure	Person 1	2	3	4	5	6	7	8	9	10
Old	40	38	37	34	39	37	42	33	39	40
New	37	37	32	40	37	35	39	32	37	38

Use a nonparametric technique to test the hypothesis that there is no difference between the old and new procedure in the incidence of errors committed. (Let $\alpha = .05$.)

9.21 There is a sequence containing 3 Ss and 17 Ts. (a) What is the minimum number of runs that can occur in this sequence? (b) What is the maximum number?

9.22 A table of one-digit random numbers begins as follows:

0 1 3 4 6 8 9 7 6 5 4 1 0 0 9 8 7 6 5 4 3 2 1

1 4 3 8 7 6 6 7 5 4 5 0 2 5 8 7 6 9 4 3 2 1 0

9 7 6 5 4 3 2 2 1 1 4 5 6 7 7 8 9 2 4 4 5

If an even number (or zero) is denoted by E and an odd number is denoted by O, how many runs of Es or Os are there in this sequence?

9.23 Based on the number of runs of even (including zero) and odd numbers in the sequence in the previous exercise, is there any evidence that the sequence of numbers given there is not random? (Let $\alpha = .05$.)

9.24 Let any number from zero to 4 be denoted by A, and any number from 5 to 9 be denoted by B. How many runs of As or Bs are there in the sequence in Exercise 9.22?

9.25. Based on the number of runs of As and Bs in the sequence in Exercise 9.22, is there any evidence that the sequence of numbers given there is not random? (In the previous exercise you found the number of such runs. Let $\alpha = .01$.)

CHAPTER REVIEW

1. The χ^2 *distribution* (with ν degrees of freedom) is the probability distribution of the sum of squares of ν independent standard normal variables. The χ^2 distribution is a continuous probability distribution and is generally skewed to the right; but as the number of degrees of freedom increases, it approaches normality. There is a zero probability that a χ^2 random variable will be negative.

2. To test whether a number of population proportions are equal (against the alternative hypothesis that they are unequal), the sample proportion (in the samples from all populations combined) should be computed. This proportion should then be multiplied by the sample size from each population to obtain the expected frequency of "successes" in this sample. The test statistic is $\Sigma(f - e)^2/e$, where f is the actual frequency of "successes" or "failures" in each sample and e is the corresponding expected frequency. If the test statistic exceeds χ^2_α (where the number of degrees of freedom is one less than the number of populations being compared), the null hypothesis should be rejected.

3. A *contingency table* contains a certain number of rows and columns, and the decision maker wants to know whether the probability distribution in one column differs from that in another column. In other words, is the variable represented by the rows independent of that represented by the columns? Assuming such independence, we can compute the expected frequencies in the table. The test statistic is $\Sigma(f - e)^2/e$, where f is an actual frequency and e is the corresponding expected frequency. If the test statistic exceeds χ^2_α (where the number of degrees of freedom equals $[r - 1][c - 1]$, r being the number of rows and c being the number of columns), the null hypothesis (of independence) should be rejected.

4. To test whether an *observed frequency distribution* conforms to a *theoretical frequency distribution*, we calculate the test statistic $\Sigma(f - e)^2/e$, where f is the observed frequency in a particular class interval and e is the theoretical, or expected, frequency in the same class interval. If the test statistic exceeds χ^2_α—where the number of degrees of freedom equals the number of class intervals minus $(1 + h)$, and h is the number of parameters estimated from the sample—the null hypothesis (that the observed frequency distribution conforms to the theoretical one) should be rejected.

5. To test whether the variance of a normal population equals σ_0^2, we compute the test statistic: $(n - 1)s^2 \div \sigma_0^2$, where n is the sample size and s^2 is the sample variance. If the test statistic exceeds $\chi^2_{\alpha/2}$ or is less than $\chi^2_{1-\alpha/2}$ (where $n - 1$ is the number of degrees of freedom), the null hypothesis should be rejected if the alternative hypothesis is two-sided.

6. The *sign test* is used to test the null hypothesis that the population median equals a certain amount. A plus sign is placed next to a sample observation exceeding this amount; a minus sign is placed next to one that

falls below it. The number of plus signs x has a binomial distribution (where $P = 1/2$) if the null hypothesis is true. For large samples, the null hypothesis should be rejected if $x < n/2 - z_{\alpha/2}\sqrt{n/4}$ or if $x > n/2 + z_{\alpha/2}\sqrt{n/4}$, assuming that a two-tailed test is appropriate.

7. The *Mann-Whitney test* is used to test the null hypothesis that two samples come from the same population. The first step is to rank all observations in the two samples combined and to compute R_1, the sum of the ranks for the first sample. One then computes the test statistic

$$\frac{n_1n_2 + \dfrac{n_1(n_1 + 1)}{2} - R_1 - \dfrac{n_1n_2}{2}}{\sqrt{n_1n_2(n_1 + n_2 + 1)/12}},$$

where n_1 is the number of observations in the first sample and n_2 is the number of observations in the second sample. If this test statistic is greater than $z_{\alpha/2}$ or less than $-z_{\alpha/2}$ the null hypothesis should be rejected. (This assumes that n_1 and n_2 are at least 10.)

8. The *runs test* is designed to test the null hypothesis that a sequence of numbers, symbols, or objects is in random order. A *run* is a sequence of *identical* numbers, symbols, objects, or events preceded and followed by *different* numbers, symbols, objects, or events, or by nothing at all. If a sequence is in random order, the expected number of runs E_r equals $[(2n_1n_2)/(n_1 + n_2)] + 1$, and the standard deviation of the number of runs σ_r equals

$$\sqrt{\frac{2n_1n_2(2n_1n_2 - n_1 - n_2)}{(n_1 + n_2)^2(n_1 + n_2 - 1)}},$$

where n_1 is the number of observations in the sequence of one type and n_2 is the number of the other type. If both n_1 and n_2 exceed 10, the null hypothesis should be rejected if the number of runs r is less than $E_r - z_{\alpha/2}\sigma_r$ or if it exceeds $E_r + z_{\alpha/2}\sigma_r$.

Getting Down to Cases:

TESTING FOR NORMALITY AT THE AMERICAN STOVE COMPANY

A number of years ago, the American Stove Company collected data concerning the heights of a metal piece that the firm manufactured. The firm measured the heights of 145 of these metal pieces, and the results are shown in the following frequency distribution:[13]

Height of metal piece (inches)	Number of metal pieces
Less than .8215	9
.8215 and under .8245	5
.8245 and under .8275	14
.8275 and under .8305	21
.8305 and under .8335	55

13. See A. J. Duncan, op. cit.

Height of metal piece (inches)	Number of metal pieces
.8335 and under .8365	23
.8365 and under .8395	7
.8395 and under .8425	6
.8425 and over	5
Total	145

For many purposes it is important to test whether a certain random variable is normally distributed. For example, as we saw in Chapter 8, some statistical tests assume normality. Suppose that the American Stove Company had given you the task of using the above data to test whether the heights of the metal pieces are normally distributed.

(a) Express the upper limit of each class interval in the frequency distribution as a deviation from the sample mean (.8314 inches), and divide this deviation by the sample standard deviation (.0059 inches).

(b) Use the results in (a)—which correspond to points on the standard normal distribution—to determine the theoretical number of metal pieces that should fall into each class interval if the heights are normally distributed.

(c) Calculate the difference between the actual and expected frequency in each class, square this difference, divide it by the theoretical frequency, and sum the results for all classes.

(d) Use the χ^2 distribution to test the hypothesis that the heights are normally distributed. (Let $\alpha = .05$.)

(e) Explain your choice of the number of degrees of freedom.

(f) Write a three-sentence memorandum summarizing your results.

10

Experimental Design and the Analysis of Variance*

10.1 Introduction

As we have stressed often in previous chapters, one of the important functions of statistics is to provide principles for the proper design and analysis of experiments. Although the word *experiment* may evoke a vision of laboratory activities, many statistical investigations are experiments which have nothing whatever to do with a laboratory. Nonlaboratory experiments of various kinds are continually being carried out by business firms and government agencies for practical reasons. To make sure that these procedures are designed effectively and analyzed properly, you must know something about experimental design and the analysis of variance, the key topics of this chapter.

10.2 Design of Industrial Experiments

Although the neophyte commonly believes that it is a simple matter to design an experiment for testing a certain hypothesis, often a considerable amount of ingenuity is required to achieve a design that really tests the hypothesis one wants to test—and does so at something near minimum cost. As we know from Chapter 1, one of the commonest pitfalls is an experimental design in which the effect of the variable one wants to estimate is inextricably entangled with the effect of some other factor. (In other words, more than one factor may be responsible for a particular observed experimental result.) When this

*Some instructors may want to skip the last sections of this chapter, which deal with the two-way analysis of variance. Others may want to take up only the first three sections, or skip this chapter entirely. Any of these options is feasible, since an understanding of this chapter is not required in subsequent chapters. Readers who skipped Chapter 9 should read Section 9.3 before reading Section 10.4.

TABLE 10.1
Incidence of Polio in Two Groups of Children

School grade	Treatment	Number of children	Number afflicted with polio	Number afflicted per 100,000
Second grade	Salk vaccine	222,000	38	17
First and third grades	No vaccine	725,000	330	46

SOURCE: W. S. Youden, "Chance, Uncertainty, and Truth in Science," *Journal of Quality Technology*, 1972.

occurs, the effect of the variable one wants to estimate is said to be *confounded* with the effect of another factor or factors.

To illustrate this, recall our discussion in Chapter 1 of the results of a nationwide test of the effectiveness of the Salk antipolio vaccine. These data (reproduced in Table 10.1) seem perfectly adequate for measuring the effects of the vaccine on the incidence of polio in children, but in fact they contain a major bias: Only those second-graders who received their parent's permission received the vaccine, whereas *all* first- and third-graders were taken into account in estimating what the incidence of polio would be without the vaccine. The problem was that the incidence of polio was substantially lower among nonvaccinated children who did not receive permission than among those who did. (Recall from Chapter 1 that children of lower-income parents are less likely to receive permission—and because they grow up in less hygienic conditons, they are less likely to get polio.) Table 10.2 shows the results of a much more adequate experimental design in which children who received permission are differentiated from those who did not. Those who received permission were assigned at random to either the group receiving the vaccine or the group receiving the placebo (similar in appearance to the vaccine, but of no medical significance). As you can see, the results indicate that the reduction in the incidence of polio due to the vaccine is much larger than is shown in Table 10.1.[1]

TABLE 10.2
Randomized Experiment with Salk Vaccine

Permission given	Treatment	Number of children	Number afflicted with polio	Number afflicted per 100,000
Yes	Salk vaccine	201,000	33	16
Yes	Placebo	201,000	115	57
No	None	339,000	121	36

SOURCE: See Table 10.1.

1. This example is from W. J. Youden, "Chance, Uncertainty, and Truth in Science," *Journal of Quality Technology*, 1972.

In general, *an experiment dictates that one group of people (or machines, materials, or other experimental units) be treated differently from another group; and the effect of this difference in treatment is estimated by comparing certain measurable characteristics of the two groups*. However, unless the people (or machines, materials, or other experimental units) are assigned to one group or the other *at random*, all sorts of confounding can occur. For example, in the case of the Salk vaccine, if doctors had decided which children were to receive the vaccine and which were not, there might have been a tendency to give the vaccine to children where the consent of the parents was easy to obtain, and such children might differ from the others in the likelihood that they would contract polio. The importance of randomization cannot be overemphasized. The experimental units or subjects should be assigned at random to the groups receiving different treatments.

SHOULD THE EFFECTS OF SEVERAL FACTORS BE STUDIED ONE AT A TIME?

In many experiments, the statistician or decision maker is interested in estimating the effects of more than one factor on a certain characteristic of the relevant experimental unit. For example, a thread manufacturer may be interested in estimating the effects of differences in raw materials and differences in types of machines on the strength of the thread produced. The traditional way of estimating these effects is the one-at-a-time method. That is, the firm would hold constant the raw material used and observe the effects of differences in the types of machine used. For example, if there are three types of raw materials that could be used (A, B, and C) and if there are three types of machines (I, II, and III), the firm might use one raw material, A, in all three types of machine in order to observe how the strength of the thread varies among the machines. Similarly, to estimate the effect of the type of raw material on the strength of the thread, the firm might use only one type of machine, I, and observe how the strength of the thread varies among the types of raw material.

An important disadvantage of the one-at-a-time approach is that the results may be too narrowly focused. For example, the experiments described in the previous paragraph will provide information concerning the effects of the type of machine on the strength of the thread *if raw material A is used*. However, the differences among machines in this regard may *not* be independent of the raw material used. Thus, although machine I may result in the strongest thread when raw material A is used, machine I may result in the weakest thread when raw material C is used. Similarly, the above experiments will provide information concerning the effects of the type of raw material on the strength of the thread when machine I is used, but the differences among raw materials in this regard may not be independent of the machine used. Thus, although raw material A may result in the strongest thread when machine I is used, material A may result in the weakest thread when machine III is used.

In general, modern statisticians tend to emphasize the advantages of not controlling an experiment too closely. Even if the effects of the factors are independent, it frequently is less expensive to conduct an experiment where the factors are allowed to vary, rather than hold one or more constant, because fewer observations often are required to obtain the same precision. For example, in the above case, a better experimental design may be to obtain data concerning the strength of the thread when each combination of raw material and machine is used. Thus, the firm might use raw material A with machine I, with machine II, and with machine III. Similarly, each of the other raw materials (B and C) might be used with machine I, with machine II, and with machine III. If the effects of the raw material are independent of the type of machine, and if the effects of the type of machine are independent of the raw material, the resulting nine observations could be used to estimate the effect of each raw material and each type of machine. (If they are not independent, more observations are needed, as we shall see below.) More will be said about this in subsequent sections of this chapter.

RANDOMIZED BLOCKS

The type of experimental design recommended in the preceding paragraph is known as a randomized block design. In a *randomized block design* there are two kinds of effects, *treatment effects* and *block effects*. These terms are derived from agricultural research , where a field may be split into several blocks, and various treatments (such as fertilizers, pesticides, or some other factor whose effects the researcher is interested in) may be randomly assigned to plots in each block. Each block is constructed so that it contains relatively homogeneous experimental conditions. In an agricultural experiment each block may be a piece of land which has relatively homogeneous soil, sunlight, rainfall, and so forth. In the case of the thread manufacturer, either the raw materials or the machines can be regarded as treatments, and the other factor can be considered blocks. If each of the machines is regarded as a treatment, each raw material can then be regarded as a block because the use of this (and only this) raw material results in relatively homogeneous experimental conditions for comparing the effects of the treatments (the machines).

In a randomized block design, the statistician obtains data concerning the effect of each treatment in each block. Thus, in this case data are obtained concerning the strength of the thread resulting from the use of each type of machine with each raw material. The results might be as shown in Table 10.3. Note that only one observation is obtained concerning the effect of each

TABLE 10.3
Results of Experiment by Thread Manufacturer

	Machine (strength of thread)		
Raw material	I	II	III
A	50	40	45
B	48	39	45
C	52	44	48

treatment in each block. That is, no attempt is made to obtain more than one measurement of the strength of the thread resulting from the use of a particular type of machine with a particular raw material. In cases of this sort, the statistician must assume that the treatment effects and block effects are independent; if this assumption is violated, the *experimental errors*—the errors due to chance variation in the effect of each treatment in each block—will tend to be overstated.

To avoid assuming that the treatment and block effects are independent, and to obtain more precise estimates of these effects, statisticians often specify that *replications* occur. In other words, they ask that the experiment be repeated so that more than one observation is obtained concerning the effect of each treatment in each block. Thus, in the case above, each type of machine might be used with each type of raw material to produce two pieces of thread, not one, and the results might be as shown in Table 10.4.

TABLE 10.4
Results of Experiment by Thread Manufacturer, where Replication Occurs

	Machine (strength of thread)		
Raw material	I	II	III
A	50,52	40,38	45,46
B	48,47	39,39	45,44
C	52,53	44,45	48,47

The following is an example of the use of randomized blocks with replications. (This is an actual case, not a hypothetical one.)

Example 10.1 A number of years ago, the claim was made by some agricultural scientists that plants grow better when watered with a dilute solution of fluorescein. To test this hypothesis, the agricultural scientists watered one group of plants with water and an adjacent group with fluorescein, as shown in panel A of Figure 10.1. The results indicated a large difference in the growth of the two groups. Unfortunately, however, these results were not supported by subsequent research, because the experimental design was faulty. In reality, the soil to which the fluorescein was applied contained more nutrients than the soil to which plain water was applied. Suggest how a randomized block design might be substituted.

Solution: A more effective test of the effect of fluorescein might have been accomplished by the distribution of plants shown in panel B of Figure 10.1.

The entire field is divided into 6 relatively homogeneous blocks, each of which contains 18 plants. Each block is then divided in half, and a coin is flipped to determine which half (that is, which nine plants) will be watered with fluorescein and which half will receive plain water. This is a randomized block with nine observations concerning the effect of each

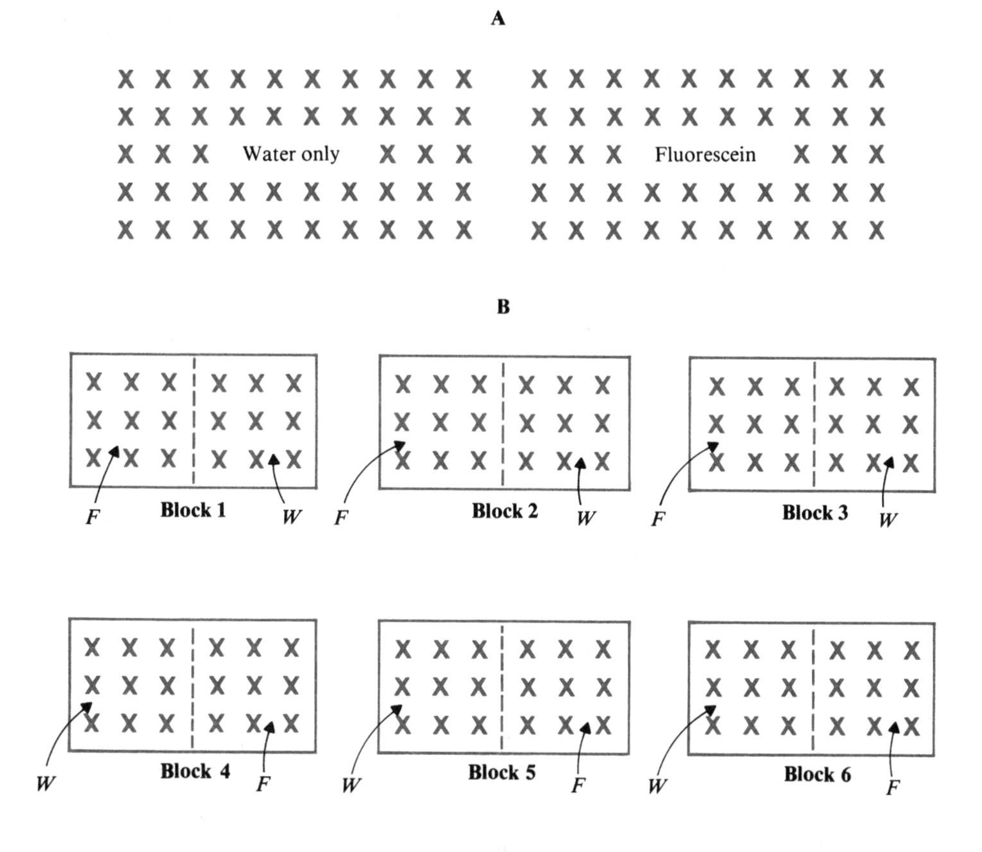

FIGURE 10.1 Spatial Distribution of Plants Receiving Water Only vs. Dilute Solution of Fluorescein

treatment in each block. The difference between the average rate of growth of the plants receiving fluorescein and the average rate of growth of those receiving plain water is a measure of the net effect of the treatment. (The net block effects—that is, the differences in rate of growth among various blocks—are of subsidiary importance in this case.)[2]

LATIN SQUARES

Still another important concept in experimental design is *balance*. In either Table 10.3 or 10.4, we can estimate the net effect of each type of machine on the strength of the thread by comparing the mean strength of the pieces of thread made by this type of machine with the mean strength of the thread made by the other types of machines. Such a comparison is meaningful because each of these means pertains to the same set of raw materials. For example, in Table 10.3, each raw material is included once in calculating such

2. Ibid.

a mean. In this sense, the design is balanced. Similarly, if we compare the average growth rate of plants receiving fluorescein with that of plants receiving water only, this comparison is balanced in the design shown in panel B of Figure 10.1, since each average pertains to the same six blocks of ground. In contrast, if the mean strength for machine I was based on the use of a different raw material than the mean strength for machine II, the design used by the thread manufacturer would not be balanced in this respect. And if the design in panel A of Figure 10.1 were used, the design would not be balanced with regard to the location and fertility of the soil.

This concept of balance is used ingeniously in a commonly employed experimental design called a *Latin square.* Suppose that the thread manufacturer wants to estimate the effect of the type of raw material, the type of machine, and the type of worker using them on the strength of the thread produced. (Note that an additional factor—type of worker—has been added to the two considered previously.) There are three types of raw material (A, B, and C), three types of machine (I, II, and III), and three types of worker (unskilled, semiskilled, and skilled). Although you may not believe it at first, the firm, by using a Latin square, can estimate the effects of all three of these factors on the basis of only nine observations.

Table 10.5 shows how this can be accomplished. Each vertical column signifies that a particular kind of raw material is used, and each horizontal row signifies that a particular type of machine is used. The letters in the body of the table indicate whether unskilled (U), semiskilled (E), or skilled (S) labor is used. There are nine entries in the table, each corresponding to an observation. For example, the entries in the first column show that in the first observation, raw material A will be used on machine I with unskilled labor; in the second observation, raw material A will be used on machine II with skilled labor; and in the third observation, raw material A will be used on machine III with semiskilled labor.

Several important points should be noted about the design shown in Table 10.5. First, each type of labor is used three times. Second, each type of labor shows up once and only once in each row, which means that each type of labor is used once and only once with each type of machine. Third, each type of labor shows up once only in each column, which means that each type of labor is used once and only once with each type of raw material. What these three points add up to is that *each type of labor is used once with each type of raw*

TABLE 10.5
A Latin-Square Experimental Design

	Raw material		
Machine	A	B	C
I	U	S	E
II	S	E	U
III	E	U	S

material and with each type of machine. Further, *each type of raw material is used once with each type of labor and with each type of machine;* and *each type of machine is used once with each type of labor and with each type of raw material.*

Because of this balance, we can compare the strength of the thread resulting from various raw materials by comparing (1) the average of the three observations where raw material A was used, with (2) the average of the three observations where raw material B was used, with (3) the average of the three observations where raw material C was used. This is a valid comparison because each of these sets of observations is based on the one-time use of each type of machine and each type of labor. Similarly, we can compare the strength of the thread resulting from various types of machines by comparing (1) the average of the three observations where machine I was used with (2) the average of the three observations where machine II was used, with (3) the average of the three observations where machine III was used. Similar comparisons can be made among various types of labor.

10.3 Testing Textile Fabrics: A Case Study[3]

Latin squares are used often in industrial experimentation. To illustrate their use, consider the actual case of a chemical firm that wanted to test the durability of four types of rubber-covered fabric. The machine which is used for tests of this sort contains four rectangular brass plates, each of which is covered with a special emery paper. A mechanical device rubs samples of fabric over each of the four plates bearing the emery paper. Although the four plates, or positions in the machine, are much the same, they differ from one another slightly. Since a different type of fabric can be tested at each position, the machine can test four types of fabric simultaneously, each such testing being called a run. It is known that because of variations in the condition of the emery paper and in temperature, humidity, and other factors, the results of each test will vary from one run to the next.

The chemical firm used a Latin-square design for comparing the four types of fabric. As shown in Table 10.6, one factor considered in the experimental design was the position of the fabric in the machine (that is, the particular brass plate on which it was tested). This factor is shown by each vertical column beneath a number. A second factor was the run, which is shown by each horizontal row. The letters in the body of Table 10.6 indicate which of the four fabrics was tested in the indicated run, in the indicated position in the machine. The four types of fabric are designated by A, B, C, and D. As you can see, each type of fabric appears only four times in the table, once only in each row and once only in each column.

The numbers in parentheses in Table 10.6 show the results of the 16 tests, each of these numbers showing the loss in weight of the fabric after a certain amount of rubbing. The average result for fabric A was 266; for fabric B, 220;

3. This example is from O. L. Davies, *The Design and Analysis of Industrial Experiments* (London: Oliver and Boyd, 1956), p. 163 ff.

TABLE 10.6
Results of a Latin-Square Design to Test Textile Fabrics

	Position in machine				
Run	4	2	1	3	Mean
2	A (251)	B (241)	D (227)	C (229)	237
3	D (234)	C (273)	A (274)	B (226)	252
1	C (235)	D (236)	B (218)	A (268)	239
4	B (195)	A (270)	C (230)	D (225)	230
Mean	229	255	237	237	

SOURCE: O. L. Davies, *The Design and Analysis of Industrial Experiments* (London: Oliver and Boyd, 1956), p. 164.

for fabric C, 242; and for fabric D, 231. These averages are comparable in the sense that each machine position and each run was included once only in each average. Based on the results, it appeared that fabric B wore best, since it experienced the lowest weight loss. Also, based on the averages in Table 10.6, it appeared that position 2 in the machine tended to result in larger weight losses than the other positions, and that run 3 tended to produce higher weight losses than the other runs. Using techniques of the sort discussed in subsequent sections of this chapter, the firm was able to test whether these differences are statistically significant.

A final point should be noted concerning Latin squares. Underlying the use of this design is the assumption that the effect of each factor is independent of the other factors. In other words, it is assumed that the difference in weight loss between fabric A and fabric B will be the same, on the average, regardless of which machine position is chosen or which run is considered. Also, the difference in weight loss between machine positions is assumed to be the same, regardless of which type of fabric or run is considered. Frequently this assumption is perfectly reasonable, but in those cases where it is not, another type of experimental design should be used.

EXERCISES

10.1 Experimental design reached the front page of the *Wall Street Journal* on August 25, 1977. The *Journal* defined a placebo as "any inactive substance or procedure used with a patient under the guise of an effective treatment." It reported that placebos often make people feel that they have recovered. Given that this is the case, suppose that a new headache remedy is administered to 1,000 patients, and 700 say that it reduces the severity of their headaches. Is this proof of the effectiveness of this remedy? Why, or why not?

10.2 In the early 1960s, "gastric freezing" was used to treat duodenal ulcers. This treatment meant lowering a small balloon into the stomach and filling it with very cold alcohol to temporarily freeze the ulcer. Ulcer patients seemed to experience dramatic improvements after receiving this treatment. Is this proof of the effectiveness of this treatment? Why, or why not?

10.3 To see whether "gastric freezing" really worked, researchers divided all patients entering a certain medical center into two groups: those who would undergo "gastric freezing," and those who would receive a placebo. (The latter patients were given the balloon treatment, but with alcohol that was not frigid enough to freeze the ulcer.) The latter group showed as much improvement as the former group. How do you account for this?

10.4 An engineer wants to test whether process A results in higher output per man-hour in his plant than process B. To find out, he produces 10 percent of the plant's output, using process A. Since process A is still experimental, he is forced to use it only on the night shift, when supervision is less strict. Comment on this experimental design.

10.5 In the 1950s, the United States Army carried out tests to determine whether large amounts of vitamins C and B complex would increase the physical performance of soldiers involved in a high-activity program in a cold environment. One way that this study could have been carried out is by picking a sample of 100 soldiers at random and sending them to a cold place to engage in calisthenics. All 100 would receive large amounts of the vitamins mentioned. Then their average calisthenics performance might be compared with the average performance of all U.S. Army recruits. What problems can you detect in this design?

10.6 Another way in which the experiment in Exercise 10.5 might be designed is as follows: Half of the 100 soldiers in the study might be given large amounts of vitamins C and B complex while the other half might receive a placebo. Then the average performance of one group might be compared with that of the other. In determining which group a soldier would join, the medical record of each individual would be inspected to see whether he had taken any vitamin supplements before. Only those who had not done so would be assigned to the group receiving the placebo. What problems can you detect in this design?

10.7 As still another way of designing the experiment in Exercise 10.5, half of the 100 soldiers in the study would be given large amounts of the vitamins, while the other half would receive nothing. Soldiers would be allocated at random to the two groups. Those receiving the vitamins would be told what they were being given, and those receiving nothing would be told that they were receiving nothing. Then the average performance of one group might be compared with that of the other. What problems can you detect in this design?

10.8 The army actually carried out the experiment in Exercise 10.5 in the following way: A random sample of soldiers was drawn from four platoons. The soldiers chosen from each platoon were randomly divided into two equal groups, one of which received large amounts of the vitamins, the other of which received a placebo. However, neither group knew that the latter was a placebo. In each platoon the average performance of one group was compared with that of the other.
(a) Is this a randomized block? Why, or why not?
(b) If so, what are the blocks, and what are the treatments?
(c) Is this a Latin square? Why, or why not?
(d) If it is a Latin square, explain the balance of the design.
(e) Does this design contain replications?

4. These data are from A. Duncan, *Quality Control and Industrial Statistics* (Homewood, Ill.: Irwin, 1959), p. 747. They are coded; that is, a constant amount is subtracted from each measurement to make the data easier to work with.

10.9 The following table shows the results of 9 different measurements of the viscosity of silicone gum rubber.[4] As can be seen, 3 of the measurements pertain to the first batch of this material, 3 pertain to the second batch, and 3 to the third batch. Also, 3 of the measurements were made by William Jones, 3 were made by John Beam, and 3 were made by Joan Read. Three of the measurements were made with type A viscosity measuring jars, 3 were made with type B measuring jars, and 3 were made with type C measuring jars.

	Measurer		
Batch	Jones	Beam	Read
I	9 (A)	8 (B)	-3 (C)
II	17 (B)	-2 (C)	7 (A)
III	-2 (C)	41 (A)	2 (B)

(a) Is this a randomized block? Why, or why not?
(b) If so, what are the blocks, and what are the treatments?
(c) Is this a Latin square? Why, or why not?
(d) If it is a Latin square, explain the balance of the design.
(e) Does the design contain replications?

10.10 (a) Based on the data in Exercise 10.9, which of the batches has the highest average rating? (b) Which type of measuring jar seems to result in the highest average rating? (c) Without carrying out the appropriate tests of significance, can we be sure that these differences among batches and among types of measuring jars are not due to chance?

10.4 The *F* Distribution

Earlier in this chapter we described particular experimental designs, but we have not yet indicated how one can test whether the observed differences among treatment means (that is, differences among the means of the observations resulting from various treatments) are statistically significant. For example, the results in Table 10.3 seem to indicate that the mean strength of thread produced by machine I is greater than that of machine II or machine III. But is this observed difference statistically significant? In other words, is it quite unlikely that it could have arisen due to chance? To answer this question, we must take up the *F* distribution, another of the most important probability distributions in statistics.

The *F* distribution is a continuous probability distribution. It was named for R. A. Fisher, the great British statistician who developed it in the early 1920s. Its definition is as follows.

> **F *distribution:*** *If a random variable* Y_1 *has a* χ^2 *distribution with* ν_1 *degrees of freedom, and if a random variable* Y_2 *has a* χ^2 *distribution with* ν_2 *degrees of freedom, then* $Y_1/\nu_1 \div Y_2/\nu_2$ *has an* F *distribution with* ν_1 *and* ν_2 *degrees of freedom (if* Y_1 *and* Y_2 *are independent).*

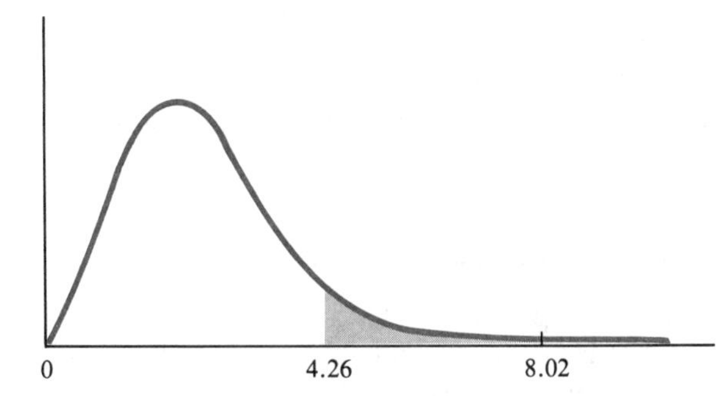

FIGURE 10.2
The *F* Probability Density Function, with 2 and 9 Degrees of Freedom

In other words, the ratio of two independent χ^2 random variables, each divided by its number of degrees of freedom, is an *F* random variable.

Like the *t* and χ^2 distributions, the *F* distribution is in reality a family of probability distributions, each corresponding to certain numbers of degrees of freedom. But unlike the *t* and χ^2 distributions, the *F* distribution has two numbers of degrees of freedom, not one. Figure 10.2 shows the *F* distribution with 2 and 9 degrees of freedom. As you can see, the *F* distribution is skewed to the right. However, as both numbers of degrees of freedom become very large, the *F* distribution tends toward normality. As in the case of the χ^2 distribution, the probability that an *F* random variable is negative is zero. This must be true since an *F* random variable is a ratio of two non-negative numbers. (Y_1/ν_1 and Y_2/ν_2 are both non-negative.) Once again, it should be emphasized that any *F* random variable has *two* numbers of degrees of freedom. Be careful to keep these numbers of degrees of freedom in *the correct order*, because an *F* distribution with ν_1 and ν_2 degrees of freedom is *not* the same as an *F* distribution with ν_2 and ν_1 degrees of freedom.

Tables are available which show the values of *F* that are exceeded with certain probabilities, such as .05 and .01. Appendix Table 9 shows, for various numbers of degrees of freedom, the value of *F* that is exceeded with probability equal to .05. For example, if the numbers of degrees of freedom are 2 and 9, the value of *F* that is exceeded with probability equal to .05 is 4.26. Similarly, Appendix Table 10 shows, for various numbers of degrees of freedom, the value of *F* that is exceeded with probability equal to .01. For example, if the numbers of degrees of freedom are 2 and 9, the value of *F* exceeded with probability equal to .01 is 8.02.

In the following sections of this chapter, we shall show how the *F* distribution can be used to test whether the differences among various treatment means are statistically significant. For now, it is important to become familiar with Appendix Tables 9 and 10. The following example shows how these tables are used.

Example 10.2 A random variable has the F distribution with 6 and 18 degrees of freedom. What is the value of this random variable that is exceeded with a probability of .05? With a probability of .01?

Solution: Appendix Table 9 shows that the answer to the first question is 2.66; Appendix Table 10 shows that the answer to the second question is 4.01.

10.5 Analysis of a Completely Randomized Design

The procedure of one of the simplest experimental designs is to divide a set of people or other experimental units into groups, and subject each group to a different treatment. (The people or other experimental units are allocated to the groups at random.) Then the differences among the mean responses of the various groups are used to measure the net effects of the various treatments. This is called a *completely randomized design*. For example, suppose that a beer manufacturer picks a random sample of 20 beer drinkers whose favorite brand of beer is known to be brand W. These beer drinkers are then divided randomly into four groups of five individuals each. The first group is asked to rate a beer in an unmarked can which is in reality their favorite (brand W). The second group is asked to rate a beer in an unmarked can which is in reality brand X. The third group is asked to rate a beer in an unmarked can which is in reality brand Y. The fourth group is asked to rate a beer in an unmarked can which is in reality brand Z. If the results are as shown in Table 10.7, the mean ratings provided by the four groups are 60, 61, 58, and 61, respectively. The beer manufacturer would like to test the hypothesis that these differences among the four means are due to chance.

In this case, the null hypothesis is that $\mu_1 = \mu_2 = \mu_3 = \mu_4$, where μ_1 is

TABLE 10.7
Results of Beer Manufacturer's Survey

	Ratings			
	Respondent's favorite beer	Brand X	Brand Y	Brand Z
	60	61	58	61
	59	60	54	57
	61	66	58	61
	55	62	58	61
	65	56	62	65
$\bar{x}_j$	60	61	58	61
$\sum_{i=1}^{5} (x_{ij} - \bar{x}_j)^2$	52	52	32	32
s_j^2	13	13	8	8

the mean rating in the population of the favorite beer, μ_2 is the mean rating in the population of brand X, μ_3 is the mean rating in the population of brand Y, and μ_4 is the mean rating in the population of brand Z. The alternative hypothesis is that μ_1, μ_2, μ_3, and μ_4 are not all equal. Clearly, the null hypothesis tends to be supported if the four sample means are close together, whereas the alternative hypothesis tends to be supported if they are far apart. A reasonable measure of how close together or far apart the four sample means are is their variance—that is, the square of their standard deviation. Specifically, their variance equals

$$s_{\bar{x}}^2 = \sum_{j=1}^{4} \frac{(\bar{x}_j - \bar{\bar{x}})^2}{3},$$

where $\bar{x}_j$ is the mean of the jth sample and $\bar{\bar{x}}$ is the mean of the four sample means.[5] If the numerical values in Table 10.7 are substituted in this equation, we have

$$s_{\bar{x}}^2 = \frac{(60 - 60)^2 + (61 - 60)^2 + (58 - 60)^2 + (61 - 60)^2}{3}$$

$$= 2.$$

Suppose that each of the four populations of ratings can be approximated by a normal distribution, and that the standard deviation of each population σ is the same. Then if the null hypothesis is true (that is, if $\mu_1 = \mu_2 = \mu_3 = \mu_4$), our four samples in Table 10.7 are four samples from the same population. Consequently, since the variance of the sample means drawn from the same population equals σ^2/n (as we know from Chapter 6), it follows that $s_{\bar{x}}^2$ is an estimate of σ^2/n where n is the size of each sample. (In this case, $n = 5$.) Thus, $ns_{\bar{x}}^2$ is an estimate of σ^2, the common variance of the four populations, if the null hypothesis and our other assumptions are true. In particular, in the case in Table 10.7, 5(2), or 10, is an estimate of σ^2 if the null hypothesis is true (since $n = 5$ and $s_{\bar{x}}^2 = 2$).

To test whether the null hypothesis is true, we compare this estimate of σ^2 with another estimate of σ^2 that is valid regardless of whether or not the null hypothesis is true. This latter estimate of σ^2 is the mean of the four variances within the samples, s_1^2, s_2^2, s_3^2, and s_4^2. As we know from Chapter 7, each of these sample variances is an unbiased estimate of the population variance (assumed to be the same in the four populations). Thus, the mean of these variances is also an unbiased estimate of the population variance; and this is true regardless of whether or not μ_1, μ_2, μ_3, and μ_4 are equal. In the case shown in Table 10.7, the mean of the sample variances equals

$$\frac{s_1^2 + s_2^2 + s_3^2 + s_4^2}{4} = \frac{13 + 13 + 8 + 8}{4} = \frac{42}{4} = 10.5.$$

The derivation of s_1^2, s_2^2, s_3^2, and s_4^2 is shown in Table 10.7.

5. The denominator is 3 because this is a sample variance, where the denominator is the sample size minus 1. There are four sample means, so the denominator is $(4 - 1) = 3$.

At this point, we have two estimates of σ^2, the first being $ns_{\bar{x}}^2$ (in this case, 10), the second being $(s_1^2 + s_2^2 + s_3^2 + s_4^2)/4$ (in this case, 10.5). To test the null hypothesis, we form the ratio of these two estimates:

$$F = \frac{ns_{\bar{x}}^2}{(s_1^2 + s_2^2 + s_3^2 + s_4^2)/4} \tag{10.1}$$

This ratio is called a *variance ratio*. If the null hypothesis is true, this ratio should be fairly close to 1, since both the numerator and the denominator should be approximately equal to σ^2. If the null hypothesis is not true, $ns_{\bar{x}}^2$ should be much greater than σ^2, since it will also reflect the variation among μ_1, μ_2, μ_3, and μ_4. Consequently, this variance ratio ,F, is used as a test statistic. High values of F are evidence that the null hypothesis should be rejected.

But how large can F be expected to be by chance if the null hypothesis is true? The answer is given by the F distribution, since the variance ratio has an F distribution with $(k - 1)$ and $k(n - 1)$ degrees of freedom (where k is the number of population means being compared and n is the sample size in each population), if the null hypothesis is true. Thus, in the case of Table 10.7, the variance ratio has an F distribution with 3 and 16 degrees of freedom, since $k = 4$ and $n = 5$. As shown in Appendix Table 9, there is a .05 chance that an F variable with 3 and 16 degrees of freedom will exceed 3.24. Thus, if the .05 significance level is appropriate, the null hypothesis should be rejected if the variance ratio exceeds 3.24. In fact, since $ns_{\bar{x}}^2 = 10$ and $(s_1^2 + s_2^2 + s_3^2 + s_4^2)/4 =$ 10.5, the variance ratio in Table 10.7 equals 0.95, and there is no reason to reject the null hypothesis.

In general, the decision rule for this test procedure is as follows.

Decision Rule: *Reject the null hypothesis that the population means are all equal if*

$$\frac{ns_{\bar{x}}^2}{\sum_{j=1}^{k} s_j^2/k} > F_{\alpha},$$

where α is the desired level of significance, and the F *distribution has degrees of freedom equal to* $(k - 1)$ *and* $k(n - 1)$*, where* k *is the number of population means being compared, and* n *is the sample size in each population. Otherwise, accept the null hypothesis.*

In the next section, we shall show how this test can be presented in another format which is customarily used in the design and analysis of experiments.

10.6 One-Way Analysis of Variance

The analysis of variance is a technique designed to divide the total variation in a set of data into its component parts, each of which can be ascribed to a particular source. For example, in the problem presented in the previous section, the total variation might be split into two parts, one representing the differences among beer brands in their average ratings, the other representing the variation among consumers in their rating of a particular brand. In this section, we will show how an analysis of variance can be carried out for this type of problem. Although this is merely another way of presenting the test described in the previous section, it is useful to know this format since it is often employed and since it can be adapted for more complex problems, as we shall see in Section 10.8.

NOTATION

In any one-way analysis of variance, the purpose is to test whether the means of k populations are equal. Since n observations are chosen from each population, the data can be arrayed as shown in Table 10.8. Each (vertical) column contains observations from the same population. Thus, x_{11} is the first observation from the first population, x_{21} is the second observation from the first population, x_{12} is the first observation from the second population, and so on. In general, x_{ij} is the ith observation from the jth population. We can apply this notation to Table 10.7, where $k = 4$ and $n = 5$. Clearly, in Table 10.7, $x_{11} = 60$, $x_{21} = 59$, $x_{12} = 61$, and so on.

TABLE 10.8
Array of Data for Analysis of Variance

	x_{11}	x_{12}	.	x_{1j}	.	x_{1k}
	x_{21}	x_{22}	.	x_{2j}	.	x_{2k}
	x_{31}	x_{32}	.	x_{3j}	.	x_{3k}
	.	.				
	x_{i1}	x_{i2}	.	x_{ij}	.	x_{ik}
	.	.	.	.	.	.
	x_{n1}	x_{n2}	.	x_{nj}	.	x_{nk}
Means	$\bar{x}_{.1}$	$\bar{x}_{.2}$	.	$\bar{x}_{.j}$	.	$\bar{x}_{.k}$

The mean of the observations in the jth column is denoted by $\bar{x}_{.j}$. Thus, in Table 10.7, $\bar{x}_{.1} = 60$, $\bar{x}_{.2} = 61$, $\bar{x}_{.3} = 58$, and $\bar{x}_{.4} = 61$. On the other hand, the mean of the observations in the ith (horizontal) row is denoted by $\bar{x}_{i.}$. The dot indicates the subscript over which the averaging takes place. Since the first subscript designates the row of the observation, $\bar{x}_{.j}$ indicates that the averaging is over the observations in all rows in the jth column. Similarly, since the second subscript designates the column of the observation, $\bar{x}_{i.}$ indicates that the averaging is over the observations in all columns in the ith row. Thus, in Table 10.7, $\bar{x}_{1.} = (60 + 61 + 58 + 61)/4$, or 60.

BASIC IDENTITY

Using this notation, we can define the total variation in a set of data as follows:

$$TSS = \sum_{i=1}^{n} \sum_{j=1}^{k} (x_{ij} - \bar{x}_{..})^2, \tag{10.2}$$

where $\bar{x}_{..}$ is the overall mean of all $n \times k$ observations. The total variation is generally referred to as the ***total sum of squares***, which accounts for the *TSS* on the left-hand side of equation (10.2). With a bit of algebraic manipulation, it can be shown that the total sum of squares is identically equal to the sum of two terms:

$$\sum_{i=1}^{n} \sum_{j=1}^{k} (x_{ij} - \bar{x}_{..})^2 = n \sum_{j=1}^{k} (\bar{x}_{.j} - \bar{x}_{..})^2 + \sum_{i=1}^{n} \sum_{j=1}^{k} (x_{ij} - \bar{x}_{.j})^2. \tag{10.3}$$

This is the basic identity underlying the one-way analysis of variance.

To understand the one-way analysis of variance, it is essential to interpret each of the terms in equation (10.3). The total sum of squares, which is on the left-hand side of this equation, is a measure of how much variation exists among all of the $n \times k$ observations in the sample. According to equation (10.3), this total variation can be split into two parts, one reflecting differences among the means of the observations taken from different populations, the other reflecting differences among the observations taken from the same population. Holding n and k constant, the larger the first part (relative to the second part), the less likely it would seem that the population means are all equal.

The first term on the right-hand side of equation (10.3) is often called the ***between-group sum of squares*** since it reflects differences between the populations in the sample means. To satisfy yourself that it does reflect such differences, note that it equals $(k - 1)$ times $ns_{\bar{x}}^2$. Another frequently used name for this term is the ***treatment sum of squares*** since the differences among the population means are often called the net effect of different treatments. *Treatment* is a completely general term used to characterize each of the populations being compared. For example, each brand of beer can be called a different treatment in Table 10.7.

The second term on the right-hand side of equation (10.3) is often referred to as the ***within-group sum of squares***, since it measures the variation within the populations. To see that it does measure such variation, note that it is equal to $k(n - 1)$ times the denominator of the variance ratio in equation (10.1). (In equation (10.1), $k = 4$ and $n = 5$.) It is also frequently called the ***error sum of squares***, since the within-group variation is often interpreted as being due to experimental error.

To illustrate that the total sum of squares is identically equal to the between-group sum of squares plus the within-group sum of squares, let's return to the data in Table 10.7. Clearly, the total sum of squares equals

$$TSS = (60 - 60)^2 + (59 - 60)^2 + (61 - 60)^2 + (55 - 60)^2$$
$$+ (65 - 60)^2 + (61 - 60)^2 + (60 - 60)^2 + (66 - 60)^2$$

$$+ (62 - 60)^2 + (56 - 60)^2 + (58 - 60)^2 + (54 - 60)^2$$
$$+ (58 - 60)^2 + (58 - 60)^2 + (62 - 60)^2 + (61 - 60)^2$$
$$+ (57 - 60)^2 + (61 - 60)^2 + (61 - 60)^2 + (65 - 60)^2$$
$$= 198.$$

The between-group sum of squares equals

$$BSS = 5[(60 - 60)^2 + (61 - 60)^2 + (58 - 60)^2 + (61 - 60)^2]$$
$$= 30.$$

And the within-group sum of squares equals

$$WSS = (60 - 60)^2 + (59 - 60)^2 + (61 - 60)^2 + (55 - 60)^2$$
$$+ (65 - 60)^2 + (61 - 61)^2 + (60 - 61)^2 + (66 - 61)^2$$
$$+ (62 - 61)^2 + (56 - 61)^2 + (58 - 58)^2 + (54 - 58)^2$$
$$+ (58 - 58)^2 + (58 - 58)^2 + (62 - 58)^2 + (61 - 61)^2$$
$$+ (57 - 61)^2 + (61 - 61)^2 + (61 - 61)^2 + (65 - 61)^2$$
$$= 168.$$

Since $198 = 30 + 168$, it is obvious that the total sum of squares does equal the sum of the between-group sum of squares and the within-group sum of squares.

ANALYSIS-OF-VARIANCE TABLE

The test procedure used in the one-way analysis of variance is precisely the same as that described in the previous section. What is different is that the computations are presented in the format of an analysis of variance table. The general form of a one-way analysis of variance table is shown in Table 10.9. The first column shows the source or type of variation, and the second column shows the corresponding sum of squares. The third column shows the number of degrees of freedom corresponding to each sum of squares, these numbers being the figures that were used in equation (10.1) to divide each sum of squares to obtain an estimate of σ^2, the variance of each population.[6] The fourth column shows each mean square, which is the sum of squares divided by the number of degrees of freedom. The last column shows the ratio of the two mean squares. If the null hypothesis is true, this ratio—which is precisely the same as the variance ratio in equation (10.1)—has an F distribution with $(k - 1)$ and $k(n - 1)$ degrees of freedom. (Note that these are the numbers of

6. In the numerator of the right-hand side of equation (10.1), the between-group sum of squares is divided by 3. (Note that $ns_{\bar{x}}^2 = BSS \div 3$.) In the denominator of the right-hand side of equation (10.1), the within-group sum of squares is divided by $4 \times (n - 1)$. (Note that $(s_1^2 + s_2^2 + s_3^2 + s_4^2) \div 4 = WSS \div [4 \times (n - 1)]$.) Since $k = 4$ in equation (10.1), the numbers of degrees of freedom are indeed the numbers used in equation (10.1) to divide the sums of squares to get estimates of σ^2.

TABLE 10.9
General One-Way Analysis-of-Variance Table

Source of variation	Sum of squares	Degrees of freedom	Mean square	F
Between groups	BSS	$k - 1$	$\frac{BSS}{k-1}$	$\frac{BSS}{k-1} \div \frac{WSS}{k(n-1)}$
Within groups	WSS	$k(n - 1)$	$\frac{WSS}{k(n-1)}$	
Total	TSS	$nk - 1$		

degrees of freedom in the first and second rows of the table.) Thus, in a one-way analysis of variance table, the decision rule is as follows.

> ***Decision Rule:*** *Reject the null hypothesis that the population means are all equal if the ratio of the between-group mean square to the within-group mean square exceeds* F_α, *where* α *is the desired significance level. Otherwise, accept the null hypothesis.*

Of course, this decision rule is essentially the same as the decision rule in Section 10.5.

To illustrate the application of a one-way analysis of variance, let's go back to the experiment carried out by the beer manufacturer. Based on the data in Table 10.7, we know from previous paragraphs that the between-group sum of squares equals 30 and that the within-group sum of squares equals 168. These figures constitute the second column of Table 10.9. And since $k = 4$ and $n = 5$, it is evident that the numbers of degrees of freedom are 3 and 16. Thus, dividing 30 by 3, we get the between-group mean square, which is 10; and dividing 168 by 16, we get the within-group mean square, which is 10.5. The ratio of the mean squares is $10 \div 10.5 = 0.95$. Since $F_{.05} = 3.24$ when there are 3 and 16 degrees of freedom, the ratio of the mean squares is not greater than $F_{.05}$, and there is no reason (at the .05 significance level) to reject the null hypothesis that the mean ratings for the four brands of beer are equal. The analysis-of-variance table is given in Table 10.10.

TABLE 10.10
One-Way Analysis-of-Variance Table, Beer Example

Source of variation	Sum of squares	Degrees of freedom	Mean square	F
Between groups	30	3	10	0.95
Within groups	168	16	10.5	
Total	198	19		

The following example illustrates further how the one-way analysis of variance is used.

Example 10.3 The thread manufacturer (encountered earlier in this chapter) wants to determine whether the mean strength of thread produced by three different types of machine are different when raw material A is used on each machine. Four pieces of thread are produced on each type of machine, the results being as follows:

Machine		
I	II	III
50	41	49
51	40	47
51	39	45
52	40	47

Use a one-way analysis of variance to test whether the mean strength of thread is equal for the three types of machine. (Let the significance level equal .05)

Solution: The mean strength for the first type of machine is 51; for the second type it is 40; and for the third type it is 47. The mean for all types of machine is 46. Thus, the between-group sum of squares equals $4 \times [(51-46)^2 + (40-46)^2 + (47-46)^2]$, or 4(62), or 248. The within-group sum of squares equals $(50-51)^2 + (51-51)^2 + (51-51)^2 + (52-51)^2 + (41-40)^2 + (40-40)^2 + (39-40)^2 + (40-40)^2 + (49-47)^2 + (47-47)^2 + (45-47)^2 + (47-47)^2 = 12$. Thus, the analysis-of-variance table is

Source of variation	Sum of squares	Degrees of freedom	Mean square	F
Between groups	248	2	124	93
Within groups	12	9	1.33	
Total	260	11		

Since there are 2 and 9 degrees of freedom, $F_{.05} = 4.26$. Since the observed value of F far exceeds this amount, the thread manufacturer should reject the null hypothesis that the mean strength is the same for the three types of machine.[7]

10.7 Confidence Intervals for Differences among Means

In the previous section, we saw how the one-way analysis of variance can be used to test whether several population means are all equal. However, in most statistical investigations the purpose is to find out the *extent* to which these means differ, not just *whether* they differ. For example, in the case of the experiment carried out by the thread manufacturer in Example 10.3, it is important to estimate the differences between the mean strengths of thread produced by various types of machines. The firm wants to answer questions

7. See Appendix 10.2 for some formulas that are useful in calculating the required sums of squares in more complicated cases.

like: What is the difference between the mean strength of thread produced on machine I and that produced on machine II? In this section we will show how confidence intervals can be constructed for *all* differences among the population means. The confidence coefficient attached to these intervals is the probability that *all* these intervals will include the respective differences among the population means *simultaneously*.

In the case of the thread manufacturer, there are three differences between the population means, since there are three population means. If μ_1 is the mean strength of thread produced by machine I, μ_2 is the mean strength of that produced by machine II, and μ_3 is the mean strength of thread produced by machine III, the probability is $(1 - \alpha)$ that *all* the following statements hold true *simultaneously*:

$$\bar{x}_{.1} - \bar{x}_{.2} - \sqrt{F_\alpha}\, s_w \sqrt{\frac{2(k-1)}{n}} < \mu_1 - \mu_2 < \bar{x}_{.1} - \bar{x}_{.2} + \sqrt{F_\alpha}\, s_w \sqrt{\frac{2(k-1)}{n}}$$

$$\bar{x}_{.1} - \bar{x}_{.3} - \sqrt{F_\alpha}\, s_w \sqrt{\frac{2(k-1)}{n}} < \mu_1 - \mu_3 < \bar{x}_{.1} - \bar{x}_{.3} + \sqrt{F_\alpha}\, s_w \sqrt{\frac{2(k-1)}{n}}$$

$$\bar{x}_{.2} - \bar{x}_{.3} - \sqrt{F_\alpha}\, s_w \sqrt{\frac{2(k-1)}{n}} < \mu_2 - \mu_3 < \bar{x}_{.2} - \bar{x}_{.3} + \sqrt{F_\alpha}\, s_w \sqrt{\frac{2(k-1)}{n}}$$

(10.4)

where s_w is the square root of the within-group mean square defined as $\text{WSS}/k(n - 1)$, F_α is the value of an F random variable (with $(k - 1)$ and $k(n - 1)$ degrees of freedom) exceeded with a probability of α, k is the number of means being compared, and n is the size of the sample taken from each population.[8]

Since $\bar{x}_{.1} = 51$, $\bar{x}_{.2} = 40$, $\bar{x}_{.3} = 47$, $F_{.05} = 4.26$, $s_w = 1.15$, $k = 3$, and $n = 4$, the 95 percent confidence interval for all of the differences between the population means is as follows:

$$6.63 < \mu_1 - \mu_2 < 11.37$$

$$1.63 < \mu_1 - \mu_3 < 6.37$$

$$-9.37 < \mu_2 - \mu_3 < -4.63.$$

Clearly, machine I seems to result in stronger thread than machine II or machine III, and machine III seems to result in stronger thread than machine II.

Expression (10.4) can be used to construct confidence intervals for the differences between the population means, no matter what the values of k and n may be. The following is an example of how these intervals are constructed.

Example 10.4 Using the data in Table 10.7, construct a 95 percent confidence interval for the six differences between the mean ratings of the four beers (the favorite, brand X, brand Y, and brand Z).

8. This result and that in Section 10.9, are from H. Scheffe, *The Analysis of Variance* (New York: Wiley, 1959).

Solution: In this case, $\bar{x}_{.1} = 60$, $\bar{x}_{.2} = 61$, $\bar{x}_{.3} = 58$, $\bar{x}_{.4} = 61$, $F_{.05} = 3.24$ (since there are 3 and 16 degrees of freedom), $s_w = \sqrt{10.5}$, $k = 4$, and $n = 5$. Thus, $\sqrt{F_\alpha}\ s_w \sqrt{(k-1)2/n} = \sqrt{3.24} \times \sqrt{10.5} \times \sqrt{(3/5) \times 2} = 6.39$, and the confidence intervals are as follows:

$$-7.39 < \mu_1 - \mu_2 < 5.39$$

$$-4.39 < \mu_1 - \mu_3 < 8.39$$

$$-7.39 < \mu_1 - \mu_4 < 5.39$$

$$-3.39 < \mu_2 - \mu_3 < 9.39$$

$$-6.39 < \mu_2 - \mu_4 < 6.39$$

$$-9.39 < \mu_3 - \mu_4 < 3.39$$

As you can see, all these confidence intervals include zero, which is what would be expected, given that the analysis of variance in the previous section concluded that none of these differences is statistically significant.

10.8 Two-Way Analysis of Variance

The one-way analysis of variance is the simplest type; there are a variety of more complicated types which are taken up in more specialized textbooks. For present purposes, it is sufficient to discuss a single extension, the two-way analysis of variance. This technique varies from the one-way analysis because two, not one, sources of variation (other than the error sum of squares) are singled out for attention. In particular, two-way analysis is the technique used to test whether the differences among treatment means in a randomized block without replications are statistically significant.

To illustrate two-way analysis of variance, let's return to the beer manufacturer discussed previously. Suppose that, when the study described in Section 10.5 is presented to the firm's managers, they suggest to the firm's statisticians that a somewhat different experimental design be used. In particular, it is suggested that the same five people be asked to rate all four of the brands of beer, rather than employing a different sample for each brand. In this way, the differences among brands in sample means will not be clouded by differences in the composition of the sample. The result is a randomized block design in which each person constitutes a block.

Suppose that the statisticians carry out the experiment suggested and the results are as shown in Table 10.11. In this new experimental design, the firm's managers are interested in the differences among brands of beer, and the differences among people. The differences among brands of beer in their mean ratings are called *differences among treatment means*, as in the previous section. The differences among persons in their mean ratings are called *differences among block means*, and they reflect the fact that some individuals may tend to rate all these brands of beer more highly than would other individuals. Our primary interest here is in whether or not the observed differences among treatment means are due to chance. The differences among the block means

TABLE 10.11
Results of Survey by Beer Manufacturer

Person	Rating				Block mean
	Favorite brand	Brand X	Brand Y	Brand Z	
Jones	63	59	62	61	61.25
Smith	61	62	57	63	60.75
Klein	61	64	60	58	60.75
Carlucci	62	62	60	62	61.50
Weill	58	63	61	61	60.75
Treatment mean	61	62	60	61	

are of only secondary interest in this case; but since they contribute to the total sum of squares, they must be included in the analysis.

As in the case of the one-way analysis of variance, the total sum of squares equals

$$TSS = \sum_{i=1}^{n} \sum_{j=1}^{k} (x_{ij} - \bar{x}_{..})^2. \tag{10.5}$$

In this case, however, the total sum of squares can be split into three parts in the following way:

$$\sum_{i=1}^{n} \sum_{j=1}^{k} (x_{ij} - \bar{x}_{..})^2 = n \sum_{j=1}^{k} (\bar{x}_{.j} - \bar{x}_{..})^2 + k \sum_{i=1}^{n} (\bar{x}_{i.} - \bar{x}_{..})^2 + \text{error sum of squares}, \tag{10.6}$$

where the error sum of squares (denoted by ESS) can be obtained by subtraction. That is,

$$ESS = \sum_{i=1}^{n} \sum_{j=1}^{k} (x_{ij} - \bar{x}_{..})^2 - n \sum_{j=1}^{k} (\bar{x}_{.j} - \bar{x}_{..})^2 - k \sum_{i=1}^{n} (\bar{x}_{i.} - \bar{x}_{..})^2. \tag{10.7}$$

The identity in equation (10.6) is the basis for the two-way analysis of variance. The first term on the right-hand side of equation (10.6) is called the ***treatment sum of squares***, as in previous sections. Clearly, it reflects differences in the treatment means (for example, differences in the average rating of various brands of beer in Table 10.11). We denote this term by BSS. The second term on the right-hand side of equation (10.6) is called the ***block sum of squares***, since it reflects differences in the block means (for example, differences among the persons in Table 10.11 in their average ratings of all beers). We denote the block sum of squares by *RSS*. The identity in equation (10.6) says that

> total sum of squares = treatment sum of squares + block sum of squares + error sum of squares.

In other words, $TSS = BSS + RSS + ESS$.

This identity is used in the two-way analysis-of-variance table shown in Table 10.12. The first column of the table shows the source of variation (treatment, block, or error), and the second column shows the corresponding sum of squares. The third column shows the number of degrees of freedom for each sum of squares—$(k - 1)$ for the treatment sum of squares, $(n - 1)$ for the block sum of squares, and $(k - 1)(n - 1)$ for the error sum of squares. The fourth column shows the mean square for treatments, blocks, and errors, each being the relevant sum of squares divided by the relevant degrees of freedom (that is, the second column divided by the third column). The last column shows (1) the treatment mean square divided by the error mean square and (2) the block mean square divided by the error mean square. As explained in the next paragraph, these two ratios are the pay dirt of the entire analysis.

TABLE 10.12
General Two-Way Analysis-of-Variance Table

Source of variation	Sum of squares	Degrees of freedom	Mean square	F
Treatments	BSS	$k - 1$	$\frac{BSS}{k-1}$	$\frac{BSS}{k-1} \div \frac{ESS}{(k-1)(n-1)}$
Blocks	RSS	$n - 1$	$\frac{RSS}{n-1}$	$\frac{RSS}{n-1} \div \frac{ESS}{(k-1)(n-1)}$
Error	ESS	$(k - 1)(n - 1)$	$\frac{ESS}{(k-1)(n-1)}$	
Total	TSS	$nk - 1$		

In a two-way analysis of variance, we can test two different null hypotheses, not just one. The first null hypothesis is that *the treatment means are all equal.* In terms of our example, this hypothesis says that the mean rating for each brand of beer is the same. To test this hypothesis, we use the ratio of the treatment mean square to the error mean square. If α is the significance level, the decision rule is the following.

> ***Decision Rule:*** *Reject the above null hypothesis if the ratio of the treatment mean square to the error mean square exceeds* F_α *where there are* $(k - 1)$ *and* $(k - 1)(n - 1)$ *degrees of freedom. Accept the above null hypothesis if this ratio does not exceed* F_α.

The second null hypothesis is that *the block means are all equal.* In terms of our example, this hypothesis says that the mean rating is the same for all people. To test this second hypothesis, we use the ratio of the block mean square to the error mean square. If α is the significance level, the decision rule is the following.

> ***Decision Rule:*** *Reject the above null hypothesis if the ratio of the block mean square to the error mean square exceeds* F_α *where there are* $(n-1)$ *and* $(k-1)(n-1)$ *degrees of freedom. Accept the above null hypothesis if this ratio does not exceed* F_α.

To illustrate the application of two-way analysis of variance, consider the data concerning the beers in Table 10.11. As shown in Table 10.13, the

TABLE 10.13
Analysis of Results of Beer Survey

Source of variation	Sum of squares	Degrees of freedom	Mean square	*F*
Treatments (beers)	10	3	3.33	$\frac{3.33}{4.50} = 0.74$
Blocks (people)	2	4	0.50	$\frac{0.50}{4.50} = 0.11$
Error	54	12	4.50	
Total	66	19		

As shown in Table 10.11, the treatment means are 61, 62, 60, and 61. Thus, since $\bar{x}_{..} = 61$,

$$\text{Treatment sum of squares} = 5[(61-61)^2+(62-61)^2+(60-61)^2+(61-61)^2]$$
$$= 5(2) = 10$$

As shown in Table 10.11, the block means are 61.25, 60.75, 60.75, 61.50, and 60.75. Thus,

$$\text{Block sum of squares} = 4[(61.25-61)^2+(60.75-61)^2+(60.75-61)^2$$
$$+(61.50-61)^2+(60.75-61)^2] = 4(.5) = 2.$$

Based on the data in Table 10.11,

$$\text{Total sum of squares} = (63-61)^2+(59-61)^2+(62-61)^2+(61-61)^2$$
$$+(61-61)^2+(62-61)^2+(57-61)^2+(63-61)^2$$
$$+(61-61)^2+(64-61)^2+(60-61)^2+(58-61)^2$$
$$+(62-61)^2+(62-61)^2+(60-61)^2+(62-61)^2$$
$$+(58-61)^2+(63-61)^2+(61-61)^2+(61-61)^2$$
$$= 66.$$

Using the previous results,

$$\text{Error sum of squares} = 66 - 10 - 2 = 54.$$

total sum of squares equals 66, the treatment sum of squares equals 10, the block sum of squares equals 2, and the error sum of squares equals 54. (Each of these numbers is derived under the table.) Since $k = 4$ and $n = 5$, the degrees of freedom for each sum of squares is as shown in the third column of the table; and dividing each sum of squares by its number of degrees of freedom, we get the mean squares in the fourth column. Finally, dividing the treatment mean square by the error mean square, we get 0.74. This is less than 3.49, which is $F_{.05}$ when there are 3 and 12 degrees of freedom. Thus, there is no reason to reject the hypothesis that the treatment means are equal. Dividing the block mean square by the error mean square, we get 0.11. This is less than 3.26, which is $F_{.05}$ when there are 4 and 12 degrees of freedom. Thus, we should not reject the hypothesis that the block means are equal. Overall, the results of this analysis indicate that the individuals in the sample do not exhibit different average ratings for all included beer brands, and there is no evidence of differences in the average ratings of the four brands of beer.

The two-way analysis of variance is of great practical importance because, as we know from earlier sections, randomized blocks are a commonly used experimental design. The following is a further example of how two-way analysis of variance is carried out.

Example 10.5 Use the data in Table 10.3 to construct a two-way analysis of variance to test whether the mean strength of thread differs among the three types of machine. Then test whether the mean strength of thread differs among the three types of raw material. Set the significance level of each test equal to .05.

Solution: Let x_{ij} be the strength of the thread made from the ith type of raw material on the jth type of machine, where A is the first type of raw material, B the second type, and so on. Based on Table 10.3, it is clear that

$$\bar{x}_{.1} = 50 \qquad \bar{x}_{1.} = 45$$

$$\bar{x}_{.2} = 41 \qquad \bar{x}_{2.} = 44$$

$$\bar{x}_{.3} = 46 \qquad \bar{x}_{3.} = 48.$$

If the various types of machines are regarded as treatments, and if the raw materials are regarded as blocks, the treatment sum of squares is

$$3 \times \left[\left(50 - 45\frac{2}{3}\right)^2 + \left(41 - 45\frac{2}{3}\right)^2 + \left(46 - 45\frac{2}{3}\right)^2\right] = 122,$$

and the block sum of squares is

$$3 \times \left[\left(45 - 45\frac{2}{3}\right)^2 + \left(44 - 45\frac{2}{3}\right)^2 + \left(48 - 45\frac{2}{3}\right)^2\right] = 26,$$

since $\bar{x}_{..} = 45\ 2/3$. Thus, the error sum of squares is

$$\left(50 - 45\frac{2}{3}\right)^2 + \left(48 - 45\frac{2}{3}\right)^2 + \left(52 - 45\frac{2}{3}\right)^2$$

$$+ \left(40 - 45\frac{2}{3}\right)^2 + \left(39 - 45\frac{2}{3}\right)^2 + \left(44 - 45\frac{2}{3}\right)^2$$

$$+ \left(45 - 45\frac{2}{3}\right)^2 + \left(45 - 45\frac{2}{3}\right)^2 + \left(48 - 45\frac{2}{3}\right)^2$$

$$- 122 - 26 = 150 - 122 - 26 = 2.$$

The analysis-of-variance table is as follows:

Source of variation	Sum of squares	Degrees of freedom	Mean square	F
Machines (treatments)	122	2	61	61/0.5 = 122
Raw materials (blocks)	26	2	13	13/0.5 = 26
Error	2	4	0.5	
Total	150	8		

Since $F_{.05} = 6.94$ when there are 2 and 4 degrees of freedom, both values of F in the table exceed $F_{.05}$. Thus, the firm should reject both the null hypothesis that the mean strength is the same for all machines and the null hypothesis that the mean strength is the same for all raw materials.[9]

10.9 Confidence Intervals for Differences among Means

In the previous section, we saw how the two-way analysis of variance can be used to test whether the treatment means are all equal. In most experiments, however, the purpose is to find out the *extent* to which the treatment means differ, not just *whether* they differ. In this section, we will show how confidence intervals can be constructed for all differences between treatment means. The confidence coefficient attached to these intervals is the probability that *all* these intervals will include the true differences among treatment means *simultaneously*.

In the case of the thread manufacturer there are three differences among the treatment (that is, machine) means. Assuming that the raw material is held constant, let μ_1 once again be the mean strength of thread produced by machine I, μ_2 the mean strength for machine II, and so on. The probability is $(1 - \alpha)$ that *all* the following statements hold true *simultaneously*:

$$\bar{x}_{.1} - \bar{x}_{.2} - \sqrt{F_\alpha}\, s_E \sqrt{\frac{2(k-1)}{n}} < \mu_1 - \mu_2 < \bar{x}_{.1} - \bar{x}_{.2} + \sqrt{F_\alpha}\, s_E \sqrt{\frac{2(k-1)}{n}}$$

$$\bar{x}_{.1} - \bar{x}_{.3} - \sqrt{F_\alpha}\, s_E \sqrt{\frac{2(k-1)}{n}} < \mu_1 - \mu_3 < \bar{x}_{.1} - \bar{x}_{.3} + \sqrt{F_\alpha}\, s_E \sqrt{\frac{2(k-1)}{n}}$$

$$\bar{x}_{.2} - \bar{x}_{.3} - \sqrt{F_\alpha}\, s_E \sqrt{\frac{2(k-1)}{n}} < \mu_2 - \mu_3 < \bar{x}_{.2} - \bar{x}_{.3} + \sqrt{F_\alpha}\, s_E \sqrt{\frac{2(k-1)}{n}} \quad (10.8)$$

where k is the number of treatments, n the number of blocks, F_α is the value

9. See Appendix 10.2 for some formulas that are useful in calculating the required sums of squares in more complicated cases.

of F with $(k - 1)$ and $(k - 1)(n - 1)$ degrees of freedom exceeded with a probability of α, and s_E is the square root of the error mean square. Since $\bar{x}_{.1} = 50$, $\bar{x}_{.2} = 41$, $\bar{x}_{.3} = 46$, $F_{.05} = 6.94$, $k = 3$, $n = 3$, and $s_E = .71$, the 95 percent confidence interval for all the differences between treatment (machine) means when the block (raw material) is held constant, is as follows:

$$6.84 < \mu_1 - \mu_2 < 11.16$$
$$1.84 < \mu_1 - \mu_3 < 6.16$$
$$-7.16 < \mu_2 - \mu_3 < -2.84.$$

It is also possible to construct confidence intervals of this sort for the differences among the block means when the treatment is held constant. Thus, in this case, when the type of machine is held constant, one can construct confidence intervals for the differences among the means corresponding to various raw materials. (Since this entails only a slight modification of the procedures covered in the previous paragraph, we describe this procedure in a footnote.)[10]

The following is a further illustration of the construction of confidence intervals of this sort.

Example 10.6 Use the data in Table 10.11 to construct a 95 percent confidence interval for the six differences between the mean ratings of the four beers, when the person making the ratings is held constant.

Solution: In this case, $\bar{x}_{.1} = 61$, $\bar{x}_{.2} = 62$, $\bar{x}_{.3} = 60$, $\bar{x}_{.4} = 61$, $F_{.05} = 3.49$, $s_E = \sqrt{4.50}$, $k = 4$, and $n = 5$. Thus, $\sqrt{F_\alpha}\, s_E \sqrt{[(k - 1)/n] \times 2} = \sqrt{3.49} \times \sqrt{4.50} \times \sqrt{6/5} = 4.34$, and the confidence intervals are as follows:

$$-5.34 < \mu_1 - \mu_2 < 3.34$$
$$-3.34 < \mu_1 - \mu_3 < 5.34$$
$$-4.34 < \mu_1 - \mu_4 < 4.34$$
$$-2.34 < \mu_2 - \mu_3 < 6.34$$
$$-3.34 < \mu_2 - \mu_4 < 5.34$$
$$-5.34 < \mu_3 - \mu_4 < 3.34.$$

All the above confidence intervals include zero, which would be expected, since the analysis of variance in the previous section concluded that none of the differences among the treatment means is statistically significant.

10.10 A Final Caution

Before concluding this chapter, it is important to note the assumptions underlying the analysis of variance. If these assumptions are not met, the use

10. To obtain confidence intervals for the block means, all that one has to do is substitute k for n and vice versa in expression (10.8). Of course, F_α must also differ from that in expression (10.8) because there are $(n - 1)$ and $(k - 1)(n - 1)$ degrees of freedom in this case.

of the analysis of variance may be misleading. One assumption is that the populations being compared are normally distributed. Fortunately, this assumption is not as stringent as it sounds, since studies have shown that the validity of the analysis of variance is not significantly affected by moderate departures from normality. In the jargon of the statistician, the analysis of variance is a "robust test" in this regard. Another assumption underlying the analysis of variance is that the variances of the populations are equal. (Appendix 10.1 shows how the F distribution can be used to test the hypothesis that the variances of two populations are equal.) If this assumption is not met, trouble can result. Statisticians have devised techniques for handling this problem in some cases; these techniques are more properly discussed in more advanced texts. Still another important assumption is that the observations are statistically independent. To repeat, the analysis of variance should not be used unless the relevant assumptions are at least approximately fulfilled.

EXERCISES

10.11 If an F random variable has 40 and 30 degrees of freedom, what is the probability that it will exceed (a) 1.79; (b) 2.30?

10.12 Suppose that X has a χ^2 distribution with 4 degrees of freedom and Y has a χ^2 distribution with 7 degrees of freedom. (a) What is the distribution of $X/4 \div Y/7$? (b) What is the distribution of $Y/7 \div X/4$?

10.13 Suppose that a random variable has an F distribution with 10 and 12 degrees of freedom. (a) What is the value of this variable that is exceeded with a probability of .05? (b) What is the value of this variable that is exceeded with a probability of .01?

10.14 An American automobile manufacturer runs an experiment in which four of its cars are chosen at random, four of another U.S. firm's cars are chosen at random, four of a German firm's cars are chosen at random, and four of a Japanese firm's cars are chosen at random. Each of the 16 cars is operated under identical conditions for a month, and the mileage per gallon of gasoline is determined. The results are as follows:

Miles per gallon

Auto firm	U.S. competitor	German firm	Japanese firm
18	22	25	29
20	21	27	28
19	24	26	24
17	20	28	25

Test the hypothesis that the mean number of miles per gallon is the same for all four firms' cars (using $\alpha = .05$).

10.15 Construct an analysis-of-variance table summarizing the results in Exercise 10.14.

10.16 Suppose that the automobile manufacturer in Exercise 10.14 designs the following experiment: As before, it picks at random four of its cars, four of its U.S. competitor's cars, four of the German firm's cars, and four of the Japanese firm's cars. This time it has one of each firm's cars drive in city traffic; one of each firm's cars drive under suburban conditions; one of each firm's cars drive under mountainous conditions; and one of each firm's cars drive in flat, open country. The results are as follows:

Manufacturer	Miles per gallon			
	City	Suburbs	Mountains	Flat country
Auto firm	14	20	21	24
U.S. competitor	15	21	25	26
German firm	18	24	26	28
Japanese firm	19	25	25	27

(a) Construct an analysis-of-variance table summarizing the results.
(b) What are the treatment effects here? What are the block effects?

10.17 (a) Based on the results in Exercise 10.16, does it appear that each make of car gets the same number of miles per gallon? (b) If not, which seems to get the most, and which seems to get the least?

10.18 (a) Based on the results in Exercise 10.16, does it appear that the number of miles per gallon is the same for all types of driving conditions? (b) If not, which type of driving condition seems to result in the largest number of miles per gallon, and which type seems to result in the smallest number of miles per gallon?

10.19 Use the data in Exercise 10.14 to construct 95 percent confidence intervals for the differences among the four types of cars with respect to the mean number of miles per gallon.

10.20 Use the data in Exercise 10.16 to construct 95 percent confidence intervals for the differences among the four types of cars with respect to the mean number of miles per gallon when the type of driving condition is held constant.

CHAPTER REVIEW

1. An *experiment* dictates that one group of people, machines, or other experimental units be treated differently from another group; and the effect of this difference in treatment is estimated by comparing certain measurable characteristics of the two groups. In an attempt to prevent confounding, statisticians recommend that the people, machines, or other experimental units be assigned to one group or the other *at random.*

2. A frequently used experimental design is the *randomized block design,* in which the experimental units are classified into blocks, and the statistician obtains data concerning the effect of each *treatment* in each *block*. In this design it is possible to estimate both the differences among the treatment means (holding the block constant) and the differences among the block means (holding the treatment constant).

3. Another commonly used experimental design is the *Latin square.* In a Latin square design there are three factors: (1) the treatments being compared; (2) a factor corresponding to the horizontal rows of the Latin square; and (3) a factor corresponding to the vertical columns of the Latin square. In each horizontal row of a Latin square, each treatment appears only once; and in each vertical column each treatment appears only once.

4. If Y_1 has a χ^2 distribution with ν_1 degrees of freedom, and if Y_2 has a

χ^2 distribution with ν_2 degrees of freedom, then $Y_1/\nu_1 \div Y_2/\nu_2$ has an F distribution with ν_1 and ν_2 degrees of freedom (if Y_1 and Y_2 are independent). The F distribution is generally skewed to the right, but as both degrees of freedom become very large, the F distribution tends toward normality.

5. In a *completely randomized design*, experimental units are classified at random into groups, and each group is subjected to a different treatment. Then the sample mean in each group is used as an estimate of the population mean, and a test is made of the null hypothesis that the population means are all equal. The null hypothesis should be rejected if $ns_{\bar{x}}^2 \div \Sigma s_j^2 / k > F_\alpha$, where k is the number of population means being compared, n is the sample size in each population, $s_{\bar{x}}^2$ is the variance of the k sample means, s_j^2 is the variance of the sample from the jth population, and F_α is the value of an F random variable with $(k - 1)$ and $k(n - 1)$ degrees of freedom that is exceeded with a probability of α.

6. An *analysis of variance* is a technique designed to analyze the total variation in a set of data, the object being to split up this total variation into component parts, each of which can be ascribed to a particular source. The *one-way analysis of variance* splits the total sum of squares into two parts, the *between-group* (or *treatment*) *sum of squares* and the *within-group* (or *error*) *sum of squares*. Using this breakdown of the total sum of squares, the one-way analysis of variance tests the null hypothesis that the population means in a completely randomized design are all equal. The test procedure is precisely the same as that given in the previous paragraph. What is different is that the computations are presented in the format of an analysis-of-variance table.

7. In most experiments, the purpose is to find out the *extent to which* the population means (corresponding to the various treatments) differ, not just *whether* they differ. Confidence intervals can be constructed for all differences among the population means. The *confidence coefficient* attached to these intervals is the probability that all these intervals will include the respective differences among the population means simultaneously.

8. The *two-way analysis of variance* recognizes not one, but two sources of variation other than the error sum of squares. The two-way technique is used to test whether the differences among treatment means in a randomized block (without replications) are statistically significant. The two-way analysis of variance splits the total sum of squares into three parts, the ***treatment sum of squares***, the ***block sum of squares***, and the ***error sum of squares***. The null hypothesis that the treatment means are equal should be rejected if the ratio of the treatment mean square to the error mean square exceeds F_α, where there are $(k - 1)$ and $(k - 1)(n - 1)$ degrees of freedom. The null hypothesis that the block means are equal should be rejected if the ratio of the block mean square to the error mean square exceeds F_α, where there are $(n - 1)$ and $(k - 1)(n - 1)$ degrees of freedom.

9. Confidence intervals can be constructed for all differences among the treatment means (holding the block constant). The confidence coefficient attached to these intervals is the probability that all these intervals will include the respective differences among the treatment means simultaneously. Confidence intervals of this sort can also be constructed for the differences among the block means (holding the treatment constant).

Getting Down to Cases:
HOW A TEXTILE MANUFACTURER USED THE ANALYSIS OF VARIANCE

A textile plant was interested in determining whether some types of cotton were more likely than others to break during the weaving process. The fewer breaks per unit-length of warp in a type of cotton, the higher its weaving quality. (A *warp* is a quantity of warp yarn that goes into one loom as a unit.) In particular, the plant wanted to know whether two growths of cotton (A and B) differed in weaving quality. Also, the plant's statisticians wanted to know whether there was a difference in this respect among the number of turns per inch in the yarn: low (L), medium (M), and high (H). Thus, there were six combinations of growth of cotton and number of turns in the yarn that the plant wanted to study: (1) low number of turns with cotton A; (2) medium number of turns with cotton A; (3) high number of turns with cotton A; (4) low number of turns with cotton B; (5) medium number of turns with cotton B; and (6) high number of turns with cotton B.

The firm chose at random nine warps of each of these six types, and the number of warp threads that broke during the weaving of each warp was counted and expressed as a particular number of breaks per unit-length of warp. The results were as follows:[11]

Low number of turns with cotton A	Medium number of turns with cotton A	High number of turns with cotton A	Low number of turns with cotton B	Medium number of turns with cotton B	High number of turns with cotton B
26	18	36	27	42	20
30	21	21	14	26	21
54	29	24	29	19	24
25	17	18	19	16	17
70	12	10	29	39	13
52	18	43	31	28	15
51	35	28	41	21	15
26	30	15	20	39	16
67	36	26	44	29	28

The firm used a one-way analysis of variance to test the observed differences among the mean breakage rates.

(a) Which of the six combinations of growth of cotton and number of turns seems to have the lowest breakage rate? Which seems to have the highest breakage rate?

(b) Are the observed differences in the mean breakage rate among the six combinations statistically significant? (Let $\alpha = .01$.)

(c) Suppose that previous experience indicates that the dispersion of breakage rates is higher for cotton A than for cotton B. Would this violate the assumptions of the statistical analysis you did in (b)?

(d) Can this experiment be viewed as a randomized block? If so, explain how.

11. See L. Tippett, *Technological Applications of Statistics* (New York: Wiley, 1950). These data come from an actual case.

(e) On the basis of your results, what advice would you give the textile manufacturer?

APPENDIX 10.1
Comparing Two Population Variances

The F distribution can be used to test the hypothesis that the variance of one normal population equals the variance of another normal population. This test is often useful because a decision maker wants to determine whether one population is more variable than another. For example, a production manager may want to determine whether the variability of the errors made by one measuring instrument is less than the variability of those made by another measuring instrument. In addition, this test is often used to determine whether the assumptions underlying other statistical tests are valid. For example, in carrying out the t test to determine whether the means of two populations are the same, we assume that the variances of the two populations are equal. (Recall Chapter 8.) This assumption can be checked by carrying out the test described below.

The null hypothesis is that the variance of one normal population σ_1^2 equals the variance of the other normal population σ_2^2. To test this hypothesis, a sample of n_1 observations is taken from the first population and a sample of n_2 observations is taken from the second population. The test statistic is $s_1^2 \div s_2^2$, where s_1^2 is the sample variance of the observations taken from the first population and s_2^2 is the sample variance of the observations taken from the second population. If the null hypothesis (that $\sigma_1^2 = \sigma_2^2$) is true, this test statistic has the F distribution with $(n_1 - 1)$ and $(n_2 - 1)$ degrees of freedom.

> **Decision Rule:** *When the alternative hypothesis*[12] *is* $\sigma_1^2 > \sigma_2^2$, *reject the null hypothesis if the test statistic exceeds* F_α. *When the alternative hypothesis is* $\sigma_1^2 \neq \sigma_2^2$, *let the population with the larger sample variance be the first population (that is, the one whose sample variance is in the numerator of the test statistic); and reject the null hypothesis if the test statistic exceeds* $F_{\alpha/2}$.

To illustrate the use of this test, suppose that a firm wants to test whether the variance of the length of life of type A light bulbs equals the variance of the length of life of type B light bulbs. (The significance level is set at .02) A random sample of 25 type A bulbs is selected, as is a random sample of 25 type B bulbs, and it is found that the sample variance is 50 for type A bulbs and 80 for type B. Since the alternative hypothesis is that the two variances are unequal (regardless of which is bigger), we let the type B bulbs constitute the first population (since its sample variance is larger than the sample variance of type A bulbs). Thus, $s_1^2 \div s_2^2 = 80 \div 50$, or 1.6. The null hypothesis should be rejected if this test statistic exceeds $F_{.01}$ (the number of degrees of freedom

12. When the alternative hypothesis is one-sided (that is, when the variance of one population is larger than that of the other population, according to the alternative hypothesis), the population with the larger variance according to the alternative hypothesis should be designated as the first population.

being 24 and 24). According to Appendix Table 10, $F_{.01} = 2.66$. Since the test statistic does not exceed $F_{.01}$, the null hypothesis should not be rejected.

APPENDIX 10.2

Formulas for Computations in the Analysis of Variance

In calculating the sums of squares that are required by the analysis of variance, it is usually best to use formulas that require finding only sums and sums of squares of the observations. Such formulas are given below for the sums of squares in Table 10.9:

$$BSS = \frac{1}{n}\sum_{j=1}^{k} T_j^2 - \frac{1}{nk}T^2$$

$$TSS = \sum_{i=1}^{n}\sum_{j=1}^{k} x_{ij}^2 - \frac{1}{kn}T^2,$$

where T_j is the total of the observations from the jth population (or the jth treatment), and T is the total of all observations. Once these two sums of squares are calculated, we can obtain WSS by subtraction. That is,

$$WSS = TSS - BSS.$$

An additional formula that applies to Table 10.12 is

$$RSS = \frac{1}{k}\sum_{i=1}^{n} T_i^2 - \frac{1}{kn}T^2,$$

where T_i is the total of the observations in the ith block.

To illustrate the use of these formulas, consider once more the data in Example 10.3. Clearly, $T = 552$, $T_1 = 204$, $T_2 = 160$, and $T_3 = 188$. Thus,

$$BSS = \frac{1}{4}(204^2 + 160^2 + 188^2) - \frac{1}{4(3)}552^2 = 25{,}640 - 25{,}392 = 248$$

$$TSS = 25{,}652 - \frac{1}{4(3)}552^2 = 25{,}652 - 25{,}392 = 260$$

$$WSS = 260 - 248 = 12.$$

Comparing these results with those in the solution to Example 10.3, we find that they are identical. The advantage of the formulas given here is that they are easier and more efficient to calculate. Although this is not obvious in this case (since the numbers were intentionally chosen so that the calculations would be simple), this is generally true.

11

Regression and Correlation Techniques

11.1 Introduction

Statisticians frequently must estimate how one variable is related to, or affected by, another variable. A firm may need to determine how its sales are related to the gross national product; or it may need to determine how its total production costs are related to its output rate. To estimate such relationships, statisticians use regression techniques; and to determine how strong such relationships are, they use correlation techniques. Regression and correlation are among the most important and most frequently used methods of statistics. In this chapter, we begin the study of regression and correlation. Subsequent chapters will provide further information concerning these topics.

11.2 Relationship among Variables, and the Scatter Diagram

DETERMINISTIC AND STATISTICAL RELATIONSHIPS

The statistical techniques presented in previous chapters were concerned with a single variable, X. For example, in Chapter 7 we described how to estimate the mean of X; and in Chapter 8 we described how to test whether this mean equals a specified value. In many important practical situations, statisticians must be concerned with more than a single variable; in particular, they must be concerned with the relationships among variables. For example, they may want to determine whether changes in one variable, X, tend to be associated with changes in another variable, Y. (For example, does Y tend to increase when X increases?) The techniques presented in previous chapters are powerless to handle such a problem.

When we say that statisticians are interested in the relationships among variables, it is important to note at the outset that these relationships seldom are deterministic. To see what we mean by a *deterministic relationship*, suppose that Y is the variable we want to estimate and X is the variable whose value will be used to make this estimate. If the relationship between Y and X is *exact*, we say that it is a *deterministic relationship*. For example, if Y is the perimeter of a square and X is the length of a side of the square, it is evident that $Y = 4X$. This is a deterministic relationship since, once we are given X, we can predict Y *exactly*. (For example, in this case if X equals 2, we know that Y must equal 8—no more, no less.)

Statisticians are generally interested in statistical, not deterministic, relationships. If a *statistical relationship* exists between Y and X, the average value of Y tends to be related to the value of X, but it is impossible to predict with certainty the value of Y on the basis of the value of X. For example, suppose that X is a family's annual income and Y is the amount the family saves per year. On the average, the amount saved by a family tends to increase as its income increases; and this relationship can be used to predict how much a family will save, if we know the amount of its income. However, this relationship is far from exact. Since families with the same income do not all save the same amount, it is impossible to predict with certainty the amount a family will save on the basis of its income alone.

REGRESSION ANALYSIS

Regression analysis describes the way in which one variable is related to another. (As we shall see in Chapter 12, regression and correlation techniques can handle more than two variables, but only two are considered in this chapter.) Regression analysis derives an equation which can be used to estimate the unknown value of one variable on the basis of the known value of the other variable. For example, suppose that a hosiery mill is scheduled to produce 4 tons of output next month and wants to estimate how much its costs will be. In this case, although the mill's output is known, its costs are unknown. Regression analysis can be used to estimate the value of the costs on the basis of the known value of output. Regression analysis can also be used to estimate the level of capital expenditures required to establish a plant with a certain capacity. In the case of the hosiery mill, if the plant's capacity were known, regression analysis could be used to predict the firm's level of expenditure.

The term *regression analysis* comes from studies carried out by the English statistician Francis Galton about 80 years ago. Galton compared the heights of parents with the heights of their offspring and found that very tall parents tended to have offspring who were shorter than their parents, while very short parents tended to have offspring who were taller than their parents. In other words, the heights of the offspring of unusually tall or unusually short parents tended to "regress" toward the mean height of the population. Because Galton used the height of the parent to predict the height of the offspring, this type of analysis came to be called regression analysis, even though subsequent applications actually had very little to do with Galton's "regression" of heights toward the mean.

SCATTER DIAGRAM

Since regression analysis is concerned with how one variable is related to another, an analysis of this sort generally begins with data concerning the two variables in question. Again, suppose that the hosiery mill wants to estimate how its monthly costs are related to its monthly output rate. Clearly, a sensible first step for the firm is to obtain data regarding its costs and output during a sample of months in the past. Suppose that the firm collects such data for a sample of 9 months, the results being shown in Table 11.1. It is convenient to plot data of this sort in a so-called ***scatter diagram***. In this diagram, the known variable—in this case the monthly output rate—is plotted along the horizontal axis and is called the *independent variable*. The unknown variable—the monthly cost, in this case—is plotted along the vertical axis and is called the *dependent variable*. Of course, during the period to which the data pertain, *both* variables are known, but when the regression analysis is used to estimate how much it will cost to produce a particular output, only output, rather than cost, will be known.

TABLE 11.1
Cost and Output of a Hosiery Mill, Sample of Nine Months

Output (tons)	Production cost (thousands of dollars)
1	2
2	3
4	4
8	7
6	6
5	5
8	8
9	8
7	6

Figure 11.1 shows the scatter diagram based on the data in Table 11.1. Clearly, this diagram provides a useful visual portrait of the relationship between the dependent and independent variables. Based on such a diagram, one can form an initial impression concerning at least the three questions which are given below.

Is the Relationship Direct or Inverse? The relationship between X and Y is *direct* if increases in X tend to be associated with increases in Y, and decreases in X tend to be associated with decreases in Y. On the other hand, the relationship between X and Y is *inverse* if increases in X tend to be associated with decreases in Y, and decreases in X tend to be associated with increases in Y. Figure 11.1 indicates that the relationship between the firm's output and its costs is direct, as would be expected. Panel A of Figure 11.2 shows a case where the relationship between X and Y is inverse. Not all scatter diagrams indicate either a direct or inverse relationship. Some, like panel C, indicate no

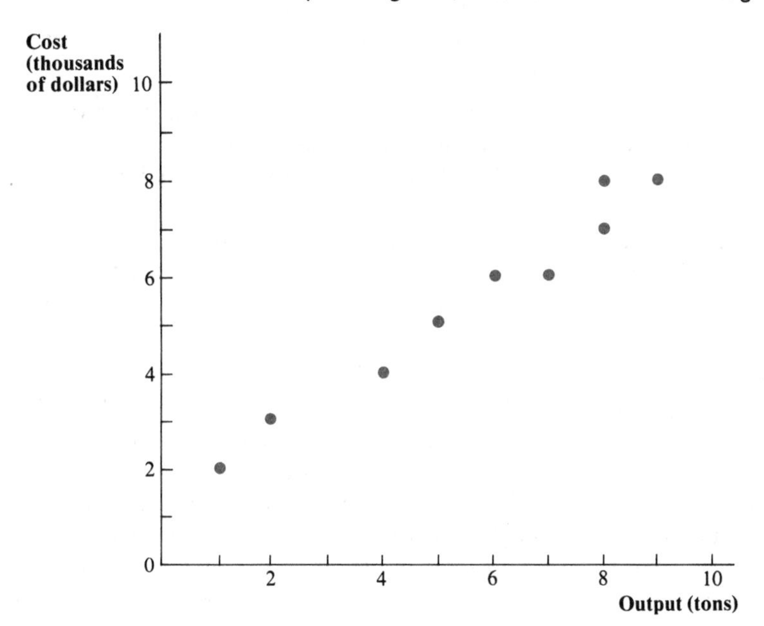

FIGURE 11.1 Scatter Diagram of Data on Cost and Output of Hosiery Mill

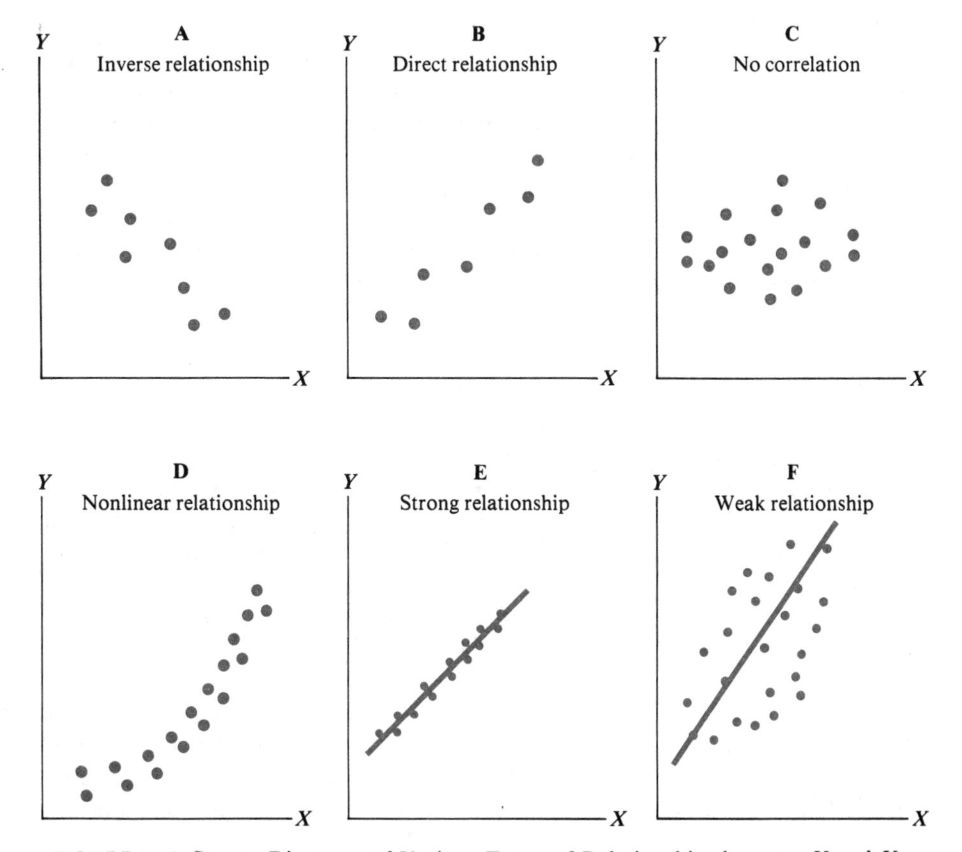

FIGURE 11.2 Scatter Diagrams of Various Types of Relationships between X and Y

correlation at all between X and Y. That is, changes in X do not seem to have any effect on the value of Y.

Is the Relationship Linear or Nonlinear? A relationship between X and Y is *linear* if a straight line provides an adequate representation of the average relationship between the two variables. On the other hand, if the points in the scatter diagram fall along a curved line or depart in some other way from a linear relationship, the relationship between X and Y is *nonlinear*. Figure 11.1 suggests that the relationship between cost and output is linear (at least in this range). Panel D of Figure 11.2 shows a case where the relationship between X and Y is nonlinear.

How Strong is the Relationship? The relationship between X and Y is relatively strong if the points in the scatter diagram lie close to the line of average relationship. For example, panel E of Figure 11.2 shows a case where the relationship is strong enough so that one can predict the value of Y quite accurately on the basis of the value of X. This is evidenced by the fact that all the points lie very close to the line. On the other hand, panel F of Figure 11.2 shows a case where the relationship is so weak that one cannot predict the value of Y at all well on the basis of the value of X. This follows from the fact that the points are scattered widely around the line.

CORRELATION ANALYSIS

Correlation analysis is concerned with the strength of the relationship between two variables. As we have seen, some relationships among variables are much stronger than others. For example, the size of a person's left foot is ordinarily very strongly related to the size of his or her right foot. On the other hand, there is some relationship between a firm's size and how rapidly it adopts new techniques, but this relationship may be rather weak. Correlation analysis is an important and useful complement to regression analysis. Whereas regression analysis describes the *type* of relationship between the two variables, correlation analysis describes the *strength* of this relationship.

11.3 Aims of Regression and Correlation Analysis

Basically, there are four principal goals of regression and correlation analysis. First, *regression analysis provides estimates of the dependent variable for given values of the independent variable.* If the hosiery mill in Figure 11.1 wants to estimate its monthly cost of producing 4 tons of output per month, regression analysis provides such an estimate, based on the *regression line.* This line (which is fitted to the data by a method described in Section 11.6) estimates the mean value of Y for each value of X. Thus, in the case of the hosiery mill the regression line would estimate the mean value of cost for each value of output.

Second, *regression analysis provides measures of the errors that are likely to be involved in using the regression line to estimate the dependent variable.* For example, in the case of the hosiery mill it clearly would be useful to know how much faith one can put in the cost estimate based on the regression line. To answer such questions, statisticians construct confidence intervals which are described in Section 11.9.

Third, *regression analysis provides an estimate of the effect on the mean value of* Y *of a one-unit change in* X. For example, in the case of the hosiery mill, the management might well be interested in the value of the marginal cost, which, of course, is the increase in the total cost due to a one-unit increase in output. If the relationship between cost and output is linear, the slope of the regression line equals the mill's marginal cost. Regression analysis enables us to estimate this slope and to test hypotheses concerning its value, as shown in Section 11.13.

Fourth, *correlation analysis provides estimates of how strong the relationship is between the two variables.* The *coefficient of correlation* and the *coefficient of determination* are two measures generally used for this purpose. These will be discussed in detail in Sections 11.10 and 11.11.

11.4 Linear Regression Model

A *model* is a simplified or idealized representation of the real world. All scientific inquiry is based to some extent on the use of models. In this section, we describe the model—that is, the set of simplifying assumptions—on which regression analysis is based. To begin with, the statistician visualizes a population of all relevant pairs of observations of the independent and dependent variables. For example, in the case of the hosiery mill, the statistician would visualize a population of pairs of observations concerning output and cost. This population would include all the levels of cost corresponding to all the output rates in the history of the mill.

Holding constant the value of X (the independent variable), the statistician assumes that each corresponding value of Y (the dependent variable) is drawn at random from the population. For example, the second pair of observations in Table 11.1 is a case where output equals 2 tons and cost equals \$3,000. The statistician views this pair of observations as arising in the following way. The value of output (the independent variable) is fixed at 2 tons. The value of cost (the dependent variable) is the result of a random choice of all levels of cost corresponding to an output of 2 tons. Thus, the value of the dependent variable is a random variable, which happens in this case to equal \$3,000.

Clearly, the probability distribution of the dependent variable Y is determined by the distribution of values of the dependent variable in the population, when the value of the independent variable is fixed at its specified value. For example, suppose that the probability distribution of cost, given that output equals 2 tons, is as shown in Figure 11.3. According to this figure, this probability distribution is bell-shaped with a mean of \$2,950. Why does the probability distribution have this shape? Because in the population as a whole the values of cost (the dependent variable), when output is fixed at 2 tons, have a bell-shaped distribution with a mean of \$2,950.

The probability distribution of Y, given a specified value of X, is called

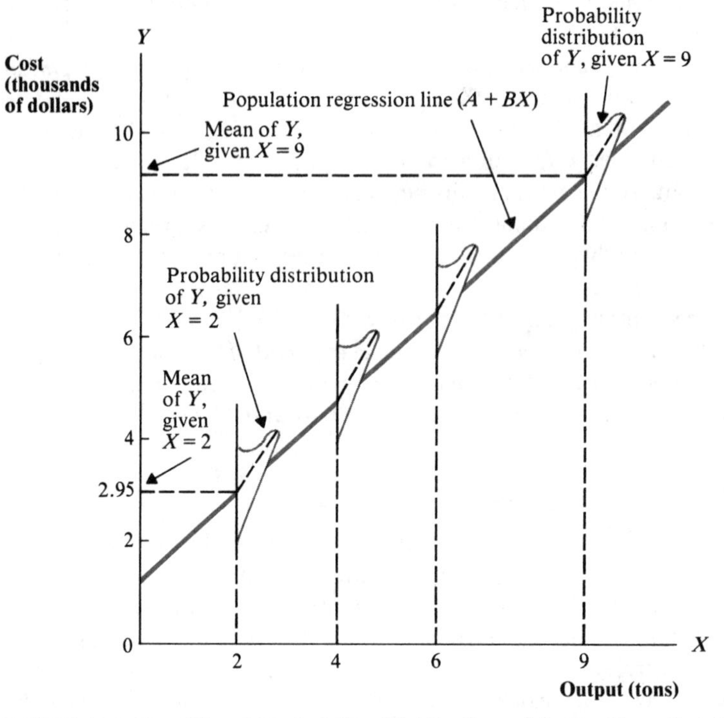

FIGURE 11.3 Conditional Probability Distribution of Cost, Given that Output Equals 2, 4, 6, or 9 Tons

the ***conditional probability distribution of*** Y. Thus, Figure 11.3 shows the conditional probability distribution of cost, given that output equals 2 tons (in other words, under the *condition* that output equals 2 tons).[1] Figure 11.3 also shows the conditional probability distribution of cost, given that output equals 4 tons, 6 tons, and 9 tons. The conditional probability distribution of Y, given the specified value of X, is denoted by

$$P(Y|X),$$

where Y is the value of the dependent variable and X is the specified value of the independent variable. The mean of this conditional probability distribution is denoted by $\mu_{Y.X}$, and the standard deviation of this probability distribution is denoted by $\sigma_{Y.X}$. For example, in Figure 11.3, $\mu_{Y.X}$ equals $2,950 if output equals 2 tons. (That is, $\mu_{Y.2} = 2.95$ thousands of dollars.)

Regression analysis makes the following assumptions about the conditional probability distribution of Y. First, it assumes that *the mean value of* Y, *given the value of* X, *is a linear function of* X. In other words, the mean value of

1. The concept of a conditional probability distribution has already been touched on in Appendix 4.3. The present discussion is self-contained and does not presume any knowledge of the material in Appendix 4.3.

the dependent variable is assumed to be a linear function of the independent variable. Put still differently, the means of the conditional probability distributions are assumed to lie on a straight line, the equation of this line being

$$\mu_{Y \cdot X} = A + BX.$$

Figure 11.3 shows a case of this sort, as evidenced by the fact that the means lie on a straight line. This straight line is called the ***population regression line*** or the ***true regression line***.

Second, regression analysis assumes that *the standard deviation of the conditional probability distribution is the same, regardless of the specified value of the independent variable.* Thus, in Figure 11.3, the spread of each of the conditional probability distributions is the same. For example, the standard deviation of the probability distribution of cost, given that output is 2 tons, is the same as the standard deviation of cost, given that output is 9 tons. This characteristic (of equal standard deviations) is called *homoscedasticity*.

Third, regression analysis assumes that *the values of* Y *are independent of one another.* For example, if one observation lies below the mean of its conditional probability distribution, it is assumed that this will not affect the chance that some other observation in the sample will lie below the mean of its conditional probability distribution. Obviously, this assumption need not be true. For example, in the case of the hosiery mill, if one month's costs are below average, the next month's costs may also be below average because the same factors may be at work for an extended period of time.

Fourth, regression analysis assumes that *the conditional probability distribution of* Y *is normal.* Actually, as pointed out below, not all aspects of regression analysis require this assumption, but some do. It is also worth noting that in regression analysis only Y is regarded as a random variable. The values of X are assumed to be fixed. Thus, when regression analysis is used to estimate Y on the basis of X, the true value of Y is subject to error, but the value of X is known. For example, if regression analysis is used to estimate the hosiery mill's costs when its output is 4 tons, the true cost of producing this output can be predicted only subject to error; but the output (4 tons) is known precisely.

The four assumptions underlying regression analysis can be stated somewhat differently. Together they imply that

$$Y_i = A + BX_i + e_i, \tag{11.1}$$

where Y_i is the ith observed value of the dependent variable, X_i is the ith observed value of the independent variable, and e_i is a normally distributed random variable with a mean of zero and a standard deviation equal to σ_e. Essentially, e_i is an *error term*, that is, a random amount that is added to $A + BX_i$ (or subtracted from it if e_i is negative). Because of the presence of this error term, the observed values of Y_i fall around the population regression line, not on it. Thus, as shown in Figure 11.4, if e_1 (the value of the error term for the first observation) is -1, Y_1 will lie 1 below the population regression

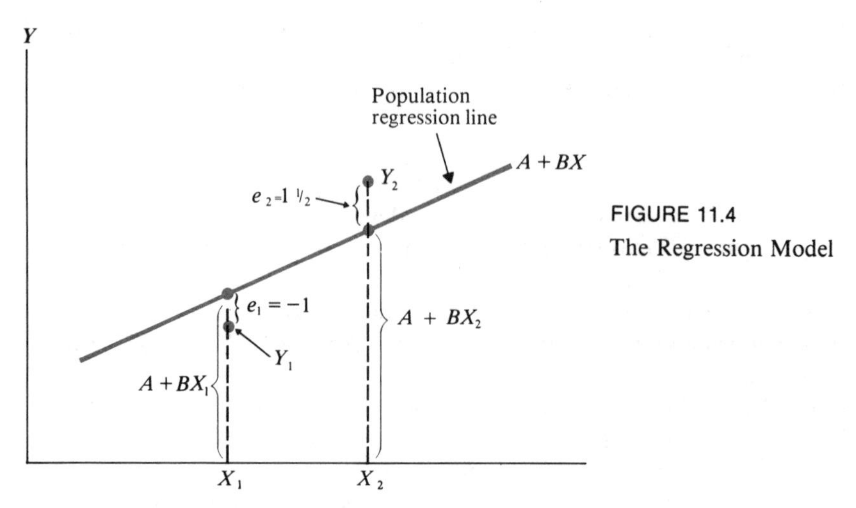

FIGURE 11.4
The Regression Model

line. And if e_2 (the value of the error term for the second observation) is $+1.50$, Y_2 will lie 1.50 above the population regression line. Regression analysis assumes that the values of e_i are independent.[2]

Although the assumptions underlying regression analysis are unlikely to be met completely, they are close enough to the truth in a sufficiently large number of cases so that regression analysis is a powerful technique. Nonetheless, it is important to recognize at the start that if these assumptions are not at least approximately valid, the results of a regression analysis can be misleading. In the next chapter we shall indicate how tests can be carried out to check these assumptions, and we will describe how violations of these assumptions may affect the results.

11.5 Sample Regression Line

To carry out a regression analysis, we must obtain the mathematical equation for a line that describes the average relationship between the dependent and independent variable. This line is calculated from the sample observations and is called the *sample* or *estimated regression line.* It should not be confused with the *population regression line* discussed in the previous section. Whereas the population regression line is based on the entire population, the sample regression line is based only on the sample.

The general expression for the sample regression line is

$$\hat{Y} = a + bX,$$

where $\hat{Y}$ is the value of the dependent variable predicted by the regression line, and a and b are estimators of A and B, respectively. Since this equation

2. In this chapter (and in Chapter 12), we use capital letters X and Y to denote both random variables and realized values. Which is referred to should be obvious from the context. (Practically all texts use the same symbol for both at this point. There seems to be general agreement that there is little danger of confusion.)

implies that $\hat{Y} = a$ when $X = 0$, it follows that a is the value of Y at which the line intersects the Y axis. Thus, a is often called the Y *intercept* of the regression line. And b, which clearly is the *slope* of the line, measures the change in the predicted value of Y associated with a one-unit increase in X.

Figure 11.5 shows the estimated regression line for the data concerning cost and output of the hosiery mill. The equation for this regression line is

$$\hat{Y} = 1.266 + 0.752X,$$

where $\hat{Y}$ is monthly cost in thousands of dollars and X is monthly output in tons. What is 1.266? It is the value of a, the estimator of A. What is 0.752? It is the value of b, the estimator of B. We are not interested here in how this equation was determined. (The methods used and their rationale are described in detail in the next section.) What we do want to consider is how this equation should be interpreted.

To begin with, note the difference between Y and $\hat{Y}$. Whereas Y denotes an *observed* value of monthly cost, $\hat{Y}$ denotes the *computed* or *estimated* value of monthly cost, based on the regression line. For example, the first row of Table 11.1 shows that in the first month the actual value of cost was $2,000 when output was 1 ton. Thus, $Y = 2.0$ thousands of dollars when $X = 1$. In contrast, the regression line indicates that $\hat{Y} = 1.266 + 0.752(1)$, or 2.018 thousands of dollars when $X = 1$. In other words, while the regression line predicts that cost will equal $2,018 when output equals 1 ton, the actual cost under these circumstances (in the first month) was $2,000.

It is important to be able to identify and interpret the Y intercept and slope of a regression line. What is the Y intercept of the regression line in the case of the hosiery mill? It is 1.266 thousands of dollars. This means that if the monthly output is zero, the estimated monthly cost would be $1,266. As shown in Figure 11.5, $1,266 is the value of the dependent variable at which the regression line intersects the vertical axis. What is the slope of the regression line in this case? It is 0.752 thousands of dollars. This means that the estimated monthly cost increases by $752 when the monthly output increases by 1 ton.

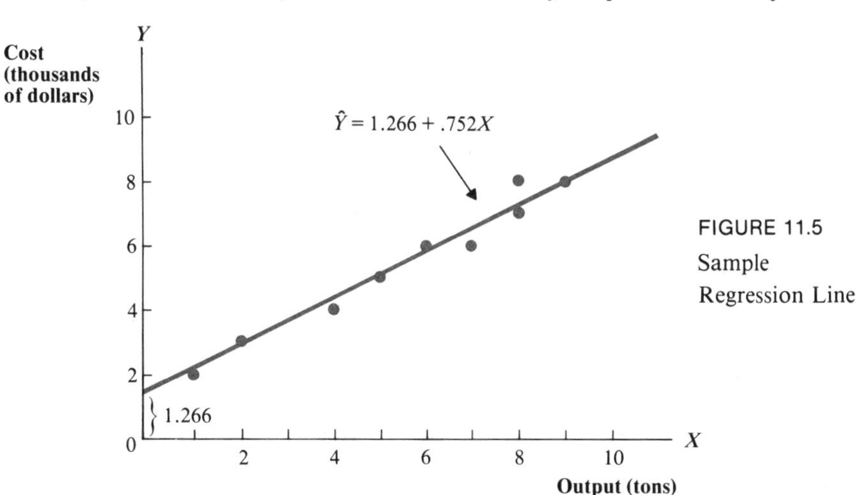

FIGURE 11.5
Sample Regression Line

11.6 Method of Least Squares

In this section, we describe how a sample regression line is calculated. To illustrate how this is done, suppose that we want to estimate a regression line to represent the relationship between cost and output in Figure 11.1. Since the equation for this line is

$$\hat{Y} = a + bX,$$

the estimation of a regression line really amounts to the choice of numerical values of a and b. There are an infinite number of possible values of a and b we could choose. One possibility, resulting in the line shown in panel A of Figure 11.6, is that $a = 1/2$ and $b = 1$. Another possibility, resulting in the

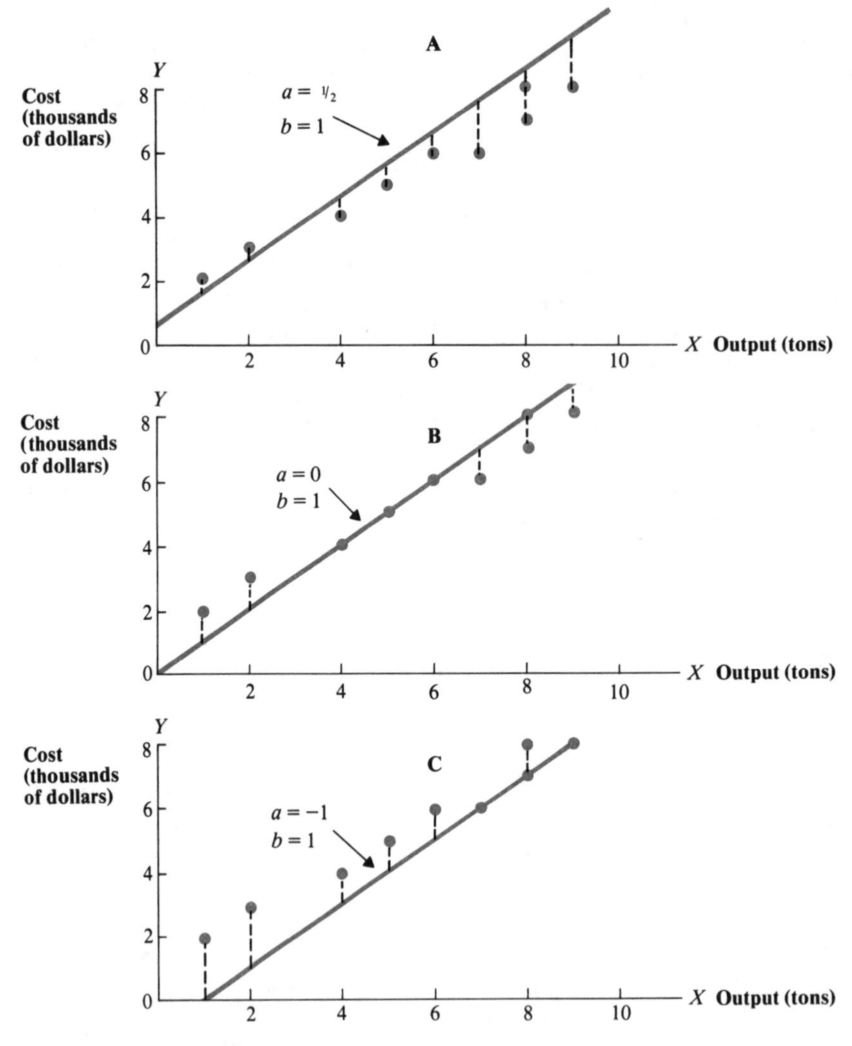

FIGURE 11.6 Alternative Values of a and b

line in panel B of Figure 11.6, is that $a = 0$ and $b = 1$. Still another possibility, resulting in the line in panel C of Figure 11.6, is that $a = -1$ and $b = 1$. How can we figure out which is best?

The *method of least squares* answers this question in the following way: We should take each possible line (that is, each possible value of a and b) and measure the deviation of each point in the sample from this line. Thus, in panel A of Figure 11.6 the deviation of each point from the line is measured by the broken vertical line from the point to the line. Then, according to the method of least squares, we square each of these deviations and add them up. Thus, in panel A of Figure 11.6 the sum of the squared deviations of the points from the line shown there is 8.25. In panel B the sum of squared deviations is 5, and in panel C the sum of squared deviations is 12.[3] Then *the method of least squares dictates that we choose the line where the sum of the squared deviations of the points from the line is a minimum.*

Certainly, on an intuitive level, it makes sense to choose the line (that is, the values of a and b) that minimizes the sum of squared deviations of the data in the sample from the line. Why? Because the bigger the sum of squared deviations of the data from the line, the poorer the line fits the data. Thus, one should minimize the sum of squared deviations if one wants to obtain a line that fits the data as well as possible. This is illustrated in Figure 11.6. Clearly, the line in panel C does not fit the data as well as the line in panel A, which in turn does not fit the data as well as the line in panel B. This fact is reflected in the differences in the sum of the squared deviations. As would be expected, the sum of squared deviations is higher for the line in panel C than for the line in panel A, and higher for the line in panel A than for the line in panel B.

Mathematically, it can be shown that if Y_i and X_i are the ith pair of observations concerning the dependent and independent variables, the values of a and b that result in the minimization of the sum of squared deviations from the regression line satisfy the following equations:

$$\sum_{i=1}^{n} Y_i = na + b \sum_{i=1}^{n} X_i$$

$$\sum_{i=1}^{n} X_i Y_i = a \sum_{i=1}^{n} X_i + b \sum_{i=1}^{n} X_i^2,$$

where n is the number of values of X_i (and Y_i) on which the calculation of the sample regression line is based. Solving for a and b, which are generally referred to as the *least-squares estimators of* A *and* B, we obtain the following important formulas:

3. In panel A the deviations of the points from the line are 1/2, 1/2, −1/2, −1/2, −1/2, −3/2, −3/2, −1/2, −3/2. Thus, the sum of squared deviations is 8.25. In panel B the deviations are 1, 1, 0, 0, 0, −1, −1, 0, −1. Thus, the sum of squared deviations is 5. In panel C the deviations are 2, 2, 1, 1, 1, 0, 1, 0, 0. Thus, the sum of squared deviations is 12. Since the deviations are in units of thousands of dollars, the squared deviations are in units of millions of dollars squared.

$$b = \frac{\sum_{i=1}^{n} (X_i - \bar{X})(Y_i - \bar{Y})}{\sum_{i=1}^{n} (X_i - \bar{X})^2}, \quad (11.2a)$$

$$a = \bar{Y} - b\bar{X}. \quad (11.2b)$$

The value of b in equation (11.2a) is often called the ***estimated regression coefficient.***

From the standpoint of computational ease, it frequently is preferable to use a somewhat different formula for b than the one given in equation (11.2a). This alternate formula, which yields the same answer as equation (11.2a), is

$$b = \frac{n \sum_{i=1}^{n} X_i Y_i - \left(\sum_{i=1}^{n} X_i\right) \left(\sum_{i=1}^{n} Y_i\right)}{n \sum_{i=1}^{n} X_i^2 - \left(\sum_{i=1}^{n} X_i\right)^2}.$$

In the case of the hosiery mill, Table 11.2 shows the calculation of $\Sigma X_i Y_i$, ΣX_i^2, ΣX_i, and ΣY_i. Based on these calculations,

$$b = \frac{9(319) - (50)(49)}{9(340) - 50^2} = \frac{2871 - 2450}{3060 - 2500}$$

$$= \frac{421}{560} = .752.$$

Thus, the value of b, the least-squares estimator of B, is .752 thousands of

TABLE 11.2

Computation of ΣX_i, ΣY_i, ΣX_i^2, ΣY_i^2, and $\Sigma X_i Y_i$.

	X_i	Y_i	X_i^2	Y_i^2	$X_i Y_i$
	1	2	1	4	2
	2	3	4	9	6
	4	4	16	16	16
	8	7	64	49	56
	6	6	36	36	36
	5	5	25	25	25
	8	8	64	64	64
	9	8	81	64	72
	7	6	49	36	42
Total	50	49	340	303	319

$$\bar{X} = \frac{50}{9} = 5.556$$

$$\bar{Y} = \frac{49}{9} = 5.444$$

dollars, which is the result given in the previous section. In other words, an increase in output of 1 ton results in an increase in estimated cost of about $752.

Having calculated b, we can readily determine the value of a, the least-squares estimator of A. According to equation (11.2b),

$$a = \bar{Y} - b\bar{X},$$

where $\bar{Y}$ is the mean of the values of Y, and $\bar{X}$ is the mean of the values of X. Since, as shown in Table 11.2, $\bar{Y} = 5.444$ and $\bar{X} = 5.556$, it follows that

$$a = 5.444 - .752(5.556)$$

$$= 1.266.$$

Thus, the least-squares estimate of A is 1.266 thousands of dollars. Recall that this is the result given in the previous section.

Given a and b, it is a simple matter to specify the average relationship in the sample between cost and output for the hosiery firm. This relationship is

$$\hat{Y} = 1.266 + 0.752X, \tag{11.3}$$

where $\hat{Y}$ is measured in thousands of dollars and X is measured in tons. As we know, this line is often called the *sample regression line* or the *regression of* Y *on* X. It is the line that we presented in the previous section and that we plotted in Figure 11.5. Now we have shown how this line is derived.

A regression line of this sort can be of great practical importance in business and economics. For example, suppose that the managers of the hosiery mill want to predict the firm's monthly costs if they decide to produce 4 tons per month. Using equation (11.3), the firm's statisticians would predict that its costs would be

$$1.266 + 0.752(4) = 4.274. \tag{11.3a}$$

Since costs are measured in thousands of dollars, this means that total costs would be expected to be 4.274 thousands of dollars, or $4,274.

The following example illustrates further how one calculates a least-squares regression line.

Example 11.1 An economist wants to estimate the relationship between a family's annual income and the amount that the family saves. The following data from nine families are obtained:

Annual income (thousands of dollars)	Annual savings (thousands of dollars)
12	0.0
13	0.1
14	0.2
15	0.2
16	0.5
17	0.5
18	0.6
19	0.7
20	0.8

Calculate the least-squares regression line, where annual savings is the dependent variable and annual income is the independent variable.

Solution: Letting X_i be the income (in thousands of dollars) of the ith family, and Y_i be the saving (in thousands of dollars) of the ith family, we find that

$$\sum_{i=1}^{9} X_i Y_i = 63.7 \quad \sum_{i=1}^{9} Y_i = 3.6 \quad \bar{Y} = 0.4$$

$$\sum_{i=1}^{9} X_i^2 = 2364 \quad \sum_{i=1}^{9} X_i = 144 \quad \bar{X} = 16.$$

Thus, substituting these values in the alternate formula for b, we obtain

$$b = \frac{9(63.7) - (144)(3.6)}{9(2364) - 144^2} = \frac{573.3 - 518.4}{21{,}276 - 20{,}736} = .1017.$$

Consequently,

$$a = \bar{Y} - b\bar{X} = 0.4 - .1017(16) = -1.2272.$$

Thus, the regression line is

$$\hat{Y} = -1.2272 + .1017X,$$

where both X and Y are measured in thousands of dollars.

11.7 Characteristics of Least-Squares Estimates

Optimally, in their calculations the decision maker and the statistician would like to know the population regression line. For example, the hosiery mill would like to know the population regression line relating its costs to its output—the regression based on *all* the possible observations of cost and output. However, this regression line cannot be calculated because the statistician has only the sample of observations to work with. Therefore, the best that the statistician can do is to calculate the sample regression line and to use it as an estimate of the population regression line. The statistics—a and b—defined in the previous section are estimators of A and B, the constants in the population regression. ***Whether or not the conditional probability distribution of the dependent variable is normal,*** these estimators have the following desirable properties:

1. *Unbiasedness.* It can be shown that a is an unbiased estimator of A, and that b is an unbiased estimator of B. In other words, if we were to draw one sample after another, and calculate the least-squares estimators a and b from each sample, the mean value of a would equal A, and the mean value of b would equal B, if we were to draw a very large number of samples. (Recall from Chapter 7 that unbiasedness is one of the criteria statisticians use for choosing among estimators.)

2. *Efficiency.* It can also be shown that of all estimators which are unbiased (and which are linear functions of the dependent variables), a and b have the smallest standard deviation. In other words, they are the most

efficient estimators of this type. This very important result was proved by the so-called *Gauss-Markov theorem*. (Recall from Chapter 7 that efficiency is one of the criteria used by statisticians to choose among estimators.)

3. *Consistency*. We also find that a is a consistent estimator of A, and that b is a consistent estimator of B. In other words, as the sample size becomes larger and larger, the value of a homes in on A, and the value of b homes in on B. (Recall from Chapter 7 that consistency is one of the criteria used by statisticians for choosing among estimators.)

Because of these very desirable properties, the least-squares estimators a *and* b *are the standard estimators used by statisticians to estimate the constants in the population regression line,* A *and* B. The fact that a and b have these desirable properties provides fundamental and strong support for the intuitive judgment in the previous section that the method of least squares is a good way to fit a sample regression line.

11.8 Standard Error of Estimate

In previous sections we have shown how regression analysis provides estimates of the dependent variable for given values of the independent variable (the first goal cited in Section 11.3). We will now describe how regression analysis provides measures of the errors that are likely to be involved in this estimation procedure (the second goal cited in Section 11.3). First, it is essential to recall that the standard deviation of the conditional probability distribution of the dependent variable is assumed to be the same, regardless of the value of the independent variable. *This standard deviation, which we denote by σ_e, is a measure of the amount of scatter about the regression line in the population.* If σ_e is large, there is much scatter; if σ_e is small, there is little scatter.

The sample statistic used to estimate σ_e is the standard error of estimate. It is defined as

$$s_e = \sqrt{\frac{\sum_{i=1}^{n} (Y_i - \hat{Y}_i)^2}{n - 2}}, \qquad (11.4)$$

where Y_i is the ith value of the dependent variable, n is the sample size, and $\hat{Y}_i$ is the estimate of Y_i from the regression line (that is, $\hat{Y}_i = a + bX_i$). Clearly, the value of s_e rises with increases in the amount of scatter about the regression line in the sample. If there is no scatter at all (which means that all the points are on the regression line), s_e equals zero, since Y_i and $\hat{Y}_i$ are always the same. But if there is much scatter, Y_i often differs greatly from $\hat{Y}_i$, with the result that s_e will be large.

Another formula used frequently for the standard error of estimate is

$$s_e = \sqrt{\frac{\sum_{i=1}^{n} Y_i^2 - a\sum_{i=1}^{n} Y_i - b\sum_{i=1}^{n} X_iY_i}{n - 2}}. \qquad (11.5)$$

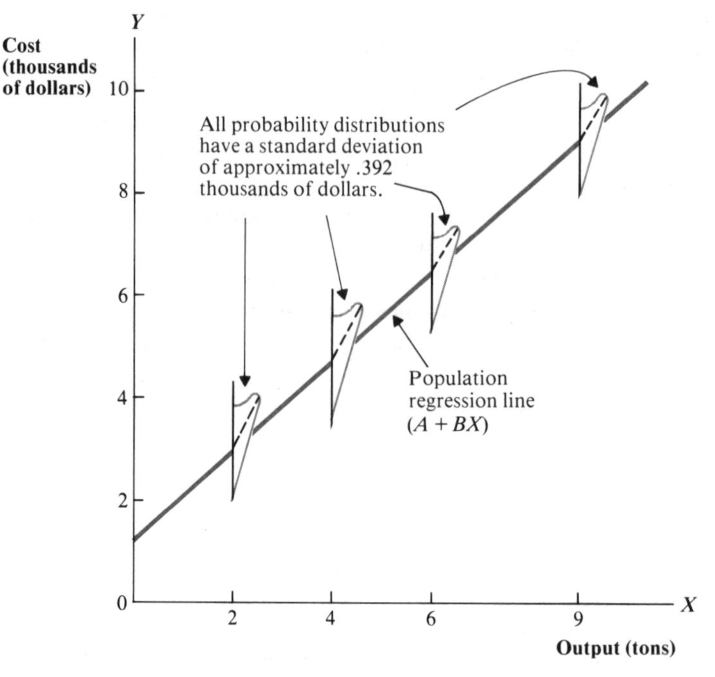

FIGURE 11.7 Conditional Probability Distribution of Cost, Given that Output Equals 2, 4, 6, or 9 Tons

This expression is often easier to calculate than the one given in equation (11.4). Of course, equations (11.4) and (11.5) will always give the same result. In each of these equations $(n - 2)$ is the denominator because this results in s_e^2 being an unbiased estimate of σ_e^2.

To illustrate the use of equation (11.5), let's return to the case of the hosiery mill. Since Table 11.2 shows that $\Sigma Y_i^2 = 303$, $\Sigma Y_i = 49$, and $\Sigma X_i Y_i = 319$, it follows that

$$s_e = \sqrt{\frac{303 - (1.266)(49) - (0.752)(319)}{7}} = \sqrt{.154} = 0.392.$$

Thus, if we knew the true values of A and B, the standard deviation of the errors in prediction based on this true regression line would be about .392 thousands of dollars, or \$392. In other words, the situation would be as shown in Figure 11.7.

The standard error of estimate will be used frequently in subsequent sections. The example below further illustrates how it is calculated.

Example 11.2 Based on the data in Example 11.1, the economist wants to calculate the standard error of estimate. What is its value?

Solution: From equation (11.5), it follows that

$$s_e = \sqrt{\frac{\sum_{i=1}^{9} Y_i^2 - a\sum_{i=1}^{9} Y_i - b\sum_{i=1}^{9} X_iY_i}{7}}$$

$$= \sqrt{\frac{2.08 + (1.2272)(3.6) - (.1017)(63.7)}{7.}}$$

$$= \sqrt{\frac{2.08 + 4.4179 - 6.4783}{7}} = \sqrt{.0028}$$

$$= .053.$$

Thus, the standard error of estimate is .053 thousands of dollars, or $53.

11.9 Estimators of (1) the Conditional Mean and (2) an Individual Value of Y

ESTIMATING THE CONDITIONAL MEAN

In this section we show how one can estimate the conditional mean of Y. In contrast to the previous section, it is not assumed that we know A and B; instead, we assume that both the regression line and the estimate are based on least-squares estimates of A and B. To be specific, let's return again to the example of the hosiery mill. Suppose that the managers of the mill are interested in predicting the *mean* monthly cost of the mill if it were to achieve an output of 4 tons per month *over and over again*. In other words, the firm's managers are interested in estimating the vertical coordinate of the point on the population regression line corresponding to an output of 4 tons per month. In Figure 11.8, this conditional mean is denoted by $\mu_{y.4}$.

Since the population regression line is unknown, the best that the mill's managers can do is to substitute the sample regression line. That is, to estimate $A + B(4)$, one uses $a + b(4)$, or $4,274. This, of course, is the point on the sample regression line corresponding to $X = 4$, as shown in Figure 11.8. In general, in order to estimate the conditional mean of Y (that is, the vertical coordinate of the point on the true regression line) when the independent variable equals X^*, one should use

$$a + bX^*.$$

A confidence interval for the conditional mean of Y (given that the independent variable equals X^*) is

$$(a + bX^*) \pm t_{\alpha/2}s_e \sqrt{\frac{1}{n} + \frac{(X^* - \bar{X})^2}{\sum_{i=1}^{n} X_i^2 - n\bar{X}^2}} \qquad (11.6)$$

If the conditional probability distribution of the dependent variable is normal, the

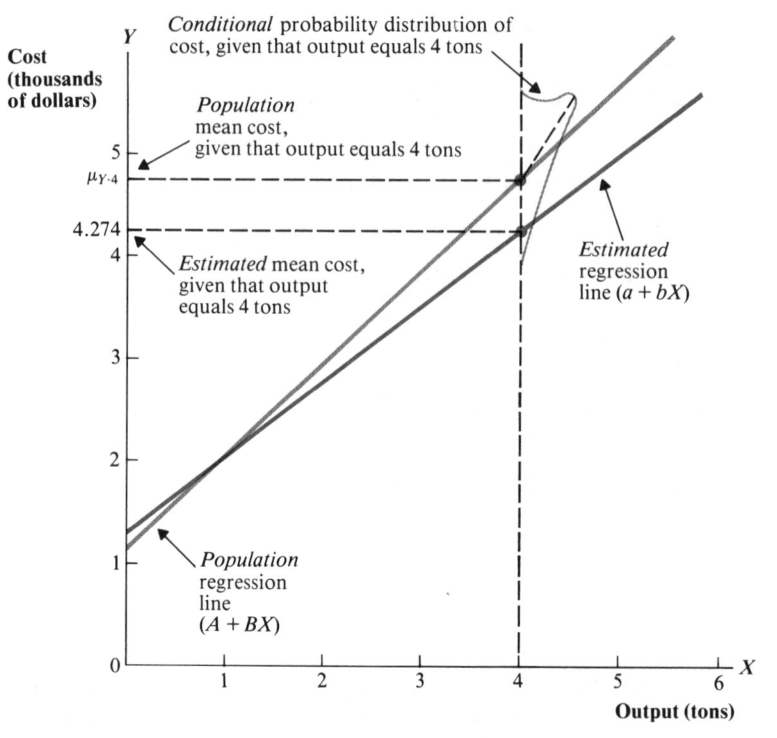

FIGURE 11.8 Estimated (and Population) Mean Cost and Conditional Probability Distribution of Cost, if Output Equals 4 Tons

probability is $(1 - \alpha)$ *that this interval will include this conditional mean.* Note that this confidence interval becomes wider as the value of X^* lies farther and farther from $\bar{X}$. This makes sense because as one moves farther and farther from $\bar{X}$, an error in the estimated slope of the regression line will result in an increasingly larger error in the location of the regression line.

To illustrate the use of the confidence interval in expression (11.6), let's consider the prediction of the mean cost that would be incurred if the hosiery mill were to achieve repeatedly an output of 4 tons per month. In this case, the 95 percent confidence interval for this mean cost would be

$$4.274 \pm 2.365(0.392)\sqrt{\frac{1}{9} + \frac{(4 - 5.556)^2}{62.2}}$$

or

$$4.274 \pm .359.$$

(Note that $t_{\alpha/2}$ is based on $(n - 2)$, or 7, degrees of freedom.) Thus, if the firm is interested in estimating the point on the population regression line corresponding to an output of 4 tons per month, the 95 percent confidence interval for this conditional mean is 3.915 to 4.633 thousands of dollars (that is, \$3,915 to \$4,633).

The following example further illustrates how a confidence interval is computed for a conditional mean.

Example 11.3 Based on the data in Example 11.1, calculate a 95 percent confidence interval for the mean amount of family savings among families with an income of \$20,000.

Solution: Substituting in equation (11.6), we obtain

$$-1.2272 + .1017(20) \pm 2.365(.053)\sqrt{\frac{1}{9} + \frac{(20 - 16)^2}{2364 - 9(16)^2}},$$

or

$$.8068 \pm 2.365(.053)(.615),$$

or

$$0.730 \text{ to } 0.884.$$

Thus, the 95 percent confidence interval for the mean amount of savings among families with an income of \$20,000 is .730 to .884 thousands of dollars, or \$730 to \$884.

PREDICTING AN INDIVIDUAL VALUE OF Y

Sample regression lines are often used to predict an individual value of Y. The hosiery mill's managers might want to predict the mill's costs for next month if it produces 4 tons of output then. This kind of prediction can be made with the use of the sample regression line. The vertical coordinate of the point on the sample regression line corresponding to $X = 4$ can be used as a predictor. As shown in Figure 11.8 (and as we already know from Section 11.6), the prediction in this case would be that $Y = 4.274$ thousands of dollars, or \$4,274. In general, to predict an individual value of Y when the independent variable equals X^*, one should use

$$a + bX^*.$$

A confidence interval for the value of Y that will occur if the independent variable is set at X^* is

$$(a + bX^*) \pm t_{\alpha/2}s_e\sqrt{\frac{n+1}{n} + \frac{(X^* - \bar{X})^2}{\sum_{i=1}^{n} X_i^2 - n\bar{X}^2}} \qquad (11.7)$$

If the conditional probability distribution of the dependent variable is normal, the probability is $(1 - \alpha)$ that this interval will include the true value of Y. A comparison of this confidence interval with the one shown in expression (11.6) indicates that this one is wider than the other. This, of course, is reasonable because the sampling error in predicting an individual value of the dependent variable will be greater than the sampling error in estimating the conditional mean value of the dependent variable.

To illustrate the use of this confidence interval, suppose that the hosiery mill is interested in predicting the cost that will be incurred if next month's output rate is set at 4 tons. Based on expression (11.7), the 95 percent confidence interval for this cost is

$$4.274 \pm 2.365(0.392)\sqrt{\frac{10}{9} + \frac{(4 - 5.556)^2}{62.2}}$$

or

$$4.274 \pm 0.994.$$

(Again, $t_{\alpha/2}$ is based on $(n - 2)$, or 7, degrees of freedom.) Thus, if the firm is interested in estimating the cost in a particular month when the output rate is 4 tons, the 95 percent confidence interval for this cost is 3.280 to 5.268 thousands of dollars, or $3,280 to $5,268.

It is important to recognize that the estimate of the conditional mean of Y covered above is quite different from the prediction of an individual value of Y being discussed here. For example, an estimate of the *average* cost that would occur if output were set *repeatedly* at 4 tons per month is not the same as a prediction of the cost that would occur if output were set at 4 tons next month only. In the latter case, we are trying to predict the value of the random variable Y whose distribution is shown in Figure 11.8; in the former case, we are attempting to estimate the conditional mean of this distribution ($\mu_{Y.4}$), also shown in Figure 11.8. Put still differently, in the latter case we are attempting to predict an individual value of Y, while in the former we are trying to estimate a point on the population regression line. Although these two things are not the same, we use the same point estimate—$a + b(4)$—for both. However, as we have seen, we use a different confidence interval for each.

Before leaving the topic of the confidence intervals in expressions (11.6) and (11.7), it is important to spell out the underlying assumptions once more. First, *it is assumed that the population regression is linear; that is, that the conditional mean of the dependent variable is a linear function of the independent variable.* Second, *it is assumed that the conditional probability distribution of the dependent variable is normal and that its standard deviation is the same, regardless of the value of the independent variable.* If any one of these assumptions is violated, the results will be in error to at least some degree. It is also worth noting that the confidence interval in expression (11.7) relates only to a single prediction. In other words, the confidence interval pertains to the entire process of gathering a sample of size n and making a single prediction. It does not pertain to more than one prediction based on a single sample.

Example 11.4 shows how a confidence interval is computed for a predicted value of the dependent variable.

Example 11.4 Based on the data in Example 11.1, predict the savings of a family chosen at random from among families with an income of $20,000, and calculate a 95 percent confidence interval for this prediction.

Solution: The prediction is $-1.2272 + .1017(20)$, or .807 thousands of dollars (that is, \$807). Using expression (11.7), the confidence interval is

$$-1.2272 + .1017(20) \pm 2.365(.053)\sqrt{\frac{10}{9} + \frac{(20-16)^2}{2364 - 9(16^2)}},$$

or

$$.8068 \pm 2.365(.053)1.174,$$

or

$$.660 \text{ to } .954.$$

Thus, the 95 percent confidence interval for the savings of a family chosen at random from among families with an income of \$20,000 is .660 to .954 thousands of dollars, or \$660 to \$954.

EXERCISES

11.1 The United States Department of Agriculture has published data concerning the strength of cotton yarn and the length of the cotton fibers that make up the yarn.[4] Results for 10 pieces of yarn are as follows:

Strength of yarn (pounds)	Fiber length (hundredths of an inch)
99	85
93	82
99	75
97	74
90	76
96	74
93	73
130	96
118	93
88	70

(a) Construct a scatter diagram of these data. (b) Based on this scatter diagram, does the relationship between these two variables seem to be direct or inverse? (c) Is this in accord with common sense? Why, or why not? (d) Does the relationship seem to be linear?

11.2 Assume that the conditional mean value of yarn strength is a linear function of fiber length. Calculate the least-squares estimates of the parameters (A and B) of this linear function, based on the data in Exercise 11.1.

11.3 (a) What is the sample regression line for the data in Exercise 11.1? (b) Use this regression line to predict the average strength of yarn made from fibers of length equal to 0.80 inches. (c) Use this regression line to predict the average strength of yarn made from fibers of length equal to 0.90 inches.

11.4 A firm examines a random sample of 10 spot welds of steel. In each case, the

4. U.S. Department of Agriculture, *Results of Fiber and Spinning Tests for Some Varieties of Upland Cotton Grown in the U.S.*, April 1945.

shear strength of the weld and the diameter of the weld are determined, the results being as follows:[5]

Shear strength (pounds)	Weld diameter (thousandths of an inch)
680	190
800	200
780	209
885	215
975	215
1,025	215
1,100	230
1,030	250
1,175	265
1,300	250

(a) Construct a scatter diagram of these data. (b) Based on this scatter diagram, does the relationship between these two variables seem to be direct or inverse? (c) Does this accord with common sense? Why, or why not? (d) Does the relationship seem to be linear?

11.5 Assume that the conditional mean value of a weld's shear strength is a linear function of its diameter. Calculate the least-squares estimates of the parameters (A and B) of this linear function, based on the data in Exercise 11.4.

11.6 (a) Plot the regression line for the data in Exercise 11.4. (b) Use this regression line to predict the average shear strength of a weld 1/5 inch in diameter. (c) Use the regression line to predict the average shear strength of a weld 1/4 inch in diameter.

11.7 Using the data in Exercise 11.1 and the calculations you have already made in Exercise 11.2, calculate the standard error of estimate. What does this number mean?

11.8 Using the data in Exercise 11.1 and the calculations in Exercises 11.2 and 11.7, compute a 90 percent confidence interval for the point on the population regression line corresponding to a fiber length of 0.80 inches.

11.9 (a) Using the results of Exercise 11.8, compute a 90 percent confidence interval for the strength of a piece of yarn, if the fiber length is 0.80 inches. (b) Why is this confidence interval wider than the confidence interval in Exercise 11.8?

11.10 Using the data in Exercise 11.4 and the calculations you made in Exercise 11.5, compute the standard error of estimate. What does this number mean?

11.11 Using the data in Exercise 11.4 and the calculations in Exercises 11.5 and 11.10, compute the 95 percent confidence interval for the mean shear strength among welds with a diameter of 1/4 inch.

11.12 Using the results of Exercise 11.11, compute the 95 percent confidence interval for the shear strength of a weld if its diameter is 1/4 inch. What assumptions underlie the calculation of this confidence interval?

11.10 Coefficient of Determination

In previous sections, we have shown how a regression line can be calculated. Once the regression line has been found, the statistician wants to know how

5. A. Duncan, *Quality Control and Industrial Statistics* (Homewood, Ill.: Irwin, 1959), p. 651.

well this line fits the data. There can be vast differences in how well a regression line fits a set of data, as shown in Figure 11.9. Clearly, the regression line in panel B of Figure 11.9 provides a better fit than the regression line in panel A of the same figure. How can we measure how well a regression line fits the data?

As a first step toward answering this question, we must discuss the concept of *variation*, which refers to a sum of squared deviations. The total variation in the dependent variable Y equals

$$\sum_{i=1}^{n} (Y_i - \bar{Y})^2.$$

In other words, the total variation equals the sum of the squared deviations of Y from its mean. (In Chapter 10, this was often called the total sum of squares.)

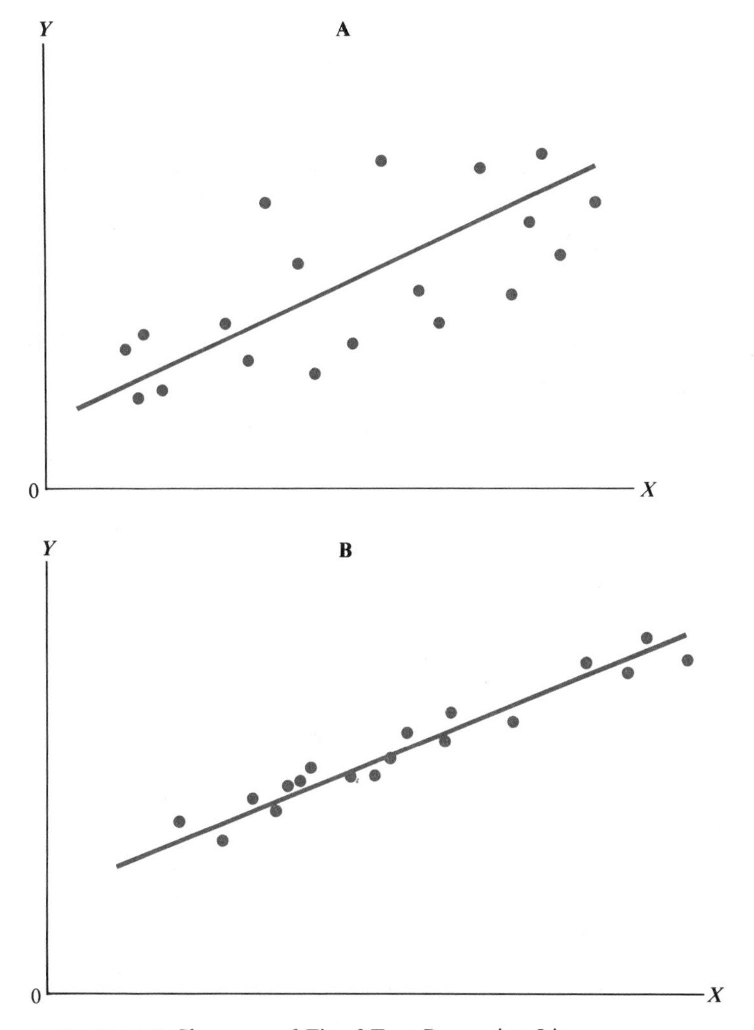

FIGURE 11.9 Closeness of Fit of Two Regression Lines

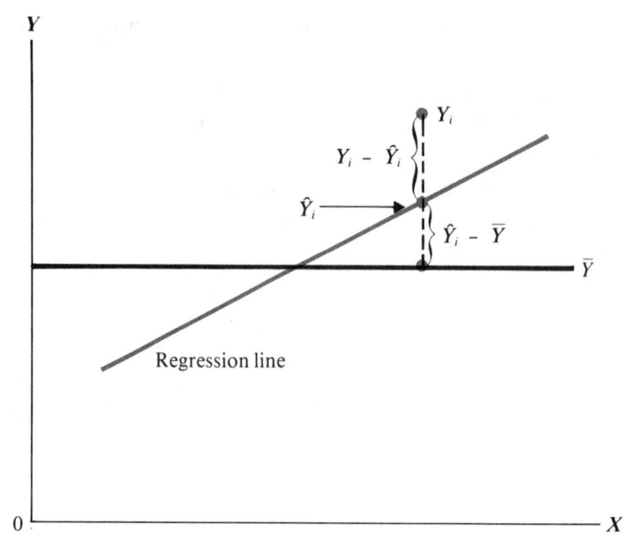

FIGURE 11.10 Division of $(Y_i - \bar{Y})$ into Two Parts: $(Y_i - \hat{Y}_i)$ and $(\hat{Y}_i - \bar{Y})$

To measure how well a regression line fits the data, we divide the total variation in the dependent variable into two parts: (1) the variation that *can* be explained by the regression line; and (2) the variation that *cannot* be explained by the regression line. To divide the total variation in this way, we must note that for the ith observation,

$$(Y_i - \bar{Y}) = (Y_i - \hat{Y}_i) + (\hat{Y}_i - \bar{Y}), \tag{11.9}$$

where $\hat{Y}_i$ is the value of Y_i that would be predicted on the basis of the regression line. In other words, as shown in Figure 11.10, the discrepancy between Y_i and the mean value of Y can be split into two parts: the discrepancy between Y_i and the point on the regression line directly below (or above) Y_i and the discrepancy between the point on the regression line directly below (or above) Y_i and $\bar{Y}$.

If we square both sides of equation (11.9) and sum the result over all values of i, we find that[6]

$$\sum_{i=1}^{n} (Y_i - \bar{Y})^2 = \sum_{i=1}^{n} (Y_i - \hat{Y}_i)^2 + \sum_{i=1}^{n} (\hat{Y}_i - \bar{Y})^2.$$

The term on the left-hand side of this equation shows the *total variation* in the

6. To derive this result, note that

$$\sum_{i=1}^{n} (Y_i - \bar{Y})^2 = \sum_{i=1}^{n} [(Y_i - \hat{Y}_i) + (\hat{Y}_i - \bar{Y})]^2$$

$$= \sum_{i=1}^{n} (Y_i - \hat{Y}_i)^2 + \sum_{i=1}^{n} (\hat{Y}_i - \bar{Y})^2 + 2\sum_{i=1}^{n} (Y_i - \hat{Y}_i)(\hat{Y}_i - \bar{Y}).$$

The last term on the right-hand side equals zero, so this result follows.

dependent variable. The first term on the right-hand side measures the *variation in the dependent variable that is not explained by the regression.* This is a reasonable interpretation of this term since it is the sum of squared deviations of the actual observations from the regression line. Clearly, the larger the value of this term, the poorer the regression equation fits the data.

The second term on the right-hand side of the equation measures the *variation in the dependent variable that is explained by the regression.* This is a reasonable interpretation of this term since it shows how much the dependent variable would be expected to vary on the basis of the regression alone. Putting it differently, this second term shows the reduction in unexplained variation due to the use of the regression instead of $\bar{Y}$ as an estimator of Y. Using $\bar{Y}$ as an estimator, the total variation is unexplained, whereas the first term on the right is unexplained when predictions based on the regression are used. Thus, the second term on the right shows the reduction in unexplained variation due to the use of predictions based on the regression instead of $\bar{Y}$.

To measure the closeness of fit of a regression line, statisticians use the ***coefficient of determination***:

$$r^2 = 1 - \frac{\sum_{i=1}^{n}(Y_i - \hat{Y}_i)^2}{\sum_{i=1}^{n}(Y_i - \bar{Y})^2}. \tag{11.10}$$

In other words,

$$r^2 = 1 - \frac{\text{Variation not explained by regression}}{\text{Total variation}} = \frac{\text{Variation explained by regression}}{\text{Total variation}}. \tag{11.11}$$

Clearly, the coefficient of determination is a reasonable measure of the closeness of fit of the regression line, since it equals *the proportion of the total variation in the dependent variable that is explained by the regression line.*

In practical work, a more convenient formula for the coefficient of determination is

$$r^2 = \frac{\left[n\sum_{i=1}^{n}X_iY_i - \left(\sum_{i=1}^{n}X_i\right)\left(\sum_{i=1}^{n}Y_i\right)\right]^2}{\left[n\sum_{i=1}^{n}X_i^2 - \left(\sum_{i=1}^{n}X_i\right)^2\right]\left[n\sum_{i=1}^{n}Y_i^2 - \left(\sum_{i=1}^{n}Y_i\right)^2\right]}. \tag{11.12a}$$

To illustrate the computation of the coefficient of determination, Table 11.2 shows the various quantities needed in equation (11.12a) in the case of the hosiery mill's cost function. Substituting these quantities into equation (11.12a), we have

$$r^2 = \frac{[9(319) - 50(49)]^2}{[9(340) - 50^2][9(303) - 49^2]} = \frac{421^2}{560(326)}$$

$$= 0.97.$$

Thus, the coefficient of determination between cost and output for the hosiery mill is 0.97. In other words, the regression line in Figure 11.5 can explain about 97 percent of the variation in cost.

Still another formula that is sometimes even more convenient is

$$r^2 = \frac{a\sum_{i=1}^{n} Y_i + b\sum_{i=1}^{n} X_iY_i - \frac{1}{n}\left(\sum_{i=1}^{n} Y_i\right)^2}{\sum_{i=1}^{n} Y_i^2 - \frac{1}{n}\left(\sum_{i=1}^{n} Y_i\right)^2}. \tag{11.12b}$$

The advantage of this formula over equation (11.12a) is that if one has already calculated the regression line, the values of a and b are already available. If this formula is used in the case of the hosiery mill, the result is

$$r^2 = \frac{1.266(49) + .752(319) - 266.778}{36.222} = \frac{62.034 + 239.888 - 266.778}{36.222}$$

$$= .97,$$

which, of course, is the answer we obtained in the previous paragraph.

11.11 The Correlation Coefficient

As pointed out at the beginning of this chapter, the purpose of correlation analysis is to measure the strength of the relationship between two variables, X and Y. The assumptions (or model) underlying correlation analysis are as follows: First, both X and Y are assumed to be normally distributed random variables. This is different from regression analysis where Y is assumed to be a random variable but X is not. Second, the standard deviation of the Ys is assumed to be constant for all values of X, and the standard deviation of the Xs is assumed to be constant for all values of Y.

The correlation coefficient is commonly used as a measure of the strength of the relationship between two variables. The ***correlation coefficient*** r is simply the square root of the coefficient of determination. That is,

$$r = \sqrt{r^2}.$$

The sign of r must equal the sign of the slope of the regression line. Thus, the positive square root of r^2 is taken if $b > 0$, and the negative square root is taken if $b < 0$. For ease of computation, the following formula is often used:[7]

$$r = \frac{n \sum_{i=1}^{n} X_i Y_i - \sum_{i=1}^{n} X_i \sum_{i=1}^{n} Y_i}{\sqrt{n \sum_{i=1}^{n} X_i^2 - \left(\sum_{i=1}^{n} X_i\right)^2} \sqrt{n \sum_{i=1}^{n} Y_i^2 - \left(\sum_{i=1}^{n} Y_i\right)^2}}. \quad (11.13)$$

The correlation coefficient cannot be greater than 1 or less than -1. *If* r $=$ *1, there is a perfect linear relationship between the independent and dependent variables, and the relationship is direct.* In other words, the situation is like that shown in panel A of Figure 11.11. On the other hand, *if* r $= -1$, *there is a perfect linear relationship between the independent and dependent variables, and the relationship is inverse.* The situation is like that shown in panel B of Figure 11.11. In each case the relationship is perfect in the sense that the regression explains all the variation in the dependent variable (since all the points fall on the regression line). Why does the correlation coefficient equal either $+1$ or -1 if all the variation in Y is explained by the regression line? Because under these circumstances, the actual value of Y must always be equal to the value computed from the regression, which means that $\Sigma(Y_i - \hat{Y}_i)^2 = 0$. In other words, none of the variation in Y is unexplained by the regression. Hence, it follows from equation (11.10) that r^2 must equal 1, which means that r must equal $+1$ or -1.

If r $=$ *0, there is zero correlation between the independent and dependent variables.* In this case, the least-squares estimate of B will turn out to be zero, indicating that on the average, changes in the independent variable have no effect on the dependent variable. Under such circumstances, the correlation coefficient is zero because the regression explains none of the variation in the dependent variable. In other words, $\Sigma(\hat{Y}_i - \bar{Y})^2 = 0$ since the mean value of Y is always equal to the value computed from the regression. Also, $\Sigma(Y_i - \hat{Y}_i)^2 = \Sigma(Y_i - \bar{Y})^2$ since $\hat{Y}_i$ always equals $\bar{Y}$. Hence, it follows from equation (11.10) that r^2 must equal 0, which means that r must equal 0. In a situation of this sort, the best estimate of the dependent variable is $\bar{Y}$; the value of the independent variable provides no additional useful information on this score. The situation is like that shown in panel C of Figure 11.11.

The calculation and interpretation of the sample correlation coefficient are illustrated below.

7. The correlation coefficient in the text is the correlation coefficient *unadjusted* for degrees of freedom, and is a biased estimate of the population correlation coefficient. (It is biased away from zero.) An unbiased estimate is the *adjusted* correlation coefficient, which is

$$r = \sqrt{1 - \frac{\sum_{i=1}^{n} (Y_i - \hat{Y}_i)^2 \div (n-2)}{\sum_{i=1}^{n} (Y_i - \bar{Y})^2 \div (n-1)}},$$

where n is the number of observations. (*Adjusted* here means adjusted for degrees of freedom.)

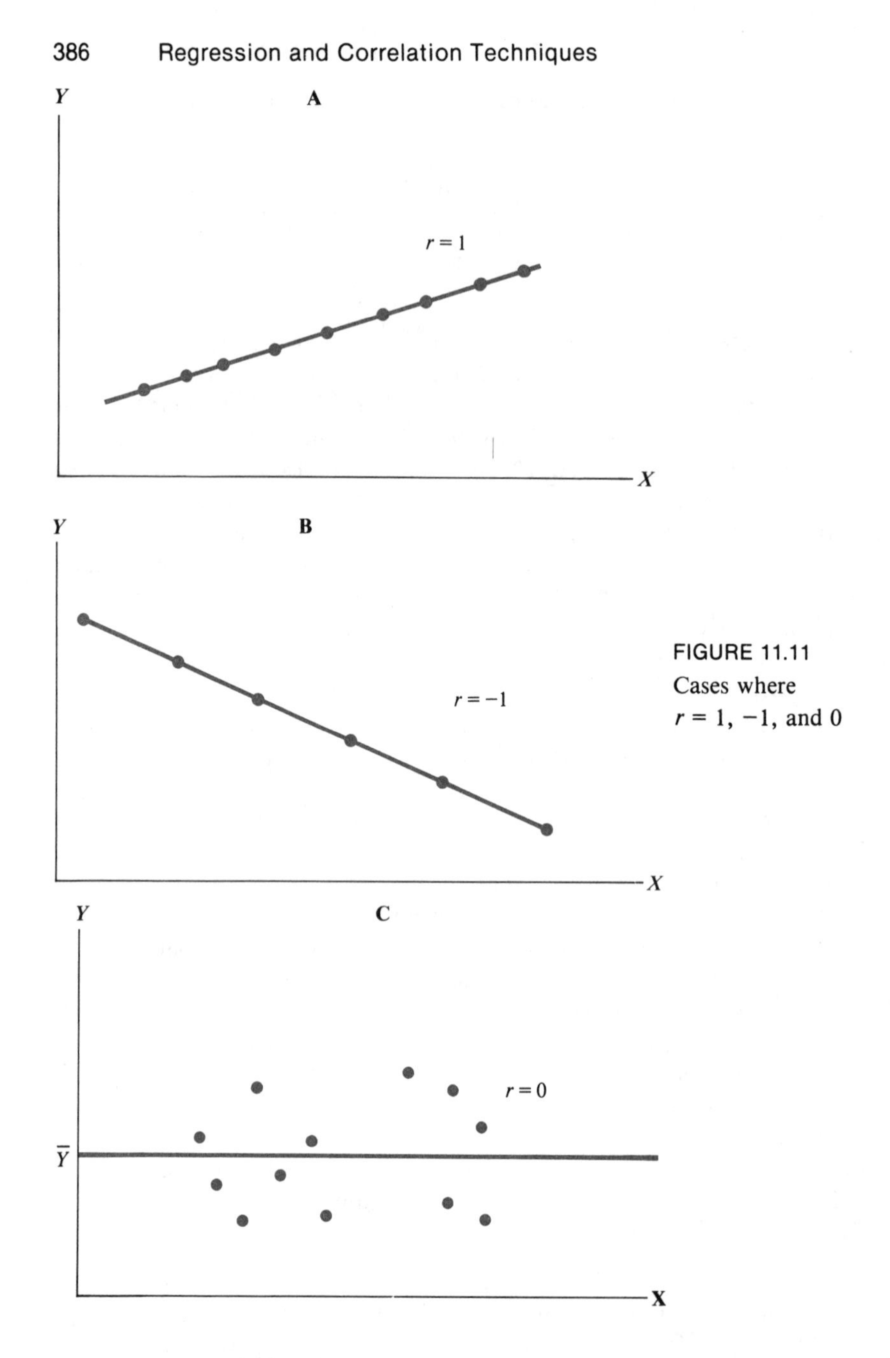

FIGURE 11.11
Cases where
$r = 1, -1$, and 0

Example 11.5 An industrial psychologist obtains the IQ score and productivity of 10 workers, the results being

IQ score	Productivity (output per hour)
110	5.2
120	6.0
130	6.3

126	5.7
122	4.8
121	4.2
103	3.0
98	2.9
80	2.7
97	3.2

Compute the correlation coefficient between IQ score and productivity. Is the relationship direct or inverse?

Solution: Letting the productivity of the ith worker equal Y_i and the IQ score of the ith worker equal X_i,

$$\sum_{i=1}^{10} X_i Y_i = 5042.6 \quad \sum_{i=1}^{10} X_i = 1107 \quad \sum_{i=1}^{10} Y_i = 44.0$$

$$\sum_{i=1}^{10} Y_i^2 = 210.84 \quad \sum_{i=1}^{10} X_i^2 = 124{,}823 \quad n = 10.$$

Thus, from equation (11.13), it follows that

$$r = \frac{10(5{,}042.6) - (1107)(44)}{\sqrt{10(124{,}823) - (1107)^2}\ \sqrt{10(210.84) - (44)^2}}$$

$$= \frac{50{,}426 - 48{,}708}{\sqrt{1{,}248{,}230 - 1{,}225{,}449}\ \sqrt{2108.4 - 1936}} = \frac{1718}{\sqrt{22{,}781}\ \sqrt{172.4}}$$

$$= \frac{1718}{(150.93)(13.13)} = \frac{1718}{1982} = .867.$$

The value of the correlation coefficient is about .87. Since this value is positive, the relationship seems to be direct.

11.12 Inference Concerning the Population Correlation Coefficient[8]

The correlation coefficient described in the previous section is a *sample* correlation coefficient, and it varies from one sample to another. In contrast, the ***population correlation coefficient,*** ρ (rho), pertains to the entire population. The definition of ρ is the same as that of r, the only difference being that r is based on the sample data whereas ρ is based on all the data in the population. In most cases, statisticians are much more interested in the correlation coefficient in the population than in the sample, and they use r as an estimate of ρ.

Frequently, statisticians are interested in testing whether the population correlation coefficient is zero. *If one variable is independent of another variable,*

8. Some instructors may prefer to take up Chapter 11 before Chapter 8. Chapter 11 has been written so this can be done; but Sections 11.12 and 11.13 should be taken up after Chapter 8 has been covered.

the population correlation coefficient equals zero. For example, if an employee's productivity on a particular job is completely unrelated to his or her age, the correlation coefficient in the population between a person's productivity and age will be zero. The personnel director of a firm may want to determine whether productivity and age really are uncorrelated for a particular job. Having measured productivity and age for a sample of workers, the director may want to test the null hypothesis that $\rho = 0$.

To carry out such a test, the first step is to specify the null hypothesis and the alternative hypothesis. The null hypothesis is that $\rho = 0$, and the alternative hypothesis is that $\rho \neq 0$. To test the null hypothesis, we compute

$$t = \frac{r}{\sqrt{(1 - r^2)/(n - 2)}}, \tag{11.14}$$

which has the t distribution with $(n - 2)$ degrees of freedom if the null hypothesis is true. Thus, the decision rule used to test this null hypothesis is: *Reject the null hypothesis that* $\rho = 0$ *if* t *is greater than* $t_{\alpha/2}$ *or less than* $-t_{\alpha/2}$; *accept the null hypothesis otherwise.*

To illustrate the way in which this test is carried out, suppose that in a sample of 18 workers the sample correlation coefficient between productivity and age is 0.42. To test the null hypothesis that the population correlation coefficient equals zero, we compute

$$t = \frac{0.42}{\sqrt{1 - .42^2)/16}} = \frac{0.42}{\sqrt{.2059}} = 0.93.$$

If the significance level of the test is set at .05, the null hypothesis should be rejected if $t > 2.12$ or if $t < -2.12$, since Appendix Table 6 shows that $t_{.025} = 2.12$ if there are 16 degrees of freedom. Because the observed value of t does not exceed 2.12 or fall below -2.12, the null hypothesis should not be rejected.

11.13 Inference Concerning the Value of B

In regression analysis, the slope of the sample regression line b varies from one sample to another. Like any sample statistic, it has a sampling distribution; and an estimate of the standard deviation of this sampling distribution is

$$s_b = s_e \div \sqrt{\sum_{i=1}^{n} X_i^2 - n\bar{X}^2},$$

which is often called the *standard error of* b. There are many occasions when the statistician wants to use the observed value of b to calculate a confidence interval for B, the slope of the population regression line. Such a confidence interval is

$$b \pm t_{\alpha/2} s_b. \tag{11.15}$$

If the conditional probability distribution of the dependent variable is normal, the probability is $(1 - \alpha)$ *that these limits will include the true value of* B. To illustrate the use of this formula, we can calculate the 95 percent confidence interval for B in the case of the hosiery mill. Since $b = .752$, $t_{.025} = 2.365$, $s_e = .392$, and $\Sigma X_i^2 - n\bar{X}^2 = 62.2$, it follows that the confidence interval is

$$.752 \pm 2.365(.392) \div \sqrt{62.2},$$

or

$$.752 \pm .118.$$

Thus, the 95 percent confidence interval for the cost of producing an extra ton of output is .634 to .870 thousands of dollars (that is, \$634 to \$870).

In addition to estimating the value of B, the statistician frequently wants to test the hypothesis that B equals zero. If this hypothesis is true, the mean of the dependent variable is the same, regardless of the value of the independent variable. (Specifically, the mean equals A, since $BX = 0$.) Thus, a knowledge of the independent variable is of no use in predicting the dependent variable since the conditional probability distribution of the dependent variable is not influenced by the value of the independent variable. In other words, if this hypothesis is true (and if the assumptions given in Section 11.4 hold), there is *no relationship* between the dependent and the independent variable.

If B equals zero, it does not follow that b must equal zero. On the contrary, it is quite likely that b will be non-zero because of random fluctuations. The decision rules for testing the null hypothesis that $B = 0$ are given below. These rules assume that the conditional probability distribution of the dependent variable is normal, and which rule is appropriate depends on the nature of the alternative hypothesis.

Alternative Hypothesis: $B > 0$***.*** *Reject the null hypothesis if* $b \div s_b > t_\alpha$*; accept the null hypothesis if* $b \div s_b \leqslant t_\alpha$*. (The number of degrees of freedom is* $n - 2$*, and* α *is the significance level.)*

Alternative Hypothesis: $B < 0$***.*** *Reject the null hypothesis if* $b \div s_b < -t_\alpha$*; accept the null hypothesis if* $b \div s_b \geqslant -t_\alpha$*.*

Alternative Hypothesis: $B \neq 0$***.*** *Reject the null hypothesis if* $b \div s_b > t_{\alpha/2}$ *or* $< -t_{\alpha/2}$*; accept the null hypothesis if* $-t_{\alpha/2} \leqslant b \div s_b \leqslant t_{\alpha/2}$*.*

As an illustration of how this test is carried out, suppose that the hosiery mill wants to test whether the slope of the population regression line relating cost to output equals zero (against the alternative hypothesis that $B \neq 0$.) The significance level is .05. From previous sections, we know that in this case $b = 0.752$, $s_e = 0.392$, $\Sigma X_i^2 - n\bar{X}^2 = 62.2$, and $n = 9$. Thus,

$$b \div s_b = \frac{.752}{.392 \div \sqrt{62.2}} = 15.1.$$

Since $n = 9$, there are 7 degrees of freedom, and $t_{.025} = 2.365$. Since the value of the test statistic (15.1) exceeds 2.365, the null hypothesis should be rejected.

Below is a further illustration of how one can test the hypothesis that $B = 0$.

Example 11.6 Based on the data in Example 11.1, test the hypothesis that the slope of the true regression line relating savings to income equals zero. The alternative hypothesis is that this slope exceeds zero. Use the .05 significance level.

Solution: From the results in Examples 11.1 and 11.2, we know that $b = .1017$, $s_e = .053$, $\Sigma X_i^2 - n\bar{X}^2 = 60$, and $n = 9$. Thus,

$$b \div s_b = \frac{.1017}{.053 \div \sqrt{60}} = 15.$$

Since $t_{.05} = 1.895$ when there are 7 degrees of freedom, the null hypothesis should be rejected. That is, the evidence seems to indicate that the slope of the true regression is positive, not zero.

11.14 Statistical Cost Functions: A Case Study

In preceding sections we have used some hypothetical data concerning the costs and output of a hosiery mill to illustrate regression and correlation techniques. Since these data are hypothetical, they may give the impression that the procedures employed are not of much practical use in estimating and analyzing cost functions. Nothing could be further from the truth. One of the classic studies carried out in industrial economics used precisely these techniques to estimate the relationship between cost and output for a hosiery mill. This study,[9] carried out by Joel Dean of the Columbia University Graduate School of Business, obtained the following regression line:

$$\hat{Y} = 2936 + 1.998X,$$

where $\hat{Y}$ is the computed monthly cost of production in dollars and X is the monthly output of the mill in dozens of pairs of stockings. The scatter diagram and the above regression line are shown in Figure 11.12. The standard error of estimate was \$6110. The correlation coefficient was 0.973.

This study is famous for several reasons, one being that it provided some early evidence concerning the shape of firms' cost functions. In particular, the study suggested that within the relevant range marginal cost (that is, the extra cost of an extra unit of output) did not vary with the output level of the firm. However, this may have been due in considerable part to the limited range of the observations. In other words, if data were obtained concerning much higher output levels than those shown in Figure 11.12, it is likely that costs would no longer conform to the linear relationship, due to capacity constraints. Since Dean's pioneer study, a large number of similar investigations have been made for other firms in a variety of industries. The results, based on

9. J. Dean, "Statistical Cost Functions of a Hosiery Mill," *Journal of Business*, 1941. Reprinted in part in E. Mansfield, *Statistics for Business and Economics: Readings and Cases* (New York: Norton, 1980).

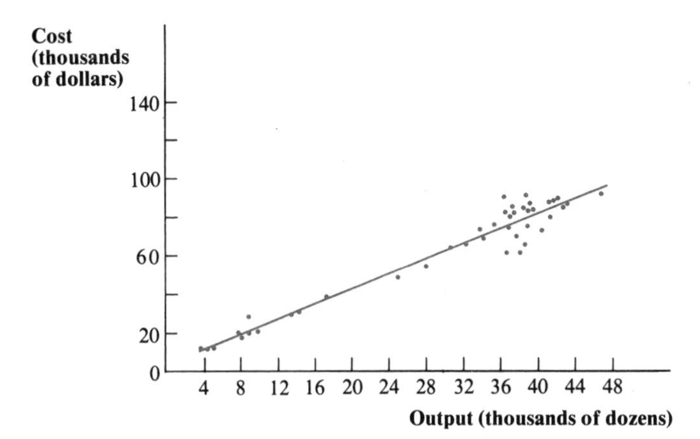

FIGURE 11.12 Actual Regression of Cost on Output, Hosiery Mill

SOURCE: J. Dean, "Statistical Cost Functions of a Hosiery Mill," *Journal of Business*, 1941.

the types of regression and correlation techniques described in this chapter, have proved valuable both to the firms themselves and to economists and government agencies.

11.15 Hazards and Problems in Regression and Correlation

There are a number of pitfalls in regression and correlation analyses that should be emphasized. First, *it is by no means true that a high coefficient of determination (or correlation coefficient) between two variables means that one variable causes the other variable to vary*. For example, if one regresses the size of a person's left foot on the size of his or her right foot, this regression is bound to fit very well, since the size of a person's left foot is closely correlated with the size of his or her right foot. But this does not mean that the size of a person's right foot *causes* a person's left foot to be as large or small as it is. Two variables can be highly correlated without causation being implied.

Second, *even if an observed correlation is due to a causal relationship, the direction of causation may be the reverse of that implied by the regression.* For example, suppose that we regress a firm's profits on its R and D (research and development) expenditure, the firm's profits being the dependent variable and its R and D expenditures being the independent variable. If the correlation between these two variables turns out to be high, does this imply that high R and D expenditures produce high profits? Obviously not. The line of causation could run the other way: High profits could result in high R and D expenditures. Thus, in interpreting the results of regression and correlation studies, it is important to ask oneself whether the line of causation assumed in the studies is correct.

Third, *regressions are sometimes used to forecast values of the dependent variable corresponding to values of the independent variable lying beyond the sample range*. For example, in Figure 11.13, the scatter diagram shows that the data for the independent variable range from about 1 to 7. But the regression may be used to forecast the dependent variable when the independent variable assumes a value of 9, which is outside the sample range. This procedure, known as ***extrapolation***, is dangerous because the available data provide no evidence that the true regression is linear beyond the range of the sample data. For example, the true regression may be as shown in Figure 11.13, in which case a forecast based on the estimated regression may be very poor.

Fourth, *one should be careful to avoid creating a spurious correlation by dividing both the independent and dependent variable by the same quantity*. For example, we might want to determine whether a country's copper output is related to its lead output. To normalize for differences among the sizes of countries, it may seem sensible to use copper output *per capita* as one variable and lead output *per capita* as the other. This procedure may result in these ratios being positively correlated (because the denominator of the ratio is the same for both variables), even if there is no relationship whatever between a country's copper output and its lead output. For example, suppose that there

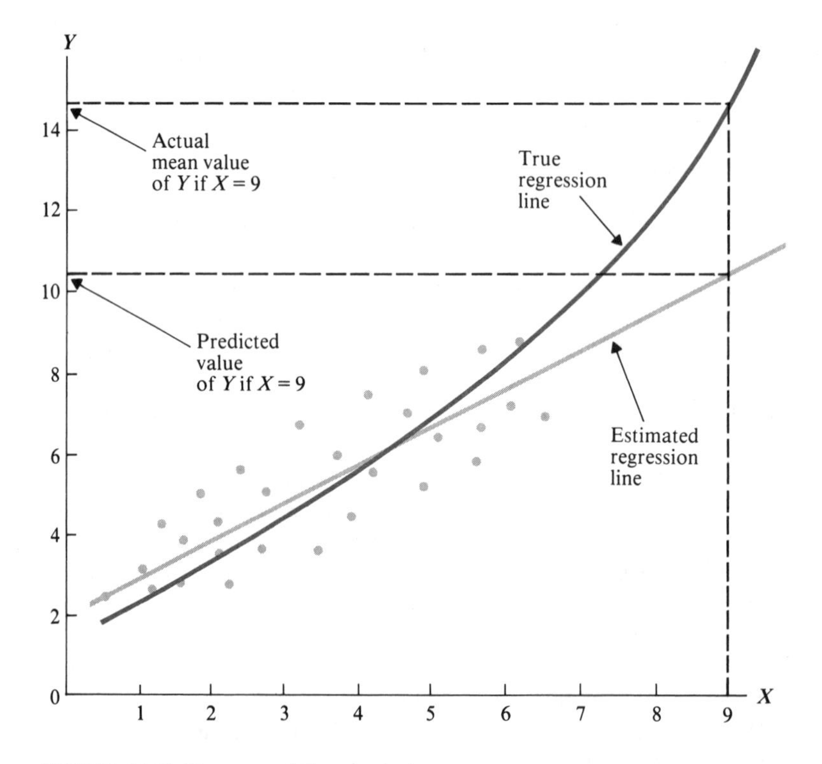

FIGURE 11.13 Dangers of Extrapolation

NOTE: If the curved line is the true (that is, the population) regression line, and if the estimated regression line is extrapolated to forecast the value of Y when $X = 9$, the result will be very inaccurate. Whereas the forecast is that Y will equal about 10.5, the mean value of Y(if $X = 9$) is really about 14.5.

are three countries, A, B, and C, and that each country's copper output and lead output are as shown in panel A of Figure 11.14. Clearly, there is no relationship between the two outputs. However, if the populations of these countries are 1 million, 10 million, and 20 million respectively, there is a strong positive correlation between copper output *per capita* and lead output *per capita*, as shown in panel B of Figure 11.14.

Fifth, *it is important to recognize that a regression based on past data may not be a good predictor, due to shifts in the regression line.* For example, suppose that the hosiery mill experiences considerable increases over time in wage rates and materials prices. If so, the regression line relating cost to output is likely to shift upward and to the left; and predictions based on historical data are likely to underestimate future costs. On the other hand, suppose that while input prices remain constant the hosiery mill experiences considerable productivity growth due to new technology. If this happens, the regression line relating cost to output is likely to shift downward and to the right. Predictions based on historical data are likely to overestimate future costs in a situation of this sort.

Finally, *when carrying out a regression, it is important to try to make sure that the assumptions in Section 11.4 are met.* Statisticians often plot the

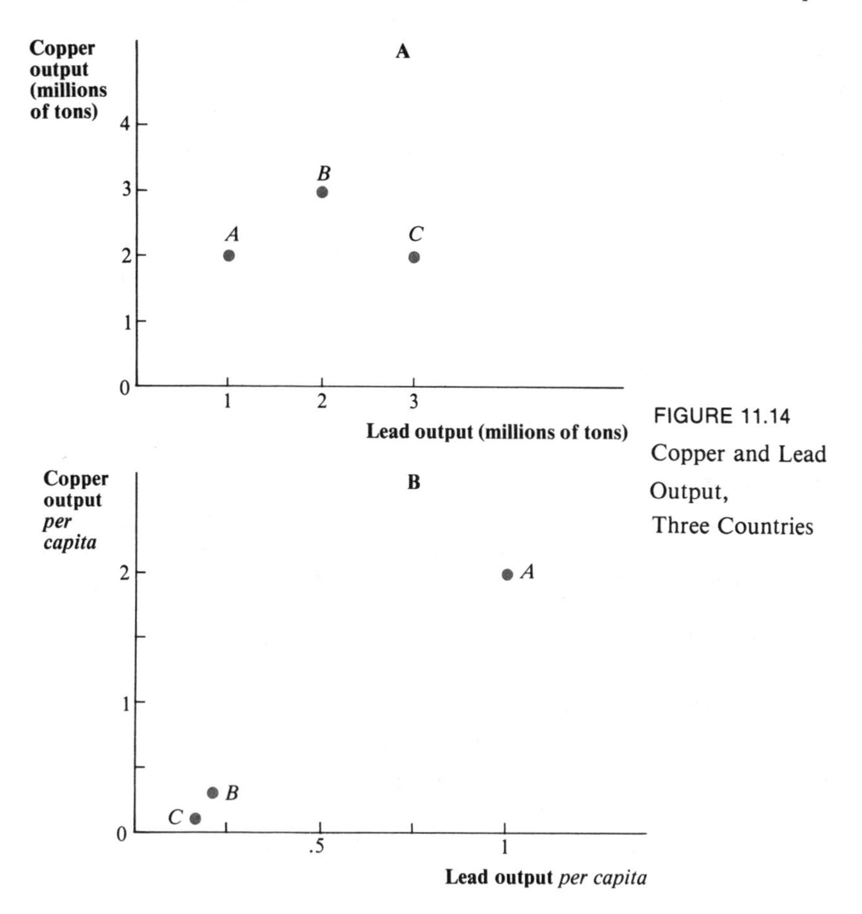

FIGURE 11.14
Copper and Lead Output, Three Countries

regression line on the scatter diagram and look for evidence of departures from the assumptions. For example, the scatter diagram may indicate that the relationship between the variables is curvilinear, not linear. (See panel D of Figure 11.2.) If this is so, logarithmic regression techniques (described in Appendix 11.1) or multiple regression (described in the next chapter) may be more appropriate than the simple linear regression techniques discussed here. There are a considerable number of tests that can be carried out to determine whether the assumptions underlying the use of regression techniques are met. Some of the most commonly used tests are presented in the next chapter.

EXERCISES

11.13 (a) Using the data in Exercise 11.1, calculate the sample correlation coefficient between fiber length and yarn strength. (b) What proportion of the variation in yarn strength in the sample can be explained by fiber length?

11.14 Based on the data in Example 11.5, test whether the population correlation coefficient between IQ score and productivity is zero. (Let $\alpha = .05$, and use a two-tailed test.)

11.15 (a) Using the data in Exercise 11.4, calculate the sample correlation coefficient between weld strength and weld diameter. (b) What proportion of the variation in weld strength in the sample can be explained by weld diameter?

11.16 If the sample correlation coefficient is .30 and n equals 20, test whether the population correlation coefficient is zero. (Let $\alpha = .05$, and use a two-tailed test.)

11.17 Using the data in Exercise 11.1, test the hypothesis that the slope of the true regression line relating yarn strength (the dependent variable) and fiber length (the independent variable) equals zero. The alternative hypothesis is that the slope exceeds zero. (Use the .01 significance level.)

11.18 Using the data in Exercise 11.4, test the hypothesis that the slope of the true regression line relating a weld's strength (the dependent variable) to its diameter (the independent variable) equals zero. The alternative hypothesis is that the slope is greater or less than zero. (Use the .05 significance level.)

11.19 On the basis of our results in previous exercises, can we be sure that variation in fiber length *causes* variation in yarn strength? Why, or why not?

11.20 On the basis of our results in previous exercises, can we be sure that variation in a weld's shear strength does *not* cause variation in its diameter? Why, or why not?

11.21 If the standard deviation of the conditional probability distribution of the dependent variable varies with the value of the independent variable, *heteroscedasticity* is said to occur. Can you think of any reasons why heteroscedasticity might occur in the case of the relationship between savings and income in Example 11.1?

11.22 What factors might cause heteroscedasticity in the case of the relationship between a firm's costs and its level of output?

11.23 On the basis of the data in Table 11.1 and the regression equation derived therefrom, the hosiery mill estimates that if output is 10 tons per month, costs should be $8,786 per month. Comment on this estimate.

11.24 If a regression line cannot explain 36 percent of the variation in the independent variable, what is the correlation coefficient?

11.25 If two variables X and Y are statistically independent, their correlation coefficient is zero. But is it true that if the population correlation coefficient between X and Y is zero they must be statistically independent? Explain.

CHAPTER REVIEW

1. *Regression analysis* indicates how one variable is related to another. A first step in describing such a relationship is to plot a scatter diagram. The *regression line* shows the average relationship between the dependent variable and the independent variable. The method of least squares is the standard technique used to fit a regression line to a set of data. If the regression line is $\hat{Y} = a + bX$, and if a and b are calculated by least squares,

$$b = \frac{\sum_{i=1}^{n} (X_i - \bar{X})(Y_i - \bar{Y})}{\sum_{i=1}^{n} (X_i - \bar{X})^2}$$

$$a = \bar{Y} - b\bar{X}.$$

This value of b is often called the *estimated regression coefficient.*

2. The regression line calculated by the method of least squares is generally based on a sample, not on the entire population. Suppose that the means of the conditional probability distributions of the dependent variable fall on a straight line: $\mu_{Y \cdot X} = A + BX$. Also, suppose that the standard deviation of these conditional probability distributions is the same for all values of X, that the observations in the sample are statistically independent, and that the values of X are known with certainty. Then a is an unbiased and consistent estimator of A, and is the most efficient estimator of A (among those unbiased estimators that are linear functions of the dependent variable). Similarly, b is an unbiased and consistent estimator of B, and is the most efficient estimator of B (among those unbiased estimators that are linear functions of the dependent variable).

3. The *standard deviation of the conditional probability distribution of the dependent variable is* σ_e. It is a measure of the amount of scatter about the regression line in the population. The sample statistic which is used to estimate σ_e is the standard error of estimate, defined as

$$s_e = \sqrt{\frac{\sum_{i=1}^{n} (Y_i - \hat{Y}_i)^2}{n - 2}},$$

where $\hat{Y}_i$ is the estimate of Y_i based on the regression line.

4. If we can assume that the conditional probability distribution of the dependent variable is normal, we can calculate a confidence interval for the conditional mean of the dependent variable when the independent variable

equals X^*. This confidence interval is

$$(a + bX^*) \pm t_{\alpha/2}s_e \sqrt{\frac{1}{n} + \frac{(X^* - \bar{X})^2}{\sum_{i=1}^{n} X_i^2 - n\bar{X}^2}}$$

if the confidence coefficient is $(1 - \alpha)$. Also, we can calculate a confidence interval for the value of the dependent variable that will occur if the independent variable is set at X^*. This confidence interval is

$$(a + bX^*) \pm t_{\alpha/2}s_e \sqrt{\frac{n+1}{n} + \frac{(X^* - \bar{X})^2}{\sum_{i=1}^{n} X_i^2 - n\bar{X}^2}}$$

if the confidence coefficient is $(1 - \alpha)$. The number of degrees of freedom is $n - 2$.

5. Assuming that the conditional probability distribution of the dependent variable is normal, a confidence interval for the slope of the population regression line B is

$$b \pm t_{\alpha/2}s_b,$$

where

$$s_b = s_e \div \sqrt{\sum_{i=1}^{n} X_i^2 - n\bar{X}^2}$$

and where $(1 - \alpha)$ is the confidence coefficient. To test the null hypothesis that $B = 0$ (against the alternative hypothesis that $B \neq 0$), reject the null hypothesis if $b \div s_b > t_{\alpha/2}$ or $< -t_{\alpha/2}$; accept the null hypothesis if $-t_{\alpha/2} \leqslant b \div s_b \leqslant t_{\alpha/2}$. The number of degrees of freedom is $(n - 2)$, and α is the significance level.

6. To measure the closeness of fit of a regression line, statisticians often use the *coefficient of determination*, defined as

$$r^2 = \frac{\left[n \sum_{i=1}^{n} X_i Y_i - \left(\sum_{i=1}^{n} X_i\right)\left(\sum_{i=1}^{n} Y_i\right)\right]^2}{\left[n \sum_{i=1}^{n} X_i^2 - \left(\sum_{i=1}^{n} X_i\right)^2\right]\left[n \sum_{i=1}^{n} Y_i^2 - \left(\sum_{i=1}^{n} Y_i\right)^2\right]}.$$

The *coefficient of determination* equals the proportion of the total variation in the dependent variable that is explained by the regression line.

7. *Correlation analysis* is concerned with measuring the strength of the relationship between two variables. The correlation coefficient, which is the square root of the coefficient of determination, is often used for this purpose. If $r = 1$, there is a perfect linear relationship between the two variables, and the relationship is direct. If $r = -1$, there is a perfect linear relationship between the two variables, and the relationship is inverse. If two variables are statistically independent, $r = 0$. To test whether the population correlation

coefficient ρ equals zero, one can calculate

$$t = \frac{r}{\sqrt{\frac{(1 - r^2)}{(n - 2)}}}$$

The null hypothesis that $\rho = 0$ should be rejected if this test statistic exceeds $t_{\alpha/2}$ or is less than $-t_{\alpha/2}$, where α is the significance level and the number of degrees of freedom equals $n - 2$. The alternative hypothesis is that $\rho \neq 0$.

8. There are many pitfalls in regression and correlation analysis. A high correlation between two variables does not necessarily mean that the variables are causally related. And even if they are causally related, the direction of causation may be different from that presumed in the analysis. It is extremely dangerous to extrapolate a regression line beyond the range of the data. One should be careful to avoid causing spurious correlation by dividing both the independent and the dependent variable by the same quantity. A regression based on past data may not be a good predictor, due to shifts in the regression line. It is important to try to determine whether the assumptions underlying regression analysis are met.

Getting Down to Cases:

PIG IRON AND LIME CONSUMPTION IN THE PRODUCTION OF STEEL

A number of years ago, the well-known British statistician L. H. C. Tippett conducted a study of the relationship between the percentage of pig iron in a cast of steel and the lime consumption per cast.[10] The basic data used were the following:

Percentage of pig iron	Lime consumption (hundred weights)	Percentage of pig iron	Lime consumption (hundred weights)	Percentage of pig iron	Lime consumption (hundred weights)
23	164	37	140	45	187
25	141	37	170	45	194
26	140	37	176	45	216
29	156	37	182	45	219
30	165	37	191	45	219
30	177	37	194	46	205
30	178	37	198	46	235
30	182	37	216	47	193
30	184	38	145	47	197
31	172	38	157	47	206
32	159	38	164	47	218
32	185	38	175	47	218
33	138	38	225	47	220
33	155	38	281	47	274

10. See L. H. C. Tippett, *Technological Application of Statistics* (New York: Wiley, 1950).

Percentage of pig iron	Lime consumption (hundred weights)	Percentage of pig iron	Lime consumption (hundred weights)	Percentage of pig iron	Lime consumption (hundred weights)
33	170	39	190	47	310
33	192	39	201	48	170
33	228	40	138	48	205
34	161	40	200	48	241
35	133	40	223	48	242
35	146	40	241	49	193
35	156	41	212	49	204
35	165	42	166	49	206
35	176	42	182	50	158
35	193	42	194	50	195
35	194	42	213	50	196
36	124	42	246	50	198
36	132	43	207	52	208
36	146	43	210	52	219
36	174	43	212	52	262
36	180	44	176	53	170
36	195	44	215	53	188
36	201	45	174	53	193
37	126	45	184	53	219
				53	240

(a) Plot the above data in a scatter diagram.

(b) Calculate the least-squares regression of lime consumption on percentage of pig iron. (That is, let lime consumption be the dependent variable and percentage of pig iron be the independent variable.)

(c) Compute the correlation coefficient between lime consumption and percentage of pig iron.

(d) Interpret the results of (a), (b), and (c) above.

APPENDIX 11.1
Logarithmic Regression

In this chapter we have dealt only with the case where the regression of Y on X is linear. That is, we have assumed that $\mu_{Y.X}$, the conditional mean of Y, is a linear function of X. That is,

$$\mu_{Y.X} = A + BX_i.$$

But in many cases, the mean value of Y_i is not a linear function of X_i. Despite this fact, the methods described in this chapter may still be serviceable in some instances. In this appendix we describe two ways in which these methods can be modified to handle cases where the relationship between Y_i and X_i is nonlinear.

Where Y_i *is an Exponential Function of* X_i. Suppose that

$$Y_i = \alpha\beta^{X_i} z_i, \tag{11.16}$$

where z_i is an error term with a positive (+) value. It is assumed that the

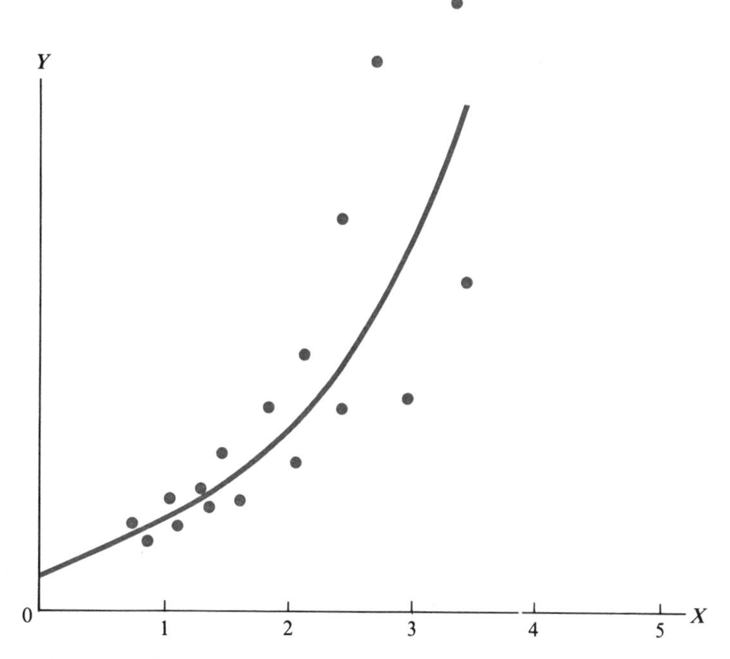

FIGURE 11.15 Exponential Function (Assuming $\beta = 2$)

expected value of log z_i equals zero, that the variance of log z_i equals σ_e^2, and that the values of log z_i are independent. Figure 11.15 shows this relationship between Y_i and X_i. In economic statistics, this sort of relationship occurs frequently. For example, if an employee's salary increases at an approximately constant percentage rate, there will be an exponential relationship between the employee's years of experience and his or her salary. By taking the logarithm of both sides of equation (11.16) we obtain

$$\log Y_i = \log \alpha + \log \beta X_i + \log z_i.$$

If we let $\log \alpha = A$, $\log \beta = B$, and $\log z_i = e_i$, we have

$$\log Y_i = A + BX_i + e_i,$$

which is an ordinary linear equation. Thus, we can estimate A ($=\log \alpha$) and B ($=\log \beta$) by regressing log Y_i on X_i. In other words, the trick here is to use log Y_i, not Y_i, as the dependent variable in the regression.

Where the Relationship Between Y_i *and* X_i *is*

$$Y_i = \alpha X_i^B z_i. \tag{11.17}$$

Figure 11.16 shows the relationship between Y_i and X_i when $B = .6$ and when $B = -1$. This sort of relationship occurs frequently in economics. For example, the relationship in panel B of Figure 11.16 is a demand curve of unit elasticity, if Y_i is quantity demanded and X_i is price. (The relationship in panel A will be explained in the following section.) Once again, we can convert

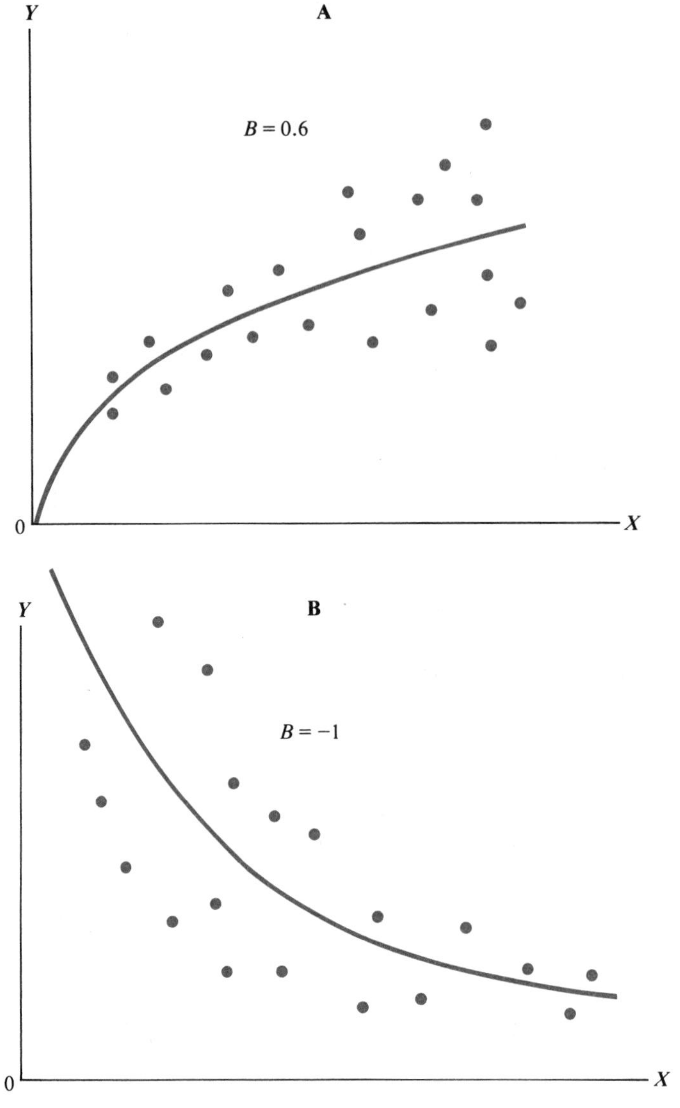

FIGURE 11.16 Graph of $Y = \alpha X_i^B$, for $B = 0.6$ and -1.0

this equation to linear form by taking logarithms of both sides, the result being

$$\log Y_i = \log \alpha + B \log X_i + \log z_i.$$

If we let $\log \alpha = A$, and $\log z_i = e_i$, we have

$$\log Y_i = A + B \log X_i + e_i,$$

which is an ordinary linear equation. We can estimate A $(=\log \alpha)$ and B by regressing $\log Y_i$ on $\log X_i$. In other words, the trick is to use $\log Y_i$, not Y_i, as the dependent variable and to use $\log X_i$, not X_i, as the independent variable in the regression.

LOGARITHMIC REGRESSION AND THE .6 RULE: A CASE STUDY

Engineers and industrial economists have often used the so-called .6 rule to estimate how capital expenditures vary with a plant's capacity. According to this rule, a 1 percent increase in capacity results in about a 0.6 percent increase in the cost of building the plant. In other words,

$$C_i = \alpha K_i^{0.6} z_i \tag{11.18}$$

where C_i is the cost of building the ith plant, K_i is the capacity of this plant, and α is a number that varies depending on the type of plant, time period, and other such variables. The relationship between C_i and K_i is like that shown in panel A of Figure 11.16.

In a well-known study,[11] Frederick Moore, then of the Rand Corporation, used regression analysis to see how well the .6 rule fit the data for a number of industries. In accord with the procedure described in the preceding section, Moore took logarithms of both sides of equation (11.18), the result being

$$\log C_i = \log \alpha + B \log K_i + e_i,$$

where $e_i = \log z_i$ and $B = 0.6$, according to the .6 rule. Among the industries Moore studied were cement and oxygen. For each industry he computed the least-squares estimate of B, the formula being exactly like equation (11.2a), except that $\log C_i$ is substituted for Y_i and $\log K_i$ is substituted for X_i.

The resulting estimate of B was 0.63 for the oxygen plants and 0.77 for the cement plants. Thus, for these two industries the estimate of B was not very different from the theoretical value predicted by the .6 rule. Indeed, it can be shown that the discrepancies between b (the estimate of B) and 0.6 could readily be due to chance.[12] Moore's study is an interesting practical illustration of the use of the techniques described in this appendix.

APPENDIX 11.2
Rank Correlation

In Chapter 9 we pointed out that nonparametric tests are often available as analogues to parametric tests. An important nonparametric technique used for measuring the degree of correlation between two variables Y and X is *rank correlation*. This technique is particularly useful when one or more of the variables in question cannot be measured in ordinary ways, but can only be ranked. For example, a psychologist may want to determine how closely a person's persistence is associated with his or her IQ score. Since there is no objective measure of persistence, the psychologist asks a professor to rank 12 students on the basis of his perception of their persistence. Each student's persistence rank and IQ score are shown in Table 11.3.

*The **rank correlation coefficient** measures the closeness of the relationship*

11. F. Moore, "Economics of Scale: Some Statistical Evidence," *Quarterly Journal of Economics*, May 1959. Reprinted in part in E. Mansfield, op. cit.

12. In each industry a confidence interval for B (of the sort discussed in Section 11.13) includes .6.

TABLE 11.3
Results of Psychologist's Tests

Person	(1) Persistence Rank	(2) IQ score	(3) IQ rank	(4) d_i = (3) − (1)	(5) d_i^2
A	1	120	8	7	49
B	9	130	3	−6	36
C	4	132	2	−2	4
D	3	127	4	1	1
E	8	126	5	−3	9
F	2	110	11	9	81
G	11	98	12	1	1
H	10	113	10	0	0
I	6	121	7	1	1
J	5	123	6	1	1
K	7	119	9	2	4
L	12	140	1	−11	121
Σd_i^2					308

$$r_r = 1 - \frac{6(308)}{12(144 - 1)} = 1 - \frac{1848}{1716} = -.077$$

between the two sets of rankings—that is, between the rankings of the one variable and the rankings of the other variable. For example, column (3) in Table 11.3 shows each person's rank with regard to IQ score and column (1) shows the corresponding rank with regard to persistence. Are these ranks correlated? In other words, do people who rank highly with regard to IQ score also tend to rank highly with regard to persistence? If so, is this relationship close? To help answer such questions, statisticians have devised the rank correlation coefficient, which is defined as follows:

$$r_r = 1 - \frac{6 \sum_{i=1}^{n} d_i^2}{n(n^2 - 1)}, \quad (11.19)$$

where d_i is the difference between the two ranks of the ith observation, and n is the number of observations. As shown in Table 11.3, the rank correlation coefficient in the case of the psychologist's study equals −.077.

The value of the rank correlation coefficient can vary from +1 to −1. A positive value of the rank correlation coefficient suggests that the two variables are directly related; a negative value suggests that they are inversely related. If the rank correlation coefficient is +1, there is a perfect direct relationship between the two rankings. If the rank correlation coefficient is −1, there is a perfect inverse relationship between the two rankings. If the two variables are statistically independent, the population rank correlation coefficient is zero. (However, due to chance, the *sample* rank correlation coefficient may not be zero under these circumstances.)

To test whether the population rank correlation coefficient is zero, we compute the following test statistic:

$$t = \frac{r_r}{\sqrt{(1 - r_r^2)/(n - 2)}}.$$

If the null hypothesis (that the population rank correlation coefficient is zero) is true, this statistic has the t distribution with $(n - 2)$ degrees of freedom. Thus, if the alternative hypothesis is that the population rank correlation coefficient is either greater or less than zero the decision rule is as follows.

Decision Rule: *Reject the null hypothesis that the population rank correlation coefficient is zero if* $t < -t_{\alpha/2}$ *or if* $t > t_{\alpha/2}$*; accept the null hypothesis if* $-t_{\alpha/2} \leqslant t \leqslant t_{\alpha/2}$.(There are $(n - 2)$ *degrees of freedom, and* α *is the significance level.)*

To illustrate the application of this decision rule, let's return to the psychologist's results in Table 11.3. Since the sample rank correlation coefficient in this case equals $-.077$,

$$t = \frac{-.077}{\sqrt{.9941/10}} = \frac{-.077}{.315} = -.24.$$

If the significance level is set at .05, $t_{.025} = 2.228$ since there are 10 degrees of freedom. Since the observed value of t is not less than -2.228 or greater than 2.228, the psychologist should not reject the null hypothesis that persistence and IQ score are unrelated.

The rank correlation coefficient is often used instead of the ordinary correlation coefficient when the dependent variable is far from normally distributed. Wide departures from normality have no effect on the test described in the two previous paragraphs because this test does not assume normality. Also, the rank correlation coefficient is sometimes preferred to the ordinary correlation coefficient because the former is easier to compute. As you can see from Table 11.3, the computations are relatively simple.

12

Multiple Regression and Correlation*

12.1 Introduction

In the previous chapter we discussed regression and correlation techniques in the case where there is only one independent variable. In practical applications of regression and correlation techniques, it frequently is necessary and desirable to include two or more independent variables. In this chapter we will extend our treatment of regression and correlation to the case in which there is more than one independent variable. In addition, we will indicate how tests can be carried out to check some of the assumptions underlying regression analysis, and we will describe how violations of these assumptions may affect the results.

12.2 Multiple Regression: Nature and Purposes

Whereas a *simple regression* includes only one independent variable, a *multiple regression* includes two or more independent variables. Basically, there are two important reasons why a multiple regression must often be used instead of a simple regression. First, *one frequently can predict the dependent variable more accurately if more than one independent variable is used.* In the case of the hosiery mill (discussed in the previous chapter), the firm's statisticians may feel that factors other than the output rate have an important effect on the firm's costs. For example, it may seem likely that costs will tend to decrease if the mill's manager is relatively experienced. Thus, holding the output rate constant, the expected value of the firm's costs may be a linear function of the number of years of experience of the mill's manager. In other words, it may be

*Some instructors may prefer to omit the latter half of this chapter, which deals with dummy variables, multicollinearity, and serial correlation. Others may wish to skip this chapter completely. Either option is feasible since an understanding of subsequent chapters does not depend on coverage of this chapter.

reasonable to assume that

$$E(Y_i) = A + B_1X_{1i} + B_2X_{2i}, \tag{12.1}$$

where Y_i is the cost (in thousands of dollars) for the ith month, X_{1i} is the output rate (in tons) for that month, and X_{2i} is the number of years of experience of the manager in charge of the mill during the ith month. Of course, if increases in the experience of the manager result in lower costs, B_2 is negative.

According to equation (12.1), the expected value of cost in a certain month is dependent on the output rate for that month and on the amount of experience of the mill's manager during that month. The equation says that if the output rate increases by 1 ton, the expected value of cost increases by B_1 thousands of dollars, and that if the amount of experience of the manager increases by 1 year, the expected value of cost increases by B_2 thousands of dollars. (Since B_2 is presumed to be negative, this means that the expected value of cost will decrease if the amount of experience increases by 1 year.) If it is true that cost depends on both the output rate and the amount of managerial experience, then the firm's statisticians can predict cost more accurately by using a multiple regression—that is, an equation relating the dependent variable (cost) to both independent variables (output rate and the amount of managerial experience)—than by using the simple regression in the previous chapter. The latter relates the dependent variable (cost) to only one of the independent variables (output rate).

A second reason for using multiple regression instead of simple regression is that *if the dependent variable depends on more than one independent variable, a simple regression of the dependent variable on a single independent variable may result in a biased estimate of the effect of this independent variable on the dependent variable*. For example, suppose that, in accord with equation (12.1), cost is dependent on both output and the amount of managerial experience. Estimating the simple regression of cost on output (as we did in the previous chapter) may result in a biased estimate of B_1, which measures the cost of an extra ton of output. To understand why such a bias may arise, suppose that the mill has tended to produce greater outputs in those months when it has had relatively experienced managers than in months when it has had relatively inexperienced managers. If this is the case, the costs of the low-output months have tended to be high because the managers have tended to be inexperienced, and the costs of the high-output months have tended to be low because the managers have tended to be experienced. Thus, the estimated simple regression will result in an estimate of B_1 that will be biased downward.[1]

When a dependent variable is a function of more than one independent variable, the observed relationship between the dependent variable and any

1. Why will the estimate of B_1 be biased downward? Because the costs of the high-output months are lower and the costs of the low-output months are higher than if the amount of managerial experience were held constant at its average level. Thus, the extra cost resulting from a certain extra amount of output is underestimated, which is the same as saying that B_1 is biased downward.

one of the independent variables may be misleading because the observed relationship may reflect the variation in the other independent variables. Since these other independent variables are totally uncontrolled, they may be varying in such a way as to make it appear that this independent variable has more effect or less effect on the dependent variable than in fact is true. To estimate the true effects of this independent variable on the dependent variable, we must include all the independent variables in the regression; that is, we must construct a multiple regression.

12.3 The Multiple-Regression Model

As pointed out in Chapter 11, the basic model underlying simple regression is

$$Y_i = A + BX_i + e_i,$$

where Y_i is the ith observed value of the dependent variable, X_i is the ith observed value of the independent variable, and e_i is a normally distributed random variable with a mean of zero and a standard deviation of σ_e. Essentially, e_i is an *error term*—that is, a random amount that is added to $A + BX_i$ (or subtracted from it, if e_i is negative). The conditional mean of Y_i is assumed to be a linear function of X_i—namely, $A + BX_i$. And the values of e_i are assumed to be statistically independent.

The model underlying multiple regression is essentially the same as that above, the only difference being that the conditional mean of the dependent variable is assumed to be a linear function of more than one independent variable. If there are two independent variables X_1 and X_2, the model is

$$Y_i = A + B_1X_{1i} + B_2X_{2i} + e_i, \tag{12.2}$$

where e_i is an error term. As in the case of simple regression, it is assumed that the expected value of e_i is zero, that e_i is normally distributed, and that the standard deviation of e_i is the same, regardless of the value of X_{1i} or X_{2i}. Also, the values of e_i are assumed to be statistically independent. In contrast to the case of simple regression, the conditional mean of Y_i is a linear function of both X_{1i} and X_{2i}. Specifically, the conditional mean equals $A + B_1X_{1i} + B_2X_{2i}$.

12.4 Least-Squares Estimates of the Regression Coefficients

The first step in multiple-regression analysis is to identify the independent variables, and to specify the mathematical form of the equation relating the expected value of the dependent variable to these independent variables. In the case of the hosiery mill this step is carried out in equation (12.1), which indicates that the independent variables are the output rate and the manager's years of experience. The relationship between the expected value of the dependent variable (costs) and these independent variables is linear. Having carried out this first step, *we next estimate the unknown constants* A, B$_1$ *and* B$_2$ *in the true regression equation.* Just as in the case of simple regression, these constants are estimated by finding the value of each that minimizes the sum of the squared deviations of the observed values of the dependent variable from the values of the dependent variable predicted by the regression equation.

To understand more precisely the nature of least-squares estimates of A, B_1, and B_2, suppose that a is an estimator of A, b_1 an estimator of B_1, and b_2 an estimator of B_2. Then the value of the dependent variable $\hat{Y}_i$ predicted by the estimated regression equation is

$$\hat{Y}_i = a + b_1X_{1i} + b_2X_{2i},$$

and the deviation of this predicted value from the actual value of the dependent variable is

$$Y_i - \hat{Y}_i = Y_i - a - b_1X_{1i} - b_2X_{2i}.$$

Just as in the case of simple regression, the closeness of fit of the estimated regression equation to the data is measured by the sum of squares of these deviations:

$$\sum_{i=1}^{n} (Y_i - \hat{Y}_i)^2 = \sum_{i=1}^{n} (Y_i - a - b_1X_{1i} - b_2X_{2i})^2, \tag{12.3}$$

where n is the number of observations in the sample. The larger this sum of squares, the less closely the estimated regression equation fits; the smaller this sum of squares, the more closely it fits. Thus, it seems reasonable to choose the values of a, b_1, and b_2 that minimize the expression in equation (12.3). These estimates are least-squares estimates, as in the case of simple regression.

It can be shown that the values of a, b_1, and b_2 that minimize the sum of squared deviations in equation (12.3) must satisfy the following (so-called normal) equations:

$$\begin{aligned} \sum_{i=1}^{n} Y_i &= na + b_1 \sum_{i=1}^{n} X_{1i} + b_2 \sum_{i=1}^{n} X_{2i} \\ \sum_{i=1}^{n} X_{1i}Y_i &= a \sum_{i=1}^{n} X_{1i} + b_1 \sum_{i=1}^{n} X_{1i}^2 + b_2 \sum_{i=1}^{n} X_{1i}X_{2i} \\ \sum_{i=1}^{n} X_{2i}Y_i &= a\sum_{i=1}^{n} X_{2i} + b_1 \sum_{i=1}^{n} X_{1i}X_{2i} + b_2 \sum_{i=1}^{n} X_{2i}^2. \end{aligned} \tag{12.4}$$

Solving these equations for a, b_1, and b_2, we obtain the following results:

$$b_1 = \frac{\sum_{i=1}^{n} (X_{2i} - \bar{X}_2)^2 \sum_{i=1}^{n} (X_{1i} - \bar{X}_1)(Y_i - \bar{Y}) - \sum_{i=1}^{n} (X_{1i} - \bar{X}_1)(X_{2i} - \bar{X}_2) \sum_{i=1}^{n} (X_{2i} - \bar{X}_2)(Y_i - \bar{Y})}{\sum_{i=1}^{n} (X_{1i} - \bar{X}_1)^2 \sum_{i=1}^{n} (X_{2i} - \bar{X}_2)^2 - \left[\sum_{i=1}^{n} (X_{1i} - \bar{X}_1)(X_{2i} - \bar{X}_2)\right]^2}$$

$$b_2 = \frac{\sum_{i=1}^{n} (X_{1i} - \bar{X}_1)^2 \sum_{i=1}^{n} (X_{2i} - \bar{X}_2)(Y_i - \bar{Y}) - \sum_{i=1}^{n} (X_{1i} - \bar{X}_1)(X_{2i} - \bar{X}_2) \sum_{i=1}^{n} (X_{1i} - \bar{X}_1)(Y_i - \bar{Y})}{\sum_{i=1}^{n} (X_{1i} - \bar{X}_1)^2 \sum_{i=1}^{n} (X_{2i} - \bar{X}_2)^2 - \left[\sum_{i=1}^{n} (X_{1i} - \bar{X}_1)(X_{2i} - \bar{X}_2)\right]^2}$$

$$a = \bar{Y} - b_1\bar{X}_1 - b_2\bar{X}_2. \tag{12.5}$$

To make these computations simpler, note that

$$\sum_{i=1}^{n}(X_{1i}-\bar{X}_1)^2=\sum_{i=1}^{n}X_{1i}^2-\frac{\left(\sum_{i=1}^{n}X_{1i}\right)^2}{n}$$

$$\sum_{i=1}^{n}(X_{2i}-\bar{X}_2)^2=\sum_{i=1}^{n}X_{2i}^2-\frac{\left(\sum_{i=1}^{n}X_{2i}\right)^2}{n}$$

$$\sum_{i=1}^{n}(X_{1i}-\bar{X}_1)(Y_i-\bar{Y})=\sum_{i=1}^{n}X_{1i}Y_i-\frac{\left(\sum_{i=1}^{n}X_{1i}\right)\left(\sum_{i=1}^{n}Y_i\right)}{n}$$

$$\sum_{i=1}^{n}(X_{2i}-\bar{X}_2)(Y_i-\bar{Y})=\sum_{i=1}^{n}X_{2i}Y_i-\frac{\left(\sum_{i=1}^{n}X_{2i}\right)\left(\sum_{i=1}^{n}Y_i\right)}{n}$$

$$\sum_{i=1}^{n}(X_{1i}-\bar{X}_1)(X_{2i}-\bar{X}_2)=\sum_{i=1}^{n}X_{1i}X_{2i}-\frac{\left(\sum_{i=1}^{n}X_{1i}\right)\left(\sum_{i=1}^{n}X_{2i}\right)}{n}.$$

The following example illustrates how least-squares estimates of A, B_1, and B_2 are calculated. Although (as we shall see in subsequent sections) electronic computers are generally used for such calculations, it is worthwhile to work through at least one sample calculation of this sort by hand.

Example 12.1 The hosiery mill's statisticians feel that equation (12.1) is true, and they want to obtain least-squares estimates of A, B_1, and B_2. Data are obtained concerning cost, output, and managerial experience for nine months, the results being shown in Table 12.1. Calculate the least-squares estimates of A, B_1, and B_2. Compare the least-squares estimate of B_1 with the estimate of B that we obtained in the previous chapter, based on the

TABLE 12.1

Cost and Output of a Hosiery Mill, with the Number of Years of Experience of Manager in Charge during the Month (Sample of Nine Months)

Output (tons)	Production cost (thousands of dollars)	Manager's years of experience
2	3	0
1	2	1
8	8	2
5	5	3
6	6	4
4	4	5
7	6	6
9	8	7
8	7	8

same data concerning cost and output. Why is the present estimate different from the latter? Which of the two estimates is likely to be the better?

Solution: Based on the data in Table 12.1,

$$\sum_{i=1}^{9} X_{1i}^2 = 340 \qquad \sum_{i=1}^{9} X_{1i} = 50 \qquad \sum_{i=1}^{9} Y_i^2 = 303$$

$$\sum_{i=1}^{9} X_{2i}^2 = 204 \qquad \sum_{i=1}^{9} X_{2i} = 36 \qquad \sum_{i=1}^{9} Y_i = 49$$

$$\sum_{i=1}^{9} X_{1i}X_{2i} = 245 \qquad \sum_{i=1}^{9} X_{1i}Y_i = 319 \qquad \sum_{i=1}^{9} X_{2i}Y_i = 225.$$

Inserting these figures into the formulas given prior to this example,

$$\sum_{i=1}^{9} (X_{1i} - \bar{X}_1)^2 = 340 - \frac{50^2}{9} = 340 - 277.78 = 62.22$$

$$\sum_{i=1}^{9} (X_{2i} - \bar{X}_2)^2 = 204 - \frac{36^2}{9} = 204 - 144 = 60$$

$$\sum_{i=1}^{9} (X_{1i} - \bar{X}_1)(Y_i - \bar{Y}) = 319 - \frac{(50)(49)}{9} = 319 - 272.22 = 46.78$$

$$\sum_{i=1}^{9} (X_{2i} - \bar{X}_2)(Y_i - \bar{Y}) = 225 - \frac{(36)(49)}{9} = 225 - 196 = 29$$

$$\sum_{i=1}^{9} (X_{1i} - \bar{X}_1)(X_{2i} - \bar{X}_2) = 245 - \frac{(50)(36)}{9} = 245 - 200 = 45.$$

Then, inserting these figures into the equations in (12.5), we obtain

$$b_1 = \frac{(60)(46.78) - (45)(29)}{(62.22)(60) - 45^2} = 0.88$$

$$b_2 = \frac{(62.22)(29) - (45)(46.78)}{(62.22)(60) - 45^2} = -0.18$$

$$a = \frac{49}{9} - .88\left(\frac{50}{9}\right) + .18\left(\frac{36}{9}\right) = 1.28.$$

Consequently, the estimated regression equation is

$$\hat{Y}_i = 1.28 + 0.88X_{1i} - 0.18X_{2i}.$$

The estimated value of B_1 is 0.88, as contrasted with the estimate of B in the previous chapter, which was 0.75. In other words, a one-ton increase in output results in an increase in estimated costs of 0.88 thousands of dollars, as contrasted with .75 thousands of dollars in the previous chapter. The reason these estimates differ is that the present estimate of the effect of output on costs holds constant the manager's years of experience, whereas the earlier estimate did not hold this factor constant. Since this factor affects costs, the earlier estimate is likely to be a biased estimate of the effect of output on cost.[2]

12.5 Confidence Intervals and Tests of Hypotheses Concerning B_1 and B_2

As in the case of simple regression, least-squares estimators have many statistically desirable properties. Specifically, a, b_1, *and* b_2 *are unbiased and consistent.* Moreover, the Gauss-Markov theorem tells us that *of all unbiased estimators that are linear functions of the dependent variables* a, b_1, *and* b_2 *have the smallest standard deviation.* As in the case of simple regression, these desirable properties hold if the observations are independent and if the standard deviation of the conditional probability distribution is the same, regardless of the value of the independent variables. It is *not* necessary that the conditional probability distribution of the dependent variable be normal. Or, stating the same thing differently, it is *not* necessary that the probability distribution of e_i be normal.

If we are willing to assume that e_i is normally distributed, we can calculate a confidence interval for B_1 or B_2. This frequently is an important purpose of a multiple-regression analysis. For example, in the case of the hosiery mill, an important purpose of the analysis may be to obtain a confidence interval for B_1, which is often called the ***true regression coefficient of*** $\mathbf{X_1}$. (The least-squares estimator of B_1—that is, b_1—is often called the *estimated regression coefficient of* X_1). This true regression coefficient is of interest because it measures the effect of a one-unit increase in X_1 (tons of output) on the expected value of Y (cost), when X_2 (the number of years of experience of the mill's manager) is held constant. Another important purpose of the analysis may be to obtain a confidence interval for B_2, which is often called the ***true regression coefficient of*** $\mathbf{X_2}$. (The least-squares estimator of B_2—that is, b_2—is often called the *estimated regression coefficient of* X_2). This true regression coefficient is of interest because it measures the effect of a one-unit increase in X_2 (the number of years of experience of the mill's manager) on the expected value of Y (cost), when X_1 (output) is held constant.[3]

As pointed out in the previous section, multiple regressions are generally carried out by electronic computer, not by hand. Because of the importance of

2. Of course, this regression is only supposed to be appropriate when X_{1i} and X_{2i} vary in a certain limited range. If X_{2i} is large and X_{1i} is small, the regression would predict a negative value of cost, which obviously is inadmissable. But as long as the regression is not used to make predictions for values of X_{1i} and X_{2i} outside the range of the data given in Table 12.1, this is no problem. For simplicity, we assume in equation (12.1) that the effect of the number of years of managerial experience on the expected value of cost (holding output constant) can be regarded as linear in the relevant range. Alternatively, we could have assumed that it was quadratic (or some other nonlinear forms could have been used). In Section 12.13 we discuss how multiple regression can be used to estimate quadratic equations.

3. Sometimes statisticians are also interested in obtaining a confidence interval for A, which is often called the intercept of the regression, and which measures the expected value of Y when both X_1 and X_2 are zero. (See note 4.)

multiple-regression techniques, standard programs have been formulated for calculation by computer of the least-squares estimates of A, B_1, and B_2. Since these estimates are sample statistics, they are obviously subject to sampling error. Besides calculating the values of the least-squares estimators of A, B_1, and B_2 (that is, a, b_1, and b_2), these programs provide an estimate of the standard deviation of b_1 (often called the *standard error of* b_1), and an estimate of the standard deviation of b_2 (often called the *standard error of* b_2).

Given the computer printout, it is relatively simple to construct a confidence interval for B_1 or B_2. As noted above, in any standard computer printout the standard error of b_1 and the standard error of b_2 are shown. If the confidence coefficient is set at $(1 - \alpha)$, *a confidence interval for* B_1 *is*

$$b_1 \pm t_{\alpha/2} s_{b_1}, \tag{12.6}$$

where s_{b_1} is the standard error of b_1 and where t has $n - k - 1$ degrees of freedom (where k is the number of independent variables included in the regression—two in this case). If the confidence coefficient is set at $(1 - \alpha)$, *a confidence interval for* B_2 *is*

$$b_2 \pm t_{\alpha/2} s_{b_2}, \tag{12.7}$$

where s_{b_2} is the standard error of b_2.[4]

Given the computer printout, it is also easy to test the null hypothesis that B_1 or B_2 equals zero, if we assume once again that e_i—in equation (12.2)—is normally distributed. The computer printout shows the *t statistic (or t value) for* b_1, this statistic being defined as $b_1 \div s_{b_1}$. If B_1 equals zero, this t statistic has the t distribution with $(n - k - 1)$ degrees of freedom. Thus, if the alternative hypothesis is two-sided and α is the significance level, the decision rule is as follows.

> **Decision rule:** *Reject the null hypothesis that* B_1 *equals zero if the* t *statistic for* b_1 *exceeds* $t_{\alpha/2}$ *or is less than* $-t_{\alpha/2}$; *otherwise accept the null hypothesis. (The number of degrees of freedom is* $n - k - 1$.*)*

Similarly, if B_2 equals zero, the t statistic (or t value) for b_2—defined as $b_2 \div s_{b_2}$—has the t distribution with $(n - k - 1)$ degrees of freedom; and the decision rule is as follows.

> **Decision rule:** *Reject the null hypothesis that* B_2 *equals zero if the* t *statistic for* b_2 *exceeds* $t_{\alpha/2}$ *or is less than* $-t_{\alpha/2}$; *otherwise accept the null hypothesis. (The number of degrees of freedom is* $n - k - 1$.*)*

The following examples illustrate how, with a computer printout of the results of a multiple regression, one can construct confidence intervals for

4. Some computer printouts also show s_a, the standard error of a. If the confidence coefficient is set at $(1 - \alpha)$, a confidence interval for A is $a \pm t_{\alpha/2} s_a$.

some of the true regression coefficients, and test whether some of these coefficients are zero.[5]

Example 12.2 The computer printout of the multiple regression of cost on (1) output and (2) the number of years of experience of the mill's manager is shown, in part, in Table 12.2. (The basic data were given in Table 12.1.) In this printout, output is designated as variable 2, and the number of years of experience of the mill's manager is designated as variable 3. As you can see, the computer prints out the value of b_1, s_{b_1}, and the t statistic for b_1 in the first row, labeled 2 (for variable 2). In the second row, labeled 3 (for variable 3), it prints out the value of b_2, s_{b_2}, and the t statistic for b_2. (Because of rounding errors, the estimates of a, b_1, and b_2 obtained by the computer will differ slightly from those obtained by hand.) Use these results to calculate a 95 percent confidence interval for B_1; then use these results to test the hypothesis that B_2 equals zero. (The alternative hypothesis is two-sided, and the significance level should be set equal to 0.5.)

TABLE 12.2

Section of Computer Printout, Showing Results of Multiple Regression of Cost on Output and Manager's Years of Experience

```
INDEPENDENT VARIABLE(S)  2  3
DEPENDENT VARIABLE          1

VAR        COEFF        STD. ERROR      T-VALUE
 2        0.8790           0.0347       25.3400
 3       -0.1759           0.0353       -4.9800

INTERCEPT    1.26472
```

Solution: Since the standard error of b_1 equals .0347, a 95 percent confidence interval for B_1 is $0.879 \pm t_{.025}(.0347)$. Because $n - k - 1 = 9 - 2 - 1$, or 6, the t distribution has 6 degrees of freedom, and $t_{.025} = 2.447$. Thus, the confidence interval for B_1 is $0.879 \pm .0849$. In other words, the 95 percent confidence interval for the increase in expected cost due to a one-ton increase in output is .794 to .964 thousands of dollars.

If B_2 were zero, the probability would be .05 that the t statistic for b_2 would be greater than 2.447 or less than -2.447. (This is because $t_{.025} = 2.447$.) Since the t statistic for b_2 equals -4.98, it is less than -2.447. Thus, we must reject the null hypothesis that B_2 equals zero.

Example 12.3 The computer printout in Table 12.3 indicates the results of a multiple regression where the dependent variable is the rate of productivity

5. It is also possible, if the printout shows the t statistic (or t value) for a, to test the null hypothesis that A equals zero. If A equals zero, the t statistic (or t value) for a—defined as $a \div s_a$—has the t distribution with $(n - k - 1)$ degrees of freedom; and the decision rule is: Reject the null hypothesis that A equals zero if the t statistic for a exceeds $t_{\alpha/2}$ or is less than $-t_{\alpha/2}$; otherwise, accept the null hypothesis. (The number of degrees of freedom is $n - k - 1$.)

TABLE 12.3
Section of Computer Printout, Showing Results of Multiple Regression of an Industry's Rate of Productivity Increase(Variable 1) on Its Percent of Union Employees (Variable 2), Percent of Value-Added Spent on Basic Research (Variable 3), and Percent of Value-Added Spent on Applied Research and Development (Variable 4)

```
INDEPENDENT VARIABLE(S)   2  3  4
DEPENDENT VARIABLE        1
```

VAR	COEFF	STD. ERROR	T-VALUE
2	-0.0559	0.0146	-3.8400
3	1.3625	0.4670	2.9200
4	0.0727	0.0276	2.6400
INTERCEPT	4.78826		

increase (in percentage points per year) in an industry during 1948–66. Unlike the case in Example 12.2, there are three, not two, independent variables: the percent of the industry's employees belonging to a union in 1953 (variable 2), the percent of the industry's value-added spent in 1958 on basic research (variable 3), and the percent of the industry's value-added spent in 1958 on applied research and development (variable 4). This multiple regression is based on data for 20 industries. Calculate a 95 percent confidence interval for the true regression coefficient of the percent of value-added spent on basic research. Using a two-tailed test, test the null hypothesis that the true regression coefficient of the percent of value-added spent on applied research and development is zero. (Set the significance level at .05.)

Solution: Since the estimated regression coefficient of the percent of value-added spent on basic research is 1.3625, and the standard error of this estimated regression coefficient is .4670, the 95 percent confidence interval is $1.3625 \pm 2.120(.4670)$, since $t_{.025} = 2.120$ when there are $20 - 3 - 1$, or 16 degrees of freedom. In other words, the confidence interval is $1.3625 \pm .990$, or 0.3725 to 2.3525. To interpret this result, suppose that the amount spent by an industry on basic research (as a percent of its value-added) increases by 1 percentage point. Based on the regression, it appears that such an increase is associated, on the average, with an increase in the rate of productivity increase of from 0.3725 to 2.3525 percentage points, according to the 95 percent confidence interval.[6]

The t value of the estimated regression coefficient of the percent of value-added spent on applied research and development is 2.64, which exceeds 2.120, the value of $t_{.025}$ when there are 16 degrees of freedom. Thus,

6. These numerical results, like many others in this text, are merely illustrative. That is, they are used to explain the nature of a statistical technique and should not be interpreted as an accurate or complete representation of the relevant economic phenomena. Indeed, some of the data used in other examples of this sort are hypothetical, as is evident from our treatment of them.

we should reject the null hypothesis that the true regression coefficient of the percent of value-added spent on applied research and development is zero.

12.6 Multiple Coefficient of Determination

In the previous chapter we described how the coefficient of determination can be used to measure how well a simple regression equation fits the data. When a multiple regression is calculated, the multiple coefficient of determination, rather than the simple coefficient of determination discussed in the previous chapter, is used for this purpose. The multiple coefficient of determination is defined as

$$R^2 = 1 - \frac{\sum_{i=1}^{n}(Y_i - \hat{Y}_i)^2}{\sum_{i=1}^{n}(Y_i - \bar{Y})^2},$$

where $\hat{Y}_i$ is the value of the dependent variable that is predicted from the regression equation.[7] Thus, as in the case of the simple coefficient of determination covered in the previous chapter,

$$R^2 = \frac{\text{Variation explained by regression}}{\text{Total variation}}, \tag{12.9}$$

which means that R^2 *measures the proportion of the total variation in the dependent variable that is explained by the regression equation.* The positive square root of the multiple coefficient of determination is called the ***multiple correlation coefficient*** and is denoted by R. It, too, is sometimes used to measure how well a multiple-regression equation fits the data.

If there are only two independent variables in a multiple regression, as in equation (12.1), a relatively simple way to compute the multiple coefficient of determination is as follows:

$$R^2 = \frac{b_1 \sum_{i=1}^{n}(X_{1i} - \bar{X}_1)(Y_i - \bar{Y}) + b_2 \sum_{i=1}^{n}(X_{2i} - \bar{X}_2)(Y_i - \bar{Y})}{\sum_{i=1}^{n} Y_i^2 - \frac{\left(\sum_{i=1}^{n} Y_i\right)^2}{n}}. \tag{12.10}$$

If there are more than two independent variables, a multiple regression is

7. This is the *unadjusted* multiple coefficient of determination, which is biased away from zero. An unbiased estimate of the population multiple coefficient of determination is the *adjusted* multiple coefficient of determination, which is

$$\bar{R}^2 = 1 - \frac{\sum_{i=1}^{n}(Y_i - \hat{Y}_i)^2 \div (n - k - 1)}{\sum_{i=1}^{n}(Y_i - \bar{Y})^2 \div (n - 1)},$$

where n is the number of observations and k is the number of independent variables. (By *adjusted*, we mean adjusted for degrees of freedom.) See Section 12.4 for a more detailed definition of $\hat{Y}_i$.

almost always carried out on an electronic computer, which is programmed to print out the value of the multiple coefficient of determination (or of the multiple correlation coefficient).

The following are examples of the calculation and interpretation of the multiple coefficient of determination.

Example 12.4 Use the data in Table 12.1 to calculate the multiple coefficient of determination between the hosiery mill's costs, on the one hand, and its output and the number of years of experience of the mill's manager, on the other. Interpret your result.

Solution: We know from Example 12.1 that $b_1 = 0.88$, $b_2 = -0.18$,

$$\sum_{i=1}^{9} (X_{1i} - \bar{X}_1)(Y_i - \bar{Y}) = 46.78, \quad \sum_{i=1}^{9} (X_{2i} - \bar{X}_2)(Y_i - \bar{Y}) = 29, \text{ and}$$

$$\sum_{i=1}^{9} Y_i^2 - \frac{\left(\sum_{i=1}^{9} Y_i\right)^2}{9} = 303 - \frac{(49)^2}{9} = 303 - 266.78 = 36.22.$$

Thus it follows from equation (12.10) that

$$R^2 = \frac{.88(46.78) - .18(29)}{36.22} = .99.$$

This means that 99 percent of the variation in the firm's monthly costs during the period covered by the data can be explained by the multiple regression equation derived in Example 12.1.

Example 12.5 Table 12.4 shows another part of the computer printout of the results of the multiple regression of an industry's rate of productivity increase on the three independent variables described in Example 12.3. Interpret the figure labeled Multiple R-Squared in this printout.

TABLE 12.4

Another Section of the Computer Printout, Showing Results of Multiple Regression Concerning Industries' Rates of Productivity Increase

```
S.E. OF ESTIMATE          0.62555
F-VALUE                   8.66
MULTIPLE R-SQUARED         .619
```

Solution: This figure—.619—is the multiple coefficient of determination. It means that the regression equation shown in the printout explains 61.9 percent of the variation among these industries in the rate of productivity increase.

12.7 Geometrical Interpretation of Results of Multiple Regression and Correlation

As we have seen, the estimated multiple regression equation shows the average relationship between the dependent variable and the independent variables. If there are only two independent variables, it is possible to represent this average relationship by a plane rather than by a line (which was used for this purpose in the previous chapter). Figure 12.1 shows the plane corresponding to the regression equation relating the hosiery mill's costs to its output rate and the number of years of experience of the mill's manager:

$$\hat{Y}_i = 1.28 + 0.88X_{1i} - 0.18X_{2i}.$$

When both X_{1i} and X_{2i} are zero, Figure 12.1 shows that the predicted value of cost is 1.28 thousands of dollars, which is the intercept of the regression. Figure 12.1 shows that when X_{1i} is held constant, if X_{2i} increases by one unit (one year in this case), the average value of the dependent variable decreases by an amount equal to the regression coefficient of X_{2i}, which is .18 thousands of dollars (or \$180) in this case. Figure 12.1 also shows that when X_{2i} is held constant, if X_{1i} is increased by one unit (one ton in this case), the average value of the dependent variable increases by an amount equal to the regression coefficient of X_{1i}, which is .88 thousands of dollars (or \$880).

Just as each observation in a simple regression can be represented as a point in a scatter diagram, each observation in a multiple regression (with two independent variables) can be represented as a point in three-dimensional

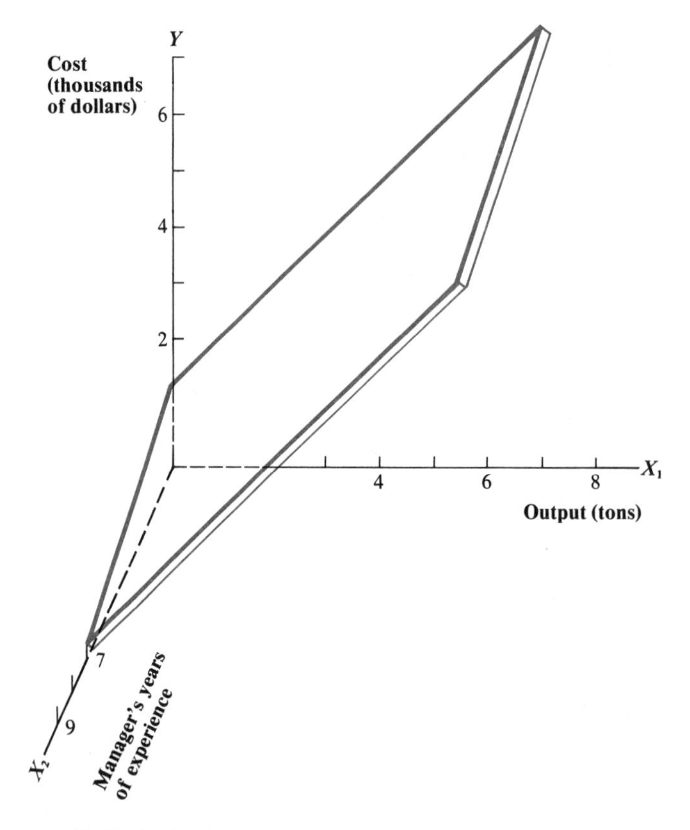

FIGURE 12.1 Plane Corresponding to Regression Equation Relating Cost to Output and Manager's Years of Experience

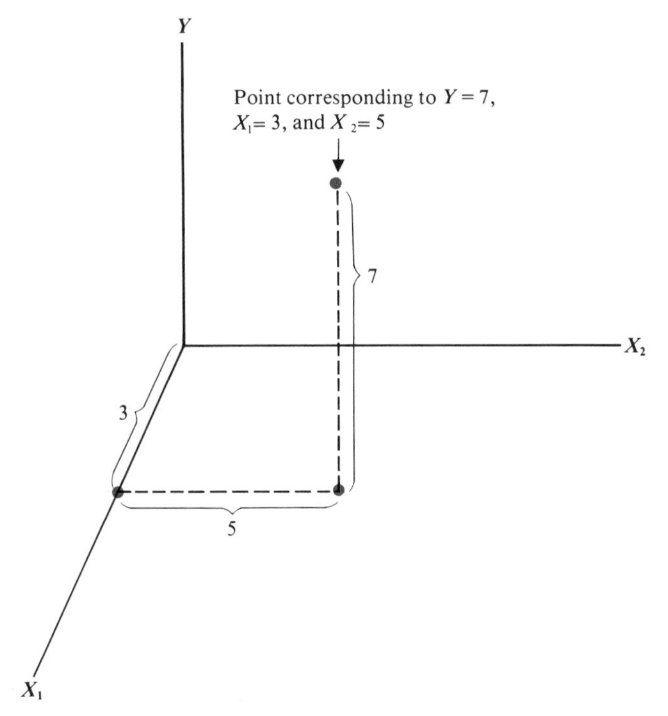

FIGURE 12.2 Geometrical Representation of an Observation (Y Is the Value of the Dependent Variable; X_1 and X_2 Are the Values of the Independent Variables)

space. Figure 12.2 shows how we can represent an observation where the dependent variable equals 7 and the independent variables equal 3 and 5. *The multiple coefficient of determination is a measure of how well the plane representing the regression equation fits the points representing the individual observations.* For example, panel A of Figure 12.3 shows a case where the regression plane does not fit the points at all well; on the other hand, panel B of Figure 12.3 shows a case where the regression plane fits the points much more closely than in panel A. Clearly, the multiple coefficient of determination (as well as the multiple correlation coefficient) is higher in panel B than in panel A.

12.8 Analysis of Variance[8]

The analysis of variance, which we discussed at length in Chapter 10, is used to test the overall statistical significance of a regression equation. That is, it is used to test whether *all* the true regression coefficients in the equation equal zero. In the case of the hosiery mill, we might want to test whether both B_1 and B_2 are zero in order to see whether there is any relationship between the

8. To understand this section, it is not essential that the reader be familiar with Chapter 10. However, it is essential that Section 10.4 (on the F distribution) be read prior to studying this section.

A: R^2 relatively low

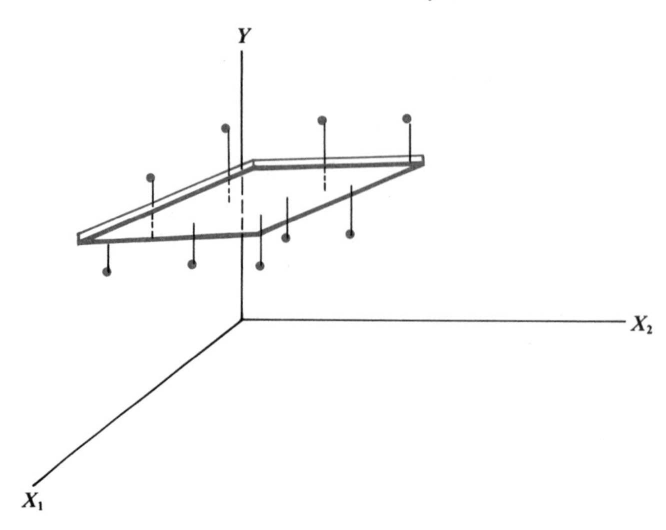

B: R^2 relatively high

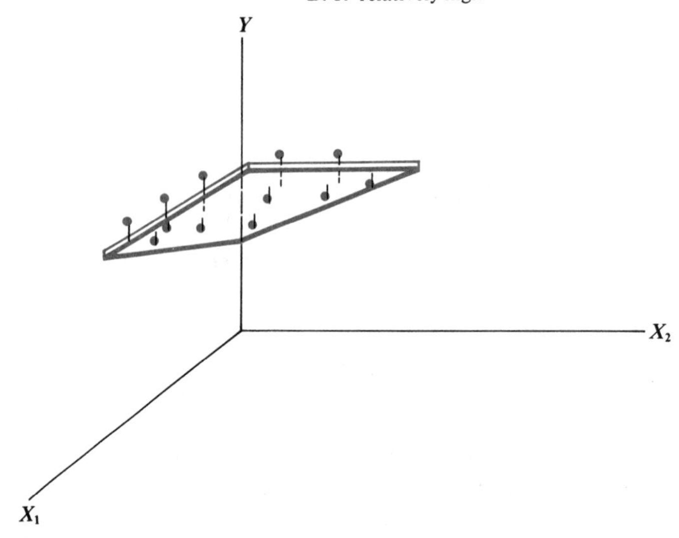

FIGURE 12.3 Closeness of Fit of Regression Plane: Two Cases

dependent variable and all the independent variables taken together. The analysis of variance can be used in this way in simple as well as multiple regressions. In simple regressions, the result is precisely the same as in the test of $B = 0$, described in Section 11.13.

In our discussion of the analysis of variance in Chapter 10, we defined the total variation in Y as the sum of the squared deviations of the values of Y from the mean of Y. Thus, the total variation in Y equals

$$\sum_{i=1}^{n} (Y_i - \bar{Y})^2.$$

As pointed out in Section 11.10,

$$\sum_{i=1}^{n}(Y_i - \bar{Y})^2 = \sum_{i=1}^{n}(\hat{Y}_i - \bar{Y})^2 + \sum_{i=1}^{n}(Y_i - \hat{Y}_i)^2.$$

The first term on the right is the variation explained by the regression, and the second term on the right is the variation unexplained by the regression (also called the *error sum of squares* or the *residual sum of squares*). Thus

Total variation = Explained variation + Unexplained variation.

To carry out the analysis of variance, we construct a table of the general form of Table 12.5. The first column shows the source or type of variation, and the second column shows the corresponding sum of squares (that is, the corresponding variation). The third column shows the number of degrees of freedom corresponding to each sum of squares. For the explained variation, the number of degrees of freedom equals the number of independent variables k. For the unexplained variation, the number of degrees of freedom equals $(n - k - 1)$. The fourth column shows the mean square, which is the sum of squares divided by the number of degrees of freedom. The fifth column shows the ratio of the two mean squares.

TABLE 12.5
Analysis of Variance for Regression

Source of variation	Sum of squares	Degrees of freedom	Mean square	F ratio
Explained by regression	$\sum_{i=1}^{n}(\hat{Y}_i - \bar{Y})^2$	k	$\dfrac{\sum_{i=1}^{n}(\hat{Y}_i - \bar{Y})^2}{k}$	$\dfrac{\sum_{i=1}^{n}(\hat{Y}_i - \bar{Y}^2)/k}{\sum_{i=1}^{n}(Y_i - \hat{Y}_i)^2/(n-k-1)}$
Unexplained by regression	$\sum_{i=1}^{n}(Y_i - \hat{Y}_i)^2$	$n - k - 1$	$\dfrac{\sum_{i=1}^{n}(Y_i - \hat{Y}_i)^2}{(n - k - 1)}$	
Total	$\sum_{i=1}^{n}(Y_i - \bar{Y})^2$	$n - 1$		

If the null hypothesis is true (that is, if all the true regression coefficients are zero), this ratio has an F distribution with k and $(n - k - 1)$ degrees of freedom. Thus, to carry out the appropriate analysis of variance, the decision rule is as follows.

Decision Rule: *Reject the null hypothesis that the true regression coefficients are all zero if the ratio of the explained mean square to the unexplained mean square exceeds* F_α, *where* α *is the desired significance level. Otherwise, accept the null hypothesis.*

Because multiple regressions are generally carried out on an electronic computer, it is seldom necessary to calculate the numbers in Table 12.5 by hand. The computer printout generally gives the analysis-of-variance table. Some printouts refer to the explained mean square as the *regression mean square* and to the unexplained mean square as the *error mean square* or the *residual mean square*. The ratio of the explained to the unexplained mean square is often referred to as the F value or the F ratio. The way in which the analysis-of-variance table is displayed on the computer printout may vary from that in Table 12.5, but the information imparted is the same. In Section 12.14 we will examine the printout of one computer program in some detail.

The following is an example of the application of the analysis of variance.

Example 12.6 Table 12.4 shows the F ratio for the multiple regression of an industry's rate of productivity increase on the three independent variables described in Example 12.3. Use this F ratio to test the hypothesis that all the true regression coefficients in this multiple regression are zero. (Set $\alpha = .05$.)

Solution: Table 12.4 shows that the F ratio equals 8.66. Since $n = 20$ and $k = 3$, the number of degrees of freedom are 3 and 16. (Why? Because $k = 3$ and $(n - k - 1) = 16$.) Thus, $F_{.05}$ equals 3.24. (See Appendix Table 9.) Since the F ratio exceeds 3.24, we should reject the null hypothesis that all the true regression coefficients in this multiple regression equal zero.

EXERCISES

12.1 The Rotunda Corporation believes that its annual sales depend on disposable income in its city of location and on the price of its product. Data for the past 15 years concerning these variables are given below:

Year	Number of units sold annually (millions)	Disposable income (billions of dollars)	Price of product (dollars per unit)
1980	8	4	9
1979	8	4	8
1978	5	3	8
1977	4	3	9
1976	6	3	7
1975	4	3	10
1974	2	2	8
1973	3	2	6
1972	4	2	5
1971	2	2	7
1970	2	2	8
1969	1	1	6
1968	1	1	5
1967	1	1	7
1966	1	1	5

Calculate the multiple regression equation, if Rotunda's sales are the dependent variable and disposable income and price of product are the independent variables.

12.2 Based on your results in the previous exercise, predict the volume of annual sales for the Rotunda Corporation under each of the following sets of circumstances:
(a) disposable income equals $3 billion and price equals $7;
(b) disposable income equals $1 billion and price equals $6;
(c) disposable income equals $2 billion and price equals $10.

12.3 (a) Based on your results in Exercise 12.1, what is the estimated effect on the expected value of Rotunda's annual sales of an increase of $1 billion in disposable income? (b) What is the estimated effect of an increase of $1 in price?

12.4 The president of the Rotunda Corporation uses the multiple regression equation you derived in Exercise 12.1 to predict the firm's annual sales for next year. Since he believes that disposable income will be $6 billion and product price will be $11, he bases his prediction on these values of the independent variables. What objections might legitimately be raised against these procedures?

12.5 (a) Use the data in Exercise 12.1 to calculate the multiple coefficient of determination between Rotunda's sales, on the one hand, and disposable income and price, on the other. (b) Interpret your results. Does it appear that the regression equation you derived in Exercise 12.1 fits the data well? Why, or why not?

12.6 Between 1966 and 1980, the population of the city in which the Rotunda Corporation is located grew steadily. By 1980, it was about four times its size in 1966. The marketing director of the Rotunda Corporation objects to the equation that you derived in Exercise 12.1 on the grounds that population, not disposable income, has been the variable that has influenced Rotunda's sales.
(a) Is the marketing director correct in saying that if population and disposable income are perfectly correlated, the observed effect of disposable income in your equation may be due to population?
(b) Can you suggest some ways of altering the multiple regression so that the marketing director's hypothesis can be tested?

12.7 An economist asks a research assistant to calculate a multiple regression on an electronic computer. The dependent variable is a particular industry's rate of productivity increase during 1948–66, and the independent variables are (1) the percent of the industry's employees belonging to unions in 1953 (variable 2); (2) the percent of the industry's value-added spent on applied research and development in 1958 (variable 4); and (3) the percentage of the firms in the industry reporting in 1958 that they expected their R and D (research and development) expenditures to pay out in no less than six years (variable 8). Only 17 industries can be included because data are lacking concerning the third independent variable for 3 of the industries. The research assistant comes back with the printout, which is as follows:

```
INDEPENDENT VARIABLE(S)   2   4   8
DEPENDENT VARIABLE        1

VAR          COEFF         STD. ERROR      T-VALUE
 2         -0.0622           0.0090       -6.9400
 4          0.0963           0.0154        6.2700
 8          0.0704           0.0098        7.2000

INTERCEPT   4.65948
```

(a) What is the estimated regression equation?
(b) Calculate a 95 percent confidence interval for the true regression coefficient of the percent of the industry's value-added spent on applied research and development.
(c) Calculate a 90 percent confidence interval for the true regression coefficient of the percent of the industry's employees belonging to unions.

12.8 The economist in the previous exercise wants to test the hypothesis that whether an industry's R and D consists mainly of long-term or short-term projects has no effect on its rate of productivity increase. If the significance level is set at .05, and if the alternative hypothesis is two-sided, use the printout in the previous exercise to test this hypothesis.

12.9 Another part of the computer printout in Exercise 12.7 is shown below. (a) What is the multiple coefficient of determination in this case? What does this result mean? (b) Test the null hypothesis that all the true regression coefficients in the multiple regression are zero. (Set $\alpha = .05$.)

```
S.E. OF ESTIMATE          0.3458
F-VALUE                   29.01
MULTIPLE R-SQUARED        .87
```

12.10 An economist believes that there is a quadratic relationship between the interest rate in a particular year and the amount spent that year by a particular industry on plant and equipment. In other words, the economist believes that

$$E(S_t) = A + B_1 i_t + B_2 i_t^2,$$

where S_t is the amount spent on plant and equipment by this industry during the tth year and i_t is the interest rate during the tth year. Can the economist use multiple regression techniques to estimate A, B_1, and B_2? If so, what assumptions must be made?

12.9 Dummy-Variable Techniques

Multiple regression can be used to analyze the effects of *qualitative* variables (that is, variables that do not assume numerical values) as well as ***quantitative*** variables. For example, suppose that an economist wants to estimate the effect on a family's saving rate of two variables: (1) the family's annual income, and (2) whether the family owns its home or rents. The second independent variable is a qualitative, not a quantitative, variable; yet it can be included in a multiple regression as a so-called dummy variable, defined below.

Dummy variable: *A dummy variable is a variable that equals zero or 1.*

In the present case, the economist can construct a dummy variable H_i, which equals 1 if the family owns its own home and zero if the ith family rents. Then, if the relationship between the dependent and independent variables is linear, it may be assumed that

$$S_i = A + B_1 I_i + B_2 H_i + e_i, \tag{12.11}$$

where S_i is the annual amount (in thousands of dollars) saved by the ith

TABLE 12.6

Annual Savings and Annual Income of 6 Home-Owning and 14 Home-Renting Families

Name	Annual savings (thousands of dollars)	Annual income (thousands of dollars)	Owns/rents	Value of H_i
Jones	1.0	20	Rents	0
Smith	1.3	24	Rents	0
Kargill	0.7	12	Rents	0
Mennon	0.8	16	Rents	0
Billings	0.5	11	Rents	0
Stratahan	2.4	32	Owns	1
Cohen	0.3	10	Rents	0
Lamb	3.2	40	Owns	1
Schmidt	2.8	32	Owns	1
Palucci	0.0	7	Rents	0
Chichester	0.3	9	Rents	0
LaRue	0.0	6	Rents	0
Liu	1.0	18	Rents	0
Armour	2.0	20	Owns	1
Christenson	0.4	12	Rents	0
Howe	0.7	14	Rents	0
Pitt	1.5	15	Owns	1
Drummond	1.6	16	Owns	1
Tracy	0.6	15	Rents	0
Holland	0.6	14	Rents	0

family, I_i is its annual income (in thousands of dollars), and e_i is the difference between S_i and $E(S_i)$. Using the procedures described in Section 12.4, least-squares estimates of A, B_1, and B_2 can be obtained in the ordinary way.

To understand more clearly what equation (12.11) means and what the use of dummy variables really amounts to, suppose that the economist obtains data from 20 families (6 of whom own their homes and 14 of whom rent), and that these data are as shown in Table 12.6. If we plot the amount that each family saves yearly against its annual income, we obtain the results shown in Figure 12.4. Clearly, the relationship between the amount saved yearly and annual income seems different for the home owners than for the renters: Specifically, the home owners seem to save more at each level of income than do the renters. In other words, there seem to be two regression lines, one for families who own their homes, and one for those who rent. These two regression lines have the same slope but different intercepts. Suppose that two regression lines (with the same slope but different intercepts) exist in the population, as well as in this sample. What equation (12.11) does is to compress these two regression lines into a single equation.

To see that equation (12.11) does this, suppose that the model representing the amounts saved by renting families is

$$S_i = A + B_1 I_i + e_i,$$

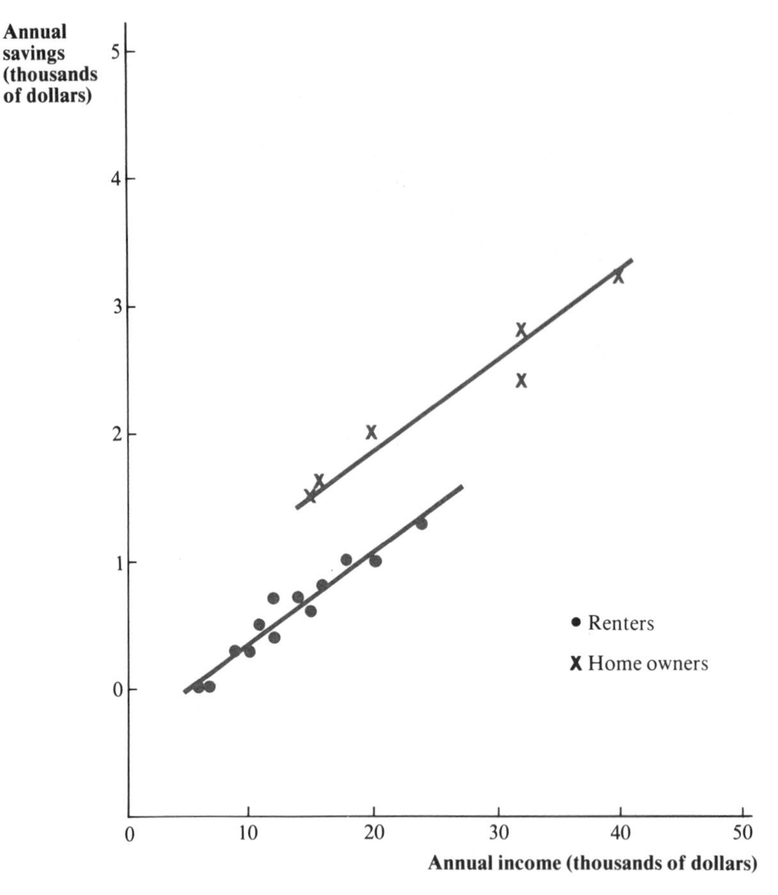

FIGURE 12.4 Relationship between Amount Saved and Income, Renters and Home Owners

and suppose that the model representing the amounts saved by home-owning families is

$$S_i = (A + B_2) + B_1 I_i + e_i,$$

where B_2 is the *extra* amount (in thousands of dollars) that home owners save, on the average. Then it follows that equation (12.11) is valid both for families who own their homes and for those who rent. (Why? Because $B_2 H_i$ in equation (12.11) equals zero for renting families, while it equals B_2 for home-owning families.)

If equation (12.11) is accepted as a reasonable model, the least-squares estimates of A, B_1, and B_2 can be calculated in the usual way, as shown in Table 12.7. The resulting regression equation, shown by the two parallel lines in Figure 12.4, is

$$\hat{S}_i = -.3207 + .0675 I_i + .827 H_i, \tag{12.12}$$

where $\hat{S}_i$ is the amount of the ith family's savings (in thousands of dollars) predicted by the regression equation. According to this equation, a $1,000 increase in income results in an increase in estimated savings of .0675 thousands of dollars, or $67.50. Holding income constant, a home-owning family is estimated to save .827 thousands of dollars (or $827) more than a renting family.

Note that if H_i were omitted from this equation and if the simple regression of S_i on I_i alone were used to estimate B_1, the result would be a biased estimate of B_1. As is evident from Figure 12.4, the simple regression of S_i on I_i would overestimate B_1 because families who rent (1) tend to have lower incomes than home owners, and (2) tend to save less than home owners even when their incomes equal those of home owners. This illustrates the fact

TABLE 12.7

Calculation of Least-Squares Estimates of A, B_1, and B_2, Based on Data in Table 12.6

$$\sum_{i=1}^{20}(H_i - \bar{H})^2 = \sum_{i=1}^{20} H_i^2 - \frac{\left(\sum_{i=1}^{20} H_i\right)^2}{20} = 6 - 1.8 = 4.2$$

$$\sum_{i=1}^{20}(I_i - \bar{I})^2 = \sum_{i=1}^{20} I_i^2 - \frac{\left(\sum_{i=1}^{20} I_i\right)^2}{20} = 7{,}377 - 5{,}882.4 = 1{,}494.6$$

$$\sum_{i=1}^{20}(S_i - \bar{S})^2 = \sum_{i=1}^{20} S_i^2 - \frac{\left(\sum_{i=1}^{20} S_i\right)^2}{20} = 39.27 - 23.544 = 15.726$$

$$\sum_{i=1}^{20}(I_i - \bar{I})(S_i - \bar{S}) = \sum_{i=1}^{20} I_i S_i - \frac{\left(\sum_{i=1}^{20} I_i\right)\left(\sum_{i=1}^{20} S_i\right)}{20} = 516.1 - 372.15 = 143.95$$

$$\sum_{i=1}^{20}(H_i - \bar{H})(S_i - \bar{S}) = \sum_{i=1}^{20} H_i S_i - \frac{\left(\sum_{i=1}^{20} H_i\right)\left(\sum_{i=1}^{20} S_i\right)}{20} = 13.5 - 6.51 = 6.99$$

$$\sum_{i=1}^{20}(I_i - \bar{I})(H_i - \bar{H}) = \sum_{i=1}^{20} I_i H_i - \frac{\left(\sum_{i=1}^{20} I_i\right)\left(\sum_{i=1}^{20} H_i\right)}{20} = 155 - 102.9 = 52.1$$

$$b_1 = \frac{(4.2)(143.95) - (52.1)(6.99)}{(1{,}494.6)(4.2) - (52.1)^2} = \frac{604.590 - 364.179}{6{,}277.32 - 2{,}714.41} = .0675$$

$$b_2 = \frac{(1{,}494.6)(6.99) - (52.1)(143.95)}{(1{,}494.6)(4.2) - (52.1)^2} = \frac{10{,}447.254 - 7{,}499.795}{6{,}277.32 - 2{,}714.41} = .827$$

$$a = \left(\frac{21.7}{20}\right) - .0675\left(\frac{343}{20}\right) - .827\left(\frac{6}{20}\right) = -.3207$$

(In deriving a, note that $\sum_{i=1}^{20} S_i = 21.7$, $\sum_{i=1}^{20} I_i = 343$, and $\sum_{i=1}^{20} H_i = 6$.)

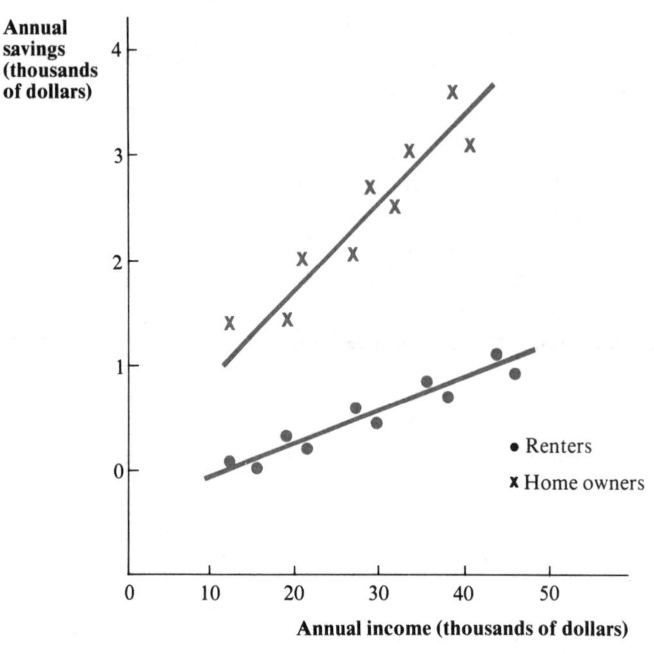

FIGURE 12.5 Case where Slope of Regression Line Depends on Value of Dummy Variable

(stressed at the beginning of this chapter) that if a dependent variable is influenced by more than one independent variable, a simple regression of the dependent variable on one of the independent variables is likely to result in a biased estimate of the effect of this independent variable on the dependent variable.

When this dummy variable technique is used, it is assumed that the values of the other regression coefficients in the regression equation are not affected by the value of the dummy variable. For example, in equation (12.11) it is assumed that the value of B_1 is the same regardless of whether H_i equals zero or 1. In other words, it is assumed that the slope of the relationship between savings and income is the same among families who rent as among those who own their homes. This assumption may or may not be true. (Figure 12.5 shows a case where it is not true.) If this is not true, separate regression equations should be estimated for families who rent and for those who own their homes.[9]

12.10 Multicollinearity

Regression analysis, like any tool, should not be applied blindly. It is important to check whether the assumptions underlying regression analysis are at least approximately correct, and to be aware of the problems that regression

9. Alternatively, other more advanced techniques can be used in a case of this sort.

analysis can encounter. One important problem that can arise in multiple regression studies is multicollinearity, which is defined below.

> ***Multicollinearity:*** *Multicollinearity is a situation in which two or more of the independent variables are very highly correlated.*

In the case of the hosiery mill, suppose that the relationship in the past between output and the number of years of experience of the mill's manager has been as shown in Figure 12.6. If so, there has been a perfect linear relationship between the two independent variables in equation (12.1). In a case of this sort, it is impossible to estimate the regression coefficients of both independent variables (X_1 and X_2) because the data provide no information concerning the effect of one independent variable, holding the other independent variable constant. All that can be observed is the effect of both independent variables together, given that they both move together in the way they have in the past.

To understand why it is impossible to estimate the regression coefficients of both independent variables in a case of this sort, let's look at the case of the hosiery mill in the situation shown in Figure 12.6. Regression analysis estimates the effect of each independent variable by seeing how much effect this one independent variable has on the dependent variable when other independent variables are held constant. However, in the situation in Figure 12.6 it is impossible for such an analysis to be carried out because one cannot separate the effects of the output rate on cost from the effects of the years of experience of the mill's manager. Whenever the output rate increased, so did the manager's number of years of experience. Given that the two independent variables move together in this rigid, lockstep fashion, there is no way to tell

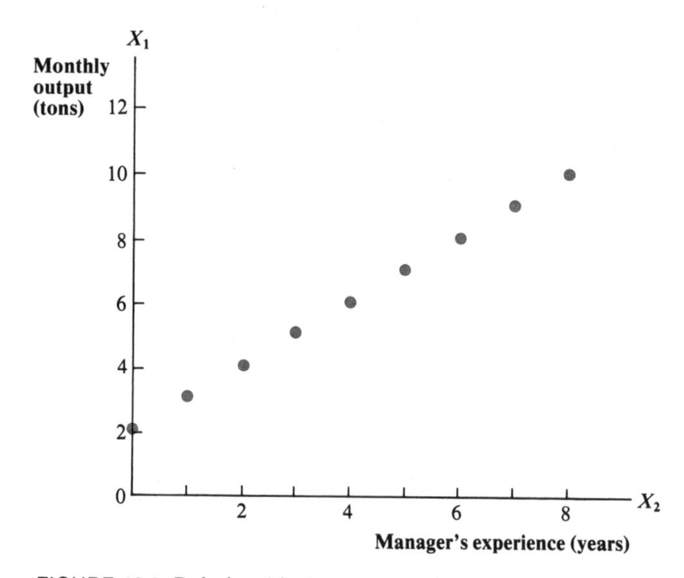

FIGURE 12.6 Relationship between Independent Variables in Equation (12.1), Case of Complete Multicollinearity

how much effect each has separately; all that we can observe is the effect of both combined. Stated in terms of our earlier discussion of experimental design, these effects are hopelessly confounded.

Another way of characterizing multicollinearity is by portraying it geometrically. Figure 12.7 shows the relationship between cost on the one hand and output and amount of managerial experience on the other, given the situation in Figure 12.6. Because the two independent variables are perfectly correlated (that is, the correlation coefficient equals 1), there is no single plane that minimizes the sum of squared deviations; instead, there are any number of planes that do equally well in this regard. Thus, the method of least squares does not result in estimates of the effect of each independent variable and can only be used to estimate their combined effects.

If there is good reason to believe that the independent variables will continue to move in lockstep in the future as they have in the past,

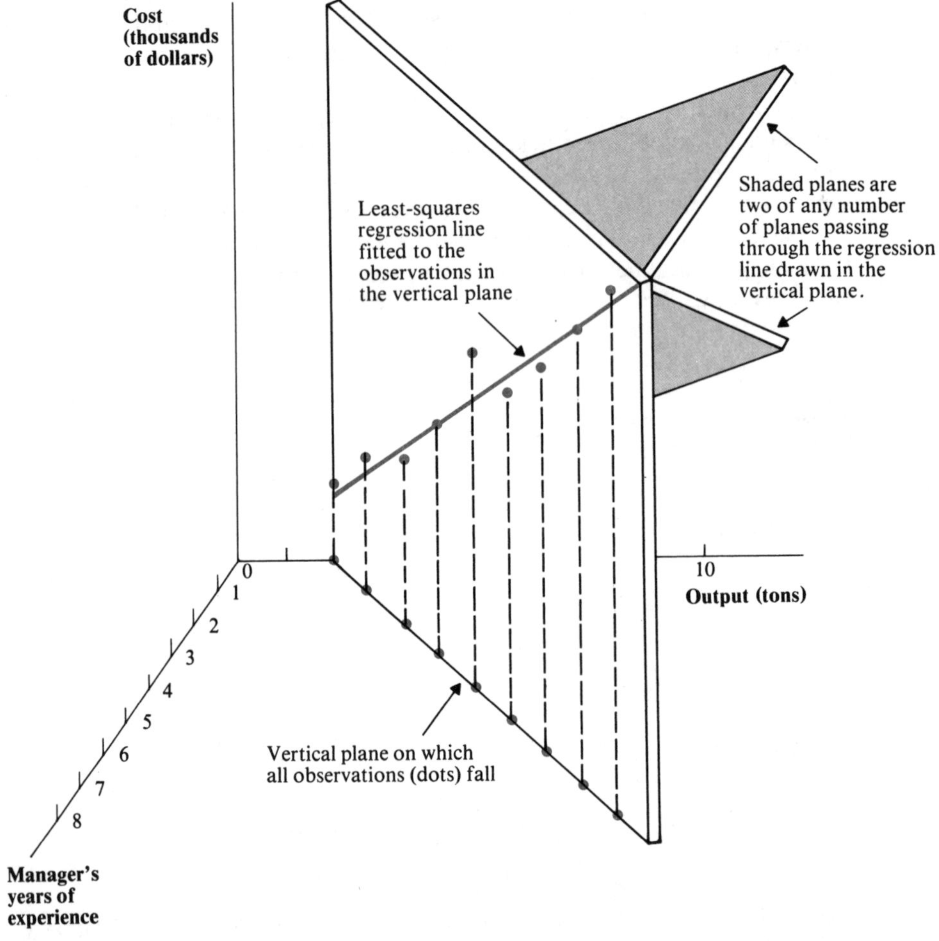

FIGURE 12.7 Geometrical Representation of Multicollinearity, Based on Situation in Figure 12.6

multicollinearity does not prevent us from using regression analysis to predict the dependent variable. Since the two independent variables are perfectly correlated, one of them in effect stands for both; and we therefore need use only one in the regression analysis. However, if the independent variables cannot be counted on to continue to move in lockstep this procedure is dangerous, since it takes no account of the variation in the excluded independent variable.

In fact, one seldom encounters cases where independent variables are *perfectly* correlated, as they are in Figure 12.6. *But one often encounters cases where independent variables are so highly correlated that, although it is possible to estimate the regression coefficient of each variable, these regression coefficients cannot be estimated at all accurately.* How can one tell whether multicollinearity is so strong that it is responsible for the large standard errors of certain estimated regression coefficients? This is done by estimating the correlation coefficients among the independent variables. If some of these correlation coefficients are close to 1 (or −1), multicollinearity is likely to be a problem.[10]

In cases where multicollinearity exists, it sometimes is possible to alter the independent variables in such a way as to reduce it. For example, suppose that a statistician wants to estimate a regression equation where the quantity demanded per year of a certain good is the dependent variable and the average price of this good and disposable income of American consumers are the independent variables. If disposable income is measured in money terms (that is, without adjustment for changes in the price level), there may be a high correlation between the independent variables. But if disposable income is measured in real terms (that is, with adjustment for changes in the price level), this correlation may be reduced considerably. Thus, this may be a good reason to measure disposable income in real rather than money terms.

If multicollinearity cannot be avoided in this way, there may be no alternative but to acquire new data which do not contain the high correlation among the independent variables. Just as bricks cannot be made without straw, there may be no way to estimate accurately the regression coefficient of a particular independent variable that is very highly correlated with some other independent variable.

12.11 Serial Correlation and the Durbin-Watson Test

SERIAL CORRELATION

Besides multicollinearity, another important problem that can arise in regression analysis is that the error terms (that is, the values of e_i) are not

10. Of course, if there are only two independent variables, there is only one simple correlation coefficient between the independent variables. But as the number of independent variables gets larger, the number of such correlation coefficients increases also. Besides looking at the simple correlation coefficients among pairs of independent variables, it is advisable to look at the multiple correlation coefficients of each of the independent variables on all of the others. See J. Johnston, *Econometric Methods*, 2d ed. (New York: McGraw-Hill, 1972), p. 163.

independent. This, of course, is a violation of one of the assumptions underlying regression analysis. (Recall Section 12.3.) Many regressions (both simple and multiple) in business and economics are based on time series; that is, the data consist of observations pertaining to various periods of time. For example, the data in Exercise 12.1 are time series because they pertain to 15 years of sales by the Rotunda Corporation. When regressions are based on time series, the error terms frequently are serially correlated.

> ***Serial correlation:*** *If the value of a time series at time* t *is correlated with its value* h *periods before, the time series exhibits serial correlation (also known as autocorrelation). If no such correlation exists, the time series is said to be serially independent.*

To illustrate serial correlation, take the case of the Dow-Jones average of stock prices. If this average is relatively high (by historical standards) today, it is likely to be relatively high tomorrow as well. Not all economic time series are serially correlated. For example, in Chapter 8 the time series of the mean height of metal pieces (part of which is shown in Figure 8.13) shows no serial correlation.

As we pointed out above, the error terms in equation (12.2) may be serially correlated. Since this may be true for either simple or multiple regressions, let's assume that there is only one independent variable and that

$$Y_i = A + BX_i + e_i,$$

where Y_i is the ith value of the dependent variable, X_i is the ith value of the independent variable, and e_i is the difference between Y_i and the expected value of Y_i. Suppose that the error terms—that is, the e_i—are serially correlated. In particular, suppose that each error term is positively correlated with the subsequent error term. Thus, if Y_1 tends to be above its expected value (which means that $e_1 > 0$), Y_2 tends also to be above its expected value (which means that $e_2 > 0$).[11] If serial correlation of this sort is present, the least-squares method may still result in unbiased estimates of A and B. *The principal problem resulting from such serial correlation is that the standard errors of the estimators of* A *and* B *are really larger than would be estimated from the formulas that assume no serial correlation.* For example, the customary formula for the standard error of the estimator of B is

$$s_b = \frac{s_e}{\sqrt{\sum_{i=1}^{n}(X_i - \bar{X})^2}} = \frac{s_e}{\sqrt{\sum_{i=1}^{n} X_i^2 - n\bar{X}^2}},$$

where s_e is the standard error of estimate. This formula assumes that the e_i (the error terms) are independent. But if serial correlation of this sort is present, this formula for the standard error of b is an underestimate of the true standard error.

11. This is a case of *positive* serial correlation. (It is the sort of situation frequently encountered in economics.) If Y_2 tends to be *below* its expected value when Y_1 is *above* its expected value, this is a case of *negative* serial correlation. More is said about this below.

Another effect of serial correlation is that *it may make it appear that cycles exist in the data.* For example, if an observation is above the mean in one period, the next period's observation may also tend to be above the mean; and if an observation is below the mean in one period, the next period's observation may also tend to be below the mean. The result gives the appearance of a cycle but is actually due to serial correlation. Still another effect of serial correlation is that *it may induce a falsely high or falsely low agreement between two variables.* If both of the variables are serially correlated, what appears to be a large number of cases where the two variables are much the same or not the same may really amount to relatively few cases because the observations are not independent. (The observations are dependent on one another and tend to measure the same thing.)

DURBIN-WATSON TEST

To test whether serial correlation is present in the error terms in a regression, we can use the Durbin-Watson test. Let $\hat{e}_i$ be the difference between Y_i and $\hat{Y}_i$, the value of Y_i predicted by the sample regression; then

$$\hat{e}_i = Y_i - a - bX_i.$$

In order to apply the Durbin-Watson test, we must compute

$$d = \frac{\sum_{i=2}^{n} (\hat{e}_i - \hat{e}_{i-1})^2}{\sum_{i=1}^{n} \hat{e}_i^2}. \tag{12.13}$$

Durbin and Watson have provided tables which show whether d is so high or so low that the null hypothesis that there is no serial correlation should be rejected.

Suppose that we want to test this null hypothesis against the alternative hypothesis that there is *positive* serial correlation. (*Positive* serial correlation would mean that e_i is *directly* related to e_{i-1}.) If so, we should reject the null hypothesis if $d < d_L$ and accept the null hypothesis if $d > d_u$. If $d_L \leqslant d \leqslant d_u$, the test is inconclusive. The values of d_L and d_u are shown in Appendix Table 11. (Note that these values depend on the sample size n and on k, the number of independent variables in the regression.) On the other hand, suppose that the alternative hypothesis is that there is *negative* serial correlation. (*Negative* serial correlation would mean that e_i is *inversely* related to e_{i-1}.) If so, we should reject the null hypothesis if $d > 4 - d_L$ and accept the null hypothesis if $d < 4 - d_u$. If $4 - d_u \leqslant d \leqslant 4 - d_L$, the test is inconclusive. Finally, for a two-tailed test of both positive and negative serial correlation, reject the null hypothesis if $d < d_L$ or if $d > 4 - d_L$, and accept the null hypothesis if $d_u < d < 4 - d_u$. Otherwise, the test is inconclusive. For a two-tailed test, note that the significance level is double that shown in Appendix Table 11.

TABLE 12.8

Computation of Durbin-Watson Statistic in Example 12.7

Day	$\hat{e}_i$	$(\hat{e}_i - \hat{e}_{i-1})^2$	$\hat{e}_i^2$	Day	$\hat{e}_i$	$(\hat{e}_i - \hat{e}_{i-1})^2$	$\hat{e}_i^2$
1	− 190	—	36,100	31	− 400	10,000	160,000
2	− 200	100	40,000	32	350	562,500	122,500
3	− 220	400	48,400	33	50	90,000	2,500
4	− 100	14,400	10,000	34	50	0	2,500
5	800	810,000	640,000	35	50	0	2,500
6	20	608,400	400	36	150	10,000	22,500
7	300	78,400	90,000	37	− 900	1,102,500	810,000
8	400	10,000	160,000	38	− 800	10,000	640,000
9	0	160,000	0	39	300	1,210,000	90,000
10	− 440	193,600	193,600	40	300	0	90,000
11	800	1,537,600	640,000	41	− 30	108,900	900
12	300	250,000	90,000	42	50	6,400	2,500
13	500	40,000	250,000	43	0	2,500	0
14	500	0	250,000	44	− 50	2,500	2,500
15	200	90,000	40,000	45	300	122,500	90,000
16	210	100	44,100	46	− 300	360,000	90,000
17	− 50	67,600	2,500	47	− 700	160,000	490,000
18	0	2,500	0	48	300	1,000,000	90,000
19	50	2,500	2,500	49	− 150	202,500	22,500
20	200	22,500	40,000	50	100	62,500	10,000
21	600	160,000	360,000	51	− 150	62,500	22,500
22	− 100	490,000	10,000	52	− 150	0	22,500
23	450	302,500	202,500	53	− 100	2,500	10,000
24	500	2,500	250,000	54	0	10,000	0
25	− 300	640,000	90,000	55	− 500	250,000	250,000
26	− 600	90,000	360,000	56	− 1000	250,000	1,000,000
27	200	640,000	40,000	57	− 900	10,000	810,000
28	700	250,000	490,000	58	− 150	562,500	22,500
29	300	160,000	90,000	59	300	202,500	90,000
30	− 300	360,000	90,000	60	− 400	490,000	160,000
				61	− 150	62,500	22,500
					Total	13,908,400	9,711,000

The following example illustrates the use of the Durbin-Watson test to detect serial correlation.

Example 12.7 An ice cream firm regresses its daily sales on two independent variables (namely, the mean temperature during the day and the price it charges). The residual from this regression for each of the 61 days included in the sample—that is, each value of $\hat{e}_i$—is shown in Table 12.8. Use the Durbin-Watson test to determine whether there is evidence of serial correlation in the error terms. (Set the significance level equal to .05, and use a two-tailed test.)

Solution: As shown in Table 12.8,

$$d = \frac{13{,}908{,}400}{9{,}711{,}000} = 1.43.$$

The alternative hypothesis here is two-tailed, because we would like to detect either positive or negative serial correlation. Since .05 is the significance level, we should look in Appendix Table 11 for $\alpha = .025$. Because $n = 61$, and $k = 2$, this table shows that $d_L = 1.44$ and $d_u = 1.57$ (since n is approximately 60), which means that we should reject the null hypothesis if $d < 1.44$ or $d > 2.56$, and that we should accept the null hypothesis if $1.57 < d < 2.43$. Since $d = 1.43$, we should reject the null hypothesis. In other words, there seems to be statistically significant evidence of serial correlation.

12.12 Analyzing the Residuals

In the previous section we defined $\hat{e}_i$ as the difference between Y_i and $\hat{Y}_i$. In other words, it is the difference between the actual value of Y_i and the value predicted by the regression. Since it is a measure of the extent to which Y_i *cannot* be explained by the regression, $\hat{e}_i$ is often called the ***residual*** for the ith observation. If the assumptions underlying regression analysis are met, the error terms should be successively independent, as we know from previous sections. If they are not independent, this is likely to show up in a plot of the residuals. For example, panel A in Figure 12.8 shows a case where successive residuals are positively correlated. (Why positively? Because if one residual is large, the next tends to be large; and if one residual is small, the next tends to be small.) Panel B of Figure 12.8 shows a case where successive residuals are negatively correlated. (Why negatively? Because if one residual is large, the next tends to be small; and if one residual is small, the next tends to be large.)

The printouts of many computer programs show the residuals, and statisticians frequently plot these to detect departures from the assumptions underlying regression analysis. To illustrate how an analysis of the residuals can be useful in detecting such departures, let's return to the ice cream firm in Example 12.7. After regressing the firm's daily sales on two independent variables (temperature and price), the firm's statistician calculates the residuals (shown in Table 12.8), plots them (Figure 12.9), and determines whether their pattern suggests that an important independent variable has been omitted. The statistician notices that many of the days when the residuals were large and positive were days when the firm had all three of its salespersons on the job, whereas many of the days when the residuals were large and negative were days when only one of the firm's salespeople worked (because the others were ill or on vacation). This pattern, shown in Figure 12.9, suggests that the regression does not contain all the important independent variables. (Statisticians often refer to this as a *specification error*.) To decide whether the number of salespeople working on a particular day should be used as an additional independent variable in the regression, the statistician can use this number as a third independent variable, and see whether its regression coefficient is statistically significant.

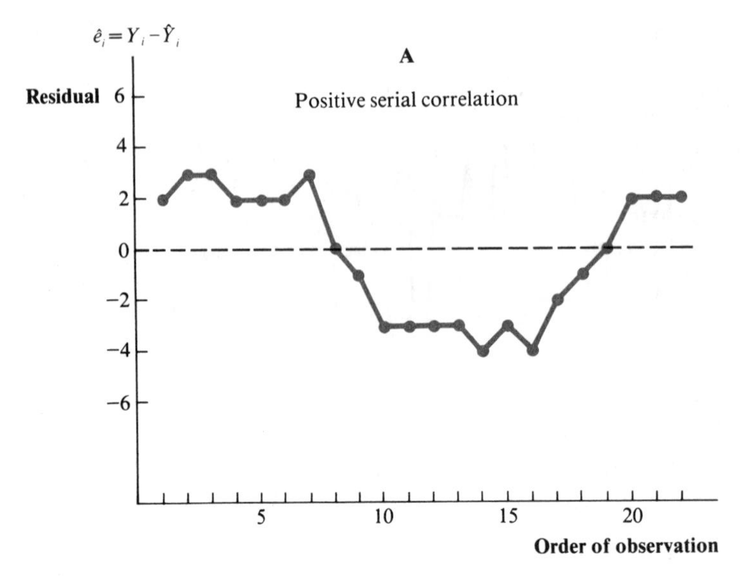

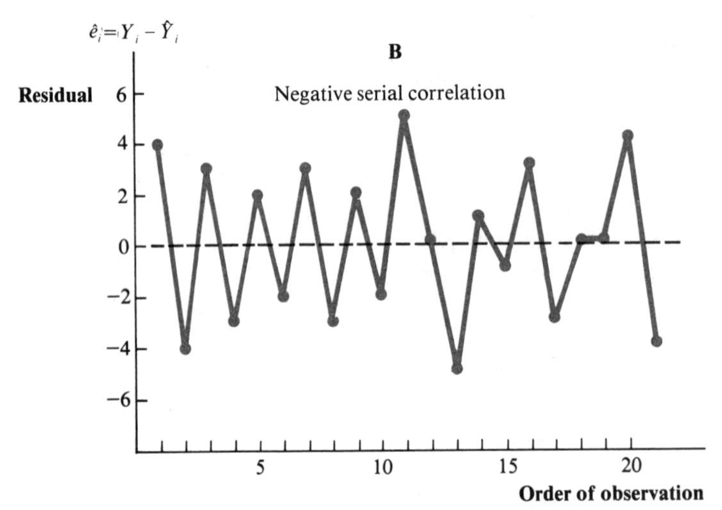

FIGURE 12.8
Serial Correlation of Residuals

Besides being useful in detecting specification errors, the residuals can also be plotted to detect departures from the assumption that the standard deviation of the error terms in the regression is the same, regardless of the values of the independent variables. For example, suppose that the ice cream firm's statistician plots each day's residual against the mean temperature for that day, the results being shown in Figure 12.10. Apparently, the standard deviation of the residuals tends to increase as the temperature increases. If this tendency is statistically significant, and if the standard deviation of the residuals is a reasonably good estimate of the standard deviation of the error terms, it indicates a departure from one of the assumptions underlying regression analysis. (As emphasized, the form of regression analysis described here assumes that the standard deviation of the error terms is constant.)

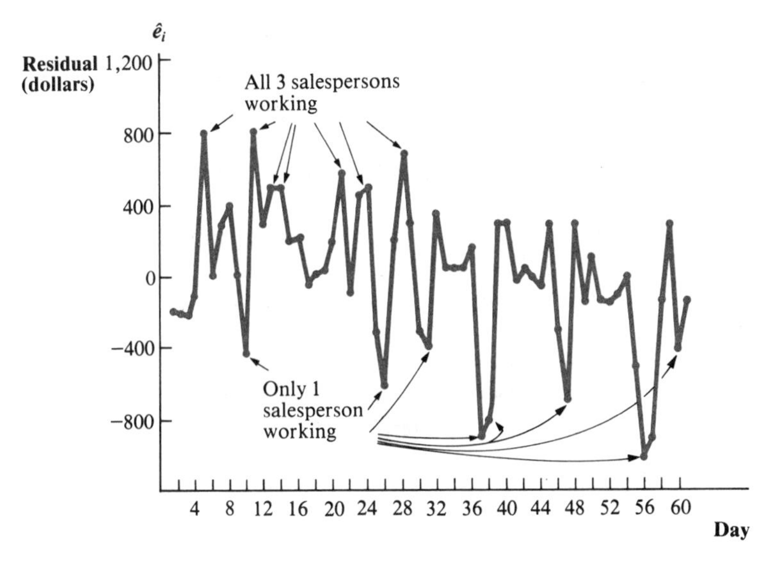

FIGURE 12.9 Residuals in Table 12.8: Detecting Specification Errors

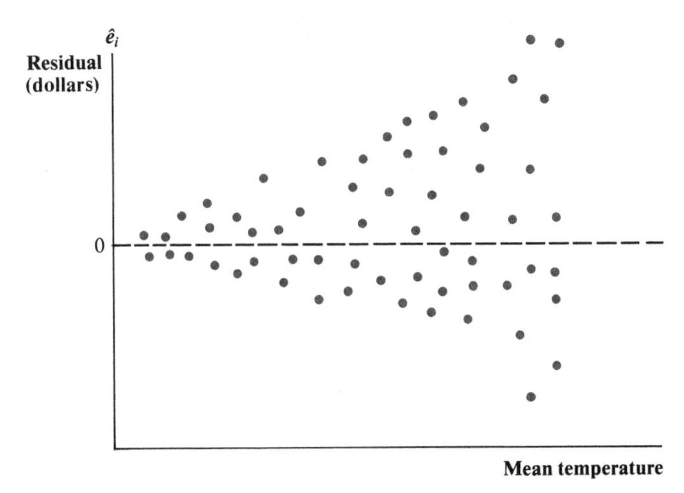

FIGURE 12.10 Scatter Diagram of Residual and Temperature: Departure from Homoscedasticity

Statisticians generally plot and analyze the residuals from their regression equations; so do economists, as is indicated by Nobel laureate Paul A. Samuelson's well-known statement: "To the scientific forecaster I say, 'Always study your residuals.'" What Samuelson meant was that in evaluating any model it is useful to calculate the difference between each observation and what the model predicts this observation will be. These differences—or residuals—are very useful in indicating whether the model excludes some important explanatory variables and whether its assumptions are valid.

12.13 Coping with Departures from the Assumptions

In previous sections and in Chapter 11, we have discussed ways of detecting certain departures from the assumptions underlying regression analysis. What can be done if a particular assumption is violated? Let's begin with the assumption that the conditional mean of the dependent variable is a linear function of the independent variables. If an examination of the data indicates that the relationship is curvilinear rather than linear, what can be done? One possibility is to use the logarithmic regression techniques described in Appendix 11.1 to fit a nonlinear relationship to the data. Another possibility is to fit a quadratic equation to the data. For example, suppose that a statistician is interested in the relationship between a family's income and the amount the family spends on food. To detect signs of nonlinearity, the statistician examines the scatter diagram and finds there is an obvious tendency for the relationship between these two variables to be curvilinear. He then assumes that

$$Y_i = A + B_1X_i + B_2X_i^2 + e_i,$$

where Y_i is the amount spent on food by the ith family and X_i is its income. Letting X_i be the first independent variable and X_i^2 be the second independent variable, A, B_1, and B_2 are estimated in the way described in Section 12.4. In this way, a quadratic relationship is fitted to the data.

In a case of this sort, the least-squares estimates of B_1, B_2, and A are as follows:

$$b_1 = \frac{\sum_{i=1}^{n}(u_i - \bar{u})^2 \sum_{i=1}^{n}(X_i - \bar{X})(Y_i - \bar{Y}) - \sum_{i=1}^{n}(X_i - \bar{X})(u_i - \bar{u}) \sum_{i=1}^{n}(Y_i - \bar{Y})(u_i - \bar{u})}{\sum_{i=1}^{n}(X_i - \bar{X})^2 \sum_{i=1}^{n}(u_i - \bar{u})^2 - \left[\sum_{i=1}^{n}(X_i - \bar{X})(u_i - \bar{u})\right]^2}$$

$$b_2 = \frac{\sum_{i=1}^{n}(X_i - \bar{X})^2 \sum_{i=1}^{n}(u_i - \bar{u})(Y_i - \bar{Y}) - \sum_{i=1}^{n}(X_i - \bar{X})(u_i - \bar{u}) \sum_{i=1}^{n}(Y_i - \bar{Y})(X_i - \bar{X})}{\sum_{i=1}^{n}(X_i - \bar{X})^2 \sum_{i=1}^{n}(u_i - \bar{u})^2 - \left[\sum_{i=1}^{n}(X_i - \bar{X})(u_i - \bar{u})\right]^2}$$

$$a = \bar{Y} - b_1\bar{X} - b_2\bar{u},$$

where $u_i = X_i^2$. To derive these formulas, one can substitute X_i for X_{1i} and u_i for X_{2i} in the equations in (12.5).

Next, let's turn to the assumption that the error terms in a regression are independent. If a statistician finds evidence of serial correlation in the error terms, what can be done about it? Specifically, suppose that a dependent variable Y is regressed on an independent variable X, and the Durbin-Watson test indicates the presence of serial correlation. What can be done? One possible way to proceed is to perform a transformation on the data. In other words, rather than calculating the regression of Y on X, one can calculate the regression of some function of the Ys on some function of the Xs. The particular function that should be used is too complicated to be presented here,[12] but for present purposes, all that we want to convey is that if serial correlation is detected, there are methods for handling the problem.

Similarly, if other complications exist, such as departures from homoscedasticity, there are techniques for coping with these problems. This does not mean that the application of these techniques is cut-and-dried, or that it is possible to deal with any and all departures from the assumptions. But anyone who uses regression techniques should appreciate the importance of testing the assumptions on which these techniques are based, while realizing that when these tests indicate that the assumptions are violated, a variety of methods exist that are aimed at handling the problem. Descriptions of these methods can be found in more advanced texts.[13]

12.14 Computer Programs and Multiple Regression

The advent of computer technology has caused a marked reduction in the amount of effort and expense required to calculate multiple regressions with large numbers of independent variables. Thirty years ago there was an enormous amount of drudgery in computing a multiple regression with more than a few independent variables; now such computations are relatively simple. It is important for you to be familiar with the kind of information printed out by computers and the form in which it appears. Since there is a wide variety of "canned" programs for calculating regressions, there is no single format or list of items which are printed out. However, the various sorts of computer printouts are sufficiently similar so that it is worthwhile looking at one illustration in some detail.

Table 12.9 shows the printout from a multiple regression of an industry's rate of productivity increase during 1948–66 on three independent variables: (1) the percent of the industry's employees who are union members (variable 2); (2) the percent of the industry's value-added spent on basic research (variable 3); and (3) the percent of the industry's value-added spent on applied research and development (variable 4). We are already familiar with the top four horizontal rows of this printout, which were reproduced in Table 12.3. As we already know, these rows show the value, standard error, and t value of each of the estimated regression coefficients, as well as the estimated intercept of the regression equation.

The next six rows in Table 12.9 present the results of an analysis of variance for the multiple regression. The *variation explained by the regression*

12. See J. Johnston, op. cit.

13. In some cases, an equation is part of a system of equations. In such cases, the unwary statistician may obtain biased estimates of the population regression equation unless special techniques are used. For a simple introduction to this so-called identification problem, see E. Working, "What Do Statistical Demand Curves Show?" in E. Mansfield, *Statistics for Business and Economics: Readings and Cases* (New York: Norton, 1980). For a much more complete and advanced discussion, see J. Johnston, op. cit.

TABLE 12.9
Computer Printout of Results of Multiple Regression concerning Industries' Rates of Productivity Increase

```
INDEPENDENT VARIABLE(S)   2  3  4
DEPENDENT VARIABLE           1

VAR          COEFF       STD. ERROR     T-VALUE
  2        -0.0559         0.0146       -3.8400
  3         1.3625         0.4670        2.9200
  4         0.0727         0.0276        2.6400

INTERCEPT   4.78826

REGRESSION
   DEGREES OF FREEDOM     3
   SUM OF SQUARES         10.17104
   MEAN SQUARE            3.39035

ERROR
   DEGREES OF FREEDOM     16
   SUM OF SQUARES         6.26096
   MEAN SQUARE            0.39131

S.E. OF ESTIMATE          0.62555
F-VALUE                   8.66
MULTIPLE R-SQUARED        .619
```

equals 10.17104. The *error sum of squares* equals 6.26096. The *regression mean square* is the variation explained by the regression divided by its degrees of freedom (which equal the number of independent variables). Thus, the regression mean square in this case equals 10.17104 ÷ 3, or 3.39035. The *error mean square* here is the error sum of squares divided by its degrees of freedom (which is one less than the number of observations minus the number of independent variables). Thus, the error mean square in this case equals 6.26096 ÷ 16, or .39131. The *F* value in Table 12.9 is shown in the next-to-last row in the table.

The only item left to be explained in Table 12.9 is the *standard error of estimate.* As in the case of simple regression, the standard error of estimate is an estimate of the standard deviation of the probability distribution of the dependent variable when the independent variables are all held constant. Thus, it is a measure of the amount of scatter of individual observations about the regression equation. In this case the standard error of estimate is 0.62555, which means that in the population as a whole, the standard deviation of the differences between an industry's rate of productivity increase and that predicted by the true regression is estimated to be 0.62555.[14]

12.15 Choosing the Best Form of a Multiple-Regression Equation

As emphasized throughout this chapter, the widespread availability and application of computer technology have facilitated greatly the use of multiple regression and correlation. Statisticians now can try various versions of a particular multiple regression equation to see which version fits best. These experiments can take various forms. For one thing, one can experiment with *various measures of a particular variable.* For example, the hosiery mill's statisticians may want to try alternative output measures which take account of the fact that all tons of output are not the same. For another thing, one can experiment with *various forms of the regression equation.* Thus, in the case of the hosiery mill, the statistician might want to try the logarithm of output, rather than output, as the independent variable.

Still another way in which statisticians can experiment with a particular regression equation is to *add and omit various independent variables to see how the results are affected.* The hosiery mill's statisticians might be interested in adding another independent variable, the price of nylon, to the multiple regression to see whether it seems to have a significant effect on the firm's costs. Also, they may want to determine (1) whether the estimated regression coefficients of the independent variables currently used are statistically significant when this new independent variable is introduced, and (2) whether the values of these regression coefficients are altered considerably by its introduction.

Statisticians sometimes use a procedure called *stepwise multiple regression* to specify which independent variables seem to provide the best explanation of the behavior of the dependent variable. Using commonly available programs, a computer can determine which of a set of independent variables is most highly correlated with the dependent variable Y. (The computer programs allow the inclusion of dozens of independent variables in the set.) Suppose that this independent variable is X. The computer then selects the independent variable from the remainder of the set that results in the greatest reduction in the variation unexplained by the regression of Y on X. Suppose that this added independent variable is R. Then the computer selects the independent variable from the remainder of this set that results in the greatest reduction in the variation unexplained by the regression of Y on X and R. If this independent variable is S, then the computer selects the independent variable from the remainder of this set that results in the greatest reduction in the varia-

14. The printout in Table 12.9 is not meant to be exhaustive. Besides the information shown there, the printout often shows the values of each of the dependent and independent variables (and their means and standard deviations); the simple correlation coefficient between the dependent variable and each independent variable and between each pair of independent variables; the values of the residuals; and the value of d (to be used in the Durbin-Watson test). Each of these items has been explained previously.

tion unexplained by the regression of Y on X, R, and S. This process continues until all the independent variables in the set have been added to the regression, or until none of the remaining independent variables in the set reduces significantly the unexplained variation.

Given the relative ease with which one can calculate alternative forms and versions of a multiple regression equation, it is extremely important that the statistician have good a priori reasons for including each of the independent variables in the regression equation being used. Using stepwise multiple regression, it is not very difficult to select some independent variables that will explain much of the variation in practically any dependent variable, even if these independent variables really have little or no effect on the dependent variable. Why is this so? Because some variables are bound to be correlated with the dependent variable by chance, or they may be influenced by the same factors as the dependent variable, with the result that they are correlated with the dependent variable even though they do not influence the dependent variable. If one uses the computer to hunt around long enough, one is likely to find a combination of independent variables that explains much of the variation in the dependent variable. But if these variables are merely the result of an indiscriminate, mechanical quest for a good-fitting regression equation, the resulting equation is likely to be useless for purposes of prediction.

In choosing the best form of a multiple regression equation, the statistician sometimes must decide whether to include an independent variable which on *a priori* grounds is almost certain to influence the dependent variable, but which has an estimated regression coefficient that is not statistically significant. If there are very strong *a priori* grounds for believing that an independent variable affects the dependent variable, it is acceptable to include this variable in the regression equation, even if its regression coefficient is not statistically significant. After all, the fact that the estimated regression coefficient is not statistically significant does not prove that the true regression coefficient is zero. And the estimated regression coefficient would be likely to constitute a better estimate of the true regression coefficient than zero (which would be the value attributed to it if the independent variable were omitted from the equation).

12.16 Railroad Costs: A Case Study

To illustrate the use of multiple regression, let's consider another study of a firm's cost function.[15] This research was carried out for one of the nation's largest railroads, the purpose being to determine how the costs incurred in a freight yard are related to the output of the yard. The two most important services performed by a yard are switching and delivery, and it seems reasonable to use the number of cuts switched and the number of cars delivered during a particular period as a measure of output. (A *cut* is a group

15. E. Mansfield and H. Wein, "A Regression Control Chart for Costs," *Applied Statistics*, March 1958.

of cars that rolls as a unit onto the same classification track; it is often used as a unit of switching output.)

The study assumed that

$$C_i = A + B_1S_i + B_2D_i + e_i,$$

where C_i is the cost incurred in this freight yard on the ith day, S_i is the number of cuts switched in this yard on the ith day, D_i is the number of cars delivered in this yard on the ith day, and e_i is the difference between C_i and $E(C_i)$. Data were obtained regarding C_i, S_i, and D_i for 61 days. Based on the procedures described in Section 12.4, these data were used to obtain estimates of A, B_1, and B_2. The resulting regression equation was

$$\hat{C}_i = 4{,}914 + 0.42S_i + 2.44D_i, \tag{12.14}$$

where $\hat{C}_i$ is the cost (in dollars) predicted by the regression equation for the ith day.

This equation was of considerable value to the railroad sponsoring the study. Particularly interesting was the fact that the marginal cost of switching a cut at this yard was about 42 cents and that of delivering a car was about \$2.44. The standard error of each of these estimated regression coefficients was also computed, and based on the material presented in Section 12.5, it was easy to obtain a confidence interval for each regression coefficient. Taking B_2 as an illustration, the standard error of the estimate of B_2 is .40. Since there are 58 degrees of freedom, $t_{.025} = 2.00$, and $t_{.025}s_{b_2} = .80$. Thus, a 95 percent confidence interval for B_2 is \$2.44 ± 0.80, or \$1.64 to \$3.24.

EXERCISES

12.11 A government agency wants to estimate the effect of a city's unemployment rate on its crime rate (as measured by the number of major crimes per million people). The agency believes that the crime rate (holding the unemployment rate constant) tended to be higher in 1980 than in 1978. Using the data below, calculate the multiple regression of the crime rate on the unemployment rate and a dummy variable representing the year to which the data pertain.

City	Crime rate	Unemployment rate (percent)	Year
A	100	5	1978
B	120	7	1978
C	140	8	1978
D	170	9	1978
E	110	5	1980
F	160	8	1980
G	200	9	1980
H	120	6	1980
I	130	7	1980
J	150	7	1980

12.12 (a) Based on your results in the previous exercise, what is the least-squares estimate of the mean difference between 1980 and 1978 in a city's crime rate, holding the unemployment rate constant? (b) What is the effect of an increase of one percentage point in the unemployment rate, holding the year constant?

12.13 Suppose that a noted criminologist were to publish a study indicating that the effect of a 1 percent increase in the unemployment rate on the crime rate was much greater in 1980 than in 1978. (a) Would this make you skeptical of the results you obtained in Exercise 12.11? Why, or why not? (b) Can the data in Exercise 12.11 be used to determine whether the criminologist's hypothesis is true? If so, how?

12.14 A firm is interested in estimating the effect of its advertising expenditures and its number of salespeople on its sales. The firm calculates a multiple regression of its sales on its advertising expenditures and its number of salespeople. Are the independent variables in this regression equation the only ones that should be included? If not, what are some other variables that might be considered?

12.15 Calculate the deviation of each of the observations in Exercise 12.1 from the multiple regression equation (when sales are regressed on disposable income and price).

12.16 Using the data in Exercise 12.1, test whether there is serial correlation in the error terms (that is, the deviations of the observations from the regression equation) when the Rotunda Corporation's sales are regressed on disposable income and price. Let $\alpha = .05$, and assume the alternative hypothesis to be that there is either positive or negative serial correlation.

12.17 A statistician regresses a dependent variable on four independent variables, based on 30 observations. The Durbin-Watson statistic equals 1.06. Test whether there is serial correlation in the deviations of the observations from the regression equation. Let $\alpha = .05$, and suppose that the alternative hypothesis is that there is positive serial correlation.

12.18 Using 40 observations, an economist regresses a dependent variable on two independent variables. The Durbin-Watson statistic equals 2.71. Test whether there is serial correlation in the deviations of the observations from the regression equation. Let $\alpha = .01$, and assume the alternative hypothesis to be that there is negative serial correlation.

12.19 The printout of the regression (based on the data in Table 12.6) of the annual amount saved by a family on its income (variable 2) and the dummy variable showing whether it owns its home or rents (variable 3) is shown (in part) below:

```
INDEPENDENT VARIABLE(S)  2  3
DEPENDENT VARIABLE       1

VAR          COEFF        STD. ERROR       T-VALUE
  2          0.0675           0.0040       16.8900
  3          0.8273           0.0754       10.9800

INTERCEPT  0.32037
```

Calculate a 95 percent confidence interval for the true regression coefficient of the family's income.

12.20 Is there any evidence from the printout in the previous exercise that multicollinearity is a problem? How would you determine whether multicollinearity is a problem in this case?

12.21 An economist wants to estimate a multiple regression equation in which the amount saved by the ith family depends on the family's income, whether the family is white or nonwhite, and whether the family is headed by a male or a female. Explain how a regression equation of this sort can be estimated. What is the dependent variable? What are the independent variables? What assumptions must be made?

CHAPTER REVIEW

1. Whereas a *simple regression* includes only one independent variable, a *multiple regression* includes more than one independent variable. An advantage of multiple regression over simple regression is that one frequently can predict the dependent variable more accurately if more than one independent variable is used. Also, if the dependent variable is influenced by more than one independent variable, a simple regression of the dependent variable on a single independent variable may result in a biased estimate of the effect of this independent variable on the dependent variable.

2. The first step in *multiple regression analysis* is to identify the independent variables, and then to specify the mathematical form of the equation relating the expected value of the dependent variable to the independent variables. For example, if Y is the dependent variable and X_1 and X_2 are identified as the independent variables, one might specify that

$$Y_i = A + B_1X_{1i} + B_2X_{2i} + e_i,$$

where e_i is the difference between Y_i and $E(Y_i)$. To estimate B_1 and B_2 (called the true regression coefficients of X_1 and X_2) as well as A (the intercept of this true regression equation), we use the values that minimize the sum of squared deviations of Y_i from $\hat{Y}_i$, the value of the dependent variable predicted by the estimated regression equation.

3. Multiple regressions are generally calculated by computers rather than by hand. The standard programs print out the *estimated standard deviations* of the least-squares estimators of B_1 and B_2 (these estimated standard deviations being called *standard errors*). Using the value of the least-squares estimator of B_1 or B_2 together with these standard errors, one can obtain a confidence interval for B_1 or B_2. For example, a confidence interval for B_1 is $b_1 \pm t_{\alpha/2}s_{b_1}$ where b_1 is the least-squares estimator of B_1 and s_{b_1} is the standard error of b_1. The standard programs also print out the t value for each estimated regression coefficient, the t value being the estimated regression coefficient divided by its standard error. To test whether a true regression coefficient equals zero, one should see whether its t value exceeds $t_{\alpha/2}$ or is less than $-t_{\alpha/2}$; if so, one should reject the hypothesis that the true regression coefficient is zero. (This is a two-tailed test.)

4. The *multiple coefficient of determination* R^2 equals the ratio of the variation explained by the multiple regression to the total variation in the dependent variable. The positive square root of the multiple coefficient of determination is called the *multiple correlation coefficient* and is denoted by R. Both R^2 and R are measures of how well the regression equation fits the data:

The closer they are to zero, the poorer the fit; the closer they are to 1, the better the fit.

5. Multiple regression can be used to analyze the effects of *qualitative variables* (that is, variables which do not assume numerical values) as well as *quantitative variables.* To represent a qualitative variable we use a *dummy variable* (that is, a variable that can equal 0 or 1). For example, to include whether or not a family owns its home in a regression, we construct a dummy variable which equals 1 if the family owns its home and zero if it rents.

6. An important problem that can occur in multiple regression is *multicollinearity*, a situation where two or more of the independent variables are highly correlated. If two independent variables are perfectly correlated, there is no way of estimating the effect of each, holding the other constant; all that we can observe is the effect of both combined. Even if two independent variables are highly (but not perfectly) correlated, there may be no way of estimating the regression coefficient of each with even minimal accuracy. In cases of this sort, it may be necessary to obtain additional data or to redefine the variables in order to reduce the correlation between the independent variables.

7. Another problem arises when the error terms in the equation (that is, the values of e_i) are serially correlated. If the basic data constitute a time series, the existence of serial correlation means that successive values of e_i are not independent. For example, if the value of the error term for one period is positive, the value of the error term for the next period may tend to be positive also. The principal problem resulting from serial correlation of this sort is that the ordinary formulas underestimate the standard errors of the regression coefficients and intercept. To test whether serial correlation of this kind is present, we can use the *Durbin-Watson test*, which is based on the value of d, where

$$d = \sum_{i=2}^{n} (\hat{e}_i - \hat{e}_{i-1})^2 \div \sum_{i=1}^{n} \hat{e}_i^2,$$

where $\hat{e}_i$ equals $Y_i - \hat{Y}_i$. (Each value of $\hat{e}_i$ is called a *residual.*) Tables are available which show whether d is so high or so low that the null hypothesis (that there is no serial correlation) should be rejected.

8. The advent of computer technology has caused a marked reduction in the amount of effort and expense required to calculate multiple regressions. We have described and explained the items generally printed out by "canned" programs, including the F test for determining whether all true regression coefficients are zero. With existing computer technology, one can experiment with various forms of a regression equation. It is important that the statistician have good *a priori* reasons for including each of the independent variables in the equation. If independent variables are chosen indiscriminately in a mechanical quest for a good-fitting equation, the resulting equation is likely to be useless for purposes of prediction.

Getting Down to Cases:

DETERMINANTS OF THE STRENGTH OF COTTON YARN

The U.S. Department of Agriculture has presented data concerning the skein strength (in pounds) of cotton yarn, which is perhaps the most important single measure of spinning quality. For a sample of varieties of upland cotton grown in the United States, the Department showed the skein strength (in pounds) as well as the fiber length (in hundredths of an inch) and fiber tensile strength (in thousands of pounds per square inch). The results for a sample of 20 pieces of cotton yarn were as follows:[16]

Skein strength	Fiber length	Fiber tensile strength
99	85	76
93	82	78
99	75	73
97	74	72
90	76	73
96	74	69
93	73	69
130	96	80
118	93	78
88	70	73
89	82	71
93	80	72
94	77	76
75	67	76
84	82	70
91	76	76
100	74	78
98	71	80
101	70	83
80	64	79

(a) Calculate the regression of skein strength on fiber length and fiber tensile strength. (That is, let skein strength be the dependent variable, and let fiber length and fiber tensile strength be the independent variables.)

(b) Based on your results in (a), what is the effect on the average strength of a piece of yarn of an increase in fiber length of .01 inches? What is the effect on the average strength of a piece of yarn of an increase in fiber tensile strength of 1,000 pounds per square inch?

(c) Based on your results in (a), estimate the strength of a piece of yarn if the fiber length is .80 inches and the fiber tensile strength is 75,000 pounds per square inch.

(d) Write a one-paragraph report summarizing your findings.

16. See U.S. Department of Agriculture, ***Results of Fiber and Spinning Tests for Some Varieties of Upland Cotton Grown in the United States***, Washington, D.C., 1945; and A. J. Duncan, *Quality Control and Industrial Statistics,* (Homewood, Ill.: Irwin, 1959).

13

Introduction to Time Series

13.1 Introduction

Business and economic data generally are of two types: ***cross-section*** data and ***time-series*** data. *Cross-section data pertain to a variety of units or entities at a given point in time.* For example, the profits data regarding the petroleum firms exhibited in Table 1.8 are cross section data. They show the profitability of various oil firms during a particular period of time, 1973. *Time series data, on the other hand, pertain to a given unit or entity at a number of points in time.* Thus, if we had obtained data concerning the profitability of Exxon during each year from 1955 to 1980, this would have been an example of a time series.

Decision makers in business and government are continually involved with time series. Month by month, business executives pore over the latest month's sales figures. They compare these figures with those for previous periods and with past forecasts in an attempt to gauge their current performance and determine their future moves. Moreover, sales figures are only one of a myriad of time series that are generated by most firms. The accounting department of the typical company generates time series on sales, costs, taxes, profits, assets, debts, dividends, and many other variables. Moreover, the engineering department, production department, marketing department, and other parts of a firm also contribute to the generation of time series.

Statisticians have developed certain techniques for describing and analyzing time series. For one thing, they have tried in various ways to break down a time series into such elements as its trend, seasonal variation, and cyclical variation. This chapter describes these procedures and indicates how they have been used in business and economic forecasting. In addition, we point out the limitations of these techniques. (As we shall see, they are based on a very rudimentary theory of the determinants of economic variables, and should be used with caution.) Finally, we will describe briefly some alternative types of forecasting techniques which are based on econometric models.

13.2 The Traditional Time-Series Model

The classical approach to the analysis of time series, devised primarily by economic statisticians, was essentially descriptive. It assumed that an economic time series could be decomposed into four components: trend, seasonal variation, cyclical variation, and irregular movements. More specifically, it assumed that the value of an economic variable at a certain point in time could be represented as the product of each of these four components. For example, the value of a company's sales in January 1980 was viewed as equal to

$$T \times S \times C \times I, \tag{13.1}$$

where T is the trend value of the firm's sales during that month, S is the seasonal variation attributable to January, C is the cyclical variation occurring that month, and I is the irregular variation that occurred then.[1] Each of these components is defined below.

Trend: *A trend is a relatively smooth long-term movement of a time series.* For example, the population of the United States increased rather steadily between 1940 and 1978, as shown in Figure 13.1. Thus, there has been an

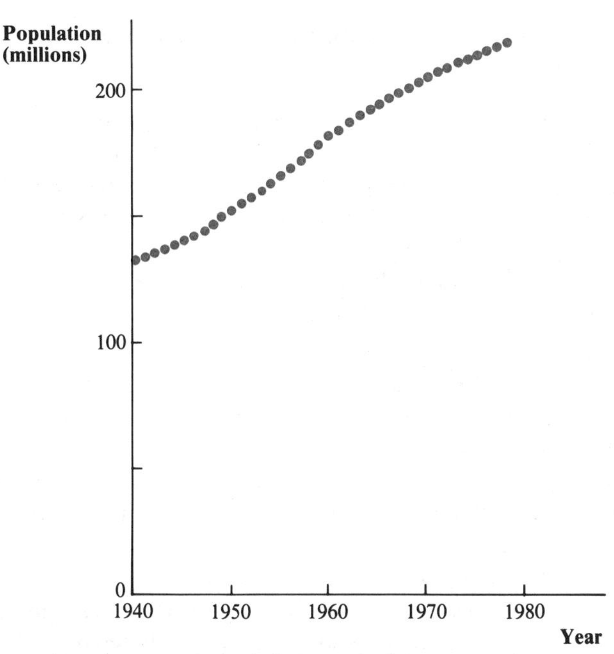

FIGURE 13.1 Population of the United States, 1940–78

SOURCE: *Economic Report of the President, 1979* (Washington, D.C.: Government Printing Office, 1979).

1. In some versions of this model, the components are added rather than multiplied. That is, it is assumed that

$$Y = T + S + C + I,$$

where Y is the value of the time series.

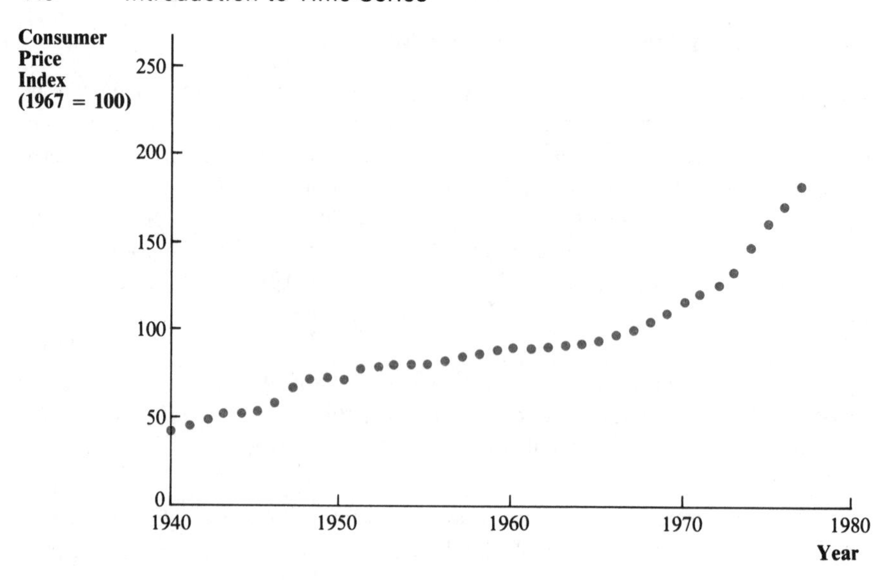

FIGURE 13.2 Consumer Price Index, 1940–77

SOURCE: *Economic Report of the President, 1979* (Washington, D.C.: Government Printing Office, 1979).

upward trend in the U.S. population. Similarly, there has been an upward trend in the price level in the United States, as shown in Figure 13.2. Of course, not all trends are upward. For example, the trend in the average length of the marketable interest-bearing public debt in the United States has been downward, as shown in Figure 13.3.[2] Whether upward or downward, the trend of a time series is represented by a smooth curve. In equation (13.1) T is the value of the firm's sales that would be predicted for January 1980, based on such a curve.

Seasonal Variation: *In a particular month, the value of an economic variable is likely to differ from what would be expected on the basis of its trend, due to seasonal factors.* For example, consider the sales of a firm that produces air conditioners. Since the demand for air conditioners is much higher in the summer than in the winter, one would expect that the monthly time series of the firm's sales would show a pronounced and predictable seasonal pattern. Specifically, sales each year would tend to be higher from June through August than during the rest of the year. As we shall see, it is possible to calculate *seasonal indexes* which estimate how much each month departs from what would be expected on the basis of its trend. In equation (13.1) we must multiply the trend value T by the seasonal index S to allow for the effect of this seasonal variation.

Cyclical Variation: *Another reason why an economic variable may differ from its trend value is that it may be influenced by the so-called business cycle.* As

2. In still other cases the trend is horizontal; that is, there is no upward tendency or downward tendency in the time series. In these cases it is often said that there is *no trend.*

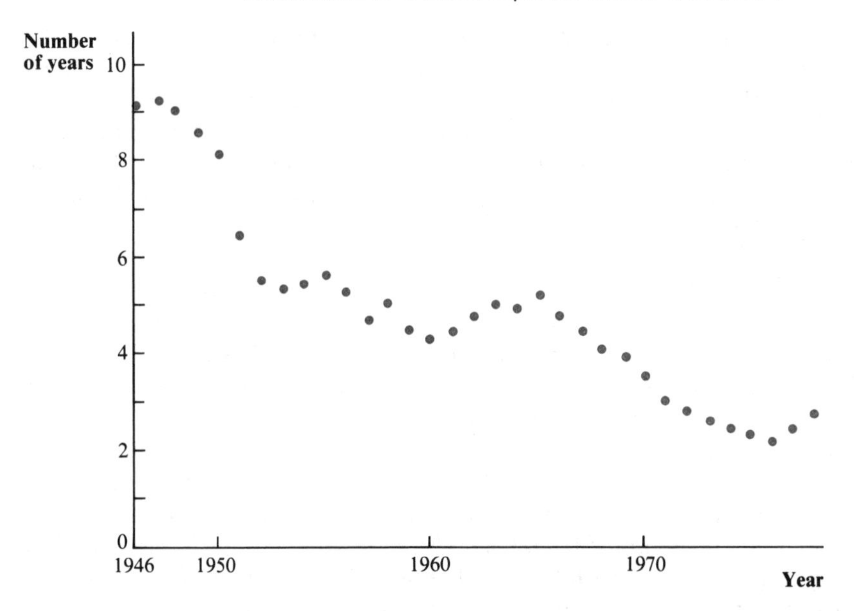

FIGURE 13.3 Average Length of Marketable Interest-Bearing U.S. Debt, 1946–78

SOURCE: *Economic Report of the President, 1979* (Washington, D.C.: Government Printing Office, 1979).

is well known, the general tempo of economic activity in our society has exhibited a cyclical nature, with booms being followed by recessions, and recessions being followed by expansions. These cycles have not been regular or consistent (which is one reason why many economists prefer the term *business fluctuations* to *business cycles*); but unquestionably there has been a certain cyclical ebb and flow of economic activity, which has been reflected in a great many time series. For this reason, $T \times S$ is multiplied by C, which is supposed to indicate the effect of cyclical variation on the firm's sales in equation (13.1).

Irregular Variation: After having been multiplied by both S and C, the trend value T has been altered to reflect seasonal and cyclical forces. However, besides these forces, *a variety of short-term, erratic forces are also at work*. Their effects are represented by I. In effect, I reflects the effects of all factors other than the trend, seasonal variation, and cyclical variation. According to the classical model, these irregular forces are too unpredictable to be useful for forecasting purposes.

13.3 Estimation of a Least-Squares Linear Trend Line

Many studies have been carried out to estimate the trend, seasonal variation, and cyclical variation in particular economic time series. In Sections 13.3 to 13.7 of this chapter we discuss the methods used to estimate a trend; in subsequent sections we shall take up seasonal and cyclical variation. First, we take up the case where the long-term overall movement of the time series seems to be linear. For example, this seems true for the profits of the American Telephone and Telegraph Company (AT and T) during 1957–69. (These profits are plotted in Figure 13.4.) In a case where the trend seems to be linear, statisticians frequently use the method of least squares to calculate the trend. In other words, they assume that if the long-term forces underlying the trend were the only ones at work the time series would be approximately linear. Specifically, they assume that

$$Y_t = A + Bt, \tag{13.2}$$

where Y_t is the trend value of the variable at time t. (Note that t assumes values like 1980 or 1981 if time is measured in years.) The *trend value* is the value of the variable that would result if only the trend were at work. The deviation of Y, the actual value of the variable, from the trend value is *the deviation from trend.*

If the deviations from trend—that is, $Y - Y_t$—can be regarded as random variables with zero mean and constant standard deviation, and if they are independent, we know from Chapter 11 that the most efficient linear unbiased estimates of A and B can be obtained by the method of least squares. Actually, because the deviations from trend often are not independent (see Section 12.11) these assumptions frequently are not met. Thus, one should be very cautious about the interpretation and use of such trend lines. As stressed in previous chapters, regression techniques can be quite misleading when the assumptions on which they are based are not valid.

To illustrate the calculation of a linear trend, let's examine American

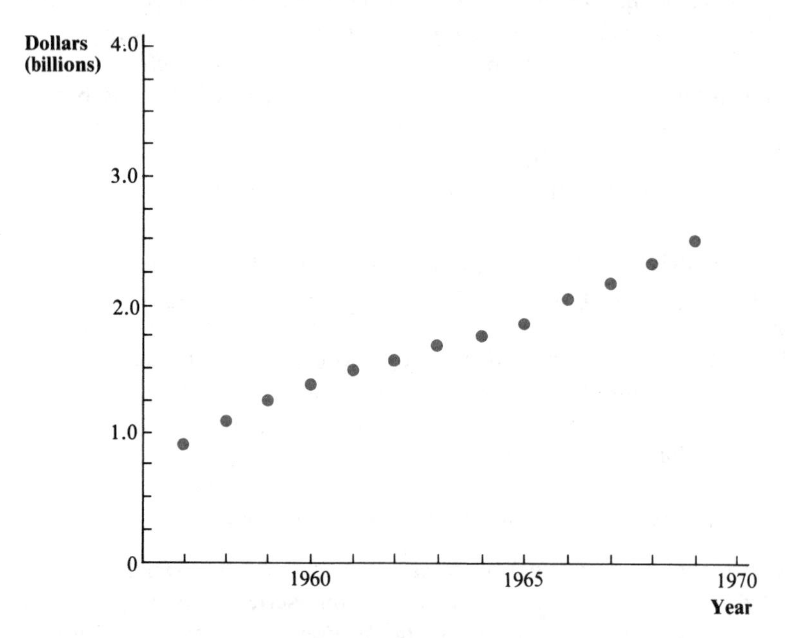

FIGURE 13.4 Net Operating Income, American Telephone and Telegraph Company, 1957–69

TABLE 13.1

Net Operating Income, American Telephone and Telegraph Company, 1957–69, and Calculation of ΣS_t, $\Sigma S_t t'$, and $\Sigma t'^2$

Year (t)	Operating income (billions of dollars) (S_t)	t'	$S_t t'$	t'^2
1957	0.88	−6	−5.28	36
1958	1.06	−5	−5.30	25
1959	1.22	−4	−4.88	16
1960	1.32	−3	−3.96	9
1961	1.43	−2	−2.86	4
1962	1.52	−1	−1.52	1
1963	1.65	0	0	0
1964	1.80	1	1.80	1
1965	1.95	2	3.90	4
1966	2.16	3	6.48	9
1967	2.32	4	9.28	16
1968	2.36	5	11.80	25
1969	2.58	6	15.48	36
Total	22.25	0	24.94	182

$$\bar{S} = \frac{22.25}{13} = 1.712$$

Telephone and Telegraph's annual profits from 1957 to 1969 as shown in Table 13.1. Since profit in year t is the dependent variable and t is the independent variable, it follows from our discussion in Chapter 11 that

$$b = \frac{\sum_{t=t_0}^{t_0+n-1} (S_t - \bar{S})(t - \bar{t})}{\sum_{t=t_0}^{t_0+n-1} (t - \bar{t})^2}, \tag{13.3}$$

$$a = \bar{S} - b\bar{t}, \tag{13.4}$$

where S_t is profit (in billions of dollars) in year t, t_0 is the earliest year in the time series (that is, 1957), and $t_0 + n - 1$ is the latest year in the time series (that is, 1969).

In equations (13.3) and (13.4), time is measured in ordinary calendar years. In other words, t varies from 1957 to 1969. To carry out the calculations, it is advisable to convert time into a coded variable which has a mean of zero. If there are an odd number of years (as in this case), let $t' = 0$ for the middle year, and let the rest of the years be . . . , $-3, -2, -1, 0, 1, 2, 3, \ldots$. If there is an even number of years, let $t' = -1$ and $t' = 1$ for the two middle years, and let the rest of the years be . . . $-5, -3, -1, 1, 3, 5, \ldots$. The advantage of this coding is that if it is carried out,[3]

$$b = \frac{\Sigma S_{t'}t'}{\Sigma t'^2} \tag{13.5}$$

$$a = \bar{S}. \tag{13.6}$$

These expressions are simpler to compute than those in the previous paragraph.

Table 13.1 shows the values of $\Sigma S_{t'}t'$, $\Sigma t'^2$, and $\bar{S}$ for the time series of AT and T's profits. Inserting these values into equations (13.5) and (13.6), we have

$$b = \frac{24.94}{182} = 0.137$$

$$a = 1.712.$$

Thus, the trend line (shown in Figure 13.5) is

$$S_{t'} = 1.712 + 0.137t'.$$

To interpret this equation, note that $t' = 0$ when $t = 1963$. Thus, this equation says that the estimated trend value of AT and T's profits was 1.712 billions of

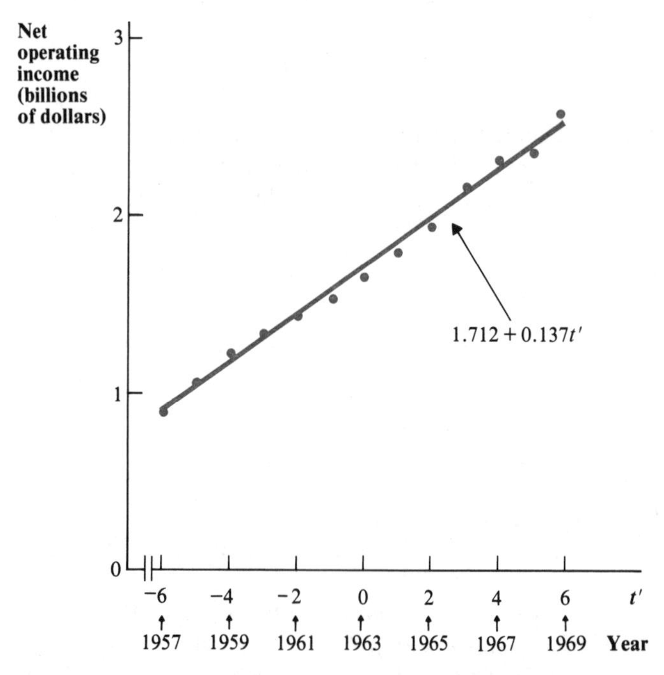

FIGURE 13.5 Linear Trend in Net Operating Income, American Telephone and Telegraph Company, 1957–69

3. Since the time scale is changed so that $\bar{t} = 0$, it is obvious that equation (13.3) simplifies to equation (13.5) and that equation (13.4) simplifies to equation (13.6).

dollars in 1963, and that it increased by 0.137 billions of dollars per year.

As pointed out in the previous paragraph, this trend line has 1963, the middle year in the time series, as its origin. This is because $t' = 0$ for 1963. Suppose that we want to obtain an equation for the trend where t, not t', is the independent variable. In other words, suppose that we want to return to the original time scale where time is measured in calendar years rather than in deviations from 1963. To change scale in this way we need only change a, since b will be unaffected. Why? Because regardless of whether time is measured in calendar years or in deviations from 1963, the annual increase in the trend value is b (=.137) billions of dollars.

How is the value of a changed if we return to the original time scale where time is measured in calendar years (like 1963 or 1964)? Letting a' be the new value of a, it must be true that

$$a' + b(1963) = 1.712.$$

Why is this true? Because if time is measured in calendar years (and the origin is at year zero), the left-hand side of the above equation equals the trend value in 1963. Since this trend value must be the same regardless of which scale is used for measuring time, it must equal 1.712, the former value of a. Thus, since $b = .137$,

$$a' = 1.712 - .137(1963) = 1.712 - 268.931 = -267.219.$$

Consequently, if the original scale is used for time—that is, if t is set equal to 1963 in 1963—the trend line is

$$S_t = -267.219 + 0.137t. \tag{13.7}$$

The following example shows how, once a trend line of this sort has been obtained, it can be used to summarize and describe the long-term behavior of a time series.

Example 13.1 Use the least-squares trend line to estimate AT and T's 1964 profits which would have been expected on the basis of this trend alone. Use this trend line to estimate the average annual increase in AT and T's profits from 1957 to 1969.

Solution: According to equation (13.7), the trend value of AT and T's profits in 1964 was $-267.219 + 0.137(1964)$, or 1.849 billions of dollars. According to this equation, the firm's profits tended to increase annually by 0.137 billions of dollars (that is, by $137 million) during this period.

13.4 Estimation of a Nonlinear Trend

Many time series exhibit nonlinear trends. In some such cases, a quadratic function of time provides an adequate trend. Such a trend can be represented as

$$Y_t = A + B_1t + B_2t^2.$$

To estimate A, B_1, and B_2, we can use the multiple-regression techniques

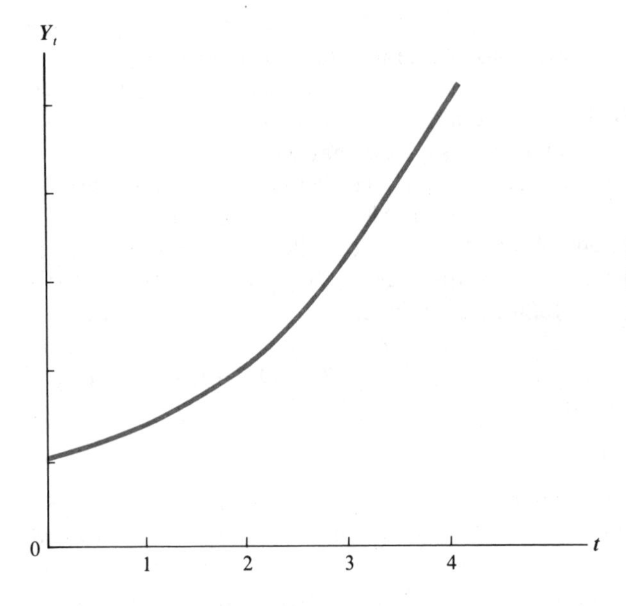

FIGURE 13.6
Exponential Trend,
Assuming $\beta = 1.5$

described in the previous chapter. (Recall Section 12.13.)[4] As noted in the previous chapter, there are standard computer programs for making these computations.

For many variables, an exponential curve provides a better-fitting trend than a quadratic curve. The equation for such a trend (shown in Figure 13.6) is

$$Y_t = \alpha\beta^t, \quad (13.8)$$

where Y_t is the trend value of the time series at time t. A trend of this sort seems to fit many business and economic time series, including the capital-labor ratio and the savings-income ratio of the American economy. If this sort of trend is present, we can take logarithms of both sides of equation (13.8), the result being

$$\log Y_t = A + Bt, \quad (13.9)$$

where $A = \log \alpha$ and $B = \log \beta$. Since equation (13.9) is linear, we can estimate A and B by the method of least squares.[5] Then we can take antilogs of A and B to estimate α and β, the unknown coefficients in equation (13.8). In this way, we can estimate the nonlinear trend shown in equation (13.8).

The following example illustrates the calculation of an exponential trend.

Example 13.2 An economist wants to estimate the trend in the ratio of U.S. consumption expenditures to net national product. Table 13.2 shows this ratio (expressed as a percentage and given at 10-year intervals) from 1900 to 1950. Calculate an exponential trend for this ratio.

Solution: Taking logarithms of the ratios, we obtain the figures in the

4. For those readers who skipped Chapter 12, the only point necessary to grasp is that it is possible to calculate a quadratic trend by the method of least squares.

5. This has been pointed out in Appendix 11.1.

TABLE 13.2

Calculation of Exponential Trend for Consumption as a Percent of Net National Product, United States, 1900–50

Year	(Y)	$\log Y$	t'	$\log Y(t')$	t'^2
1900	84.2	1.9253	−25	−48.1325	625
1910	87.3	1.9410	−15	−29.1150	225
1920	87.1	1.9400	−5	− 9.7000	25
1930	94.3	1.9745	5	9.8725	25
1940	90.6	1.9571	15	29.3565	225
1950	93.8	1.9722	25	49.3050	625
Total		11.7101	0	1.5865	1750

SOURCE: S. Kuznets, *Capital in the American Economy* (New York: National Bureau of Economic Research, 1959).

third column of the table. The least-squares estimator of B is

$$b = \frac{\Sigma \log Y(t')}{\Sigma t'^2} \tag{13.10}$$

and the least-squares estimator of A is

$$a = \frac{\Sigma \log Y}{n}, \tag{13.11}$$

if time is coded as shown in Table 13.2. (That is, time equals −25, −15, −5, 5, 15, 25, which means that 1925 is the new origin.) Inserting the figures from Table 13.2 into equations (13.10) and (13.11), we have

$$b = \frac{1.5865}{1750} = 0.0009066$$

$$a = \frac{11.7101}{6} = 1.9517.$$

Since the antilog of b is 1.002 and the antilog of a is 89.5, the regression line in equation (13.8) is estimated to be

$$Y_t = 89.5\,(1.002)^{t'},$$

where t' is measured in years from 1925.

13.5 The Trend in the Capital-Labor Ratio: A Case Study

An important characteristic of any economy is the ratio of capital to labor. Since capital includes the various means of production such as buildings, equipment, and inventories, the capital-labor ratio shows the average amount of such equipment and related items (in constant dollars) that can be utilized by a member of the labor force. All other things equal, the higher the capital-labor ratio, the more productive a laborer will be. There has been considerable interest in the changes over time in the capital-labor ratio in the

TABLE 13.3

Capital per Worker, United States, Selected Years 1900–53

Year	Capital per worker (dollars)	Year	Capital per worker (dollars)
1900	4,820	1927	6,157
1903	4,843	1930	6,666
1906	4,970	1933	7,326
1909	5,231	1936	6,202
1912	5,309	1939	6,393
1915	5,580	1942	5,736
1918	5,090	1945	5,499
1921	5,995	1948	6,066
1924	5,877	1951	6,049
		1953	6,182

American economy. Nobel laureate Simon Kuznets of Harvard University analyzed this ratio in some of his studies; his estimates are shown in part in Table 13.3. Apparently, there was a steady growth in this ratio from 1900 to 1953; it was about $5000 per person in 1900 and about $6,000 per person after World War II.[6]

Lawrence Klein of the University of Pennsylvania and Richard Kosobud of Wayne State University estimated the trend in this ratio, using Kuznets's data (part of which is shown in Table 13.3).[7] Klein and Kosobud assumed that the trend line conformed to equation (13.8). That is, they assumed that

$$Y_t = \alpha\beta^{t'},$$

where Y_t in this case is the trend value of the capital-labor ratio at time t', and t' is measured in units of six months from the beginning of 1927. In accord with our discussion in the previous section, they regressed log Y on t' to obtain the least-squares estimates of log α and log β, the result being

$$\log Y_t = 3.76126 + 0.0010t'.$$

Taking antilogs of both sides, Klein and Kosobud obtained the following trend line:

$$Y_t = 5771\,(1.0023)^{t'}, \qquad (13.12)$$

since the antilog of 3.76126 is 5771 and the antilog of .0010 is 1.0023. The trend line in equation (13.12) indicates that on the average, the capital-labor ratio has grown at a compound rate of slightly under 1/4 of 1 percent semi-annually. This result is of considerable interest to economists concerned with the long-term economic growth of the United States.

6. The data in Table 13.3 are from S. Kuznets, *Capital in the American Economy* (New York: National Bureau of Economic Research, 1959) and J. Kendrick, *Productivity Trends in the United States* (New York: National Bureau of Economic Research, 1960).

7. L. Klein and R. Kosobud, "Great Ratios of Economics," *Quarterly Journal of Economics*, May 1961; reprinted in E. Mansfield, *Statistics for Business and Economics: Readings and Cases* (New York: Norton, 1980).

13.6 Moving Averages

In some cases, there is no relatively simple mathematical function that can adequately portray the long-term movement of a particular time series. For example, consider the annual price of the common stock of the General Dynamics Corporation during 1956–73. Clearly, this time series, shown in Figure 13.7, does not exhibit a simple linear, exponential, or quadratic trend. Instead, the price of the stock reached a peak in 1958, declined until 1963, increased again until 1967, and fell again during the late 1960s. In cases of this sort, statisticians frequently use moving averages to ***smooth*** the time series —that is, *to generate a smooth curve showing the long-term movements of the series*. Thus, *moving averages are sometimes used to estimate trends where the trends do not lend themselves to the treatment described in previous sections.*

To illustrate what we mean by a moving average, consider Table 13.4 which shows the annual price of General Dynamics common stock from 1956 to 1973. *To obtain a* ***smoothed*** *or trend value for a particular year, we average the figures for an interval of time centered on this year.* For example, suppose that we were to compute a *five-year moving average.* In this case, we would use the mean price during 1956–60 as the trend value for 1958; the mean price during 1957–61 as the trend value for 1959; the mean price during 1958–62 as the trend value for 1960; and so on. The computation of the five-year moving average is shown in Table 13.4, and the moving average itself is plotted in Figure 13.8.

Of course, a five-year moving average is not the only kind. Suppose that we want to compute a *seven-year moving average* for the data in Table 13.4. In this case, we would use the mean price during 1956–62 as the trend value for 1959; the mean price during 1957–63 as the trend value for 1960; the mean price during 1958–64 as the trend value for 1961; and so on. The computation of the seven-year moving average is shown in Table 13.4, and the moving

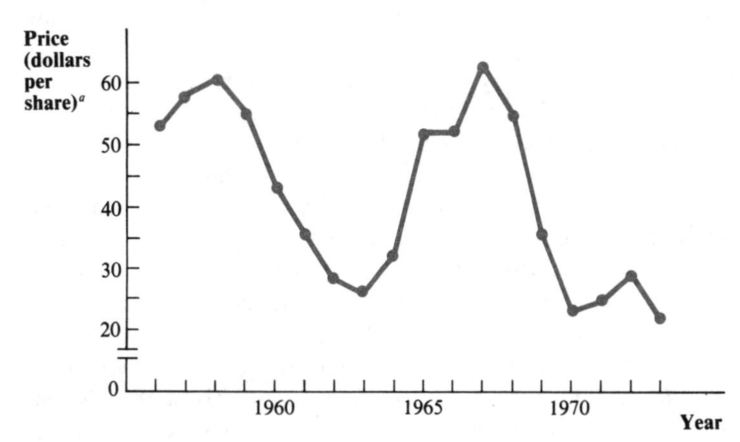

FIGURE 13.7 Annual Price of the Common Stock of General Dynamics Corporation, 1956–73

[a]Mean of annual high and low price.

TABLE 13.4

Five-year and Seven-year Moving Averages of the Price of General Dynamics Common Stock

Year	Price (dollars per share)[a]	Five-year moving total	Five-year moving average	Seven-year moving total	Seven-year moving average
1956	52				
1957	58				
1958	61	269	53.8		
1959	55	252	50.4	332	47.4
1960	43	222	44.4	306	43.7
1961	35	187	37.4	280	40.0
1962	28	164	32.8	271	38.7
1963	26	173	34.6	268	38.3
1964	32	190	38.0	288	41.1
1965	52	225	45.0	308	44.0
1966	52	254	50.8	316	45.1
1967	63	258	51.6	313	44.7
1968	55	229	45.8	306	43.7
1969	36	202	40.4	283	40.4
1970	23	168	33.6	253	36.1
1971	25	135	27.0		
1972	29				
1973	22				

[a]Mean of annual high and low price.

average itself is plotted in Figure 13.8. Other possible kinds of moving averages are nine-year moving averages, ten-year moving averages, and so on.

The basic idea underlying the use of moving averages is that, *if the time series contains certain fluctuations or cycles that tend to recur, the effect of these cycles can be eliminated by taking a moving average where the number of years in the average equals the period of the cycle.* For example, Figure 13.9 shows a time series where there is a simple four-year cycle superimposed on a linear trend. In this case, if a moving average is taken of the time series (the number of years in the average being equal to the period of the cycle, which is four years), the result will be the linear trend only. Obviously, however, practically no business or economic time series contains as simple or regular a cycle as that shown in Figure 13.9. Thus, all that the use of a moving average can do is to smooth a time series, not eliminate the short-term fluctuations entirely.

Finally, several points should be noted concerning moving averages. First, *if a time series is a purely random sequence of numbers, a moving average of this time series will tend to exhibit cyclical fluctuations.* (In terms of our discussion in the previous chapter, this is due to the fact that a moving average is serially correlated.) Thus, one must be careful to recognize that many apparent cycles in moving averages may be spurious. Second, *the peaks and troughs in the moving average may occur at different times than the peaks and*

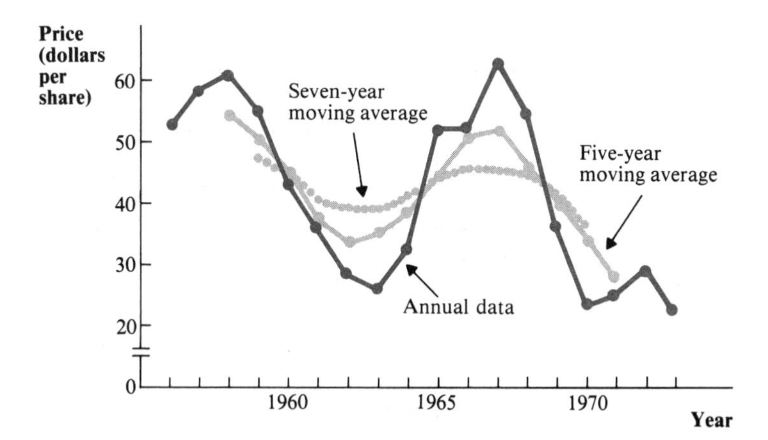

FIGURE 13.8 Five-year and Seven-year Moving Averages of Price of Common Stock of General Dynamics Corporation, 1956–73

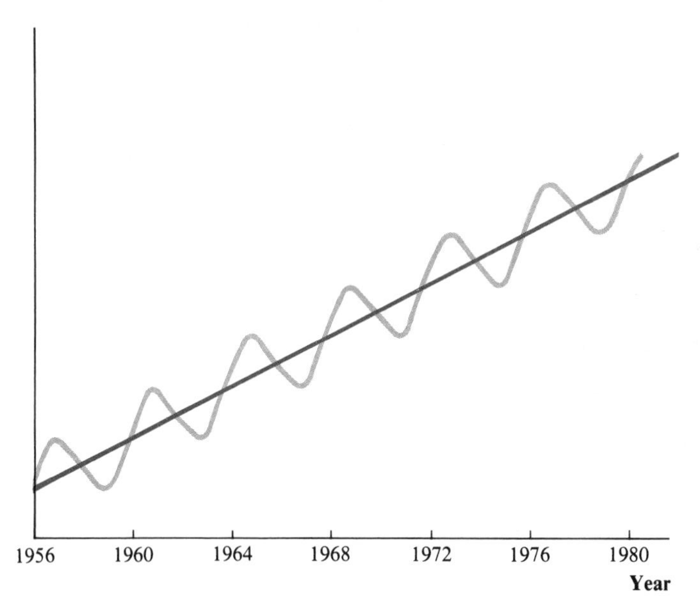

FIGURE 13.9 Time Series Composed of a Linear Trend plus a Simple Four-Year Cycle

troughs in the original time series. (For example, as shown in Figure 13.8, the price of General Dynamics common stock reached a trough in 1963, but the five-year moving average reached a trough in 1962.) Third, *a moving average of this type cannot be calculated for the latest or earliest years in a time series*, since the average depends on numbers that precede or occur after the time series.

13.7 Exponential Smoothing

Another method of smoothing a time series is to use *exponential smoothing.* According to this method, *the trend value at time* t *is a weighted average of all available previous values, where the weights decline geometrically as one goes backward in time.* Using the moving average techniques described in the previous section, the trend value at time t is obtained by calculating an unweighted average of the observations centered at time t, the number of observations being 5, 7, or some other fixed number. In contrast, exponential smoothing uses *all* the previous observations to obtain the trend value at time t, and the weight attached to each observation declines *geometrically* with the age of the observation.

To illustrate the nature of exponential smoothing, let's assume that a firm has been in existence for five years and that its sales have been \$1 million, \$3 million, \$3 million, \$2 million, and \$4 million. (See Figure 13.10.) Then, the trend value in the fifth year would be a weighted average of \$1 million, \$3 million, \$3 million, \$2 million, and \$4 million, where the weights decline geometrically as we go backward in time. Specifically, the weight attached to the observation at time t equals θ, the weight attached to the observation at time $t - 1$ equals $(1 - \theta)\theta$, the weight attached to the observation at time $t - 2$ equals $(1 - \theta)^2\theta$, the weight attached to the observation at time $t - 3$ equals $(1 - \theta)^3\theta, \ldots$, and the weight attached to the observation at the earliest relevant point in time (time 0) equals $(1 - \theta)^t$. Clearly, the weights decline geometrically as one goes backward in time; that is, the weight attached to the observation at time $t - 1$ is $(1 - \theta)$ times the weight attached to the observation at time t; the weight attached to the observation at time $t - 2$ is $(1 - \theta)$ times the weight attached to the observation at time $t - 1$; and so on.

To calculate an exponentially smoothed time series, it is necessary to choose a value of θ, which is designated the ***smoothing constant.*** If we choose a value of 0.5 for θ, the exponentially smoothed value of the firm's sales in each of the five years is as follows:

$$S_0 = 1$$

$$S_1 = (.5)(3) + (1 - .5)(1) = 2$$

$$S_2 = (.5)(3) + (1 - .5)(.5)(3) + (1 - .5)^2(1) = 2.5$$

$$S_3 = (.5)(2) + (1 - .5)(.5)(3) + (1 - .5)^2(.5)(3) + (1 - .5)^3(1) = 2.25$$

$$S_4 = (.5)(4) + (1 - .5)(.5)(2) + (1 - .5)^2(.5)(3) + (1 - .5)^3(.5)(3) + (1 - .5)^4(1) = 3.125,$$

where S_0 is the exponentially smoothed value of the firm's sales in the first year of its existence, S_1 is this value in the second year, S_2 the value in the third year, and so on. Figure 13.10 shows both the original time series and the exponentially smoothed time series.

A noteworthy characteristic of an exponentially smoothed time series is

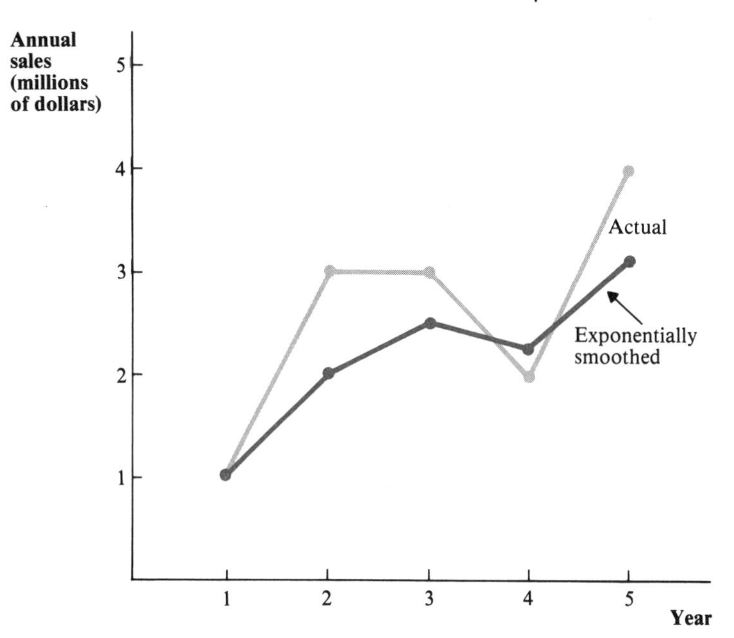

FIGURE 13.10 Sales of Firm, Actual and Exponentially Smoothed[a]
[a]Year 1 is the firm's first year in existence, Year 2 is its second year, and so on.

that in order to calculate the value of such a smoothed time series at time t all we need is the value of the smoothed time series at time $t - 1$ and the actual value of the time series at time t. This is because the smoothed value of the time series at time t is a simple weighted average of the smoothed value at time $t - 1$ and the actual value at time t. If S_t is the smoothed value at time t,

$$\boxed{S_t = \theta Y(t) + (1 - \theta)S_{t-1},} \qquad (13.13)$$

where $Y(t)$ is the value of the time series at time t.[8] This means that in order to calculate an exponentially smoothed time series, one does not need to keep all the previous values of the actual time series. For example, the firm in the previous paragraph does not need to keep all its previous sales figures; *it only*

8. It is easy to prove that equation (13.13) is true . If $Y(t)$ is the actual value of the time series at time t, then equation (13.13) implies that

$$\begin{aligned}
S_t &= \theta Y(t) + (1 - \theta)S_{t-1} \\
&= \theta Y(t) + (1 - \theta)[\theta Y(t - 1) + (1 - \theta)S_{t-2}] \\
&= \theta Y(t) + (1 - \theta)\theta Y(t - 1) + (1 - \theta)^2[\theta Y(t - 2) + (1 - \theta)S_{t-3}] \\
&= \theta Y(t) + (1 - \theta)\theta Y(t - 1) + (1 - \theta)^2\theta Y(t - 2) + \ldots + (1 - \theta)^t Y(0).
\end{aligned}$$

Since the last expression on the right-hand side is equivalent to the definition of an exponentially smoothed time series in the first paragraph of this section, it follows that equation (13.13) is true.

needs to keep the value of the exponentially smoothed sales in the previous year. From this information alone (together with the present value of sales and the smoothing constant), the firm can calculate the smoothed current sales figure. For example, if the firm's sales in its sixth year of existence are $5 million, the smoothed value of sales for the sixth year is

$$(0.5)(5) + (1 - 0.5)(3.125) = 4.062,$$

or $4.062 million.

In choosing the value of the smoothing constant θ, it is essential that a number between 0 and 1 be picked. In other words, it is essential that $0 \leqslant \theta \leqslant 1$. If θ is close to 1, past values of the time series are given relatively little weight (compared to recent values) in calculating smoothed values. If θ is close to 0, past values of the time series are given considerable weight (as compared to recent values) in calculating smoothed values. If the time series contains a great deal of random variation, it is often advisable to choose a relatively small value of θ since this results in relatively little weight being put on $Y(t)$, which is more influenced than S_{t-1} by this variation. On the other hand, if one wants the smoothed time series to reflect relatively quickly whatever changes occur in the average level of the time series, the value of θ should be set at a high level.

Exponential smoothing has both theoretical and practical advantages. If the time series conforms to certain assumptions concerning its probability distribution, it can be shown (with more advanced mathematical techniques than are appropriate here) that exponential smoothing has desirable theoretical properties.[9] As for its practical advantages, it is obvious that exponential smoothing can be carried out very simply if the computations are based on equation (13.13). For these reasons, exponential smoothing is sometimes preferred over the use of simple moving averages. (For a discussion of the use of exponential smoothing for forecasting purposes, see Appendix 13.2.)

The following example illustrates the construction of an exponentially smoothed time series.

Example 13.3 A bicycle shop sells the following number of bicycles from 1968 to 1979. Calculate an exponentially smoothed time series from these data. (Set θ equal to 1/3.)

Year	Number sold (thousands)	Year	Number sold (thousands)
1968	3	1974	6
1969	3	1975	6
1970	3	1976	6
1971	3	1977	9
1972	6	1978	10
1973	6	1979	12

Solution: Let S_0 be the exponentially smoothed sales (in thousands) in

9. For example, see R. G. Brown, *Smoothing, Forecasting, and Prediction of Discrete Time Series* (Englewood Cliffs: Prentice-Hall, 1963).

1968, S_1 the exponentially smoothed sales in 1969, and so on. Then, from equation (13.13),

$$S_0 = 3.00$$
$$S_1 = 1/3(3) + 2/3(3) = 3.00$$
$$S_2 = 1/3(3) + 2/3(3) = 3.00$$
$$S_3 = 1/3(3) + 2/3(3) = 3.00$$
$$S_4 = 1/3(6) + 2/3(3) = 4.00$$
$$S_5 = 1/3(6) + 2/3(4) = 4.67$$
$$S_6 = 1/3(6) + 2/3(4.67) = 5.11$$
$$S_7 = 1/3(6) + 2/3(5.11) = 5.41$$
$$S_8 = 1/3(6) + 2/3(5.41) = 5.61$$
$$S_9 = 1/3(9) + 2/3(5.61) = 6.74$$
$$S_{10} = 1/3(10) + 2/3(6.74) = 7.83$$
$$S_{11} = 1/3(12) + 2/3(7.83) = 9.22$$

EXERCISES

13.1 "Trends are useless. By itself time does not cause a variable to change." Comment and evaluate.

13.2 "Although a trend may be a useful summary description of the long-term movement in a time series, it tells us little or nothing about why the time series behaved in this way." Comment and evaluate.

13.3 "Trends are inexorable, if they fit the data well. Once you find a trend that fits past data, you can be sure it will continue into the future." Comment and evaluate.

13.4 "You can fit a trend by plotting the time series against time and by drawing a freehand curve that seems to fit well. You don't have to bother with least squares techniques, since the assumptions underlying the use of least squares are not met." Comment and evaluate.

13.5 The Union Carbide Corporation's sales during 1960–1975 are given below:

Year	Sales (billions of dollars)	Year	Sales (billions of dollars)
1960	1.5	1968	2.7
1961	1.6	1969	2.9
1962	1.6	1970	3.0
1963	1.7	1971	3.0
1964	1.9	1972	3.3
1965	2.1	1973	3.9
1966	2.2	1974	5.3
1967	2.5	1975	5.7

Fit a linear (least-squares) trend line to these data.

13.6 (a) In Exercise 13.5, what is the equation for the trend line if the origin is set at 1968? (b) What is the equation for the trend line if the origin is set at 1965?

13.7 Fit an exponential trend line to the data in Exercise 13.5.

13.8 (a) Plot the linear trend line (from Exercise 13.5) against the time series in Exercise 13.5. (b) Plot the exponential trend line (from Exercise 13.7) on the same graph. (c) Which trend line—the linear or the exponential—seems to provide a better fit to the data?

13.9 The capacity utilization rate in all U.S. manufacturing is shown below for 1947 to 1974.[10]

Year	Percent utilized of all manufacturing capacity
1947	95
1948	92
1949	81
1950	90
1951	91
1952	88
1953	92
1954	83
1955	91
1956	90
1957	86
1958	76
1959	82
1960	80
1961	78
1962	81
1963	83
1964	86
1965	91
1966	96
1967	93
1968	94
1969	96
1970	88
1971	85
1972	90
1973	96
1974	91

Calculate a three-year moving average of this time series.

13.10 Calculate a five-year moving average of the time series in Exercise 13.9.

13.11 Calculate a seven-year moving average of the time series in Exercise 13.9.

13.12 (a) Plot the time series in Exercise 13.9, together with the three-year moving average (derived in Exercise 13.9). (b) Based on this diagram, does it appear that this time series has a strong linear trend?

13.13 Suppose that a statistician decides to change the smoothing constant from 1/3 to 1/5. He has the exponentially smoothed value of a time series for the previous period, S_{t-1}, based on the old smoothing constant. Is it true that $S_t = 1/5Y + (1 - 1/5)S_{t-1}$, where S_{t-1} is based on $\theta = 1/3$? Why, or why not?

10. See *Economic Report of the President, 1976* (Washington, D.C.: Government Printing Office, 1976), p. 211. This is the Wharton series.

13.14 Using the data in Exercise 13.5, calculate the exponentially smoothed value of Union Carbide's annual sales during 1960–1964, if the smoothing constant equals 1/5.

13.15 Using the data in Exercise 13.5, calculate the exponentially smoothed value of Union Carbide's annual sales during 1960–1964, if the smoothing constant equals 1/10.

13.8 Seasonal Variation

Many time series are composed of monthly or quarterly rather than annual data. For such time series, decision makers and statisticians must recognize that seasonal variation is likely to be present in the series. As pointed out in Section 13.2, seasonal variation in many economic time series is due to the weather. However, this is not always the case, as illustrated by a firm that sells ornaments for Christmas trees. There is likely to be a pronounced seasonal variation in this firm's sales, but this is due to the location of a specific holiday (Christmas) on the calendar, not to the weather. Still other reasons for seasonal variation are the fact that some industries tend to grant vacations at a particular time of year, or that taxes have to be paid at particular times of the year, or that schools tend to open at particular times of the year.

As we stated early in this chapter, statisticians have devised methods for estimating the pattern of seasonal variation in a particular time series. In other words, they can determine the extent to which a particular month or quarter is likely to differ from what would be expected on the basis of the trend and cyclical variation in the time series. (In terms of the traditional model in equation (13.1), statisticians can determine the value of S for each month or quarter.) For example, the statistician for a manufacturer of soft drinks may tell the company's managers that U.S. production of soft drinks tends in June to be 5.9 percent higher than what the trend and cyclical variation in soft drink production would indicate. Or the statistician may find that U.S. production of soft drinks in December tends to be 7.0 percent lower than the trend and cyclical variation would indicate.

The seasonal variation in a particular time series is described by a figure for each month, the **seasonal index,** *which shows the way in which that month tends to depart from what would be expected on the basis of the trend and cyclical variation in the time series.* For example, Table 13.5 shows the seasonal variation in U.S. production of soft drinks. January's production tends to be about 93.4 percent of the amount expected on the basis of trend and cyclical variation; February's production tends to be about 89.3 percent of this amount; March's production tends to be about 90.7 percent of this amount; and so on. Figures of this sort can be used in a number of ways. *One important application is to forecast what the time series will be in the future.* For example, suppose that, based on the trend and cyclical variation, it appears likely that about 30 million gallons of soft drinks will be produced next January. If this is the case, a reasonable forecast of actual January production is .934 (30 million) = 28.02 million gallons, since January's production tends to be 93.4 percent of the amount expected on the basis of trend and cyclical variation.

In addition, *a knowledge of seasonal variation is often useful in interpreting*

TABLE 13.5
Seasonal Variation in Production of Soft Drinks in the United States

Month	Seasonal index
January	93.4
February	89.3
March	90.7
April	94.9
May	99.0
June	105.9
July	112.4
August	113.4
September	108.3
October	103.9
November	95.8
December	93.0

current or recent developments. For example, suppose that it was announced in June 1980 that soft drink output in the United States had increased by about 4 percent between April and May. Firms and government agencies interested in the soft drink industry were likely to ask: To what extent is this increase due to seasonal factors? In other words, even if no increase occurred between April and May in the trend value, and if cyclical factors remained constant, would one expect an increase of this magnitude on the basis of seasonal factors alone? Looking at Table 13.5, since May's output is generally about 99.0 percent of what would be expected on the basis of trend and cyclical variation, and since April's output is generally about 94.9 percent of this amount, one would expect May's output to be 4.3 percent higher than April's, even if the trend value and cyclical variation were the same for each month.[11] Thus, the observed increase was no more than would be expected on the basis of seasonal factors alone.

Another way of handling this problem is to **deseasonalize** the data, a procedure that is often carried out in business and government. *By deseasonalizing we mean removing the seasonal element from the data.* Recall that according to the traditional model, the value of a time series at time t is equal to the seasonal variation at t times a combination of trend, cyclical, and irregular factors. *By dividing the value of the time series at time* t *by the seasonal index for time* t *(divided by 100), we obtain the deseasonalized value at time* t. This deseasonalized value should be purged of the effects of seasonal variation.

For example, if each month's soft drink output in 1980 was as shown in Table 13.6, we can use the seasonal indexes in Table 13.5 to remove the seasonal element from each month. For example, since January is generally

11. If May's output is 99.0 percent of what would be expected on the basis of trend and cyclical variation, and if April's output is 94.9 percent of the same amount, May's output is 99.0/94.9 × April's output, or 104.3 percent of April's output.

TABLE 13.6
Deseasonalizing Production Figures for Soft Drinks in 1980

Month	Output (millions of gallons) (1)	Seasonal index (2)	Deseasonalized output (millions of gallons) (1) ÷ [(2)/100]
January	25	93.4	26.8
February	24	89.3	26.9
March	24	90.7	26.5
April	25	94.9	26.3
May	26	99.0	26.3
June	28	105.9	26.4
July	30	112.4	26.7
August	30	113.4	26.5
September	30	108.3	27.7
October	31	103.9	29.8
November	30	95.8	31.3
December	30	93.0	32.3

93.4 percent of the deseasonalized amount, the deseasonalized output in January 1980 was 25 million ÷ .934, or 26.8 million gallons. Similarly, since July is generally 112.4 percent of the deseasonalized amount, the deseasonalized output in July 1980 was 30 million ÷ 1.124, or 26.7 million gallons. Table 13.6 shows the result for each month of 1980. To see whether the 4 percent increase between April and May of 1980 was greater than would be expected on the basis of seasonal variation, we can compare the deseasonalized figures for April and May. Since May's deseasonalized output was no greater than April's, the increase between April and May was no greater than would be expected due to seasonal factors alone. Of course, this is precisely the same result that we obtained in the next-to-last paragraph.

The example below further illustrates the construction and use of deseasonalized data.

Example 13.4 A statistician estimates the seasonal index for each month for a firm's sales, the result being

January	90	May	100	September	110
February	90	June	100	October	110
March	90	July	100	November	110
April	90	August	100	December	110

If actual sales in January of this year are $20 million, and if each month's sales rose this year by $100,000, what was the deseasonalized value of sales for each month? Taking account of seasonal factors, did sales generally tend to rise or fall during this year?

Solution: The deseasonalized value of sales each month (in millions of dollars) was

January	20.0 ÷ .90 = 22.2	July	20.6 ÷ 1.00 = 20.6
February	20.1 ÷ .90 = 22.3	August	20.7 ÷ 1.00 = 20.7
March	20.2 ÷ .90 = 22.4	September	20.8 ÷ 1.10 = 18.9
April	20.3 ÷ .90 = 22.6	October	20.9 ÷ 1.10 = 19.0
May	20.4 ÷ 1.00 = 20.4	November	21.0 ÷ 1.10 = 19.1
June	20.5 ÷ 1.00 = 20.5	December	21.1 ÷ 1.10 = 19.2

It is therefore evident that when seasonal factors are taken into account, sales generally tended to fall during this year.

13.9 Calculation of a Seasonal Index: Ratio-to-Moving-Average Method

How can we calculate a seasonal index like that shown in Table 13.5? In other words, how do we know that January's output tends to be 93.4 percent of the amount that would be expected on the basis of trend and cyclical variation, that February's output tends to be 89.3 percent of this amount, and so on? The most commonly used method is to begin by computing a 12-month moving average of the monthly time series. Table 13.7 shows the 12-month moving average computed for the time series for the U.S. output of soft drinks. Note, however, that this moving average, shown in the second column of Table 13.7, is centered between two months rather than at the middle of a month. Thus, to center the moving average at the *middle* of each month, we take the average of two moving averages: (1) the one centered at the *beginning* of the month in question, and (2) the one centered at the *end* of the month in question. The results are shown in the third column of Table 13.7.

The moving average pertaining to the middle of a given month is viewed as a good approximation of what would be expected on the basis of trend and cyclical variation. Why? Because, as pointed out in Section 13.6, a moving average should eliminate the effect of a cycle if the length of the moving average is equal to the period of the cycle. Clearly, the period of the seasonal variation is 12 months; thus, a 12-month moving average should eliminate the seasonal variation if this variation is completely regular from year to year. Stated differently, the 12-month moving average, since it is 1/12 of the annual total (for the year centered at the middle of the month), should be essentially free of seasonal variation, because seasonal variation is defined so that it averages out over a one-year period. Besides eliminating most of the seasonal variation, the 12-month moving average eliminates much of the irregular variation, with the result that it is viewed as a good approximation of what would be expected on the basis of trend and cyclical variation.

The next step in constructing the monthly seasonal index is to *divide the actual value for each month by the moving average.* According to the traditional model in equation (13.1), the actual value equals $T \times S \times C \times I$, and the moving average is approximately equal to $T \times C$. Thus, the ratio of the actual value to the moving average, shown in the last column of Table 13.7, is an estimate of $S \times I$. Because these ratios reflect both seasonal and irregular variation, we want to eliminate the irregular variation to the extent possible.

TABLE 13.7

Computation of Centered 12-Month Moving Average for Production of Soft Drinks, U.S., 1967–71

Month	Soft drink output (1967 = 100.0)	12-Month moving average[a]	Centered 12-month moving average[b]	Actual output as a ratio of centered moving average[c]
January 1967	88.0			
February	86.1			
March	91.8			
April	96.4			
May	100.1			
June	107.5			
		100.0		
July	111.0		100.4	1.106
		100.8		
August	114.1		101.2	1.127
		101.6		
September	108.9		101.8	1.070
		102.0		
October	103.4		102.3	1.011
		102.5		
November	95.7		102.9	.930
		103.2		
December	96.7		103.7	.932
		104.1		
January 1968	97.9		104.8	.934
		105.6		
February	95.5		106.2	.899
		106.8		
March	97.0		107.5	.902
		108.2		
April	102.7		108.9	.943
		109.6		
May	108.5		110.3	.984
		111.0		
June	118.0		111.4	1.059
		111.8		
July	128.5		112.4	1.143
		113.0		
August	129.3		113.5	1.139
		114.1		
September	125.4		114.6	1.094
		115.1		
October	120.0		115.7	1.037
		116.4		
November	112.9		117.0	.965
		117.7		
December	106.2		118.3	.898
		118.9		
January 1969	111.5		119.5	.933
		120.2		
February	108.9		120.9	.901
		121.7		
March	108.7		122.5	.887
		123.4		
April	118.7		124.0	.957
		124.6		
May	124.2		125.0	.994
		125.4		
June	132.1		126.1	1.048
		126.7		
July	144.7		127.0	1.139
		127.3		
August	146.5		127.6	1.148
		128.0		
September	145.9		128.5	1.135
		129.0		
October	134.3		129.3	1.039
		129.5		
November	123.5		130.0	.950
		130.4		
December	121.4		130.9	.927
		131.4		
January 1970	118.9		131.6	.903
		131.8		
February	117.0		132.0	.886
		132.1		
March	121.2		132.1	.917
		132.1		
April	124.7		132.5	.941
		132.9		
May	134.3		133.3	1.008
		133.6		
June	144.9		134.0	1.081
		134.4		

(table continues)

TABLE 13.7 (continued)

Month	Soft drink output (1967 = 100.0)	12-month moving average[a]	Centered 12-month moving average[b]	Actual output as a ratio of centered moving average[c]
July	149.3	135.3	134.8	1.108
August	150.2	135.7	135.5	1.108
September	145.7	136.3	136.0	1.071
October	143.6	137.2	136.7	1.050
November	132.4	137.6	137.4	.964
December	130.3	138.2	137.9	.945
January 1971	129.9	139.5	138.8	.936
February	121.4	140.3	139.9	.868
March	128.4	141.8	141.0	.911
April	135.7	142.8	142.3	.954
May	139.1	144.0	143.4	.970
June	152.9	145.2	144.6	1.057
July	163.9			
August	159.8			
September	164.3			
October	156.0			
November	146.8			
December	144.3			

SOURCE: Federal Reserve Board, *Industrial Production,* 1971 edition.

[a] To obtain each figure in this column, the first step is to calculate the 12-month moving sum of the output figures. Thus, the sum of the first 12 output figures on page 469 is 88.0 + 86.1 + . . . + 95.7 + 96.7 = 1199.7. Dividing this sum by 12, we get 100.0, the first figure in this column. The sum of the second through thirteenth output figures is 86.1 + 91.8 + . . . + 96.7 + 97.9 = 1209.6. Dividing this sum by 12, we get 100.8, the second figure in this column. Continuing in this way, we derive each figure in this column.

[b] To obtain each figure in this column, average (1) the figure that is 1/2 line ***above*** it in the previous column and (2) the figure that is 1/2 line ***below*** it in the previous column. Thus, the top figure on page 469 is the average of 100.0 and 100.8, or 100.4. And the next figure is the average of 100.8 and 101.6, or 101.2.

[c] To obtain each figure in this column, divide the figure in the first column by the figure in the third column.

To do this, *we calculate the median of the ratios for each month,* which should be relatively free from the effects of irregular variation. Table 13.8 shows that (after all ratios are multiplied by 100) the median of the January ratios is 93.35, the median of the February ratios is 89.25, and so on.

Finally, *we adjust each month's median so that the mean value of the 12 monthly indexes equals 100.* To understand why, suppose that the deseasonalized value of a time series is the same for each month of the year. Due to seasonal variation, the actual value varies from month to month. However, the seasonal variation cancels out *for the year as a whole.* In this case, if we sum up the values for the year as a whole, the sum of the actual values must equal the sum of the deseasonalized values. That is, the mean value of the monthly indexes must be 100, and the sum of the indexes for all 12 months must be

TABLE 13.8
Computation of Seasonal Index for Soft Drink Production

Month	Ratios (times 100)				Median	Seasonal index
January	93.4	93.3	90.3	93.6	93.35	93.4
February	89.9	90.1	88.6	86.8	89.25	89.3
March	90.2	88.7	91.7	91.1	90.65	90.7
April	94.3	95.7	94.1	95.4	94.85	94.9
May	98.4	99.4	100.8	97.0	98.90	99.0
June	105.9	104.8	108.1	105.7	105.80	105.9
July	110.6	114.3	113.9	110.8	112.35	112.4
August	112.7	113.9	114.8	110.8	113.30	113.4
September	107.0	109.4	113.5	107.1	108.25	108.3
October	101.1	103.7	103.9	105.0	103.80	103.9
November	93.0	96.5	95.0	96.4	95.70	95.8
December	93.2	89.8	92.7	94.5	92.95	93.0
Total					1199.15	1200.0

1200. To be sure that this holds true for the monthly indexes we are constructing, we must multiply each median (in the next-to-last column of Table 13.8) by 1200 and divide the result by the sum of the medians, 1199.15. The results are shown in the last column of Table 13.8.

The figures in the last column are the monthly indexes we want. As pointed out in the previous section, these indexes have a variety of uses in business and government. However, now that we have shown how they are calculated, it is evident that they are based on a number of simplifying assumptions. In particular, we have assumed implicitly that the pattern of seasonal variation has and will remain relatively constant over time. This can be a dangerous assumption. For example, the seasonal variation in electric power consumption has been affected by the advent of air conditioning. There are ways to take account of changes of this sort in seasonal variation, but the simple methods described here make no attempt to do so. Nevertheless, these procedures are used very extensively in business and economics, and are a standard part of the statistician's equipment.

13.10 Calculation of Seasonal Variation: Dummy-Variable Method[12]

Another way of calculating the seasonal variation in a time series is to use the dummy-variable technique discussed in the previous chapter on multiple regression. Suppose, for example, that a statistician has a time series composed of quarterly values; that is, each observation pertains to the first, second, third, or fourth quarter of a year. If the statistician believes that the time series has a

12. This section should be skipped if the reader has not covered the latter part of Chapter 12.

linear trend, he or she may assume that the value of the observation at time t equals

$$Y(t) = A + B_1t + B_2Q_1 + B_3Q_2 + B_4Q_3 + e_t, \tag{13.14}$$

where Q_1 equals 1 if time t is the first quarter and 0 otherwise, Q_2 equals 1 if time t is the second quarter and zero otherwise, Q_3 equals 1 if time t is the third quarter and 0 otherwise, and e_t equals $Y_t - E(Y(t))$.

It is important to understand the meaning of B_1, B_2, B_3, and B_4 in equation (13.14). Clearly, B_1 is the slope of the linear trend, but what are B_2, B_3, and B_4? The answer is that B_2 *is the difference between the expected value of an observation in the first quarter and the expected value of an observation in the fourth quarter, holding the trend value constant.* To see that this is true, note that if an observation pertains to the first quarter, its expected value equals

$$A + B_1t + B_2,$$

according to equation (13.14). Similarly, if an observation pertains to the fourth quarter, its expected value equals

$$A + B_1t,$$

according to equation (13.14). Thus, if the effect of the trend could be held constant (and if t were the same), the difference between the expected value of an observation in the first quarter and the expected value of an observation in the fourth quarter would be

$$(A + B_1t + B_2) - (A + B_1t) = B_2,$$

which is what we set out to prove. Holding constant the trend value, one can show in the same way that B_3 *is the difference between the expected value of an observation in the second quarter and the expected value of an observation in the fourth quarter; and* B_4 *is the difference between the expected value of an observation in the third quarter and the expected value of an observation in the fourth quarter.*

Thus, if equation (13.14) is valid, the statistician can represent the seasonal variation in the time series by the three numbers B_2, B_3, and B_4. To estimate each of these numbers, ordinary multiple regression techniques can be used.[13] The dependent variable is $Y(t)$, and the independent variables are t, Q_1, Q_2, and Q_3. The latter three independent variables—Q_1, Q_2, and Q_3—are dummy variables (as we know from the previous chapter); that is, each of these can assume only two values: 0 or 1. As indicated in the previous chapter, the constants in equation (13.14)—A, B_1, B_2, B_3, and B_4—can be estimated by the ordinary least-squares procedure.

When using this dummy-variable technique, the statistician assumes that seasonal effects are *added* to the trend value, as shown in equation (13.14). This differs from the traditional model in equation (13.1), where it is assumed

13. Of course, it should be recalled that multiple regression assumes that the error terms are independent and that the standard deviation of the error terms is constant. These assumptions may not be met. For example, the error terms may be serially correlated. See Chapter 12.

that seasonal effects *multiply* the trend value. (See footnote 1.) The application of the appropriate assumption to the appropriate case is facilitated by knowing that either assumption can be made.

The following example shows how the dummy-variable technique can be used to estimate the seasonal variation in monthly data.

Example 13.5 A statistician has monthly data concerning the sales of a particular firm. Indicate how he can estimate the seasonal variation (from month to month) in sales if there is a linear trend.

Solution: The statistician can assume that

$$S(t) = A + B_1t + B_2M_1 + B_3M_2 + \dots + B_{12}M_{11} + e_t, \qquad (13.15)$$

where $S(t)$ is the firm's sales in the month t, M_1 equals 1 if month t is January and 0 otherwise, . . . , M_{11} equals 1 if month t is November and 0 otherwise, and e_t equals $S(t) - E(S(t))$. Using ordinary multiple regression techniques, the statistician can estimate $A, B_1, B_2, \dots, B_{12}$. The estimates of $B_2, B_3, \dots, B_{11}$, and B_{12} indicate the seasonal variation in the firm's sales. In particular, B_2 is the difference between the expected value of sales for January and December, B_3 is the difference between the expected value of sales for February and December, and so on, until B_{12} is the difference between the expected value of sales for November and December (holding constant the trend value for each case).

13.11 Cyclical Variation

As pointed out at the beginning of this chapter, time series in business and economics frequently exhibit cyclical variation, such variation often being termed the *business cycle*. To illustrate what we mean by the *business cycle* or *business fluctuations*, let's look at how national output has grown in the United States since World War I. Figure 13.11 shows the behavior of gross national product (GNP) in constant dollars in the United States since 1919. It is clear that output has grown considerably during this period; indeed, GNP is more than five times what it was 50 years ago. It is also clear that this growth has not been steady. On the contrary, although the long-term trend has been upward, there have been periods—1919–21, 1929–33, 1937–38, 1944–46, 1948–49, 1953–54, 1957–58, 1969–70, and 1973–75—when national output has declined.

Let's define the *full-employment level* of GNP as the total amount of goods and services that could have been produced if there had been full employment. Figure 13.11 shows that national output tends to rise and approach its full-employment level for a while, then falter and fall below this level, then rise to approach it once more, then fall below it again, and so on. For example, output remained close to the full-employment level in the prosperous mid-1920s, fell far below this level in the depressed 1930s, and rose again to this level once we entered World War II. This movement of national output is sometimes called the business cycle, but it must be recognized that these "cycles" are far from regular or consistent. (On the contrary, they are very irregular.)

FIGURE 13.11 Gross National Product (1958 Dollars), United States 1918–78 (Excluding World War II)

Each cycle can be divided by definition into four phases, as shown in Figure 13.12. *The* ***trough*** *is the point where national output is lowest relative to its full-employment level.* ***Expansion*** *is the subsequent phase during which national output rises. The* ***peak*** *occurs when national output is highest relative to its full-employment level. Finally,* ***recession*** *is the subsequent phase during which national output falls.*[14] Besides these four phases, two other terms are frequently used to describe stages of the business cycle. A *depression* is a period when national output is well below its full-employment level; it is a severe recession. (Depressions are, of course, periods of excessive unemployment.) *Prosperity* is

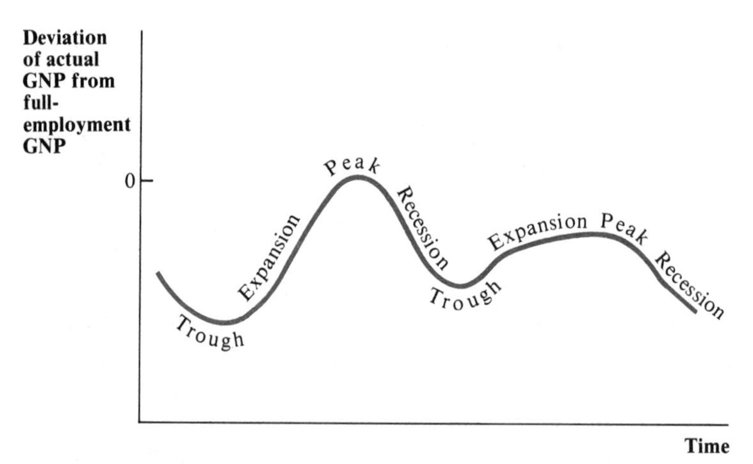

FIGURE 13.12 Four Phases of Business Fluctuations

14. The peak and trough may also be defined in terms of deviations from the long-term trend of GNP rather than in terms of deviations from the full-employment level of GNP.

a period when national output is close to its full-employment level. Prosperity, if total spending is too high relative to potential output, can be a time of rapid inflation. Of course, in some business cycles the peak may not be a period of prosperity because output may be below its full-employment level; in the same way, the trough may not be a period of depression because output may not be far below its full-employment level.

Many business and economic time series go up and down with the business cycle. For example, industrial output tends to be above its trend line at the peak of the business cycle and tends to fall below its trend line at the trough. Similarly, such diverse series as the money supply, industrial employment, and stock prices reflect the business cycle. However, not all series go up and down at exactly the same time. Some turn upward before others at a trough, and some turn downward before others at a peak. As we shall see in Section 13.13, the fact that some time series tend to precede others in cyclical variation sometimes is used to forecast the pace of economic activity.

13.12 Elementary Forecasting Techniques

One of the most difficult and important tasks that statisticians confront is forecasting. (As a well-known politician once pointed out with extraordinary incisiveness: "Today the real problem is the future.") In general, all forecasting techniques are extremely fallible, and all forecasts should be treated with caution. Nonetheless, businesses and government agencies have no choice but to make forecasts, however crude. Since governments, firms, and private individuals must continually make decisions that hinge on what they expect will happen, they must make implicit forecasts even if they do not make explicit ones. Thus, the central question is how best to forecast, not whether to forecast. In this section we shall present some elementary forecasting techniques; but these should be viewed as crude first approximations rather than as highly sophisticated methods. (More sophisticated techniques are taken up briefly in Section 13.14.)

The simplest type of forecasting method is a straightforward extrapolation of a trend. For example, let's return to the American Telephone and Telegraph Company. At the end of 1969, suppose that AT and T wanted to forecast its 1970 profits. During 1957–69, we know from Section 13.3 that the firm's profits could be represented by the following trend line:

$$S_{t'} = 1.712 + 0.137t',$$

where t' = the year in question minus 1963. To forecast its 1970 sales, AT and T could simply insert 7 (that is, 1970 − 1963) for t' in the above equation. Thus, the forecast for 1970 is

$$1.712 + 0.137(7) = 2.671,$$

or 2.671 billions of dollars. As shown in Figure 13.13, this forecast is a simple extension, or extrapolation, of the trend line into the future.[15]

15. How accurate would this forecast have been? American Telephone and Telegraph's profits for 1970 were $2.82 billion, so this forecast was in error by about 6 percent.

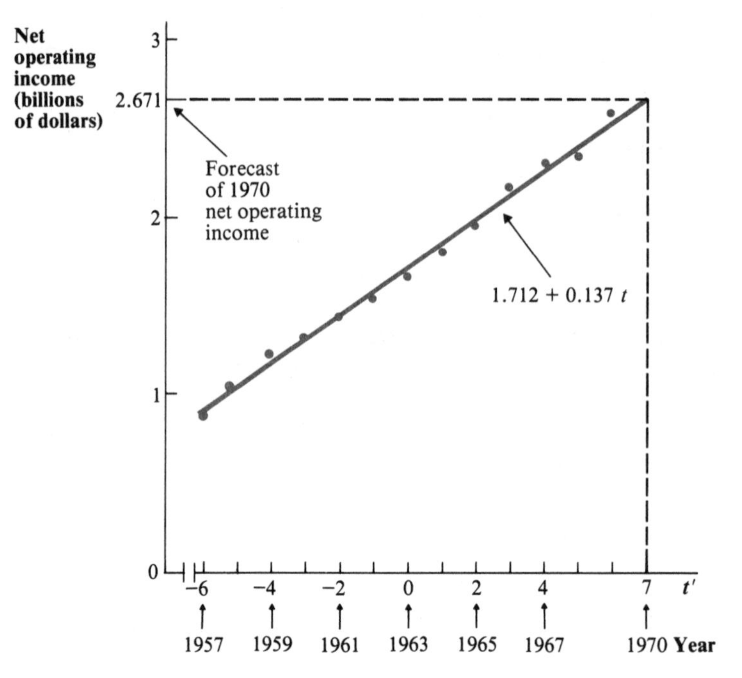

FIGURE 13.13 Simple Trend Extrapolation to Forecast 1970 Net Operating Income of American Telephone and Telegraph Company

In many cases, firms and government agencies want to forecast monthly rather than annual amounts. In such cases, it is necessary to recognize that seasonal variation, as well as trend, is likely to affect the value for a particular month. To see how a forecast can be made under such circumstances consider a beer manufacturer that wants to forecast its sales during each month of 1981. On the basis of data for each month during 1960–80, the firm determines that its sales seem to conform to the following trend:

$$S_t = 1{,}410 + 33t,$$

where S_t is the trend value of the firm's monthly sales (in thousands of dollars) and t is time measured in months from January 1980. Thus, if this trend continues, the expected sales for each month in 1981 would be as shown in the second column of Table 13.9. But this ignores whatever seasonal variation exists in the firm's sales. In order to include seasonal variation, suppose that the beer manufacturer's statisticians analyze past sales data and, using the method described in Section 13.9, find that the monthly seasonal index for sales is as shown in the third column of Table 13.9. Thus, if this seasonal pattern continues in 1981 as in the past, we would expect that actual sales each month would equal the trend value (in the second column) times the seasonal index (in the third column) divided by 100. The result, which is shown in the

TABLE 13.9

Forecasted Trend Value of Sales, Seasonal Index, and Forecasted Monthly Sales of Beer Manufacturer, 1981

Month	Forecasted trend value of sales[a]	Seasonal index	Forecasted sales (reflecting both trend and seasonal variation)[a]
January	1,806	80	1,445
February	1,839	80	1,471
March	1,872	90	1,685
April	1,905	90	1,715
May	1,938	110	2,132
June	1,971	120	2,365
July	2,004	120	2,405
August	2,037	120	2,444
September	2,070	110	2,277
October	2,103	100	2,103
November	2,136	100	2,136
December	2,169	80	1,735

[a] Expressed in units of $1,000.

fourth column of Table 13.9, is a forecast that includes both the trend and the seasonal variation.

Of course, it must be recognized that this entire procedure is simply a mechanical extrapolation of the firm's sales data into the future. Essentially, the assumption is made that the past trend and the past seasonal variation will continue. The hazards involved in simple extrapolation of this sort have been emphasized in the previous two chapters. Moreover, it is assumed that the trend and seasonal variation are the predominantly important factors that will determine sales in the coming months. The validity of this assumption depends on many considerations, including the extent to which the time series in question (in this case, sales) is affected by cyclical factors and the extent to which the economy is likely to change its cyclical position. In the next section, we will turn our attention to a particular method of forecasting business fluctuations.

13.13 Leading Indicators

Firms often want to modify their forecasts in order to take account of prospective overall changes in economic activity. For example, if the beer manufacturer in Table 13.9 is convinced that a serious depression will occur in 1981, it is likely to modify the forecasts in Table 13.9 accordingly. But how does the beer manufacturer—or anyone else—predict whether there is going to be a depression? There are a variety of ways of doing this, all of which are very imperfect. In this section, we discuss an essentially empirical approach, reserving a discussion of more sophisticated techniques for the next section.

Perhaps the simplest way to forecast business fluctuations is to use leading indicators, which are certain economic series that typically go down or up before gross national product does. The National Bureau of Economic Research, founded by Wesley C. Mitchell (1874–1948), has carried out detailed and painstaking examinations of the behavior of various economic variables over a long period of time, in some cases over as long as 100 years. The Bureau has attempted to find out whether each variable turns downward before, at, or after the peak of the business cycle, and whether it turns upward before, at, or after the trough. *Variables that go down before the peak and up before the trough are called* **leading series**. *Variables that go down at the peak and up at the trough are called* **coincident series**. *Variables that go down after the peak and up after the trough are called* **lagging series**.

It is worthwhile examining the kinds of variables that fall into each of these three categories, since they provide important facts about the anatomy of the cycle. According to the Bureau, some important leading series are business failures, new orders for durable goods, the average work week, building contracts, stock prices, certain wholesale prices, and new incorporations. These are the variables that tend to turn downward before the peak and upward before the trough.[16] Coincident series include employment, industrial production, corporate profits, and gross national product, among many others. Some typical lagging series are retail sales, manufacturers' inventories, and personal income.

Economists sometimes use leading series as forecasting devices. There are sound economic reasons why these series turn downward before a peak or upward before a trough: In some cases leading series indicate changes in spending in strategic areas of the economy, while in others they indicate changes in businessmen's and investors' expectations. In order to guide the government in setting economic policies and to guide firms in their planning, it is important to try to spot turning points—peaks and troughs—in advance. (This, of course, is the toughest part of economic forecasting.) Economists sometimes use leading indicators as evidence that a turning point is about to occur. If a large number of leading indicators turn down, this is viewed as a sign of a coming peak. The upturn of a large number of leading indicators is thought to signal an impending trough.

Unfortunately, leading indicators are not very reliable. It is true that the economy has seldom turned downward in recent years without a warning from these indicators, but unfortunately these indicators have turned down on several occasions—in 1952 and 1962, for example—when the economy did *not* turn down subsequently. Thus, leading indicators sometimes provide false signals. Also, in periods of expansion they sometimes turn downward too far ahead of the real peak. And in periods of recession they sometimes turn upward only a very short while before the trough, so that we've turned the corner before anything can be done. Nonetheless, leading indicators are not worthless; they are watched closely and used to supplement other more sophisticated forecasting techniques.

16. Of course, business failures turn upward before the peak and downward before the trough.

13.14 Econometric Models as Forecasting Techniques

In recent years, statisticians and economists have tended to base their forecasts less on simple trend extrapolation of the sort described in Section 13.12 and more on multiple regression techniques and multi-equation models of the sort described in Chapter 12. *The emphasis has shifted toward the construction and estimation of an equation or system of equations that will show the effects of various independent variables on the variable or variables one wants to forecast.* For example, we may want to estimate the quantity of onions that will be supplied by American producers next year. Based on elementary economic theory, one would expect this quantity to depend on this year's price of onions and this year's level of production costs. To make such a forecast, modern statisticians and economists would be likely to estimate an equation relating these independent variables to the quantity of onions supplied; then they would use this equation to make the forecast.

To illustrate this, let's consider the study by D. B. Suits and S. Koizumi[17] which determined that the quantity of onions supplied in year t (designated as Y_t) was related in the following way to the price of onions in year $t - 1$ (designated as P_{t-1}) and the level of costs of producing onions in year $t - 1$ (designated as C_{t-1}):

$$\log Y_t = 0.134 + 0.0123(t - 1924) + 0.324 \log P_{t-1} - 0.512 \log C_{t-1}$$

Given the known price of onions this year and the known level of this year's production costs, this equation may be used to forecast the quantity of onions that will be supplied next year. If we insert the relevant values of P, C, and t into the right-hand side of this equation, the resulting value of Y is the desired forecast. (The procedure for estimating an equation of this sort has already been described in considerable detail in Chapter 12.)[18]

To forecast many economic variables (such as gross national product) statisticians often use multi-equation models. One of these is the Wharton Econometric Model, described in Appendix 13.1. The Wharton Model contains over 200 equations variously intended to explain the level of expenditures by households, the level of business investment, aggregate output and employment, and wages, prices, and interest rates. The forecasts produced by the Wharton Model and other large ones like it are followed closely by major business firms and government agencies. Indeed, some firms and agencies have constructed their own multi-equation models. Of course, this does not mean that these large models have an unblemished forecasting record; on the contrary, they, like all other forecasting techniques, are quite fallible. However, these models continue to be widely used in business and

17. D. B. Suits and S. Koizumi, "The Dynamics of the Onion Market," *Journal of Farm Economics*, 38, 1956, pp. 475–84. This study is cited for illustrative purposes only. It was carried out a number of years ago and would have to be updated if used today. However, this does not reduce its usefulness in the present context, since we are citing it only as a well-known example of equations of this sort.

18. Readers who skipped Chapter 12 can simply take on faith the fact that equations of this sort can be estimated by statisticians.

TABLE 13.10

Average Error (in percentage points) of the Forecasts of Six Econometric Models, 1970–74[a]

Econometric model	Rate of increase of GNP (in money terms)	Rate of increase of GNP (in real terms)	Unemployment rate	Rate of price increase
	Average error (in percentage points)			
Bureau of Economic Analysis	1.2	1.5	0.3	1.7
Chase Econometrics	1.0	1.2	0.3	1.6
Data Resources, Inc.	1.3	1.2	0.3	1.7
Ray C. Fair	1.5	1.9	0.4	2.1
General Electric	1.0	1.2	0.3	1.3
Wharton	1.3	1.0	0.4	1.8

SOURCE: Stephen McNees, "How Accurate Are Economic Forecasts?" *New England Economic Review*, November 1974.

[a] All forecasts included here were for one year ahead.

government. (Table 13.10 describes the accuracy of six leading econometric models during the 1970s.)

Both the single-equation model used to forecast the quantity of onions supplied and the Wharton Model, with its more than 200 equations, are examples of econometric models. Although it would be inappropriate in an elementary course to go far into econometrics, it is important that you know the definition of an econometric model.

> ***Econometric model:*** *An econometric model is a system of equations (or a single equation) estimated from past data that is used to forecast economic and business variables.*

The essence of any econometric model is that it blends economic theory with statistical techniques. Since econometric models have come to play such an important role in business and economic forecasting, students who want to delve more deeply into forecasting procedures should consider taking the specialized econometrics courses available at many colleges and universities.

13.15 Forecasting Company Sales: A Case Study[19]

To illustrate the ways in which the simple forecasting techniques described in previous sections have been used, let's consider the case of the Long Island

19. This case study is from the Conference Board, *Studies in Business Policy* and is reprinted in E. Mansfield, *Managerial Economics and Operations Research*, 3d ed. (New York: Norton, 1975). Needless to say, it pertains only to the firm's practices at the time of the Conference Board's publication. For an account of the sales forecasting procedures of the Timken Company, Cummins Engine Company, and RCA Corporation, see the Conference Board, *Sales Forecasting*

Lighting Company, a major New York public utility. This company forecasts monthly sales for each of the next 18 months and annual sales for each of the next 5 years. Obviously, these forecasts play an important role in the firm's planning and decision making.

The first step in Long Island Lighting's forecasting procedure is to use the trend of population in Nassau and Suffolk Counties to forecast the population in the area served by the firm in the relevant month or year in the future. The next step is to multiply the forecasted population by the number of meters per person (which remains relatively constant) in order to obtain a forecast of the number of meters in the relevant month or year in the future. Next, the 12-month moving average of energy use per meter is projected into the future on the basis of a trend line of the sort discussed in previous sections. The relationship between energy use per meter and sales per meter is then used to forecast the sales per meter corresponding to the forecasted energy use per meter. Finally, to obtain a sales forecast, the predicted level of sales per meter is multiplied by the forecasted number of meters.

This first sales forecast is checked against another which is derived in a different way, but which is also produced in a number of steps. The firm's engineering department begins by estimating the peak demand for the previous year and then estimates the increments in peak demand for the next five years. Next, the trend in the ratio of average kilowatt demand to peak demand is projected into the future. (According to the firm, this is reasonably accurate, since the ratio moves slowly.) Then the forecast of the peak demand is multiplied by the forecast of the ratio of average demand to peak demand, and the product is multiplied by 8,760 (the number of hours in a year). The result (after the amounts of energy lost and company consumption are deducted) represents another forecast of the company's sales.

A planning committee reconciles these two forecasts, and the result is the forecast used in the firm's planning. This method has been used by Long Island Lighting for a number of years. During an eight-year period, "the margin of error has been slightly above 1% for forecasts made one year in advance and between 5% and 10% for the most distant years."[20] Of course, it should not be assumed that the Long Island Lighting Company's techniques will work this well for other firms. But because there was a rather well-defined and stable trend in energy use per meter, in the ratio of average kilowatt demand to peak demand, and in other key variables, these techniques seem to have been relatively effective. For firms facing quite different conditions, such techniques might give much less accurate results.

EXERCISES

13.16 The Bona Fide Washing Machine Company's statistician calculates a seasonal index for the firm's sales, the results being shown in the second column on the next page. The firm's monthly 1978 sales are shown in the third column on the next page.

(New York: Conference Board, 1978); reprinted in E. Mansfield, "Sales Forecasting: Three Case Studies," *Statistics for Business and Economics: Readings and Cases* (New York: Norton, 1980).

20. Ibid.

Month	Seasonal index	1978 Sales (millions of dollars)
January	97	2.5
February	96	2.4
March	97	2.7
April	98	2.9
May	99	3.0
June	100	3.1
July	101	3.2
August	103	3.1
September	103	3.2
October	103	3.1
November	102	3.0
December	101	2.9

Calculate deseasonalized sales figures for 1978.

13.17 The Acme Corporation calculates a 12-month moving average of its sales. After centering this moving average at the middle of the month, Acme divides each month's sales by this centered moving average. The results (multiplied by 100) are as follows:

	Ratios			
Month	1975	1976	1977	1978
January	91.0	90.0	90.5	94.0
February	93.0	92.0	92.5	95.0
March	95.0	94.0	94.5	96.0
April	97.0	96.0	96.5	97.0
May	99.0	98.0	98.5	98.0
June	101.0	100.0	100.5	99.0
July	103.0	102.0	102.5	101.0
August	105.0	104.0	104.5	102.0
September	107.0	106.0	106.5	103.0
October	109.0	108.0	108.5	104.0
November	100.0	110.0	105.0	105.0
December	100.0	100.0	100.0	106.0

Calculate a seasonal index for the Acme Corporation's sales.

13.18 A firm claims that the seasonal index for its sales is 110 for each of the first six months of the year, and 95 for each of the last six months of the year. Comment on this claim.

13.19 (a) Calculate the difference between each year's sales of the Union Carbide Corporation and the least-squares trend you derived in Exercise 13.5. (b) Do these deviations from trend show a cyclical pattern? If so, when are some of the most pronounced peaks and troughs?

13.20 (a) Does the time series of capacity utilization rates in Exercise 13.9 show cyclical variation? (b) Do the peaks and troughs tend to coincide with those in the economy as a whole?

13.21 (a) Use a simple extrapolation of the linear trend in Exercise 13.5 to forecast the Union Carbide Corporation's sales in 1985. (b) Use a simple extrapolation of the exponential trend in Exercise 13.7 to do the same thing. (c) How close to one another are the results?

13.22 On the basis of a simple extrapolation of the linear trend, the predicted value of the Bona Fide Washing Machine Company's sales in February 1982 is \$4 million. Adjust this figure to include seasonal variation, on the basis of the figures in Exercise 13.16.

13.23 On the basis of a simple extrapolation of a linear trend, the predicted value of the Acme Corporation's sales in February 1981 is \$25 million. Adjust this figure to include seasonal variation on the basis of the figures in Exercise 13.17.

13.24 Explain why new orders for durable goods and building contracts would be expected to be leading series rather than lagging series.

13.25 Explain why the level of stock prices would be expected to be a leading series rather than a lagging series.

13.26 In equation (13.14), suppose that the statistician estimates B_2 to be 2, B_3 to be 3, and B_4 to be -2. Holding constant the trend value, what is the expected difference in the value of Y between the first and third quarters?

13.27 In equation (13.14), suppose that the statistician believes the trend to be exponential rather than linear. How should equation (13.14) be changed to reflect this?

13.28 In equation (13.15), suppose that the statistician estimates B_2 to be 2 and B_3 to be 3. Holding constant the trend value, what is the expected difference between January and February in the value of S?

13.29 Suits and Koizumi estimated an equation to explain the quantity of onions supplied. Do you think that an equation could be estimated to explain the quantity of onions demanded? If so, what independent variables would you propose to include in such an equation?

CHAPTER REVIEW

1. The traditional approach to the analysis of time series, devised primarily by economic statisticians, was essentially descriptive. It assumed that an economic time series can be decomposed into four components: trend, seasonal variation, cyclical variation, and irregular movements. A *trend* is a relatively smooth long-term movement of a time series. *Seasonal variation*, which results in one month's (or one season's) being consistently higher or lower than another, is frequently due to the weather, holidays, and other such factors. *Cyclical variation* reflects the effects of business fluctuations, or the so-called business cycle. *Irregular variation* is attributable to short-term, erratic forces that are too unpredictable to be of use for forecasting purposes.

2. If the trend in a time series is *linear*, simple regression may be used to estimate an equation representing the trend. If it seems to be *nonlinear*, a quadratic equation may be estimated by multiple regression, or an exponential trend may be fitted. In some cases, there is no relatively simple mathematical function that can adequately portray the long-term movement of a particular time series. In cases of this sort, statisticians frequently use *moving averages* to smooth the time series. If the time series contains certain fluctuations or cycles that tend to recur, the effect of these cycles can be eliminated by taking a moving average where the number of years in the average is equal to the period of the cycle. Another method of smoothing a time series is to use *exponential smoothing*. According to this method, the smoothed value at a

given point in time is a weighted average of all previous values where the weights decline geometrically as one goes backward in time.

3. The seasonal variation in a particular time series is described by a figure for each month (the *seasonal index*) that shows the extent to which that month's value typically departs from what would be expected on the basis of trend and cyclical variation. Such seasonal indexes can be used to *deseasonalize* a time series, that is, to remove the seasonal element from the data. To calculate a seasonal index for a given month, we compute a 12-month moving average centered at the beginning of this month, as well as a 12-month moving average centered at the end of this month. Then we express this month's actual value as a ratio of the mean of these two moving averages (and multiply by 100). The median value of this ratio (adjusted so that the mean value of all months' seasonal indexes is 100) is the seasonal index for this month. Another way to calculate a seasonal index is to use dummy variables in a multiple regression.

4. Many business and economic time series go up and down with the fluctuations of the economy as a whole. This cyclical variation, as well as trend and seasonal variation, is reflected in many time series. It is customary to divide business fluctuations into four phases: *trough, expansion, peak,* and *recession*. Variables that go down before the peak and up before the trough are called *leading series*. Some important leading series are business failures, new orders for durable goods, average work week, building contracts, stock prices, certain wholesale prices, and new incorporations. Economists sometimes use these series, which are often called *leading indicators*, to forecast whether a turning point is about to occur. If a large number of leading indicators turn downward, this is viewed as a sign of a coming peak. If a large number turn upward, this is thought to signal an impending trough. Although these indicators are not very reliable, they are watched closely and are used to supplement other more sophisticated forecasting techniques.

5. One of the most difficult and important tasks that statisticians confront is forecasting. In general, all forecasting techniques are extremely fallible, and all forecasts should be treated with caution. The simplest kind of forecasting method is a straightforward extrapolation of a trend. To allow for seasonal variation, such an extrapolation can be multiplied by the seasonal index (divided by 100) for the month to which the forecast applies. This entire procedure is simply a mechanical extrapolation of the time series into the future. In recent years, statisticians and economists have tended to base their forecasts less on simple extrapolations of this sort and more on equations (or systems of equations) showing the effects of various independent variables on the variable (or variables) one wants to forecast. These equations (or systems of equations), which are estimated using the techniques described in the previous chapter, are often called *econometric models*.

Getting Down to Cases:

INDUSTRIAL PRODUCTION IN SEVEN MAJOR COUNTRIES

The Organization for Economic Cooperation and Development (OECD), which includes the major non-communist industrialized nations like France, Germany, Japan, the United Kingdom, and the United States, analyzes and publishes data concerning economic developments in all its member countries. In July 1976, OECD published charts (given below and on the next page) showing changes in industrial production in seven major countries.[21]

(a) How useful are these data in estimating the trend of industrial production in each country?

(b) Would you expect the trend of industrial production in each of these countries to be much the same? Why, or why not?

(c) How useful are these data in estimating cyclical variation in each country?

(d) During the period covered by the data, when did a major peak occur in each country? When was the trough? When was the recession? When was the expansion?

(e) In this period, was the business cycle in one country synchronized rather closely with that in another country? Explain.

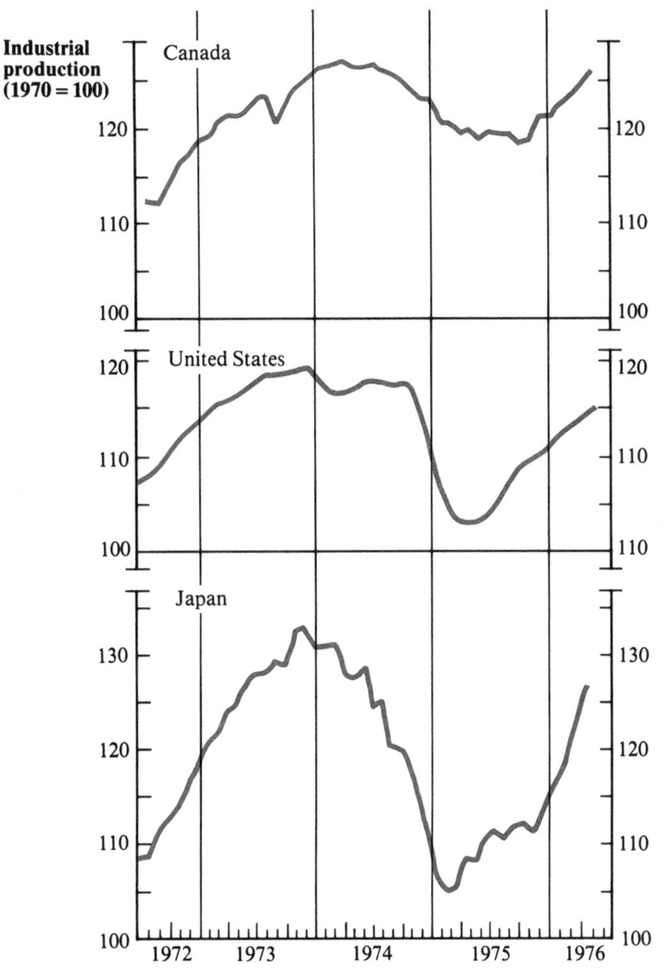

21. Organization of Economic Cooperation and Development, *Economic Outlook*, Paris, July 1976, p. 12.

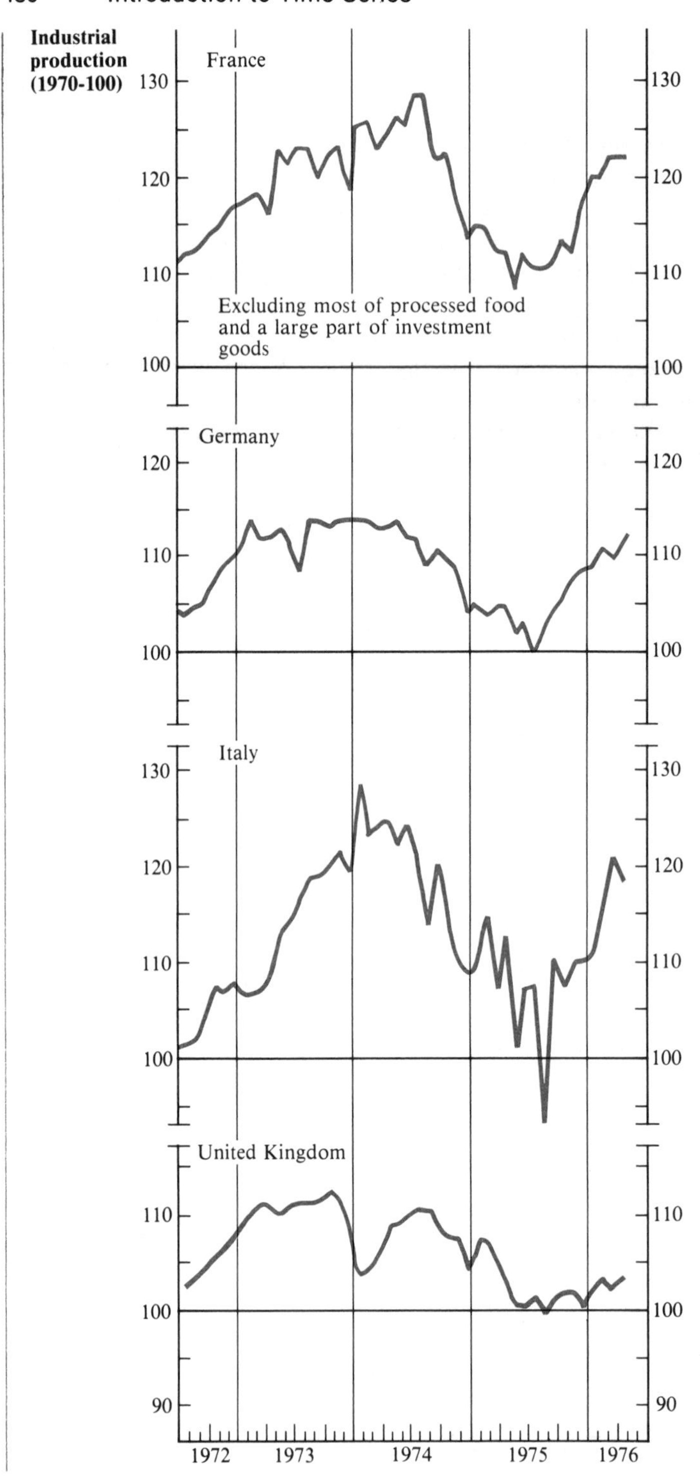
Industrial
production
(1970-100)
France
130
120
110
100
Excluding most of processed food
and a large part of investment
goods
Germany
120
110
100
Italy
130
120
110
100
United Kingdom
110
100
90
1972
1973
1974
1975
1976

APPENDIX 13.1
The Wharton Econometric Model

One of the most widely used econometric models is the Wharton model. Composed of over 200 equations, this model has been in operation for over 10 years. In addition to being used to forecast gross national product, it attempts to forecast the composition of gross national product, the price level, unemployment, and other major variables pertaining to the national economy. With the help of high-speed electronic computers, the coefficients of the many equations in this model can be estimated and revised from time to time.

Due to the model's large size it is not feasible to list all its equations. But it is possible to describe the nature of some of the more important equations, thus providing some feel for the structure of the model. First, there are several equations to explain the level of personal consumption expenditures, one equation pertaining to autos, one to nonautomotive durables, one to nondurables, and one to services. In each of these equations, the level of expenditure is dependent on disposable income and prices, among other things. This, of course, is in keeping with basic economic theory.

Second, there are several equations which are used to explain the level of investment, one equation pertaining to business purchases of plant and equipment, one pertaining to construction of houses and apartments, and one pertaining to changes in inventories. In the case of expenditures on plant and equipment, the relevant equation makes these expenditures dependent on changes in output or the extent of capacity utilization as well as on the cost of capital. In the case of residential investment, the tightness of monetary markets is an important explanatory variable in the relevant equation.

Third, there are equations to explain output and employment. These variables are determined in the model primarily by the level of aggregate demand, which depends on consumption expenditure, investment, imports and exports, and government expenditures. Government expenditures are an exogenous variable in the model. Equations are included to explain imports and exports, these variables being functions of things like the prices of imported goods relative to domestic goods and the competitiveness of U.S. export prices relative to world prices. To estimate employment, production functions are estimated.[22] Given the estimates of output, these production functions can be used to forecast employment. Unemployment can also be forecasted, based on information concerning the size of the labor force.

Fourth, there are equations to explain wages, prices, and interest rates. The wage equations make the rate of increase of wages depend on the rate of unemployment and the rate of increase of prices. The price level is determined by the extent of capacity utilization, among other things. A number of equations relate to the monetary sector of the economy. These equations attempt to explain interest rates, the money supply, and other financial variables.

22. A *production function* is defined by economists as the relationship between input and output for a firm, industry, or the economy as a whole.

Every three months the Wharton Model is used to produce forecasts of gross national product, the price level, the unemployment rate, and other such variables for the next two years. During the 1960s its forecasting performance was particularly impressive; the average error of its forecasts of gross national product during 1959–67 was only about $3 billion (less than 1/2 of 1 percent). During the mid-1970s, its forecasting performance was less impressive, due in part to political events such as the behavior of the OPEC oil producers. (Obviously, such events are difficult to predict and incorporate into such a model.) Nonetheless, this and other multi-equation models are widely used by business firms and government agencies. They have an important influence on both government policy and business decision making here and abroad.

APPENDIX 13.2
Forecasting Based on Exponential Smoothing

In Section 13.7 we described how exponential smoothing can be used to smooth a time series. It is worth noting that exponential smoothing can also be used to forecast a time series. When used for forecasting purposes, the basic equation for exponential smoothing is

$$F_t = \theta A(t - 1) + (1 - \theta)F_{t-1}, \tag{13.16}$$

where $A(t - 1)$ is the actual value of the time series at time $(t - 1)$, and F_t is the forecast for time t. We suppose that the forecast is being made at time $(t - 1)$, so the actual value of the time series at this time is known. Given the smoothing constant θ, the forecast for time t is simply a weighted average of the *actual* value at time $(t - 1)$ and the *forecasted* value for time $(t - 1)$, where the actual value is weighted by θ and the forecasted value is weighted by $(1 - \theta)$. It can easily be shown that the forecast for time t is the weighted sum of the actual values prior to time t, where the weight attached to each value declines geometrically with the age of the observation.

To illustrate the use of exponential smoothing for forecasting purposes, let's return to the firm in Section 13.7 that had been in existence for five years. Sales during the first year were $1 million, and we assume that the firm's sales forecast for the first year was also $1 million. What will be its sales forecast for the second year? To make such a forecast, the firm begins by choosing a value for the smoothing constant θ. Suppose that a value of .2 is chosen. Then the forecast for the second year is .2(1) + .8(1) = 1, or $1 million. The firm's actual sales in the second year turn out to be $3 million. Hence, its sales forecast for the third year will be .2(3) + .8(1) = 1.4, or $1.4 million. The firm's actual sales in the third year turn out to be $3 million. Hence, its sales forecast for the fourth year will be .2(3) + .8(1.4) = 1.72, or $1.72 million. And so on.

Exponential smoothing is often used in this way to make forecasts, particularly where there is a need for a cheap, fast, and rather mechanical method to make forecasts for a large number of items. For example, to implement various kinds of inventory control models, demand forecasts for hundreds or thousands of items may be required. A major advantage of this technique is that the only number that must be stored until the next time period is the forecast for the next period. Why? Because this number, together with the actual value for the next period and θ, determines the forecast for the period following the next period.

In accord with our discussion in Section 13.7, it is essential that θ be between zero and 1. If θ is close to 1, past values of the time series are given relatively little weight (compared to recent values) in determining the forecast. If θ is close to zero, past values are given relatively heavy weight (compared to recent values) in determining the forecast. In practice, values of .3 or less are often used. Finally, a variety of smoothing techniques can be used as the basis for forecasting methods. The technique given in this Appendix is certainly among the simplest, and thus has been chosen as the logical introduction to methods of this kind.

14

Index Numbers*

14.1 Introduction

Decision makers and analysts in business and government must frequently try to summarize the differences that have occurred in a number of economic variables over a particular period of time. For example, between 1970 and 1980 changes occurred in the prices of practically all goods and services; and these price changes were by no means uniform. For many purposes, it is useful to construct a price index to summarize this variety of price changes into a single number which indicates the extent to which the cost of living in 1980 exceeded that in 1970. Similarly, the price of a given item is different in Los Angeles than in Chicago, and this intercity difference varies from one item to another. Again, it is useful to summarize this variety of price differences into a single number which tells whether (and to what extent) the cost of living is higher in Los Angeles than in Chicago. Statisticians and economists have devoted considerable attention to the formulation of index numbers of this kind; the purpose of this chapter is to describe and illustrate the methods used to construct such numbers.

At the outset, let's define somewhat more precisely what we mean by an index number.

> ***Index number:*** *An index number is a ratio (generally expressed in percentage terms) of one quantity to another where one of the quantities summarizes a given group of items and the other quantity summarizes a base group of items. The "base" group is used as a basis of comparison with the "given" group.*

For example, a government statistician might be asked to construct an index number showing changes in steel prices between 1970 and 1980. In such an index number, 1970 would be the *base* year, and 1980 would be the *given* year. The 1980 prices of the various types of steel (sheets, bars, tin plate, stainless, and so on) would be compared with those in 1970, and this index number would indicate how much higher these prices were in 1980 than in 1970. Thus,

*Some instructors may wish to skip Sections 14.8 and 14.9 which deal with quantity indexes. These sections can be omitted without loss of continuity.

if this index number equals 140, this means that on the average, steel prices were 40 percent higher in 1980 than in 1970.

14.2 Unweighted Index Numbers

As a first step toward understanding how index numbers are constructed, we consider so-called unweighted index numbers. Although unweighted index numbers are very seldom used (because of flaws discussed below), they are the simplest kind and thus provide a good beginning for our discussion. To see what we mean by unweighted index numbers, suppose that we want to construct an index number showing changes over time in the price of major food crops. The price per bushel of four major crops—corn, wheat, oats, and soybeans—is shown in Table 14.1 for 1960, 1965, 1970, and 1972. Using 1960 as the base period, an index of these prices can be constructed by summing up the prices of each of these commodities in a given year and then dividing the sum by the corresponding sum in the base year. For example, since \$1.00 is the price of corn in 1960, since \$1.74 is the price of wheat in 1960, and so on, the sum for the base year is

$$\$1.00 + 1.74 + 0.60 + 2.13 = \$5.47.$$

Similarly, the corresponding sum for 1972 is

$$\$1.29 + 1.67 + 0.67 + 3.49 = \$7.12.$$

Thus, the value of the index number for 1972 is

$$\frac{\$7.12}{\$5.47}(100) = 130,$$

which indicates that these crop prices rose by 30 percent between 1960 and 1972.

TABLE 14.1

Price per Bushel of Four Major Food Crops, United States, 1960–72

Crop	1960 (dollars)	1965 (dollars)	1970 (dollars)	1972 (dollars)
Corn	1.00	1.16	1.33	1.29
Wheat	1.74	1.35	1.33	1.67
Oats	0.60	0.62	0.62	0.67
Soybeans	2.13	2.54	2.85	3.49

SOURCE: *Statistical Abstract of the United States* (Washington, D.C.: Government Printing Office, 1972 and 1973).

In general, the simple unweighted index we have constructed can be expressed as follows: *If* P_{01} *is the price of the first commodity (corn) in the base period and* P_{11} *is its price in year* 1, P_{02} *is the price of the second commodity (wheat) in the base period and* P_{12} *is its price in year* 1, *and so forth, then the simple unweighted price index for year* 1 *is*

$$I_1 = \frac{P_{11} + P_{12} + \ldots + P_{1m}}{P_{01} + P_{02} + \ldots + P_{0m}}(100),$$

where m *is the number of commodities included in the index.* Thus, we can express this index as

$$I_1 = \frac{\sum_{i=1}^{m} P_{1i}}{\sum_{i=1}^{m} P_{0i}}(100). \tag{14.1}$$

One major difficulty with this ***simple aggregative index***, as it is called, is that *the result one obtains depends on the units in which each of the commodities is expressed.* For example, suppose that we measure soybeans in units of 100 bushels rather than bushels. Then the price of soybeans would be $213 in 1960 and $349 in 1972. Thus, the simple aggregative price index for 1972 would be

$$\frac{\$1.29 + 1.67 + 0.67 + 349.00}{\$1.00 + 1.74 + 0.60 + 213.00}(100) = \frac{352.63}{216.34}(100) = 163,$$

which indicates that these crop prices rose by 63 percent between 1960 and 1972. Clearly, this result is quite different from that obtained when the price of a single bushel of soybeans was used.

One way of avoiding this difficulty is to use an index which is an ***average of relatives.*** For example, to obtain an index of food crop prices, one could begin by obtaining relatives for each price, a *relative* being *the ratio of the price in the given year to that in the base year.* Table 14.2 shows the price relatives for each commodity. Then to obtain the index, we can compute the mean of these relatives. For example, the value of the index number for 1972 can be

TABLE 14.2
Price Relatives for Four Major Food Crops, United States, 1960–72

Crop	1965 price / 1960 price	1970 price / 1960 price	1972 price / 1960 price
Corn	$\frac{1.16}{1.00} = 1.16$	$\frac{1.33}{1.00} = 1.33$	$\frac{1.29}{1.00} = 1.29$
Wheat	$\frac{1.35}{1.74} = 0.78$	$\frac{1.33}{1.74} = 0.76$	$\frac{1.67}{1.74} = 0.96$
Oats	$\frac{0.62}{0.60} = 1.03$	$\frac{0.62}{0.60} = 1.03$	$\frac{0.67}{0.60} = 1.12$
Soybeans	$\frac{2.54}{2.13} = 1.19$	$\frac{2.85}{2.13} = 1.34$	$\frac{3.49}{2.13} = 1.64$

computed from the relatives shown in the last column of Table 14.2:

$$\frac{1.29 + 0.96 + 1.12 + 1.64}{4}(100) = 125.$$

Note that we use the mean of the price relatives; the median or mode could have been used instead, but the mean is the kind of average that is typically employed.

In general, an index number based on a simple average of relatives can be expressed in the following way. Using the notation described above, the price index for year 1 equals

$$I_1 = \frac{\frac{P_{11}}{P_{01}} + \frac{P_{12}}{P_{02}} + \frac{P_{13}}{P_{03}} + \ldots + \frac{P_{1m}}{P_{0m}}}{m}(100).$$

Or using the summation symbol,

$$I_1 = \frac{\sum_{i=1}^{m} \frac{P_{1i}}{P_{0i}}}{m}(100). \tag{14.2}$$

14.3 Weighted Index Numbers

Unweighted relative price indexes have a very important disadvantage which results in their seldom being used: *The various prices included in the index are given equal importance.* Consider the four commodities in Tables 14.1 and 14.2. As shown in Table 14.3, corn is produced and consumed in much greater quantities than wheat, oats, or soybeans. Consequently, it seems obvious that changes in the price of corn should have a greater weight in a price index than

TABLE 14.3

Production and Value of Output of Major Food Crops, United States, 1960–72

Crop	1960	1965	1970	1972
	Production (millions of bushels)			
Corn	3,907	4,084	4,099	5,553
Wheat	1,355	1,316	1,370	1,545
Oats	1,153	927	909	695
Soybeans	555	846	1,124	1,283
	Value of output (millions of dollars)			
Corn	3,907	4,737	5,452	7,163
Wheat	2,358	1,777	1,822	2,580
Oats	692	575	564	466
Soybeans	1,182	2,149	3,203	4,478

changes in the price of the other three crops. However, unweighted relative price indexes would treat them all as equally important.

To remedy this deficiency, statisticians generally use weighted index numbers. In the case of price indexes, the weights most often used for price relatives are value weights. For example, in the case of Table 14.3, the price relative for corn would be weighted by the value of corn produced, the price relative for wheat would be weighted by the value of wheat produced, and so on. Using the values in the base year as weights, it is clear from Table 14.2 (which contains the price relatives) and Table 14.3 (which contains the value weights) that the value of the price index for 1972 is

$$\frac{3{,}907(1.29) + 2{,}358(0.96) + 692(1.12) + 1{,}182(1.64)}{3{,}907 + 2{,}358 + 692 + 1{,}182}(100)$$

$$= \frac{10{,}017}{8{,}139}(100) = 123.$$

In other words, the price relative for each commodity is weighted by its value in the base year, and the weighted sum of the price relatives is divided by the sum of the weights.

In symbols, this weighted relative price index can be expressed as

$$\frac{\sum_{i=1}^{m} W_i \frac{P_{1i}}{P_{0i}}(100)}{\sum_{i=1}^{m} W_i}, \tag{14.3}$$

where W_i is the weight attached to the price relative of the ith commodity. When, as in this case, the weight equals $Q_{0i}P_{0i}$, where Q_{0i} is the amount consumed of the ith commodity in the base period, the resulting price index is

$$\frac{\sum_{i=1}^{m} (Q_{0i}P_{0i})\left(\frac{P_{1i}}{P_{0i}}\right)}{\sum_{i=1}^{m} Q_{0i}P_{0i}}(100).$$

This is the Laspeyres index, defined below.

Laspeyres index: *The Laspeyres price index is the ratio (expressed as a percentage) of the total cost in the given year of the quantity of each commodity consumed in the base year to what was the total cost of these quantities in the base year. In symbols, the Laspeyres index is*

$$\boxed{\frac{\sum_{i=1}^{m} Q_{0i}P_{1i}}{\sum_{i=1}^{m} Q_{0i}P_{0i}}(100).} \tag{14.4}$$

Another frequently encountered type of weighted relative price index is the Paasche index, defined as follows.

Paasche index: *The Paasche price index is the ratio (expressed as a percentage) of the total cost in the given year of the quantity of each commodity consumed in the given year to what would have been the total cost of these quantities in the base year. In symbols, the Paasche index is*

$$\frac{\sum_{i=1}^{m} Q_{1i}P_{1i}}{\sum_{i=1}^{m} Q_{1i}P_{0i}}(100). \tag{14.5}$$

Both the Laspeyres and Paasche indexes measure the change, from the base year to the given year, in the total cost of the commodities consumed; but while the Laspeyres index uses the quantities consumed in the *base year*, the Paasche index uses the quantities consumed in the *given year* (that is, year 1).

Table 14.4 shows the value of the Paasche index for our four commodities in 1970 and 1972. In 1972, for example, since the amounts

TABLE 14.4
1970 and 1972 Paasche Price Indexes

Computation of 1970 Paasche price index		
Crop	1970 output × 1970 price	1970 output × 1960 price
Corn	5,452	4,099
Wheat	1,822	2,384
Oats	564	545
Soybeans	3,203	2,394
Total	11,041	9,422

Value of index $= \frac{11{,}041}{9{,}422}(100) = 117.$

Computation of 1972 Paasche price index		
Crop	1972 output × 1972 price	1972 output × 1960 price
Corn	7,163	5,553
Wheat	2,580	2,688
Oats	466	417
Soybeans	4,478	2,733
Total	14,687	11,391

Value of index $= \frac{14{,}687}{11{,}391}(100) = 129.$

consumed of each commodity were as shown in the last column of Table 14.3, it would have cost

$$5{,}553(\$1.00) + 1{,}545(\$1.74) + 695(\$0.60) + 1{,}283(\$2.13),$$

or $11,391 million to buy these amounts in the base period. Thus, since it actually cost $14,687 million to buy these amounts in 1972,[1] the value of the Paasche index for 1972 is

$$\frac{14{,}687}{11{,}391}(100) = 129.$$

Although the Paasche index is potentially useful, it has important practical disadvantages in cases where an index is calculated for one period after another. For example, the Consumer Price Index and the Producer Price Index (formerly the Wholesale Price Index) are calculated on a monthly and annual basis. To use the Paasche index for a series of this sort, one would have to obtain fresh data each period concerning the quantity weights, Q_{1i}. This would be expensive and time-consuming, making the use of the Paasche index inappropriate on practical grounds.

14.4 Other Weighting Schemes and Base Periods

Although the Laspeyres index weights individual prices by quantity consumed or produced in the *base period*, and the Paasche index weights them by quantity consumed or produced in the *given period* (that is, the period for which the index is computed), these two periods are not the only ones which could be used. For example, consider our index of crop prices. The Laspeyres 1972 index calculated in the previous section weighted individual prices by output in 1960; and the Paasche index for 1972 weighted individual prices by output in 1972. *But there is nothing to prevent us from using as weights output in some year other than 1960 or 1972.* Indeed, many important price indexes published by the federal government do just that.

One reason for using output for some period other than the base year or current year is that it is desirable for the weights to refer to a "normal" year. Suppose that 1960 and 1972 were both rather abnormal years for agriculture, whereas 1965 seemed much freer of major abnormalities. Then rather than using 1960 quantity weights (as in the Laspeyres index) or 1972 weights (as in the Paasche index), we might use 1965 quantity weights. In other words, to obtain the index for 1972, we might calculate the amount it would cost in 1972 to buy the amount of each commodity consumed in 1965. The result is

$$4{,}084(\$1.29) + 1{,}316(\$1.67) + 927(\$0.67) + 846(\$3.49) = \$11{,}040.$$

Then we might calculate the amount it would have cost in 1960 to buy the

1. Table 14.3 shows that the value of output in 1972 was 7,163 + 2,580 + 466 + 4,478 = 14,687 millions of dollars. The same thing is shown in the latter part of Table 14.4.

amount of each commodity consumed in 1965. The result is

$$4{,}084(\$1.00) + 1{,}316(\$1.74) + 927(\$0.60) + 846(\$2.13) = \$8{,}732.$$

Thus, the index for 1972 is

$$\frac{11{,}040}{8{,}732}(100) = 126.$$

In symbols, this index, sometimes called a ***fixed-weight aggregative index***, can be expressed as

$$I_1 = \frac{\sum_{i=1}^{m} Q_{wi}P_{1i}}{\sum_{i=1}^{m} Q_{wi}P_{oi}}(100), \tag{14.6}$$

where Q_{wi} is the amount consumed of the ith commodity in year w. Year w is, of course, the period (chosen because it is freer of abnormalities or for some other reason) to which the weights pertain. Note that the only difference between this index and the Laspeyres index or the Paasche index is that Q_{wi} is used as a weight here, whereas Q_{0i} is used in the Laspeyres index and Q_{1i} is used in the Paasche index.

Going a step further, there is no reason why the quantity weights must pertain to a single year. Instead, *one can use as weights the average level of output or consumption in a number of years.* For example, in the case of the index of crop prices, we might feel that the best set of weights is an average of 1965 and 1970 quantities, as given in column three of Table 14.5. Based on these weights, the index for 1972 is computed (in Table 14.5) in the ordinary way. The resulting index number for 1972 is 128. Among the many

TABLE 14.5

Fixed-Weight Aggregative Index of Crop Prices, Quantity Weights Being Averages of 1965 and 1970

Crop	(1) 1965 output	(2) 1970 output	(3) Quantity weight [average of (1) and (2)]	(4) Quantity weight × 1960 price (dollars)	(5) Quantity weight × 1972 price (dollars)
Corn	4,084	4,099	4,092	4,092	5,279
Wheat	1,316	1,370	1,343	2,337	2,243
Oats	927	909	918	551	615
Soybeans	846	1,124	985	2,098	3,438
Total				9,078	11,575

Index for 1972 $= \frac{11{,}575}{9{,}078}(100) = 128.$

TABLE 14.6

Fixed-Weight Aggregative Index of Crop Prices, Using 1965 Quantity Weights and Average of 1960 and 1965 as Price Base

Crop	1965 output	1960 price (dollars)	1965 price (dollars)	Average of 1960 and 1965 prices (dollars)	1965 output × average price (dollars)	1965 output × 1972 price (dollars)
Corn	4,084	1.00	1.16	1.08	4,411	5,268
Wheat	1,316	1.74	1.35	1.55	2,040	2,198
Oats	927	0.60	0.62	0.61	565	621
Soybeans	846	2.13	2.54	2.34	1,980	2,953
Total					8,996	11,040

$$\text{Index for 1972} = \frac{11{,}040}{8{,}996}(100) = 123.$$

well-known indexes which have used quantity weights of this sort is the U.S. Department of Agriculture's index of prices received by farmers. During the 1960s this index used as quantity weights the average quantities during 1953–57.

Finally, *the base period for prices (as well as output) can also be an average of several years*. Suppose that we decide that an average of 1960 and 1965 is a good base period for crop prices. Assuming that we use 1965 quantity weights, Table 14.6 shows how the index for 1972 is computed. Government price indexes frequently use an average of several years as the base price. For example, during the 1960s, the Producer Price Index (then called the Wholesale Price Index) used 1957–59 as the base period for prices. Of course, there is no reason why the period to which the quantity weights pertain (the *weight base*) must be the same as the period from which price changes are measured (the *reference base*). For example, the weight base in Table 14.6 is 1965, whereas the reference base is the average of 1960 and 1965.

14.5 Chain Index Numbers

To complete our survey of the various techniques of index number construction, we must describe chain index numbers. To understand the basic principles behind chain index numbers, one need only consider Table 14.7, which shows the wholesale price of aluminum ingot during 1968–72. Based on these data, it is a simple matter to make up an index of the price of aluminum during this period. All we need to do is divide the price for each year by the price in 1968 and multiply by 100.

The resulting index in column (2) of Table 14.7 shows how the price of aluminum has varied relative to the 1968 base price. Suppose, however, that we are interested in year-to-year changes in price. In this case, it makes sense to use as a base the price in the previous year, not the price in 1968. In other words, *we can use a shifting price base rather than a fixed base*. For example, the

TABLE 14.7
Index of Wholesale Price of Aluminum Ingot, 1968–72

Year	(1) Price of aluminum ingot (cents per pound)	(2) Price index (1968 = 100)	(3) Price index (Previous year = 100)
1968	25.6	$\frac{25.6}{25.6}(100) = 100$	—
1969	27.1	$\frac{27.1}{25.6}(100) = 106$	$\frac{27.1}{25.6}(100) = 106$
1970	28.8	$\frac{28.8}{25.6}(100) = 112$	$\frac{28.8}{27.1}(100) = 106$
1971	22.7	$\frac{22.7}{25.6}(100) = 89$	$\frac{22.7}{28.8}(100) = 79$
1972	22.4	$\frac{22.4}{25.6}(100) = 88$	$\frac{22.4}{22.7}(100) = 99$

SOURCE: *Statistical Abstract of the United States* (Washington, D.C.: Government Printing Office, 1971 and 1973).

index for 1970 relative to a 1969 base is

$$I_{69}^{70} = \frac{28.8}{27.1}(100) = 106,$$

and the index for 1971 relative to a 1970 base is

$$I_{70}^{71} = \frac{22.7}{28.8}(100) = 79.$$

These are chain index numbers.

From chain index numbers one can easily obtain the ordinary fixed-base index numbers described in previous sections. To show this, let P_{72} be the price of aluminum in 1972, P_{71} its price in 1971, and so on. Then if the 1968 price is used as a base, the ordinary index number for 1972 is

$$I_{72} = \frac{P_{72}}{P_{68}}(100).$$

But this can be written

$$I_{72} = \left(\frac{P_{72}}{P_{71}}\right)\left(\frac{P_{71}}{P_{70}}\right)\left(\frac{P_{70}}{P_{69}}\right)\left(\frac{P_{69}}{P_{68}}\right)(100).$$

Since P_{72}/P_{71} is the chain index (divided by 100) indicating the change from 1971 to 1972, P_{71}/P_{70} is the chain index (divided by 100) indicating the change from 1970 to 1971, and so forth, we can substitute $I_{71}^{72} \div 100$ for P_{72}/P_{71},

$I_{70}^{71} \div 100$ for P_{71}/P_{70}, and so on, the result being

$$I_{72} = \left(\frac{I_{71}^{72}}{100}\right)\left(\frac{I_{70}^{71}}{100}\right)\left(\frac{I_{69}^{70}}{100}\right)\left(\frac{I_{68}^{69}}{100}\right)(100).$$

Clearly, the chain index technique can be used to compute the value of an index number, given that one knows the value of the index number in the previous period. To see how this can be done, let's return to our example of the aluminum price index. It is apparent that the value of the index for 1971 was

$$I_{71} = \left(\frac{P_{71}}{P_{70}}\right)\left(\frac{P_{70}}{P_{69}}\right)\left(\frac{P_{69}}{P_{68}}\right)(100).$$

Thus, if we substitute I_{71} for its above equivalent expression in the formula for I_{72} at the bottom of the previous page we have

$$I_{72} = \frac{P_{72}}{P_{71}} I_{71}.$$

The meaning of this formula is as follows: To get the value of the price index for 1972, all that we have to do is multiply the 1971 value of the price index by the ratio of 1972 price to 1971 price. This latter ratio is, of course, the chain index (divided by 100) indicating the price change from 1971 to 1972. This is the way that many important price indexes, such as the Consumer Price Index, are calculated.

EXERCISES

14.1 According to the Department of Agriculture, the prices received by farmers can be represented by the following index numbers:

Year	All farm products	Livestock	Cotton	Tobacco	Fruit
1910–14	100	100	100	100	100
1930	125	134	104	140	149
1950	258	280	282	402	194
1970	280	326	183	604	233
1974	467	453	433	821	349

(a) For each of these index numbers, specify the base period and the given period.
(b) Did the price of farm products increase at a more rapid or less rapid annual rate during 1970–74 than during 1950–70?
(c) Did the price of farm products increase at a more rapid or less rapid annual rate during 1970–74 than during 1930–50?
(d) Which of the specific types of farm products (livestock, cotton, tobacco, and fruit) in the table above experienced the greatest percentage price increase from 1910–14 to 1974?

14.2 The Acme Corporation manufactures three types of machine tools. The price of each of these tools in 1960, 1970, and 1980 is given below:

Year	Type A	Type B	Type C
1960	$1,000	$ 500	$2,000
1970	1,800	1,000	4,200
1980	2,500	1,500	6,100

(a) Calculate a simple aggregative index—that is, a simple unweighted index—of this firm's prices. (Use 1960 as the base year, and use 1970 and 1980 as the given years.) (b) Indicate the limitations of this kind of price index, and state how it can be improved.

14.3 (a) Using the data in Exercise 14.2, calculate index numbers based on a simple average of relatives to represent the changes over time in the Acme Corporation's prices. (b) Indicate the limitations of this kind of price index, and state how it can be improved.

14.4 Suppose that the quantity of each type of machine sold each year by the Acme Corporation was as follows:

	Number of tools sold		
Year	Type A	Type B	Type C
1960	100	200	10
1970	200	300	20
1980	300	400	20

Calculate a Laspeyres price index for 1970 and 1980 for this firm. (Again, use 1960 as the base period.)

14.5 Using the data in Exercises 14.2 and 14.4, calculate a Paasche price index for this firm, assuming that 1980 is the given period and 1960 is the base period.

14.6 Suppose that the Acme Corporation considers 1970 as a much more normal year than either 1960 or 1980. (a) Use the data in Exercises 14.2 and 14.4 to calculate a fixed-weight aggregative price index for 1960, where the weights pertain to 1970, with 1960 as the base period. (b) Calculate a fixed-weight aggregative price index for 1980, where the weights pertain to 1970, with 1960 as the base period.

14.7 Let I_{60}^{70} be the 1970 index (relative to a 1960 base) of the price of type A machines. (a) Based on the data in Exercise 14.2, what is the value of I_{60}^{70}? (b) If I_{70}^{71} was 110, what was I_{60}^{71}?

14.8 Irving Fisher, a well-known American economist, suggested that an "ideal" index number would equal

$$\sqrt{(\text{Laspeyres index})(\text{Paasche index})}.$$

(Some reasons why Fisher felt this way are touched on in subsequent exercises.) Using the data for the Acme Corporation, calculate the "ideal" price index for 1980, using 1960 as the base period.

14.9 Using the symbols in equations (14.4) and (14.5), construct a general formula for Fisher's "ideal" price index.

14.6 Basic Considerations in Index Number Construction

Thus far we have focused exclusively on the types of formulas used to construct an index number. However, in many cases the practical problems of index number construction outweigh in importance the choice of a formula. This section discusses four fundamental questions that must be faced in constructing any index number.

First, *what is the purpose of the index number? What is it intended to represent or measure*? Unless this question is answered properly, there is a danger that the index number may not really measure what it is supposed to measure. In our example of crop prices, does the index number represent changes in the price level of *all* crops, or just food crops—in a number of countries, or only in the United States? Does the index number measure short-term changes—weekly or monthly price movements—or long-term changes? All these questions must be answered before the statistician can construct a proper index number.

Second, assuming that a particular index number is a price index, the statistician must decide *which prices should be included.* In our index of crop prices, should potatoes be included? Should rice be included? Such questions can only be answered on the basis of the *objectives* of the index, and the statistician must specify very carefully the nature and characteristics of the items involved. In establishing our index of crop prices, the analyst must be careful to specify precisely the *kind* of wheat being included. If such specifications are lacking, the price data may not be comparable, thereby rendering the index quite erroneous. Thus, if the 1972 wheat price was lower than the 1960 price because it pertained to a poorer grade of wheat, our index is misleading. One must constantly be on guard against lack of comparability, and the problems involved in obtaining comparable data should not be underestimated. Since concepts and definitions as well as techniques of measurement frequently differ from source to source and from period to period, data that seem comparable frequently turn out not to be.

Third, the statistician must determine *the time period to which the index number should pertain.* Given the objectives of our index number of crop prices, should we be comparing recent price levels with 1960, or should we be comparing them with 1965, or 1970? In other words, what period should be used as the base period? Naturally, the answer depends on the purposes of the index number. Often, a base period is chosen in order to avoid a period of abnormality such as war, severe depression, hyperinflation, and so on. The reasons for this are evident. For example, an index of prices with 1933 (the bottom of the Great Depression) as a base year is bound to exaggerate the extent of recent price increases. The federal government has established a standard base period for most indexes computed by government agencies. During the early 1970s, this base period was 1967; during the late 1970s, it was 1972 for some (but not all) indexes.

Fourth, the statistician must determine *the weights to be used.* As we have seen, there is a considerable amount of choice in this regard. The weights may pertain to the base period (as in the Laspeyres index), to the given period (as in the Paasche index), or to some other period (as in the fixed-weight aggregative index). But this is only part of the range of choice. In addition, there is often some question about the nature of the weights. For example, in a price index, should the quantity weights be the amounts consumed or the amounts produced? And the amounts consumed or produced by whom? Like so many of the questions taken up in this section, these questions must be answered by the good judgment of the statistician in response to the particular objectives of the index number under construction. There are no mechanical rules or formulas for solving these problems.

14.7 The Consumer Price Index: A Case Study

Perhaps the most widely heralded and most influential index in the United States is the Consumer Price Index, computed by the U.S. Department of Labor's Bureau of Labor Statistics. Until 1978, the purpose of this index was to measure changes in prices of goods and services purchased by urban wage earners and clerical workers and their families. In 1978 the index was expanded to include all urban consumers (although the narrower index was not discontinued). The index includes the price of food, automobiles, clothing, homes, furniture, home supplies, drugs, fuel, medical fees, legal fees, rents, repairs, transportation fares, recreational goods, and so forth. Prices, as defined in the index, include sales taxes and excise taxes. Also, real estate taxes—but not income taxes or personal property taxes—are included in the index. As the *New York Times* put it, "... of all the torrent of statistics pouring out of Washington, none exceeds in importance the monthly Consumer Price Index issued by the Bureau of Labor Statistics."[2]

The base period for the Consumer Price Index has been changed from time to time. During the 1960s, the average of 1957–59 was used as a base period; during the 1970s, the base period was 1967. Undoubtedly, the base period will be changed again in the future. The Consumer Price Index is computed monthly. A separate index is calculated for each of 28 Standard Metropolitan Statistical Areas as well as for all the urban places in the United States. Thus, the index provides information on inter-city differences in the rate of price increases. Table 14.8 shows the behavior of the Consumer Price Index during 1978. Table 14.9 provides index numbers for prices of various kinds of goods and services included in the Consumer Price Index during 1960–78. Clearly, the rate of increase of prices has varied from one kind of

TABLE 14.8
Consumer Price Index, January 1978–December 1978

Month	Consumer Price Index (1967 = 100)
January	187.2
February	188.4
March	189.8
April	191.5
May	193.3
June	195.3
July	196.7
August	197.8
September	199.3
October	200.9
November	202.0
December	202.9

SOURCE: Bureau of Labor Statistics.

2. *The New York Times*, June 22, 1974.

TABLE 14.9

Consumer Price Index, by Commodity and Service Groups, 1960–78

Year	All items	Food	Durable commodities	Nondurable commodities[a]	Rent	Services (less rent)
			(1967 = 100)			
1960	88.7	88.0	96.7	90.7	91.7	81.9
1961	89.6	89.1	96.6	91.2	92.9	83.9
1962	90.6	89.9	97.6	91.8	94.0	85.5
1963	91.7	91.2	97.9	92.7	95.0	87.3
1964	92.9	92.4	98.8	93.5	95.9	89.2
1965	94.5	94.4	98.4	94.8	96.9	91.5
1966	97.2	99.1	98.5	97.0	98.2	95.3
1967	100.0	100.0	100.0	100.0	100.0	100.0
1968	104.2	103.6	103.1	104.1	102.4	105.7
1969	109.8	108.9	107.0	108.8	105.7	113.8
1970	116.3	114.9	111.8	113.1	110.1	123.7
1971	121.3	118.4	116.5	117.0	115.2	130.8
1972	125.3	123.5	118.9	119.8	119.2	135.9
1973	133.1	141.4	121.9	124.8	124.3	141.8
1974	147.7	161.7	130.6	140.9	130.6	156.0
1975	161.2	175.4	145.5	151.7	137.3	171.9
1976	170.5	180.8	154.3	158.3	144.7	186.8
1977	181.5	192.2	163.2	166.5	153.5	201.6
1978 (Nov.)	202.0	217.8	180.0	179.1	168.5	227.8

SOURCE: *Economic Report of the President, 1979* (Washington, D.C.: Government Printing Office, 1979).

[a] Food is excluded from this and the previous column.

good to another. For example, note the large jump in food prices from 1972 to 1974.

The Consumer Price Index is widely used by industry and government. Labor contracts often stipulate that wages must increase in accord with changes in the Consumer Price Index. Also, other contracts, such as long-term leases, sometimes call for automatic adjustments based on the Index. In addition, it is often used by economists and government officials to measure changes in the purchasing power of the dollar. Pensions, welfare payments, royalties, and even alimony payments are often related to the Consumer Price Index.

The formula used to compute the Consumer Price Index is of the fixed-weight aggregative type. During more than 60 years of existence, the Consumer Price Index has experienced several revisions of its quantity weights. There has also been some change in the list of items in the Index in order to include new products like synthetic fibers and television sets and drop obsolete products such as buggy whips. At present the Consumer Price Index contains approximately 400 items. Prices are obtained by representatives of the Bureau of Labor Statistics from a sample of about 25,000 retail stores and service establishments. To make sure that quality and quantity do not vary over time, the Bureau has formulated detailed specifications for each item.

Finally, it is important to recognize the limitations of the Consumer Price Index. Despite warnings to the contrary, many people seem to believe that this index measures a family's living costs. This clearly is not so, since a family's living costs depend on how much the family consumes as well as on the level of prices it pays. All that the Consumer Price Index measures is the change in prices. Another important misconception is that this index applies to all Americans, whereas it applies only to urban consumers. Also, it is important to recognize that the Index does not take into account changes in income taxes and personal property taxes; nor is it free from errors due to sampling or incorrect reporting of prices. Still and all, the Consumer Price Index provides a reasonably accurate indication of changes over time in the prices paid by urban consumers, although it may not be a very good measure for a particular worker or occupation.[3]

The following example illustrates how the Consumer Price Index is used by industry and labor.

Example 14.1 In many situations, ranging from the formulation of government economic policy to a particular labor negotiation between management and a union, it is important to distinguish between *money* wages and *real* wages. Money wages are wages expressed in ordinary dollars, whereas real wages are corrected to take into account changes in the purchasing power of the dollar. In other words, real wages are expressed in dollars of constant purchasing power. Suppose that your money wage during selected years from 1967 to 1977 was as shown in Table 14.10, and that the Consumer Price Index (divided by 100) during each year was as shown in the final column of this table. What was your real wage in 1967 dollars in each year?

Solution: To compute real wages, one must divide money wages by the appropriate price index (divided by 100). Assuming that the Consumer

TABLE 14.10

Your Annual Wage and the Consumer Price Index

Year	Your money wage (dollars)	Consumer Price Index ÷ 100 (1967 = 1.00)
1967	15,000	1.000
1968	16,000	1.042
1969	17,000	1.098
1970	18,000	1.163
1971	19,000	1.213
1972	20,000	1.253
1973	21,000	1.331
1977	25,000	1.815

3. This discussion omits many important considerations that are too detailed to be taken up here. For further details, see Bureau of Labor Statistics, "The Consumer Price Index," in E. Mansfield, *Statistics for Business and Economics: Readings and Cases* (New York: Norton, 1980).

Price Index is appropriate, your real wage was

1967: \$15,000 ÷ 1.000 = \$15,000
1968: 16,000 ÷ 1.042 = 15,355
1969: 17,000 ÷ 1.098 = 15,482
1970: 18,000 ÷ 1.163 = 15,477
1971: 19,000 ÷ 1.213 = 15,663
1972: 20,000 ÷ 1.253 = 15,961
1973: 21,000 ÷ 1.331 = 15,777
1977: 25,000 ÷ 1.815 = 13,774

14.8 Quantity Indexes

Our discussion up to this point has been concerned solely with price indexes. In the next two sections we consider another important kind of index number, quantity indexes. Just as price indexes measure the change over time in prices, quantity indexes measure the change over time in the quantity produced. For example, to construct a quantity index for the production of food crops in the United States we can use formulas that are entirely analogous to those used for price indexes. Basically, we only need to substitute price terms for quantity terms, and vice versa, since prices are used as weights in quantity indexes.

Specifically, a *Laspeyres quantity index* can be computed, the formula being:

$$I_1 = \frac{\sum_{i=1}^{m} P_{0i}Q_{1i}}{\sum_{i=1}^{m} P_{0i}Q_{0i}}(100). \tag{14.7}$$

TABLE 14.11

Computation of Laspeyres Quantity Index for 1972

Crop	(1) 1972 quantity	(2) 1960 quantity	(3) Weight (1960 price)	(3) × (1)	(3) × (2)
Corn	5,553	3,907	1.00	5,553	3,907
Wheat	1,545	1,355	1.74	2,688	2,358
Oats	695	1,153	0.60	417	692
Soybeans	1,283	555	2.13	2,733	1,182
Total				11,391	8,139

$$\text{Index} = \frac{11,391}{8,139}(100) = 140.$$

The weights are the prices in the base year (year 0). To illustrate the computations, Table 14.11 shows how a Laspeyres quantity index for 1972 can be computed for the food crops, given that 1960 is the base year.

Similarly, a *Paasche quantity index* can be computed, the formula being

$$I_1 = \frac{\sum_{i=1}^{m} P_{1i} Q_{1i}}{\sum_{i=1}^{m} P_{1i} Q_{0i}} (100). \tag{14.8}$$

The weights are the prices in the given year (year 1). To illustrate the computations, Table 14.12 shows how a Paasche quantity index for 1972 can be computed for the food crops, given that 1960 is the base year.

TABLE 14.12

Computation of Paasche Quantity Index for 1972

Crop	(1) 1972 quantity	(2) 1960 quantity	(3) Weight (1972 price)	(3) × (1)	(3) × (2)
Corn	5,553	3,907	1.29	7,163	5,040
Wheat	1,545	1,355	1.67	2,580	2,263
Oats	695	1,153	0.67	466	773
Soybeans	1,283	555	3.49	4,478	1,937
Total				14,687	10,013[a]

$$\text{Index} = \frac{14{,}687}{10{,}013}(100) = 147.$$

[a] Due to rounding errors, this differs from 10,017 in the first equation in Section 14.3.

Also, *a fixed-weight aggregative quantity index* can be computed, the formula being

$$I_1 = \frac{\sum_{i=1}^{m} P_{wi} Q_{1i}}{\sum_{i=1}^{m} P_{wi} Q_{0i}} (100). \tag{14.9}$$

where P_{wi} is the price of the ith commodity in year w. The weights are the prices in year w. To illustrate the computations, Table 14.13 shows how an index of this sort can be computed for 1972 food crops, given that 1960 is the base (reference) year and the weights are 1965 prices.

TABLE 14.13

Computation of Fixed-Weight Aggregative Quantity Index for 1972, with 1965 Prices as Weights

Crop	(1) 1972 quantity	(2) 1960 quantity	(3) Weight (1965 price)	(3) × (1)	(3) × (2)
Corn	5,553	3,907	1.16	6,441	4,532
Wheat	1,545	1,355	1.35	2,086	1,829
Oats	695	1,153	0.62	431	715
Soybeans	1,283	555	2.54	3,259	1,410
Total				12,217	8,486

$$\text{Index} = \frac{12{,}217}{8{,}486}(100) = 144.$$

14.9 The Index of Industrial Production: A Case Study

Analysts in business and economics pay close attention to the Index of Industrial Production, computed by the Federal Reserve Board. This is a quantity index which measures changes in the output of American manufacturing, mining, and electric and gas industries, as well as in individual parts of each of these three industrial sectors. For example, in manufacturing, individual indexes are computed for such industries as primary metals, fabricated metal products, machinery, transportation equipment, instruments, ordnance, lumber, furniture, textiles, paper, chemicals, and food.

Although the Index of Industrial Production pertains to only a part of the economy (such activities as wholesale and retail trade, finance, other services, transportation, construction, and agriculture being omitted), it nevertheless is an important indicator of the pace of economic activity in the United States. Table 14.14 shows the values of the Index during 1947–78. In general, the Index rose during this period, but, as one would expect, it fell during recessions. (For example, it dropped from 1948 to 1949, from 1953 to 1954, from 1957 to 1958, from 1969 to 1970, and from 1973 to 1975.)

The Index of Industrial Production, which is computed monthly, used 1967 as a base (reference) period during the early and middle 1970s. As with the Consumer Price Index and other government indexes, its base period is revised periodically, and the weights used are also updated from time to time. (During the early and middle 1970s they, too, referred to 1967.) These weights are somewhat different from those described in the previous section. Instead of using the price of a commodity as a weight, the value-added per unit of output is used since this provides a better measure of the worth of a unit of an industry's output.

To understand why this is so, let us consider what is meant by *value-added*. In briefest terms, it can be defined as *the amount of value added by the industry to the total value of the product*. For example, suppose that $170

TABLE 14.14
Index of Industrial Production, 1947–78 (1967 = 100)

Year	Index	Year	Index
1947	39.4	1963	76.5
1948	41.0	1964	81.7
1949	38.8	1965	89.2
1950	44.9	1966	97.9
1951	48.7	1967	100.0
1952	50.6	1968	105.7
1953	54.8	1969	110.7
1954	51.9	1970	106.6
1955	58.5	1971	106.8
1956	61.1	1972	115.2
1957	61.9	1973	125.6
1958	57.9	1974	124.8
1959	64.8	1975	117.8
1960	66.2	1976	129.8
1961	66.7	1977	137.1
1962	72.2	1978[a]	145.1

SOURCE: Federal Reserve Board.
[a] Preliminary estimate by Council of Economic Advisers.

million of bread was produced in the United States in 1980. In order to produce it, flour mills turned out \$90 million worth of flour. Thus, the value of the bread industry's production was the extra value it added to the \$90 million of flour it bought—that is, \$80 million. (In other words, the bread industry's value-added is \$80 million.) Similarly, in the automobile industry the total value of the automobile is not attributable to this industry alone since the steel, tires, glass, and many other components of cars were not produced by the auto manufacturers. What the auto makers produce is the extra value that is added to the value of the steel, tires, glass, and other products that they purchase.

If we let $V_{67,i}$ be the value-added per unit of output of the ith commodity in 1967, the formula for the Index of Industrial Production is

$$I_1 = \frac{\sum_{i=1}^{m} V_{67,i} Q_{1i}}{\sum_{i=1}^{m} V_{67,i} Q_{67,i}} (100).$$

From this formula, it is clear that (as we have pointed out) the value-added per unit of output is used to weight each commodity's output.

EXERCISES

14.10 On the next page are three important price indexes for the years, 1970–75. (a) How do the objectives of these index numbers differ? (b) Which includes the widest range of goods and services? (c) Which of these index numbers do most economists prefer as a measure of general inflation? Why?

Year	Consumer Price Index (1967 = 100)	Producer Price Index (formerly Wholesale Price Index) (1967 = 100)	Price deflator for GNP (1972 = 100)
1970	116.3	110.4	91.36
1971	121.3	113.9	96.02
1972	125.3	119.1	100.00
1973	133.1	134.7	105.92
1974	147.7	160.1	116.20
1975	161.2	174.9	126.35

14.11 "Since the data in Exercise 14.10 show that the Consumer Price Index for 1975 was 161.2, while the price deflator for gross national product (GNP) for 1975 was 126.35, it follows that consumer prices rose more rapidly during 1970–75 than did the prices of gross national product generally." Do you agree? Why, or why not?

14.12 The average weekly earnings in manufacturing during 1970–75 are given below:

Year	Average weekly earnings
1970	$133.73
1971	142.44
1972	154.69
1973	166.06
1974	176.40
1975	189.51

Using the Consumer Price Index (given in Exercise 14.10), compute real weekly earnings.

14.13 Based on your results in Exercise 14.12, by what percentage did real average weekly earnings in manufacturing increase between 1970 and 1975? Describe in detail what the word *real* means in this context.

14.14 A vegetarian couple of ages 72 and 70 are considering moving to Atlanta, Georgia from Boston, Massachusetts. They read that the Consumer Price Index is 1 percent higher for Boston than for Atlanta. From this they conclude that it will cost them 1 percent less to maintain their existing standard of living in Atlanta than in Boston. Do you agree? Why, or why not?

14.15 Suppose that the price of a trip from New York to London changed as follows during 1900–1980:

Year	Price
1900	$100
1940	200
1980	400

Are these prices of a ship or airplane ticket an adequate measure of the cost of the trip? Why, or why not? If not, what additional factors should be taken into account? (Hint: What quality changes have occurred?)

14.16 The Acme Corporation wants to construct a quantity index for its production of machine tools. Using the data in Exercises 14.2 and 14.4, construct a Laspeyres index of this kind for 1970 and 1980. (Use 1960 as the base period.)

14.17 Using the data in Exercises 14.2 and 14.4, construct a Paasche quantity index for the Acme Corporation, where 1980 is the given year and 1960 is the base period.

14.18 It is frequently asserted that the Laspeyres quantity index overstates changes, whereas the Paasche quantity index understates them. Can you think of an important reason why this is likely to be true?

14.19 It seems desirable that when a price index (divided by 100) is multiplied by a quantity index (divided by 100), the result should equal the value index, which is

$$\sum_{i=1}^{m} Q_{1i}P_{1i} \div \sum_{i=1}^{m} Q_{0i}P_{0i}.$$

Basically, the reason why this is desirable is that

Price × Quantity = Value.

Consequently, if the price index for a commodity doubles and if the quantity index for the commodity also doubles, the value index should quadruple. (This is the so-called *factor-reversal* test.)
(a) If both the price index and the quantity index are Laspeyres indexes, does this desirable result occur?
(b) If both the price index and the quantity index are Paasche indexes, does this desirable result occur?
(c) If both the price index and the quantity index are of Fisher's "ideal" type (see Exercise 14.8), does this desirable result occur?

14.20 It also seems desirable that if an index for year 1 with base year 0 equals 40, then an index for year 0 with base year 1 should equal 250. After all, if the item in question is 40 percent as big in year 1 as in year 0, it should be 250 percent as big in year 0 as in year 1. Whether or not an index number has this property determines whether it passes the *time-reversal* test.
(a) Does the Laspeyres index pass this test?
(b) Does Fisher's "ideal" index pass this test?

CHAPTER REVIEW

1. An *index number* is a ratio (generally expressed as a percentage) of one quantity to another, where one of the quantities summarizes a given group of items and the other quantity summarizes a base group of items. The *base* group is used as a standard of comparison with the *given* group. For example, an index number might be constructed to indicate the level of steel prices in 1980 (the given period) relative to 1970 (the base period).

2. If P_{0i} is the price of the ith commodity in the base period and P_{1i} is its price in the given period, the *simple aggregative price index* is

$$\frac{\sum_{i=1}^{m} P_{1i}}{\sum_{i=1}^{m} P_{0i}}(100),$$

and the *average of relatives* is

$$\frac{\sum_{i=1}^{m} \frac{P_{1i}}{P_{0i}}}{m}(100).$$

These indexes are seldom used because the simple aggregative price index is influenced by the units in which the quantities of various commodities are expressed, and the average of relatives does not weight prices by their importance.

3. In constructing price indexes, statisticians generally weight each commodity's *price relative* by its value. If the value in the base period is used as a weight, the result is the *Laspeyres price index*, which is

$$\frac{\sum_{i=1}^{m} Q_{0i}P_{1i}}{\sum_{i=1}^{m} Q_{0i}P_{0i}}(100),$$

where Q_{0i} is the amount consumed of the ith commodity in the base period. If the value in the given period is used as a weight, the result is the *Paasche price index*, which is

$$\frac{\sum_{i=1}^{m} Q_{1i}P_{1i}}{\sum_{i=1}^{m} Q_{1i}P_{0i}}(100),$$

where Q_{1i} is the amount consumed of the ith commodity in the given period. If the quantity weights refer to a period other than the base year or the given year, the result is a *fixed-weight aggregative index*, which is

$$\frac{\sum_{i=1}^{m} Q_{wi}P_{1i}}{\sum_{i=1}^{m} Q_{wi}P_{0i}}(100),$$

where Q_{wi} is the amount consumed of the ith commodity in the period to which the quantity weights pertain.

4. There are a host of practical problems that must be met in the construction of an index number. It is particularly important that the purpose of the index number be kept in mind, and that a proper decision be made concerning which prices should be included, the time period to which the index number should pertain (including which base period to use), and the nature of the weights to be used. To a considerable extent proper decisions depend on the good judgment and integrity of the statistician, since there are no mechanical rules insuring good results.

5. Just as price indexes measure the change over time (or from place to place) in prices, so *quantity indexes* measure the change over time (or from place to place) in the quantity produced. Among the possible indexes of this sort are the *Laspeyres quantity index* (which uses base-period prices as weights), the *Paasche quantity index* (which uses given-period prices as weights), and the *fixed-weight aggregative quantity* index (which uses prices in a period other than the base or given period as weights).

6. Probably the most famous and important price index is the Consumer Price Index, which is issued monthly by the Bureau of Labor Statistics. It is an important measure of the rate of inflation, and is used widely as a basis for escalator clauses that raise wages, pensions, and other payments when the price level rises. An important quantity index is the Index of Industrial Production, computed monthly by the Federal Reserve Board. Although it excludes such parts of the economy as wholesale and retail trade, services, finance, transportation, construction, and agriculture, it is regarded as an important indicator of the pace of economic activity in the United States.

Getting Down to Cases:

SEASONAL VARIATION IN INDEXES OF INDUSTRIAL PRODUCTION

As noted previously, one of the most carefully watched indexes is the Federal Reserve Board's Index of Industrial Production. This index is broken into separate indexes for various types of goods. As the Federal Reserve Board points out in the following quotation, the seasonal variation for these various types of goods can be quite different.

> [Page 514] shows the seasonal patterns of production for several major market groups . . . . Output of consumer goods reaches a seasonal peak in the autumn months primarily as a result of the processing of food crops and a seasonal accumulation of business inventories of non-food products to supply cold weather and holiday demands. There is a less pronounced seasonal high in consumer goods output in February and March to meet spring demands for clothing and home goods. The subsequent decline and a marked rise in June, as shown in the top panel in the chart, are followed by sharp curtailments in July production schedules. In recent years these reductions have amounted to about 7.5 per cent.
>
> Total equipment output shows less seasonal variation than consumer goods, and the defense and space component shows virtually none. Output of business equipment in the new index reaches an advanced level before seasonal adjustment during the first half of the year as production of construction and farm machinery reaches a peak in the spring months. A second high period is evident during the autumn.
>
> Output of construction products has a marked seasonal rise—about 12 percent—during the spring and early summer, as shown in the bottom panel [on page 514]. The seasonal decline in such output during the . . . month of July is more than that for business equipment and about the same as that for consumer goods.
>
> Output of materials is at an advanced level from February to June. This has usually been true for total products too.[4]

(a) Why do you think that output of all these types of goods is relatively low in December?

(b) Why do you think that output of all these types of goods is relatively low in July?

(c) Does the seasonal variation in construction products seem more severe than for business equipment? If so, why?

4. Federal Reserve Board, *Industrial Production, 1971* (Washington, D.C.: Government Printing Office, 1972), pp. 47–48.

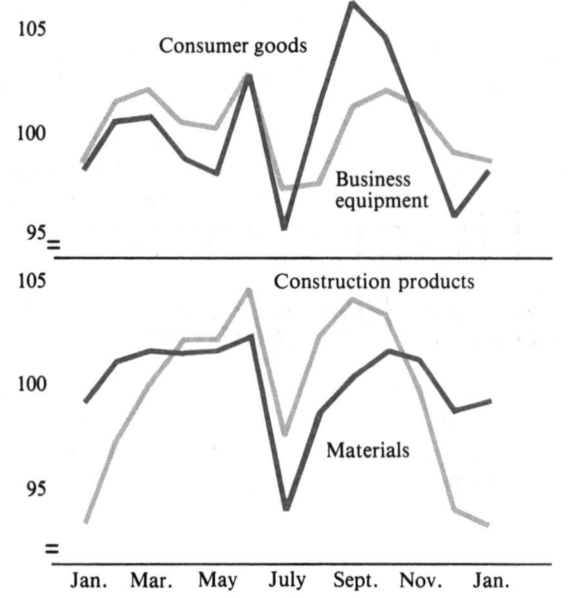

Implied indexes derived by dividing indexes without seasonal adjustment by seasonally adjusted indexes.

(d) If the output of consumer goods is 3 percent lower in November than in September, is this a decrease when seasonal variation is taken into account? Why, or why not?

(e) In recent years, provisions for a longer Christmas holiday have been written into labor contracts in the auto industry. What effect do you think this will have on the seasonal variation in the auto industry?

15

Decision Theory: Prior Analysis and Posterior Analysis

15.1 Introduction

One of the most interesting developments in statistics over the past 25 years has been the growing application of decision theory. Decision makers in business and government are continually faced with the problem of making choices under conditions of uncertainty. For example, a business frequently must decide whether or not to introduce a new product while still uncertain of the size of the potential market for the product. And firms must decide on the location, size, and staffing of new plants when they are uncertain of the future pattern of demand, costs, and a variety of other relevant factors. Statistical decision theory provides useful ways of analyzing such problems and helps decision makers arrive at rational choices. In this and the following chapter we will describe this theory and its application.

15.2 Decision Trees

Any problem of decision making under uncertainty has the following two characteristics. First, *the decision maker must make a choice, or perhaps a series of choices, among alternative courses of action*. Second, *this choice leads to some consequence, but the decision maker cannot tell in advance the exact nature of this consequence because it depends on some unpredictable event, or series of events, as well as on the choice itself*. For example, let's consider the case of a marketing manager who must decide whether or not to adopt a new label for a product. In this case, the choice is between two alternatives: adopt the new label or stick with the old one. And the consequence of adopting the new label is uncertain since the marketing manager cannot be sure whether the new label is superior to the old one. If the new label is a better label the firm will gain \$800,000, but if it is not the firm will lose \$500,000. The marketing manager believes that if he adopts the new label there is a 50-50 chance that it will be superior and a 50-50 chance that it will not.

To represent any such problem of decision making under uncertainty, a decision tree is useful. Its definition is as follows.

> **Decision tree:** *A decision tree represents a decision problem as a series of choices, each of which is depicted by a fork (sometimes called a juncture or branching point). A* **decision fork** *is a juncture representing a choice where the decision maker is in control of the outcome; a* **chance fork** *is a juncture where "chance" controls the outcome.*

To differentiate between a decision fork and a chance fork, we shall place a small square at the former juncture but not at the latter. Figure 15.1 shows the decision tree for the problem facing the marketing manager. Beginning at the left-hand side of the diagram, the first choice is up to the marketing manager, who can either follow the branch representing the adoption of the new label or the branch representing the continued use of the old label. Since this fork is a decision fork, it is represented by a square. If the branch representing the continued use of the old label is followed, the consequence is certain: The firm will have no extra profits or losses. Thus, zero extra profit is shown at the end of this branch. If the branch representing the adoption of the new label is followed, we come to a chance fork since it is uncertain whether the new label is superior to the old one. The upper branch following this chance fork represents the consequence that the new label is superior, in which case the extra profit to the firm is \$800,000, shown at the end of this branch. The lower branch following this chance fork represents the consequence that the new label is not superior, in which case the outcome is − \$500,000 (a loss), shown at the end of this branch. The probability that "chance" will choose each of these branches is shown above the branch.

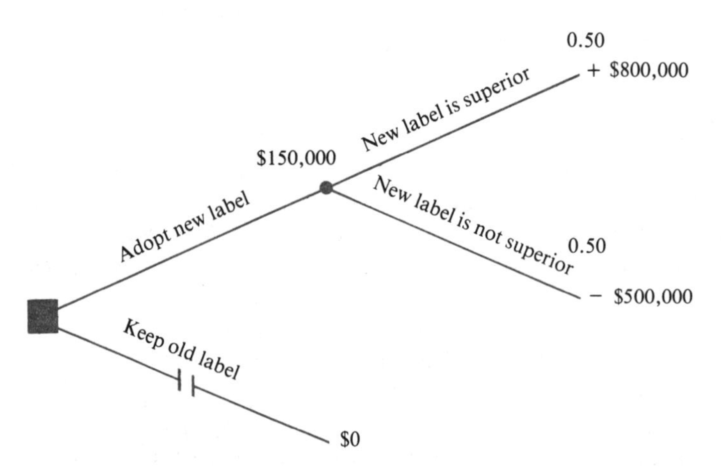

FIGURE 15.1 Decision Tree for Marketing Manager's Problem of Whether or Not to Adopt New Label

Given such a decision tree, it is easy to determine which branch the marketing manager should choose in order to maximize expected extra profit to the firm. The process by which we solve this problem, known as *backward induction*, requires that we begin at the right-hand side of the decision tree, where the monetary payoff figures are located. The first step is to calculate the expected monetary value of being situated at the chance fork immediately to the left of these payoff figures. In other words, this is the expected extra profit to the firm given that "chance" will choose which subsequent branch will be followed. Since there is a 0.50 probability that the branch culminating in a profit increase of $800,000 will be followed, and a 0.50 probability that the branch culminating in a profit decrease of $500,000 will be followed, the expected monetary value of being situated at this chance fork is

$$0.50(\$800{,}000) + 0.50(-\$500{,}000) = \$150{,}000.$$

This number is written above the chance fork in question to show that this is the expected monetary outcome of being located at that fork. Moving further along the decision tree to the left, it is evident that the marketing manager has a choice of two branches, one of which leads to an expected extra profit for the firm of $150,000, the other of which leads to a zero extra profit. If the marketing manager wants to maximize expected monetary value (in this case, extra profit), he should choose the former branch. In other words, he should adopt the new label. Since the latter branch (Keep old label) is nonoptimal, we place two vertical lines through it.

At this point, it is worth noting that this graphic procedure for analyzing the marketing manager's problem amounts to precisely the same thing as the calculations we made in Example 4.6 where this same problem was presented. Recall that in order to solve this problem then, we compared the expected extra profit if the new label was adopted ($150,000) with the extra profit if it was not adopted ($0) and followed the course of action that resulted in the larger of the two. Our procedure in Figure 15.1 is exactly the same.

15.3 Decision Trees: A Case Study[1]

To illustrate how decision trees can be used in actual circumstances, let's consider the problem of whether or not the federal government should seed hurricanes. In the early 1970s statisticians at Stanford Research Institute (now SRI International) analyzed this decision for the United States Department of Commerce, and used decision trees in their work. To construct the decision

1. This case study is based on R. Howard, J. Matheson, and D. North, "The Decision to Seed Hurricanes," *Science*, June 1972. It is reprinted in part in E. Mansfield, *Statistics for Business and Economics: Readings and Cases* (New York: Norton, 1980). Note that only a small portion of the authors' findings can be discussed here, and that we must necessarily oversimplify their methods and results. The interested reader is referred to the original paper for a fuller discussion. Also see Roscoe Braham, "Field Experimentation in Weather Modification" (and the comments by William Kruskal, Frederick Mosteller, Jerzy Neyman, and others), *Journal of the American Statistical Association*, 74, March 1979, pp. 52–104. For further discussion, see Section 5.9.

tree they began by noting that the government (the decision maker in this case) has a choice of two courses of action—to seed or not to seed. Thus, in Figure 15.2 at the left-hand end of the decision tree, there is a decision fork with two branches, one representing seeding, the other representing not seeding.

If the government follows the branch corresponding to seeding, it moves to a chance fork where there are five branches corresponding respectively to a large increase, a moderate increase, no change, a moderate decrease, and a large decrease in sustained wind speed. The property damage (in millions of dollars) corresponding to each of these consequences (plus the cost of seeding, which is $250,000) is shown at the right-hand end of each branch. (For example, the total cost is $336.05 million if there is a large increase in sustained wind speed.) The choice of which of these consequences will occur is up to "chance." The probability that each one will occur is given above each branch.

Continuing the method above, if the government follows the lower branch in Figure 15.2—the one corresponding to *not* seeding—it moves to a chance fork where there also are five branches corresponding respectively to a large increase, a moderate increase, no change, a moderate decrease, and a

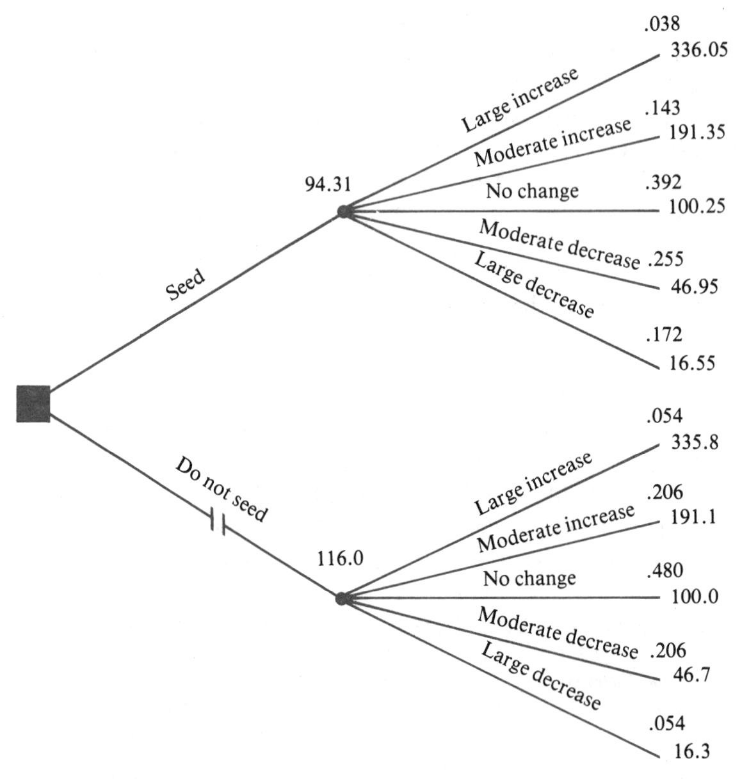

FIGURE 15.2 Decision Tree for the Government's Decision to Seed or Not to Seed a Hurricane

large decrease in sustained wind speed. The property damage resulting from each of these consequences is shown at the right-hand end of the relevant branch, and the probability that each will occur is given above its branch.

To determine whether the government should decide to seed or not to seed under these circumstances, the Stanford statisticians went through the procedure described in the previous section. That is, they calculated the expected cost (in millions of dollars) of being at the chance fork resulting from seeding, which is

$$.038(336.05) + .143(191.35) + .392(100.25)$$
$$+ .255(46.95) + .172(16.55) = 94.31.$$

They entered this figure (94.31) above the relevant chance fork in Figure 15.2. They then calculated the expected cost (in millions of dollars) of being at the chance fork resulting from not seeding, which is

$$.054(335.8) + .206(191.1) + .480(100.0) + .206(46.7) + .054(16.3)$$
$$= 116.0.$$

And they then entered the result (116.0) above its chance fork in Figure 15.2. If the government wants to minimize the expected cost, it should choose the branch that leads to the situation where the expected cost is lower.[2] Since the branch corresponding to seeding leads to an expected cost of $94.31 million and the one corresponding to not seeding leads to an expected cost of $116.00 million, it follows that the optimal decision is to seed. (And two vertical lines are placed through the nonoptimal branch.)

The following example, based on the data in Figure 15.2, provides further practice in using decision trees.

Example 15.1 Suppose that the property damage corresponding to a large decrease in sustained wind speed is zero, rather than $16.3 million (the amount in Figure 15.2). If the government wants to minimize expected cost, should it decide to seed?

Solution: The expected cost (in millions of dollars) of being at the chance fork resulting from seeding is

$$.038(336.05) + .143(191.35) + .392(100.25)$$
$$+ .255(46.95) + .172(.25) = 91.51.$$

(Under these circumstances, the cost when there is a large decrease in sustained wind speed is .25, since this is the cost of seeding.) The expected cost (in millions of dollars) of being at the chance fork resulting from not seeding is

2. Whereas the monetary values in the marketing manager's problem were profits, the monetary values in this case are costs. Consequently, the decision maker wants to *minimize* the expected value of the monetary values in this case, whereas he wants to *maximize* the expected value of the monetary values in the previous section.

$$.054(335.8) + .206(191.1) + .480(100.0)$$
$$+ .206(46.7) + .054(0) = 115.11.$$

Since seeding leads to an expected cost of $91.51 million and not seeding leads to an expected cost of $115.11 million, the government should seed in order to minimize expected cost.

15.4 Maximization of Expected Utility

In discussing both the marketing manager's labeling decision and the U.S. Government's problem of whether or not to seed hurricanes, we have assumed that the decision maker wants to maximize expected monetary gain (or to minimize expected monetary loss). In this and the following section we will explain in greater detail why this may not be the right criterion, and will discuss how a more appropriate criterion can be formulated.[3] To understand why a decision maker may not want to maximize expected monetary gain, consider a situation where you are given a choice between (1) receiving $1,000,000 for certain, and (2) a gamble in which a fair coin is tossed, and you will receive $2,100,000 if it comes up heads, or you will lose $50,000 if it comes up tails. The expected monetary gain for the gamble is

$$0.50(\$2{,}100{,}000) + 0.50(-\$50{,}000) = \$1{,}025{,}000,$$

so you should choose the gamble over the certainty of $1,000,000 if you want to maximize expected monetary gain. However, it seems likely that many persons would prefer the certainty of $1,000,000 since the gamble entails a 50-50 chance that you will lose $50,000, a very substantial sum. Moreover, many people may feel that they can do almost as much with $1,000,000 as with $2,100,000, and therefore the extra amount is not worth the risk of losing $50,000.

Clearly, *whether or not you will want to maximize expected monetary gain in this situation depends on your attitude toward risk.* If you are a widow of modest means, you will probably be overwhelmed at the thought of taking a 50-50 chance of losing $50,000. On the other hand, if you are the president of a big corporation the prospect of a $50,000 loss may be not the least bit unsettling, and you may prefer the gamble to the certainty of a mere $1,000,000. And if you are the sort of person who enjoys danger and risk, you may prefer the gamble even though a $50,000 loss may wipe you out completely.

Fortunately, there is no need to assume that the decision maker wants to maximize expected monetary gain. Instead, as John von Neumann and Oskar Morgenstern pointed out several decades ago,[4] we can construct a ***utility function*** for the decision maker based on his or her attitudes toward risk; from

3. We assume here that the relevant probabilities can be formulated. If not, still other criteria of the sort discussed in more advanced texts on decision theory may have to be used.

4. J. von Neumann and O. Morgenstern, *The Theory of Games and Economic Behavior* (Princeton: Princeton University Press, 1944).

this, we can then go on to choose the alternative that offers the decision maker the maximum expected utility. The procedure used to construct a utility function is described in the following section, but before proceeding to this discussion, we must present the underlying assumptions of a utility function.

First, we must assume that *the decision maker's preferences are transitive.* That is, if he or she prefers Budweiser beer to Pabst beer, and Pabst to Coors, he or she must therefore prefer Budweiser to Coors. Similarly, if the decision maker is indifferent between a hot dog and a hamburger and indifferent between a hamburger and a salami sandwich, he or she must therefore be indifferent between a hot dog and a salami sandwich. The assumption of transitivity plays an important role in the theory of consumer behavior. Students who have taken a course in microeconomics are likely to have encountered this assumption before.

Second, we must assume that *if there are three outcomes,* A, B, *and* C, *and if the decision maker prefers* A *to* B *and* B *to* C, *then there must be some probability* P *such that the decision maker will be indifferent between the certainty of* B *and a gamble where there is a probability of* P *that* A *will occur and a probability of (*1 − P*) that* C *will occur.* This probability may be big or small: that doesn't matter. What is important is that some value of P exists so that the decision maker is indifferent between the gamble that A or C will occur and the certainty of B.

Third, we must assume that *if the decision maker is indifferent between a hot dog and a hamburger, then he or she will be indifferent between two lottery tickets that are identical except that one offers a hot dog as a prize while the other offers a hamburger.* This is known as the *independence axiom.*

Fourth, we must assume that *the decision maker, faced with two lottery tickets for identical prizes, will always choose the one with the higher probability of winning.* Also, we must assume that if the decision maker is offered a lottery ticket whose prize is another lottery ticket, his attitude toward it will be the same as if he had computed the *ultimate* odds of winning or losing that are involved in this compound lottery ticket.

Although questions have been raised by some statisticians and economists concerning a few of these assumptions, most people seem to regard them as quite reasonable foundations on which to build a theory of choice under uncertainty. It is important to note, however, that we are not assuming that individuals actually conform to all these assumptions in their actual decision-making processes. Even if a person agreed with all the axioms involved, he or she might make mistakes or act irrationally at times. Our theory is designed to indicate how people *should* make choices if their decisions are to be in accord with their own preferences. This does not necessarily indicate how they *do* make choices.

15.5 Construction of a Utility Function

If the four assumptions listed in the previous section are met, it can be shown that *a rational decision maker will maximize expected utility.* In other words, the decision maker should choose the course of action with the highest expected utility. But what is a *utility*? It is a number that is attached to a possible outcome of the decision. Each outcome has a utility. The decision maker's *utility function* shows the utility that he or she attaches to each possible outcome. This utility function, as we shall see, shows the decision maker's preferences with respect to risk.

How can we know the utility that the decision maker attaches to each possible outcome? In other words, how can we construct a utility function for the decision maker? In order for us to construct such a function the decision maker must respond to a series of questions which indicate his or her preferences with regard to risk. For example, consider the case of the decision to seed or not to seed a hurricane. As indicated in Figure 15.2, the property damages corresponding to a large increase, a moderate increase, no change, a moderate decrease, and a large decrease in sustained wind speed are 335.8, 191.1, 100.0, 46.7, and 16.3 millions of dollars, respectively. To tell whether the expected utility resulting from seeding exceeds the expected utility of not seeding a hurricane, we must know the utility that the decision maker attaches to a monetary loss of $335.8 million—which we will represent by $U(-335.8)$—and the utility that he or she attaches to a monetary loss of $191.1 million—that is, $U(-191.1)$—and $U(-100.0)$, $U(-46.7)$, and $U(-16.3)$. Why? Because the expected utility of not seeding equals $.054U(-335.8) + .206U(-191.1) + .480U(-100.0) + .206U(-46.7) + .054U(-16.3)$. Also, we must know $U(-336.05)$, $U(-191.35)$, $U(-100.25)$, $U(-46.95)$, and $U(-16.55)$, since the expected utility of seeding equals

$$.038U(-336.05) + .143U(-191.35) + .392U(-100.25)$$
$$+ .255U(-46.95) + .172U(-16.55).$$

(For the source of these numbers, see Figure 15.2.)

We can find each of the required utilities in two steps. The first step is simple: *We set the utility attached to two monetary values arbitrarily*. The utility of the better consequence is set higher than the utility of the worse one. Often, the worst consequence involved is given a utility of zero, and the best consequence is given a utility of 1. Consequently, let's set $U(-336.05)$ equal to zero and $U(-16.3)$ equal to 1. It turns out that the ultimate results of the analysis do not depend on which two numbers we choose, as long as the utility of the better consequence is set higher than the utility of the worse one. Thus, we could set $U(-336.05)$ equal to 4 and $U(-16.3)$ equal to 10. It would make no difference to the ultimate outcome of the analysis.

The second step is somewhat more complicated: *In this step we present the decision maker with a choice between the certainty of one of the other monetary values and a gamble where the possible outcomes are the two monetary values whose utilities we set arbitrarily*. For example, suppose that we want to find $U(-191.1)$. To do so, we ask the decision maker whether he or she would prefer the certainty of a $191.1 million loss to a gamble where there is a probability of P that the loss is $16.3 million and a probability of $(1 - P)$ that the loss is $336.05 million. We then try various values of P until we find the one where the decision maker is indifferent between the certainty of a $191.1 million loss and this gamble. Suppose that this value of P is 0.6.

If the decision maker is indifferent between the certain loss of \$191.1 million and this gamble, it must be that the expected utility of the certain loss of \$191.1 million equals the expected utility of the gamble. (Why? Because under the assumptions in the previous section, the decision maker maximizes expected utility.) Thus,

$$U(-191.1) = 0.4U(-336.05) + 0.6U(-16.3).$$

And since we set $U(-336.05)$ equal to zero and $U(-16.3)$ equal to 1, it follows that $U(-191.1)$ must equal 0.6. In other words, the utility attached to a loss of \$191.1 million is 0.6.

Similarly, we can find $U(-100.0)$, $U(-46.7)$, and the other utilities needed to compute the expected utility of seeding and the expected utility of not seeding. For example, to obtain $U(-100.0)$, we ask the decision maker whether he or she would prefer the certainty of a \$100.0 million loss to a gamble where there is a probability of P that the loss is \$16.3 million and a probability of $(1 - P)$ that the loss is \$336.05 million. Then we try various values of P until we find the one where the decision maker is indifferent between the certainty of a \$100.0 million loss and this gamble. Suppose that this value of P is .85. Then the expected utility of a certain loss of \$100.0 million must equal the expected utility of this gamble, which means that

$$U(-100.0) = .15U(-336.05) + .85U(-16.3).$$

And since $U(-336.05)$ equals zero and $U(-16.3)$ equals one, it follows that $U(-100.0)$ equals .85.

The decision maker's utility function is the graph or relationship showing the utility he or she attaches to each amount of monetary gain or loss. For example, in the case of the seeding example, the decision maker's utility function may be as shown in Figure 15.3. Through the repeated use of the procedure described

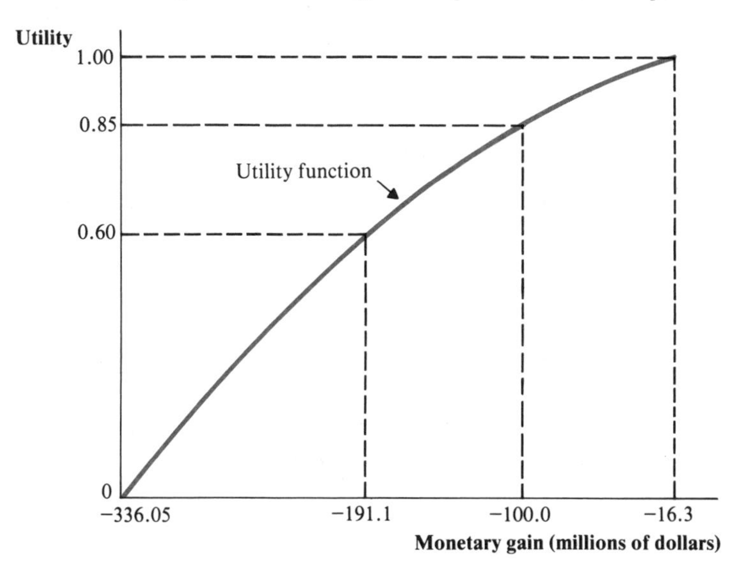

FIGURE 15.3 Decision Maker's Utility Function, Hurricane-Seeding Example

above, we can obtain as many points on this utility function as we like—or as the decision maker's patience permits. But to see whether the decision maker, in the light of his or her own preferences toward risk, should seed or not, all that we need are the utilities shown in Table 15.1. Based on these utilities, each of which is obtained by the above procedure, it is clear that the decision maker should decide to seed, since the expected utility if a hurricane is seeded (.832288) exceeds that if it is not seeded (.781354).

TABLE 15.1

Expected Utility from Seeding or Not Seeding a Hurricane

Seed				Do not seed			
Monetary loss (millions of dollars)	(1) Utility	(2) Probability	(1) × (2)	Monetary loss (millions of dollars)	(1) Utility	(2) Probability	(1) × (2)
336.05	0.000	.038	.000000	335.8	0.001	.054	.000054
191.35	0.599	.143	.085657	191.1	0.600	.206	.123600
100.25	0.849	.392	.332808	100.0	0.850	.480	.408000
46.95	0.949	.255	.241995	46.7	0.950	.206	.195700
16.55	0.999	.172	.171828	16.3	1.000	.054	.054000
	Expected utility:		.832288		Expected utility:		.781354

15.6 Characteristics of Utility Functions

Not all utility functions look like the one in Figure 15.3. Although one can expect that utility increases with monetary gain, the shape of the utility function can vary greatly, depending on the preferences of the decision maker. Figure 15.4 shows three general types of utility functions. The one in panel A is like that in Figure 15.3 in the sense that utility increases with monetary value, but *at a decreasing rate.* In other words, an increase in monetary gain of $1 is associated with *smaller and smaller* increases in utility as the monetary gain increases in size. People with utility functions of this sort are ***risk averters.*** That is, when confronted with gambles with equal expected monetary gains, they prefer a gamble with a more certain outcome to one with a less certain outcome.

The utility function in panel B is one where utility increases with monetary value, but *at an increasing rate.* In other words, an increase in monetary gain of $1 is associated with *larger and larger* increases in utility as the monetary gain increases in size. People with utility functions of this sort are ***risk lovers.*** That is, when confronted with gambles with equal expected monetary gains, they prefer a gamble with a less certain outcome to one with a more certain outcome.

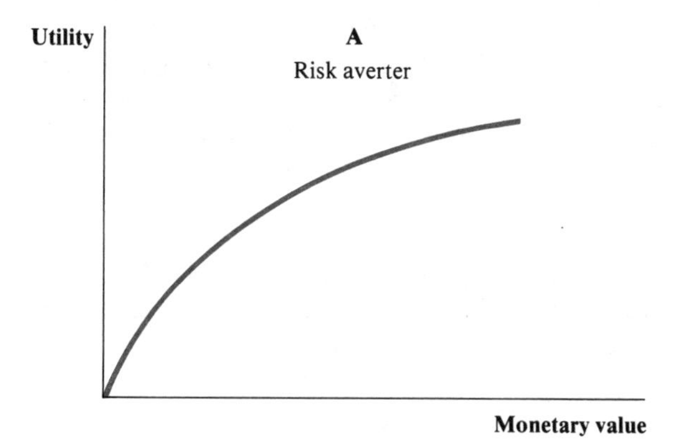

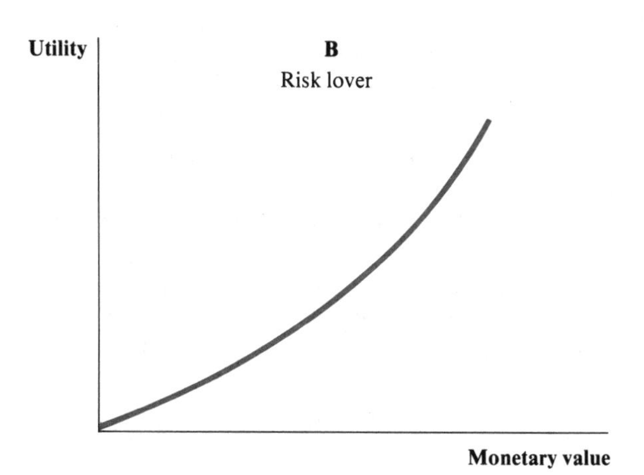

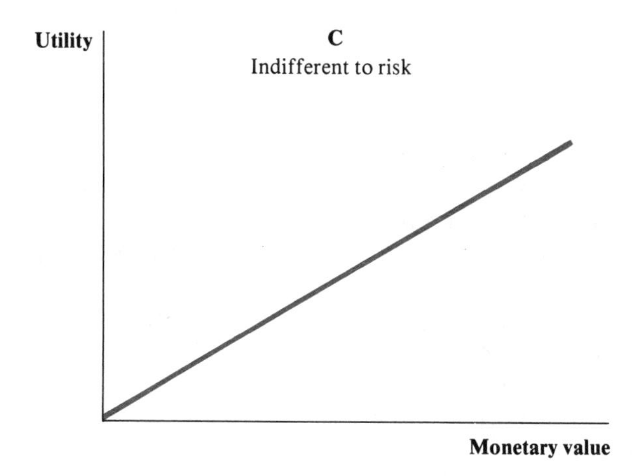

FIGURE 15.4 Three Types of Utility Functions

Finally, the utility function in panel C is one where utility increases with monetary value and *at a constant rate*. In other words, an increase of $1 in monetary gain is associated with a *constant* increase in utility as the monetary gain grows larger and larger. Stated differently, utility in this case is a linear function of monetary gain:

$$U = a + bM, \tag{15.1}$$

where U is utility, M is monetary gain, and a and b are constants. (Of course, $b > 0$.) People with utility functions of this sort are ***indifferent to risk***. That is, they maximize expected monetary gain or value, regardless of risk. It is easy to show that this is true. Clearly, if equation (15.1) holds,

$$E(U) = a + bE(M), \tag{15.2}$$

where $E(U)$ is expected utility and $E(M)$ is expected monetary gain. (For a proof, see Appendix 4.1.) Thus, since expected utility is directly related to expected monetary gain, it can only be a maximum when expected monetary gain is a maximum.

In the remainder of this book, we shall assume that the decision maker's utility function is of the sort shown in panel C. If this is so, we can assume that the decision maker wants to maximize expected monetary gain. The reason for this assumption is that it simplifies the discussion. (Also, in many actual cases it may be a satisfactory approximation). In cases where this assumption does not hold, it is easy to alter our results accordingly. All that is required is that we substitute the decision maker's utilities for the monetary values in all subsequent sections.

The following example provides a simple illustration of the concepts discussed in this section.

Example 15.2 John Jones and Cynthia Brown are each offered a choice of a certain gain of $200 or a gamble where there is a 50-50 chance of gaining $400 and a 50-50 chance of losing $100. John Jones chooses the gamble; Cynthia Brown chooses the certain gain of $200. Can we deduce from this that Jones is a risk lover and that Brown is a risk averter?

Solution: Since the expected value of the gamble equals 0.5($400) − 0.5($100), or $150, John Jones must be a risk lover because he chose the alternative with the less certain outcome, even though it had a lower expected monetary value than the certain gain of $200. On the other hand, Cynthia Brown may or may not be a risk averter. If she were a risk averter, she would choose the certain gain because it has both a higher expected monetary gain and less uncertainty than the gamble. But if she were indifferent to risk, she would also choose the certain $200 gain, since it has a higher expected monetary gain than the gamble.

EXERCISES

15.1 The Bona Fide Washing Machine Company is considering the purchase of a small firm that produces clocks. Bona Fide's management feels that there is a 50-50 chance, if Bona Fide buys the firm, that it can mold the firm into an effective producer

of washing machine parts. If the firm can be transformed in this way, Bona Fide believes that it will make $500,000 if it buys the firm; if it cannot be transformed in this way, Bona Fide believes that it will lose $400,000.
(a) Construct a decision tree to represent Bona Fide's problem.
(b) What are the decision forks? (Are there more than one?)
(c) What are the chance forks? (Are there more than one?)

15.2 Use the decision tree you constructed in Exercise 15.1 to solve Bona Fide's problem. In other words, assuming that the firm wants to maximize expected extra profit, should Bona Fide buy the firm or not?

15.3 Before Bona Fide makes a decision concerning the purchase of the firm, Bona Fide's president learns that if the clock producer cannot be made into an effective producer of washing machine parts, there is a 0.2 probability that it can be resold to a Saudi Arabian syndicate at a profit of $100,000. (If the firm cannot be resold, Bona Fide will lose $400,000.)
(a) Does this information alter the decision tree?
(b) Can you think of three mutually exclusive outcomes if Bona Fide buys the firm?
(c) What is the probability of each of these outcomes?
(d) What is the monetary value to Bona Fide of each of these outcomes?

15.4 Use your results in Exercise 15.3 to solve Bona Fide's problem under this new set of conditions. In other words, on the basis of this new information, should Bona Fide buy the firm or not?

15.5 Bona Fide's executive vice president discovers an error in the estimate of how much Bona Fide will gain if it buys the clock manufacturer and turns it into an effective producer of washing machine parts.
(a) Under the circumstances in Exercises 15.1 and 15.2, how big would this error have to be to reverse the indicated decision?
(b) Under the circumstances in Exercises 15.3 and 15.4, how big would the error have to be to reverse the indicated decision?

15.6 The president of the Uphill Corporation, a maker of bicycle pedals, is indifferent between a certain gain of $10,000 and a gamble where there is a 0.50 chance of winning $100,000 and a 0.50 chance of losing $90,000. Draw three points on his utility function.

15.7 Based on the results in Exercise 15.6, does it appear that Uphill's president is (a) indifferent to risk; (b) a risk lover; (c) a risk averter?

15.8 (a) Based on the results in Exercise 15.6, does it appear that Uphill's president's utility function is linear? (b) Suppose that his utility function is linear between – $90,000 and $10,000 and linear between $10,000 and $100,000 (but that the slopes of these two line segments are not necessarily equal). Does this mean he maximizes expected monetary gain?

15.9 A local bank president says that he is indifferent to risk. Suppose that we let 0 be the utility he attaches to $100,000 and 1 be the utility he attaches to $200,000. If what he says is true, what is the utility he attaches to (a) $300,000; (b) $50,000; (c) – $10,000?

15.10 Suppose that the bank president in Exercise 15.9 must decide whether or not to make a loan where he perceives the probability to be 0.6 that he will gain $20,000 and 0.4 that he will lose $25,000. If he does not make the loan, he will incur neither a loss nor a gain. If what he says (in Exercise 15.9) is true, will he make the loan?

15.11 In contrast to the bank president in Exercises 15.9 and 15.10, the president of the

Crooked Arrow National Bank is a risk lover. (a) Is it possible that he would make the loan in Exercise 15.10? (b) Can you be sure he would make such a loan? Why, or why not?

15.7 Posterior Analysis

Up to this point, we have been concerned with ***prior analysis***—that is, decision making based solely on the basis of prior probabilities. A *prior probability* is a probability based on whatever information is available prior to the gathering of new experimental or sample evidence. As you will recall from Chapter 3, a prior probability may be a subjective probability that a certain hypothesis is true. For example, the probability that the new label (in Section 15.2) will be superior to the old one is a prior probability, just as the probability of a large increase in sustained wind speed if a hurricane is seeded (in Section 15.3) is a prior probability. In the next few sections, we turn to ***posterior analysis***, in which decision making is based both on prior probabilities and on new experimental or sample evidence. Using Bayes' theorem, which we discussed at length in Chapter 3, we revise the prior probabilities to take account of this new, additional evidence.[5]

To illustrate the nature of posterior analysis, let's return to the marketing manager who must decide whether or not to adopt a new label for the firm's product. In contrast to Section 15.2 (and Example 4.6), suppose that the marketing manager does not want to rely solely on his prior probabilities that the new label will or will not be superior to the old one.[6] Instead, he sets in motion a survey of consumers. This survey is designed so that *if the new label is really superior*, the probability that the survey results will be favorable to the new label is 0.8, the probability that they will be neither favorable nor unfavorable to the new label is 0.1, and the probability that they will be unfavorable to the new label is 0.1. On the other hand, *if the new label is really not superior*, the probability that the survey results will be unfavorable to the new label is 0.7, the probability that they will be neither favorable nor unfavorable to the new label is 0.1, and the probability that they will be favorable to the new label is 0.2.[7]

Suppose that the survey results turn out to be favorable to the new label. Using Bayes' theorem, we can compute the posterior probabilities that the new label is or is not superior to the old one. These posterior probabilities, which take account of the survey evidence, are

$$P(B \mid b) = \frac{P(b \mid B)P(B)}{P(b \mid B)P(B) + P(b \mid N)P(N)} \tag{15.3}$$

5. Note that *prior* and *posterior* are relative terms. For example, a prior probability may be derived, at least in part, from a previous sample. The essential point is that it is the probability prior to the availability of the evidence in question, not all evidence.

6. Recall that the marketing manager's prior probabilities were .5 that the new label would be superior to the old one and .5 that it would not be superior to the old one.

7. The reader may recall that this survey was discussed in Section 3.8, where we derived the posterior probabilities used in this section.

and

$$P(N|b) = \frac{P(b|N)P(N)}{P(b|B)P(B) + P(b|N)P(N)}, \tag{15.4}$$

where $P(B|b)$ is the posterior probability that the new label is really superior and $P(N|b)$ is the posterior probability that it is not. The prior probability that the new label is superior is $P(B)$, and the prior probability that it is not is $P(N)$. The probability that the survey results are favorable to the new label when it in fact is superior is $P(b|B)$, while the probability that they are favorable to the new label when it in fact is not superior is $P(b|N)$.

Inserting the numerical values into equations (15.3) and (15.4), we obtain

$$P(B|b) = \frac{0.8(0.5)}{0.8(0.5) + 0.2(0.5)} = 0.8$$

$$P(N|b) = \frac{0.2(0.5)}{0.8(0.5) + 0.2(0.5)} = 0.2.$$

Using these posterior probabilities, we can calculate the expected increase in profit if the marketing manager adopts the new label and if he does not. If he adopts the new label, the expected increase in profit, based on these posterior probabilities, is

$$0.8(\$800{,}000) + 0.2(-\$500{,}000) = \$540{,}000.$$

If he does not adopt the new label, the increase in profit is zero. Thus, if he wants to maximize expected profit, he should adopt the new label.

The effect of the evidence obtained from the survey has been to enlarge the expected increase in profit, should the new label be adopted. Before the survey the expected profit increase (if the new label were adopted) was \$150,000. This figure was based only on the marketing manager's prior probabilities, and is called the *prior expected value*. After the survey the expected profit increase was \$540,000, should the new label be adopted. This was based on the posterior probabilities and is called the *posterior expected value*. The reason why the posterior expected value exceeded the prior expected value in this case was that the sample produced evidence that increased the probability that the new label is in fact superior to the old one. Needless to say, the posterior expected value does not always exceed the prior expected value.

15.8 Posterior Analysis in a Drug Firm: A Case Study[8]

To illustrate the use of posterior analysis, let's examine an actual situation. In 1968, the Canadian subsidiary of an American drug firm was considering introducing a certain kind of throat lozenge to the Canadian market. John Stonier of the firm's Consumer Products Division was charged with the

8. See C. H. von Lanzenauer, *Cases in Operations Research* (London and Canada: University of Western Ontario, 1975). This example is based on an actual case, but the names and numbers have been changed, and the circumstances and outcome have been altered and simplified considerably.

responsibility of recommending whether or not to introduce the new product. To simplify the analysis, he used three profit levels—$200,000 per year, zero, and −$200,000 per year—to represent the possible profits from the new product. The prior probabilities he attached to these outcomes were 0.5, 0.3, and 0.2, respectively. Thus, on the basis of the prior probabilities, the expected profit increase resulting from the introduction of the new product was

$$0.5(\$200{,}000) + 0.3(0) + 0.2(-\$200{,}000) = \$60{,}000.$$

And since if the new product was not introduced the profit increase would be zero, Stonier was inclined to recommend introducing the product.

But Stonier also had the opportunity of carrying out a test-market study before making the final decision. Table 15.2 shows his estimate that the test market would indicate each level of increased profit from the introduction of the throat lozenges, when the actual increased profit from their introduction was as given at the top of each column in the table. For example, the first column of this table shows that if the actual profit increase would be $200,000, he felt that there was a 0.6 probability that the test-market study would indicate this fact, a 0.3 probability that it would indicate a zero profit increase, and a 0.1 probability that it would indicate a $200,000 loss.

TABLE 15.2

Probability of Each Outcome of the Test-Market Study, Given Actual Level of Increased Profit from the New Product

Increased profit, test-market study (dollars)	Actual increased profit from new product (dollars)		
	200,000	0	−200,000
200,000	.6	.2	.1
0	.3	.6	.3
−200,000	.1	.2	.6
	1.00	1.00	1.00

If Stonier's test-market study indicated that the profit increase was −$200,000, should he recommend the introduction of this new product? Let $P_1(200{,}000)$ be the posterior probability that the profit increase will be $200,000, $P_1(0)$ be the posterior probability that it will be zero, and $P_1(-200{,}000)$ be the posterior probability that it will be −$200,000. Under these circumstances,[9]

$$P_1(200{,}000) = \frac{.1(.5)}{.1(.5) + .2(.3) + .6(.2)} = 5/23,$$

$$P_1(0) = \frac{.2(.3)}{.1(.5) + .2(.3) + .6(.2)} = 6/23,$$

$$P_1(-200{,}000) = \frac{.6(.2)}{.1(.5) + .2(.3) + .6(.2)} = 12/23.$$

9. To illustrate how the posterior probabilities (5/23, 6/23, 12/23) are derived, consider

Thus, if the new product is introduced, the posterior expected value of the increased profit is

$$(5/23)(\$200{,}000) + (6/23)(0) + (12/23)(-\$200{,}000) = -\$60{,}870.$$

Since the expected increased profit is negative (and consequently less than the expected increased profit of zero if the new product is not introduced), it appears that Stonier should recommend that the product not be introduced. In contrast to the situation in the previous section, this is a case where the sample evidence recommends a change in the decision.

The following example is a further illustration of posterior analysis, based on the data in Table 15.2.

Example 15.3 Suppose that the test-market study indicated an increased profit of zero. Should Stonier recommend the introduction of the new throat lozenge?

Solution: Let $P_1(200{,}000)$ be the posterior probability that the profit increase will be \$200,000, $P_1(0)$ be the posterior probability that it will be zero, and $P_1(-200{,}000)$ be the posterior probability that it will be $-$\$200,000. Since the test-market study indicates that the profit increase will be zero, these posterior probabilities are as follows:[10]

$$P_1(200{,}000) = \frac{.3(.5)}{.3(.5) + .6(.3) + .3(.2)} = .384,$$

$P_1(200{,}000)$. Based on Bayes' theorem, it equals

$$\frac{P(-200{,}000 \mid 200{,}000)P_0(200{,}000)}{P(-200{,}000 \mid 200{,}000)P_0(200{,}000) + P(-200{,}000 \mid 0)P_0(0) + P(-200{,}000 \mid -200{,}000)P_0(-200{,}000)},$$

where $P_0(200{,}000)$, $P_0(0)$, and $P_0(-200{,}000)$ are the prior probabilities of an increased profit of \$200,000, 0, and $-$\$200,000, respectively. $P(-200{,}000 \mid 200{,}000)$ is the probability that the test-market study will indicate an increased profit of $-$\$200,000, given that the profit increase will be \$200,000; $P(-200{,}000 \mid 0)$ is the probability that the test-market study will indicate an increased profit of $-$\$200,000, given that the profit increase will be zero; and so forth. Inserting into this expression the prior probabilities in the first paragraph of this section and the conditional probabilities in Table 15.2, we get the equation for $P_1(200{,}000)$ in the text.

The equations in the text for $P_1(0)$ and $P_1(200{,}000)$ can be derived in an analogous way. (Do this as an exercise.)

10. To illustrate how the posterior probabilities (.384, .462, and .154) are derived, consider $P_1(200{,}000)$. Based on Bayes' theorem, it equals

$$\frac{P(0 \mid 200{,}000)P_0(0)}{P(0 \mid 200{,}000)P_0(200{,}000) + P(0 \mid 0)P_0(0) + P(0 \mid -200{,}000)P_0(-200{,}000)},$$

where $P_0(200{,}000)$, $P_0(0)$, and $P_0(-200{,}000)$ are the prior probabilities of an increased profit of \$200,000, 0, and $-$\$200,000, respectively. $P(0 \mid 200{,}000)$ is the probability that the test-market study will indicate an increased profit of zero, given that the profit increase will be \$200,000; $P(0 \mid 0)$ is the probability that the test-market study will indicate an increased profit of zero, given that the profit increase will be zero; and so forth. Inserting the prior probabilities given in the first paragraph of this section and the conditional probabilities in Table 15.2 into this expression, we get the equation for $P_1(200{,}000)$ in the text.

The equations in the text for $P_1(0)$ and $P_1(-200{,}000)$ can be derived in an analogous way. (Do this as an exercise.)

$$P_1(0) = \frac{.6(.3)}{.3(.5) + .6(.3) + .3(.2)} = .462,$$

$$P_1(-200{,}000) = \frac{.3(.2)}{.3(.5) + .6(.3) + .3(.2)} = .154.$$

Thus, if the new product is introduced, the posterior expected value of increased profit is

$$.384(\$200{,}000) + .462(0) + .154(-\$200{,}000) = \$46{,}000.$$

Since this exceeds zero, it appears that Stonier should recommend the introduction of the new product.

15.9 Sample Size and the Relationship between Prior and Posterior Probability Distributions

It is important to recognize that the relationship between the prior and posterior probability distributions of a random variable is influenced by the size of the sample. *As the sample size increases, the posterior distribution depends more and more on the sample results and less and less on the prior distribution.* To illustrate this, suppose that the marketing manager in Section 15.7 takes a second sample to indicate whether the new label is superior and that it, like the first sample, is favorable to the new label. Given that the surveys are independent and designed so that if the new label is superior, the probability that *each survey* will be favorable to it is 0.8, the probability that *both* surveys will be favorable to it under these circumstances is 0.8(0.8) = 0.64. And given that the surveys are designed so that if the new label is not superior, the probability that *each* will be favorable to it is 0.2, the probability that *both* surveys will be favorable to the new label when in fact it is not superior is 0.2(0.2) = 0.04.

Using Bayes' theorem, we can compute the posterior probabilities that the new label is or is not superior to the old one. These posterior probabilities are

$$P(B\,|\,bb) = \frac{P(bb|B)P(B)}{P(bb|B)P(B) + P(bb|N)P(N)} \tag{15.5}$$

and

$$P(N\,|\,bb) = \frac{P(bb|N)P(N)}{P(bb|B)P(B) + P(bb|N)P(N)}, \tag{15.6}$$

where $P(bb|B)$ is the probability that both surveys are favorable to the new label, given that it is in fact superior, and $P(bb|N)$ is the probability that both surveys are favorable to it, given that it is in fact not superior.

Inserting the numerical values into equations (15.5) and (15.6), we obtain

$$P(B \mid bb) = \frac{0.64(0.5)}{0.64(0.5) + 0.04(0.5)} = 16/17 = .941,$$

$$P(N \mid bb) = \frac{0.04(0.5)}{0.64(0.5) + 0.04(0.5)} = 1/17 = .059.$$

Given the larger sample size, the posterior probability that the new label is superior is much higher than it was after the first sample (0.941 vs. 0.8). And the posterior probability that the new label is not superior is much lower than after the first sample (0.059 vs. 0.2). This is what we would expect. As the sample size increases, the sample becomes increasingly trustworthy, and purely subjective feelings must be given relatively less weight. Thus, in this case, although the decision maker's prior probability that the new label was superior was only 0.5, the posterior probability increased to 0.8 after the first sample indicated that it was superior, and moved even further from 0.5 (to 0.941) after the second sample also indicated that it was superior.

15.10 Expected Value of Perfect Information, Prior and Posterior

An important concept in decision theory is the expected value of perfect information, which is defined as follows.

> ***Expected Value of Perfect Information:*** *The expected value of perfect information is the increase in expected profit if the decision maker could obtain completely accurate information concerning the random variable in question (but if he or she does not yet know what this information will be).*

In the case of the marketing manager, this expected value is the increase in expected profit if the manager could obtain perfectly accurate information indicating whether or not the new label is superior.

To illustrate how the expected value of perfect information can be computed, consider the marketing manager (before the survey described in Section 15.7 was carried out). To determine the expected value of perfect information, we must evaluate the expected gain to the company if he can obtain access to perfectly accurate information. Then we must calculate the extent to which this expected gain exceeds the expected gain based on the information actually available to the marketing manager.

If the manager obtains perfect information, he will be able to make the correct decision, regardless of whether the new label is superior. In other words, if the new label is superior the manager will be aware of this fact, and he will adopt the label. If the new label is not superior he will be aware of this fact also, and he will not adopt the label. Thus, given that the manager has access to perfect information, the expected value of the gain to the company is

$$0.5(\$800{,}000) + 0.5(0) = \$400{,}000.$$

To see why this is the expected gain if he has access to perfect information, it is important to recognize that although it is assumed that he has access to perfect information, *he does not yet know what this information will be*. Based on his prior probabilities, there is a .5 probability that this information will show that the new label really is superior, in which case he will adopt it and the gain will be $800,000. There is also a .5 prior probability that the information will show that the new label is really not superior, in which case he will not adopt it and the gain will be zero. Thus, as shown above, the expected value of the company's gain if the manager has access to perfect information (that is not yet revealed to him) is $400,000.

In contrast, the expected value of the company's gain if he bases his decision on existing information is $150,000, as we saw in Section 15.2. The difference between these two figures—$400,000 minus $150,000, or $250,000—is the expected value of perfect information. It is a measure of the value of perfect information. *It shows the amount by which the expected value of the firm's gains increases as a consequence of having access to perfect information.* Put differently, *it is the maximum amount that the decision maker should pay to obtain perfect information.*

In the case above, the $250,000 is the *prior* expected value of perfect information. In other words, it is the expected value of perfect information before the survey in Section 15.7 was carried out, and is based entirely on the decision maker's prior probabilities. After the survey has been carried out, we can compute the *posterior* expected value of perfect information—that is, the expected value of perfect information based on posterior probabilities. In the case of the marketing manager, the posterior expected value of perfect information is

$$0.8(\$800{,}000) + 0.2(0) - \$540{,}000 = \$100{,}000.$$

The posterior expected value of perfect information differs from the prior expected value of perfect information for two reasons. First, the posterior probabilities that the new label is or is not superior (0.8 and 0.2) are used in place of the prior probabilities (0.5 and 0.5). Second, the expected profit based on the posterior probabilities ($540,000) is used in place of the expected profit based on the prior probabilities ($150,000).

It is interesting to compare the prior and posterior expected values of perfect information. Before the survey was taken, the marketing manager should have been willing to pay up to $250,000 for perfect information concerning whether or not the new label was superior since this was the expected value of perfect information. But after the survey was taken and the result indicated that the new label was in fact superior, the manager should have been willing to pay only up to $100,000 for perfect information since this was then the expected value of perfect information. The survey, by reducing the decision maker's doubts concerning the superiority of the new label, has consequently reduced the expected value of perfect information.

Note, however, that the posterior expected value of perfect information can be either greater or less than the prior expected value of perfect information. For example, in the case of the Canadian drug executive in

Section 15.8, the prior expected value of perfect information was[11]

$$0.5(\$200{,}000) + 0.3(0) + 0.2(0) - \$60{,}000 = \$40{,}000,$$

while the posterior expected value of perfect information was[12]

$$(5/23)(\$200{,}000) + (6/23)(0) + (12/23)(0) - (0) = \$43{,}478.$$

The reason why the latter exceeds the former is that the posterior probabilities were quite different from the prior probabilities, and the optimal action was altered by the test-market results. In effect, the test-market results increased doubt concerning the proper decision and increased the expected value of perfect information.

15.11 Expected Opportunity Loss

In both prior analysis and posterior analysis, a frequently used concept is that of opportunity loss, defined as follows.

> ***Opportunity loss:*** *Given that the decision maker takes a particular action, the opportunity loss if a particular event (or state of nature) occurs is the difference between the profit actually achieved and the profit that would have been achieved if the action had been the best possible one for this event.*

For each event that can occur, one can compute the opportunity loss. For example, in the case of the marketing manager, there are two possible states of nature: that the new label is superior, or that it is not. If the former is the true state of nature, the best action that can be taken is to adopt the new label. Since the profit resulting from this act is $800,000 higher than if the old label is kept, the opportunity loss of keeping the old label is $800,000, and the opportunity loss of adopting the new label is zero. If the latter is the true state of nature (that is, if the new label is not superior), the best action that can be taken is to keep the old label. Since the profit resulting from this act is

11. According to Section 15.8, there is a 0.5 prior probability that the profit increase will be $200,000, in which case the new product will be introduced (and the profit increase will be $200,000). There is a 0.3 prior probability that the profit increase will be zero, in which case the new product will not be introduced (and the profit increase will be zero). And there is a 0.2 prior probability that the profit increase would be −$200,000, in which case the new product will not be introduced (and the profit increase will be zero). Thus, the expected value of the firm's increased profit if it has access to perfect information is 0.5 ($200,000) + 0.3(0) + 0.2(0). Subtracting $60,000, which (as we know from Section 15.8) is the expected profit increase with existing information, the expected value of perfect information is as shown in the equation in the text.

12. According to Section 15.8, there is a 5/23 posterior probability that the profit increase will be $200,000, in which case the new product will be introduced (and the profit increase will be $200,000). There is a 6/23 posterior probability that the profit increase will be zero, in which case the new product will not be introduced (and the profit increase will be zero). And there is a 12/23 posterior probability that the profit increase would be −$200,000, in which case the new product will not be introduced (and the profit increase will be zero). Thus, the expected value of the firm's increased profit if it has access to perfect information is (5/23)($200,000) + (6/23)(0) + (12/23)(0). Subtracting zero, which is the expected profit increase with existing information (since, as we know from Section 15.8, the new product will not be introduced), the expected value of perfect information is as shown in the text.

$500,000 higher than if the new label is adopted, the opportunity loss of adopting the new label is $500,000, and the opportunity loss of keeping the old label is zero. Table 15.3 shows, for each state of nature, both the profit and the opportunity loss resulting from each action the decision maker can take.

TABLE 15.3

Profit and Opportunity Loss for Each Action and Each Possible State of Nature[a]

	Profit (dollars)		Opportunity loss (dollars)	
State of nature	Adopt new label	Keep old label	Adopt new label	Keep old label
New label is superior	800,000*	0	0	800,000
New label is not superior	−500,000	0*	500,000	0

[a]The profit for the best act corresponding to each state of nature is marked with an asterisk. This informs the reader that for each state of nature opportunity losses are measured as differences from the figure marked with an asterisk.

If the decision maker wants to maximize expected monetary value, it can be shown that he or she should choose the action that minimizes the expected opportunity loss. This is an alternative way of taking the action that maximizes expected monetary value. To demonstrate that the minimization of expected opportunity loss leads to the same choice as the maximization of expected monetary gain, consider the case of the marketing manager. Before the survey, the expected opportunity loss *if the marketing manager adopts the new label* is

$$0.5(0) + 0.5(\$500{,}000) = \$250{,}000.$$

Why? Because there is a 0.5 prior probability that the new label will be superior to the old one, in which case there is a zero opportunity loss (see Table 15.3); and because there is a 0.5 prior probability that the new label will not be superior to the old one, in which case there is a $500,000 opportunity loss (see Table 15.3). The expected opportunity loss *if he does not adopt the new label* is

$$0.5(\$800{,}000) + 0.5(0) = \$400{,}000.$$

Why? Because there is a 0.5 prior probability that the new label will be superior to the old one, in which case there is an $800,000 opportunity loss (see Table 15.3); and because there is a 0.5 prior probability that the new label will not be superior to the old one, in which case there is a zero opportunity loss (see Table 15.3).

Thus, if he minimizes the expected opportunity loss, he will choose the same action—that is, he will adopt the new label—as if he maximized expected profit. As stated above, this will be true in general.

Before concluding this section, two additional points should be noted. First, if you compare the results of this section with those of the last one, you will find that *the expected opportunity loss corresponding to the best action ($250,000) equals the expected value of perfect information (also $250,000).* This, too, will always be true. Thus, an alternative way of calculating the expected value of perfect information is to find the expected opportunity loss corresponding to the best action.

Second, as stated at the beginning of this section, *the concepts discussed here are relevant for posterior as well as prior analysis.* To illustrate their application in a posterior analysis, consider the case of the marketing manager after he received the survey results indicating that the new label was in fact superior. Using the resulting posterior probabilities, the expected opportunity loss if the marketing manager adopts the new label is[13]

$$.8(0) + .2(\$500{,}000) = \$100{,}000.$$

The expected opportunity loss if he does not adopt it is[14]

$$.8(\$800{,}000) + .2(0) = \$640{,}000.$$

In accord with our previous discussion, the action that minimizes expected opportunity loss—namely, adopting the new label—is the same as the action that maximizes expected profit. Also, the expected opportunity loss corresponding to the best action ($100,000) equals what was shown in the previous section to be the expected value of perfect information.

15.12 Two-Action Problems with Linear Profit Functions

In the final sections of this chapter we will look in some detail at a special kind of problem which is encountered rather frequently and which can be handled simply and effectively by the methods discussed above. *This kind of problem is characterized by only two possible actions that the decision maker can take, and by the fact that the profit received is a linear function of the random variable representing the relevant chance event.* As an illustration, suppose that a firm is considering whether or not to establish a plant for producing and selling television sets in a particular foreign country. There are 5 million potential buyers in that country annually, and p is the proportion of these potential buyers who would buy this firm's television set if it were available. The fixed cost of the new plant would be $1 million per year and the gross profit from each television set sold—that is, the price less the unit variable cost—is $20.

The firm must choose between two possible actions: to establish or not to establish the plant. If it established the plant, the firm's profit can easily be shown to be a linear function of p, which is a random variable. To see that this is the case, note that the number of television sets sold (in millions per year)

13. According to Section 15.7, under these circumstances there is a 0.8 posterior probability that the new label will be superior to the old, in which case there is a zero opportunity loss (see Table 15.3); and there is a 0.2 posterior probability that the new label will not be superior to the old, in which case there is a $500,000 opportunity loss (see Table 15.3). Thus, the expected opportunity loss is given by the equation in the text.

14. According to Section 15.7, under these circumstances there is a 0.8 posterior probability that the new label will be superior to the old, in which case there is an $800,000 opportunity loss (see Table 15.3); and there is a 0.2 posterior probability that the new label will not be superior to the old, in which case there is a zero opportunity loss (see Table 15.3). Thus, the expected opportunity loss is given by the equation in the text.

will equal $5p$ so that profit (in millions of dollars per year) will be

$$\pi = -1 + 20(5)(p),$$

since annual fixed cost equals $1 million, and each television set sold results in a gross profit of $20. Simplifying terms, we have

$$\pi = -1 + 100p, \tag{15.7}$$

which shows that profit is a linear function of p. If the firm does not establish the plant, the profit clearly is zero.

In a problem of this sort, there is a break-even value of the random variable which makes the decision maker indifferent between the two actions. If the random variable falls below this value, one action is optimal; if it falls above this value, the other action is optimal. In the case at hand, the break-even value of p is .01, since if $p = .01$ the profit obtained by establishing the plant is

$$-1 + 100(.01) = 0,$$

which is the same as the profit obtained by not establishing the plant. Figure 15.5 shows the profit functions if the firm establishes the plant and if it does not. As you can see, these functions intersect at the break-even value of p, which is .01. If p is less than .01, the optimal action is not to establish the plant; if it is greater than .01, the optimal action is to establish it.

Suppose that the firm has prior probabilities concerning p of the following sort: It believes that there is a .2 probability that p will equal .006, a .3 probability that it will equal .009, a .4 probability that it will equal .012, and

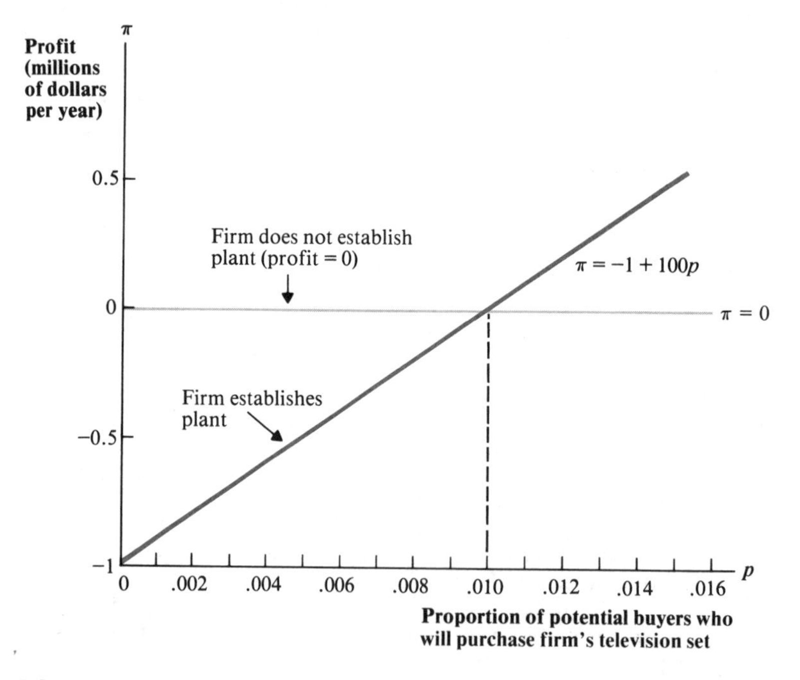

FIGURE 15.5 Profit Functions if Firm Does or Does Not Establish Plant

a .1 probability that it will equal .015. Under these circumstances, if the firm has a linear utility function (and thus wants to maximize expected profit) which action should it take? As we shall see in the following sections, there is a simple way of answering this question.

15.13 Solving a Two-Action Problem

Before we describe the simple way to solve this problem, let's solve it in the way discussed in previous sections. Figure 15.6 shows the decision tree in this case. Based on the firm's prior probabilities, the expected profit if it establishes the plant is

$$.2(-\$400{,}000) + .3(-\$100{,}000) + .4(\$200{,}000)$$
$$+ .1(\$500{,}000) = \$20{,}000.$$

In other words, the profit[15] is $-\$400{,}000$ if $p = .006$ (and the probability that this is the case is .2), the profit is $-\$100{,}000$ if $p = .009$ (and the probability that this is the case is .3), the profit is \$200,000 if $p = .012$ (and the probability that this is the case is .4), and the profit is \$500,000 if $p = .015$ (and the probability that this is the case is .1). Because the result—\$20,000—exceeds zero (the expected profit if the firm does not establish the plant), the firm should establish the plant.

A very simple way of solving this type of problem is to recognize that if profit is a linear function of p (as shown in equation 15.7), then the expected

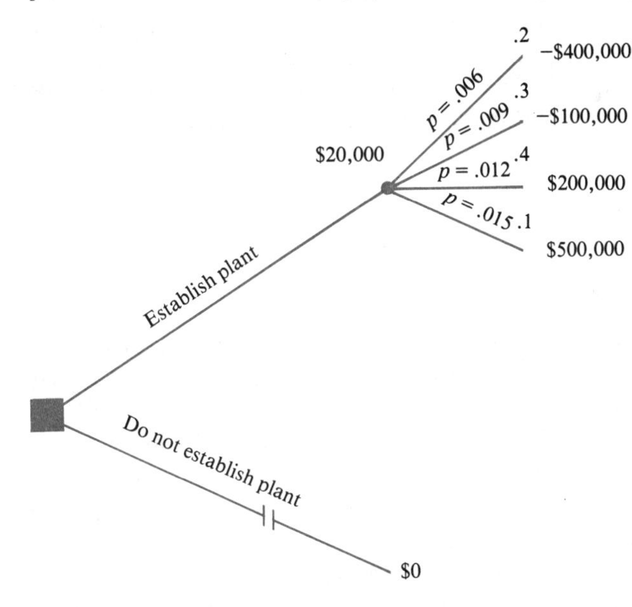

FIGURE 15.6
Decision Tree for Firm's Decision to Establish or Not Establish Plant

15. To obtain each of these profit figures, we insert the relevant value of p into equation (15.7). For example, when $p = .006$, equation (15.7) tells us that $\pi = -1 + 100(.006) = -.4$. Since π is expressed in millions of dollars, this means that the profit is $-\$400{,}000$ if $p = .006$.

value of profit is the same linear function of the expected value of p. In other words, if A and B are any two constants, and if

$$\pi = A + Bp,$$

then

$$E(\pi) = A + BE(p).$$

(The proof of this is given in Appendix 4.1.) Using this proposition, if follows from equation (15.7) that

$$E(\pi) = -1 + 100E(p) \tag{15.8}$$

if the firm establishes the plant. Thus, all we have to do to find the expected profit under these circumstances is determine the expected value of p, which is

$$.2(.006) + .3(.009) + .4(.012) + .1(.015) = .0102,$$

and substitute it in equation (15.8) to get

$$E(\pi) = -1 + 100(.0102) = .02,$$

which equals $20,000, the same result we got at the beginning of this section.[16]

An important point is that *the only essential information required to solve this type of problem is the expected value of the random variable—$E(p)$ in this* case. We do not need to know the variance of p, or whether the probability distribution of p is skewed, or the shape of the probability distribution. We only need to know the expected value. This is fortunate because prior probability distributions frequently cannot be specified with a great deal of confidence, and the decision maker may be more confident of the expected value than of the entire distribution. Based on equation (15.8), it is clear that the expected profit from establishing the plant will exceed zero (the expected profit if it does *not* establish the plant) if

$$-1 + 100E(p) > 0,$$

or if

$$E(p) > .01.$$

Thus, the decision boils down to whether or not the expected value of p *exceeds .01 (the break-even value of* p *discussed in the previous section). If it does, the firm should establish the plant; if not, it shouldn't.*

To gain additional insight into two-action problems of this sort, it is useful to consider the same situation from a somewhat different point of view.

Example 15.4 Using the data provided in this section concerning the firm that is considering whether or not to establish a plant in a particular foreign country, find the course of action that minimizes the expected opportunity loss of the firm. Show that (in keeping with Section 15.11) the result is the same as that obtained by maximizing expected profit.

16. Since π is expressed in millions of dollars, we must multiply .02 by $1 million to get the expected profit, which is $20,000. This explains how the figure in the text was derived.

Solution: The first step is to specify the opportunity loss corresponding to each chance event—that is, to each value of p. If p is less than its break-even value (.01 in this case), the optimal action is to not establish the plant, and the opportunity loss (in millions of dollars per year) if the firm establishes the plant is $100\ (.01 - p)$. Why? Because if the firm establishes the plant, its losses, according to equation (15.7), will equal

$$1 - 100p,$$

which is the same as $100\ (.01 - p)$. For example, if $p = .006$, the firm's losses will equal $100\ (.01 - .006) = .4$—that is, \$400,000.

On the other hand, if p is greater than its break-even value (.01 in this case), the optimal action is to establish the plant, and the opportunity loss (in millions of dollars per year) if the firm does not establish it is $100\ (p - .01)$. Why? Because if the firm does not establish the plant, it forgoes profits, according to equation (15.7), equal to

$$-1 + 100p,$$

which is the same as $100\ (p - .01)$. For example, if $p = .015$, it forgoes profits of $100\ (.015 - .01) = .5$—that is, \$500,000.

Table 15.4 summarizes these results; it contains the opportunity losses corresponding to each action and each possible value of p. (Figure 15.7 plots the opportunity loss corresponding to each possible action and each value of p.) Based on these results, it is a simple matter to calculate the expected opportunity loss resulting from each action. If the firm establishes the plant, the expected opportunity loss is

$$.2(\$400{,}000) + .3(\$100{,}000) + .4(0) + .1(0) = \$110{,}000.$$

In other words, the opportunity loss is \$400,000 if $p = .006$ (and the probability that this is the case is .2), the opportunity loss is \$100,000 if $p = .009$ (and the probability that this is the case is .3), the opportunity loss is zero if $p = .012$ (and the probability that this is the case is .4), and the opportunity loss is zero if $p = .015$ (and the probability that this is the case is .1). If the firm does not establish the plant, the expected opportunity loss is

$$.2(0) + .3(0) + .4(\$200{,}000) + .1(\$500{,}000) = \$130{,}000.$$

In other words, the opportunity loss is zero if $p = .006$ (and the probability

TABLE 15.4

Opportunity Loss for Each Action and Each Value of p

	Opportunity loss (millions of dollars)	
Value of p	Establish plant	Do not establish plant
$p \leq .01$	$100\ (.01 - p)$	0
$p > .01$	0	$100\ (p - .01)$

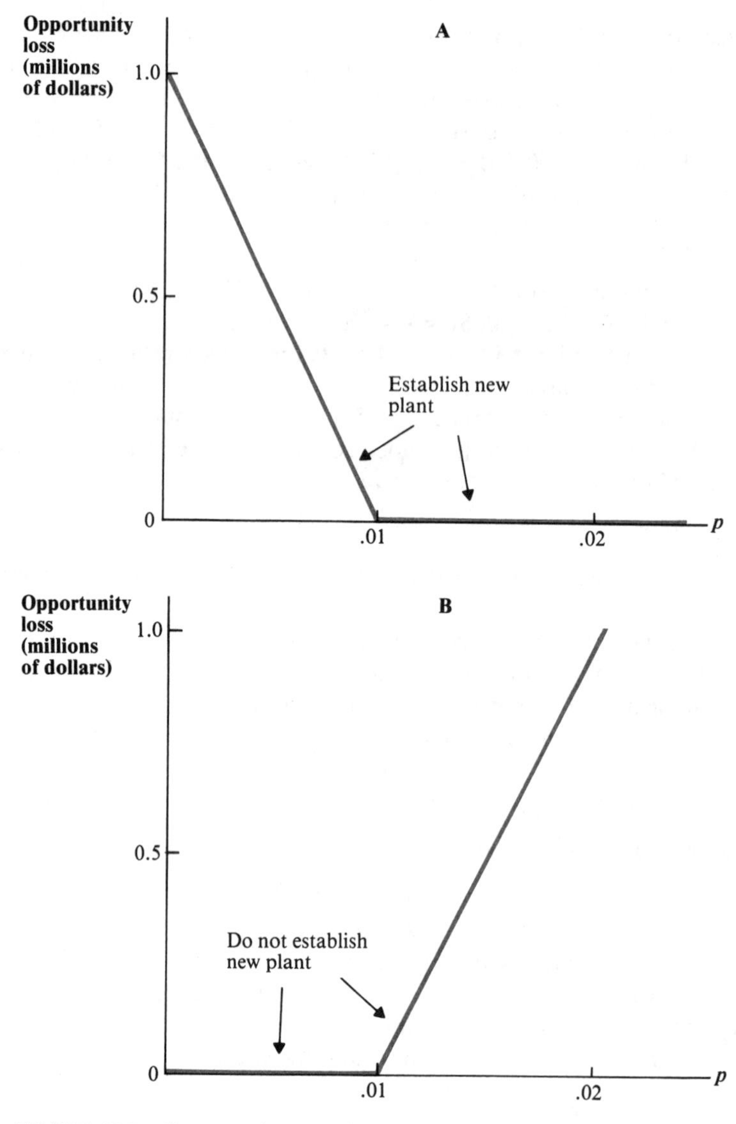

FIGURE 15.7 Opportunity-Loss Function for Each Action

that this is the case is .2), the opportunity loss is zero if $p = .009$ (and the probability that this is the case is .3), the opportunity loss is $200,000 if $p = .012$ (and the probability that this is the case is .4), and the opportunity loss is $500,000 if $p = .015$ (and the probability that this is the case is .1).

The firm should establish the plant because the expected opportunity loss ($110,000) if it does so is less than the expected opportunity loss ($130,000) if it does not do so. Thus, the result is the same as we obtained by maximizing expected profit, which is what we set out to prove. As we emphasized in Section 15.11, this will always be true.

15.14 Posterior Analysis of the Two-Action Problem

Up to this point, we have been carrying out a prior analysis of the television firm's problem. If the firm decides to carry out a survey of potential television buyers in the relevant country in order to estimate the value of p, it is possible (after the survey results are known) to carry out a posterior analysis. For example, suppose that the firm conducts a random sample of 100 potential buyers, and that none says that he or she would buy the firm's product. We can use the binomial distribution to obtain the probability of this result, given that the true value of p is .006, .009, .012, or .015. Since the probability that each potential buyer would not purchase the firm's product is $1 - p$, the probability that 100 potential buyers would not purchase it is $(1 - p)^{100}$. The third column of Table 15.5 shows the value of $(1 - p)^{100}$, given that the true value of p is .006, .009, .012, or .015. Applying Bayes' theorem, we get the posterior probabilities in the last column of Table 15.5. That is, the posterior probability is .29 that $p = .006$, the posterior probability is .33 that $p = .009$, and so on.[17]

TABLE 15.5
Calculation of Posterior Probabilities[a]

(1) Value of p	(2) Prior probability of this value of p	(3) $(1 - p)^{100}$	(2) × (3)	Posterior probability
.006	.20	.549	.1098	.29
.009	.30	.407	.1221	.33
.012	.40	.301	.1204	.32
.015	.10	.223	.0223	.06
Total			.3746	1.00

[a] The values in column (3) were obtained by using the Poisson distribution as an approximation to the binomial distribution. Readers who skipped the Poisson distribution can evaluate $(1 - p)^{100}$ by straightforward means.

To conduct a posterior analysis of the problem, it is necessary only to find the expected value of p based on the posterior probabilities. Why? Because, as we emphasized in Section 15.13, the decision boils down to whether or not the expected value of p exceeds .01, the break-even value. If the expected value does exceed .01, the firm should establish the plant; if not, it shouldn't. Based on the posterior probabilities (derived and shown in Table 15.5), the expected value of p is

$$.29(.006) + .33(.009) + .32(.012) + .06(.015) = .00945.$$

Thus, on the basis of this new information, the firm should not establish the

17. To obtain each posterior probability in Table 15.5, the number in the next-to-last column must be divided by .3746, the sum of the numbers in this column. Thus, the posterior probability that p equals .006 is .1098 ÷ .3746, or .29. This procedure is an application of Bayes' theorem.

plant. Here, as in the case of the Canadian drug firm, the optimal action is changed by the survey results.

EXERCISES

15.12 The Acme Corporation is considering the introduction of a new, lighter type of bicycle. The firm thinks that there is a 30 percent probability that the bicycle will be a major success (in which case the firm will make $1 million), a 40 percent probability that it will be no better (and no worse) than the firm's existing line of bicycles (in which case the firm will make nothing), and a 30 percent probability that it will be a flop (in which case the firm will lose $800,000). Acme hires a market-research firm to study the new bicycle's potential market, and this firm concludes that the new bike will be no better and no worse than the firm's existing line. Based on previous experience, the probability of coming to this conclusion, given that the new bicycle would be a major success, is thought to be 0.3. The probability of coming to this conclusion, given that the new bicycle is no better and no worse than the existing line, is thought to be 0.5. And the probability of coming to this conclusion, given that the new bicycle would be a flop, is thought to be 0.2.
(a) Calculate the posterior probabilities that the new bicycle will be a major success; no better or worse than the existing line; a flop.
(b) Based on these posterior probabilities, construct a decision tree representing Acme's decision problem.
(c) If Acme wants to maximize expected monetary gain, should it introduce the new bicycle?

15.13 In the posterior analysis in Exercise 15.12, what is the expected value of perfect information?

15.14 Suppose that Acme had based its decision solely on the prior probabilities in Exercise 15.12. (a) What was the prior expected value of perfect information? (b) How does it compare with the posterior expected value of perfect information? (c) Why isn't the prior expected value of perfect information equal to the posterior expected value of perfect information?

15.15 Construct a table showing the opportunity loss corresponding to each action, given each state of nature in the situation described in Exercise 15.12.

15.16 (a) Do a prior analysis based on the minimization of expected opportunity loss, using the data in Exercise 15.12. (b) Show that the expected opportunity loss corresponding to the best action is equal to the expected value of perfect information.

15.17 (a) Do a posterior analysis based on the minimization of expected opportunity loss, using the data in Exercise 15.12. (b) Show that the expected opportunity loss corresponding to the best action is equal to the expected value of perfect information.

15.18 The Merriwether Auto Company is offered the opportunity to buy 1,000 used cars from a concern that is going out of business. The price for the entire group of cars is $1 million. Each car with no major defects can be sold by Merriwether for $2,000. Each car with major defects will have to be scrapped, and Merriwether will get only $50 for it.
(a) In this situation, is Merriwether's profit a linear function of a random variable?
(b) If so, what is the random variable? What is the linear function?

15.19 Suppose that Merriwether's owner thinks that there is a .30 probability that 40 percent of the cars in this group have no major defects, that there is a .30 probability that 50 percent of the cars in this group have no major defects, and that there is a .40 probability that 55 percent of the cars in this group have no major defects. If he wants to maximize expected profit, should he buy the group of 1,000 used cars? Construct a decision tree to answer this question.

15.20 Prove that the answer to the question in Exercise 15.19 depends only on the expected value of the percentage of cars in the group with major defects.

15.21 Construct a table showing the opportunity loss corresponding to each action and to each number of cars with major defects in Exercise 15.8.

15.22 Construct a decision tree based on opportunity losses in order to solve the problem posed in Exercise 15.19.

15.23 Merriwether's owner decides, before entering into the purchase, to take a random sample of four of the cars in the group in order to inspect each one carefully for major defects. He finds that three of the four cars have major defects. Based on this result, would you advise him to buy the group of cars? Why, or why not?

CHAPTER REVIEW

1. Any problem of decision making under uncertainty has two characteristics. First, the decision maker must make a choice (or perhaps a series of choices) among alternative courses of action. Second, this choice (or series of choices) leads to some consequence, but the decision maker cannot tell in advance the exact nature of this consequence because it depends on some unpredictable event (or series of events) as well as on his or her choice (or series of choices). A *decision tree* represents such a problem as a series of choices, each of which is depicted by a fork. A *decision fork* is a juncture representing a choice where the decision maker is in control; a *chance fork* is a juncture where "chance" is in control of the outcome. By the process of *backward induction*, one can work one's way from the right-hand end of a decision tree to the left-hand end to solve the problem.

2. Whether a decision maker wants to maximize expected monetary gain depends on his or her preferences with regard to risk. To reflect these preferences, we can construct a *utility function* for the decision maker (if the necessary assumptions are met). The first step in constructing a utility function is to establish arbitrarily the utilities attached to two monetary values. The second step is to present the decision maker with a choice between a gamble where the possible outcomes are the two monetary values whose utilities were arbitrarily set and the certainty of some third monetary value. After finding the probabilities in this gamble that will make the decision maker indifferent between these two choices, we can calculate the utility of the third monetary value. Next, the latter steps can be repeated over and over in order to obtain the utilities attached to as many monetary values as we need. If the assumptions discussed in this chapter are met, *the decision maker, if he or she is rational, will maximize expected utility*.

3. *Posterior analysis* is concerned with decisions based both on prior probabilities and on new experimental or sample evidence, whereas *prior analysis* is based on prior probabilities alone. To carry out a posterior analysis, we use Bayes' theorem to compute posterior probabilities which reflect the decision maker's prior probabilities and the sample results. As the sample size increases, the posterior probabilities depend more and more on the sample results and less and less on the prior probabilities. Once the posterior probabilities have been computed, the optimal choice is the one that maximizes expected monetary value if the decision maker is indifferent toward

risk. In both posterior and prior analyses, we can compute the *expected value of perfect information.*

4. An *opportunity loss* is the loss incurred by the decision maker if he or she fails to take the best action possible. For each event that can occur, one can compute the opportunity loss. If the decision maker wants to maximize expected monetary value, it can be shown that he or she should choose the action that minimizes the expected opportunity loss. This is an alternative way of finding the action that maximizes expected monetary value. It can be applied in either a prior or a posterior analysis. An interesting point concerning this approach is that the expected opportunity loss corresponding to the best action is equal to the expected value of perfect information.

5. A special kind of problem that can be handled particularly easily is one where there are only two possible actions that the decision maker can take, and where the profit received is a linear function of a random variable representing the relevant chance event. In a problem of this sort, there is a break-even value of the random variable which makes the decision maker indifferent between the two actions. If the random variable falls below this value, one action is optimal; if it falls above this value, the other action is optimal. But, of course, the decision maker does not know what value the random variable will assume. All that is required to solve this type of problem is to compute the expected value of the random variable. If this expected value falls below the break-even value, the first action is optimal; if it falls above it, the second action is optimal. This is true regardless of whether one is carrying out a prior or a posterior analysis.

Getting Down to Cases:

MAXWELL HOUSE'S PRICING DECISION CONCERNING THE QUICK-STRIP CAN[18]

In October 1962, Folger's began test-marketing its coffee in Stockton, California in a keyless container. Whereas earlier coffee cans had to be opened with a key, the new can did not. In November 1963, Maxwell House, the nation's largest producer of coffee, surveyed 125 Stockton users of Folger's coffee in the new can, and found that 86 percent of the consumers who had tried it preferred the new can to the old can. Other pieces of marketing intelligence also indicated that a keyless can would be popular among consumers.

Maxwell House, together with the American Can Company, had helped to develop its own keyless container which operated on the tear-strip opening principle. Based on the reception of Folger's new can, Maxwell House executives were optimistic about consumer reaction to their own new can. One important decision that had to be made before introducing this new can was whether or not to raise the per-pound price of coffee in the new can by 2 cents. Coffee in the quick-strip can was expected to cost an average of 0.7 cents per pound more than that in the old container. According to Joseph Newman who studied this case, if Maxwell House

18. This case is based on a section from Joseph Newman's *Management Applications of Decision Theory* (New York: Harper and Row, 1971).

raised its price by 2 cents per pound it might have been reasonable to expect (1) a .25 probability that its market share would decline by 1.5 percentage points; (2) a .25 probability that its market share would remain constant; (3) a .25 probability that its market share would increase by 1.0 percentage points; and (4) a .25 probability that its market share would increase by 2.5 percentage points. The change in Maxwell Houses's profits corresponding to each change in its market share is given in Table 15.6.

TABLE 15.6

Changes in Profit Corresponding to Selected Changes in Market Share

Price per pound held constant		Price per pound increased by 2 cents	
Change in market share (percentage points)	Change in profit (thousands of dollars)	Change in market share (percentage points)	Change in profit (thousands of dollars)
+2.8	4,104	2.5	11,939
+1.0	− 591	1.0	6,489
0	− 840	0	2,856
−0.6	−1,218	−1.5	−1,050

According to Newman, if Maxwell House did not raise its price, it might have been reasonable to expect (1) a 0.1 probability that its market share would decline by 0.6 percentage points; (2) a 0.2 probability that its market share would remain constant; (3) a 0.5 probability that its market share would increase by 1.0 percentage points; and (4) a 0.2 probability that its market share would increase by 2.8 percentage points. The change in Maxwell House's profits corresponding to each of these market-share changes is provided in Table 15.6.

(a) Construct a decision tree representing Maxwell House's pricing problem.

(b) If Maxwell House wanted to maximize expected monetary value, should it have increased the price per pound of coffee in its new can by 2 cents?

(c) Suppose that Maxwell House was certain that its market share would increase by 2.8 percentage points if it did not raise its price. If all other aspects of the situation were unchanged, should Maxwell House have increased the price by 2 cents? Explain.

16

Decision Theory: Preposterior Analysis and Sequential Analysis*

16.1 Introduction

In the previous chapter, we discussed how decisions can be made on the basis of prior probabilities alone (prior analysis) and how they can be made after additional evidence has been obtained through sampling or experimentation (posterior analysis). However, nothing has yet been said about how the decision maker can determine whether or not it is worthwhile to obtain additional evidence of this sort. Preposterior analysis and sequential analysis are concerned with this latter problem, as well as with the problem of how much additional evidence to obtain and what action to take on the basis of it.

Whereas prior analysis and posterior analysis are concerned with the final choice of a best course of action, preposterior analysis and sequential analysis recognize that the decision maker may decide not to make a final choice until additional information has been obtained. For this reason, preposterior analysis and sequential analysis are richer and more interesting than prior analysis and posterior analysis. The former techniques are also more realistic, since executives in business and government continually must make choices in a situation where an important option is to postpone a final decision and buy further information.

16.2 Preposterior Analysis of the Marketing Manager's Problem

To show the nature of preposterior analysis, let's return to the case of the marketing manager who must decide whether or not to put a new label on his

*Some instructors may wish to skip the latter part of this chapter, which deals with sequential analysis. Others may want to skip this chapter entirely. The book is written so that this can easily be done.

firm's product. At the beginning of the previous chapter we described how the manager could make this decision on the basis of his prior probabilities alone. However, this prior analysis assumed that the marketing manager could not—or did not want to—carry out a survey to obtain empirical evidence concerning the likelihood that the proposed new label was superior (or not superior) to the old label. Later, we described how the manager could make his decision after he had obtained evidence of this sort—but we did not indicate how he could decide whether it was worthwhile to carry out a survey to provide such evidence.

In contrast to the previous chapter, we now must recognize that the marketing manager really has the choice of whether or not to carry out a survey before making a final decision to use the new label or keep the old one. If he decides *not to carry out a survey,* the problem boils down to the one handled by the prior analysis in the previous chapter. Figure 16.1 shows the decision tree for this analysis. You will recall from the previous chapter that the monetary gain is \$800,000 if the new label is introduced and is successful, and is − \$500,000 if it is introduced and is unsuccessful. The prior probability is 0.50 that the new label is superior and 0.50 that it is not superior. As is evident from the decision tree, the optimal act, under these circumstances, is to introduce the new label since the expected profit is \$150,000 if the manager does so and zero if he does not.

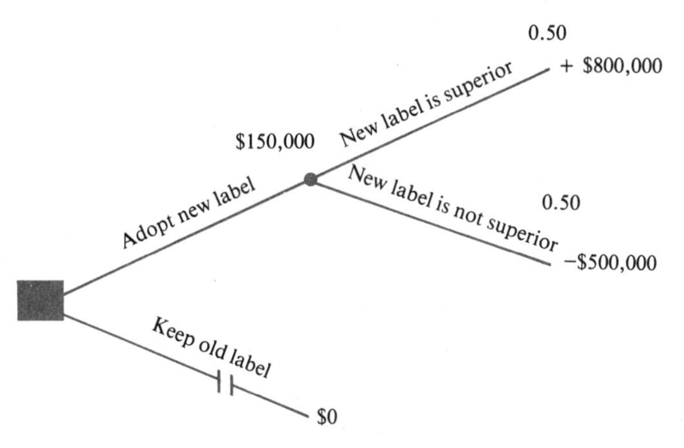

FIGURE 16.1 Decision Tree for the Marketing Manager's Problem if He Decides Not to Carry Out a Survey

But what if the manager decides to *carry out a survey* in order to obtain additional information concerning whether or not the new label is superior? In accord with the previous chapter, suppose that he can carry out a survey which will result in three possible outcomes: (1) favorable to the new label; (2) ambiguous; and (3) unfavorable to the new label. Suppose that the probability that each outcome will occur, given that the new label is superior, is as shown in the first horizontal row of Table 16.1. (For example, the probability that the survey results will be favorable to the new label, given that it is really superior, is 0.8.) The probability that each outcome will occur, given that the new label is not superior, is as shown in the second horizontal row in Table 16.1. (For example, the probability that the survey results will be unfavorable to the new label, given that it really is not superior, is 0.7.) In addition, suppose that the cost of carrying out this survey is \$100,000.

TABLE 16.1
Probability of Each Survey Outcome, Given the Actual State of Nature

	Sample outcome		
State of nature	Favorable to new label	Neither favorable nor unfavorable	Unfavorable to new label
New label is superior	0.8	0.1	0.1
New label is not superior	0.2	0.1	0.7

Given these characteristics of the sample, it is clear that if the survey is carried out the manager incurs a cost of \$100,000, and "chance" determines which of the three possible outcomes of the survey will result. Can the marketing manager estimate the probability of each of these three outcomes? Based on the information provided above, the answer is yes. *The probability that the sample results will be favorable to the new label equals*

$$P(b) = P(b|B)P(B) + P(b|N)P(N),$$

where $P(b|B)$ is the probability that the sample results are favorable, given that the new label is in fact superior, $P(b|N)$ is the probability that the sample results are favorable, given that the new label is not superior, $P(B)$ is the prior probability that the new label is superior, and $P(N)$ is the prior probability that it is not superior. Inserting numerical values into this equation,[1] we find that

$$P(b) = 0.8(0.5) + 0.2(0.5) = 0.5.$$

The probability that the sample results are ambiguous—that is, neither favorable nor unfavorable to the new label—is

$$P(a) = P(a|B)P(B) + P(a|N)P(N),$$

where $P(a|B)$ is the probability that the sample results are neither favorable nor unfavorable to the new label, given that the new label is in fact superior, and $P(a|N)$ is the probability that the sample results are neither favorable nor unfavorable to the new label, given that the new label is not superior. Inserting numerical values into this equation,[2] we find that

$$P(a) = 0.1(0.5) + 0.1(0.5) = 0.1.$$

Similarly, *the probability that the sample results are unfavorable to the new*

1. Recall from an earlier paragraph in this section that both $P(B)$ and $P(N)$ equal 0.5. Table 16.1 shows that $P(b|B) = 0.8$ and $P(b|N) = 0.2$. These are the required numerical values.
2. As pointed out earlier in this section, both $P(B)$ and $P(N)$ equal 0.5. Table 16.1 shows that $P(a|B) = 0.1$ and $P(a|N) = 0.1$. These are the required numerical values.

label is

$$P(n) = P(n|B)P(B) + P(n|N)P(N),$$

where $P(n|B)$ is the probability that the sample results are unfavorable to the new label, given that the new label is in fact superior, and $P(n|N)$ is the probability that the sample results are unfavorable to the new label, given that the new label is in fact not superior. Inserting numerical values into this equation,[3] we find that

$$P(n) = 0.1(0.5) + 0.7(0.5) = 0.4.$$

Thus, if the marketing manager decides to carry out a survey, the cost will be $100,000, and the probability is 0.5 that the survey results will be favorable to the new label, 0.1 that they will be ambiguous (neither favorable nor unfavorable), and 0.4 that they will be unfavorable to the new label.

16.3 Completing the Decision Tree

Up to this point what we have said can be summarized by the decision tree in Figure 16.2. As shown there, the marketing manager has the choice of carrying out a survey or not, this decision being shown at the left-hand end of the tree. *If he takes the upper branch* (that is, if he does not carry out the survey), he proceeds to a decision fork where he must choose whether or not to introduce the new label. The expected profit resulting from introducing the new label is $150,000 and the expected profit resulting from not doing so is zero; thus, the expected monetary value of taking the upper branch is $150,000. *If the manager takes the lower branch* (that is, if he does carry out the survey), he proceeds to a chance fork where "chance" is in control. From that point, the probability that survey results will be favorable to the new label is 0.5, the probability that ambiguous survey results will occur is 0.1, and the probability that unfavorable results will occur is 0.4.

To complete this tree we must trace out the consequences of each of the three possible survey results. First, let's take *the case where the survey results are favorable to the new label.* If this is the case, "chance" has taken the upper branch from the chance fork at the bottom of Figure 16.2. Now the next move is up to the marketing manager who must decide whether to adopt the new label or keep the old one. If he keeps the old label, the profit is zero. If he adopts the new label, the expected profit is

$$0.8(\$800{,}000) + 0.2(-\$500{,}000) = \$540{,}000$$

because the posterior probability that the new label is superior is 0.8, and the posterior probability that the new label is not superior is 0.2. (Recall the discussion in Section 15.7 of the previous chapter.) Thus, in this case the best action is to adopt the new label.

Next, let's take *the case where the survey results are neither favorable nor*

3. As we have already pointed out, both $P(B)$ and $P(N)$ equal 0.5. Table 16.1 shows that $P(n|B) = 0.1$ and $P(n|N) = 0.7$. These are the required numerical values.

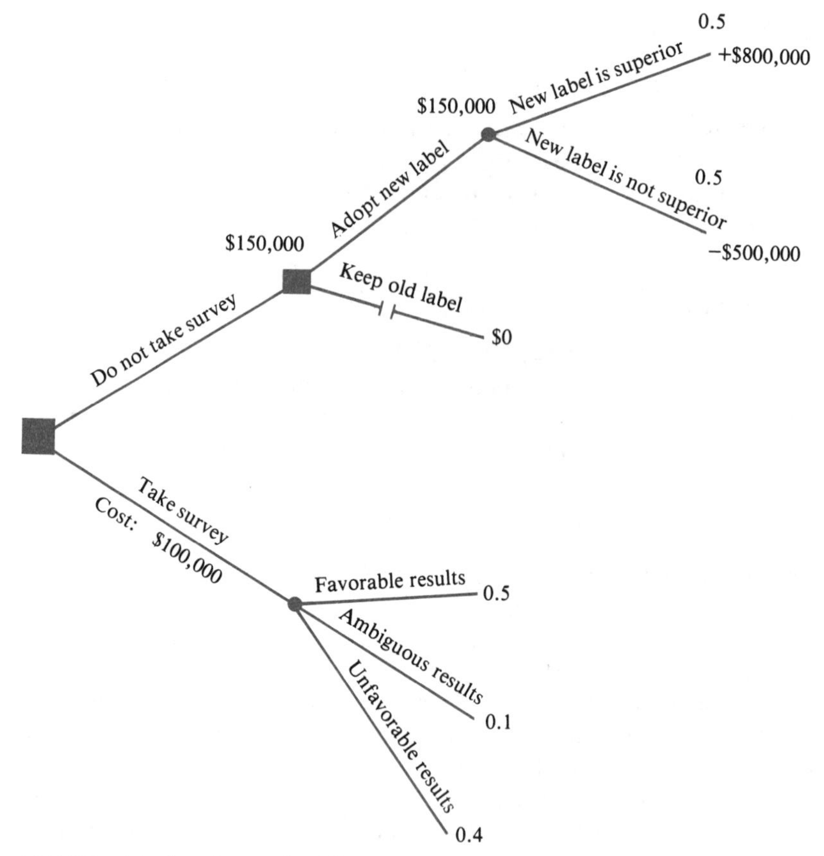

FIGURE 16.2 Portion of Decision Tree for Marketing Manager's Problem

unfavorable to the new label. If this is the case, "chance" has taken the middle branch from the chance fork at the bottom of Figure 16.2. Again, the next move is up to the marketing manager who must decide whether to adopt the new label or keep the old one. If he keeps the old label, the profit is zero. If he adopts the new label, the expected profit is

$$0.5(\$800{,}000) + 0.5(-\$500{,}000) = \$150{,}000,$$

because the posterior probability that the new label is superior is 0.5, and the posterior probability that the new label is not superior is 0.5.[4] Thus, in this case, the best action is to adopt the new label.

4. The posterior probability that the new label is superior is

$$P(B \mid a) = \frac{P(a \mid B)P(B)}{P(a \mid B)P(B) + P(a \mid N)P(N)} = \frac{(0.1)(0.5)}{(0.1)(0.5) + (0.1)(0.5.)} = 0.5.$$

The posterior probability that the new label is not superior is

$$P(N \mid a) = \frac{P(a \mid N)P(N)}{P(a \mid N)P(N) + P(a \mid B)P(B)} = \frac{(0.1)(0.5)}{(0.1)(0.5) + (0.1)(0.5)} = 0.5.$$

The values of $P(a \mid B)$ and $P(a \mid N)$ come from Table 16.1.

Finally, let's take *the case where the survey results are unfavorable to the new label.* If this is the case, "chance" has taken the bottom branch from the chance fork at the bottom of Figure 16.2, and once again the next move is up to the marketing manager, who must decide whether to adopt the new label or keep the old one. If he keeps the old label the profit is zero, whereas if he adopts the new label the expected profit is

$$(1/8)(\$800{,}000) + (7/8)(-\$500{,}000) = -\$337{,}500,$$

because the posterior probability that the new label is superior is 1/8, and the posterior probability that the new label is not superior is 7/8.[5] Thus, in this case, the best action is not to adopt the new label.

Having analyzed the consequences of each of the three possible survey results, we can complete the decision tree which is shown in its entirety in Figure 16.3. Working backward from the right-hand side of the tree, we can use *backward induction* (described in Chapter 15) to solve the problem. As noted above, the expected monetary value of being situated at decision fork (a) is \$540,000, the expected monetary value of being at decision fork (b) is \$150,000, and the expected monetary value of being at decision fork (c) is zero. Thus, since, if the marketing manager decides to carry out the survey, the probabilities of arriving at decision forks (a), (b), and (c) are 0.5, 0.1, and 0.4, respectively, it follows that the expected profit, if he carries out the survey, is

$$0.5(\$540{,}000) + 0.1(\$150{,}000) + 0.4(0) = \$285{,}000.$$

Comparing this amount with the \$150,000 expected profit that results when no survey is taken, it is clear that the marketing manager would be warranted in spending up to \$285,000 − \$150,000, or \$135,000, for the survey. This is the *expected value of the sample information.* Note, however, that this is a gross figure since it takes no account of the amount that the firm must pay for the survey—\$100,000 in this case. The *expected net gain of sample information* is the difference between the expected value of the sample information and the cost of the survey. The expected net gain shows the amount by which the survey increases or reduces the firm's expected profit when account is taken of the cost of the survey. Since the expected net gain of sample information is \$135,000 − \$100,000, or \$35,000, it follows that the marketing manager should carry out the survey. Why? Because, even after we deduct the cost of the survey, the expected profit is greater if the survey is taken than if it is not.

5. The posterior probability that the new label is superior is

$$P(B \mid n) = \frac{P(n \mid B)P(B)}{P(n \mid B)P(B) + P(n \mid N)P(N)} = \frac{(0.1)(0.5)}{(0.1)(0.5) + (0.7)(0.5)} = \frac{1}{8}.$$

The posterior probability that the new label is not superior is

$$P(N \mid n) = \frac{P(n \mid N)P(N)}{P(n \mid N)P(N) + P(n \mid B)P(B)} = \frac{(0.7)(0.5)}{(0.7)(0.5) + (0.1)(0.5)} = \frac{7}{8}.$$

The values of $P(n \mid B)$ and $P(n \mid N)$ come from Table 16.1.

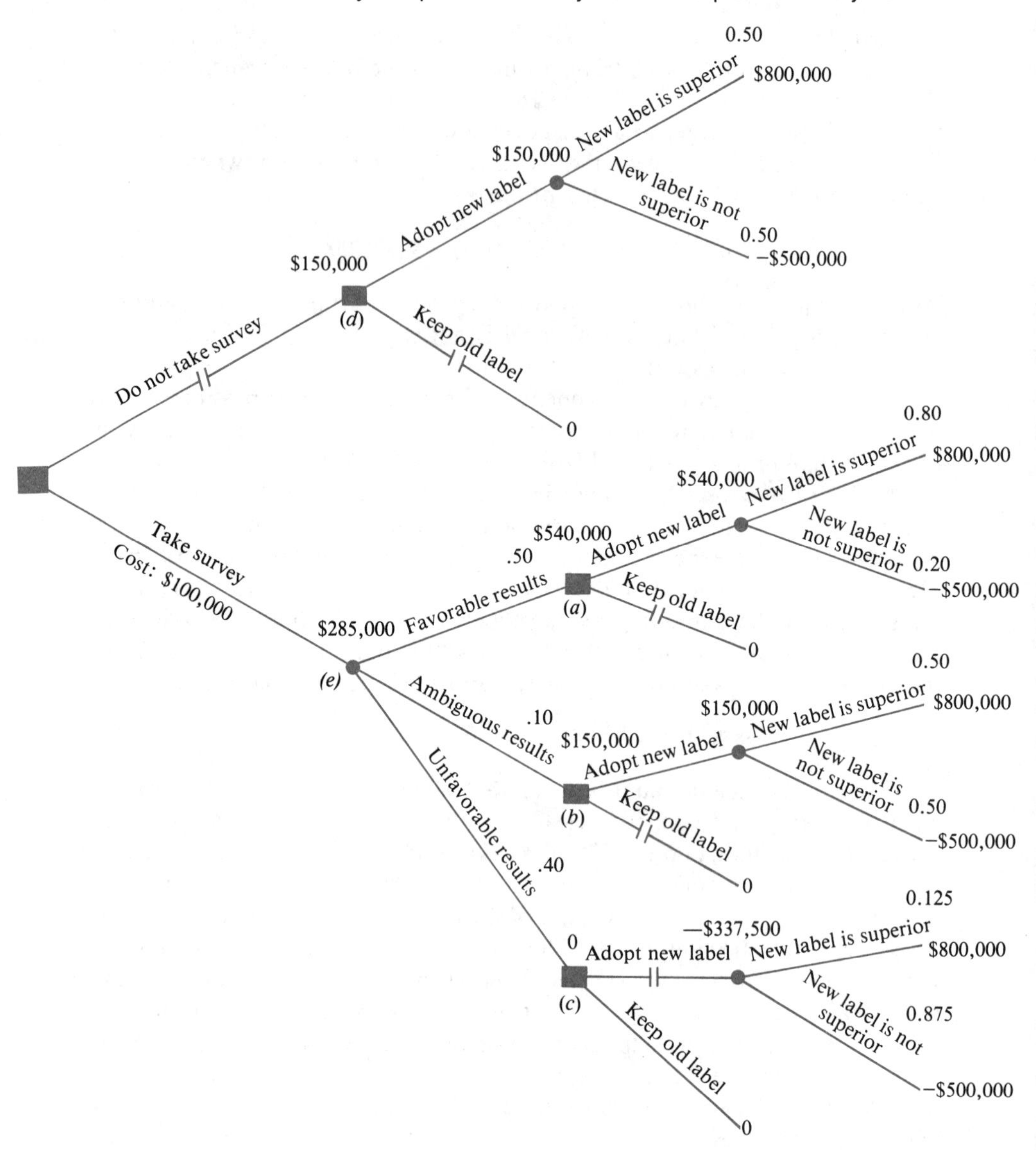

FIGURE 16.3 Complete Decision Tree for Marketing Manager's Problem

16.4 Extensive Form Analysis and Normal Form Analysis

In the previous two sections we have carried out a type of preposterior analysis known as *extensive form analysis*. More specifically, in terms of the decision tree in Figure 16.3, the upper part of the tree represents a prior analysis which results in the prior expected profit shown at fork (d), and the lower part of the tree represents an extensive form analysis which results in the preposterior expected profit shown at fork (e). To carry out an extensive form analysis, one begins by constructing a decision tree, by characterizing the states of nature at the right-hand ends of the tree, and by assigning payoffs to each action assuming that each state of nature is true. To determine whether the survey should be carried out, the preposterior expected payoff should be reduced by the cost of the survey, and the result should be compared with the prior expected payoff if no survey were carried out. If the former exceeds the latter, the survey is worthwhile.

Another type of preposterior analysis is *normal form analysis*. In this type of analysis we break down the problem in quite a different way, the first step being to list all the possible *strategies* that might be employed.

> ***Strategy:*** *A strategy is a decision rule that specifies which action the decision maker will take if each survey result occurs.*

For example, one strategy is the following: Adopt the new label if the survey results are favorable to it; keep the old label if the survey results are ambiguous or unfavorable to the new label. There are eight possible strategies that the marketing manager could choose, all of which are listed in Table 16.2. *In normal form analysis, one compares the value of all these strategies to find the one that is optimal. The result, of course, is the same as that arrived at by extensive form analysis, but, for reasons explained in a later section, it is sometimes advantageous to use a normal form analysis rather than an extensive form analysis.* (On the other hand, it is sometimes advantageous to use an extensive form analysis; the advantages of each are given in Section 16.5.)

To illustrate how a normal form analysis is carried out, let's use this type of analysis to solve the marketing manager's problem. To add some variety to the illustration, let's minimize expected opportunity loss rather than maximize expected profit. You will recall from Section 15.11 of the previous chapter that a decision maker, if he or she minimizes expected opportunity loss, will maximize expected profit, so that the result should be the same. For each strategy, we must calculate the expected opportunity loss, given that a particular state of nature prevails. Then we can use these conditional expected losses to compute the expected opportunity loss from the strategy.

Let's begin with ***strategy 1*** in Table 16.2. *Given that the new label in fact is superior,* the expected opportunity loss with this strategy is

$$0.8(0) + 0.1(0) + 0.1(0) = 0.$$

Why? Because, if the new label really is superior, the probability that the sample results will be favorable to the new label is 0.8, the probability that they will be ambiguous is 0.1, and the probability that they will be unfavorable is 0.1. (See Table 16.1.) Regardless of which of these sample outcomes occurs, strategy 1 dictates that the new label should be accepted. If the new label is in fact superior the opportunity loss associated with accepting it is zero (as shown in Table 16.3). Thus, the expected opportunity loss in this case is zero.

TABLE 16.2
Possible Strategies in Marketing Manager's Problem

Survey outcome	Strategy[a]							
	1	2	3	4	5	6	7	8
Favorable to new label	Adopt new	Keep old	Adopt new	Adopt new	Keep old	Keep old	Adopt new	Keep old
Ambiguous	Adopt new	Adopt new	Keep old	Adopt new	Keep old	Adopt new	Keep old	Keep old
Unfavorable to new label	Adopt new	Adopt new	Adopt new	Keep old	Adopt new	Keep old	Keep old	Keep old

[a] Each strategy specifies the course of action to be taken if each survey outcome occurs. Thus, strategy 1 specifies that the new label be adopted regardless of the survey outcome. Strategy 4 specifies that the new label be adopted if the survey outcome is favorable to the new label or ambiguous, but that the old label be kept if the survey outcome is unfavorable to the new label. Strategy 8 specifies that the old label be kept regardless of the survey outcome.

If the new label is in fact not superior, the expected opportunity loss with this strategy is

0.2($500,000) + 0.1($500,000) + 0.7($500,000) = $500,000.

The probability of each sample outcome is as given in Table 16.1. Regardless of which of these sample outcomes occurs, strategy 1 dictates that the new label should be accepted. If the new label really is not superior, the opportunity loss associated with accepting it is $500,000 (as shown in Table 16.3). Thus, the expected opportunity loss in this case is $500,000.

TABLE 16.3

Opportunity Loss for Each Action and Each State of Nature

State of nature	Action	
	Adopt new label (dollars)	Keep old label (dollars)
New label is superior	0	800,000
New label is not superior	500,000	0

SOURCE: Table 15.3

To find the expected opportunity loss from this strategy, we must multiply each of these conditional expected opportunity losses—zero and $500,000—by their probability of occurrence and sum the results. Since the prior probability that the new label is superior is 0.5 and the prior probability that it is not superior is 0.5, the expected opportunity loss for this strategy is

0.5(0) + 0.5($500,000) = $250,000.

Next, let's consider ***strategy*** *7.* What is the expected opportunity loss with this strategy? *If the new label is in fact superior,* the expected opportunity loss with this strategy is

0.8(0) + 0.1($800,000) + 0.1($800,000) = $160,000.

To see why, note the following. If the sample results are favorable to the new label (a probability of 0.8), this strategy calls for acceptance of the new label, which results in an opportunity loss of zero if the new label is really superior. If the sample results are ambiguous (a probability of 0.1), this strategy calls for keeping the old label, which results in an opportunity loss of $800,000 if the new label is really superior. And if the sample results are unfavorable to the new label (a probability of 0.1), this strategy calls for keeping the old label, which results in an opportunity loss of $800,000 if the new label is really superior.

If the new label really is not superior, the expected opportunity loss with this strategy is

0.2($500,000) + 0.1(0) + 0.7(0) = $100,000.

To see why, note the following. If the sample results are favorable to the new label (a probability of 0.2), this strategy calls for acceptance of the new label, which results in an opportunity loss of $500,000 if the new label is in fact not superior. If the sample results are ambiguous (a probability of 0.1), this strategy calls for keeping the old label, which results in an opportunity loss of zero if the new label is in fact not superior. And if the sample results are unfavorable to the new label (a probability of 0.7), this strategy calls for keeping the old label, which results in an opportunity loss of zero if the new label is in fact not superior.

To find the expected opportunity loss with this strategy, we must multiply each of these conditional expected opportunity losses—$160,000 and $100,000—by their probability of occurrence and sum the results. Since the prior probability that the new label is superior is 0.5, and the prior probability that it is not superior is 0.5, the expected opportunity loss with this strategy is

$$0.5(\$160{,}000) + 0.5(\$100{,}000) = \$130{,}000. \tag{16.1}$$

Since this is considerably less than the expected opportunity loss with strategy 1, strategy 7 seems preferable to strategy 1.

Finally, let's consider ***strategy 4***. What is the expected opportunity loss with this strategy? *If the new label is in fact superior,* the expected opportunity loss with this strategy is

$$0.8(0) + 0.1(0) + 0.1(\$800{,}000) = \$80{,}000.$$

This is the same as for strategy 7, except that zero is substituted for $800,000 in the second term on the left-hand side because this strategy, unlike strategy 7, calls for accepting the new label when the sample results are ambiguous, and the opportunity loss associated with doing so is zero, not $800,000. *If the new label really is not superior,* the expected opportunity loss is

$$0.2(\$500{,}000) + 0.1(\$500{,}000) + 0.7(0) = \$150{,}000.$$

This is the same as for strategy 7, except that $500,000 is substituted for zero in the second term on the left-hand side because this strategy, unlike strategy 7, calls for accepting the new label when the sample results are ambiguous, and the opportunity loss associated with doing so is $500,000, not zero.

To find the expected opportunity loss with strategy 4, we must multiply each of these conditional expected opportunity losses—$80,000 and $150,000—by their probability of occurrence and sum the results. We obtain

$$0.5(\$80{,}000) + 0.5(\$150{,}000) = \$115{,}000, \tag{16.2}$$

which is lower than the expected opportunity loss with strategy 7. Table 16.4 lists the expected opportunity loss associated with each of the eight possible strategies that the marketing manager could adopt in handling this problem. As you can see, strategy 4 has the lowest expected opportunity loss and thus is the *optimal strategy*.

TABLE 16.4
Expected Opportunity Loss for Each Possible Strategy Listed in Table 16.2

Strategy	Expected opportunity loss (dollars)
1	250,000
2	520,000
3	265,000
4	115,000
5	535,000
6	385,000
7	130,000
8	400,000

16.5 Comparison of Results, and Advantages of Each Type of Preposterior Analysis

To carry out a normal form analysis, one begins by listing all possible strategies (as in Table 16.2), each strategy being a specification of the action to be taken if each sample outcome arises. The next step is to calculate the expected opportunity loss associated with each strategy. (To carry out this step, it is convenient to compute first the conditional expected opportunity loss that corresponds to each state of nature, and then to weight each of these conditional figures by the prior probability that the relevant state of nature will occur.) Finally, one selects the strategy which has the minimum expected opportunity loss. This strategy is called ***Bayes' strategy***.

Let's compare the results of the normal form analysis carried out in the previous section with the results of the extensive form analysis of the same problem in Sections 16.2 and 16.3. The first thing to note is that both types of analysis lead to the choice of the same strategy. As you can see from the bottom part of the decision tree in Figure 16.3, the extensive form analysis resulted in the decision to accept the new label if the survey results are favorable to the new label or if they are ambiguous, and to keep the old label if they are unfavorable to the new label. This is precisely the same as strategy 4 which, as we saw in the previous section, is the optimal strategy derived through the normal form analysis.

Another important fact is that if we had carried out the extensive form analysis in terms of opportunity losses rather than profits, the expected opportunity loss of the optimal strategy would have been the figure at fork (e) in Figure 16.3. To prove that this is the case, Figure 16.4 shows the decision tree for this problem if opportunity losses rather than profits are used. As you can see, the figure at fork (e) is $115,000, the expected opportunity loss corresponding to strategy 4. To decide whether it is worthwhile to carry out the survey, this figure must be compared with the expected opportunity loss if no survey were taken—which, as Figure 16.4 shows, is $250,000. If we subtract the former figure from the latter, we get the expected value of sample information: $250,000 − $115,000 = $135,000. Reassuringly, this is precisely

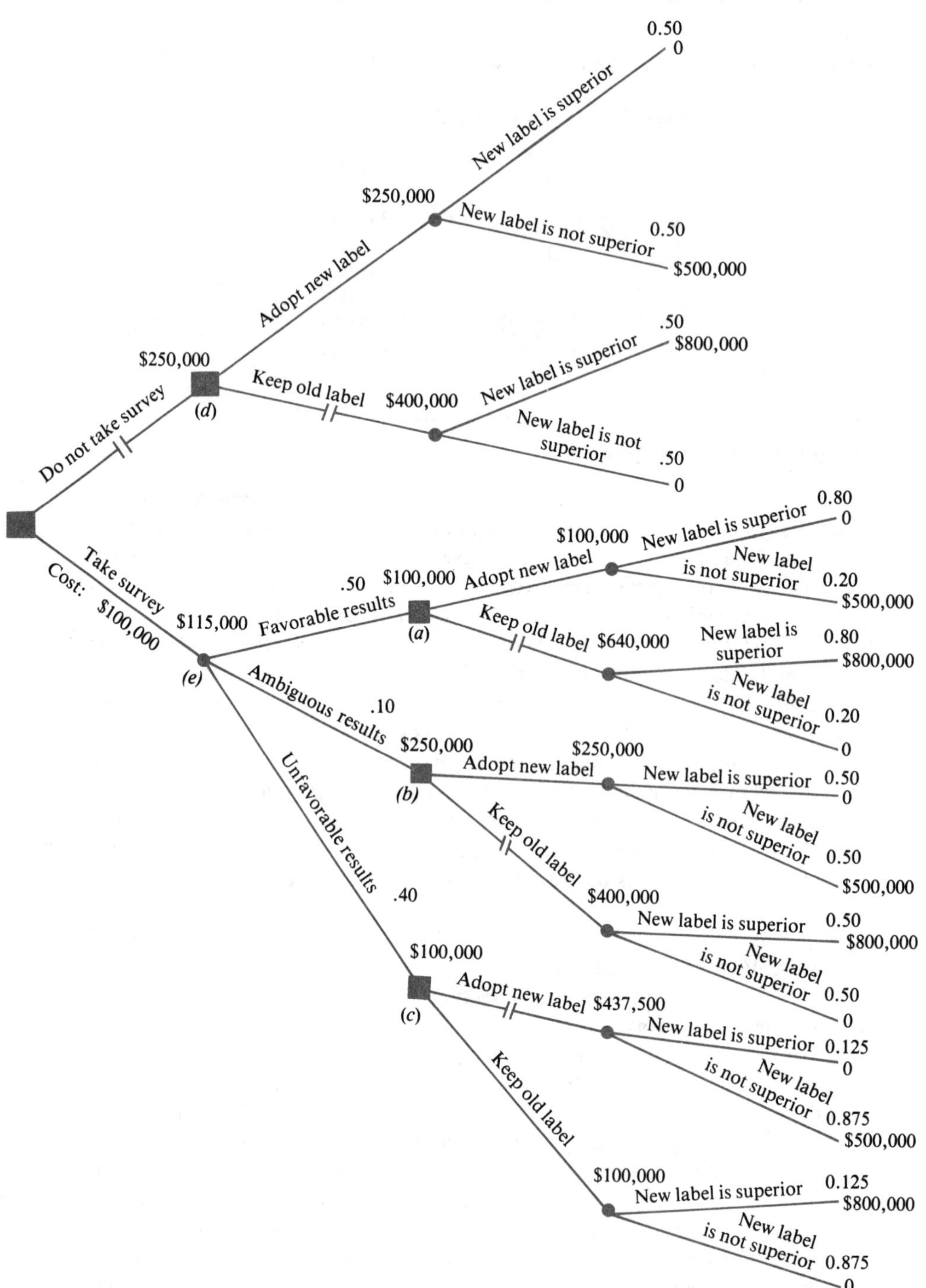

FIGURE 16.4 Decision Tree for Marketing Manager's Problem, Based on Expected Opportunity Loss

the amount that we obtained in Section 16.3 when we derived the expected value of sample information based on profits rather than on opportunity losses.

Since the extensive form analysis and the normal form analysis result in precisely the same answer, it may not seem a matter of importance as to which one is used. But this overlooks the fact that each type of analysis has certain advantages; for this reason it is worthwhile to learn both. *Perhaps the most important advantage of extensive form analysis is that the computations are much simpler and quicker than in normal form analysis.* In extensive form analysis, it is not necessary to compute the expected opportunity loss for each possible strategy. Through the process of backward induction we recognize that certain strategies are nonoptimal, and thus only the expected opportunity loss for the optimal strategy has to be calculated. This is an important advantage because real problems often involve a very large number of possible strategies, and it would be expensive and time-consuming to have to compute the expected opportunity loss for each of them.

On the other hand, *normal form analysis has the advantage of making it easier to see how sensitive the results are to the prior probabilities that are used.* As we have pointed out before, prior probabilities frequently are based on judgment or on the subjective feelings of the decision maker. It is understandable that both decision makers and analysts may be somewhat uncomfortable about these probabilities, and that they may want to see how much the results depend upon them. (Since the probabilities of survey results, given the state of nature, are often based on relative frequencies, there is usually less concern about them.)

To illustrate how normal form analysis can be used to determine the sensitivity of the results to the prior probabilities, let's return to the marketing manager's problem in the previous section. A close inspection of Table 16.4 shows that two strategies—strategies 4 and 7—are the leading contenders, given that the prior probability that the new label is superior is 0.5 (and that the prior probability that it is not superior is 0.5). Most of the other strategies in Table 16.2 do not seem very sensible. For example, strategy 5 would dictate that the decision maker keep the old label if the survey results are favorable to the new label and adopt the new label if they are unfavorable to the new label. (One hardly needs a course in statistics to reject such a strategy!) But the choice between strategies 4 and 7 is by no means obvious. As we know from Table 16.4, strategy 4 beats out strategy 7 if the prior probability that the new label is superior is 0.5. However, what if this prior probability is somewhat higher or lower than 0.5? Will strategy 4 continue to beat out strategy 7? Based on the results of a normal form analysis, this question is easy to answer.

The first thing to note is that the expected opportunity loss with strategy 4 is

$$P(\$80{,}000) + (1 - P)(\$150{,}000), \qquad (16.3)$$

where P is the prior probability that the new label is superior. This follows directly from equation (16.2), where P was set equal to 0.5. The next thing to note is that the expected opportunity loss with strategy 7 is

$$P(\$160{,}000) + (1 - P)(\$100{,}000). \tag{16.4}$$

This follows directly from equation (16.1), where P was set equal to 0.5. Clearly, the expected opportunity loss with strategy 7 goes up as P increases, while the expected opportunity loss with strategy 4 goes down as P increases.[6] To see whether there is some value of P which makes the expected opportunity losses of these two strategies equal, we can set the expressions in equation (16.3) and equation (16.4) equal to one another:

$$P(\$80{,}000) + (1 - P)(\$150{,}000) = P(\$160{,}000) + (1 - P)(\$100{,}000).$$

Then we can solve for P, the result being

$$\begin{aligned} 150{,}000 - 70{,}000P &= 100{,}000 + 60{,}000P \\ 130{,}000P &= 50{,}000 \\ P &= 5/13. \end{aligned}$$

Thus, if P is greater than 5/13, the expected opportunity loss with strategy 4 is less than that with strategy 7, while if P is less than 5/13, the reverse is true.

What does this result tell the marketing manager? It tells him that strategy 4, which he knows to be better than strategy 7 when $P = 0.5$, will continue to be better than strategy 7 when P is greater than 0.5. Also, it tells him that strategy 4 will beat out strategy 7 when P is between 5/13 and 0.5, but that strategy 4 will be poorer than strategy 7 if P is less than 5/13. On the basis of this information, the marketing manager knows precisely how sensitive his choice of a strategy is to variations in P.

Finally, it should be recognized that some strategies are better than others, regardless of the value of P. *One strategy is said to* ***dominate*** *another if its conditional expected opportunity loss is less than that of the other strategy for some possible states of nature, and is no greater than that of the other strategy for the other possible states of nature.* The following example shows how this concept of domination can be applied to the marketing manager's problem.

Example 16.1 Using the results in this section, show that strategies 4 and 7 dominate strategy 5. What implications does this have for the marketing manager?

6. It follows from equation (16.4) that the expected opportunity loss with strategy 7 equals

$$\$100{,}000 + \$60{,}000(P).$$

Thus, the expected opportunity loss increases with increases in P. Similarly, it follows from equation (16.3) that the expected opportunity loss with strategy 4 equals

$$\$150{,}000 - \$70{,}000(P).$$

Thus, the expected opportunity loss decreases with increases in P.

Solution: The expected opportunity loss with strategy 5 is[7]

$$P(\$720{,}000) + (1 - P)(\$350{,}000) = \$350{,}000 + \$370{,}000(P).$$

It follows from equation (16.3) that the expected opportunity loss with strategy 4 equals

$$\$150{,}000 - \$70{,}000(P).$$

From equation (16.4) it follows that the expected opportunity loss with strategy 7 equals

$$\$100{,}000 + \$60{,}000(P).$$

Thus, if P is the same for all strategies, the expected loss with strategy 5 is greater than with strategy 4 or strategy 7, *regardless of the value of* P.[8] This means that strategy 5 must be dominated by both strategy 4 and strategy 7. To the marketing manager, this means that strategy 5 is a poor strategy, regardless of the value of P. In other words, there is no value of P which would make this strategy a good one.

EXERCISES

16.1 The Energetic Corporation is trying to decide whether or not to establish a plant in Europe. Its management believes that if there is no resistance by Europeans to the firm's products, the firm will make $2 million by this action. The management believes that the probability of no such resistance equals 0.6. On the other hand, there is a 0.4 probability that foreign regulations and tastes will result in substantial resistance by Europeans to the firm's products. If this substantial resistance occurs, the firm will lose

7. To see why this is the expected opportunity loss with strategy 5, note that if the new label is in fact superior, the expected opportunity loss with this strategy is

$$0.8(\$800{,}000) + 0.1(\$800{,}000) + 0.1(0) = \$720{,}000.$$

If the new label is in fact not superior, the expected opportunity loss with this strategy is

$$0.2(0) + 0.1(0) + 0.7(\$500{,}000) = \$350{,}000.$$

Thus, if P is the probability that the new label is in fact superior, the expected opportunity loss with strategy 5 is

$$P(\$720{,}000) + (1 - P)(\$350{,}000),$$

as indicated above.

8. To prove that this is the case, note that the difference between the expected opportunity loss with strategy 5 and that with strategy 4 equals

$$[\$350{,}000 + \$370{,}000(P)] - [\$150{,}000 - \$70{,}000(P)]$$

$$= \$200{,}000 + \$440{,}000(P).$$

Since $P \geqslant 0$, this difference must be positive regardless of the value of P. Similarly, the difference between the expected opportunity loss with strategy 5 and that with strategy 7 equals

$$[\$350{,}000 + \$370{,}000(P)] - [\$100{,}000 + \$60{,}000(P)]$$

$$= \$250{,}000 + \$310{,}000(P).$$

Since $P \geqslant 0$ this difference, too, must be positive, regardless of the value of P.

$2.5 million. One possibility is that the firm may carry out a survey of European consumers to see what their response to the firm's products might be. A polling organization says that for a fee of $75,000 it can carry out a survey which will have the following probabilities of indicating various outcomes, given the actual state of nature:

State of nature	Sample outcome: No resistance (probability)	Sample outcome: Substantial resistance (probability)
No resistance to Energetic's products	0.7	0.3
Substantial resistance to Energetic's products	0.2	0.8

Construct the decision tree representing the firm's decision if no survey is carried out.

16.2 (a) Based on the data in Exercise 16.1, what is the probability that the survey will indicate no resistance? (b) What is the probability that it will indicate substantial resistance?

16.3 Assuming that the survey indicates no resistance, construct the subsequent branches of the decision tree.

16.4 Assuming that the survey indicates substantial resistance, construct the subsequent branches of the decision tree.

16.5 (a) Using the results obtained in Exercises 16.1–16.4, construct the entire decision tree representing the Energetic Corporation's problem. (b) Should the firm elect to have the survey carried out? (c) Given that it elects to do so, what decision should the firm take if the results indicate no resistance? (d) What decision should it take if the results indicate substantial resistance?

16.6 (a) Under the conditions described in Exercise 16.1, what is the expected value of the sample information? (b) What is the expected net gain of sample information?

16.7 List all possible strategies that might be employed in the problem described in Exercise 16.1.

16.8 For each of the strategies listed in Exercise 16.7, calculate the expected opportunity loss.

16.9 Is the strategy with the lowest expected opportunity loss (as shown by your calculations in Exercise 16.8) the same as the result obtained in Exercise 16.5? If so, explain why. If not, explain why the differences exist.

16.10 One of the possible strategies is the following: Establish the plant if the survey indicates no resistance, and do not establish it if the survey indicates substantial resistance. Letting P be the prior probability of no resistance, express the expected opportunity loss with this strategy as a function of P.

16.11 (a) For the other strategies besides that presented in Exercise 16.10, express the expected opportunity loss with each one as a function of P, the prior probability of no resistance. (b) Do some strategies dominate others? If so, which ones are dominated by which others?

16.12 Are there any values of P for which the strategy in Exercise 16.10 is not as good as some other strategies? If so, what are these values of P, and why are the other strategies superior?

16.6 Sequential Sampling

In previous sections of this chapter we have been concerned with the choice of whether or not to gather information before making a decision. Now we turn to sequential analysis, which is closely related to the sort of analysis presented in the previous sections. *In sequential analysis, it is recognized that if the decision maker chooses to gather information before making a decision, he or she has the option, after this information has been gathered, of obtaining still more information before making the decision.* For example, the marketing manager, once he obtains the results of the survey, has the option of carrying out another survey of perhaps a different type before deciding whether to adopt the new label or stick with the old one.

Before describing how one can solve sequential decision-making problems of this sort, we must consider briefly the use of sequential techniques to test hypotheses. Abraham Wald, a well-known statistician at Columbia University, did much of the pioneering work in this field during World War II. He showed how, given certain fixed probabilities of Type I and Type II errors, one can establish a sequential procedure for testing a particular hypothesis. In contrast to the test procedures in Chapters 8, 9, and 10, *no fixed sample size is specified* in sequential procedures. Instead, one computes a test statistic after each observation is chosen, and when this statistic falls beyond certain predetermined limits, the null hypothesis is accepted or rejected. Thus, it is possible that the statistician might either accept or reject the null hypothesis after only a few observations have been chosen; or it might take many observations, depending on what the observations turn out to be. *The advantage of sequential sampling is that on the average, it requires a smaller number of observations than a sample of fixed size in order to attain the same probability of Type I error (and of Type II error) as a fixed-size sample.*

To illustrate how such a sequential sampling scheme works, suppose that a firm wants to test whether the proportion of defective items in an incoming lot of goods is 0.01, the alternative hypothesis being that this proportion is 0.08. Suppose that the firm wants the probability of Type I error—that is, the probability that it will reject the hypothesis that the proportion defective is .01 when in fact this is true—to equal 0.05. Suppose also that it wants the probability of Type II error—that is, the probability that it will accept the hypothesis that the proportion defective is .01 when in fact it is .08—to equal 0.10. Then Table 16.5 shows a sequential sampling scheme that will satisfy these requirements.[9] In this table, n is the number of items that have been sampled. If x_n is the number of these items that are defective, then the firm should accept the null hypothesis if $x_n \leqslant A_n$, and it should reject the null hypothesis in favor of the alternative hypothesis if $x_n \geqslant R_n$. If $A_n < x_n < R_n$, the firm should sample another item from the shipment.

For example, suppose that the first 20 items sampled by the firm are not

9. Table 16.5, although complete enough for present purposes, only provides information up to $n = 60$. For a more complete table, see A. Duncan, *Quality Control and Industrial Statistics* (Homewood, Ill.: Irwin, 1959), p. 154.

TABLE 16.5

Sequential Sampling Plan[a]

n	A_n	R_n	n	A_n	R_n	n	A_n	R_n	n	A_n	R_n
	*		16	*	2	31	0	3	46	0	3
2	*	2	17	*	2	32	0	3	47	0	3
3	*	2	18	*	2	33	0	3	48	0	3
4	*	2	19	*	2	34	0	3	49	0	4
5	*	2	20	*	3	35	0	3	50	0	4
6	*	2	21	*	3	36	0	3	51	0	4
7	*	2	22	*	3	37	0	3	52	0	4
8	*	2	23	*	3	38	0	3	53	0	4
9	*	2	24	*	3	39	0	3	54	0	4
10	*	2	25	*	3	40	0	3	55	0	4
11	*	2	26	*	3	41	0	3	56	0	4
12	*	2	27	*	3	42	0	3	57	0	4
13	*	2	28	*	3	43	0	3	58	0	4
14	*	2	29	*	3	44	0	3	59	0	4
15	*	2	30	*	3	45	0	3	60	0	4

[a]See footnotes 9 and 10.

*A_n is less than zero.

SOURCE: A. Duncan, *Quality Control and Industrial Statistics* (Homewood, Ill.: Irwin, 1959), p. 154.

defective. Since Table 16.5 indicates that this result does not call for a decision one way or the other, the firm should sample a twenty-first item. If this item is defective, no decision is yet possible. If the next 10 items are not defective one still cannot come to a decision since the cumulative number of defectives (1) is above A_{31} (zero) and below R_{31} (which is 3). Suppose that the thirty-second item chosen is also defective. Still no decision can be made since the cumulative number of defectives (which is 2) is above A_{32} (which is zero) and below R_{32} (which is 3). Thus, another item must be sampled. Finally, if the thirty-third item to be chosen is also defective, the firm should reject the hypothesis that only 1 percent of the items in the shipment are defective since the cumulative number of defectives (3) equals R_{33} (also 3).

To repeat, the principal advantage of a sequential sampling scheme of this sort is that it allows the decision maker to test fewer items, on the average, before coming to a decision. This can be a very important advantage, particularly where testing involves the destruction of the item that is being tested. In some cases of this sort, each item tested is very expensive, and the decision maker can save large amounts of money by using sequential sampling rather than sampling based on a fixed sample size.[10]

10. Tables are available for various kinds of sequential sampling plans. For example, in the case of plans designed to test the proportion defective, one can choose the null hypothesis, the alternative hypothesis, the probability of Type I error, and the probability of Type II error. Given these choices, the tables show the sorts of information contained in Table 16.5. See A. Duncan, ibid., and A. Wald, *Sequential Analysis* (New York: Wiley, 1947).

The following example is a further illustration of how sequential sampling is carried out.

Example 16.2 A manufacturer of rockets wants to test whether the probability of failure of its rockets equals .01, the alternative hypothesis being that it equals .08. The manufacturer uses the sampling scheme shown in Table 16.5. The first rocket tested is not defective, but the next two are. Should the firm continue sampling? If not, should it accept or reject the null hypothesis?

Solution: Since $n = 3$, Table 16.5 shows that $R_n = 2$. Thus, since two out of the three rockets are defective, $x_n = 2$, which means that $x_n = R_n$. Consequently, the firm should reject the null hypothesis that the probability of failure equals .01.

16.7 Sequential Decision Making under Uncertainty

Let's return now to sequential decision-making procedures viewed from the point of view of Bayesian decision theory. To show how such procedures work, suppose that the Ellsworth Manufacturing Company is considering whether or not to install a major new process. If the process is introduced and is successful, the firm will reap a \$5 million profit. On the other hand, if the process is introduced and is not successful, the firm will incur a \$2 million loss. The firm attaches a prior probability of 0.5 to success and a prior probability of 0.5 to its being unsuccessful. Thus, on the basis of a prior analysis the firm would conclude, if it wants to maximize expected profit, that it should introduce the new process since the expected profit resulting from this action is

$$0.5(\$5,000,000) + 0.5(-\$2,000,000) = \$1,500,000,$$

while the profit resulting from its not introducing the new process is zero.

However, this prior analysis is based on the supposition that the firm gathers no evidence before making its decision. In fact, let us assume that the firm can carry out a research project to obtain information concerning the new process, and that the cost of this project is \$100,000. If the new process is in fact successful, the probability that the research project will indicate this is 0.7, while the probability that it will indicate that the new process is unsuccessful is 0.3. If the new process is in fact unsuccessful, the probability that the research project will indicate this is 0.8, while the probability that it will indicate that the new process is successful is 0.2.

If the firm decides to carry out the research project, then it can decide at the end of the project whether or not to install the new process, or it can gather still more information by building and studying a pilot plant for the new process. If the new process is in fact successful, the probability that the pilot plant will indicate this is 0.8, while the probability that it will indicate that the new process is unsuccessful is 0.2. If the new process is in fact unsuccessful, the probability that the pilot plant will indicate this is 0.7, while the probability that it will indicate that the new process is successful is 0.3. The cost of building and studying the pilot plant is \$100,000.

Figure 16.5 shows the decision tree for this problem. As you can see, the first decision the Ellsworth Manufacturing Company must make is whether or not to carry out the research project. If it decides not to do so, it must decide whether or not to install the new process. If it decides to install it, "chance"

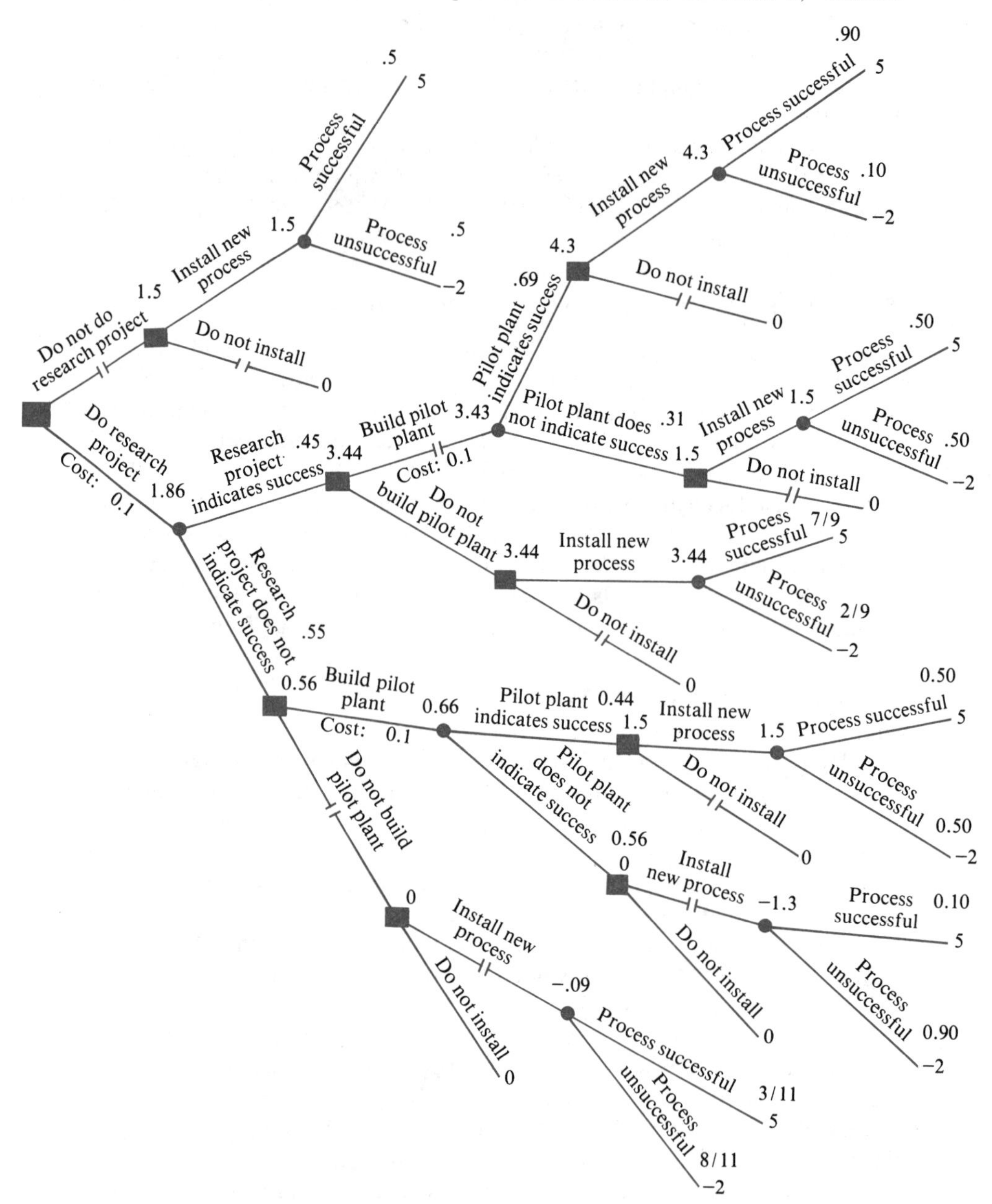

FIGURE 16.5 Decision Tree for Problem Facing Ellsworth Manufacturing Company[a]

[a]All monetary values are expressed in millions of dollars.

decides whether or not it is successful. On the other hand, if the company decides to carry out the research project, "chance" decides whether the project indicates success or not. Clearly, the probability that the research project will indicate success is $0.5(0.7) + 0.5(0.2) = 0.45$ since the prior probability that the process is successful is 0.5 and the probability that the research project will indicate that it is successful when this in fact is the case is 0.7, while the prior probability that the process is not successful is 0.5 and the probability that the research project will indicate that it is successful when in fact it is not successful is 0.2. Similarly, the probabiliity that the research project will not indicate success is 0.55.

16.8 Computation of Posterior Probabilities, Based on Results of Research Project Alone

If the research project above is carried out and *the results indicate that the new process is a success*, what is the posterior probability that it will be a success? Applying Bayes' theorem, this probability equals

$$P(S \mid s) = \frac{P(S)P(s \mid S)}{P(S)P(s \mid S) + P(N)P(s \mid N)},$$

where $P(S)$ is the prior probability that the new process is a success, $P(s \mid S)$ is the probability that the research project indicates success when the new process is in fact successful, $P(N)$ is the prior probability that the new process is not a success, and $P(s \mid N)$ is the probability that the research project indicates success when the new process is in fact not successful. Inserting numerical values in this expression,[11] we have

$$P(S \mid s) = \frac{0.5(0.7)}{0.5(0.7) + 0.5(0.2)} = 7/9.$$

Letting $P(N \mid s)$ be the posterior probability that the new process will not be a success, given that the research project indicates that it is a success, it follows that

$$P(N \mid s) = 1 - P(S \mid s) = 2/9.$$

If the research project is carried out and *the results indicate that the new process is not a success*, what is the posterior probability that it will be a success? Applying Bayes' theorem, this probability equals

$$P(S \mid n) = \frac{P(S)P(n \mid S)}{P(S)P(n \mid S) + P(N)P(n \mid N)},$$

where $P(n \mid S)$ is the probability that the research project does not indicate success when the new process is in fact successful, and $P(n \mid N)$ is the

11. Recall from the previous section that both $P(S)$ and $P(N)$ equal 0.5, and that $P(s \mid S) = 0.7$ and $P(s \mid N) = 0.2$. These are the required numerical values.

probability that the research project does not indicate success when the new process is in fact not successful. Inserting numerical values in this expression,[12] we have

$$P(S \mid n) = \frac{0.5(0.3)}{0.5(0.3) + 0.5(0.8)} = 3/11.$$

Letting $P(N \mid n)$ be the posterior probability that the new process will not be a success, given that the research project indicates that it is not a success, it follows that

$$P(N \mid n) = 1 - P(S \mid n) = 8/11.$$

With these results in hand, we can derive the expected profit if the firm carries out the research project and it indicates success, whereupon the firm decides to install the new process without building a pilot plant. This expected profit is[13]

$$(7/9)(5) + (2/9)(-2) = 3.44,$$

the figure shown in Figure 16.5. Since this expected profit exceeds zero, it is better to install the new process under these circumstances than not to do so. Also, we can derive the expected profit if the firm carries out the research project and it does not indicate success, whereupon the firm decides to install the new process without building a pilot plant. This expected profit is

$$(3/11)(5) + (8/11)(-2) = -.09,$$

which is the figure shown in Figure 16.5. Since this expected profit is less than zero, it is better not to install the new process under these circumstances.

16.9 Computation of Posterior Probabilities, Based on Results of Research Project and Pilot Plant Together

To derive the rest of the figures in Figure 16.5, we must compute the posterior probabilities that the new process will or will not be a success, based on the results of *both* the research project and the pilot plant. First, suppose that *both the research project and the pilot plant indicate success.* What is the posterior probability that the new process will be successful? Clearly, this probability

12. As pointed out in the previous section, both $P(S)$ and $P(N)$ equal 0.5, $P(n|S) = 0.3$, and $P(n|N) = 0.8$. These are the required numerical values.

13. To obtain the expected profit, we multiply the posterior probability of success (7/9) times 5, the profit (in millions of dollars) if the process is successful. To this we add the posterior probability of lack of success (2/9) times -2, the profit (in millions of dollars) if the process is not successful. The same procedure is used in the next equation in the text, but the posterior probabilities are different because they are based on the assumption that the research project does not indicate success.

equals

$$P(S|ss) = \frac{P(S)P(ss|S)}{P(S)P(ss|S) + P(N)P(ss|N)},$$

where $P(ss|S)$ is the probability that both the research project and the pilot plant indicate that the new process will be a success, given that the new process in fact is a success, and $P(ss|N)$ is the probability that both the research project and the pilot plant indicate that the new process is a success, given that the new process is not a success. If the results of the pilot plant and the research project are independent, it follows that[14]

$$P(S|ss) = \frac{0.5[(0.7)(0.8)]}{0.5[(0.7)(0.8)] + 0.5[(0.3)(0.2)]} = 0.90.$$

And we can also conclude that

$$P(N|ss) = 1 - 0.90 = 0.10,$$

where $P(N|ss)$ is the posterior probability that the new process is not a success, given that both the research project and the pilot plant indicate that it is a success.

Next, let's consider the case where *the research project indicates that the new process is a success and the pilot plant indicates that it is not a success.* Under these circumstances, what is the posterior probability that the new process is a success? Clearly, this probability equals

$$P(S|sn) = \frac{P(S)P(sn|S)}{P(S)P(sn|S) + P(N)P(sn|N)},$$

where $P(sn|S)$ is the probability that the research project will indicate that the new process is a success and that the pilot plant will indicate that it is not a success, given that the new process in fact is a success, and $P(sn|N)$ is the probability that the research project will indicate that the new process is a success and that the pilot plant will indicate that it is not a success, given that the new process in fact is not a success. Inserting numerical values,[15] we have

14. Since we assume that the results of the research project and the pilot plant are independent, the probability that the research project and the pilot plant will both indicate success is the product of the probability that the research project indicates success and the probability that the pilot plant indicates success. According to Section 16.7, the probability that the research project will indicate success (if the process is successful) is 0.7, and the probability that the pilot plant will indicate success (if the process is successful) is 0.8. Thus, $P(ss|S) = 0.7(0.8)$. According to Section 16.7, the probability that the research project will indicate success (if the process is not successful) is 0.2, and the probability that the pilot plant will indicate success (if the process is not successful) is 0.3. Thus, $P(ss|N) = 0.3(0.2)$.

15. Recall from Section 16.7 that both $P(S)$ and $P(N)$ equal 0.5. Also $P(sn|S) = 0.7(0.2)$ since the probability that the research project will indicate success (if the process is successful) is 0.7 and the probability that the pilot plant will not indicate success (if the process is successful) is 0.2, according to Section 16.7. Further, $P(sn|N) = 0.2(0.7)$ since the probability that the research project will indicate success (if the process is not successful) is 0.2, and the probability that the pilot plant will not indicate success (if the process is not successful) is 0.7 according to Section 16.7. These are the required numerical values.

$$P(S\,|\,sn) = \frac{0.5[(0.7)(0.2)]}{0.5[(0.7)(0.2)] + 0.5[(0.2)(0.7)]} = 0.5.$$

And we can also conclude that

$$P(N\,|\,sn) = 1 - 0.5 = 0.5,$$

where $P(N\,|\,sn)$ is the posterior probability that the new process is not a success, given that the research project indicates that it is a success although the pilot plant indicates that it is not a success.

Next, let's consider the case where *the research project indicates that the new process is not a success although the pilot plant indicates that it is a success.* Under these circumstances, what is the posterior probability that the new process is a success? Clearly, this probability equals

$$P(S\,|\,ns) = \frac{P(S)P(ns|S)}{P(S)P(ns|S) + P(N)P(ns|N)},$$

where $P(ns|S)$ is the probability that the research project will indicate that the new process is not a success and that the pilot plant will indicate that it is a success, given that the new process is a success, and $P(ns|N)$ is the probability that the research project will indicate that the new process is not a success and that the pilot plant will indicate that it is a success, given that in fact it is not a success. Inserting numerical values,[16] we have

$$P(S\,|\,ns) = \frac{0.5[(0.3)(0.8)]}{0.5[(0.3)(0.8)] + 0.5[(0.8)(0.3)]} = 0.5.$$

And we can also conclude that

$$P(N\,|\,ns) = 1 - 0.5 = 0.5,$$

where $P(N\,|\,ns)$ is the posterior probability that the new process is not a success if the research project indicates it is not a success and the pilot plant indicates it is a success.

Finally, suppose that *both the research project and the pilot plant indicate that the new process is not a success.* What is the posterior probability that the new process is successful? This probability equals

$$P(S\,|\,nn) = \frac{P(S)P(nn|S)}{P(S)P(nn|S) + P(N)P(nn|N)},$$

where $P(nn|S)$ is the probability that the research project and the pilot plant both indicate that the new process is not a success when in fact it is a success,

16. From Section 16.7, we know that both $P(S)$ and $P(N)$ equal 0.5. Also, $P(ns|S) = 0.3(0.8)$ since the probability that the research project will not indicate success (if the process is successful) is 0.3 and the probability that the pilot plant will indicate success (if the process is successful) is 0.8, according to Section 16.7. Further, $P(ns|N) = 0.8(0.3)$ since the probability that the research project will not indicate success (if the process is not successful) is 0.8 and the probability that the pilot plant will indicate success (if the process is not successful) is 0.3, according to Section 16.7. These are the required numerical values.

and $P(nn|N)$ is the probability that the research project and the pilot plant both indicate that the new process is not a success when in fact it is not a success. Inserting numerical values,[17] we have

$$P(S|nn) = \frac{0.5[(0.3)(0.2)]}{0.5[(0.3)(0.2)] + 0.5[(0.8)(0.7)]} = .10.$$

And we can also conclude that

$$P(N|nn) = 1 - .10 = 90,$$

where $P(N|nn)$ is the posterior probability that the new process is not a success if the research project and the pilot plant both indicate that this is the case.

16.10 Solution to the Problem

Having derived the posterior probabilities, based on the results of the research project alone and on the basis of both the research project and the pilot plant, we must now compute still another set of probabilities before being able to determine what the Ellsworth Manufacturing Company should decide. The first probability we need is the probability that the pilot plant will indicate success, given that the research project has indicated success. This probability is[18]

$$P(s|s) = \frac{P(ss)}{P(s)} = \frac{P(ss|S)P(S) + P(ss|N)P(N)}{P(s|S)P(S) + P(s|N)P(N)},$$

where $P(ss)$ is the probability that both the research project and the pilot plant indicate that the new process will be a success, and $P(s)$ is the probability that the research project indicates success. Inserting numerical values,[19] we have

$$P(s|s) = \frac{[(0.7)(0.8)](0.5) + [(0.3)(0.2)](0.5)}{(0.7)(0.5) + (0.2)(0.5)} = 0.69.$$

From this we can also conclude that the probability that the pilot plant will not indicate success, given that the research project has indicated success, equals

$$P(n|s) = 1 - .69 = .31.$$

17. As pointed out above, both $P(S)$ and $P(N)$ equal 0.5. Also, $P(nn|S) = 0.3(0.2)$ since the probability that the research project will not indicate success (if the process is successful) is 0.3 and the probability that the pilot plant will not indicate success (if the process is successful) is 0.2, according to Section 16.7. Further, $P(nn|N) = 0.8(0.7)$ since the probability that the research project will not indicate success (if the process is not successful) is 0.8 and the probability that the pilot plant will not indicate success (if the process is not successful) is 0.7, according to Section 16.7. These are the required numerical values.
18. To prove that $P(s|s) = P(ss) \div P(s)$, recall the definition of a conditional probability, given in equation (3.3).
19. The values of $P(ss|S)$ and $P(ss|N)$ are derived in footnote 14 above. $P(s|S) = 0.7$ and $P(s|N) = 0.2$, according to Section 16.7.

The next probability that we need is the probability that the pilot plant will indicate success, given that the research project has indicated that the new process will not be a success. This probability is

$$P(s|n) = \frac{P(ns)}{P(n)} = \frac{P(ns|S)P(S) + P(ns|N)P(N)}{P(n|S)P(S) + P(n|N)P(N)},$$

where $P(ns)$ is the probability that the research project will not indicate success but that the pilot plant will indicate success, and $P(n)$ is the probability that the research project will not indicate success. Inserting numerical values,[20] we have

$$P(s|n) = \frac{[(0.3)(0.8)](0.5) + [(0.8)(0.3)](0.5)}{(0.3)(0.5) + (0.8)(0.5)} = 0.44.$$

And from this it follows that the probability that the pilot plant will not indicate success, given that the research project has not indicated success, equals

$$P(n|n) = 1 - 0.44 = .56.$$

Now we have all the probabilities set forth in Figure 16.5. To solve the problem, we begin by calculating the expected profit if the firm does not carry out the research project and goes ahead and installs the new process. Since (as shown in Figure 16.5) the expected profit is $1.5 million, the firm, if it does not carry out the research project, is better off installing the new process than not installing it. Thus, *the expected profit if the firm does not carry out the research project is $1.5 million* (as we saw in Section 16.7).

Next, suppose that the firm carries out the research project and builds the pilot plant, and that both indicate the new process will be successful. Under these circumstances, the expected profit is $4.3 million if the firm introduces the new process, which compares with a zero profit if it does not introduce it. (See Figure 16.5.)[21] Thus, assuming that the research project indicates success, the expected profit if the pilot plant is built and indicates success is $4.3 million. On the other hand, suppose that whereas the research project indicates success, the pilot plant does not. In this case the expected

20. The values of $P(ns|S)$ and $P(ns|N)$ are derived in footnote 16 above. $P(n|S) = 0.3$ and $P(n|N) = 0.8$, according to Section 16.7.

21. In Figure 16.5, the expected profit (in millions of dollars) when the decision maker is situated at each fork is shown above the fork, in accord with our discussion in Section 15.2. Thus, 4.3 is shown above the fork corresponding to the situation where both the research project and the pilot plant indicate that the new process will be successful, and the firm installs the new process. Why is the expected profit equal to $4.3 million? Because (as derived in Section 16.9 and shown in Figure 16.5) there is a 0.9 probability of success (in which case the profit is $5 million) and a 0.1 probability of lack of success (in which case the profit is −$2 million). Thus, the expected profit is $0.9(5) + 0.1(-2) = 4.3$.

Each of the expected profit figures in Figure 16.5 can be calculated in a similar way. For example, what is the expected profit if the firm introduces the new process after the research project indicates success and the pilot plant does not indicate success? Using the values of $P(S|sn)$ and $P(N|sn)$ derived in Section 16.9 and shown in Figure 16.5, the answer is $0.5(5) + 0.5(-2) = 1.5$. And as you can see, this is the number shown above this fork in Figure 16.5.

profit if the new process is introduced is $1.5 million, which is obviously greater than the zero profit to be obtained if the new process is not introduced. (See Figure 16.5.) Thus, the better decision under these circumstances is to install the new process. Since the probability is 0.69 that the pilot plant will indicate success and 0.31 that it will not do so, *the expected profit if the firm builds the pilot plant (after the research project indicates success) is $3.43 million.* To compare this with the expected profit if the firm does *not* build a pilot plant (after the research project indicates success), look at the relevant section of Figure 16.5, which shows that the expected profit under the latter circumstances is $3.44 million. Thus, *if the research project indicates success, the firm will do better not to build the pilot plant. (The expected profit if it is built is $.01 million less than if it is not built, and there is an additional $.10 million cost of building it.)*

Now suppose that the firm carries out the research project and it does not indicate success. If the firm builds a pilot plant under these circumstances, and if the plant indicates success, the expected profit if the firm installs the new process is $1.5 million. Since this is more than the profit from not installing the process, the firm's best action if the pilot plant indicates success is to install the new process. If the pilot plant does not indicate success under these circumstances, the expected profit if the firm installs the new process is −$1.3 million. Thus, the firm's best action under the latter circumstances is not to introduce the new process. Since the probability is 0.44 that the pilot plant will indicate success and 0.56 that it will not do so under these circumstances, *the expected profit if the firm builds the pilot plant (after the research project does not indicate success) is $0.66 million.* If the firm does not build the pilot plant the expected profit is zero since its best action then is not to install the new process. (The expected profit if it installs the new process would be − $.09 million). Thus, *if the research project does not indicate success, the firm's best action is to build the pilot plant since (after subtracting the pilot plant's cost of $100,000) the expected profit is $0.56 million if the pilot plant is built and zero if it is not.*

At this point we can calculate the expected profit if the firm carries out the research project. Since there is a 0.45 probability that the research project will indicate success and a 0.55 probability that it will not, the expected profit is $1.86 million, as shown in Figure 16.5. Allowing for the $100,000 cost of the research project, the expected profit is nevertheless higher than $1.5 million, the expected profit if the research project is not carried out. Thus, *the firm should carry out the research project.*

To summarize, the solution to the problem is as follows: The Ellsworth Manufacturing Company, if it wants to maximize expected profit, should carry out the research project. If the research project indicates that the new process will be successful, the firm should not build the pilot plant, but go ahead and introduce the new process. If the research project does not indicate success, the firm should build the pilot plant. If the pilot plant indicates success, the firm should install the new process, but if the pilot plant does not indicate success, it should not install the new process.

16.11 Conclusion

In the problem faced by the Ellsworth Manufacturing Company, the two possible junctures at which the firm could obtain additional information were (1) before carrying out the research project, and (2) before building and studying the pilot plant. There is no reason, of course, why a third or fourth option for obtaining additional information could not be built into this analysis. For example, after the results of the pilot plant have been obtained, the firm might consider the possibility of experimenting with a larger semiworks plant; and it might then introduce the new process on a small scale in only some of its plants. Indeed, from a purely analytical point of view, there is no reason why the statistician or the decision maker cannot include as many information-gathering stages as he or she likes.

From a practical point of view, however, there are advantages in not considering too many such stages. Even with only two stages, the decision tree in Figure 16.5 becomes rather complex. If the number of stages is increased considerably, the analysis becomes increasingly difficult, even with the aid of computers. For this reason, it is generally desirable to truncate the decision tree—that is, to cut off and simplify the analysis.

Finally, a warning is in order. While preposterior analysis and sequential analysis are powerful procedures for helping to solve problems faced by business firms and government agencies, they are not magic. In particular, the statistician should recognize that if analyses of this sort are based on distorted, unreliable values of the prior probabilities, the results are likely to be incorrect and misleading. For example, in the case of the Ellsworth Manufacturing Company, if the prior probabilities of the success or failure of the new process are optimistic figments of the inventor's imagination, and if the probabilities of the outcomes of the research project and of the pilot plant (given the success or failure of the new process) reflect major biases on the part of the firm's engineers, the results may be worse than useless. Unless these and other basic data are meaningful, the results are likely to be misleading, no matter how scientific the analysis may appear. Although preposterior analysis and sequential analysis are useful they, like all statistical techniques, must be applied with proper caution.

EXERCISES

16.13 The Uphill Corporation establishes the sequential sampling scheme shown in Table 16.5 to test whether the bolts it receives from a particular supplier are of the proper size. The supplier maintains that only 1 percent of the bolts Uphill receives in any given shipment should be more than .01 inches larger or smaller than the size specified by the design. Uphill uses the sequential sampling scheme to test whether this is true.

(a) Uphill samples a given shipment and finds that 1 out of the first 30 bolts is more than .01 inches larger or smaller than specified. Should Uphill reject the shipment?

(b) Uphill continues sampling and finds that 3 out of the next 30 bolts are more than .01 inches bigger or smaller than specified. Should Uphill reject the shipment?

16.14 The Energetic Corporation is studying the desirability of buying some very large and expensive pieces of equipment in order to be eligible to bid on a particular government contract. Based on its current information, the firm believes that if it buys the equipment the probability of getting the contract is 0.55 and the probability of not getting it is 0.45. If it buys the equipment and gets the contract, the firm foresees additional profits of $2 million; if it buys the equipment and does not get the contract, the firm foresees a loss of $1.5 million. If the firm does not buy the equipment profits will be unaffected.

Before deciding whether to buy these pieces of equipment, Energetic can hire a consulting firm to look closely at the factors determining whether or not it will get the contract. Based on past experience, Energetic feels that the probability that the consulting firm will conclude that Energetic will get the contract (if it has the equipment), given that this actually is the case, is 0.9, while the probability that it will conclude that Energetic will not get the contract under these circumstances is 0.1. Energetic also feels that the probability that the consulting firm will conclude that it will get the contract (if it has the equipment), given that this actually is not the case, is 0.3, while the probability of the consultant's concluding that Energetic will not get the contract under these circumstances is 0.7. The consulting firm will charge Energetic $150,000 for its study.

Once the study has been carried out (if it is carried out) by the consulting firm, Energetic can have its report gone over by a panel of experts in order to judge whether Energetic will or will not get the contract. If Energetic will get the contract, the probability that this panel will conclude that this is so is 0.9, while the probability that it will reach the reverse conclusion is 0.1. If Energetic will not get the contract, the probability that the panel will conclude that this is so is 0.9, while the probability that it will reach the reverse conclusion is 0.1. Given each state of nature, the results of the panel and of the consulting firm are regarded as being statistically independent. The panel of experts will charge Energetic $50,000 for providing a judgment of this sort.

(a) If Energetic hires the consulting firm, and if the firm concludes that Energetic will get the contract, what is the probability that Energetic will in fact get the contract? Under these circumstances, what is the probability that Energetic will in fact not get the contract?

(b) If Energetic hires the consulting firm, and if the firm concludes that Energetic will not get the contract, what is the probability that Energetic will in fact get the contract? Under these circumstances, what is the probability that Energetic will in fact not get the contract?

(c) If both the consulting firm and the panel conclude that Energetic will get the contract, what is the probability that it will in fact get the contract? Under these circumstances, what is the probability that it will in fact not get the contract?

(d) If both the consulting firm and the panel conclude that Energetic will not get the contract, what is the probability that it will in fact get the contract? Under these circumstances, what is the probability that it will in fact not get the contract?

(e) If the consulting firm concludes that Energetic will get the contract whereas the panel of experts concludes the reverse, what is the probability that it will in fact get the contract? Under these circumstances, what is the probability that it will in fact not get the contract?

(f) If the panel of experts concludes that Energetic will get the contract whereas the consulting firm concludes the reverse, what is the probability that it will in fact get the contract? Under these circumstances, what is the probability that it will in fact not get the contract?

16.15 (a) Given the situation in Exercise 16.14, what is the probability that if the consulting firm concludes that Energetic will get the contract the panel of experts will conclude the same thing? (b) What is the probability that if the consulting firm concludes that Energetic will get the contract, the panel of experts will come to the opposite conclusion? (c) Suppose that the consulting firm concludes that Energetic will not get the contract. Under these circumstances, what is the probability that the panel of experts will come to the opposite conclusion? What is the probability that they will reach the same conclusion?

16.16 Construct the decision tree representing the Energetic Corporation's problem in Exercise 16.14. Based on your results, answer the following questions:
(a) If Energetic does not hire the consulting firm, is it better to buy the equipment or not? Why, or why not?
(b) If Energetic hires the consulting firm, and if the consulting firm says Energetic will get the contract, should Energetic hire the panel of experts? Why, or why not?
(c) If Energetic hires the consulting firm, and if the consulting firm says Energetic will not get the contract, should Energetic hire the panel of experts? Why, or why not?
(d) What is the complete solution to Energetic's problem?

CHAPTER REVIEW

1. In a *preposterior analysis* the decision maker has the option of gathering new information rather than making a final decision immediately. If it is decided not to obtain this information, the resulting expected profit can be calculated by the sort of prior analysis described in the previous chapter. If the decision maker chooses to obtain additional information, the probability distribution of the outcome of the survey or experiment can be computed. Also, using Bayes' theorem, the probability distribution of the states of nature, given the outcome of the survey or experiment, can be computed. The expected profit for each outcome can then be calculated, weighted by its probability of occurrence, and summed in order to obtain the expected profit. The difference between this value and the expected profit if it is decided not to obtain this additional information is the *expected value of the sample information.* The *expected net gain of sample information* is the expected value of the sample information less the cost of the survey or experiment. If the expected net gain of sample information is greater than zero, it is worthwhile obtaining the information.

2. There are two types of preposterior analysis: *extensive form analysis* and *normal form analysis.* To carry out an *extensive form analysis* one begins by constructing a decision tree, by characterizing the states of nature at the right-hand ends of the tree, and by assigning payoffs to each action, assuming that each state of nature is true. The next step is to calculate the probabilities that the survey or experiment will result in various outcomes, and to carry out a posterior analysis, assuming each outcome occurs, in order to find the expected payoff corresponding to each outcome. Then, as described in the previous paragraph, the preposterior expected payoff is compared with the prior expected payoff to see whether the survey or experiment is worthwhile. The principal advantage of this type of preposterior analysis is that it is less expensive and time-consuming than a normal form analysis.

3. To carry out a *normal form analysis*, the first step is to list all the possible strategies that might be employed. A *strategy* is a decision rule that

specifies which action the decision maker will take if each survey result occurs. In a normal form analysis, one compares the values of all strategies to find which one is optimal. The result, of course, is the same as that arrived at by extensive form analysis, but normal form analysis has the advantage of making it easier to see how sensitive the results are to the prior probabilities being used. For example, one can determine the range of values of the prior probabilities over which the optimal strategy remains optimal. Also, one can see whether some strategies are *dominated* by others. One strategy dominates another if (1) its conditional expected opportunity loss is less than that of the other strategy for some possible states of nature and (2) if this expected opportunity loss is no greater than that of the other strategy for the other possible states of nature.

4. In *sequential analysis* it is recognized that if the decision maker chooses to gather information before making a decision, he or she has the option, after this information has been gathered, of obtaining still more information before making the decision. *Sequential sampling techniques* can be used to test hypotheses, the hallmark of such tests being that no fixed sample size is specified. Instead, after each observation is chosen, a decision is made to accept the null hypothesis, reject the null hypothesis, or take another observation. A major advantage of sequential sampling is that on the average, a smaller sample is required than in a sample of fixed size.

5. From the point of view of Bayesian decision theory, sequential decision-making procedures are a logical extension of preposterior analysis. To illustrate such procedures, we presented a case where the decision maker could obtain information first from a research project and then from the operation of a pilot plant which tested the characteristics of a new process. In principle there is no reason why the number of information-gathering stages cannot be made very large, but because the computational problems become difficult, the statistician often truncates the analysis.

Getting Down to Cases:

MAXWELL HOUSE'S DECISION CONCERNING A THREE-MONTH SALES TEST OF THE QUICK-STRIP CAN[22]

As indicated in the case at the end of the previous chapter, Maxwell House Coffee had to decide in 1963 how quickly to introduce its new quick-strip can. One possible action at that time was to conduct a three-month sales test of the quick-strip can in Muncie and Stockton, rather than go ahead immediately with adoption of the new can (as visualized in our earlier case). A three-month sales test of this sort would have cost about $6,000. The results of such a test might indicate that the new can was (1) very successful; (2) successful; (3) no better than the old; or (4) worse than the old. According to Joseph Newman's study of this case, the probability that each market test result would occur, given the true effect of the quick-strip

22. This case is based on a section of Joseph Newman's *Management Applications of Decision Theory* (New York: Harper and Row, 1971).

can on Maxwell House's market share, is as follows:

True effect of new can on Maxwell House's market share (percentage points)	Results of three-month sales test (probabilities)			
	Very successful	Successful	No better than old	Worse than old
+2.5	0.5	0.3	0.2	0
+1.0	0.2	0.5	0.2	0.1
0	0.1	0.2	0.5	0.2
−1.5	0	0.2	0.3	0.5

Since it was assumed in this analysis that the price per pound would be increased by 2 cents, the change in Maxwell House's profits corresponding to each of these changes in market share is as follows:

Change in market share (percentage points)	Change in profit (thousands of dollars)
+2.5	11,939
+1.0	6,489
0	2,856
−1.5	−1,050

According to Newman, if Maxwell House adopted the new can without the three-month sales test, it might have been reasonable to expect a 0.25 probability that its market share would increase by 2.5 percentage points, a 0.25 probability that its market share would increase by 1.0 percentage points, a 0.25 probability that its market share would remain constant, and a 0.25 probability that its market share would decrease by 1.5 percentage points.

(a) Construct a decision tree to indicate whether or not Maxwell House should have carried out the three-month sales test.

(b) What was the posterior probability of each state of nature if the sales test results indicated that the new can was very successful? Successful? No better than the old? Worse than the old?

(c) Should Maxwell House have carried out the three-month sales test? Why, or why not?

Appendix

APPENDIX TABLE 1

Binomial Probability Distribution

This table shows the value of

$$P(x) = \frac{n!}{(n - x)!x!}P^x(1 - P)^{n-x}$$

for selected values of P and for $n = 1$ to 20. For values of P exceeding 0.5, the value of $P(x)$ can be obtained by substituting $(1 - P)$ for P and by finding $P(n - x)$. (See Section 4.7.)

		P									
n	*x*	.05	.10	.15	.20	.25	.30	.35	.40	.45	.50
1	0	.9500	.9000	.8500	.8000	.7500	.7000	.6500	.6000	.5500	.5000
	1	.0500	.1000	.1500	.2000	.2500	.3000	.3500	.4000	.4500	.5000
2	0	.9025	.8100	.7225	.6400	.5625	.4900	.4225	.3600	.3025	.2500
	1	.0950	.1800	.2550	.3200	.3750	.4200	.4550	.4800	.4950	.5000
	2	.0025	.0100	.0225	.0400	.0625	.0900	.1225	.1600	.2025	.2500
3	0	.8574	.7290	.6141	.5120	.4219	.3430	.2746	.2160	.1664	.1250
	1	.1354	.2430	.3251	.3840	.4219	.4410	.4436	.4320	.4084	.3750
	2	.0071	.0270	.0574	.0960	.1406	.1890	.2389	.2880	.3341	.3750
	3	.0001	.0010	.0034	.0080	.0156	.0270	.0429	.0640	.0911	.1250
4	0	.8145	.6561	.5220	.4096	.3164	.2401	.1785	.1296	.0915	.0625
	1	.1715	.2916	.3685	.4096	.4219	.4116	.3845	.3456	.2995	.2500
	2	.0135	.0486	.0975	.1536	.2109	.2646	.3105	.3456	.3675	.3750
	3	.0005	.0036	.0115	.0256	.0469	.0756	.1115	.1536	.2005	.2500
	4	.0000	.0001	.0005	.0016	.0039	.0081	.0150	.0256	.0410	.0625

APPENDIX TABLE 1 (Continued)

		P									
n	x	.05	.10	.15	.20	.25	.30	.35	.40	.45	.50
5	0	.7738	.5905	.4437	.3277	.2373	.1681	.1160	.0778	.0503	.0312
	1	.2036	.3280	.3915	.4096	.3955	.3602	.3124	.2592	.2059	.1562
	2	.0214	.0729	.1382	.2048	.2637	.3087	.3364	.3456	.3369	.3125
	3	.0011	.0081	.0244	.0512	.0879	.1323	.1811	.2304	.2757	.3125
	4	.0000	.0004	.0022	.0064	.0146	.0284	.0488	.0768	.1128	.1562
	5	.0000	.0000	.0001	.0003	.0010	.0024	.0053	.0102	.0185	.0312
6	0	.7351	.5314	.3771	.2621	.1780	.1176	.0754	.0467	.0277	.0156
	1	.2321	.3543	.3993	.3932	.3560	.3025	.2437	.1866	.1359	.0938
	2	.0305	.0984	.1762	.2458	.2966	.3241	.3280	.3110	.2780	.2344
	3	.0021	.0146	.0415	.0819	.1318	.1852	.2355	.2765	.3032	.3125
	4	.0001	.0012	.0055	.0154	.0330	.0595	.0951	.1382	.1861	.2344
	5	.0000	.0001	.0004	.0015	.0044	.0102	.0205	.0369	.0609	.0938
	6	.0000	.0000	.0000	.0001	.0002	.0007	.0018	.0041	.0083	.0516
7	0	.6983	.4783	.3206	.2097	.1335	.0824	.0490	.0280	.0152	.0078
	1	.2573	.3720	.3960	.3670	.3115	.2471	.1848	.1306	.0872	.0547
	2	.0406	.1240	.2097	.2753	.3115	.3177	.2985	.2613	.2140	.1641
	3	.0036	.0230	.0617	.1147	.1730	.2269	.2679	.2903	.2918	.2734
	4	.0002	.0026	.0109	.0287	.0577	.0972	.1442	.1935	.2388	.2734
	5	.0009	.0002	.0012	.0043	.0115	.0250	.0466	.0774	.1172	.1641
	6	.0000	.0000	.0001	.0004	.0013	.0036	.0084	.0172	.0320	.0547
	7	.0000	.0000	.0000	.0000	.0001	.0002	.0006	.0016	.0037	.0078
8	0	.6634	.4305	.2725	.1678	.1001	.0576	.0319	.0168	.0084	.0039
	1	.2793	.3826	.3847	.3355	.2670	.1977	.1373	.0896	.0548	.0312
	2	.0515	.1488	.2376	.2936	.3115	.2965	.2587	.2090	.1569	.1094
	3	.0054	.0331	.0839	.1468	.2076	.2541	.2786	.2787	.2568	.2188
	4	.0004	.0046	.0815	.0459	.0865	.1361	.1875	.2322	.2627	.2734
	5	.0000	.0004	.0026	.0092	.0231	.0467	.0808	.1239	.1719	.2188
	6	.0000	.0000	.0002	.0011	.0038	.0100	.0217	.0413	.0703	.1094
	7	.0000	.0000	.0000	.0001	.0004	.0012	.0033	.0079	.0164	.0312
	8	.0000	.0000	.0000	.0000	.0000	.0001	.0002	.0007	.0017	.0039
9	0	.6302	.3874	.2316	.1342	.0751	.0404	.0207	.0101	.0046	.0020
	1	.2985	.3874	.3679	.3020	.2253	.1556	.1004	.0605	.0339	.0176
	2	.0629	.1722	.2597	.3020	.3003	.2668	.2162	.1612	.1110	.0703
	3	.0077	.0446	.1069	.1762	.2336	.2668	.2716	.2508	.2119	.1641
	4	.0006	.0074	.0283	.0661	.1168	.1715	.2194	.2508	.2600	.2461

APPENDIX TABLE 1 (Continued)

		P									
n	x	.05	.10	.15	.20	.25	.30	.35	.40	.45	.50
9	5	.0000	.0008	.0050	.0165	.0389	.0735	.1181	.1672	.2128	.2461
	6	.0000	.0001	.0006	.0028	.0087	.0210	.0424	.0743	.1160	.1641
	7	.0000	.0000	.0000	.0003	.0012	.0039	.0098	.0212	.0407	.0703
	8	.0000	.0000	.0000	.0000	.0001	.0004	.0013	.0035	.0083	.0716
	9	.0000	.0000	.0000	.0000	.0000	.0000	.0001	.0003	.0008	.0020
10	0	.5987	.3487	.1969	.1074	.0563	.0282	.0135	.0060	.0025	.0010
	1	.3151	.3874	.3474	.2684	.1877	.1211	.0725	.0403	.0207	.0098
	2	.0746	.1937	.2759	.3020	.2816	.2335	.1757	.1209	.0763	.0439
	3	.0105	.0574	.1298	.2013	.2503	.2668	.2522	.2150	.1665	.1172
	4	.0010	.0112	.0401	.0881	.1460	.2001	.2377	.2508	.2384	.2051
	5	.0001	.0015	.0085	.0264	.0584	.1029	.1536	.2007	.2340	.2461
	6	.0000	.0001	.0012	.0055	.0162	.0368	.0689	.1115	.1596	.2051
	7	.0000	.0000	.0001	.0008	.0031	.0090	.0212	.0425	.0746	.1172
	8	.0000	.0000	.0000	.0001	.0004	.0014	.0043	.0106	.0229	.0439
	9	.0000	.0000	.0000	.0000	.0000	.0001	.0005	.0016	.0042	.0098
	10	.0000	.0000	.0000	.0000	.0000	.0000	.0000	.0001	.0003	.0010
11	0	.5688	.3138	.1673	.0859	.0422	.0198	.0088	.0036	.0014	.0005
	1	.3293	.3835	.3248	.2362	.1549	.0932	.0518	.0266	.0125	.0054
	2	.0867	.2131	.2866	.2953	.2581	.1998	.1395	.0887	.0513	.0269
	3	.0137	.0710	.1517	.2215	.2581	.2568	.2254	.1774	.1259	.0806
	4	.0014	.0158	.0536	.1107	.1721	.2201	.2428	.2365	.2060	.1611
	5	.0001	.0025	.0132	.0388	.0803	.1321	.1830	.2207	.2360	.2256
	6	.0000	.0003	.0023	.0097	.0268	.0566	.0985	.1471	.1931	.2256
	7	.0000	.0000	.0003	.0017	.0064	.0173	.0379	.0701	.1128	.1611
	8	.0000	.0000	.0000	.0002	.0011	.0037	.0102	.0234	.0462	.0806
	9	.0000	.0000	.0000	.0000	.0001	.0005	.0018	.0052	.0126	.0269
	10	.0000	.0000	.0000	.0000	.0000	.0000	.0002	.0007	.0021	.0054
	11	.0000	.0000	.0000	.0000	.0000	.0000	.0000	.0000	.0002	.0005
12	0	.5404	.2824	.1422	.0687	.0317	.0138	.0057	.0022	.0008	.0002
	1	.3413	.3766	.3012	.2062	.1267	.0712	.0368	.0174	.0075	.0029
	2	.0988	.2301	.2924	.2835	.2323	.1678	.1088	.0639	.0339	.0161
	3	.0173	.0852	.1720	.2362	.2581	.2397	.1954	.1419	.0923	.0537
	4	.0021	.0213	.0683	.1329	.1936	.2311	.2367	.2128	.1700	.1208
	5	.0002	.0038	.0193	.0532	.1032	.1585	.2039	.2270	.2225	.1934
	6	.0000	.0005	.0040	.0155	.0401	.0792	.1281	.1766	.2124	.2256
	7	.0000	.0000	.0006	.0033	.0115	.0291	.0591	.1009	.1489	.1934
	8	.0000	.0000	.0001	.0005	.0024	.0078	.0199	.0420	.0762	.1208
	9	.0000	.0000	.0000	.0001	.0004	.0015	.0048	.0125	.0277	.0537

APPENDIX TABLE 1 (Continued)

		P									
n	x	.05	.10	.15	.20	.25	.30	.35	.40	.45	.50
12	10	.0000	.0000	.0000	.0000	.0000	.0002	.0008	.0025	.0068	.0161
	11	.0000	.0000	.0000	.0000	.0000	.0000	.0001	.0003	.0010	.0029
	12	.0000	.0000	.0000	.0000	.0000	.0000	.0000	.0000	.0001	.0002
13	0	.5133	.2542	.1209	.0550	.0238	.0097	.0037	.0013	.0004	.0001
	1	.3512	.3672	.2774	.1787	.1029	.0540	.0259	.0113	.0045	.0016
	2	.1109	.2448	.2937	.2680	.2059	.1388	.0836	.0453	.0220	.0095
	3	.0214	.0997	.1900	.2457	.2517	.2181	.1651	.1107	.0660	.0349
	4	.0028	.0277	.0838	.1535	.2097	.2337	.2222	.1845	.1350	.0873
	5	.0003	.0055	.0266	.0691	.1258	.1803	.2154	.2214	.1989	.1571
	6	.0000	.0008	.0063	.0230	.0559	.1030	.1546	.1968	.2169	.2095
	7	.0000	.0001	.0011	.0058	.0186	.0442	.0833	.1312	.1775	.2095
	8	.0000	.0000	.0001	.0011	.0047	.0142	.0336	.0656	.1089	.1571
	9	.0000	.0000	.0000	.0001	.0009	.0034	.0101	.0243	.0495	.0873
	10	.0000	.0000	.0000	.0000	.0001	.0006	.0022	.0065	.0162	.0349
	11	.0000	.0000	.0000	.0000	.0000	.0001	.0003	.0012	.0036	.0095
	12	.0000	.0000	.0000	.0000	.0000	.0000	.0000	.0001	.0005	.0016
	13	.0000	.0000	.0000	.0000	.0000	.0000	.0000	.0000	.0000	.0001
14	0	.4877	.2288	.1028	.0440	.0178	.0068	.0024	.0008	.0002	.0001
	1	.3593	.3559	.2539	.1539	.0832	.0407	.0181	.0073	.0027	.0009
	2	.1229	.2570	.2912	.2501	.1802	.1134	.0634	.0317	.0141	.0056
	3	.0259	.1142	.2056	.2501	.2402	.1943	.1366	.0845	.0462	.0222
	4	.0037	.0348	.0998	.1720	.2202	.2290	.2022	.1549	.1040	.0611
	5	.0004	.0078	.0352	.0860	.1468	.1963	.2178	.2066	.1701	.1222
	6	.0000	.0013	.0093	.0322	.0734	.1262	.1759	.2066	.2088	.1833
	7	.0000	.0002	.0019	.0092	.0280	.0618	.1082	.1574	.1952	.2095
	8	.0000	.0000	.0003	.0020	.0082	.0232	.0510	.0918	.1398	.1833
	9	.0000	.0000	.0000	.0003	.0018	.0066	.0183	.0408	.0762	.1222
	10	.0000	.0000	.0000	.0000	.0003	.0014	.0049	.0136	.0312	.0611
	11	.0000	.0000	.0000	.0000	.0000	.0002	.0010	.0033	.0093	.0222
	12	.0000	.0000	.0000	.0000	.0000	.0000	.0001	.0005	.0019	.0056
	13	.0000	.0000	.0000	.0000	.0000	.0000	.0000	.0001	.0002	.0009
	14	.0000	.0000	.0000	.0000	.0000	.0000	.0000	.0000	.0000	.0001
15	0	.4633	.2059	.0874	.0352	.0134	.0047	.0016	.0005	.0001	.0000
	1	.3658	.3432	.2312	.1319	.0668	.0305	.0126	.0047	.0016	.0005
	2	.1348	.2669	.2856	.2309	.1559	.0916	.0476	.0219	.0090	.0032
	3	.0307	.1285	.2184	.2501	.2252	.1700	.1110	.0634	.0318	.0139
	4	.0049	.0428	.1156	.1876	.2252	.2186	.1792	.1268	.0780	.0417

APPENDIX TABLE 1 (Continued)

		P									
n	x	.05	.10	.15	.20	.25	.30	.35	.40	.45	.50
15	5	.0006	.0105	.0449	.1032	.1651	.2061	.2123	.1859	.1404	.0916
	6	.0000	.0019	.0132	.0430	.0917	.1472	.1906	.2066	.1914	.1527
	7	.0000	.0003	.0030	.0138	.0393	.0811	.1319	.1771	.2013	.1964
	8	.0000	.0000	.0005	.0035	.0131	.0348	.0710	.1181	.1647	.1964
	9	.0000	.0000	.0001	.0007	.0034	.0116	.0298	.0612	.1048	.1527
	10	.0000	.0000	.0000	.0001	.0007	.0030	.0096	.0245	.0515	.0916
	11	.0000	.0000	.0000	.0000	.0001	.0006	.0024	.0074	.0191	.0417
	12	.0000	.0000	.0000	.0000	.0000	.0001	.0004	.0016	.0052	.0139
	13	.0000	.0000	.0000	.0000	.0000	.0000	.0001	.0003	.0010	.0032
	14	.0000	.0000	.0000	.0000	.0000	.0000	.0000	.0000	.0001	.0005
	15	.0000	.0000	.0000	.0000	.0000	.0000	.0000	.0000	.0000	.0000
16	0	.4401	.1853	.0743	.0281	.0100	.0033	.0010	.0003	.0001	.0000
	1	.3706	.3294	.2097	.1126	.0535	.0228	.0087	.0030	.0009	.0002
	2	.1463	.2745	.2775	.2111	.1336	.0732	.0353	.0150	.0056	.0018
	3	.0359	.1423	.2285	.2463	.2079	.1465	.0888	.0468	.0215	.0085
	4	.0061	.0514	.1311	.2001	.2252	.2040	.1553	.1014	.0572	.0278
	5	.0008	.0137	.0555	.1201	.1802	.2099	.2008	.1623	.1123	.0667
	6	.0001	.0028	.0180	.0550	.1101	.1649	.1982	.1983	.1684	.1222
	7	.0000	.0004	.0045	.0197	.0524	.1010	.1524	.1889	.1969	.1746
	8	.0000	.0001	.0009	.0055	.0197	.0487	.0923	.1417	.1812	.1964
	9	.0000	.0000	.0001	.0012	.0058	.0185	.0442	.0840	.1318	.1746
	10	.0000	.0000	.0000	.0002	.0014	.0056	.0167	.0392	.0755	.1222
	11	.0000	.0000	.0000	.0000	.0002	.0013	.0049	.0142	.0337	.0667
	12	.0000	.0000	.0000	.0000	.0000	.0002	.0011	.0040	.0115	.0278
	13	.0000	.0000	.0000	.0000	.0000	.0000	.0002	.0008	.0029	.0085
	14	.0000	.0000	.0000	.0000	.0000	.0000	.0000	.0001	.0005	.0018
	15	.0000	.0000	.0000	.0000	.0000	.0000	.0000	.0000	.0001	.0002
	16	.0000	.0000	.0000	.0000	.0000	.0000	.0000	.0000	.0000	.0000
17	0	.4181	.1668	.0631	.0225	.0075	.0023	.0007	.0002	.0000	.0000
	1	.3741	.3150	.1893	.0957	.0426	.0169	.0060	.0019	.0005	.0001
	2	.1575	.2800	.2673	.1914	.1136	.0581	.0260	.0102	.0035	.0010
	3	.0415	.1556	.2359	.2393	.1893	.1245	.0701	.0341	.0144	.0052
	4	.0076	.0605	.1457	.2093	.2209	.1868	.1320	.0796	.0411	.0182

APPENDIX TABLE 1 (Continued)

		P									
n	*x*	.05	.10	.15	.20	.25	.30	.35	.40	.45	.50
17	5	.0010	.0175	.0668	.1361	.1914	.2081	.1849	.1379	.0875	.0472
	6	.0001	.0039	.0236	.0680	.1276	.1784	.1991	.1839	.1432	.0944
	7	.0000	.0007	.0065	.0267	.0668	.1201	.1685	.1927	.1841	.1484
	8	.0000	.0001	.0014	.0084	.0279	.0644	.1134	.1606	.1883	.1855
	9	.0000	.0000	.0003	.0021	.0093	.0276	.0611	.1070	.1540	.1855
	10	.0000	.0000	.0000	.0004	.0025	.0095	.0263	.0571	.1008	.1484
	11	.0000	.0000	.0000	.0001	.0005	.0026	.0090	.0242	.0525	.0944
	12	.0000	.0000	.0000	.0000	.0001	.0006	.0024	.0021	.0215	.0472
	13	.0000	.0000	.0000	.0000	.0000	.0001	.0005	.0021	.0068	.0182
	14	.0000	.0000	.0000	.0000	.0000	.0000	.0001	.0004	.0016	.0052
	15	.0000	.0000	.0000	.0000	.0000	.0000	.0000	.0001	.0003	.0010
	16	.0000	.0000	.0000	.0000	.0000	.0000	.0000	.0000	.0000	.0001
	17	.0000	.0000	.0000	.0000	.0000	.0000	.0000	.0000	.0000	.0000
18	0	.3972	.1501	.0536	.0180	.0056	.0016	.0004	.0001	.0000	.0000
	1	.3763	.3002	.1704	.0811	.0338	.0126	.0042	.0012	.0003	.0001
	2	.1683	.2835	.2556	.1723	.0958	.0458	.0190	.0069	.0022	.0006
	3	.0473	.1680	.2406	.2297	.1704	.1046	.0547	.0246	.0095	.0031
	4	.0093	.0700	.1592	.2153	.2130	.1681	.1104	.0614	.0291	.0117
	5	.0014	.0218	.0787	.1507	.1988	.2017	.1664	.1146	.0666	.0327
	6	.0002	.0052	.0301	.0816	.1436	.1873	.1941	.1655	.1181	.0708
	7	.0000	.0010	.0091	.0350	.0820	.1376	.1792	.1892	.1657	.1214
	8	.0000	.0002	.0022	.0120	.0376	.0811	.1327	.1734	.1864	.1669
	9	.0000	.0000	.0004	.0033	.0139	.0386	.0794	.1284	.1694	.1855
	10	.0000	.0000	.0001	.0008	.0042	.0149	.0385	.0771	.1248	.1669
	11	.0000	.0000	.0000	.0001	.0010	.0046	.0151	.0374	.0742	.1214
	12	.0000	.0000	.0000	.0000	.0002	.0012	.0047	.0145	.0354	.0708
	13	.0000	.0000	.0000	.0000	.0000	.0002	.0012	.0044	.0134	.0327
	14	.0000	.0000	.0000	.0000	.0000	.0000	.0002	.0011	.0039	.0117
	15	.0000	.0000	.0000	.0000	.0000	.0000	.0000	.0002	.0009	.0031
	16	.0000	.0000	.0000	.0000	.0000	.0000	.0000	.0000	.0001	.0006
	17	.0000	.0000	.0000	.0000	.0000	.0000	.0000	.0000	.0000	.0001
	18	.0000	.0000	.0000	.0000	.0000	.0000	.0000	.0000	.0000	.0000
19	0	.3774	.1351	.0456	.0144	.0042	.0011	.0003	.0001	.0000	.0000
	1	.3774	.2852	.1529	.0685	.0268	.0093	.0029	.0008	.0002	.0000
	2	.1787	.2852	.2428	.1540	.0803	.0358	.0138	.0046	.0013	.0003
	3	.0533	.1796	.2428	.2182	.1517	.0869	.0422	.0175	.0062	.0018
	4	.0112	.0798	.1714	.2182	.2023	.1491	.0909	.0467	.0203	.0074

APPENDIX TABLE 1 (Continued)

		P									
n	x	.05	.10	.15	.20	.25	.30	.35	.40	.45	.50
19	5	.0018	.0266	.0907	.1636	.2023	.1916	.1468	.0933	.0497	.0222
	6	.0002	.0069	.0374	.0955	.1574	.1916	.1844	.1451	.0949	.0518
	7	.0000	.0014	.0122	.0443	.0974	.1525	.1844	.1797	.1443	.0961
	8	.0000	.0002	.0032	.0166	.0487	.0981	.1489	.1797	.1771	.1442
	9	.0000	.0000	.0007	.0051	.0198	.0514	.0980	.1464	.1771	.1762
	10	.0000	.0000	.0001	.0013	.0066	.0220	.0528	.0976	.1449	.1762
	11	.0000	.0000	.0000	.0003	.0018	.0077	.0233	.0532	.0970	.1442
	12	.0000	.0000	.0000	.0000	.0004	.0022	.0083	.0237	.0529	.0961
	13	.0000	.0000	.0000	.0000	.0001	.0005	.0024	.0085	.0233	.0518
	14	.0000	.0000	.0000	.0000	.0000	.0001	.0006	.0024	.0082	.0222
	15	.0000	.0000	.0000	.0000	.0000	.0000	.0001	.0005	.0022	.0074
	16	.0000	.0000	.0000	.0000	.0000	.0000	.0000	.0001	.0005	.0018
	17	.0000	.0000	.0000	.0000	.0000	.0000	.0000	.0000	.0001	.0003
	18	.0000	.0000	.0000	.0000	.0000	.0000	.0000	.0000	.0000	.0000
	19	.0000	.0000	.0000	.0000	.0000	.0000	.0000	.0000	.0000	.0000
20	0	.3585	.1216	.0388	.0115	.0032	.0008	.0002	.0000	.0000	.0000
	1	.3774	.2702	.1368	.0576	.0211	.0068	.0020	.0005	.0001	.0000
	2	.1887	.2852	.2293	.1369	.0669	.0278	.0100	.0031	.0008	.0002
	3	.0596	.1901	.2428	.2054	.1339	.0716	.0323	.0123	.0040	.0011
	4	.0133	.0898	.1821	.2182	.1897	.1304	.0738	.0350	.0139	.0046
	5	.0022	.0319	.1028	.1746	.2023	.1789	.1272	.0746	.0365	.0148
	6	.0003	.0089	.0454	.1091	.1686	.1916	.1712	.1244	.0746	.0370
	7	.0000	.0020	.0160	.0545	.1124	.1643	.1844	.1659	.1221	.0739
	8	.0000	.0004	.0046	.0222	.0609	.1144	.1614	.1797	.1623	.1201
	9	.0000	.0001	.0011	.0074	.0271	.0654	.1158	.1597	.1771	.1602
	10	.0000	.0000	.0002	.0020	.0099	.0308	.0686	.1171	.1593	.1762
	11	.0000	.0000	.0000	.0005	.0030	.0120	.0336	.0710	.1185	.1602
	12	.0000	.0000	.0000	.0001	.0008	.0039	.0136	.0355	.0727	.1201
	13	.0000	.0000	.0000	.0000	.0002	.0010	.0045	.0146	.0366	.0739
	14	.0000	.0000	.0000	.0000	.0000	.0002	.0012	.0049	.0150	.0370
	15	.0000	.0000	.0000	.0000	.0000	.0000	.0003	.0013	.0049	.0148
	16	.0000	.0000	.0000	.0000	.0000	.0000	.0000	.0003	.0013	.0046
	17	.0000	.0000	.0000	.0000	.0000	.0000	.0000	.0000	.0002	.0011
	18	.0000	.0000	.0000	.0000	.0000	.0000	.0000	.0000	.0000	.0002
	19	.0000	.0000	.0000	.0000	.0000	.0000	.0000	.0000	.0000	.0000
	20	.0000	.0000	.0000	.0000	.0000	.0000	.0000	.0000	.0000	.0000

SOURCE: This table is taken from National Bureau of Standards, *Tables of the Binomial Probability Distribution*, Applied Mathematics Series, (U.S. Department of Commerce, 1950).

APPENDIX TABLE 2

Areas under the Standard Normal Curve

This table shows the area between zero (the mean of a standard normal variable) and z. For example, if $z = 1.50$, this is the shaded area shown below which equals .4332.

0 1.50

z	.00	.01	.02	.03	.04	.05	.06	.07	.08	.09
0.0	.0000	.0040	.0080	.0120	.0160	.0199	.0239	.0279	.0319	.0359
0.1	.0398	.0438	.0478	.0517	.0557	.0596	.0636	.0675	.0714	.0753
0.2	.0793	.0832	.0871	.0910	.0948	.0987	.1026	.1064	.1103	.1141
0.3	.1179	.1217	.1255	.1293	.1331	.1368	.1406	.1443	.1480	.1517
0.4	.1554	.1591	.1628	.1664	.1700	.1736	.1772	.1808	.1844	.1879
0.5	.1915	.1950	.1985	.2019	.2054	.2088	.2123	.2157	.2190	.2224
0.6	.2257	.2291	.2324	.2357	.2389	.2422	.2454	.2486	.2517	.2549
0.7	.2580	.2611	.2642	.2673	.2704	.2734	.2764	.2794	.2823	.2852
0.8	.2881	.2910	.2939	.2967	.2995	.3023	.3051	.3078	.3106	.3133
0.9	.3159	.3186	.3212	.3238	.3264	.3289	.3315	.3340	.3365	.3389
1.0	.3413	.3438	.3461	.3485	.3508	.3531	.3554	.3577	.3599	.3621
1.1	.3643	.3665	.3686	.3708	.3729	.3749	.3770	.3790	.3810	.3830
1.2	.3849	.3869	.3888	.3907	.3925	.3944	.3962	.3980	.3997	.4015
1.3	.4032	.4049	.4066	.4082	.4099	.4115	.4131	.4147	.4162	.4177
1.4	.4192	.4207	.4222	.4236	.4251	.4265	.4279	.4292	.4306	.4319
1.5	.4332	.4345	.4357	.4370	.4382	.4394	.4406	.4418	.4429	.4441
1.6	.4452	.4463	.4474	.4484	.4495	.4505	.4515	.4525	.4535	.4545
1.7	.4554	.4564	.4573	.4582	.4591	.4599	.4608	.4616	.4625	.4633
1.8	.4641	.4649	.4656	.4664	.4671	.4678	.4686	.4693	.4699	.4706
1.9	.4713	.4719	.4726	.4732	.4738	.4744	.4750	.4756	.4761	.4767
2.0	.4772	.4778	.4783	.4788	.4793	.4798	.4803	.4808	.4812	.4817
2.1	.4821	.4826	.4830	.4834	.4838	.4842	.4846	.4850	.4854	.4857
2.2	.4861	.4864	.4868	.4871	.4875	.4878	.4881	.4884	.4887	.4890
2.3	.4893	.4896	.4898	.4901	.4904	.4906	.4909	.4911	.4913	.4916
2.4	.4918	.4920	.4922	.4925	.4927	.4929	.4931	.4932	.4934	.4936
2.5	.4938	.4940	.4941	.4943	.4945	.4946	.4948	.4949	.4951	.4952
2.6	.4953	.4955	.4956	.4957	.4959	.4960	.4961	.4962	.4963	.4964
2.7	.4965	.4966	.4967	.4968	.4969	.4970	.4971	.4972	.4973	.4974
2.8	.4974	.4975	.4976	.4977	.4977	.4978	.4979	.4979	.4980	.4981
2.9	.4981	.4982	.4982	.4983	.4984	.4984	.4985	.4985	.4986	.4986
3.0	.4987	.4987	.4987	.4988	.4988	.4989	.4989	.4989	.4990	.4990

SOURCE: This table is adapted from National Bureau of Standards, *Tables of Normal Probability Functions*, Applied Mathematics Series 23 (U.S. Department of Commerce, 1953).

APPENDIX TABLE 3
Poisson Probability Distribution

This table shows the value of

$$P(x) = \frac{\mu^x e^{-\mu}}{x!}$$

for selected values of x and for μ = .005 to 8.0.

	μ									
x	.005	.01	.02	.03	.04	.05	.06	.07	.08	.09
0	.9950	.9900	.9802	.9704	.9608	.9512	.9418	.9324	.9231	.9139
1	.0050	.0099	.0192	.0291	.0384	.0476	.0565	.0653	.0738	.0823
2	.0000	.0000	.0002	.0004	.0008	.0012	.0017	.0023	.0030	.0037
3	.0000	.0000	.0000	.0000	.0000	.0000	.0000	.0001	.0001	.0001

	μ									
x	0.1	0.2	0.3	0.4	0.5	0.6	0.7	0.8	0.9	1.0
0	.9048	.8187	.7408	.6703	.6065	.5488	.4966	.4493	.4066	.3679
1	.0905	.1637	.2222	.2681	.3033	.3293	.3476	.3595	.3659	.3679
2	.0045	.0164	.0333	.0536	.0758	.0988	.1217	.1438	.1647	.1839
3	.0002	.0011	.0033	.0072	.0126	.0198	.0284	.0383	.0494	.0613
4	.0000	.0001	.0002	.0007	.0016	.0030	.0050	.0077	.0111	.0153
5	.0000	.0000	.0000	.0001	.0002	.0004	.0007	.0012	.0020	.0031
6	.0000	.0000	.0000	.0000	.0000	.0000	.0001	.0002	.0003	.0005
7	.0000	.0000	.0000	.0000	.0000	.0000	.0000	.0000	.0000	.0001

	μ									
x	1.1	1.2	1.3	1.4	1.5	1.6	1.7	1.8	1.9	2.0
0	.3329	.3012	.2725	.2466	.2231	.2019	.1827	.1653	.1496	.1353
1	.3662	.3614	.3543	.3452	.3347	.3230	.3106	.2975	.2842	.2707
2	.2014	.2169	.2303	.2417	.2510	.2584	.2640	.2678	.2700	.2707
3	.0738	.0867	.0998	.1128	.1255	.1378	.1496	.1607	.1710	.1804
4	.0203	.0260	.0324	.0395	.0471	.0551	.0636	.0723	.0812	.0902
5	.0045	.0062	.0084	.0111	.0141	.0176	.0216	.0260	.0309	.0361
6	.0008	.0012	.0018	.0026	.0035	.0047	.0061	.0078	.0098	.0120
7	.0001	.0002	.0003	.0005	.0008	.0011	.0015	.0020	.0027	.0034
8	.0000	.0000	.0001	.0001	.0001	.0002	.0003	.0005	.0006	.0009
9	.0000	.0000	.0000	.0000	.0000	.0000	.0001	.0001	.0001	.0002

APPENDIX TABLE 3 (Continued)

	μ									
x	2.1	2.2	2.3	2.4	2.5	2.6	2.7	2.8	2.9	3.0
0	.1225	.1108	.1003	.0907	.0821	.0743	.0672	.0608	.0550	.0498
1	.2572	.2438	.2306	.2177	.2052	.1931	.1815	.1703	.1596	.1494
2	.2700	.2681	.2652	.2613	.2565	.2510	.2450	.2384	.2314	.2240
3	.1890	.1966	.2033	.2090	.2138	.2176	.2205	.2225	.2237	.2240
4	.0992	.1082	.1169	.1254	.1336	.1414	.1488	.1557	.1622	.1680
5	.0417	.0476	.0538	.0602	.0668	.0735	.0804	.0872	.0940	.1008
6	.0146	.0174	.0206	.0241	.0278	.0319	.0362	.0407	.0455	.0504
7	.0044	.0055	.0068	.0083	.0099	.0118	.0139	.0163	.0188	.0216
8	.0011	.0015	.0019	.0025	.0031	.0038	.0047	.0057	.0068	.0081
9	.0003	.0004	.0005	.0007	.0009	.0011	.0014	.0018	.0022	.0027
10	.0001	.0001	.0001	.0002	.0002	.0003	.0004	.0005	.0006	.0008
11	.0000	.0000	.0000	.0000	.0000	.0001	.0001	.0001	.0002	.0002
12	.0000	.0000	.0000	.0000	.0000	.0000	.0000	.0000	.0000	.0001

	μ									
x	3.1	3.2	3.3	3.4	3.5	3.6	3.7	3.8	3.9	4.0
0	.0450	.0408	.0369	.0334	.0302	.0273	.0247	.0224	.0202	.0183
1	.1397	.1304	.1217	.1135	.1057	.0984	.0915	.0850	.0789	.0733
2	.2165	.2087	.2008	.1929	.1850	.1771	.1692	.1615	.1539	.1465
3	.2237	.2226	.2209	.2186	.2158	.2125	.2087	.2046	.2001	.1954
4	.1734	.1781	.1823	.1858	.1888	.1912	.1931	.1944	.1951	.1954
5	.1075	.1140	.1203	.1264	.1322	.1377	.1429	.1477	.1522	.1563
6	.0555	.0608	.0662	.0716	.0771	.0826	.0881	.0936	.0989	.1042
7	.0246	.0278	.0312	.0348	.0385	.0425	.0466	.0508	.0551	.0595
8	.0095	.0111	.0129	.0148	.0169	.0191	.0215	.0241	.0269	.0298
9	.0033	.0040	.0047	.0056	.0066	.0076	.0089	.0102	.0116	.0132
10	.0010	.0013	.0016	.0019	.0023	.0028	.0033	.0039	.0045	.0053
11	.0003	.0004	.0005	.0006	.0007	.0009	.0011	.0013	.0016	.0019
12	.0001	.0001	.0001	.0002	.0002	.0003	.0003	.0004	.0005	.0006
13	.0000	.0000	.0000	.0000	.0001	.0001	.0001	.0001	.0002	.0002
14	.0000	.0000	.0000	.0000	.0000	.0000	.0000	.0000	.0000	.0001

	μ									
x	4.1	4.2	4.3	4.4	4.5	4.6	4.7	4.8	4.9	5.0
0	.0166	.0150	.0136	.0123	.0111	.0101	.0091	.0082	.0074	.0067
1	.0679	.0630	.0583	.0540	.0500	.0462	.0427	.0395	.0365	.0337
2	.1393	.1323	.1254	.1188	.1125	.1063	.1005	.0948	.0894	.0842
3	.1904	.1852	.1798	.1743	.1687	.1631	.1574	.1517	.1460	.1404
4	.1951	.1944	.1933	.1917	.1898	.1875	.1849	.1820	.1789	.1755

APPENDIX TABLE 3 (Continued)

x	μ 4.1	4.2	4.3	4.4	4.5	4.6	4.7	4.8	4.9	5.0
5	.1600	.1633	.1662	.1687	.1708	.1725	.1738	.1747	.1753	.1755
6	.1093	.1143	.1191	.1237	.1281	.1323	.1362	.1398	.1432	.1462
7	.0640	.0686	.0732	.0778	.0824	.0869	.0914	.0959	.1002	.1044
8	.0328	.0360	.0393	.0428	.0463	.0500	.0537	.0575	.0614	.0653
9	.0150	.0168	.0188	.0209	.0232	.0255	.0280	.0307	.0334	.0363
10	.0061	.0071	.0081	.0092	.0104	.0118	.0132	.0147	.0164	.0181
11	.0023	.0027	.0032	.0037	.0043	.0049	.0056	.0064	.0073	.0082
12	.0008	.0009	.0011	.0014	.0016	.0019	.0022	.0026	.0030	.0034
13	.0002	.0003	.0004	.0005	.0006	.0007	.0008	.0009	.0011	.0013
14	.0001	.0001	.0001	.0001	.0002	.0002	.0003	.0003	.0004	.0005
15	.0000	.0000	.0000	.0000	.0001	.0001	.0001	.0001	.0001	.0002

x	μ 5.1	5.2	5.3	5.4	5.5	5.6	5.7	5.8	5.9	6.0
0	.0061	.0055	.0050	.0045	.0041	.0037	.0033	.0030	.0027	.0025
1	.0311	.0287	.0265	.0244	.0225	.0207	.0191	.0176	.0162	.0149
2	.0793	.0746	.0701	.0659	.0618	.0580	.0544	.0509	.0477	.0446
3	.1348	.1293	.1239	.1185	.1133	.1082	.1033	.0985	.0938	.0892
4	.1719	.1681	.1641	.1600	.1558	.1515	.1472	.1428	.1383	.1339
5	.1753	.1748	.1740	.1728	.1714	.1697	.1678	.1656	.1632	.1606
6	.1490	.1515	.1537	.1555	.1571	.1584	.1594	.1601	.1605	.1606
7	.1086	.1125	.1163	.1200	.1234	.1267	.1298	.1326	.1353	.1377
8	.0692	.0731	.0771	.0810	.0849	.0887	.0925	.0962	.0998	.1033
9	.0392	.0423	.0454	.0486	.0519	.0552	.0586	.0620	.0654	.0688
10	.0200	.0220	.0241	.0262	.0285	.0309	.0334	.0359	.0386	.0413
11	.0093	.0104	.0116	.0129	.0143	.0157	.0173	.0190	.0207	.0225
12	.0039	.0045	.0051	.0058	.0065	.0073	.0082	.0092	.0102	.0113
13	.0015	.0018	.0021	.0024	.0028	.0032	.0036	.0041	.0046	.0052
14	.0006	.0007	.0008	.0009	.0011	.0013	.0015	.0017	.0019	.0022
15	.0002	.0002	.0003	.0003	.0004	.0005	.0006	.0007	.0008	.0009
16	.0001	.0001	.0001	.0001	.0001	.0002	.0002	.0002	.0003	.0003
17	.0000	.0000	.0000	.0000	.0000	.0001	.0001	.0001	.0001	.0001

x	μ 6.1	6.2	6.3	6.4	6.5	6.6	6.7	6.8	6.9	7.0
0	.0022	.0020	.0018	.0017	.0015	.0014	.0012	.0011	.0010	.0009
1	.0137	.0126	.0116	.0106	.0098	.0090	.0082	.0076	.0070	.0064
2	.0417	.0390	.0364	.0340	.0318	.0296	.0276	.0258	.0240	.0223
3	.0848	.0806	.0765	.0726	.0688	.0652	.0617	.0584	.0552	.0521
4	.1294	.1249	.1205	.1162	.1118	.1076	.1034	.0992	.0952	.0912

APPENDIX TABLE 3 (Continued)

					μ					
x	6.1	6.2	6.3	6.4	6.5	6.6	6.7	6.8	6.9	7.0
5	.1579	.1549	.1519	.1487	.1454	.1420	.1385	.1349	.1314	.1277
6	.1605	.1601	.1595	.1586	.1575	.1562	.1546	.1529	.1511	.1490
7	.1399	.1418	.1435	.1450	.1462	.1472	.1480	.1486	.1489	.1490
8	.1066	.1099	.1130	.1160	.1188	.1215	.1240	.1263	.1284	.1304
9	.0723	.0757	.0791	.0825	.0858	.0891	.0923	.0954	.0985	.1014
10	.0441	.0469	.0498	.0528	.0558	.0588	.0618	.0649	.0679	.0710
11	.0245	.0265	.0285	.0307	.0330	.0353	.0377	.0401	.0426	.0452
12	.0124	.0137	.0150	.0164	.0179	.0194	.0210	.0227	.0245	.0264
13	.0058	.0065	.0073	.0081	.0089	.0098	.0108	.0119	.0130	.0142
14	.0025	.0029	.0033	.0037	.0041	.0046	.0052	.0058	.0064	.0071
15	.0010	.0012	.0014	.0016	.0018	.0020	.0023	.0026	.0029	.0033
16	.0004	.0005	.0005	.0006	.0007	.0008	.0010	.0011	.0013	.0014
17	.0001	.0002	.0002	.0002	.0003	.0003	.0004	.0004	.0005	.0006
18	.0000	.0001	.0001	.0001	.0001	.0001	.0001	.0002	.0002	.0002
19	.0000	.0000	.0000	.0000	.0000	.0000	.0000	.0001	.0001	.0001

					μ					
x	7.1	7.2	7.3	7.4	7.5	7.6	7.7	7.8	7.9	8.0
0	.0008	.0007	.0007	.0006	.0006	.0005	.0005	.0004	.0004	.0003
1	.0059	.0054	.0049	.0045	.0041	.0038	.0035	.0032	.0029	.0027
2	.0208	.0194	.0180	.0167	.0156	.0145	.0134	.0125	.0116	.0107
3	.0492	.0464	.0438	.0413	.0389	.0366	.0345	.0324	.0305	.0286
4	.0874	.0836	.0799	.0764	.0729	.0696	.0663	.0632	.0602	.0573
5	.1241	.1204	.1167	.1130	.1094	.1057	.1021	.0986	.0951	.0916
6	.1468	.1445	.1420	.1394	.1367	.1339	.1311	.1282	.1252	.1221
7	.1489	.1486	.1481	.1474	.1465	.1454	.1442	.1428	.1413	.1396
8	.1321	.1337	.1351	.1363	.1373	.1382	.1388	.1392	.1395	.1396
9	.1042	.1070	.1096	.1121	.1144	.1167	.1187	.1207	.1224	.1241
10	.0740	.0770	.0800	.0829	.0858	.0887	.0914	.0941	.0967	.0993
11	.0478	.0504	.0531	.0558	.0585	.0613	.0640	.0667	.0695	.0722
12	.0283	.0303	.0323	.0344	.0366	.0388	.0411	.0434	.0457	.0481
13	.0154	.0168	.0181	.0196	.0211	.0227	.0243	.0260	.0278	.0296
14	.0078	.0086	.0095	.0104	.0113	.0123	.0134	.0145	.0157	.0169
15	.0037	.0041	.0046	.0051	.0057	.0062	.0069	.0075	.0083	.0090
16	.0016	.0019	.0021	.0024	.0026	.0030	.0033	.0037	.0041	.0045
17	.0007	.0008	.0009	.0010	.0012	.0013	.0015	.0017	.0019	.0021
18	.0003	.0003	.0004	.0004	.0005	.0006	.0006	.0007	.0008	.0009
19	.0001	.0001	.0001	.0002	.0002	.0002	.0003	.0003	.0003	.0004
20	.0000	.0000	.0001	.0001	.0001	.0001	.0001	.0001	.0001	.0002
21	.0000	.0000	.0000	.0000	.0000	.0000	.0000	.0000	.0001	.0001

APPENDIX TABLE 4

Values of e^{-x}

x	e^{-x}	x	e^{-x}	x	e^{-x}
.0	1.000	1.5	.223	3.0	.050
.1	.905	1.6	.202	3.1	.045
.2	.819	1.7	.183	3.2	.041
.3	.741	1.8	.165	3.3	.037
.4	.670	1.9	.150	3.4	.033
.5	.607	2.0	.135	3.5	.030
.6	.549	2.1	.122	3.6	.027
.7	.497	2.2	.111	3.7	.025
.8	.449	2.3	.100	3.8	.022
.9	.407	2.4	.091	3.9	.020
1.0	.368	2.5	.082	4.0	.018
1.1	.333	2.6	.074	4.5	.011
1.2	.301	2.7	.067	5.0	.007
1.3	.273	2.8	.061	6.0	.002
1.4	.247	2.9	.055	7.0	.001

APPENDIX TABLE 5

Random Numbers

11850	53535	04260	77609	93799	92171	45524	10968	30231	70864	29908
11851	41292	15201	66342	59155	46163	69248	31029	62034	21855	27863
11852	07320	22682	09595	44805	54593	53350	61354	14029	10195	18644
11853	77676	67772	45072	08940	02592	45976	82099	90739	77072	42081
11854	43227	20568	16309	23841	53173	39475	27282	82699	00022	96419
11855	90712	41695	67474	27567	93269	10163	94190	36188	41491	71217
11856	88103	21514	60787	33170	58215	89951	01634	98155	05154	08971
11857	72252	35791	84125	31962	81093	93068	41197	57779	88515	48002
11858	51702	49516	69510	19678	47298	11355	68459	96360	13436	66314
11859	63055	86998	22187	59898	96371	61370	35937	34292	00678	33505
11860	32373	57889	85880	66515	37489	37854	72926	23437	62233	38651
11861	71996	16525	25618	56577	69130	25035	93551	54394	81572	90624
11862	26912	70619	22576	22780	99118	18487	58801	36063	32886	60453
11863	74589	82677	13353	67658	17080	43212	34585	17179	86980	81899
11864	56041	53072	19912	47466	32585	41414	07564	80712	27286	07966
11865	09286	68067	84883	10023	78195	84711	85988	31545	39904	14984
11866	33610	84843	07145	38437	06148	06094	89601	96751	49124	55092
11867	14113	06396	59084	02534	09360	81918	77118	91640	92978	24815
11868	56302	89765	63857	42747	28592	41784	00822	60356	96389	11728
11869	06362	94540	29532	09994	55277	43897	63268	40481	00312	46039
11870	48568	34412	84939	54850	84317	92032	60430	49071	68962	28953
11871	65975	60965	77679	95782	67541	50654	09482	56111	98710	35803
11872	66686	32977	48472	30226	54226	72490	18395	37338	88279	79089
11873	51610	13000	73849	46654	30324	78000	72852	28934	83197	59003
11874	47600	86103	25788	08774	72020	04543	25849	88887	41159	30131
11875	34860	67572	83116	99579	81303	41889	56577	64142	51596	25329
11876	76649	50908	67006	29332	29689	68786	98987	34815	53512	20620
11877	78321	54309	85956	04976	37863	06711	72679	03405	28770	08515
11878	35775	21295	39621	02339	16537	42246	06571	81193	94930	05376
11879	06783	21338	89886	78826	02303	37886	70453	11021	62887	36855
11880	25887	53024	71881	51208	95739	98572	01903	68043	62661	71273
11881	37784	42100	70838	78963	10927	05448	25759	74051	47577	30196
11882	02120	59536	82996	22671	89267	65924	46725	69179	15182	59158
11883	55292	03836	28883	71134	08547	93204	09656	11671	29735	59573
11884	66186	43648	97926	80469	66412	73647	36779	84688	96862	51937
11885	55010	11479	55036	82146	37120	62328	56276	28906	45311	61818
11886	02322	18679	18478	30052	05666	84405	47513	09244	78978	91819
11887	78056	67836	82582	25809	20198	37222	62629	75733	77420	58746
11888	69812	88260	83519	10062	60865	35038	14665	18163	59351	25794
11889	84904	66864	26982	37928	32988	87652	81415	24416	93778	20391

APPENDIX TABLE 5 (Continued)

11890	83143	47631	79772	08576	10311	17597	71049	63326	47168	05737
11891	44423	71197	91081	40781	72403	76245	31881	55716	89255	71997
11892	59882	58479	59609	80115	91569	23152	51781	85744	78640	80172
11893	74890	90405	75945	31645	61008	24448	42249	84909	29013	12529
11894	52174	64334	77631	19855	17723	02897	80427	20700	92210	92091
11895	41361	24347	53420	33639	83765	97935	83630	33765	21502	15589
11896	94585	84798	98480	08335	08728	60428	22282	76784	37316	08624
11897	36020	71966	61443	12554	67446	08676	46177	22422	87471	27283
11898	08112	59807	28404	60316	49676	52901	90604	48379	85233	52060
11899	05853	69681	52034	77617	78644	57321	14162	01849	94684	14628

SOURCE: The Rand Corporation, *A Million Random Digits with 100,000 Normal Deviates* (New York: Free Press, 1955), p. 238.

APPENDIX TABLE 6

Values of *t* That Will Be Exceeded with Specified Probabilities

This table shows the value of *t* where the area under the *t* distribution exceeding this value of *t* equals the specified amount. For example, the probability that a *t* variable with 14 degrees of freedom will exceed 1.345 equals .10.

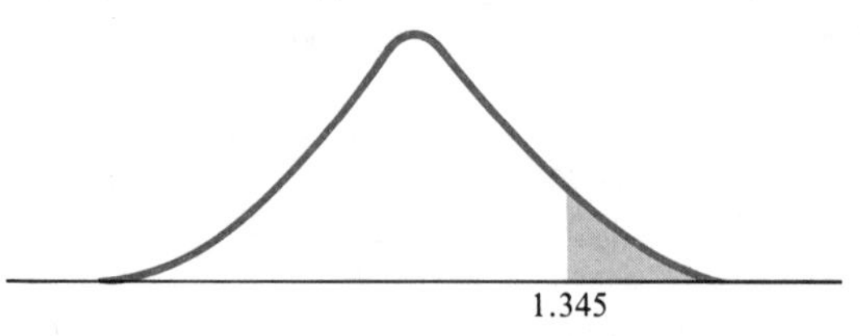

Degrees of freedom	Probability						
	0.40	0.25	0.10	0.05	0.025	0.01	0.005
1	0.325	1.000	3.078	6.314	12.706	31.821	63.657
2	.289	0.816	1.886	2.920	4.303	6.965	9.925
3	.277	.765	1.638	2.353	3.182	4.541	5.841
4	.271	.741	1.533	2.132	2.776	3.747	4.604
5	0.267	0.727	1.476	2.015	2.571	3.365	4.032
6	.265	.718	1.440	1.943	2.447	3.143	3.707
7	.263	.711	1.415	1.895	2.365	2.998	3.499
8	.262	.706	1.397	1.860	2.306	2.896	3.355
9	.261	.703	1.383	1.833	2.262	2.821	3.250
10	0.260	0.700	1.372	1.812	2.228	2.764	3.169
11	.260	.697	1.363	1.796	2.201	2.718	3.106
12	.259	.695	1.356	1.782	2.179	2.681	3.055
13	.259	.694	1.350	1.771	2.160	2.650	3.012
14	.258	.692	1.345	1.761	2.145	2.624	2.977
15	0.258	0.691	1.341	1.753	2.131	2.602	2.947
16	.258	.690	1.337	1.746	2.120	2.583	2.921
17	.257	.689	1.333	1.740	2.110	2.567	2.898
18	.257	.688	1.330	1.734	2.101	2.552	2.878
19	.257	.688	1.328	1.729	2.093	2.539	2.861
20	0.257	0.687	1.325	1.725	2.086	2.528	2.845
21	.257	.686	1.323	1.721	2.080	2.518	2.831
22	.256	.686	1.321	1.717	2.074	2.508	2.819
23	.256	.685	1.319	1.714	2.069	2.500	2.807
24	.256	.685	1.318	1.711	2.064	2.492	2.797

APPENDIX TABLE 6 (Continued)

Degrees of freedom	Probability 0.40	0.25	0.10	0.05	0.025	0.01	0.005
25	0.256	0.684	1.316	1.708	2.060	2.485	2.787
26	.256	.684	1.315	1.706	2.056	2.479	2.779
27	.256	.684	1.314	1.703	2.052	2.473	2.771
28	.256	.683	1.313	1.701	2.048	2.467	2.763
29	.256	.683	1.311	1.699	2.045	2.462	2.756
30	0.256	0.683	1.310	1.697	2.042	2.457	2.750
40	.255	.681	1.303	1.684	2.021	2.423	2.704
60	.254	.679	1.296	1.671	2.000	2.390	2.660
120	.254	.677	1.289	1.658	1.980	2.358	2.617
∞	.253	.674	1.282	1.645	1.960	2.326	2.576

SOURCE: *Biometrika Tables for Statisticians* (Cambridge, England: Cambridge University, 1954).

APPENDIX TABLE 7a

Chart Providing 95 Percent Confidence Interval for Population Proportion, Based on Sample Proportion[a]

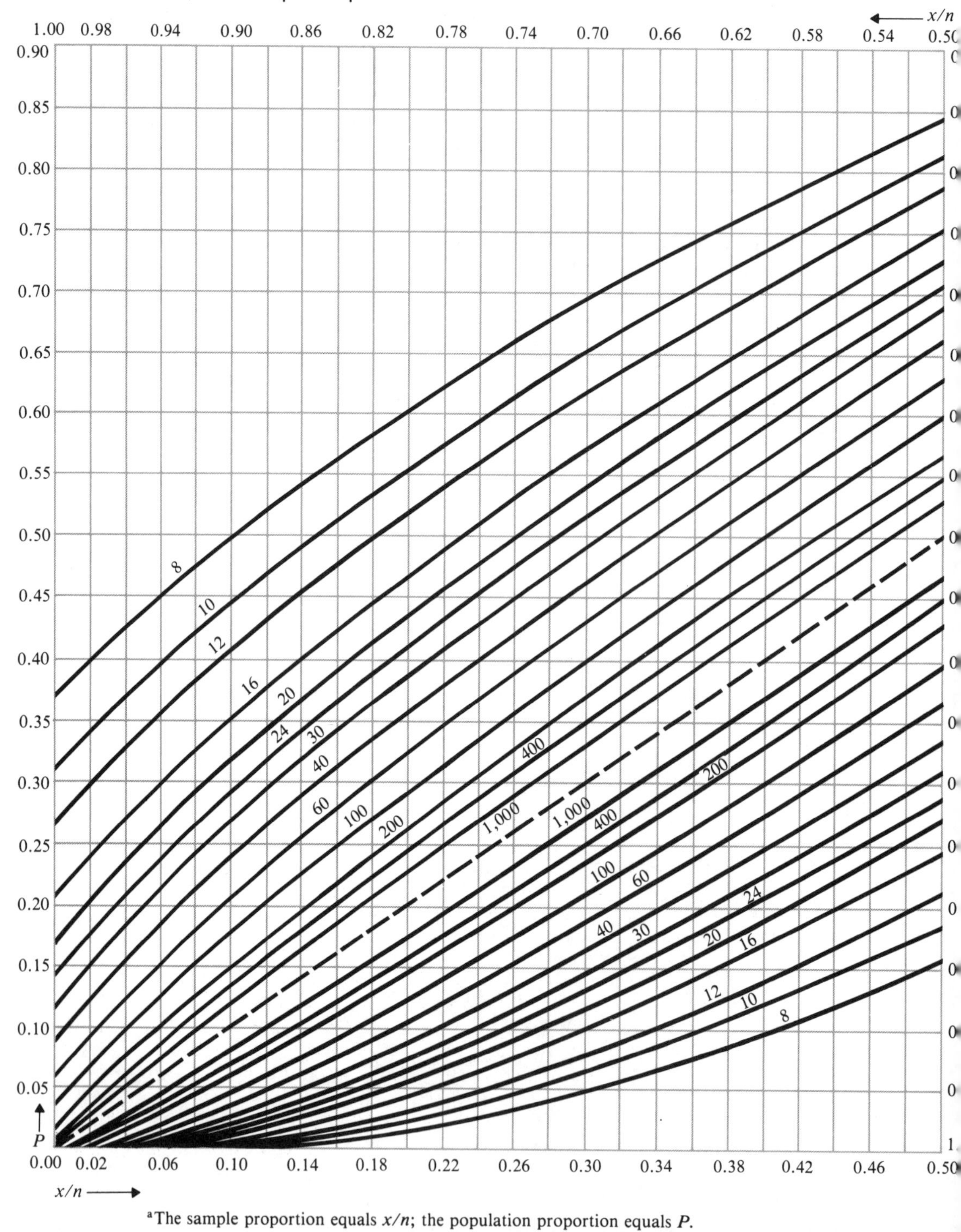

[a]The sample proportion equals x/n; the population proportion equals P.

APPENDIX TABLE 7b

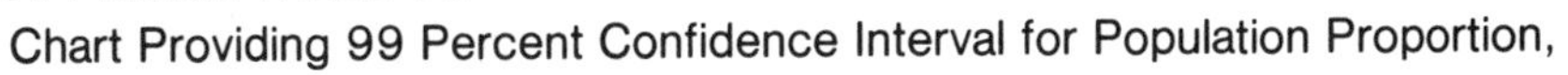

Chart Providing 99 Percent Confidence Interval for Population Proportion, Based on Sample Proportion[a]

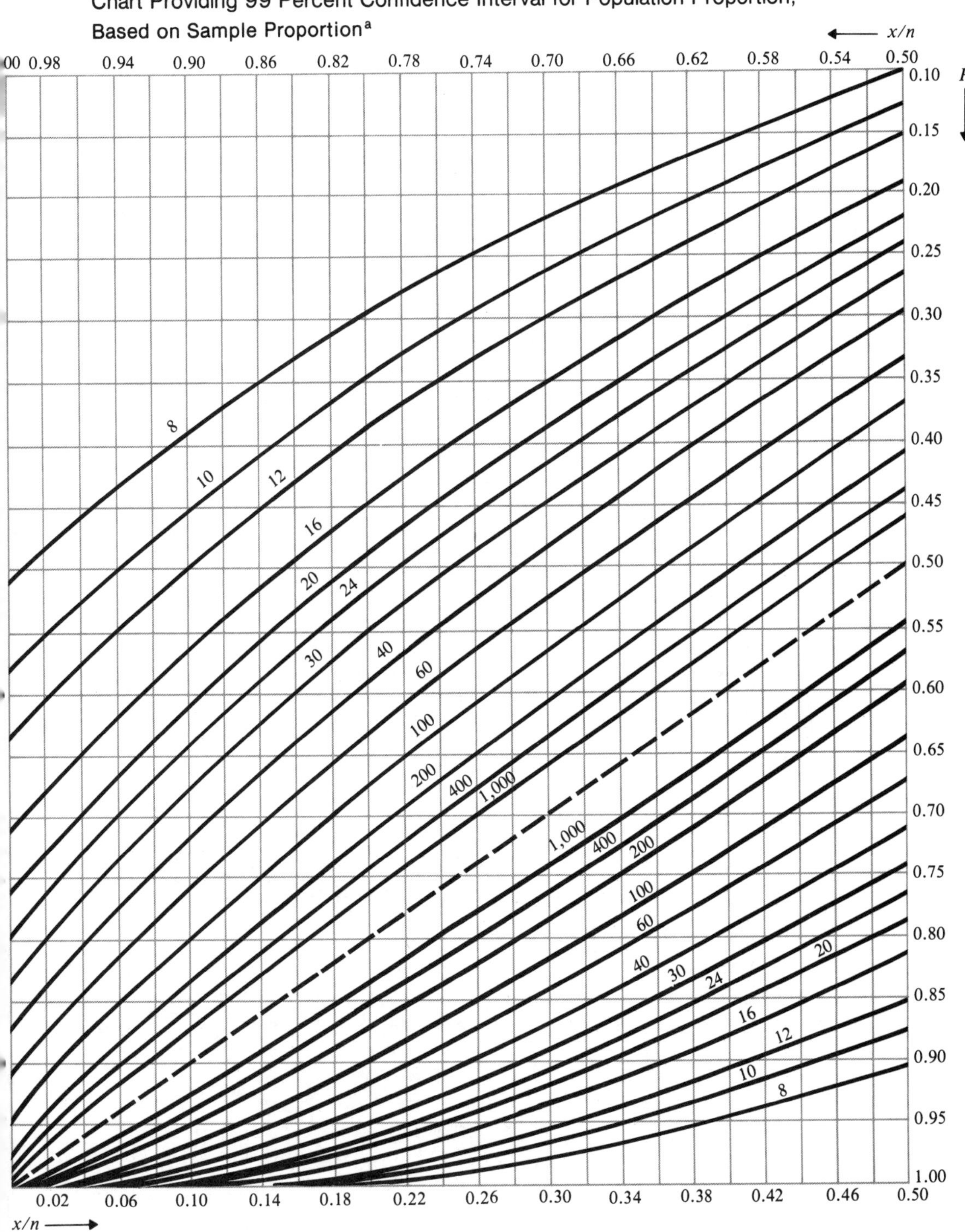

SOURCE: See Appendix Table 6.

[a]See footnote a to Appendix Table 7a.

APPENDIX TABLE 8

Values of χ^2 That Will Be Exceeded with Specified Probabilities

This table shows the value of χ^2 where the area under the χ^2 distribution exceeding this value of χ^2 equals the specified amount. For example, the probability that a χ^2 variable with 8 degrees of freedom will exceed 13.3616 equals 0.10.

0 13.3616

Degrees of freedom	Probabilities							
	0.990	0.975	0.950	0.900	0.100	0.050	0.025	0.010
1	$157088 \cdot 10^{-9}$	$982069 \cdot 10^{-9}$	$393214 \cdot 10^{-8}$	0.0157908	2.70554	3.84146	5.02389	6.63490
2	0.0201007	0.0506356	0.102587	0.210720	4.60517	5.99147	7.37776	9.21034
3	0.114832	0.215795	0.351846	0.584375	6.25139	7.81473	9.34840	11.3449
4	0.297110	0.484419	0.710721	1.063623	7.77944	9.48773	11.1433	13.2767
5	0.554300	0.831211	1.145476	1.61031	9.23635	11.0705	12.8325	15.0863
6	0.872085	1.237347	1.63539	2.20413	10.6446	12.5916	14.4494	16.8119
7	1.239043	1.68987	2.16735	2.83311	12.0170	14.0671	16.0128	18.4753
8	1.646482	2.17973	2.73264	3.48954	13.3616	15.5073	17.5346	20.0902
9	2.087912	2.70039	3.32511	4.16816	14.6837	16.9190	19.0228	21.6660
10	2.55821	3.24697	3.94030	4.86518	15.9871	18.3070	20.4831	23.2093
11	3.05347	3.81575	4.57481	5.57779	17.2750	19.6751	21.9200	24.7250
12	3.57056	4.40379	5.22603	6.30380	18.5494	21.0261	23.3367	26.2170
13	4.10691	5.00874	5.89186	7.04150	19.8119	22.3621	24.7356	27.6883
14	4.66043	5.62872	6.57063	7.78953	21.0642	23.6848	26.1190	29.1413

APPENDIX TABLE 8 (Continued)

Degrees of freedom	Probabilities 0.990	0.975	0.950	0.900	0.100	0.050	0.025	0.010
15	5.22935	6.26214	7.26094	8.54675	22.3072	24.9958	27.4884	30.5779
16	5.81221	6.90766	7.96164	9.31223	23.5418	26.2962	28.8454	31.9999
17	6.40776	7.56418	8.67176	10.0852	24.7690	27.5871	30.1910	33.4087
18	7.01491	8.23075	9.39046	10.8649	25.9894	28.8693	31.5264	34.8053
19	7.63273	8.90655	10.1170	11.6509	27.2036	30.1435	32.8523	36.1908
20	8.26040	9.59083	10.8508	12.4426	28.4120	31.4104	34.1696	37.5662
21	8.89720	10.28293	11.5913	13.2396	29.6151	32.6705	35.4789	38.9321
22	9.54249	10.9823	12.3380	14.0415	30.8133	33.9244	36.7807	40.2894
23	10.19567	11.6885	13.0905	14.8479	32.0069	35.1725	38.0757	41.6384
24	10.8564	12.4011	13.8484	15.6587	33.1963	36.4151	39.3641	42.9798
25	11.5240	13.1197	14.6114	16.4734	34.3816	37.6525	40.6465	44.3141
26	12.1981	13.8439	15.3791	17.2919	35.5631	38.8852	41.9232	45.6417
27	12.8786	14.5733	16.1513	18.1138	36.7412	40.1133	43.1944	46.9630
28	13.5648	15.3079	16.9279	18.9392	37.9159	41.3372	44.4607	48.2782
29	14.2565	16.0471	17.7083	19.7677	39.0875	42.5569	45.7222	49.5879
30	14.9535	16.7908	18.4926	20.5992	40.2560	43.7729	46.9792	50.8922
40	22.1643	24.4331	26.5093	29.0505	51.8050	55.7585	59.3417	63.6907
50	29.7067	32.3574	34.7642	37.6886	63.1671	67.5048	71.4202	76.1539
60	37.4848	40.4817	43.1879	46.4589	74.3970	79.0819	83.2976	88.3794
70	45.4418	48.7576	51.7393	55.3290	85.5271	90.5312	95.0231	100.425
80	53.5400	57.1532	60.3915	64.2778	96.5782	101.879	106.629	112.329
90	61.7541	65.6466	69.1260	73.2912	107.565	113.145	118.136	124.116
100	70.0648	74.2219	77.9295	82.3581	118.498	124.342	129.561	135.807

SOURCE: See Appendix Table 6.

APPENDIX TABLE 9

Value of an *F* Variable That Is Exceeded with Probability Equal to .05

Degrees of freedom for numerator (columns); Degrees of freedom for denominator (rows)

	1	2	3	4	5	6	7	8	9
1	161.4	199.5	215.7	224.6	230.2	234.0	236.8	238.9	240.5
2	18.51	19.00	19.16	19.25	19.30	19.33	19.35	19.37	19.38
3	10.13	9.55	9.28	9.12	9.01	8.94	8.89	8.85	8.81
4	7.71	6.94	6.59	6.39	6.26	6.16	6.09	6.04	6.00
5	6.61	5.79	5.41	5.19	5.05	4.95	4.88	4.82	4.77
6	5.99	5.14	4.76	4.53	4.39	4.28	4.21	4.15	4.10
7	5.59	4.74	4.35	4.12	3.97	3.87	3.79	3.73	3.68
8	5.32	4.46	4.07	3.84	3.69	3.58	3.50	3.44	3.39
9	5.12	4.26	3.86	3.63	3.48	3.37	3.29	3.23	3.18
10	4.96	4.10	3.71	3.48	3.33	3.22	3.14	3.07	3.02
11	4.84	3.98	3.59	3.36	3.20	3.09	3.01	2.95	2.90
12	4.75	3.89	3.49	3.26	3.11	3.00	2.91	2.85	2.80
13	4.67	3.81	3.41	3.18	3.03	2.92	2.83	2.77	2.71
14	4.60	3.74	3.34	3.11	2.96	2.85	2.76	2.70	2.65
15	4.54	3.68	3.29	3.06	2.90	2.79	2.71	2.64	2.59
16	4.49	3.63	3.24	3.01	2.85	2.74	2.66	2.59	2.54
17	4.45	3.59	3.20	2.96	2.81	2.70	2.61	2.55	2.49
18	4.41	3.55	3.16	2.93	2.77	2.66	2.58	2.51	2.46
19	4.38	3.52	3.13	2.90	2.74	2.63	2.54	2.48	2.42
20	4.35	3.49	3.10	2.87	2.71	2.60	2.51	2.45	2.39
21	4.32	3.47	3.07	2.84	2.68	2.57	2.49	2.42	2.37
22	4.30	3.44	3.05	2.82	2.66	2.55	2.46	2.40	2.34
23	4.28	3.42	3.03	2.80	2.64	2.53	2.44	2.37	2.32
24	4.26	3.40	3.01	2.78	2.62	2.51	2.42	2.36	2.30
25	4.24	3.39	2.99	2.76	2.60	2.49	2.40	2.34	2.28
26	4.23	3.37	2.98	2.74	2.59	2.47	2.39	2.32	2.27
27	4.21	3.35	2.96	2.73	2.57	2.46	2.37	2.31	2.25
28	4.20	3.34	2.95	2.71	2.56	2.45	2.36	2.29	2.24
29	4.18	3.33	2.93	2.70	2.55	2.43	2.35	2.28	2.22
30	4.17	3.32	2.92	2.69	2.53	2.42	2.33	2.27	2.21
40	4.08	3.23	2.84	2.61	2.45	2.34	2.25	2.18	2.12
60	4.00	3.15	2.76	2.53	2.37	2.25	2.17	2.10	2.04
120	3.92	3.07	2.68	2.45	2.29	2.17	2.09	2.02	1.96
∞	3.84	3.00	2.60	2.37	2.21	2.10	2.01	1.94	1.88

APPENDIX TABLE 9 (Continued)

Degrees of freedom for numerator

Degrees of freedom for denominator	10	12	15	20	24	30	40	60	120	∞
1	241.9	243.9	245.9	248.0	249.1	250.1	251.1	252.2	253.3	254.3
2	19.40	19.41	19.43	19.45	19.45	19.46	19.47	19.48	19.49	19.50
3	8.79	8.74	8.70	8.66	8.64	8.62	8.59	8.57	8.55	8.53
4	5.96	5.91	5.86	5.80	5.77	5.75	5.72	5.69	5.66	5.63
5	4.74	4.68	4.62	4.56	4.53	4.50	4.46	4.43	4.40	4.36
6	4.06	4.00	3.94	3.87	3.84	3.81	3.77	3.74	3.70	3.67
7	3.64	3.57	3.51	3.44	3.41	3.38	3.34	3.30	3.27	3.23
8	3.35	3.28	3.22	3.15	3.12	3.08	3.04	3.01	2.97	2.93
9	3.14	3.07	3.01	2.94	2.90	2.86	2.83	2.79	2.75	2.71
10	2.98	2.91	2.85	2.77	2.74	2.70	2.66	2.62	2.58	2.54
11	2.85	2.79	2.72	2.65	2.61	2.57	2.53	2.49	2.45	2.40
12	2.75	2.69	2.62	2.54	2.51	2.47	2.43	2.38	2.34	2.30
13	2.67	2.60	2.53	2.46	2.42	2.38	2.34	2.30	2.25	2.21
14	2.60	2.53	2.46	2.39	2.35	2.31	2.27	2.22	2.18	2.13
15	2.54	2.48	2.40	2.33	2.29	2.25	2.20	2.16	2.11	2.07
16	2.49	2.42	2.35	2.28	2.24	2.19	2.15	2.11	2.06	2.01
17	2.45	2.38	2.31	2.23	2.19	2.15	2.10	2.06	2.01	1.96
18	2.41	2.34	2.27	2.19	2.15	2.11	2.06	2.02	1.97	1.92
19	2.38	2.31	2.23	2.16	2.11	2.07	2.03	1.98	1.93	1.88
20	2.35	2.28	2.20	2.12	2.08	2.04	1.99	1.95	1.90	1.84
21	2.32	2.25	2.18	2.10	2.05	2.01	1.96	1.92	1.87	1.81
22	2.30	2.23	2.15	2.07	2.03	1.98	1.94	1.89	1.84	1.78
23	2.27	2.20	2.13	2.05	2.01	1.96	1.91	1.86	1.81	1.76
24	2.25	2.18	2.11	2.03	1.98	1.94	1.89	1.84	1.79	1.73
25	2.24	2.16	2.09	2.01	1.96	1.92	1.87	1.82	1.77	1.71
26	2.22	2.15	2.07	1.99	1.95	1.90	1.85	1.80	1.75	1.69
27	2.20	2.13	2.06	1.97	1.93	1.88	1.84	1.79	1.73	1.67
28	2.19	2.12	2.04	1.96	1.91	1.87	1.82	1.77	1.71	1.65
29	2.18	2.10	2.03	1.94	1.90	1.85	1.81	1.75	1.70	1.64
30	2.16	2.09	2.01	1.93	1.89	1.84	1.79	1.74	1.68	1.62
40	2.08	2.00	1.92	1.84	1.79	1.74	1.69	1.64	1.58	1.51
60	1.99	1.92	1.84	1.75	1.70	1.65	1.59	1.53	1.47	1.39
120	1.91	1.83	1.75	1.66	1.61	1.55	1.50	1.43	1.35	1.25
∞	1.83	1.75	1.67	1.57	1.52	1.46	1.39	1.32	1.22	1.00

SOURCE: See Appendix Table 6.

APPENDIX TABLE 10
Value of an F Variable Exceeded with Probability Equal to .01

Degrees of freedom for numerator

Degrees of freedom for denominator	1	2	3	4	5	6	7	8	9
1	4052	4999.5	5403	5625	5764	5859	5928	5982	6022
2	98.50	99.00	99.17	99.25	99.30	99.33	99.36	99.37	99.39
3	34.12	30.82	29.46	28.71	28.24	27.91	27.67	27.49	27.35
4	21.20	18.00	16.69	15.98	15.52	15.21	14.98	14.80	14.66
5	16.26	13.27	12.06	11.39	10.97	10.67	10.46	10.29	10.16
6	13.75	10.92	9.78	9.15	8.75	8.47	8.26	8.10	7.98
7	12.25	9.55	8.45	7.85	7.46	7.19	6.99	6.84	6.72
8	11.26	8.65	7.59	7.01	6.63	6.37	6.18	6.03	5.91
9	10.56	8.02	6.99	6.42	6.06	5.80	5.61	5.47	5.35
10	10.04	7.56	6.55	5.99	5.64	5.39	5.20	5.06	4.94
11	9.65	7.21	6.22	5.67	5.32	5.07	4.89	4.74	4.63
12	9.33	6.93	5.95	5.41	5.06	4.82	4.64	4.50	4.39
13	9.07	6.70	5.74	5.21	4.86	4.62	4.44	4.30	4.19
14	8.86	6.51	5.56	5.04	4.69	4.46	4.28	4.14	4.03
15	8.68	6.36	5.42	4.89	4.56	4.32	4.14	4.00	3.89
16	8.53	6.23	5.29	4.77	4.44	4.20	4.03	3.89	3.78
17	8.40	6.11	5.18	4.67	4.34	4.10	3.93	3.79	3.68
18	8.29	6.01	5.09	4.58	4.25	4.01	3.84	3.71	3.60
19	8.18	5.93	5.01	4.50	4.17	3.94	3.77	3.63	3.52
20	8.10	5.85	4.94	4.43	4.10	3.87	3.70	3.56	3.46
21	8.02	5.78	4.87	4.37	4.04	3.81	3.64	3.51	3.40
22	7.95	5.72	4.82	4.31	3.99	3.76	3.59	3.45	3.35
23	7.88	5.66	4.76	4.26	3.94	3.71	3.54	3.41	3.30
24	7.82	5.61	4.72	4.22	3.90	3.67	3.50	3.36	3.26
25	7.77	5.57	4.68	4.18	3.85	3.63	3.46	3.32	3.22
26	7.72	5.53	4.64	4.14	3.82	3.59	3.42	3.29	3.18
27	7.68	5.49	4.60	4.11	3.78	3.56	3.39	3.26	3.15
28	7.64	5.45	4.57	4.07	3.75	3.53	3.36	3.23	3.12
29	7.60	5.42	4.54	4.04	3.73	3.50	3.33	3.20	3.09
30	7.56	5.39	4.51	4.02	3.70	3.47	3.30	3.17	3.07
40	7.31	5.18	4.31	3.83	3.51	3.29	3.12	2.99	2.89
60	7.08	4.98	4.13	3.65	3.34	3.12	2.95	2.82	2.72
120	6.85	4.79	3.95	3.48	3.17	2.96	2.79	2.66	2.56
∞	6.63	4.61	3.78	3.32	3.02	2.80	2.64	2.51	2.41

APPENDIX TABLE 10 (Continued)

Degrees of freedom for numerator

Degrees of freedom for denominator	10	12	15	20	24	30	40	60	120	∞
1	6056	6106	6157	6209	6235	6261	6287	6313	6339	6366
2	99.40	99.42	99.43	99.45	99.46	99.47	99.47	99.48	99.49	99.50
3	27.23	27.05	26.87	26.69	26.60	26.50	26.41	26.32	26.22	26.13
4	14.55	14.37	14.20	14.02	13.93	13.84	13.75	13.65	13.56	13.46
5	10.05	9.89	9.72	9.55	9.47	9.38	9.29	9.20	9.11	9.02
6	7.87	7.72	7.56	7.40	7.31	7.23	7.14	7.06	6.97	6.88
7	6.62	6.47	6.31	6.16	6.07	5.99	5.91	5.82	5.74	5.65
8	5.81	5.67	5.52	5.36	5.28	5.20	5.12	5.03	4.95	4.86
9	5.26	5.11	4.96	4.81	4.73	4.65	4.57	4.48	4.40	4.31
10	4.85	4.71	4.56	4.41	4.33	4.25	4.17	4.08	4.00	3.91
11	4.54	4.40	4.25	4.10	4.02	3.94	3.86	3.78	3.69	3.60
12	4.30	4.16	4.01	3.86	3.78	3.70	3.62	3.54	3.45	3.36
13	4.10	3.96	3.82	3.66	3.59	3.51	3.43	3.34	3.25	3.17
14	3.94	3.80	3.66	3.51	3.43	3.35	3.27	3.18	3.09	3.00
15	3.80	3.67	3.52	3.37	3.29	3.21	3.13	3.05	2.96	2.87
16	3.69	3.55	3.41	3.26	3.18	3.10	3.02	2.93	2.84	2.75
17	3.59	3.46	3.31	3.16	3.08	3.00	2.92	2.83	2.75	2.65
18	3.51	3.37	3.23	3.08	3.00	2.92	2.84	2.75	2.66	2.57
19	3.43	3.30	3.15	3.00	2.92	2.84	2.76	2.67	2.58	2.49
20	3.37	3.23	3.09	2.94	2.86	2.78	2.69	2.61	2.52	2.42
21	3.31	3.17	3.03	2.88	2.80	2.72	2.64	2.55	2.46	2.36
22	3.26	3.12	2.98	2.83	2.75	2.67	2.58	2.50	2.40	2.31
23	3.21	3.07	2.93	2.78	2.70	2.62	2.54	2.45	2.35	2.26
24	3.17	3.03	2.89	2.74	2.66	2.58	2.49	2.40	2.31	2.21
25	3.13	2.99	2.85	2.70	2.62	2.54	2.45	2.36	2.27	2.17
26	3.09	2.96	2.81	2.66	2.58	2.50	2.42	2.33	2.23	2.13
27	3.06	2.93	2.78	2.63	2.55	2.47	2.38	2.29	2.20	2.10
28	3.03	2.90	2.75	2.60	2.52	2.44	2.35	2.26	2.17	2.06
29	3.00	2.87	2.73	2.57	2.49	2.41	2.33	2.23	2.14	2.03
30	2.98	2.84	2.70	2.55	2.47	2.39	2.30	2.21	2.11	2.01
40	2.80	2.66	2.52	2.37	2.29	2.20	2.11	2.02	1.92	1.80
60	2.63	2.50	2.35	2.20	2.12	2.03	1.94	1.84	1.73	1.60
120	2.47	2.34	2.19	2.03	1.95	1.86	1.76	1.66	1.53	1.38
∞	2.32	2.18	2.04	1.88	1.79	1.70	1.59	1.47	1.32	1.00

SOURCE: See Appendix Table 6.

APPENDIX TABLE 11

Values of d_L and d_U for the Durbin-Watson Test

A. Where α = .05

	$k = 1$		$k = 2$		$k = 3$		$k = 4$		$k = 5$	
n	d_L	d_U	d_L	d_U	d_L	d_U	d_L	d_U	d_L	d_U
15	1.08	1.36	0.95	1.54	0.82	1.75	0.69	1.97	0.56	2.21
16	1.10	1.37	0.98	1.54	0.86	1.73	0.74	1.93	0.62	2.15
17	1.13	1.38	1.02	1.54	0.90	1.71	0.78	1.90	0.67	2.10
18	1.16	1.39	1.05	1.53	0.93	1.69	0.82	1.87	0.71	2.06
19	1.18	1.40	1.08	1.53	0.97	1.68	0.86	1.85	0.75	2.02
20	1.20	1.41	1.10	1.54	1.00	1.68	0.90	1.83	0.79	1.99
21	1.22	1.42	1.13	1.54	1.03	1.67	0.93	1.81	0.83	1.96
22	1.24	1.43	1.15	1.54	1.05	1.66	0.96	1.80	0.86	1.94
23	1.26	1.44	1.17	1.54	1.08	1.66	0.99	1.79	0.90	1.92
24	1.27	1.45	1.19	1.55	1.10	1.66	1.01	1.78	0.93	1.90
25	1.29	1.45	1.21	1.55	1.12	1.66	1.04	1.77	0.95	1.89
26	1.30	1.46	1.22	1.55	1.14	1.65	1.06	1.76	0.98	1.88
27	1.32	1.47	1.24	1.56	1.16	1.65	1.08	1.76	1.01	1.86
28	1.33	1.48	1.26	1.56	1.18	1.65	1.10	1.75	1.03	1.85
29	1.34	1.48	1.27	1.56	1.20	1.65	1.12	1.74	1.05	1.84
30	1.35	1.49	1.28	1.57	1.21	1.65	1.14	1.74	1.07	1.83
31	1.36	1.50	1.30	1.57	1.23	1.65	1.16	1.74	1.09	1.83
32	1.37	1.50	1.31	1.57	1.24	1.65	1.18	1.73	1.11	1.82
33	1.38	1.51	1.32	1.58	1.26	1.65	1.19	1.73	1.13	1.81
34	1.39	1.51	1.33	1.58	1.27	1.65	1.21	1.73	1.15	1.81
35	1.40	1.52	1.34	1.58	1.28	1.65	1.22	1.73	1.16	1.80
36	1.41	1.52	1.35	1.59	1.29	1.65	1.24	1.73	1.18	1.80
37	1.42	1.53	1.36	1.59	1.31	1.66	1.25	1.72	1.19	1.80
38	1.43	1.54	1.37	1.59	1.32	1.66	1.26	1.72	1.21	1.79
39	1.43	1.54	1.38	1.60	1.33	1.66	1.27	1.72	1.22	1.79
40	1.44	1.54	1.39	1.60	1.34	1.66	1.29	1.72	1.23	1.79
45	1.48	1.57	1.43	1.62	1.38	1.67	1.34	1.72	1.29	1.78
50	1.50	1.59	1.46	1.63	1.42	1.67	1.38	1.72	1.34	1.77
55	1.53	1.60	1.49	1.64	1.45	1.68	1.41	1.72	1.38	1.77
60	1.55	1.62	1.51	1.65	1.48	1.69	1.44	1.73	1.41	1.77
65	1.57	1.63	1.54	1.66	1.50	1.70	1.47	1.73	1.44	1.77
70	1.58	1.64	1.55	1.67	1.52	1.70	1.49	1.74	1.46	1.77
75	1.60	1.65	1.57	1.68	1.54	1.71	1.51	1.74	1.49	1.77
80	1.61	1.66	1.59	1.69	1.56	1.72	1.53	1.74	1.51	1.77
85	1.62	1.67	1.60	1.70	1.57	1.72	1.55	1.75	1.52	1.77
90	1.63	1.68	1.61	1.70	1.59	1.73	1.57	1.75	1.54	1.78
95	1.64	1.69	1.62	1.71	1.60	1.73	1.58	1.75	1.56	1.78
100	1.65	1.69	1.63	1.72	1.61	1.74	1.59	1.76	1.57	1.78

APPENDIX TABLE 11 (Continued)

B. Where $\alpha = .025$

	$k = 1$		$k = 2$		$k = 3$		$k = 4$		$k = 5$	
n	d_L	d_U	d_L	d_U	d_L	d_U	d_L	d_U	d_L	d_U
15	0.95	1.23	0.83	1.40	0.71	1.61	0.59	1.84	0.48	2.09
16	0.98	1.24	0.86	1.40	0.75	1.59	0.64	1.80	0.53	2.03
17	1.01	1.25	0.90	1.40	0.79	1.58	0.68	1.77	0.57	1.98
18	1.03	1.26	0.93	1.40	0.82	1.56	0.72	1.74	0.62	1.93
19	1.06	1.28	0.96	1.41	0.86	1.55	0.76	1.72	0.66	1.90
20	1.08	1.28	0.99	1.41	0.89	1.55	0.79	1.70	0.70	1.87
21	1.10	1.30	1.01	1.41	0.92	1.54	0.83	1.69	0.73	1.84
22	1.12	1.31	1.04	1.42	0.95	1.54	0.86	1.68	0.77	1.82
23	1.14	1.32	1.06	1.42	0.97	1.54	0.89	1.67	0.80	1.80
24	1.16	1.33	1.08	1.43	1.00	1.54	0.91	1.66	0.83	1.79
25	1.18	1.34	1.10	1.43	1.02	1.54	0.94	1.65	0.86	1.77
26	1.19	1.35	1.12	1.44	1.04	1.54	0.96	1.65	0.88	1.76
27	1.21	1.36	1.13	1.44	1.06	1.54	0.99	1.64	0.91	1.75
28	1.22	1.37	1.15	1.45	1.08	1.54	1.01	1.64	0.93	1.74
29	1.24	1.38	1.17	1.45	1.10	1.54	1.03	1.63	0.96	1.73
30	1.25	1.38	1.18	1.46	1.12	1.54	1.05	1.63	0.98	1.73
31	1.26	1.39	1.20	1.47	1.13	1.55	1.07	1.63	1.00	1.72
32	1.27	1.40	1.21	1.47	1.15	1.55	1.08	1.63	1.02	1.71
33	1.28	1.41	1.22	1.48	1.16	1.55	1.10	1.63	1.04	1.71
34	1.29	1.41	1.24	1.48	1.17	1.55	1.12	1.63	1.06	1.70
35	1.30	1.42	1.25	1.48	1.19	1.55	1.13	1.63	1.07	1.70
36	1.31	1.43	1.26	1.49	1.20	1.56	1.15	1.63	1.09	1.70
37	1.32	1.43	1.27	1.49	1.21	1.56	1.16	1.62	1.10	1.70
38	1.33	1.44	1.28	1.50	1.23	1.56	1.17	1.62	1.12	1.70
39	1.34	1.44	1.29	1.50	1.24	1.56	1.19	1.63	1.13	1.69
40	1.35	1.45	1.30	1.51	1.25	1.57	1.20	1.63	1.15	1.69
45	1.39	1.48	1.34	1.53	1.30	1.58	1.25	1.63	1.21	1.69
50	1.42	1.50	1.38	1.54	1.34	1.59	1.30	1.64	1.26	1.69
55	1.45	1.52	1.41	1.56	1.37	1.60	1.33	1.64	1.30	1.69
60	1.47	1.54	1.44	1.57	1.40	1.61	1.37	1.65	1.33	1.69
65	1.49	1.55	1.46	1.59	1.43	1.62	1.40	1.66	1.36	1.69
70	1.51	1.57	1.48	1.60	1.45	1.63	1.42	1.66	1.39	1.70
75	1.53	1.58	1.50	1.61	1.47	1.64	1.45	1.67	1.42	1.70
80	1.54	1.59	1.52	1.62	1.49	1.65	1.47	1.67	1.44	1.70
85	1.56	1.60	1.53	1.63	1.51	1.65	1.49	1.68	1.46	1.71
90	1.57	1.61	1.55	1.64	1.53	1.66	1.50	1.69	1.48	1.71
95	1.58	1.62	1.56	1.65	1.54	1.67	1.52	1.69	1.50	1.71
100	1.59	1.63	1.57	1.65	1.55	1.67	1.53	1.70	1.51	1.72

APPENDIX TABLE 11 (Continued)

C. Where α = .01

	$k = 1$		$k = 2$		$k = 3$		$k = 4$		$k = 5$	
n	d_L	d_U	d_L	d_U	d_L	d_U	d_L	d_U	d_L	d_U
15	0.81	1.07	0.70	1.25	0.59	1.46	0.49	1.70	0.39	1.96
16	0.84	1.09	0.74	1.25	0.63	1.44	0.53	1.66	0.44	1.90
17	0.87	1.10	0.77	1.25	0.67	1.43	0.57	1.63	0.48	1.85
18	0.90	1.12	0.80	1.26	0.71	1.42	0.61	1.60	0.52	1.80
19	0.93	1.13	0.83	1.26	0.74	1.41	0.65	1.58	0.56	1.77
20	0.95	1.15	0.86	1.27	0.77	1.41	0.68	1.57	0.60	1.74
21	0.97	1.16	0.89	1.27	0.80	1.41	0.72	1.55	0.63	1.71
22	1.00	1.17	0.91	1.28	0.83	1.40	0.75	1.54	0.66	1.69
23	1.02	1.19	0.94	1.29	0.86	1.40	0.77	1.53	0.70	1.67
24	1.04	1.20	0.96	1.30	0.88	1.41	0.80	1.53	0.72	1.66
25	1.05	1.21	0.98	1.30	0.90	1.41	0.83	1.52	0.75	1.65
26	1.07	1.22	1.00	1.31	0.93	1.41	0.85	1.52	0.78	1.64
27	1.09	1.23	1.02	1.32	0.95	1.41	0.88	1.51	0.81	1.63
28	1.10	1.24	1.04	1.32	0.97	1.41	0.90	1.51	0.83	1.62
29	1.12	1.25	1.05	1.33	0.99	1.42	0.92	1.51	0.85	1.61
30	1.13	1.26	1.07	1.34	1.01	1.42	0.94	1.51	0.88	1.61
31	1.15	1.27	1.08	1.34	1.02	1.42	0.96	1.51	0.90	1.60
32	1.16	1.28	1.10	1.35	1.04	1.43	0.98	1.51	0.92	1.60
33	1.17	1.29	1.11	1.36	1.05	1.43	1.00	1.51	0.94	1.59
34	1.18	1.30	1.13	1.36	1.07	1.43	1.01	1.51	0.95	1.59
35	1.19	1.31	1.14	1.37	1.08	1.44	1.03	1.51	0.97	1.59
36	1.21	1.32	1.15	1.38	1.10	1.44	1.04	1.51	0.99	1.59
37	1.22	1.32	1.16	1.38	1.11	1.45	1.06	1.51	1.00	1.59
38	1.23	1.33	1.18	1.39	1.12	1.45	1.07	1.52	1.02	1.58
39	1.24	1.34	1.19	1.39	1.14	1.45	1.09	1.52	1.03	1.58
40	1.25	1.34	1.20	1.40	1.15	1.46	1.10	1.52	1.05	1.58
45	1.29	1.38	1.24	1.42	1.20	1.48	1.16	1.53	1.11	1.58
50	1.32	1.40	1.28	1.45	1.24	1.49	1.20	1.54	1.16	1.59
55	1.36	1.43	1.32	1.47	1.28	1.51	1.25	1.55	1.21	1.59
60	1.38	1.45	1.35	1.48	1.32	1.52	1.28	1.56	1.25	1.60
65	1.41	1.47	1.38	1.50	1.35	1.53	1.31	1.57	1.28	1.61
70	1.43	1.49	1.40	1.52	1.37	1.55	1.34	1.58	1.31	1.61
75	1.45	1.50	1.42	1.53	1.39	1.56	1.37	1.59	1.34	1.62
80	1.47	1.52	1.44	1.54	1.42	1.57	1.39	1.60	1.36	1.62
85	1.48	1.53	1.46	1.55	1.43	1.58	1.41	1.60	1.39	1.63
90	1.50	1.54	1.47	1.56	1.45	1.59	1.43	1.61	1.41	1.64
95	1.51	1.55	1.49	1.57	1.47	1.60	1.45	1.62	1.42	1.64
100	1.52	1.56	1.50	1.58	1.48	1.60	1.46	1.63	1.44	1.65

SOURCE: J. Durbin and G. S. Watson, "Testing for Serial Correlation in Least Squares Regression," *Biometrika*, 38, June 1951.

APPENDIX TABLE 12
Common Logarithms

N	0	1	2	3	4	5	6	7	8	9
10	0000	0043	0086	0128	0170	0212	0253	0294	0334	0374
11	0414	0453	0492	0531	0569	0607	0645	0682	0719	0755
12	0792	0828	0864	0899	0934	0969	1004	1038	1072	1106
13	1139	1173	1206	1239	1271	1303	1335	1367	1399	1430
14	1461	1492	1523	1553	1584	1614	1644	1673	1703	1732
15	1761	1790	1818	1847	1875	1903	1931	1959	1987	2014
16	2041	2068	2095	2122	2148	2175	2201	2227	2253	2279
17	2304	2330	2355	2380	2405	2430	2455	2480	2504	2529
18	2553	2577	2601	2625	2648	2672	2695	2718	2742	2765
19	2788	2810	2833	2856	2878	2900	2923	2945	2967	2989
20	3010	3032	3054	3075	3096	3118	3139	3160	3181	3201
21	3222	3243	3263	3284	3304	3324	3345	3365	3385	3404
22	3424	3444	3464	3483	3502	3522	3541	3560	3579	3598
23	3617	3636	3655	3674	3692	3711	3729	3747	3766	3784
24	3802	3826	3838	3856	3874	3892	3909	3927	3945	3962
25	3979	3997	4014	4031	4048	4065	4082	4099	4116	4133
26	4150	4166	4183	4200	4216	4232	4249	4205	4281	4298
27	4314	4330	4346	4362	4378	4393	4409	4425	4440	4456
28	4472	4487	4502	4518	4533	4548	4564	4579	4594	4609
29	4624	4639	4654	4669	4683	4698	4713	4728	4742	4757
30	4771	4786	4800	4814	4829	4843	4857	4871	4886	4900
31	4914	4928	4942	4955	4969	4983	4997	5011	5024	5038
32	5051	5065	5079	5092	5105	5119	5132	5145	5159	5172
33	5185	5198	5211	5224	5237	5250	5263	5276	5289	5302
34	5315	5328	5340	5353	5366	5378	5391	5403	5416	5423
35	5441	5453	5465	5478	5490	5502	5514	5527	5539	5551
36	5563	5575	5587	5599	5611	5623	5635	5647	5658	5670
37	5682	5694	5705	5717	5729	5740	5752	5763	5775	5786
38	5798	5809	5821	5832	5843	5855	5866	5877	5888	5899
39	5911	5922	5933	5944	5955	5966	5977	5988	5999	6010
40	6021	6031	6042	6053	6064	6075	6085	6096	6107	6117
41	6128	6138	6149	6160	6170	6180	6191	6201	6212	6222
42	6232	6243	6253	6263	6274	6284	6294	6304	6314	6325
43	6335	6345	6355	6365	6375	6385	6395	6405	6415	6425
44	6435	6444	6454	6464	6474	6484	6493	6503	6513	6522
N	0	1	2	3	4	5	6	7	8	9

APPENDIX TABLE 12 (Continued)

N	0	1	2	3	4	5	6	7	8	9
45	6532	6542	6551	6561	6571	6580	6590	6599	6609	6618
46	6628	6637	6646	6656	6665	6675	6684	6693	6702	6712
47	6721	6730	6739	6749	6758	6767	6776	6785	6794	6803
48	6812	6821	6830	6839	6848	6857	6866	6875	6884	6893
49	6902	6911	6920	6928	6937	6946	6955	6964	6972	6981
50	6990	6998	7007	7016	7024	7033	7042	7050	7059	7067
51	7076	7084	7093	7101	7110	7118	7126	7135	7143	7152
52	7160	7168	7177	7185	7193	7202	7210	7218	7226	7235
53	7243	7251	7259	7267	7275	7284	7292	7300	7308	7316
54	7324	7332	7340	7348	7356	7364	7372	7380	7388	7396
55	7404	7412	7419	7427	7435	7443	7451	7459	7466	7474
56	7482	7490	7497	7505	7513	7520	7528	7536	7543	7551
57	7559	7566	7574	7582	7589	7597	7604	7612	7619	7627
58	7634	7642	7649	7657	7664	7672	7679	7686	7694	7701
59	7709	7716	7723	7731	7738	7745	7752	7760	7767	7774
60	7782	7789	7796	7803	7810	7818	7825	7832	7839	7846
61	7853	7860	7868	7875	7882	7889	7896	7903	7910	7917
62	7924	7931	7938	7945	7952	7959	7966	7973	7980	7987
63	7993	8000	8007	8014	8021	8028	8035	8041	8048	8055
64	8062	8069	8075	8082	8089	8096	8102	8109	8116	8122
65	8129	8136	8142	8149	8156	8162	8169	8176	8182	8189
66	8195	8202	8209	8215	8222	8228	8235	8241	8248	8254
67	8261	8267	8274	8280	8287	8293	8299	8306	8312	8319
68	8325	8331	8338	8344	8351	8357	8363	8370	8376	8382
69	8388	8395	8401	8407	8414	8420	8426	8432	8439	8445
70	8451	8457	8463	8470	8476	8482	8488	8494	8500	8506
71	8513	8519	8525	8531	8537	8543	8549	8555	8561	8567
72	8573	8579	8585	8591	8597	8603	8609	8615	8621	8627
73	8633	8639	8645	8651	8637	8663	8669	8675	8681	8686
74	8692	8698	8704	8710	8716	8722	8727	8733	8739	8745
75	8751	8756	8762	8768	8774	8779	8785	8791	8797	8802
76	8808	8814	8820	8825	8831	8837	8842	8848	8854	8859
77	8865	8871	8876	8882	8887	8893	8899	8904	8910	8915
78	8921	8927	8932	8938	8943	8949	8954	8960	8965	8971
79	8976	8982	8987	8993	8998	9004	9009	9015	9020	9025
N	0	1	2	3	4	5	6	7	8	9

APPENDIX TABLE 12 (Continued)

N	0	1	2	3	4	5	6	7	8	9
80	9031	9036	9042	9047	9053	9058	9063	9069	9074	9079
81	9085	9090	9096	9101	9106	9112	9117	9122	9128	9133
82	9138	9143	9149	9154	9159	9165	9170	9175	9180	9186
83	9191	9190	9201	9206	9212	9217	9222	9227	9232	9238
84	9243	9248	9253	9258	9263	9269	9274	9279	9284	9289
85	9294	9299	9304	9309	9315	9320	9325	9330	9335	9340
86	9345	9350	9355	9360	9365	9370	9375	9380	9385	9390
87	9395	9400	9405	9410	9415	9420	9425	9430	9435	9440
88	9445	9450	9455	9460	9465	9469	9474	9479	9484	9489
89	9494	9499	9504	9509	9513	9518	9523	9528	9533	9538
90	9542	9547	9552	9557	9562	9566	9571	9576	9581	9586
91	9590	9595	9600	9605	9609	9614	9619	9624	9628	9633
92	9638	9643	9647	9652	9657	9661	9666	9671	9675	9680
93	9685	9689	9694	9699	9703	9708	9713	9717	9722	9727
94	9731	9736	9741	9745	9750	9754	9759	9763	9768	9773
95	9777	9782	9786	9791	9795	9800	9805	9808	9814	9818
96	9823	9827	9832	9836	9841	9845	9850	9854	9859	9863
97	9868	9872	9877	9881	9886	9890	9894	9899	9903	9908
98	9912	9917	9921	9926	9930	9934	9939	9943	9948	9952
99	9956	9961	9965	9969	9974	9978	9983	9987	9991	9996
N	0	1	2	3	4	5	6	7	8	9

SOURCE: Adapted from National Bureau of Standards, *Tables of* 10^x, Applied Mathematics Series 27 (U.S. Department of Commerce, 1953).

APPENDIX TABLE 13
Squares and Square Roots

n	n^2	$\sqrt{n}$	$\sqrt{10n}$	n	n^2	$\sqrt{n}$	$\sqrt{10n}$
				35	1 225	5.916 080	18.708 29
1	1	1.000 000	3.162 278	36	1 296	6.000 000	18.973 67
2	4	1.414 214	4.472 136	37	1 369	6.082 763	19.235 38
3	9	1.732 051	5.477 226	38	1 444	6.164 414	19.493 59
4	16	2.000 000	6.324 555	39	1 521	6.244 998	19.748 42
5	25	2.236 068	7.071 068	40	1 600	6.324 555	20.000 00
6	36	2.449 490	7.745 967	41	1 681	6.403 124	20.248 46
7	49	2.645 751	8.366 600	42	1 764	6.480 741	20.493 90
8	64	2.828 427	8.944 272	43	1 849	6.557 439	20.736 44
9	81	3.000 000	9.486 833	44	1 936	6.633 250	20.976 18
10	100	3.162 278	10.000 00	45	2 025	6.708 204	21.213 20
11	121	3.316 625	10.488 09	46	2 116	6.782 330	21.447 61
12	144	3.464 102	10.954 45	47	2 209	6.855 655	21.679 48
13	169	3.605 551	11.401 75	48	2 304	6.928 203	21.908 90
14	196	3.741 657	11.832 16	49	2 401	7.000 000	22.135 94
15	225	3.872 983	12.247 45	50	2 500	7.071 068	22.360 68
16	256	4.000 000	12.649 11	51	2 601	7.141 428	22.583 18
17	289	4.123 106	13.038 40	52	2 704	7.211 103	22.803 51
18	324	4.242 641	13.416 41	53	2 809	7.280 110	23.021 73
19	361	4.358 899	13.784 05	54	2 916	7.348 469	23.237 90
20	400	4.472 136	14.142 14	55	3 025	7.416 198	23.452 08
21	441	4.582 576	14.491 38	56	3 136	7.483 315	23.664 32
22	484	4.690 416	14.832 40	57	3 249	7.549 834	23.874 67
23	529	4.795 832	15.165 75	58	3 364	7.615 773	24.083 19
24	576	4.898 979	15.491 93	59	3 481	7.681 146	24.289 92
25	625	5.000 000	15.811 39	60	3 600	7.745 967	24.494 90
26	676	5.099 020	16.124 52	61	3 721	7.810 250	24.698 18
27	729	5.196 152	16.431 68	62	3 844	7.874 008	24.899 80
28	784	5.291 503	16.733 20	63	3 969	7.937 254	25.099 80
29	841	5.385 165	17.029 39	64	4 096	8.000 000	25.298 22
30	900	5.477 226	17.320 51	65	4 225	8.062 258	25.495 10
31	961	5.567 764	17.606 82	66	4 356	8.124 038	25.690 47
32	1 024	5.656 854	17.888 54	67	4 489	8.185 353	25.884 36
33	1 089	5.744 563	18.165 90	68	4 624	8.246 211	26.076 81
34	1 156	5.830 952	18.439 09	69	4 761	8.306 624	26.267 85

APPENDIX TABLE 13 (Continued)

n	n^2	$\sqrt{n}$	$\sqrt{10n}$	n	n^2	$\sqrt{n}$	$\sqrt{10n}$
70	4 900	8.366 600	26.457 51	110	12 100	10.488 09	33.166 25
71	5 041	8.426 150	26.645 83	111	12 321	10.535 65	33.316 66
72	5 184	8.485 281	26.832 82	112	12 544	10.583 01	33.466 40
73	5 329	8.544 004	27.018 51	113	12 769	10.630 15	33.615 47
74	5 476	8.602 325	27.202 94	114	12 996	10.677 08	33.763 89
75	5 625	8.660 254	27.386 13	115	13 225	10.723 81	33.911 65
76	5 776	8.717 798	27.568 10	116	13 456	10.770 33	34.058 77
77	5 929	8.774 964	27.748 87	117	13 689	10.816 65	34.205 26
78	6 084	8.831 761	27.928 48	118	13 924	10.862 78	34.351 13
79	6 241	8.888 194	28.106 94	119	14 161	10.908.71	34.496 38
80	6 400	8.944 272	28.284 27	120	14 400	10.954 45	34.641 02
81	6 561	9.000 000	28.460 50	121	14 641	11.000 00	34.785 05
82	6 724	9.055 385	28.635 64	122	14 884	11.045 36	34.928 50
83	6 889	9.110 434	28.809 72	123	15 129	11.090 54	35.071 36
84	7 056	9.165 151	28.982 75	124	15 376	11.135 53	35.213 63
85	7 225	9.219 544	29.154 76	125	15 625	11.180 34	35.355 34
86	7 396	9.273 618	29.325 76	126	15 876	11.224 97	35.496 48
87	7 569	9.327 379	29.495 76	127	16 129	11.269 43	35.637 06
88	7 744	9.380 832	29.664 79	128	16 384	11.313 71	35.777 09
89	7 921	9.433 981	29.832 87	129	16 641	11.357 82	35.916 57
90	8 100	9.486 833	30.000 00	130	16 900	11.401 75	36.055 51
91	8 281	9.539 392	30.166 21	131	17 161	11.445 52	36.193 92
92	8 464	9.591 663	30.331 50	132	17 424	11.489 13	36.331 80
93	8 649	9.643 651	30.495 90	133	17 689	11.532 56	36.469 17
94	8 836	9.695 360	30.659 42	134	17 956	11.575 84	36.606 01
95	9 025	9.746 794	30.822 07	135	18 225	11.618 95	36.742 35
96	9 216	9.797 959	30.983 87	136	18 496	11.661 90	36.878 18
97	9 409	9.848 858	31.144 82	137	18 769	11.704 70	37.013 51
98	9 604	9.899 495	31.304 95	138	19 044	11.747 34	37.148 35
99	9 801	9.949 874	31.464 27	139	19 321	11.789 83	37.282 70
100	10 000	10.000 00	31.622 78	140	19 600	11.832 16	37.416 57
101	10 201	10.049 88	31.780 50	141	19 881	11.874 34	37.549 97
102	10 404	10.099 50	31.937 44	142	20 164	11.916 38	37.682 89
103	10 609	10.148 89	32.093 61	143	20 449	11.958 26	37.815 34
104	10 816	10.198 04	32.249 03	144	20 736	12.000 00	37.947 33
105	11 025	10.246 95	32.403 70	145	21 025	12.041 59	38.078 87
106	11 236	10.295 63	32.557 64	146	21 316	12.083 05	38.209 95
107	11 449	10.344 08	32.710 85	147	21 609	12.124 36	38.340 58
108	11 664	10.392 30	32.863 35	148	21 904	12.165 53	38.470 77
109	11 881	10.440 31	33.015 15	149	22 201	12.206 56	38.600 52

APPENDIX TABLE 13 (Continued)

n	n^2	$\sqrt{n}$	$\sqrt{10n}$	n	n^2	$\sqrt{n}$	$\sqrt{10n}$
150	22 500	12.247 45	38.729 83	190	36 100	13.784 05	43.588 99
151	22 801	12.288 21	38.858 72	191	36 481	13.820 27	43.703 55
152	23 104	12.328 83	38.987 18	192	36 864	13.856 41	43.817 80
153	23 409	12.369 32	39.115 21	193	37 249	13.892 44	43.931 77
154	23 716	12.409 67	39.242 83	194	37 636	13.928 39	44.045 43
155	24 025	12.449 90	39.370 04	195	38 025	13.964 24	44.158 80
156	24 336	12.490 00	39.496 84	196	38 416	14.000 00	44.271 89
157	24 649	12.529 96	39.623 23	197	38 809	14.035 67	44.384 68
158	24 964	12.569 81	39.749 21	198	39 204	14.071 25	44.497 19
159	25 281	12.609 52	39.874 80	199	39 601	14.106 74	44.609 42
160	25 600	12.649 11	40.000 00	200	40 000	14.142 14	44.721 36
161	25 921	12.688 58	40.124 81	201	40 401	14.177 45	44.833 02
162	26 244	12.727 92	40.249 22	202	40 804	14.212 67	44.944 41
163	26 569	12.767 15	40.373 26	203	41 209	14.247 81	45.055 52
164	26 896	12.806 25	40.496 91	204	41 616	14.282 86	45.166 36
165	27 225	12.845 23	40.620 19	205	42 025	14.317 82	45.276 93
166	27 556	12.884 10	40.743 10	206	42 436	14.352 70	45.387 22
167	27 889	12.922 85	40.865 63	207	42 849	14.387 49	45.497 25
168	28 224	12.961 48	40.987 80	208	43 264	14.422 21	45.607 02
169	28 561	13.000 00	41.109 61	209	43 681	14.456 83	45.716 52
170	28 900	13.038 40	41.231 06	210	44 100	14.491 38	45.825 76
171	29 241	13.076 70	41.352 15	211	44 521	14.525 84	45.934 74
172	29 584	13.114 88	41.472 88	212	44 944	14.560 22	46.043 46
173	29 929	13.152 95	41.593 27	213	45 369	14.594 52	46.151 92
174	30 276	13.190 91	41.713 31	214	45 796	14.628 74	46.260 13
175	30 625	13.228 76	41.833 00	215	46 225	14.662 88	46.368 09
176	30 976	13.266 50	41.952 35	216	46 656	14.696 94	46.475 80
177	31 329	13.304 13	42.071 37	217	47 089	14.730 92	46.583 26
178	31 684	13.341 66	42.190 05	218	47 524	14.764 82	46.690 47
179	32 041	13.379 09	42.308 39	219	47 961	14.798 65	46.797 44
180	32 400	13.416 41	42.426 41	220	48 400	14.832 40	46.904 16
181	32 761	13.453 62	42.544 09	221	48 841	14.866 07	47.010 64
182	33 124	13.490 74	42.661 46	222	49 284	14.899 66	47.116 88
183	33 489	13.527 75	42.778 50	223	49 729	14.933 18	47.222 88
184	33 856	13.564 66	42.895 22	224	50 176	14.966 63	47.328 64
185	34 225	13.601 47	43.011 63	225	50 625	15.000 00	47.434 16
186	34 596	13.638 18	43.127 72	226	51 076	15.033 30	47.539 46
187	34 969	13.674 79	43.243 50	227	51 529	15.066 52	47.644 52
188	35 344	13.711 31	43.358 97	228	51 984	15.099 67	47.749 35
189	35 721	13.747 73	43.474 13	229	52 441	15.132 75	47.853 94

APPENDIX TABLE 13 (Continued)

n	n^2	$\sqrt{n}$	$\sqrt{10n}$	n	n^2	$\sqrt{n}$	$\sqrt{10n}$
230	52 900	15.165 75	47.958 32	270	72 900	16.431 68	51.961 52
231	53 361	15.198 68	48.062 46	271	73 441	16.462 08	52.057 66
232	53 824	15.231 55	48.166 38	272	73 984	16.492 42	52.153 62
233	54 289	15.264 34	48.270 07	273	74 529	16.522 71	52.249 40
234	54 756	15.297 06	48.373 55	274	75 076	16.552 95	52.345 01
235	55 225	15.329 71	48.476 80	275	75 625	16.583 12	52.440 44
236	55 696	15.362 29	48.579 83	276	76 176	16.613 25	52.535 70
237	56 169	15.394 80	48.682 65	277	76 729	16.643 32	52.630 79
238	56 644	15.427 25	48.785 24	278	77 284	16.673 33	52.725 71
239	57 121	15.459 62	48.887 63	279	77 841	16.703 29	52.820 45
240	57 600	15.491 93	48.989 79	280	78 400	16.733 20	52.915 03
241	58 081	15.524 17	49.091 75	281	78 961	16.763 05	53.009 43
242	58 564	15.556 35	49.193 50	282	79 524	16.792 86	53.103 67
243	59 049	15.588 46	49.295 03	283	80 089	16.822 60	53.197 74
244	59 536	15.620 50	49.396 36	284	80 656	16.852 30	53.291 65
245	60 025	15.652 48	49.497 47	285	81 225	16.881 94	53.385 39
246	60 516	15.684 39	49.598 39	286	81 796	16.911 53	53.478 97
247	61 009	15.716 23	49.699 09	287	82 369	16.941 07	53.572 38
248	61 504	15.748 02	49.799 60	288	82 944	16.970 56	53.665 63
249	62 001	15.779 73	49.899 90	289	83 521	17.000 00	53.758 72
250	62 500	15.811 39	50.000 00	290	84 100	17.029 39	53.851 65
251	63 001	15.842 98	50.099 90	291	84 681	17.058 72	53.944 42
252	63 504	15.874 51	50.199 60	292	85 264	17.088 01	54.037 02
253	64 009	15.905 97	50.299 11	293	85 849	17.117 24	54.129 47
254	64 516	15.937 38	50.398 41	294	86 436	17.146 43	54.221 77
255	65 025	15.968 72	50.497 52	295	87 025	17.175 56	54.313 90
256	65 536	16.000 00	50.596 44	296	87 616	17.204 65	54.405 88
257	66 049	16.031 22	50.695 17	297	88 209	17.233 69	54.497 71
258	66 564	16.062 38	50.793 70	298	88 804	17.262 68	54.589 38
259	67 081	16.093 48	50.892 04	299	89 401	17.291 62	54.680 89
260	67 600	16.124 52	50.990 20	300	90 000	17.320 51	54.772 26
261	68 121	16.155 49	51.088 16	301	90 601	17.349 35	54.863 47
262	68 644	16.186 41	51.185 94	302	91 204	17.378 15	54.954 53
263	69 169	16.217 27	51.283 53	303	91 809	17.406 90	55.045 44
264	69 696	16.248 08	51.380 93	304	92 416	17.435 60	55.136 20
265	70 225	16.278 82	51.478 15	305	93 025	17.464 25	55.226 81
266	70 756	16.309 51	51.575 19	306	93 636	17.492 86	55.317 27
267	71 289	16.340 13	51.672 04	307	94 249	17.521 42	55.407 58
268	71 824	16.370 71	51.768 72	308	94 864	17.549 93	55.497 75
269	72 361	16.401 22	51.865 21	309	95 481	17.578 40	55.587 77

APPENDIX TABLE 13 (Continued)

n	n^2	$\sqrt{n}$	$\sqrt{10n}$	n	n^2	$\sqrt{n}$	$\sqrt{10n}$
310	96 100	17.606 82	55.677 64	350	122 500	18.708 29	59.160 80
311	96 721	17.635 19	55.767 37	351	123 201	18.734 99	59.245 25
312	97 344	17.663 52	55.856 96	352	123 904	18.761 66	59.329 59
313	97 969	17.691 81	55.946 40	353	124 609	18.788 29	59.413 80
314	98 596	17.720 05	56.035 70	354	125 316	18.814 89	59.497 90
315	99 225	17.748 24	56.124 86	355	126 025	18.841 44	59.581 88
316	99 856	17.776 39	56.213 88	356	126 736	18.867 96	59.665 74
317	100 489	17.804 49	56.302 75	357	127 449	18.894 44	59.749 48
318	101 124	17.832 55	56.391 49	358	128 164	18.920 89	59.833 10
319	101 761	17.860 57	56.480 08	359	128 881	18.947 30	59.916 61
320	102 400	17.888 54	56.568 54	360	129 600	18.973 67	60.000 00
321	103 041	17.916 47	56.656 86	361	130 321	19.000 00	60.083 28
322	103 684	17.944 36	56.745 04	362	131 044	19.026 30	60.166 44
323	104 329	17.972 20	56.833 09	363	131 769	19.052 56	60.249 48
324	104 976	18.000 00	56.921 00	364	132 496	19.078 78	60.332 41
325	105 625	18.027 76	57.008 77	365	133 225	19.104 97	60.415 23
326	106 276	18.055 47	57.096 41	366	133 956	19.131 13	60.497 93
327	106 929	18.083 14	57.183 91	367	134 689	19.157 24	60.580 52
328	107 584	18.110 77	57.271 28	368	135 424	19.183 33	60.663 00
329	108 241	18.138 36	57.358 52	369	136 161	19.209 37	60.745 37
330	108 900	18.165 90	57.445 63	370	136 900	19.235 38	60.827 63
331	109 561	18.193 41	57.532 60	371	137 641	19.261 36	60.909 77
332	110 224	18.220 87	57.619 44	372	138 384	19.287 30	60.991 80
333	110 889	18.248 29	57.706 15	373	139 129	19.313 21	61.073 73
334	111 556	18.275 67	57.792 73	374	139 876	19.339 08	61.155 54
335	112 225	18.303 01	57.879 18	375	140 625	19.364 92	61.237 24
336	112 896	18.330 30	57.965 51	376	141 376	19.390 72	61.318 84
337	113 569	18.357 56	58.051 70	377	142 129	19.416 49	61.400 33
338	114 244	18.384 78	58.137 77	378	142 884	19.442 22	61.481 70
339	114 921	18.411 95	58.223 71	379	143 641	19.467 92	61.562 98
340	115 600	18.439 09	58.309 52	380	144 400	19.493 59	61.644 14
341	116 281	18.466 19	58.395 21	381	145 161	19.519 22	61.725 20
342	116 964	18.493 24	58.480 77	382	145 924	19.544 82	61.806 15
343	117 649	18.520 26	58.566 20	383	146 689	19.570 39	61.886 99
344	118 336	18.547 24	58.651 51	384	147 456	19.595 92	61.967 73
345	119 025	18.574 18	58.736 70	385	148 225	19.621 42	62.048 37
346	119 716	18.601 08	58.821 76	386	148 996	19.646 88	62.128 90
347	120 409	18.627 94	58.906 71	387	149 769	19.672 32	62.209 32
348	121 104	18.654 76	58.991 52	388	150 544	19.697 72	62.289 65
349	121 801	18.681 54	59.076 22	389	151 321	19.723 08	62.369 86

APPENDIX TABLE 13 (Continued)

n	n^2	$\sqrt{n}$	$\sqrt{10n}$	n	n^2	$\sqrt{n}$	$\sqrt{10n}$
390	152 100	19.748 42	62.449 98	430	184 900	20.736 44	65.574 39
391	152 881	19.773 72	62.529 99	431	185 761	20.760 54	65.650 59
392	153 664	19.798 99	62.609 90	432	186 624	20.784 61	65.726 71
393	154 449	19.824 23	62.689 71	433	187 489	20.808 65	65.802 74
394	155 236	19.849 43	62.769 42	434	188 356	20.832 67	65.878 68
395	156 025	19.874 61	62.849 03	435	189 225	20.856 65	65.954 53
396	156 816	19.899 75	62.928 53	436	190 096	20.880 61	66.030 30
397	157 609	19.924 86	63.007 94	437	190 969	20.904 54	66.105 98
398	158 404	19.949 94	63.087 24	438	191 844	20.928 45	66.181 57
399	159 201	19.974 98	63.166 45	439	192 721	20.952 33	66.257 08
400	160 000	20.000 00	63.245 55	440	193 600	20.976 18	66.332 50
401	160 801	20.024 98	63.324 56	441	194 481	21.000 00	66.407 83
402	161 604	20.049 94	63.403 47	442	195 364	21.023 80	66.483 08
403	162 409	20.074 86	63.482 28	443	196 249	21.047 57	66.558 25
404	163 216	20.099 75	63.560 99	444	197 136	21.071 31	66.633 32
405	164 025	20.124 61	63.639 61	445	198 025	21.095 02	66.708 32
406	164 836	20.149 44	63.718 13	446	198 916	21.118 71	66.783 23
407	165 649	20.174 24	63.796 55	447	199 809	21.142 37	66.858 06
408	166 464	20.199 01	63.874 88	448	200 704	21.166 01	66.932 80
409	167 281	20.223 75	63.953 11	449	201 601	21.189 62	67.007 46
410	168 100	20.248 46	64.031 24	450	202 500	21.213 20	67.082 04
411	168 921	20.273 13	64.109 28	451	203 401	21.236 76	67.156 53
412	169 744	20.297 78	64.187 23	452	204 304	21.260 29	67.230 95
413	170 569	20.322 40	64.265 08	453	205 209	21.283 80	67.305 27
414	171 396	20.346 99	64.342 83	454	206 116	21.307 28	67.379 52
415	172 225	20.371 55	64.420 49	455	207 025	21.330 73	67.453 69
416	173 056	20.396 08	64.498 06	456	207 936	21.354 16	67.527 77
417	173 889	20.420 58	64.575 54	457	208 849	21.377 56	67.601 78
418	174 724	20.445 05	64.652 92	458	209 764	21.400 93	67.675 70
419	175 561	20.469 49	64.730 21	459	210 681	21.424 29	67.749 54
420	176 400	20.493 90	64.807 41	460	211 600	21.447 61	67.823 30
421	177 241	20.518 28	64.884 51	461	212 521	21.470 91	67.896 98
422	178 084	20.542 64	64.961 53	462	213 444	21.494 19	67.970 58
423	178 929	20.566 96	65.038 45	463	214 369	21.517 43	68.044 10
424	179 776	20.591 26	65.115 28	464	215 296	21.540 66	68.117 55
425	180 625	20.615 53	65.192 02	465	216 225	21.563 86	68.190 91
426	181 476	20.639 77	65.268 68	466	217 156	21.587 03	68.264 19
427	182 329	20.663 98	65.345 24	467	218 089	21.610 18	68.337 40
428	183 184	20.688 16	65.421 71	468	219 024	21.633 31	68.410 53
429	184 041	20.712 32	65.498 09	469	219 961	21.656 41	68.483 57

APPENDIX TABLE 13 (Continued)

n	n^2	$\sqrt{n}$	$\sqrt{10n}$	n	n^2	$\sqrt{n}$	$\sqrt{10n}$
470	220 900	21.679 48	68.556 55	510	260 100	22.583 18	71.414 28
471	221 841	21.702 53	68.629 44	511	261 121	22.605 31	71.484 26
472	222 784	21.725 56	68.702 26	512	262 144	22.627 42	71.554 18
473	223 729	21.748 56	68.775 00	513	263 169	22.649 50	71.624 02
474	224 676	21.771 54	68.847 66	514	264 196	22.671 57	71.693 79
475	225 625	21.794 49	68.920 24	515	265 225	22.693 61	71.763 50
476	226 576	21.817 42	68.992 75	516	266 256	22.715 63	71.833 14
477	227 529	21.840 33	69.065 19	517	267 289	22.737 63	71.902 71
478	228 484	21.863 21	69.137 54	518	268 324	22.759 61	71.972 22
479	229 441	21.886 07	69.209 83	519	269 361	22.781 57	72.041 65
480	230 400	21.908 90	69.282 03	520	270 400	22.803 51	72.111 03
481	231 361	21.931 71	69.354 16	521	271 441	22.825 42	72.180 33
482	232 324	21.954 50	69.426 22	522	272 484	22.847 32	72.249 57
483	233 289	21.977 26	69.498 20	523	273 529	22.869 19	72.318 74
484	234 256	22.000 00	69.570 11	524	274 576	22.891 05	72.387 84
485	235 225	22.022 72	69.641 94	525	275 625	22.912 88	72.456 88
486	236 196	22.045 41	69.713 70	526	276 676	22.934 69	72.525 86
487	237 169	22.068 08	69.785 39	527	277 729	22.956 48	72.594 77
488	238 144	22.090 72	69.857 00	528	278 784	22.978 25	72.663 61
489	239 121	22.113 34	69.928 53	529	279 841	23.000 00	72.732 39
490	240 100	22.135 94	70.000 00	530	280 900	23.021 73	72.801 10
491	241 081	22.158 52	70.071 39	531	281 961	23.043 44	72.869 75
492	242 064	22.181 07	70.142 71	532	283 024	23.065 13	72.938 33
493	243 049	22.203 60	70.213 96	533	284 089	23.086 79	73.006 85
494	244 036	22.226 11	70.285 13	534	285 156	23.108 44	73.075 30
495	245 025	22.248 60	70.356 24	535	286 225	23.130 07	73.143 69
496	246 016	22.271 06	70.427 27	536	287 296	23.151 67	73.212 02
497	247 009	22.293 50	70.498 23	537	288 369	23.173 26	73.280 28
498	248 004	22.315 91	70.569 12	538	289 444	23.194 83	73.348 48
499	249 001	22.338 31	70.639 93	539	290 521	23.216 37	73.416 62
500	250 000	22.360 68	70.710 68	540	291 600	23.237 90	73.484 69
501	251 001	22.383 03	70.781 35	541	292 681	23.259 41	73.552 70
502	252 004	22.405 36	70.851 96	542	293 764	23.280 89	73.620 65
503	253 009	22.427 66	70.922 49	543	294 849	23.302 36	73.688 53
504	254 016	22.449 94	70.992 96	544	295 936	23.323 81	73.756 36
505	255 025	22.472 21	71.063 35	545	297 025	23.345 24	73.824 12
506	256 036	22.494 44	71.133 68	546	298 116	23.366 64	73.891 81
507	257 049	22.516 66	71.203 93	547	299 209	23.388 03	73.959 45
508	258 064	22.538 86	71.274 12	548	300 304	23.409 40	74.027 02
509	259 081	22.561 03	71.344 24	549	301 401	23.430 75	74.094 53

APPENDIX TABLE 13 (Continued)

n	n^2	$\sqrt{n}$	$\sqrt{10n}$	n	n^2	$\sqrt{n}$	$\sqrt{10n}$
550	302 500	23.452 08	74.161 98	590	348 100	24.289 92	76.811 46
551	303 601	23.473 39	74.229 37	591	349 281	24.310 49	76.876 52
552	304 704	23.494 68	74.296 70	592	350 464	24.331 05	76.941 54
553	305 809	23.515 95	74.363 97	593	351 649	24.351 59	77.006 49
554	306 916	23.537 20	74.431 18	594	352 836	24.372 12	77.071 40
555	308 025	23.558 44	74.498 32	595	354 025	24.392 62	77.136 24
556	309 136	23.579 65	74.565 41	596	355 216	24.413 11	77.201 04
557	310 249	23.600 85	74.632 43	597	356 409	24.433 58	77.265 78
558	311 364	23.622 02	74.699 40	598	357 604	24.454 04	77.330 46
559	312 481	23.643 18	74.766 30	599	358 801	24.474 48	77.395 09
560	313 600	23.664 32	74.833 15	600	360 000	24.494 90	77.459 67
561	314 721	23.685 44	74.899 93	601	361 201	24.515 30	77.524 19
562	315 844	23.706 54	74.966 66	602	362 404	24.535 69	77.588 66
563	316 969	23.727 62	75.033 33	603	363 609	24.556 06	77.653 07
564	318 096	23.748 68	75.099 93	604	364 816	24.576 41	77.717 44
565	319 225	23.769 73	75.166 48	605	366 025	24.596 75	77.781 75
566	320 356	23.790 75	75.232 97	606	367 236	24.617 07	77.846 00
567	321 489	23.811 76	75.299 40	607	368 449	24.637 37	77.910 20
568	322 624	23.832 75	75.365 77	608	369 664	24.657 66	77.974 35
569	323 761	23.853 72	75.432 09	609	370 881	24.677 93	78.038 45
570	324 900	23.874 67	75.498 34	610	372 100	24.698 18	78.102 50
571	326 041	23.895 61	75.564 54	611	373 321	24.718 41	78.166 49
572	327 184	23.916 52	75.630 68	612	374 544	24.738 63	78.230 43
573	328 329	23.937 42	75.696 76	613	375 769	24.758 84	78.294 32
574	329 476	23.958 30	75.762 79	614	376 996	24.779 02	78.358 15
575	330 625	23.979 16	75.828 75	615	378 225	24.799 19	78.421 94
576	331 776	24.000 00	75.894 66	616	379 456	24.819 35	78.485 67
577	332 929	24.020 82	75.960 52	617	380 689	24.839 48	78.549 35
578	334 084	24.041 63	76.026 31	618	381 924	24.859 61	78.612 98
579	335 241	24.062 42	76.092 05	619	383 161	24.879 71	78.676 55
580	336 400	24.083 19	76.157 73	620	384 400	24.899 80	78.740 08
581	337 561	24.103 94	76.223 36	621	385 641	24.919 87	78.803 55
582	338 724	24.124 68	76.288 92	622	386 884	24.939 93	78.866 98
583	339 889	24.145 39	76.354 44	623	388 129	24.959 97	78.930 35
584	341 056	24.166 09	76.419 89	624	389 376	24.979 99	78.993 67
585	342 225	24.186 77	76.485 29	625	390 625	25.000 00	79.056 94
586	343 396	24.207 44	76.550 64	626	391 876	25.019 99	79.120 16
587	344 569	24.228 08	76.615 93	627	393 129	25.039 97	79.183 33
588	345 744	24.248 71	76.681 16	628	394 384	25.059 93	79.246 45
589	346 921	24.269 32	76.746 34	629	395 641	25.079 87	79.309 52

APPENDIX TABLE 13 (Continued)

n	n^2	$\sqrt{n}$	$\sqrt{10n}$	n	n^2	$\sqrt{n}$	$\sqrt{10n}$
630	396 900	25.099 80	79.372 54	670	448 900	25.884 36	81.853 53
631	398 161	25.119 71	79.435 51	671	450 241	25.903 67	81.914 59
632	399 424	25.139 61	79.498 43	672	451 584	25.922 96	81.975 61
633	400 689	25.159 49	79.561 30	673	452 929	25.942 24	82.036 58
634	401 956	25.179 36	79.624 12	674	454 276	25.961 51	82.097 50
635	403 225	25.199 21	79.686 89	675	455 625	25.980 76	82.158 38
636	404 496	25.219 04	79.749 61	676	456 976	26.000 00	82.219 22
637	405 769	25.238 86	79.812 28	677	458 329	26.019 22	82.280 01
638	407 044	25.258 66	79.874 90	678	459 684	26.038 43	82.340 76
639	408 321	25.278 45	79.937 48	679	461 041	26.057 63	82.401 46
640	409 600	25.298 22	80.000 00	680	462 400	26.076 81	82.462 11
641	410 881	25.317 98	80.062 48	681	463 761	26.095 98	82.522 72
642	412 164	25.337 72	80.124 90	682	465 124	26.115 13	82.583 29
643	413 449	25.357 44	80.187 28	683	466 489	26.134 27	82.643 81
644	414 736	25.377 16	80.249 61	684	467 856	26.153 39	82.704 29
645	416 025	25.396 85	80.311 89	685	469 225	26.172 50	82.764 73
646	417 316	25.416 53	80.374 13	686	470 596	26.191 60	82.825 12
647	418 609	25.436 19	80.436 31	687	471 969	26.210 68	82.885 46
648	419 904	25.455 84	80.498 45	688	473 344	26.229 75	82.945 77
649	421 201	25.475 48	80.560 54	689	474 721	26.248 81	83.006 02
650	422 500	25.495 10	80.622 58	690	476 100	26.267 85	83.066 24
651	423 801	25.514 70	80.684 57	691	477 481	26.286 88	83.126 41
652	425 104	25.534 29	80.746 52	692	478 864	26.305 89	83.186 54
653	426 409	25.553 86	80.808 42	693	480 249	26.324 89	83.246 62
654	427 716	25.573 42	80.870 27	694	481 636	26.343 88	83.306 66
655	429 025	25.592 97	80.932 07	695	483 025	26.362 85	83.366 66
656	430 336	25.612 50	80.993 83	696	484 416	26.381 81	83.426 61
657	431 649	25.632 01	81.055 54	697	485 809	26.400 76	83.486 53
658	432 964	25.651 51	81.117 20	698	487 204	26.419 69	83.546 39
659	434 281	25.671 00	81.178 81	699	488 601	26.438 61	83.606 22
660	435 600	25.690 47	81.240 38	700	490 000	26.457 51	83.666 00
661	436 921	25.709 92	81.301 91	701	491 401	26.476 40	83.725 74
662	438 244	25.729 36	81.363 38	702	492 804	26.495 28	83.785 44
663	439 569	25.748 79	81.424 81	703	494 209	26.514 15	83.845 10
664	440 896	25.768 20	81.486 20	704	495 616	26.533 00	83.904 71
665	442 225	25.787 59	81.547 53	705	497 025	26.551 84	83.964 28
666	443 556	25.806 98	81.608 82	706	498 436	26.570 66	84.023 81
667	444 889	25.826 34	81.670 07	707	499 849	26.589 47	84.083 29
668	446 224	25.845 70	81.731 27	708	501 264	26.608 27	84.142 74
669	447 561	25.865 03	81.792 42	709	502 681	26.627 05	84.202 14

APPENDIX TABLE 13 (Continued)

n	n^2	$\sqrt{n}$	$\sqrt{10n}$	n	n^2	$\sqrt{n}$	$\sqrt{10n}$
710	504 100	26.645 83	84.261 50	750	562 500	27.386 13	86.602 54
711	505 521	26.664 58	84.320 82	751	564 001	27.404 38	86.660 26
712	506 944	26.683 33	84.380 09	752	565 504	27.422 62	86.717 93
713	508 369	26.702 06	84.439 33	753	567 009	27.440 85	86.775 57
714	509 796	26.720 78	84.498 52	754	568 516	27.459 06	86.833 17
715	511 225	26.739 48	84.557 67	755	570 025	27.477 26	86.890 74
716	512 656	26.758 18	84.616 78	756	571 536	27.495 45	86.948 26
717	514 089	26.776 86	84.675 85	757	573 049	27.513 63	87.005 75
718	515 524	26.795 52	84.734 88	758	574 564	27.531 80	87.063 20
719	516 961	26.814 18	84.793 87	759	576 081	27.549 95	87.120 61
720	518 400	26.832 82	84.852 81	760	577 600	27.568 10	87.177 98
721	519 841	26.851 44	84.911 72	761	579 121	27.586 23	87.235 31
722	521 284	26.870 06	84.970 58	762	580 644	27.604 35	87.292 61
723	522 729	26.888 66	85.029 41	763	582 169	27.622 45	87.349 87
724	524 176	26.907 25	85.088 19	764	583 696	27.640 55	87.407 09
725	525 625	26.925 82	85.146 93	765	585 225	27.658 63	87.464 28
726	527 076	26.944 39	85.205 63	766	586 756	27.676 71	87.521 43
727	528 529	26.962 94	85.264 29	767	588 289	27.694 76	87.578 54
728	529 984	26.981 48	85.322 92	768	589 824	27.712 81	87.635 61
729	531 441	27.000 00	85.381 50	769	591 361	27.730 85	87.692 65
730	532 900	27.018 51	85.440 04	770	592 900	27.748 87	87.749 64
731	534 361	27.037 01	85.498 54	771	594 441	27.766 89	87.806 61
732	535 824	27.055 50	85.557 00	772	595 984	27.784 89	87.863 53
733	537 289	27.073 97	85.615 42	773	597 529	27.802 88	87.920 42
734	538 756	27.092 43	85.673 80	774	599 076	27.820 86	87.977 27
735	540 225	27.110 88	85.732 14	775	600 625	27.838 82	88.034 08
736	541 696	27.129 32	85.790 44	776	602 176	27.856 78	88.090 86
737	543 169	27.147 74	85.848 70	777	603 729	27.874 72	88.147 60
738	544 644	27.166 16	85.906 93	778	605 284	27.892 65	88.204 31
739	546 121	27.184 55	85.965 11	779	606 841	27.910 57	88.260 98
740	547 600	27.202 94	86.023 25	780	608 400	27.928 48	88.317 61
741	549 081	27.221 32	86.081 36	781	609 961	27.946 38	88.374 20
742	550 564	27.239 68	86.139 42	782	611 524	27.964 26	88.430 76
743	552 049	27.258 03	86.197 45	783	613 089	27.982 14	88.487 29
744	553 536	27.276 36	86.255 43	784	614 656	28.000 00	88.543 77
745	555 025	27.294 69	86.313 38	785	616 225	28.017 85	88.600 23
746	556 516	27.313 00	86.371 29	786	617 796	28.035 69	88.656 64
747	558 009	27.331 30	86.429 16	787	619 369	28.053 52	88.713 02
748	559 504	27.349 59	86.486 99	788	620 944	28.071 34	88.769 36
749	561 001	27.367 86	86.544 79	789	622 521	28.089 14	88.825 67

APPENDIX TABLE 13 (Continued)

n	n^2	$\sqrt{n}$	$\sqrt{10n}$	n	n^2	$\sqrt{n}$	$\sqrt{10n}$
790	624 100	28.106 94	88.881 94	830	688 900	28.809 72	91.104 34
791	625 681	28.124 72	88.938 18	831	690 561	28.827 07	91.159 20
792	627 264	28.142 49	88.994 38	832	692 224	28.844 41	91.214 03
793	628 849	28.160 26	89.050 55	833	693 889	28.861 74	91.268 83
794	630 436	28.178 01	89.106 68	834	695 556	28.879 06	91.323 60
795	632 025	28.195 74	89.162 77	835	697 225	28.896 37	91.378 33
796	633 616	28.213 47	89.218 83	836	698 896	28.913 66	91.433 04
797	635 209	28.231 19	89.274 86	837	700 569	28.930 95	91.487 70
798	636 804	28.248 89	89.330 85	838	702 244	28.948 23	91.542 34
799	638 401	28.266 59	89.386 80	839	703 921	28.965 50	91.596 94
800	640 000	28.284 27	89.442 72	840	705 600	28.982 75	91.651 51
801	641 601	28.301 94	89.498 60	841	707 281	29.000 00	91.706 05
802	643 204	28.319 60	89.554 45	842	708 964	29.017 24	91.760 56
803	644 809	28.337 25	89.610 27	843	710 649	29.034 46	91.815 03
804	646 416	28.354 89	89.666 05	844	712 336	29.051 68	91.869 47
805	648 025	28.372 52	89.721 79	845	714 025	29.068 88	91.923 88
806	649 636	28.390 14	89.777 50	846	715 716	29.086 08	91.978 26
807	651 249	28.407 75	89.833 18	847	717 409	29.103 26	92.032 60
808	652 864	28.425 34	89.888 82	848	719 104	29.120 44	92.086 92
809	654 481	28.442 93	89.944 43	849	720 801	29.137 60	92.141 20
810	656 100	28.460 50	90.000 00	850	722 500	29.154 76	92.195 44
811	657 721	28.478 06	90.055 54	851	724 201	29.171 90	92.249 66
812	659 344	28.495 61	90.111 04	852	725 904	29.189 04	92.303 85
813	660 969	28.513 15	90.166 51	853	727 609	29.206 16	92.358 00
814	662 596	28.530 69	90.221 95	854	729 316	29.223 28	92.412 12
815	664 225	28.548 20	90.277 35	855	731 025	29.240 38	92.466 21
816	665 856	28.565 71	90.332 72	856	732 736	29.257 48	92.520 27
817	667 489	28.583 21	90.388 05	857	734 449	29.274 56	92.574 29
818	669 124	28.600 70	90.443 35	858	736 164	29.291 64	92.628 29
819	670 761	28.618 18	90.498 62	859	737 881	29.308 70	92.682 25
820	672 400	28.635 64	90.553 85	860	739 600	29.325 76	92.736 18
821	674 041	28.653 10	90.609 05	861	741 321	29.342 80	92.790 09
822	675 684	28.670 54	90.664 22	862	743 044	29.359 84	92.843 96
823	677 329	28.687 98	90.719 35	863	744 769	29.376 86	92.897 79
824	678 976	28.705 40	90.774 45	864	746 496	29.393 88	92.951 60
825	680 625	28.722 81	90.829 51	865	748 225	29.410 88	93.005 38
826	682 276	28.740 22	90.884 54	866	749 956	29.427 88	93.059 12
827	683 929	28.757 61	90.939 54	867	751 689	29.444 86	93.112 83
828	685 584	28.774 99	90.994 51	868	753 424	29.461 84	93.166 52
829	687 241	28.792 36	91.049 44	869	755 161	29.478 81	93.220 17

APPENDIX TABLE 13 (Continued)

n	n^2	$\sqrt{n}$	$\sqrt{10n}$	n	n^2	$\sqrt{n}$	$\sqrt{10n}$
870	756 900	29.495 76	93.273 79	910	828 100	30.166 21	95.393 92
871	758 641	29.512 71	93.327 38	911	829 921	30.182 78	95.446 32
872	760 384	29.529 65	93.380 94	912	831 744	30.199 34	95.498 69
873	762 129	29.546 57	93.434 47	913	833 569	30.215 89	95.551 03
874	763 876	29.563 49	93.487 97	914	835 396	30.232 43	95.603 35
875	765 625	29.580 40	93.541 43	915	837 225	30.248 97	95.655 63
876	767 376	29.597 30	93.594 87	916	839 056	30.265 49	95.707 89
877	769 129	29.614 19	93.648 28	917	840 889	30.282 01	95.760 12
878	770 884	29.631 06	93.701 65	918	842 724	30.298 51	95.812 32
879	772 641	29.647 93	93.755 00	919	844 561	30.315 01	95.864 49
880	774 400	29.664 79	93.808 32	920	846 400	30.331 50	95.916 63
881	776 161	29.681 64	93.861 60	921	848 241	30.347 98	95.968 74
882	777 924	29.698 48	93.914 86	922	850 084	30.364 45	96.020 83
883	779 689	29.715 32	93.968 08	923	851 929	30.380 92	96.072 89
884	781 456	29.732 14	94.021 27	924	853 776	30.397 37	96.124 92
885	783 225	29.748 95	94.074 44	925	855 625	30.413 81	96.176 92
886	784 996	29.765 75	94.127 57	926	857 476	30.430 25	96.228 89
887	786 769	29.782 55	94.180 68	927	859 329	30.446 67	96.280 84
888	788 544	29.799 33	94.233 75	928	861 184	30.463 09	96.332 76
889	790 321	29.816 10	94.286 80	929	863 041	30.479 50	96.384 65
890	792 100	29.832 87	94.339 81	930	864 900	30.495 90	96.436 51
891	793 881	29.849 62	94.392 80	931	866 761	30.512 29	96.488 34
892	795 664	29.866 37	94.445 75	932	868 624	30.528 68	96.540 15
893	797 449	29.883 11	94.498 68	933	870 489	30.545 05	96.591 93
894	799 236	29.899 83	94.551 57	934	872 356	30.561 41	96.643 68
895	801 025	29.916 55	94.604 44	935	874 225	30.577 77	96.695 40
896	802 816	29.933 26	94.657 28	936	876 096	30.594 12	96.747 09
897	804 609	29.949 96	94.710 08	937	877 969	30.610 46	96.798 76
898	806 404	29.966 65	94.762 86	938	879 844	30.626 79	96.850 40
899	808 201	29.983 33	94.815 61	939	881 721	30.643 11	96.902 01
900	810 000	30.000 00	94.868 33	940	883 600	30.659 42	96.953 60
901	811 801	30.016 66	94.921 02	941	885 481	30.675 72	97.005 15
902	813 604	30.033 31	94.973 68	942	887 364	30.692 02	97.056 68
903	815 409	30.049 96	95.026 31	943	889 249	30.708 31	97.108 19
904	817 216	30.066 59	95.078 91	944	891 136	30.724 58	97.159 66
905	819 025	30.083 22	95.131 49	945	893 025	30.740 85	97.211 11
906	820 836	30.099 83	95.184 03	946	894 916	30.757 11	97.262 53
907	822 649	30.116 44	95.236 55	947	896 809	30.773 37	97.313 93
908	824 464	30.133 04	95.289 03	948	898 704	30.789 61	97.365 29
909	826 281	30.149 63	95.341 49	949	900 601	30.805 84	97.416 63

APPENDIX TABLE 13 (Continued)

n	n^2	$\sqrt{n}$	$\sqrt{10n}$	n	n^2	$\sqrt{n}$	$\sqrt{10n}$
950	902 500	30.822 07	97.467 94	975	950 625	31.224 99	98.742 09
951	904 401	30.838 29	97.519 23	976	952 576	31.241 00	98.792 71
952	906 304	30.854 50	97.570 49	977	954 529	31.257 00	98.843 31
953	908 209	30.870 70	97.621 72	978	956 484	31.272 99	98.893 88
954	910 116	30.886 89	97.672 92	979	958 441	31.288 98	98.944 43
955	912 025	30.903 07	97.724 10	980	960 400	31.304 95	98.994 95
956	913 936	30.919 25	97.775 25	981	962 361	31.320 92	99.045 44
957	915 849	30.935 42	97.826 38	982	964 324	31.336 88	99.095 91
958	917 764	30.951 58	97.877 47	983	966 289	31.352 83	99.146 36
959	919 681	30.967 73	97.928 55	984	968 256	31.368 77	99.196 77
960	921 600	30.983 87	97.979 59	985	970 225	31.384 71	99.247 17
961	923 521	31.000 00	98.030 61	986	972 196	31.400 64	99.297 53
962	925 444	31.016 12	98.081 60	987	974 169	31.416 56	99.347 87
963	927 369	31.032 24	98.132 56	988	976 144	31.432 47	99.398 19
964	929 296	31.048 35	98.183 50	989	978 121	31.448 37	99.448 48
965	931 225	31.064 45	98.234 41	990	980 100	31.464 27	99.498 74
966	933 156	31.080 54	98.285 30	991	982 081	31.480 15	99.548 98
967	935 089	31.096 62	98.336 16	992	984 064	31.496 03	99.599 20
968	937 024	31.112 70	98.386 99	993	986 049	31.511 90	99.649 39
969	938 961	31.128 76	98.437 80	994	988 036	31.527 77	99.699 55
970	940 900	31.144 82	98.488 58	995	990 025	31.543 62	99.749 69
971	942 841	31.160 87	98.539 33	996	992 016	31.559 47	99.799 80
972	944 784	31.176 91	98.590 06	997	994 009	31.575 31	99.849 89
973	946 729	31.192 95	98.640 76	998	996 004	31.591 14	99.899 95
974	948 676	31.208 97	98.691 44	999	998 001	31.606 96	99.949 99

Answers to Odd-Numbered Exercises*

Chapter 1

). 13–14) **1.1** (a) The population consists of all people walking on the town's boardwalk between 10 A.M. and noon during the second week in August.

(b) Finite.

(c) Qualitative. Each person in the population feels either that an elementary school with less than 200 students should remain in operation, or that it should not.

(d) The population is different from the one that is really relevant. What the mayor should be concerned with is the population of townspeople, since they are the ones who pay the taxes and whose children are affected. Many of the people on the boardwalk are visitors, not permanent residents of the town.

(e) A sample might be taken of the permanent residents of the town, not of the people on the boardwalk.

1.3 Yes. The alumni who answered are likely to be different in relevant ways from those who did not. For example, those who feel strongly that the university's policies are wrong are probably more likely to reply than those who do not feel strongly in this regard.

1.5 No. This comparison is foolish, because it takes no account of the fact that there are many more flights for longer distances in 1980 than in 1940.

). 23–24) **1.7** (a) It is possible to construct a frequency distribution, but the advantages derived from doing so are not considerable when there are so few observations.

(b) One possibility might be

Income per farm (dollars)	Number of states
5,000 and under 12,000	3
12,000 and under 19,000	3
19,000 and under 26,000	3
26,000 and under 33,000	0
33,000 and under 40,000	1
Total	10

*Numerical answers depend on how calculations are rounded off, and on the order of operations; thus your answers may not agree exactly with the answers given here. I am indebted to Professor Warren Boe of the University of Iowa for supervising the checking of these answers.

1.9 You would be able to answer (a), (b), and (e).

1.11

Error (dollars)	Number of customers
Less than − 1.00	0
Less than − 0.75	1
Less than − 0.50	3
Less than − 0.25	7
Less than 0.00	37
Less than + 0.25	43
Less than + 0.50	45
Less than + 0.75	47
Less than + 1.00	49
Less than + 1.25	50

Chapter 2

(pp. 36–39) **2.1** (a) 45.5/9 = 5.06 percent
(b) 41.9/9 = 4.66 percent
(c) The mean profit rate for durable manufacturing industries was about 4/10 of a percentage point higher than for nondurable manufacturing industries in 1974.

2.3 The mean profit rate for all manufacturing industries is the mean of the mean profit rate for nondurable manufacturing industries and the mean profit rate for durable manufacturing industries.

2.5 An estimate of the mean is [60($10) + 30($30) + 10($50)] ÷ 100 = $20. An estimate of the total is $2,000.

2.7 The mode is 1/2 acre. The mean is [100(1/4) + 500(1/2) + 50(1) + 20(2)] ÷ 670, or 365 ÷ 670 = 0.54. Thus, the mode is not greater than the mean. The median is 1/2 acre, so the mode and median are equal.

2.9 Zero.

(pp. 47–48) **2.11** (a) If these lots are regarded as a sample, the standard deviation equals

$$\sqrt{\frac{300(1/4)^2 + 400(1/2)^2 + 200(1^2) + 100(2)^2 - \frac{1}{1,000}(675)^2}{999}},$$

based on equation (2.13). Thus, it equals

$$\sqrt{\frac{18.75 + 100 + 200 + 400 - 455.625}{999}} = \sqrt{.2634} = .51 \text{ acres.}$$

(b) From the above, it is obvious that the variance equals .2634 (acres)^2.
(c) 1.75 acres
(d) 1 acre − 1/4 acre, or 3/4 acres

2.13 No. The range must be less than $60, since the smallest observation is no less than zero and the largest observation is less than $60. The range must be more than $20, since the smallest observation is less than $20 and the largest observation is $40 or more.

2.15 Since we know from Exercise 2.5 that the mean is $20, the standard deviation (if the data in Exercise 2.5 are the entire population) equals

$$\sqrt{\frac{60(10 - 20)^2 + 30(30 - 20)^2 + 10(50 - 20)^2}{100}} = \sqrt{180} = \$13.42.$$

If we calculate the standard deviation in cents, we have

$$\sqrt{\frac{60(1000 - 2000)^2 + 30(3000 - 2000)^2 + 10(5000 - 2000)^2}{100}} = \sqrt{1,800,000}$$

$$= 1,342 \text{ cents.}$$

The ratio of the latter standard deviation to the former is 100, since each observation expressed in cents is 100 times the same observation expressed in dollars.

2.17 It is skewed to the right. If you plot the histogram, this is obvious.

(p. 56) **2.19** This evidence is biased, because farmers constitute a large segment of the population in rural Illinois. For this reason alone one might expect that a great many farmers would be treated for alcoholism (or practically any disease) by health organizations in this area.

2.21 He is using the mean. None.

2.23 Although it may be true that most patents are invalidated, those that are not invalidated may have very great importance and value. The variation about the average is neglected. Also, many patents are never contested before the Supreme Court. Thus, this may not be the relevant population.

Chapter 3

69–70) **3.1**

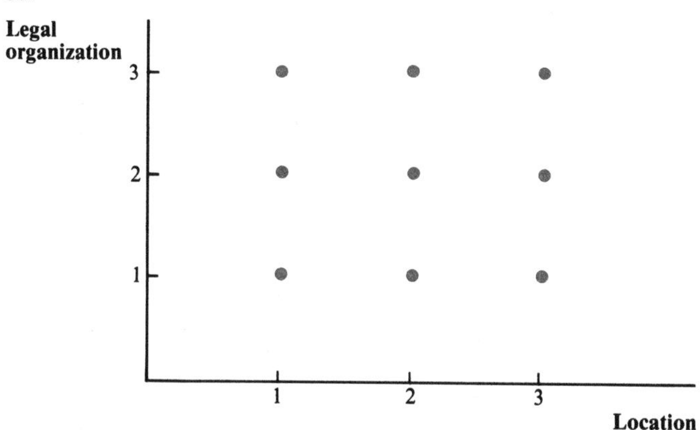

3.3 (a) It buys a Michigan proprietorship, or a Michigan partnership, or a Michigan corporation.

(b) It buys a Michigan corporation, or an Illinois corporation, or a New York corporation.

(c) It buys a Michigan proprietorship, or an Illinois proprietorship, or a New York proprietorship; or a Michigan corporation, or an Illinois corporation, or a New York corporation.

3.5 (a) Two events must occur: (1) Bona Fide must buy a proprietorship; (2) Bona Fide must buy a New York firm.

(b) Two events must occur: (1) Bona Fide must buy a corporation; (2) Bona Fide must buy a Michigan firm.

(pp. 80–82) **3.7** (a) 0.7
(b) 0.3
(c) Yes, it would decrease the answer (if the probability that they will buy a Ford is 0.3, if the probability that they will buy a Chevrolet is 0.4, and if they will buy only one of these three cars.)

3.9 (a) .70
(b) .40

3.11 (a) 100 ÷ 900
(b) 100 ÷ 1100
(c) 400 ÷ 2100
(d) 300 ÷ 2500

3.13 $P(\text{supplier I}|\text{defective}) = P(\text{defective and supplier I}) \div P(\text{defective}) = 0.10 \div .20 = 0.50$.

3.15 (a) No, because $P(A$ and $B)$ does not equal zero.
(b) No. If A and B were statistically independent, $P(A$ and $B)$ would equal $P(A)$ times $P(B)$.

3.17 (a) From Figure 3.1 it is clear that there are five points in the sample space where the sum equals 6: (5,1), (4,2), (3,3), (2,4), and (1,5). Thus, the probability is 5/36.
(b) From Figure 3.1 it is clear that there are 10 points in the sample space where the sum equals less than 6: (1,1), (1,2), (1,3), (1,4), (2,1), (2,2), (2,3), (3,1), (3,2), (4,1). Thus, the probability equals 10/36.
(c) Since $P(7 \text{ or more}) = 1 - P(6) - P(\text{less than } 6)$, it follows from parts (a) and (b) that $P(7 \text{ or more}) = 1 - 5/36 - 10/36 = 21/36$.

(pp. 87–88) **3.19** For Tom the probability is

$$\frac{(.001)(.50)}{(.001)(.50) + (.002)(.25) + (.001)(.25)} = \frac{.00050}{.00050 + .00050 + .00025} = \frac{50}{125}$$

$$= .40.$$

For Dick the probability is

$$\frac{(.002)(.25)}{(.001)(.50) + (.002)(.25) + (.001)(.25)} = \frac{.00050}{.00050 + .00050 + .00025} = \frac{50}{125}$$

$$= .40.$$

For Jane the probability is

$$\frac{(.001)(.25)}{(.001)(.50) + (.002)(.25) + (.001)(.25)} = \frac{.00025}{.00050 + .00050 + .00025} = \frac{25}{125}$$

$$= .20.$$

3.21 There are (4)(3), or 12, possible routes. Specifically, they are *Aa, Ab, Ac, Ba, Bb, Bc, Ca, Cb, Cc, Da, Db,* and *Dc*. If each route is equally likely, there is a 1/12 probability that he is right.

3.23 2/3

3.25 Most statisticians do not regard Bayes' theorem and subjective probability as useless for either scientific analysis or business decision making. However, the accuracy of the results depends on the accuracy of the subjective probabilities. If the latter are worthless, the results are likely to be the same.

(p. 97–98) **3.27** One-fifth of all the firm's motors must come from each supplier. Since each supplier ships 1/4 of its motors to each plant, (1/4)(1/5) or 1/20 of all the firm's motors will have an A on them.

3.29 There are 7! ÷ (4!3!) different groups of three that can be chosen. Since 7! ÷ (4!3!) = (7)(6)(5) ÷ (3)(2)(1), the answer is 35.

3.31 Since

$$\binom{n}{x} = \frac{n!}{(n-x)!x!},$$

it follows that

$$\binom{n}{n-x} = \frac{n!}{[n-(n-x)]!(n-x)!} = \frac{n!}{x!(n-x)!}$$
$$= \binom{n}{x}$$

3.33 The probability that the first item is not defective is 45/50. The probability that the second item is not defective (given that the first item is not defective) is 44/49. Thus, the answer is

$$\left(\frac{45}{50}\right)\left(\frac{44}{49}\right) = \frac{1{,}980}{2{,}450} = 0.808.$$

Chapter 4

(p. 108–9) **4.1** (a) Yes.
(b) 0, 1, 2, 3, or 4.
(c)

Number of bicycles	Probability
0	1/5
1	1/5
2	1/5
3	1/5
4	1/5

4.3 1/5
3/5
4/5
1
1

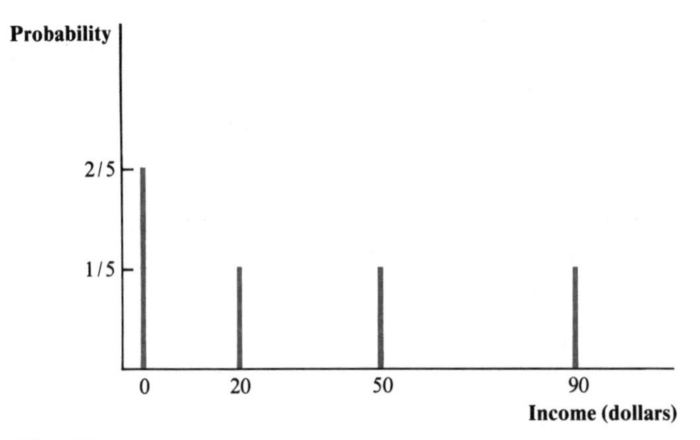

4.7 (a)

Number of bicycles	Probability
0	1/25
1	2/25
2	3/25
3	4/25
4	5/25
5	4/25
6	3/25
7	2/25
8	1/25

(b) 19/25
10/25
3/25
0

4.9

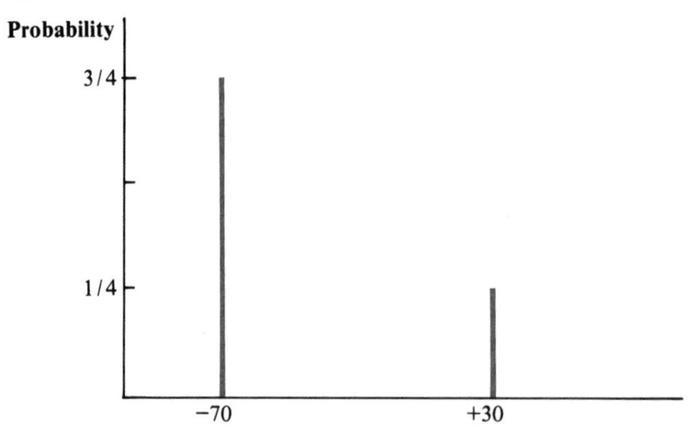

(pp. 115–16) **4.11** (a) The expected value equals

$$(1/5)(0) + (1/5)(1) + (1/5)(2) + (1/5)(3) + (1/5)(4) = 2.$$

(b) Yes.

4.13 (a) The salesman's income under each possible set of circumstances is shown below:

Number of bicycles sold			
First day	Second day	Probability	Income (dollars)
0	0	1/25	0
0	1	1/25	0
0	2	1/25	20
0	3	1/25	50
0	4	1/25	90
1	0	1/25	0
1	1	1/25	0
1	2	1/25	20
1	3	1/25	50
1	4	1/25	90
2	0	1/25	20
2	1	1/25	20
2	2	1/25	40
2	3	1/25	70
2	4	1/25	110
3	0	1/25	50
3	1	1/25	50
3	2	1/25	70
3	3	1/25	100
3	4	1/25	140
4	0	1/25	90
4	1	1/25	90
4	2	1/25	110
4	3	1/25	140
4	4	1/25	180

Thus, the probability distribution of his income in this two-day period is

Income (dollars)	Probability
0	4/25
20	4/25
40	1/25
50	4/25
70	2/25
90	4/25
100	1/25
110	2/25
140	2/25
180	1/25

Thus, the expected value equals

$$(4/25)(0) + (4/25)(\$20) + (1/25)(\$40) + (4/25)(\$50) + (2/25)(\$70) +$$

$$(4/25)(\$90) + (1/25)(\$100) + (2/25)(\$110) + (2/25)(\$140) + (1/25)(\$180) =$$

$$\frac{1600}{25} = \$64.$$

(b) Yes. Under the stated circumstances, one would certainly expect the expected value of his income in two days to be double his expected income in one day. (See Appendix 4.3.)

4.15 The expected value of this gamble is (.9)(\$5,000) + (.1)(– \$10,000) = \$3,500. He should therefore purchase the bicycles, since the expected value exceeds zero.

4.17 (a) The variance equals

$$(2/5)(0 - 32)^2 + 1/5(20 - 32)^2 + 1/5(50 - 32)^2 + 1/5(90 - 32)^2$$

$$= 2/5(1024) + 1/5(144) + 1/5(324) + 1/5(3364)$$

$$= (1/5)(5880) = 1{,}176.$$

(b) The standard deviation equals $\sqrt{1176}$ dollars, or \$34.29.

4.19 If the diameter of a rod is less than 8.25 or greater than 8.50 inches, it is more than 5 standard deviations from its mean. Using Chebyshev's inequality, we know that the probability of this occurring is less than 1/25. Thus, the maximum probability of rejection is 1/25.

(pp. 129–30) **4.21** (a) Since $n = 20$ and $P = .20$, Appendix Table 1 indicates that this probability equals .0005 + .0001 + .0000 = .0006.

(b) According to Appendix Table 1, this probability equals .0115 + .0576 + .1369 = 0.206.

(c) According to Appendix Table 1, the probability that the number expressing dissatisfaction is more than 20 percent of the sample equals 1 – (.0115 + .0576 + .1369 + .2054 + .2182) = 1 – .6296 = .3704. Thus, the expected value of the gamble to the personnel director is (.3704)(– \$100) + (.6296)(\$100) = \$25.92. This is not a fair bet, since the expected value does not equal zero.

(d) The expected number equals $nP = (20)(.2)$, or 4 employees. The standard deviation equals $\sqrt{nP(1 - P)} = \sqrt{20(.2)(.8)} = \sqrt{3.2}$, or 1.79 employees.

(e) This upper bound is 1/9. The true probability that the number will be more than 3 standard deviations (that is, 3 times 1.79, or 5.37) from the mean (which equals 4) is the probability that the number exceeds 9.37. According to Appendix Table 1, this probability equals .0020 + .0005 + .0001, or .0026.

4.23 The expected value is $nP = (20)(.15)$, or 3 people. The standard deviation is $\sqrt{nP(1 - P)} = \sqrt{(20)(.15)(.85)} = \sqrt{2.55}$, or 1.6 people.

4.25 Since the expected value is nP and the variance is $nP(1 - P)$, it follows that the ratio of the variance to the expected value is $nP(1 - P) \div nP$, or $(1 - P)$. Thus, if this ratio equals 1/2, P must equal 1/2. One cannot tell what n is.

4.27 This is proved in Appendix 4.2. Substitute 20 for N, 3 for A, and 3 for n in equation (4.12). The result is the expression given here.

Chapter 5

(pp. 158–59) **5.1** (a) .5 – .4893 = .0107
(b) .5 – .4987 = .0013
(c) .5 – .2580 = .2420
(d) .4772 – .3413 = .1359
(e) .3413 + .4772 = .8185

5.3 No. This is true only if the variable's mean equals zero.

5.5 (a) 1/2

(b) Since the mean is .01 inches greater than the design, all pedals of widths .01 inches less than the mean or greater exceeded the design. The point on the standard normal distribution corresponding to .01 inches less than the mean is $-1/2$, since the standard deviation equals .02 inches. Using Appendix Table 2, the probability that the standard normal variable exceeds $-1/2$ equals .1915 + .5000, or .6915.

(c) Since the average width equaled the design, if a pedal was .04 inches wider than the design, its width was .04 inches above the mean. Since the standard deviation was .02 inches, the point on the standard normal distribution corresponding to .04 inches above the mean is 2.0. Using Appendix Table 2, the probability that the standard normal variable exceeds 2.0 is .0228.

(d) Since the average width was .01 inches greater than the design, if a pedal was .04 inches wider than the design, its width was .03 inches above the mean. The point on the standard normal distribution corresponding to .03 inches above the mean is 1.5. Using Appendix Table 2, the probability that the standard normal variable exceeds 1.5 is .0668.

5.7 (a) X/σ is the standard normal variable if $\mu = 0$.
(b) $X - \mu$ is the standard normal variable if $\sigma = 1$.

5.9 According to Appendix Table 2, the point on the standard normal distribution below which there is a probability of .10 is -1.28. The weight that corresponds to this point on the standard normal distribution is $170 - (1.28)(20)$, or 144.4 pounds.

165-66) **5.11** (a) .2707
(b) .2707
(c) .1804

5.13 The Poisson distribution.

5.15 Since the standard deviation of a Poisson variable equals $\sqrt{\mu}$, its coefficient of variation equals $\sqrt{\mu} \div \mu$, or $1 \div \sqrt{\mu}$. Thus, if its coefficient of variation equals 2,

$$\frac{1}{\sqrt{\mu}} = 2,$$

which implies that $\sqrt{\mu} = 1/2$, or $\mu = 1/4$.

5.17 If $P = .001$, the probability that none of the sample is defective is $(.999)^n$. If n is at least 20, the Poisson approximation should be useful. According to Appendix Table 3, the probability that $x = 0$ (that is, that there are no defectives) equals .9512 (which is very close to .95) when $\mu = .05$. Since $\mu = nP$, n must equal 50 because $P = .001$. Thus, 50 pedals should be examined from each day's output.

5.19 According to Appendix Table 3, the probability distribution of the number of hits per area would be as follows, if it has a Poisson distribution with mean equal to 1.0.

Number of hits per area	Probability	Expected number of areas	Actual number of areas
0	.3679	212	229
1	.3679	212	211
2	.1839	106	93
3	.0613	35	35
4	.0153	9	7
5 or more	.0037	2	1

Multiplying each probability by 576 (the total number of areas), we get the expected number of areas with each number of hits, shown above. Finally, we can compare the *expected* and *actual* frequency distributions, shown in the last two columns of the above Table. The comparison shows that these distributions are reasonably similar. In Chapter 9, we shall see how one can test whether the differences are due to chance.

(p. 172) **5.21** (a) Since $\lambda\Delta = 3$, the probability of zero arrivals is .0498, according to Appendix Table 3.

(b) The probability that the time interval between arrivals lies between 1/2 hour and 1 hour equals $e^{-(3)(1/2)} - e^{-(3)(1)}$, or $e^{-1.5} - e^{-3}$. Using Appendix Table 4, this probability equals .223 − .050 = .173.

5.23 Since $\lambda = 3$ and $m = 6$, this probability equals 3/6 or 1/2.

5.25 (a) $\dfrac{(1/2)^2}{1 - 1/2} = \dfrac{1/4}{1/2} = 1/2$ customers

(b) 1/2 customer-hours

Chapter 6

(pp. 183–84) **6.1** AB, AC, AD, AE, BC, BD, BE, CD, CE, DE. Of course, this assumes that sampling is without replacement.

6.3 5, 10.

6.7 Families where both husband and wife work are unlikely to be included in the sample. Single persons who work are unlikely to be included. Children of school age are unlikely to be included.

6.9 (a) No, it is a systematic sample.
(b) Yes.

6.11 If proportional allocation is used, the number sampled in each stratum is proportional to the total number in the stratum. Since the total sample size is 100, 1/30 of the farms in each stratum will be chosen. Thus, the number chosen from each stratum is as follows:

Farm size (acres)	Number of farms in sample
0–50	1,000/30 = 33
51–100	500/30 = 17
101–150	500/30 = 17
151–200	400/30 = 13
201–250	400/30 = 13
Over 250	200/30 = 7
Total	100

6.13 No, because the farms contained in a particular cluster are not close together (unless people with names beginning with the same initial live close to one another, which is unlikely).

(pp. 201–2) **6.15** The point on the standard normal distribution corresponding to 3.01 inches is (3.01 − 3.00) ÷ .05 = 0.20. According to Appendix Table 2, the probability that the standard normal variable will exceed 0.20 is .4207.

6.17 \$50,000

$$\frac{\$5{,}000}{\sqrt{25}}\sqrt{\frac{100-25}{100-1}} = \$1{,}000\sqrt{\frac{75}{99}} = \$870$$

6.19 (a) The standard deviation of the sample mean is cut in half.

(b) The ratio of the new standard deviation of the sample mean to the old one is $\sqrt{4/6}$, or .82.

(c) The ratio of the new standard deviation of the sample mean to the old one is $\sqrt{100/102}$, or .99.

6.21 Suppose that the median is based on a sample size of n_1, and the mean is based on a sample size of n_2. If

$$\sqrt{\frac{\pi}{2}}\frac{\sigma}{\sqrt{n_1}} = \frac{\sigma}{\sqrt{n_2}},$$

it follows that $n_1/n_2 = \pi/2$. Since $\pi = 3.1416$, $\pi/2 = 1.57$. Thus, n_1 must be 1.57 times n_2.

6.23 The standard deviation of the sample proportion equals $.3 \div \sqrt{n}$, since the standard deviation of the population equals .3. Thus, if $.3 \div \sqrt{n} = .06$, $\sqrt{n} = 5$, or $n = 25$. Thus, the sample size must be 25.

6.25
$$\sigma_{\bar{x}} = \frac{\sigma}{\sqrt{n}}\sqrt{\frac{N-n}{N-1}}$$

$$= \frac{\sigma}{\sqrt{n}}\sqrt{\left(1-\frac{n}{N}\right) \div \left(1-\frac{1}{N}\right)}$$

If N is large, $(1 - 1/N)$ is approximately equal to 1. Thus,

$$\sigma_{\bar{x}} \doteq \frac{\sigma}{\sqrt{n}}\sqrt{1-\frac{n}{N}}.$$

Chapter 7

. 215-16) **7.1** The sample mean is 4.448 thousands of hours.

7.3 A sample of this size need not be unreliable; whether it is unreliable depends on the required accuracy. The sample percentage is not a biased estimate of the population percentage.

7.5 Because it is the expected difference between the value of the estimator and θ.

7.7 Yes. As the sample size approaches the size of the entire population, the sample proportion approaches the population proportion. The standard deviation of the sample proportion equals $\sqrt{P(1-P)/n}$. As n becomes larger and larger, it tends to zero, which means that the probability distribution of the sample proportion becomes concentrated ever more tightly about the population proportion.

. 225-26) **7.9** 4,376 to 4,624 hours

7.11 The width of a 95 percent confidence interval is $(2)(1.96)(\sigma/\sqrt{n})$, whereas the width of a 90 percent confidence interval is $(2)(1.64)(\sigma/\sqrt{n})$. Thus, the ratio of the former to the latter is 1.96/1.64, or 1.20. Therefore, he is correct.

7.13 (a) 3,964.8 to 4,279.6 hours
(b) 3,927.1 to 4,317.4 hours
(c) We assume that the length of life of motors received from supplier IV is normally distributed.

7.15 The answer is 9.40 to 10.20 days. Because the population is larger, the sample is a smaller proportion of the population, and the confidence interval is wider than that in Exercise 7.14.

7.17 If s is (approximately) distributed normally with mean equal to σ and standard deviation equal to $\sigma \div \sqrt{2n}$, then $(s - \sigma) \div \sigma/\sqrt{2n}$ has (approximately) the standard normal distribution. Thus,

$$Pr\left\{-z_{\alpha/2} < \frac{s-\sigma}{\sigma/\sqrt{2n}} < z_{\alpha/2}\right\} = 1 - \alpha$$

$$Pr\left\{\frac{-z_{\alpha/2}}{\sqrt{2n}} < \frac{s}{\sigma} - 1 < \frac{z_{\alpha/2}}{\sqrt{2n}}\right\} = 1 - \alpha$$

$$Pr\left\{\frac{-z_{\alpha/2}}{\sqrt{2n}} + 1 < \frac{s}{\sigma} < \frac{z_{\alpha/2}}{\sqrt{2n}} + 1\right\} = 1 - \alpha$$

$$Pr\left\{\frac{1}{1 + \frac{z_{\alpha/2}}{\sqrt{2n}}} < \frac{\sigma}{s} < \frac{1}{1 - \frac{z_{\alpha/2}}{\sqrt{2n}}}\right\} = 1 - \alpha$$

$$Pr\left\{\frac{s}{1 + \frac{z_{\alpha/2}}{\sqrt{2n}}} < \sigma < \frac{s}{1 - \frac{z_{\alpha/2}}{\sqrt{2n}}}\right\} = 1 - \alpha.$$

(pp. 233–34) **7.19** −0.735 to −.265 points

7.21 .262 to .418

7.23 According to Appendix Table 7(a), the confidence interval is 5 percent to 28 percent.

7.25 \$7,202 to \$9,519

(p. 240) **7.27** 1,076

7.29 If the proportion is about .4, n should be about 9,220.

Chapter 8

(pp. 267–68) **8.1** (a) The null hypothesis H_0 is that P, the proportion of deposit slips filled out incorrectly, equals .01.
(b) If it rejects the hypothesis that $P = .01$ when it is true, the bank incurs a Type I error.
(c) If it accepts the hypothesis that $P = .01$ when it is false, the bank incurs a Type II error.
(d) The relative costs of a Type I and Type II error.

8.3 (a) The alternative hypothesis is that the applicant is qualified for the job.
(b) The consequence of a Type I error is that the bank accepts an unqualified

applicant as being qualified. The consequence of a Type II error is that the bank turns away a qualified applicant.

(c) The relative costs of a Type I and Type II error.

8.5 Since $\bar{x}$ is less than 26.824 miles per gallon, the company should reject the hypothesis that the population mean is 28 miles per gallon, based on the test given in Exercise 8.4.

8.7 Yes, because the sample mean is less than 27.016 miles per gallon.

8.9 The null hypothesis (that $P = .50$) should be rejected if $p > .50 + 1.64\sqrt{(.5)(.5)/100}$ or if $p < .50 - 1.64\sqrt{(.5)(.5)/100}$. In other words, it should be rejected if $p > .582$ or if $p < .418$. Since the observed value of p is .34, it should be rejected.

8.11 (a) Reject the hypothesis that the teller is performing adequately if 1 or more out of 10 randomly chosen transactions per day contain an error.

(b) The probability that 1 or more out of 10 randomly chosen transactions contain an error equals

$$1 - (.99)^{10} = 1 - (.9606)^2(.9801) = 1 - (.9228)(.9801) = 1 - .904 = .096,$$

if 1 percent contain errors.

8.13

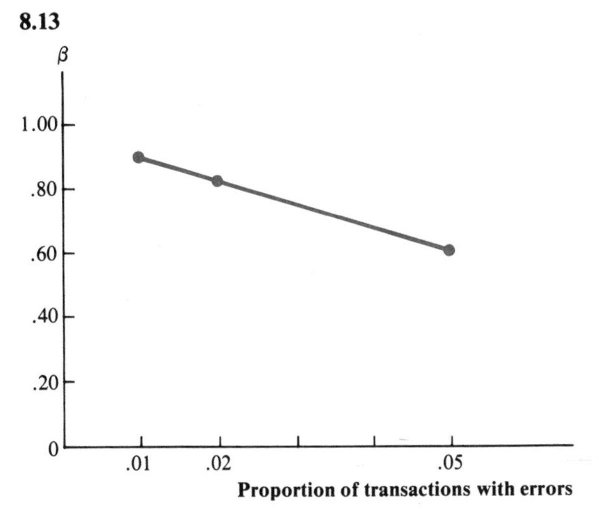

275–76) **8.15** (a) The null hypothesis is that the population mean score of the engineering graduates equals that of the business graduates. The alternative hypothesis is that these population mean scores are not equal.

(b) Reject the null hypothesis if $(\bar{x}_1 - \bar{x}_2) \div \sqrt{(100/100) + (100/100)}$ is greater than $z_{\alpha/2}$ or less than $-z_{\alpha/2}$. Otherwise, accept the null hypothesis.

(c) Since $(\bar{x}_1 - \bar{x}_2) \div \sqrt{2} = (80 - 78) \div 1.414 = 1.414$, it follows that it does not exceed $z_{.05}$, which equals 1.64. Thus, the null hypothesis should not be rejected.

(d) Since $z_{.025} = 1.96$, the null hypothesis should not be rejected.

(e) Since $z_{.005} = 2.576$, the null hypothesis should not be rejected.

8.17 (a) The null hypothesis is that the population mean score of the engineering graduates equals that of the business graduates. The alternative hypothesis is that the latter exceeds the former.

(b) Reject the null hypothesis if $(\bar{x}_1 - \bar{x}_2) \div \sqrt{(100/100) + (100/100)} < -z_\alpha$, where the scores of the engineering graduates are population 1 and those of the business graduates are population 2. Otherwise, accept the null hypothesis.

(c) Since 1.414 is not less than -1.28, the null hypothesis should not be rejected.

(d) Since 1.414 is not less than -1.64, the null hypothesis should not be rejected.

(e) Since 1.414 is not less than -2.33, the null hypothesis should not be rejected.

8.19 (a) The null hypothesis is that P_1 equals P_2 (where P_1 is the proportion in the U.S. and P_2 is the proportion in Canada). The alternative hypothesis is that $P_2 > P_1$.

(b) Reject the null hypothesis if $(p_1 - p_2) \div \sqrt{p(1 - p)[(1/81) + (1/100)]} < -z_\alpha$. Otherwise, accept the null hypothesis.

(c) He should accept it.

(d) He should accept it.

(e) He should accept it.

(pp. 285-87) **8.21** Since $n = 25$, $\bar{x} = 4{,}448$ hours, and $s = 341.7$ hours,

$$t = \frac{4{,}448 - 4{,}900}{341.7/5} = \frac{-452}{68.34} = -6.61.$$

Since $t_{.05} = 1.711$, the observed value of t is less than $-t_{.05}$. Thus, the null hypothesis should be rejected.

8.23 (a) Since $n = 9$, $\bar{x} = \$15{,}078$, and $s = \$5{,}624$,

$$t = \frac{15{,}078 - 20{,}000}{5{,}624 \div 3} = \frac{-4{,}922}{1{,}875} = -2.625.$$

Since $t_{.05} = 1.860$, the null hypothesis should be rejected since t is less than -1.860.

(b) It should not reject the null hypothesis since t is not less than -2.896.

8.25 Since $n = 9$, $\bar{x} = 4122.22$ hours, and $s = 253.9$ hours,

$$t = \frac{4122.2 - 4{,}200}{253.9 \div 3} = \frac{-77.8}{84.6} = -0.92.$$

Since $t_{.025} = 2.306$, the null hypothesis should not be rejected, because the observed value of t is not less than -2.306. Similarly, since $t_{.05} = 1.860$, the null hypothesis should not be rejected, because the observed value of t is not less than -1.860.

8.27 Since $(3)(.0004 \div \sqrt{4}) = .0006$ inches, the upper and lower control limits are at $.1020 + .0006$ inches and at $.1020 - .0006$ inches:

Chapter 9

(p. 307-9) **9.1** (a) 11.0705
(b) 18.3070
(c) 31.4104

9.3 Since the mean equals ν, and the standard deviation equals $\sqrt{2\nu}$, the coefficient of variation equals $\sqrt{2\nu} \div \nu$, or $\sqrt{2/\nu}$, where ν is the number of degrees of freedom. If $\sqrt{2/\nu}$ equals 1, ν must equal 2.

9.5 (a) It has the χ^2 distribution with 15 degrees of freedom.
(b) 15
(c) 30

9.7 The proportion containing falsified information is $(6 + 8 + 9 + 12) \div 800 = .04375$. Thus, the expected number containing falsified information at each branch is 8.75, and the expected number containing no falsified information at each branch is 191.25. Thus,

$$\Sigma \frac{(f - e)^2}{e} = \frac{(6 - 8.75)^2}{8.75} + \frac{(194 - 191.25)^2}{191.25} + \frac{(8 - 8.75)^2}{8.75} + \frac{(192 - 191.25)^2}{191.25}$$

$$+ \frac{(9 - 8.75)^2}{8.75} + \frac{(191 - 191.25)^2}{191.25} + \frac{(12 - 8.75)^2}{8.75} + \frac{(188 - 191.25)^2}{191.25}$$

$$= (1/8.75)(7.5625 + .5625 + .0625 + 10.5625)$$

$$+ (1/191.25)(7.5625 + .5625 + .0625 + 10.5625)$$

$$= (.1143 + .0052)(18.7500) = (.1195)(18.7500) = 2.2406.$$

Since $\chi^2_{.05} = 7.81473$, the null hypothesis (that the proportion was the same at each branch) should not be rejected.

9.9 The expected frequencies, if there are no regional differences in these probabilities, are

Region	Never heard of Alpha's bicycle	Thought Alpha's bicycle was overpriced	Thought Alpha's bicycle was not overpriced
South	$\frac{(131)(100)}{400} = 32.75$	$\frac{(52)(100)}{400} = 13$	$\frac{(217)(100)}{400} = 54.25$
North	$\frac{(131)(100)}{400} = 32.75$	$\frac{(52)(100)}{400} = 13$	$\frac{(217)(100)}{400} = 54.25$
East	$\frac{(131)(100)}{400} = 32.75$	$\frac{(52)(100)}{400} = 13$	$\frac{(217)(100)}{400} = 54.25$
West	$\frac{(131)(100)}{400} = 32.75$	$\frac{(52)(100)}{400} = 13$	$\frac{(217)(100)}{400} = 54.25$

Thus,

$$\Sigma \frac{(f - e)^2}{e} = \frac{(30 - 32.75)^2}{32.75} + \frac{(41 - 32.75)^2}{32.75} + \frac{(28 - 32.75)^2}{32.75} + \frac{(32 - 32.75)^2}{32.75}$$

$$+ \frac{(10 - 13)^2}{13} + \frac{(21 - 13)^2}{13} + \frac{(7 - 13)^2}{13} + \frac{(14 - 13)^2}{13}$$

$$+ \frac{(60 - 54.25)^2}{54.25} + \frac{(38 - 54.25)^2}{54.25} + \frac{(65 - 54.25)^2}{54.25} + \frac{(54 - 54.25)^2}{54.25}$$

$$= (1/32.75)(7.5625 + 68.0625 + 22.5625 + .5625)$$

$$+ (1/13)(9 + 64 + 36 + 1)$$

$$+ (1/54.25)(33.0625 + 264.0625 + 115.5625 + .0625)$$

$$= .0305(98.75) + .0769(110) + .0184(412.75)$$

$$= 3.0119 + 8.4590 + 7.5946 = 19.0655.$$

Since $\chi^2_{.01} = 16.8119$, the null hypothesis (that there are no regional differences in these probabilities) should be rejected.

9.11 The expected frequency of each number is as follows:

2	3	4	5	6	7
$\frac{200}{36} = 5.56$	$\frac{400}{36} = 11.11$	$\frac{600}{36} = 16.67$	$\frac{800}{36} = 22.22$	$\frac{1000}{36} = 27.78$	$\frac{1200}{36} = 33.33$

8	9	10	11	12
$\frac{1000}{36} = 27.78$	$\frac{800}{36} = 22.22$	$\frac{600}{36} = 16.67$	$\frac{400}{36} = 11.11$	$\frac{200}{36} = 5.56$

Thus,

$$\Sigma \frac{(f - e)^2}{e} = \frac{(0 - 5.56)^2}{5.56} + \frac{(2 - 11.11)^2}{11.11} + \frac{(2 - 16.67)^2}{16.67} + \frac{(16 - 22.22)^2}{22.22} + \frac{(30 - 27.78)^2}{27.78}$$

$$+ \frac{(100 - 33.33)^2}{33.33} + \frac{(30 - 27.78)^2}{27.78} + \frac{(14 - 22.22)^2}{22.22} + \frac{(2 - 16.67)^2}{16.67}$$

$$+ \frac{(4 - 11.11)^2}{11.11} + \frac{(0 - 5.56)^2}{5.56}$$

$$= 11.12 + \frac{1}{11.11}(9.11^2 + 7.11^2) + \frac{1}{16.67}(14.67^2 + 14.67^2) +$$

$$\frac{1}{22.22}(6.22^2 + 8.22^2) + \frac{1}{27.78}(2.22^2 + 2.22^2)$$

$$+ \frac{66.67^2}{33.33}$$

$$= 11.12 + \frac{(82.99 + 50.55)}{11.11} + \frac{2(215.21)}{16.67} + \frac{(38.69 + 67.57)}{22.22} + \frac{2(4.93)}{27.78} + \frac{4{,}444.89}{33.33}$$

$$= 11.12 + \frac{133.54}{11.11} + \frac{430.42}{16.67} + \frac{106.26}{22.22} + \frac{9.86}{27.78} + 133.36$$

$$= 11.12 + 12.02 + 25.82 + 4.78 + 0.35 + 133.36$$

$$\doteq 187.$$

Since $\chi^2_{.01} = 23.2093$, the gambler should reject the null hypothesis that the dice are true. (It might be noted that they seem to be particularly likely to show a 7.)

9.13 The confidence interval is

$$\frac{(29)(10)}{\chi^2_{.01}} < \sigma^2 < \frac{(29)(10)}{\chi^2_{.99}},$$

or

$$\frac{290}{49.5879} < \sigma^2 < \frac{290}{14.2565}$$

or

$$5.85 < \sigma^2 < 20.34.$$

319–21) **9.15** $\frac{n}{2} - z_\alpha\sqrt{\frac{n}{4}} = 50 - 1.64\sqrt{25} = 50 - 8.2 = 41.8.$

Since the number of times the new technique results in an increase of more than 10 units per hour is 41, the null hypothesis (that the researcher's claim is correct) should be rejected, because this number (41) is less than 41.8.

9.17 (a) Yes.

(b) No. Exercise 9.16 tests a hypothesis about the median, whereas Exercise 8.22 tests a hypothesis about the mean.

(c) No. The test in Exercise 8.22 assumes normality, whereas the test in Exercise 9.16 does not. However, since $n > 30$, the test in Exercise 8.22 should be dependable even if the population is not normal.

9.19 (a) Let population 1 be the distribution of the number of errors committed per day before the installation of the new procedure. Let population 2 be the distribution of the number of errors committed per day after the installation of the new procedure. Then

$$t = \frac{\bar{x}_1 - \bar{x}_2}{\sqrt{s^2\left(\frac{1}{n_1} + \frac{1}{n_2}\right)}} = \frac{8.18 - 6.55}{\sqrt{3.018(.0909 + .0909)}} = \frac{1.63}{\sqrt{.5487}} = \frac{1.63}{0.74} = 2.20,$$

since

$$s^2 = \frac{33.636 + 26.727}{20} = \frac{60.363}{20} = 3.018.$$

Since $t_{.025} = 2.086$, the null hypothesis should be rejected, because 2.20 exceeds 2.086.

(b) Yes.

(c) The Mann-Whitney test is aimed at testing whether the two populations are identical, whereas the t test is aimed at testing whether the two means are the same. However, the t test assumes that the populations have the same variance and that both are normal.

(d) No. The t test assumes that the populations are normal and have the same variance, but these assumptions are not made by the Mann-Whitney test.

9.21 (a) 2

(b) 7

9.23 Since n_1 (the number of even numbers, including zeros) equals 34 and n_2 (the number of odd numbers) equals 33,

$$E_r = \frac{2(34)(33)}{67} + 1 = \frac{2244}{67} + 1 = 33.49 + 1 = 34.49$$

$$\sigma_r = \sqrt{\frac{2(34)(33)[2(34)(33) - 67]}{67^2(66)}} = \sqrt{\frac{2244(2244 - 67)}{4489(66)}}$$

$$= \sqrt{\frac{(2244)(2177)}{296{,}274}} = \sqrt{\frac{4{,}885{,}188}{296{,}274}} = \sqrt{16.49} = 4.06.$$

We should reject the null hypothesis (of randomness) if the number of runs exceeds 34.49 + 1.96(4.06) = 34.49 + 7.96 = 42.45. Since the number of runs is 52, the null hypothesis should be rejected.

9.25 Since n_1 (the number of A's) equals 33 and n_2 (the number of B's) equals 34,

$$E_r = \frac{2(33)(34)}{67} + 1 = 34.49$$

$$\sigma_r = \sqrt{\frac{2(33)(34)[2(33)(34) - 67]}{67^2(66)}} = 4.06.$$

We should reject the null hypothesis (of randomness) if the number of runs is less than 34.49 − 2.576(4.06) = 34.49 − 10.46 = 24.03, or greater than 34.49 + 10.46 = 44.95. Since the number of runs equals 16, the null hypothesis should be rejected.

Chapter 10

(pp. 332–34) **10.1** No, because it may perform no better than a placebo.

10.3 As pointed out in Exercise 10.1, placebos can make people feel better.

10.5 It would not tell anything about how these men would have performed in the absence of vitamin supplementation.

10.7 It would be better not to tell the soldiers whether they are receiving the vitamins or the placebo.

10.9 (a) No.
(b) Not applicable.
(c) Yes.
(d) Each batch occurs only once with each viscosity jar and each measurer. Each viscosity jar occurs only once with each batch and each measurer. Each measurer occurs only once with each batch and each viscosity jar.
(e) No.

(pp. 352–53) **10.11** (a) .05
(b) .01

10.13 (a) 2.75
(b) 4.30

10.15 The table is as follows:

Source of variation	Sum of squares	Degrees of freedom	Mean square	F
Between groups	183.6875	3	61.229	20.55
Within groups	35.7500	12	2.979	
Total	219.4375	15		

10.17 (a) No, the value of F (that is, 20.78) far exceeds $F_{.05}$, which is 3.86.

(b) The means are as follows:

This firm: 19.75 miles per gallon
U.S. competitor: 21.75 miles per gallon
German firm: 24.00 miles per gallon
Japanese firm: 24.00 miles per gallon

10.19 Let μ_1 be the mean for this firm, μ_2 be the mean for its U.S. competitor, μ_3 be the mean for the German firm, and μ_4 be the mean for the Japanese firm. Then

$$\sqrt{3.49}\,(1.726)\sqrt{\frac{(3)(2)}{4}} = 3.95.$$

Consequently,

$$18.5 - 26.5 - 3.95 < \mu_1 - \mu_4 < 18.5 - 26.5 + 3.95$$

$$\boxed{-11.95 < \mu_1 - \mu_4 < -4.05}$$

$$18.5 - 21.75 - 3.95 < \mu_1 - \mu_2 < 18.5 - 21.75 + 3.95$$

$$\boxed{-7.20 < \mu_1 - \mu_2 < 0.70}$$

$$18.5 - 26.5 - 3.95 < \mu_1 - \mu_3 < 18.5 - 26.5 + 3.95$$

$$\boxed{-11.95 < \mu_1 - \mu_3 < -4.05}$$

$$21.75 - 26.5 - 3.95 < \mu_2 - \mu_3 < 21.75 - 26.5 + 3.95$$

$$\boxed{-8.70 < \mu_2 - \mu_3 < -0.80}$$

$$21.75 - 26.5 - 3.95 < \mu_2 - \mu_4 < 21.75 - 26.5 + 3.95$$

$$\boxed{-8.70 < \mu_2 - \mu_4 < -0.80}$$

$$26.5 - 26.5 - 3.95 < \mu_3 - \mu_4 < 26.5 - 26.5 + 3.95$$

$$\boxed{-3.95 < \mu_3 - \mu_4 < 3.95}$$

Chapter 11

(pp. 379–80) **11.1** (a)

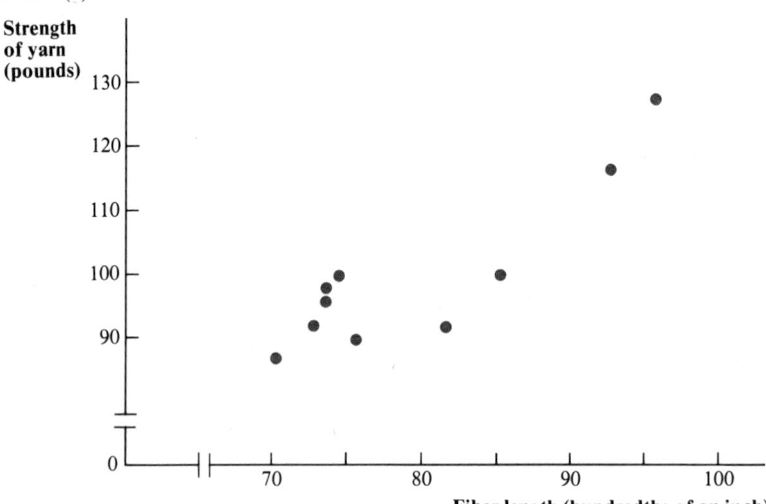

(b) Direct.

(c) Yes, since one would expect longer fibers to be associated with greater strength.

(d) It is hard to tell, but it may be linear.

11.3 (a) The sample regression line is

$$\hat{Y} = -5.8 + 1.33x.$$

(b) $\hat{Y} = -5.8 + (1.33)(80) = -5.8 + 106.4 = 100.6$ pounds.

(c) $\hat{Y} = -5.8 + (1.33)(90) = -5.8 + 119.7 = 113.9$ pounds.

11.5

	X	Y	X^2	Y^2	XY
	190	680	36,100	462,400	129,200
	200	800	40,000	640,000	160,000
	209	780	43,681	608,400	163,020
	215	885	46,225	783,225	190,275
	215	975	46,225	950,625	209,625
	215	1,025	46,225	1,050,625	220,375
	230	1,100	52,900	1,210,000	253,000
	250	1,030	62,500	1,060,900	257,500
	265	1,175	70,225	1,380,625	311,375
	250	1,300	62,500	1,690,000	325,000
Total	2239	9750	506,581	9,836,800	2,219,370
Mean	223.9	975.0			

$$b = \frac{10(2{,}219{,}370) - (2239)(9750)}{10(506{,}581) - (2239)^2} = \frac{22{,}193{,}700 - 21{,}830{,}250}{5{,}065{,}810 - 5{,}013{,}121}$$

$$= \frac{363{,}450}{52{,}689} = 6.898$$

$$a = 975 - (6.898)(223.9) = 975 - 1{,}544.46 = -569.46$$

11.7 $s_e = \sqrt{\dfrac{102{,}193 + (5.818)(1003) - (1.3298)(80{,}991)}{8}}$

$= \sqrt{40.80} = 6.387$ pounds.

The standard deviation of the conditional probability distribution of the strength of yarn is estimated to be 6.387 pounds. (Note that a and b are taken out to more decimal places than in Exercise 11.3.)

11.9 The confidence interval is

$$100.566 \pm 11.8798\sqrt{1.100056},$$

or

$$100.566 \pm (11.8798)(1.049),$$

or

$$100.566 \pm 12.461.$$

That is, it is 88.1 to 113.0 pounds. This confidence interval is wider than that in Exercise 11.8 because the sampling error in predicting the strength of a particular piece of yarn is greater than the sampling error in estimating the conditional mean strength.

11.11 The confidence interval is

$$-569.46 + 6.898(250) \pm (2.306)(99.9)\sqrt{\frac{1}{10} + \frac{(250 - 223.9)^2}{5{,}268.9}},$$

or

$$1{,}155.04 \pm 230.37\sqrt{.10 + \frac{681.21}{5{,}268.9}},$$

or

$$1{,}155.04 \pm 110.35.$$

That is, it equals 1044.7 to 1265.4 pounds.

394-95) **11.13** (a) $r^2 = \dfrac{9{,}516^2}{(7{,}156)(15{,}921)} = \dfrac{90{,}554{,}256}{113{,}930{,}676} = .7948.$

Thus, $r = .89$.

(b) 79.48 percent of the variation can be explained.

11.15 (a) $r^2 = \dfrac{363{,}450^2}{(52{,}689)(3{,}305{,}500)}$

$= (6.898)(.10995) = .7584.$

Thus, $r = .87$.

(b) 75.84 percent of the variation is explained.

11.17 $s_b = \dfrac{6.387}{\sqrt{715.6}} = \dfrac{6.387}{26.75} = .239.$

Thus, $b \div s_b = 1.330/.239 = 5.565$. Since $t_{.01} = 2.896$, the null hypothesis (that $B = 0$) should be rejected because $b \div s_b$ exceeds 2.896.

11.19 No. Correlation does not prove causation.

11.21 One might expect the standard deviation of the amount saved to increase as income increases.

11.23 This estimate involves extrapolation of the regression line beyond the range of the data. For reasons given in Section 11.15, this is a very hazardous procedure.

11.25 No. There may be a relationship like that shown below:

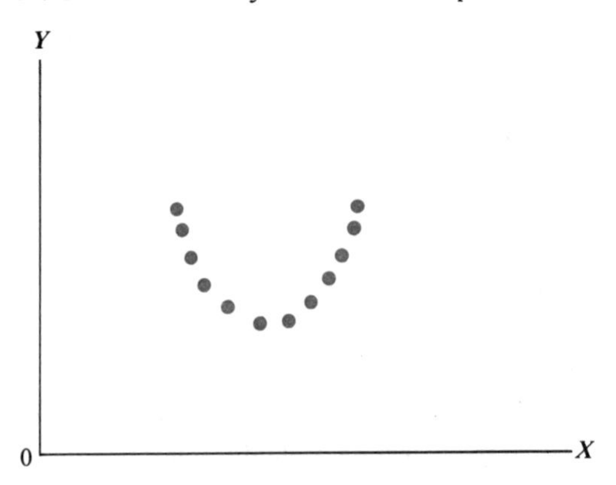

In this case the population correlation coefficient between X and Y is zero, but X and Y obviously are not independent. What the correlation coefficient measures is the closeness of the *linear* relationship between two variables.

Chapter 12

(pp. 420–22) **12.1**

	X_1	X_2	Y	X_1^2	X_2^2	Y^2	X_1Y	X_2Y	X_1X_2
	4	9	8	16	81	64	32	72	36
	4	8	8	16	64	64	32	64	32
	3	8	5	9	64	25	15	40	24
	3	9	4	9	81	16	12	36	27
	3	7	6	9	49	36	18	42	21
	3	10	4	9	100	16	12	40	30
	2	8	2	4	64	4	4	16	16
	2	6	3	4	36	9	6	18	12
	2	5	4	4	25	16	8	20	10
	2	7	2	4	49	4	4	14	14
	2	8	2	4	64	4	4	16	16
	1	6	1	1	36	1	1	6	6
	1	5	1	1	25	1	1	5	5
	1	7	1	1	49	1	1	7	7
	1	5	1	1	25	1	1	5	5
Total	34	108	52	92	812	262	151	401	261

Using equation (12.5),

$$b_1 = \frac{[812 - (108)(7.2)][151 - (34)(3.4667)] - [261 - (34)(7.2)][401 - (52)(7.2)]}{[92 - (34)(2.2667)][812 - (108)(7.2)] - [261 - (34)(7.2)]^2},$$

since $\dfrac{\left(\sum_{i=1}^{n} X_{2i}\right)^2}{n} = (108)(7.2)$, $\dfrac{\left(\sum_{i=1}^{n} X_{1i}\right)\left(\sum_{i=1}^{n} Y_i\right)}{n} = (34)(3.4667)$ and so on. Thus,

$$b_1 = \frac{(812 - 777.6)(151 - 117.8678) - (261 - 244.8)(401 - 374.4)}{(92 - 77.0678)(812 - 777.6) - (261 - 244.8)^2}$$

$$= \frac{(34.4)(33.1322) - (16.2)(26.6)}{(14.9322)(34.4) - 16.2^2}$$

$$= \frac{1139.75 - 430.92}{513.67 - 262.44} = \frac{708.83}{251.23} = 2.82.$$

$$b_2 = \frac{[92 - (34)(2.2667)][401 - (52)(7.2)] - [261 - (34)(7.2)][151 - (34)(3.4667)]}{251.23}$$

$$= \frac{(92 - 77.0678)(401 - 374.4) - (261 - 244.8)(151 - 117.8678)}{251.23}$$

$$= \frac{(14.9322)(26.6) - (16.2)(33.1322)}{251.23}$$

$$= \frac{397.197 - 536.742}{251.23} = \frac{-139.55}{251.23} = -0.555.$$

$$a = 3.4667 - (2.82)(2.2667) + (.555)(7.2) = 3.4667 - 6.3921 + 3.9960$$

$$= 1.07.$$

Thus, the multiple regression equation is

$$\hat{Y} = 1.07 + 2.82X_1 - 0.555X_2.$$

12.3 (a) The estimated effect is to increase the expected value of Rotunda's annual sales by 2.82 millions of units.

(b) The estimated effect is to reduce the expected value of Rotunda's annual sales by .555 millions of units.

12.5 (a) $R^2 = \dfrac{(2.82)(33.13) - (0.555)(26.6)}{262 - (52)(3.4667)}$

$$= \frac{93.4266 - 14.7630}{262 - 180.27} = \frac{78.6636}{81.73} = 0.96.$$

(b) The multiple regression equation explains 96 percent of the observed variation in Rotunda's annual sales. Thus, it fits the data very well.

12.7 (a) $\hat{Y} = 4.659 - 0.062X_2 + 0.096X_4 + 0.070X_8$.

(b) $.0963 \pm (2.16)(.0154)$. In other words, the confidence interval is .063 to .130.

(c) $-0.062 \pm (1.771)(.009)$. In other words, the confidence interval is $-.078$ to $-.046$.

12.9 (a) 0.87. It means that the regression equation explains 87 percent of the variation in the dependent variable.

(b) With 3 and 13 degrees of freedom, $F_{.05}$ equals 3.41. Thus, the null hypothesis should be rejected.

(pp. 441–43) **12.11** Let $X_2 = 1$ if the year is 1980 and 0 if it is 1978. Express Y in units of 10.

	Y	X_1	X_2	Y^2	X_1^2	X_2^2	X_1Y	X_2Y	X_1X_2
	10	5	0	100	25	0	50	0	0
	12	7	0	144	49	0	84	0	0
	14	8	0	196	64	0	112	0	0
	17	9	0	289	81	0	153	0	0
	11	5	1	121	25	1	55	11	5
	16	8	1	256	64	1	128	16	8
	20	9	1	400	81	1	180	20	9
	12	6	1	144	36	1	72	12	6
	13	7	1	169	49	1	91	13	7
	15	7	1	225	49	1	105	15	7
Total	140	71	6	2044	523	6	1030	87	42

Using equation (12.5),

$$b_1 = \frac{[6 - (6)(.6)][1030 - (71)(14)] - [42 - (71)(0.6)][87 - (6)(14)]}{[523 - (71)(7.1)][6 - (6)(.6)] - [42 - (71)(0.6)]^2}$$

$$= \frac{(2.4)(36) - (-0.6)(3)}{(18.9)(2.4) - (0.6)^2} = \frac{86.4 + 1.8}{45.36 - .36} = \frac{88.2}{45}$$

$$= 1.96$$

$$b_2 = \frac{[523 - (71)(7.1)][87 - (14)(6)] - [42 - (71)(0.6)][1030 - (71)(14)]}{45}$$

$$= \frac{(18.9)(3) - (-0.6)(36)}{45} = \frac{56.7 + 21.6}{45} = \frac{78.3}{45}$$

$$= 1.74$$

$$a = 14.0 - (1.96)(7.1) - (1.74)(0.6) = 14.0 - 13.916 - 1.044$$

$$= -.96.$$

Thus, the regression equation is

$$\hat{Y} = -.96 + 1.96X_1 + 1.74X_2.$$

But it should be recalled that Y is measured in units of 10 in this equation. To put the equation into the original units, a, b_1, and b_2 must be multiplied by 10. Thus, the regression equation is

$$\hat{Y} = -9.6 + 19.6X_1 + 17.4X_2.$$

12.13 (a) Yes, because these results assume that the effect of an increase of one percentage point in the unemployment rate on the crime rate is the same in both years.

(b) Yes. One can compute two separate regressions of the crime rate on the unemployment rate, one for 1978 and one for 1980. Then one can test whether the slopes are equal. However, such a test will be quite weak due to the small number of observations.

12.15

Year	Deviation of sales from multiple regression equation
1980	$8 - 1.07 - 2.82(4) + 0.555(9) = 0.65$
1979	$8 - 1.07 - 2.82(4) + 0.555(8) = 0.09$
1978	$5 - 1.07 - 2.82(3) + 0.555(8) = -0.09$
1977	$4 - 1.07 - 2.82(3) + 0.555(9) = -0.53$
1976	$6 - 1.07 - 2.82(3) + 0.555(7) = 0.35$
1975	$4 - 1.07 - 2.82(3) + 0.555(10) = 0.02$
1974	$2 - 1.07 - 2.82(2) + 0.555(8) = -0.27$
1973	$3 - 1.07 - 2.82(2) + 0.555(6) = -0.38$
1972	$4 - 1.07 - 2.82(2) + 0.555(5) = 0.07$
1971	$2 - 1.07 - 2.82(2) + 0.555(7) = -0.83$
1970	$2 - 1.07 - 2.82(2) + 0.555(8) = -0.27$
1969	$1 - 1.07 - 2.82(1) + 0.555(6) = 0.44$
1968	$1 - 1.07 - 2.82(1) + 0.555(5) = -0.11$
1967	$1 - 1.07 - 2.82(1) + 0.555(7) = 0.99$
1966	$1 - 1.07 - 2.82(1) + 0.555(5) = -0.11$

12.17 Since $n = 30$ and $k = 4$, Appendix Table 11 shows that for $\alpha = .05$, $d_L = 1.14$ and $d_u = 1.74$. Since the alternative hypothesis is that there is positive serial correlation, we should reject the null hypothesis (of no serial correlation) if $d < 1.14$ and accept it if $d > 1.74$. Since $d = 1.06$, the null hypothesis should be rejected.

12.19 Since $n = 20$, there are 17 degrees of freedom, and $t_{.025} = 2.11$. Thus, the confidence interval is

$$.0675 \pm (2.11)(.0040),$$

or

$$.0675 \pm .0084.$$

In other words, the confidence interval is .0591 to .0759.

12.21 The dependent variable is the amount saved by the family. There are three independent variables, two of which are dummy variables. The first independent variable is the family's income; the second is a dummy variable that is 0 if the family is white and 1 if the family is nonwhite; and the third is a dummy variable that is 0 if the family is headed by a male and 1 if the family is headed by a female. We assume that the effects of extra income on extra saving are the same regardless of whether a family is white or nonwhite and regardless of whether it is headed by a male or a female.

Chapter 13

(463–65) **13.1** Trends are not useless, since it is interesting and important to recognize and take account of long-run movements over time of relevant variables. Trends are useful in predicting the future and understanding the past.

13.3 No. You cannot be sure that a trend will continue into the future.

13.5 Let $t = 0$ when the year is 1960.

Y	t	Y^2	t^2	Yt
1.5	0	2.25	0	0
1.6	1	2.56	1	1.6

(Table continues)

	Y	t	Y^2	t^2	Yt
	1.6	2	2.56	4	3.2
	1.7	3	2.89	9	5.1
	1.9	4	3.61	16	7.6
	2.1	5	4.41	25	10.5
	2.2	6	4.84	36	13.2
	2.5	7	6.25	49	17.5
	2.7	8	7.29	64	21.6
	2.9	9	8.41	81	26.1
	3.0	10	9.00	100	30.0
	3.0	11	9.00	121	33.0
	3.3	12	10.89	144	39.6
	3.9	13	15.21	169	50.7
	5.3	14	28.09	196	74.2
	5.7	15	32.49	225	85.5
Total	44.9	120	149.75	1240	419.4

Using the formulas in Section 11.6, we obtain

$$b = \frac{16(419.4) - (44.9)(120)}{16(1{,}240) - 120^2} = \frac{6{,}710.4 - 5{,}388}{19{,}840 - 14{,}400} = \frac{1{,}322.4}{5{,}440} = 0.243$$

$$a = 2.806 - (.243)(7.5) = 2.806 - 1.823 = 0.983.$$

Thus, the equation for the trend line is

$$Y_t = 0.983 + 0.243t,$$

where the origin is set at 1960.

13.7 Let $t = 0$ when the year is 1960.

Y	$u = \log Y$	t	u^2	t^2	ut
1.5	.1761	0	.0310112	0	0
1.6	.2041	1	.0416568	1	.2041
1.6	.2041	2	.0416568	4	.4082
1.7	.2304	3	.0530842	9	.6912
1.9	.2788	4	.0777294	16	1.1152
2.1	.3222	5	.1038128	25	1.6110
2.2	.3424	6	.1172378	36	2.0544
2.5	.3979	7	.1583244	49	2.7853
2.7	.4314	8	.1861060	64	3.4512
2.9	.4624	9	.2138138	81	4.1616
3.0	.4771	10	.2276244	100	4.7710
3.0	.4771	11	.2276244	121	5.2481
3.3	.5185	12	.2688422	144	6.2220
3.9	.5911	13	.3493992	169	7.6843
5.3	.7243	14	.5246105	196	10.1402
5.7	.7559	15	.5713848	225	11.3385
Total	6.5938	120	3.1939187	1240	61.8863
Mean	.4121	7.5			

$$b = \frac{16(61.8863) - (6.5938)(120)}{16(1{,}240) - 120^2} = \frac{990.1808 - 791.256}{19{,}840 - 14{,}400} = \frac{198.9248}{5{,}440}$$

$$= .036567$$

$$a = .4121 - (7.5)(.036567) = .4121 - .2743 = .1378.$$

Thus, the equation for the trend line is

$$\log Y_t = .1378 + .03657t,$$

where the origin is set at 1960.

13.9 to **13.11**

Year	Moving total 3-year	Moving total 5-year	Moving total 7-year	Moving average 3-year	Moving average 5-year	Moving average 7-year
1948	268			89.3		
1949	263	449		87.7	89.8	
1950	262	442	629	87.3	88.4	89.9
1951	269	442	617	89.7	88.4	88.1
1952	271	444	616	90.3	88.8	88.0
1953	263	445	625	87.7	89.0	89.3
1954	266	444	621	88.7	88.8	88.7
1955	264	442	606	88.0	88.4	86.6
1956	267	426	600	89.0	85.2	85.7
1957	252	425	588	84.0	85.0	84.0
1958	244	414	583	81.3	82.8	83.3
1959	238	402	573	79.3	80.4	81.9
1960	240	397	566	80.0	79.4	80.9
1961	239	404	566	79.7	80.8	80.9
1962	242	408	581	80.7	81.6	83.0
1963	250	419	595	83.3	83.8	85.0
1964	260	437	608	86.7	87.4	86.9
1965	273	449	624	91.0	89.8	89.1
1966	280	460	639	93.3	92.0	91.3
1967	283	470	644	94.3	94.0	92.0
1968	283	467	643	94.3	93.4	91.9
1969	278	456	642	92.7	91.2	91.7
1970	269	453	642	89.7	90.6	91.7
1971	263	455	640	87.7	91.0	91.4
1972	271	450		90.3	90.0	
1973	277			92.3		

13.13 No, because S_{t-1} is not based on $\theta = 1/5$; it is based on the old smoothing constant.

13.15 $S_0 = 1.5$

$S_1 = (.1)(1.6) + (.9)(1.5) = .16 + 1.35 = 1.510$

$S_2 = (.1)(1.6) + (.9)(1.51) = .16 + 1.359 = 1.519$

$S_3 = (.1)(1.7) + (.9)(1.519) = .17 + 1.367 = 1.537$

$S_4 = (.1)(1.9) + (.9)(1.537) = .19 + 1.383 = 1.573.$

(pp. 482–83) **13.17** The medians are as follows:

January	90.75	May	98.25	September	106.25
February	92.75	June	100.25	October	108.25
March	94.75	July	102.25	November	105.00
April	96.75	August	104.25	December	100.00

Total 1199.50

Since the total of these medians is 1199.50, each median should be multiplied by 1200 ÷ 1199.50, or 1.0004. The results are as follows:

January	90.79	May	98.29	September	106.29
February	92.79	June	100.29	October	108.29
March	94.79	July	102.29	November	105.04
April	96.79	August	104.29	December	100.04

13.19 (a)

Year	Deviation from trend
1960	1.5 − 0.983 = 0.517
1961	1.6 − 1.226 = .374
1962	1.6 − 1.469 = .131
1963	1.7 − 1.712 = −.012
1964	1.9 − 1.955 = −.055
1965	2.1 − 2.198 = −.098
1966	2.2 − 2.441 = −.241
1967	2.5 − 2.684 = −.184
1968	2.7 − 2.927 = −.227
1969	2.9 − 3.170 = −.270
1970	3.0 − 3.413 = −.413
1971	3.0 − 3.656 = −.656
1972	3.3 − 3.899 = −.599
1973	3.9 − 4.142 = −.242
1974	5.3 − 4.385 = .915
1975	5.7 − 4.628 = 1.072

(b) In general, the deviations seem to reflect the nonlinearity of the trend rather than cyclical variation. The trough in 1971 does not correspond to a business cycle trough for the economy as a whole. The downward movement from 1960 to 1971 and the upward movement from 1971 to 1975 do not correspond to a recession and expansion for the economy as a whole.

13.21 (a) 0.983 + (.243)(25) = 0.983 + 6.075 = $7.058 billion.

(b) 0.1378 + (.03657)(25) = .1378 + .9142 = 1.052. The antilog is 11.27 billions of dollars.

(c) The results are not close to one another at all. Specifically they differ by over 50 percent.

13.23 (25)(.9279) = $23.2 millions.

13.25 Among other reasons, it is related to the business community's (and the general public's) expectations concerning the economic future. Changes in expectations can play an important role in influencing business conditions.

13.27 The logarithm of Y might be used, rather than Y.

13.29 Yes. The price of onions, the prices of goods that are substitutes for onions or that are complements to onions, and the income level of consumers.

Chapter 14

(500–501) **14.1** (a) The base period for all these index numbers is 1910–14. The given period for each of them is the year shown at the left end of the row in which the number is located. For example, the given period for all index numbers in the last row is 1974.

(b) A more rapid rate of price increase occurred during 1970–74 than during 1950–70.

(c) A more rapid annual rate of price increase occurred during 1970–74 than during 1930–50.

(d) Tobacco.

14.3 (a) If 1970 is the given year, the index number is

$$\left(\frac{\dfrac{P_{11}}{P_{01}} + \dfrac{P_{12}}{P_{02}} + \dfrac{P_{13}}{P_{03}}}{3}\right)(100) = \frac{1.8 + 2.0 + 2.1}{3}(100) = 197.$$

If 1980 is the given year, the index number is

$$\frac{\dfrac{P_{11}}{P_{01}} + \dfrac{P_{12}}{P_{02}} + \dfrac{P_{13}}{P_{03}}}{3}(100) = \frac{2.50 + 3.00 + 3.05}{3}(100) = 285.$$

(b) The various prices included are unweighted. This deficiency is remedied by the Laspeyres or Paasche indexes, discussed in subsequent exercises.

14.5 The index number is

$$\frac{300(2500) + 400(1500) + 20(6100)}{300(1000) + 400(500) + 20(2000)}(100)$$

$$= \frac{750{,}000 + 600{,}000 + 122{,}000}{300{,}000 + 200{,}000 + 40{,}000}(100) = 273.$$

14.7 (a) $\dfrac{1800}{1000}(100) = 180.$

(b) $I_{60}^{71} = \dfrac{I_{60}^{70}}{100}\dfrac{I_{70}^{71}}{100}(100) = (1.80)(1.10)(100) = 198.$

14.9 This index equals

$$\sqrt{\left[\frac{\sum_{i=1}^{m} Q_{0i}P_{1i}}{\sum_{i=1}^{m} Q_{0i}P_{0i}}(100)\right]\left[\frac{\sum_{i=1}^{m} Q_{1i}P_{1i}}{\sum_{i=1}^{m} Q_{1i}P_{0i}}(100)\right]}.$$

(510–11) **14.11** No, because the Consumer Price Index was 116.3 in 1970, whereas the price deflator for the gross national product was 91.36 in 1970.

14.13 Real average weekly earnings increased by about 2.2 percent between 1970 and 1975. By real earnings, we mean earnings adjusted for changes in the price level. Obviously, however, the adjustments we make are rough in many respects.

14.15 These prices are only a partial and incomplete measure of the costs. There is also the time involved in making the trip. Further, one should consider the amount of discomfort involved. Also, the value of the dollar has fallen.

14.17 $$\frac{2{,}500(300) + 1{,}500(400) + 6{,}100(20)}{2{,}500(100) + 1{,}500(200) + 6{,}100(10)}(100)$$

$$= \frac{750{,}000 + 600{,}000 + 122{,}000}{250{,}000 + 300{,}000 + 61{,}000}(100) = 241.$$

14.19 (a) No.
(b) No.
(c) Yes.

Chapter 15

(pp. 526–28) **15.1** (a)

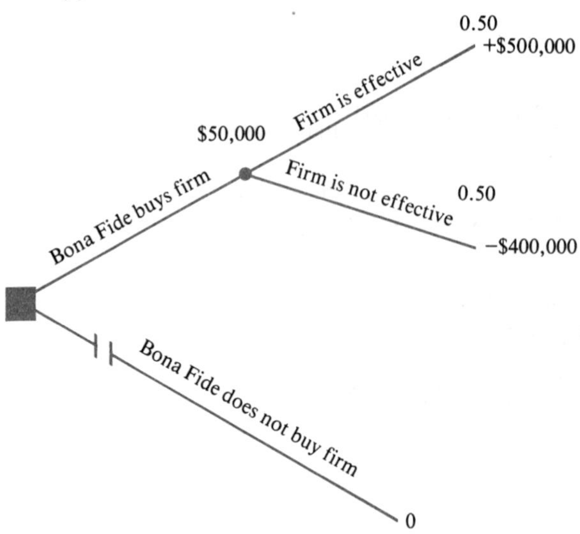

(b) There is only one: whether or not to buy the firm.
(c) There is only one: whether the firm becomes an effective producer or not.

15.3 (a) Yes.
(b) (1) The firm becomes an effective producer of washing machine parts.
(2) The firm does not become an effective producer of washing machine parts, and is sold to the Saudis.
(3) The firm does not become an effective producer of washing machine parts, and cannot be sold to the Saudis.
(c) The probability of the first outcome in (b) is 0.5; the probability of the

second outcome is (0.5)(0.2), or 0.1; and the probability of the third outcome is (0.5)(0.8), or 0.4.

(d) The extra profit to Bona Fide from the first outcome is $500,000; the extra profit from the second outcome is $100,000; the extra profit from the third outcome is − $400,000.

15.5 (a) If the extra profit if the firm is made into an effective producer of washing machine parts is $400,000 or less, the decision will be reversed. Stated differently, if the *error* was an *overstatement* of this extra profit by $100,000 or more, the decision will be reversed.

(b) If the extra profit if the firm is made into an effective producer of washing machine parts is $300,000 or less, the decision will be reversed. Stated differently, if the *error* was an *overstatement* of this extra profit by $200,000 or more, the decision will be reversed.

15.7 (a) No.
(b) Yes.
(c) No.

15.9 (a) 2.0
(b) − 0.5
(c) − 1.1

15.11 (a) Yes.

(b) Yes, because making the loan results in both a higher expected value of monetary gain and a greater risk than not making the loan.

544-45) **15.13** The expected value of perfect information is (.26)($1 million) + (.57)(0) + (.17)(0) − $124,000 = $136,000.

15.15

State of nature	Opportunity loss (dollars): Introduce new bicycle	Opportunity loss (dollars): Do not introduce new bicycle
New bicycle is major success	0	1 million
New bicycle is no better and no worse	0	0
New bicycle is flop	800,000	0

15.17 (a) If Acme introduces the new bicycle, the expected opportunity loss is (.17)($800,000) = $136,000. If Acme does not introduce the new bicycle, the expected opportunity loss is (.26)($1 million) = $260,000. Thus, Acme should introduce the new bicycle.

(b) Part (a) of this question shows that the expected opportunity loss corresponding to the best action equals $136,000. Exercise 15.13 shows that the expected value of perfect information equals $136,000. Thus, the two are equal.

15.19 If 40 percent have no major defects, 600 have major defects, and π = $1,000,000 − ($1950)(600) = − $170,000. If 50 percent have no major defects, 500 have major defects, and π = $1,000,000 − ($1950)(500) = $25,000. If 55 percent have no major defects, 450 have major defects, and π = $1,000,000 − ($1950)(450) = $122,500. Thus, the decision tree is as shown below, and he should buy the cars.

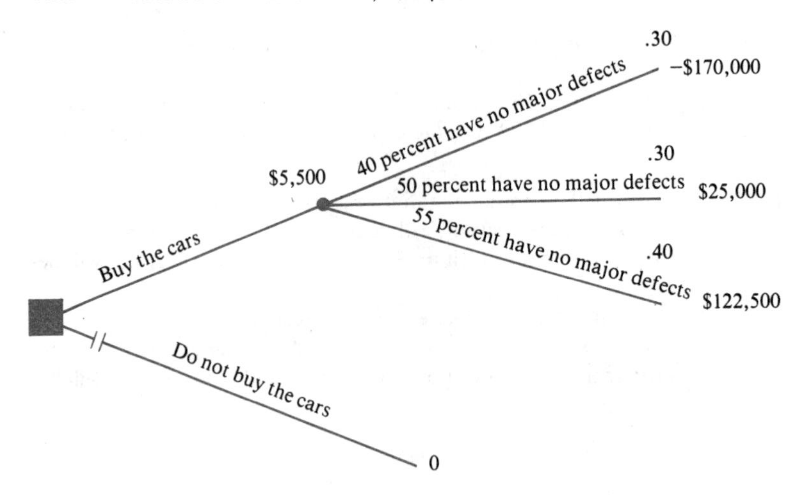

15.21 Let X be the number of cars in the group with major defects.

	Opportunity loss (dollars)	
Value of X	Buy cars	Do not buy cars
$X < 512.8$	0	1,000,000 − 1950X
$X > 512.8$	1950X − 1,000,000	0

15.23 According to Appendix Table 1, the probability that 1 out of 4 do not have major defects is .3456 if 40 percent have no major defects and is .2500 if 50 percent have no major defects. If 55 percent have no major defects, this means that 45 percent have major defects. According to Appendix Table 1, the probability that 3 out of 4 do have major defects is .2005 if 45 percent have major defects. Thus, the posterior probability that 40 percent have no major defects is

$$\frac{(.3456)(.3)}{(.3456)(.3) + (.2500)(.30) + (.2005)(.40)} = \frac{.10368}{.10368 + .07500 + .08020} = .40.$$

The posterior probability that 50 percent have no major defects is

$$\frac{(.2500)(.30)}{(.3456)(.3) + (.2500)(.30) + (.2005)(.40)} = \frac{.07500}{.10368 + .07500 + .08020} = .29.$$

The posterior probability that 55 percent have no major defects is

$$\frac{(.2005)(.40)}{(.3456)(.30) + (.2500)(.30) + (.2005)(.40)} = \frac{.08020}{.10368 + .07500 + .08020} = .31.$$

Thus, the expected profit if he buys the cars is

$$\begin{aligned}(.40)(-\$170{,}000) &+ (.29)(\$25{,}000) + (.31)(\$122{,}500)\\ &= -\$68{,}000 + \$7{,}250 + \$37{,}975\\ &= -\$22{,}775.\end{aligned}$$

Since the expected profit is negative, he should not buy the cars, if he wants to maximize expected profit.

Chapter 16

. 563–64) **16.1**

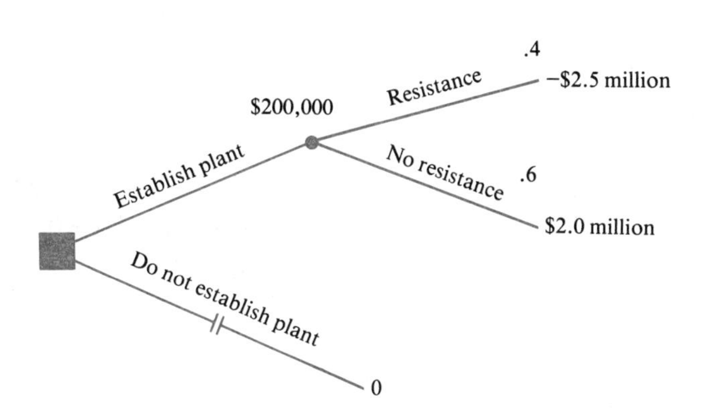

16.3 The posterior probability of resistance equals

$$\frac{(0.2)(0.4)}{(0.2)(0.4) + (0.7)(0.6)} = \frac{.08}{.50} = .16.$$

The posterior probability of no resistance equals

$$\frac{(0.7)(0.6)}{(0.2)(0.4) + (0.7)(0.6)} = \frac{.42}{.50} = .84.$$

Thus, the subsequent branches of the decision tree are as follows:

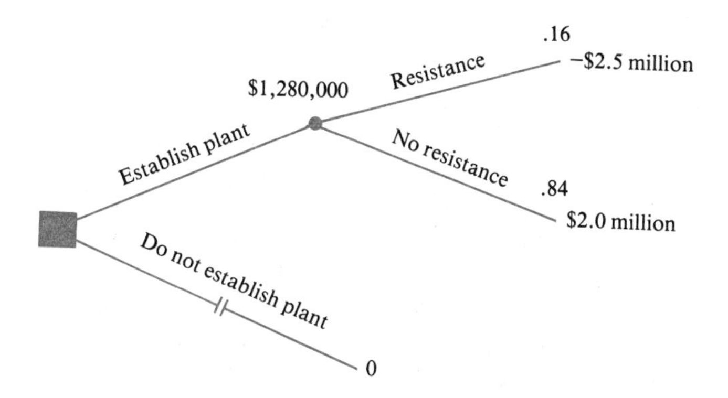

16.5 (a)

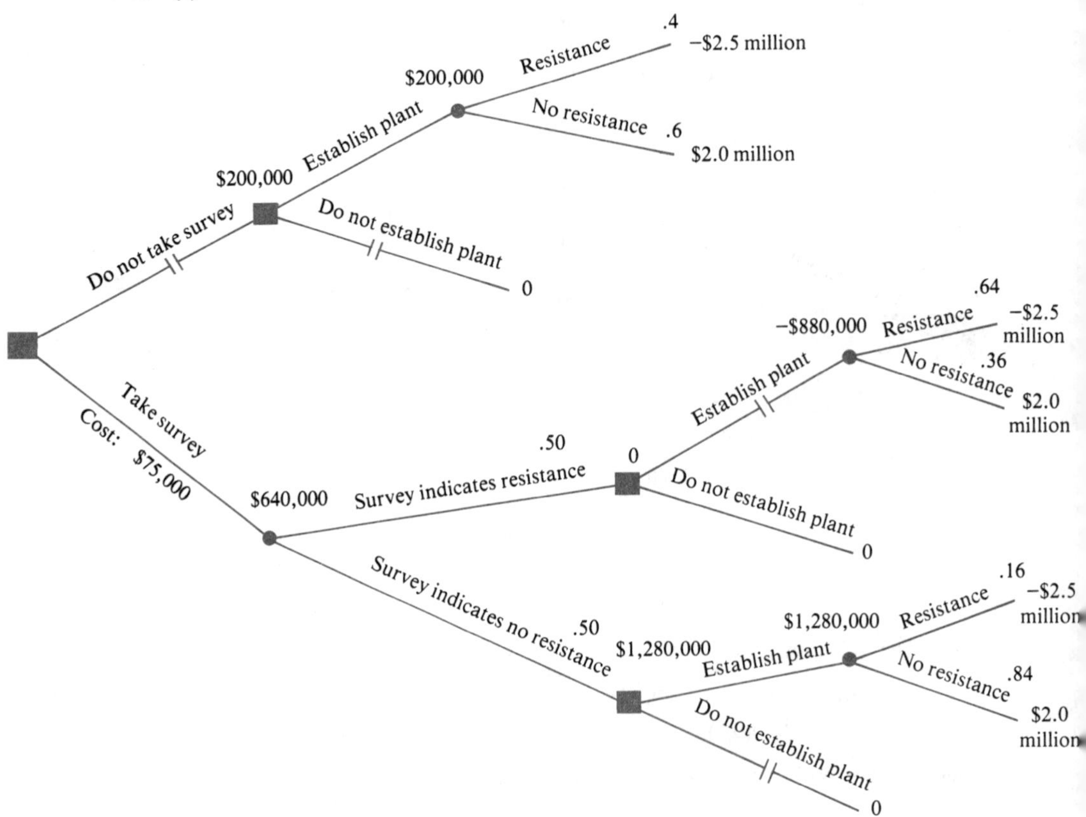

(b) Yes.
(c) Establish the plant.
(d) Do not establish the plant.

16.7 There are four possible strategies, as indicated in the table below:

	Strategy			
Survey outcome	1	2	3	4
Indicates substantial resistance	Do not establish plant	Do not establish plant	Establish plant	Establish plant
Indicates no resistance	Establish plant	Do not establish plant	Establish plant	Do not establish plant

16.9 Yes. As pointed out in the text, normal form analysis and extensive form analysis will always lead to the choice of the same strategy.

16.11 (a) For strategy 2 in the answer to Exercise 16.7, the expected opportunity loss equals P($2.0 million). For strategy 3, it equals $(1 - P)$($2.5 million). For strategy 4, it equals P($1.4 million) + $(1 - P)$($2.0 million).

(b) Clearly, strategy 4 is dominated by strategy 1.

. 576-77) **16.13** (a) No.

(b) Yes.

16.15 (a) $$\frac{(0.9)(0.9)(0.55) + (0.3)(0.1)(0.45)}{(0.9)(0.55) + (0.3)(0.45)} = \frac{.4455 + .0135}{.495 + .135} = \frac{.459}{.630} = .73$$

(b) $1 - .73 = .27$

(c) $$\frac{(0.1)(0.9)(0.55) + (0.7)(0.1)(0.45)}{(0.1)(0.55) + (0.7)(0.45)} = \frac{.0495 + .0315}{.055 + .315} = \frac{.081}{.370} = .22.$$

The probability that the panel comes to the same conclusion is $1 - 0.22 = 0.78$.

Index

GLOSSARY OF SYMBOLS

(Continued from inside the front cover)

Symbol	Meaning*
P_{oi}	Price of the *i*th commodity in the base year (14.2)
P_{1i}	Price of the *i*th commodity in year 1 (14.2)
p	Sample proportion (6.7)
p_1, p_2	Sample proportions from populations 1 and 2 (7.7)
$\hat{P}$	Estimate of population proportion (7.8)
$Pr\{a<X<b\}$	Probability that random variable *X* lies between *a* and *b* (7.5)
$Pr\{A\}$	Probability of event *A* (5.9)
Q_i	Dummy variable for *i*th quarter (13.10)
Q_{oi}	Quantity of *i*th commodity in base period (14.3)
Q_{1i}	Quantity of *i*th commodity in year 1 (14.3)
Q_{wi}	Quantity of *i*th commodity in year *W* (14.4)
R	Sample multiple correlation coefficient (12.6)
R_1	Sum of ranks in Mann-Whitney test (9.11)
R_n	Critical limit in sequential sampling plan (16.6)
R^2	Sample multiple coefficient of determination (12.6)
RSS	Block sum of squares (10.8)
r	Sample correlation coefficient (11.11)
r^2	Sample coefficient of determination (11.10)
r	Number of rows (9.5)
r	Number of population proportions being compared (9.4)
r_r	Sample rank correlation coefficient (Appendix 11.2)
ρ	Population correlation coefficient (11.12)
S	Sample space (3.2)

Symbol	Meaning*
S	Seasonal variation (13.2)
S_t	Smoothed value of time series at time *t* (13.7)
s	Sample standard deviation (2.5)
s^2	Sample variance (2.5)
s_1, s_2	Sample standard deviations of populations 1 and 2 (7.7)
s^2_j	Sample variance from *j*th population (10.5)
s_e	Sample standard error of estimate (11.8)
s_b	Standard error of sample regression coefficient (11.13)
s_{b1}, s_{b2}	Standard errors of b_1 and b_2 (12.5)
s_E	Square root of error mean square (10.9)
s_w	Square root of within-group mean square (10.7)
σ	Population standard deviation (2.5)
σ^2	Population variance (2.5)
σ_0^2	Population variance if null hypothesis is true (9.8)
$\sigma(X)$	Standard deviation of random variable *X* (4.5)
$\sigma^2(X)$	Variance of random variable *X* (4.5)
$\sigma_{\bar{x}}$	Standard deviation of sampling distribution of sample mean (6.8)
σ_1, σ_2	Standard deviations of populations 1 and 2 (8.6)
σ_e	Population standard error of estimate (11.8)
σ_b	Population standard error of *b* (11.13)
σ_p	Standard deviation of the prior distribution of the sample mean (Appendix 7.1)
σ_r	Standard deviation of number of runs (9.12)
σ_u	Standard deviation of *U* in Mann-Whitney test (9.11)

*Number in parenthesis indicates the section where the symbol is introduced.